JSTICE SYSTEM

CORRECTIONS

»» Trial   »» Sentencing   »» Probation   »» Prison   »» Parole

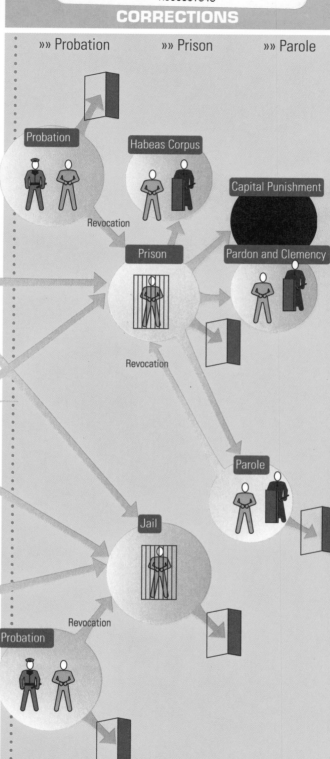

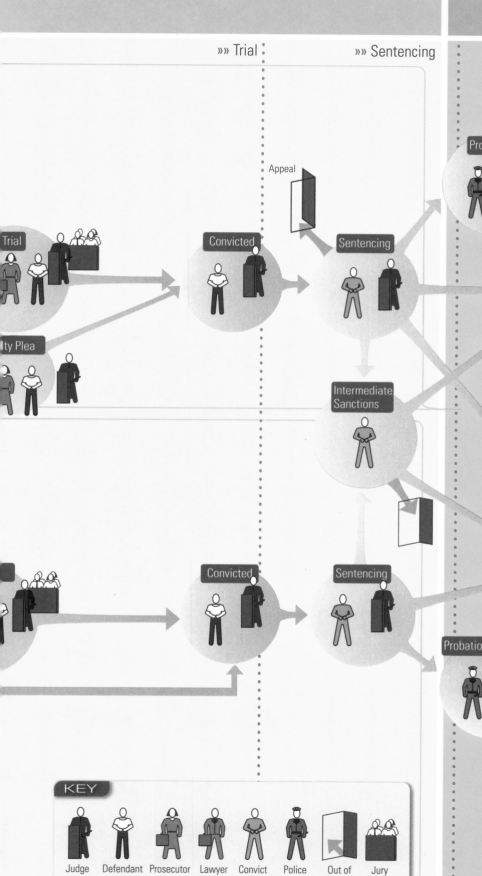

Appeal

Trial

ty Plea

Convicted

Sentencing

Intermediate Sanctions

Convicted

Sentencing

Probation

Revocation

Probation

Habeas Corpus

Prison

Revocation

Jail

Revocation

Capital Punishment

Pardon and Clemency

Parole

KEY

Judge | Defendant | Prosecutor | Lawyer | Convict | Police | Out of System | Jury

Exploring
Criminal
Justice

Robert M. Regoli, PhD
University of Colorado at Boulder

John D. Hewitt, PhD
Grand Valley State University

JONES AND BARTLETT PUBLISHERS
Sudbury, Massachusetts
BOSTON   TORONTO   LONDON   SINGAPORE

World Headquarters
Jones and Bartlett Publishers
40 Tall Pine Drive
Sudbury, MA 01776
978-443-5000
info@jbpub.com
www.jbpub.com

Jones and Bartlett Publishers
Canada
6339 Ormindale Way
Mississauga, Ontario L5V 1J2
Canada

Jones and Bartlett Publishers
International
Barb House, Barb Mews
London W6 7PA
United Kingdom

Jones and Bartlett's books and products are available through most bookstores and online booksellers. To contact Jones and Bartlett Publishers directly, call 800-832-0034, fax 978-443-8000, or visit our website www.jbpub.com.

Substantial discounts on bulk quantities of Jones and Bartlett's publications are available to corporations, professional associations, and other qualified organizations. For details and specific discount information, contact the special sales department at Jones and Bartlett via the above contact information or send an email to specialsales@jbpub.com.

Production Credits

Chief Executive Officer: Clayton E. Jones
Chief Operating Officer: Donald W. Jones, Jr.
President, Higher Education and Professional Publishing: Robert W. Holland, Jr.
V.P., Sales and Marketing: William J. Kane
V.P., Production and Design: Anne Spencer
V.P., Manufacturing and Inventory Control: Therese Connell
Publisher, Public Safety Group: Kimberly Brophy
Acquisitions Editor: Jeremy Spiegel
Associate Managing Editor: Robyn Schafer
Production Supervisor: Jenny L. Corriveau
Associate Photo Researcher and Photographer: Christine McKeen
Director of Marketing: Alisha Weisman
Marketing Manager: Wendy Thayer
Interior Design: Anne Spencer
Cover Design: Kristin E. Ohlin
Manufacturing and Inventory Coordinator: Amy Bacus
Composition: NK Graphics
Cover Image: Man behind bars: © Comstock Images/age fotostock; gavel: © Creativ Studio Heinem/age fotostock; police badge: © Corbis/age fotostock; Court: © Michael G. Smith/ShutterStock, Inc.
Text Printing and Binding: Courier Kendallville
Cover Printing: Courier Kendallville

Library of Congress Cataloging-in-Publication Data

Regoli, Robert M.
  Exploring criminal justice / Robert M. Regoli and John D. Hewitt.
    p. cm.
  ISBN-13: 978-0-7637-4284-3 (hardcover)
  ISBN-10: 0-7637-4284-8 (hardcover)
  1. Criminal justice, Administration of—United States. I. Hewitt, John D., 1945– II. Title.
  HV9950.R445 2007
  364.973—dc22

                                                    2007023930

6048

Printed in the United States of America
11 10 09 08 07   10 9 8 7 6 5 4 3 2 1

To Debbie, the one who keeps the candle glowing in the depths of darkness.

— Robert M. Regoli

To my wife, Avis, my children, Eben and Sara, and my grandchildren, Zoë, Henry, and Hugh. Thank you for always being there.

— John D. Hewitt

BRIEF CONTENTS

CONTENTS

Chapter 7

Chapter 8

SECTION 3

Courts   218

Chapter 9

Chapter 10

Chapter 11

Chapter 12

SECTION 5

Special Issues   450

Chapter 16

Chapter 17

RESOURCE PREVIEW

Exploring Criminal Justice provides students with a clear, complete, and credible introduction to the U.S. criminal justice system. It effectively informs students about the criminal justice system, its processes, and its various components while encouraging them to think critically about criminal justice issues. Features that reinforce and expand on essential information include:

Thinking about Criminal Justice: Challenges readers to think like practicing criminal justice professionals with real-world case studies and accompanying discussion questions.

Full-color photos: Probes students to consider photographs and critical thinking questions relevant to chapter topics.

Around the Globe: Offers a window into law and law enforcement around the world considering subjects such as drug enforcement policies in The Netherlands, community policing in Japan, and crime and punishment under Islamic law.

Focus on Criminal Justice: Explores the nature of crime and the criminal justice response concerning issues such as police crackdowns in San Diego, fast track prosecution of serious offenders, and modern forms of cruel and unusual punishment.

Headline Crime: Highlights recent high-profile cases, such as sentencing terrorists, probation officer misconduct, and throwing out false confessions.

Links: Connects the current topic to related subjects in other chapters.

Voice of the Court: Provides accessible summaries and excerpts from court rulings which have been instrumental in shaping the criminal justice system and process, including those affecting the actions of police officers, due process in criminal prosecutions, and the rights of prisoners.

- The *Texas Syndicate* was formed in the early 1970s in California's Folsom prison as a direct response to other prison gangs, especially the Mexican Mafia and Aryan Brotherhood.
- The *Aryan Brotherhood*, also known as the Brand, was formed in 1967 in the prison at San Quentin in California. This white supremacist group organized to serve the interests of white inmates and has since spread to prisons around the country. Much of its activities have been centered on drug trafficking, prostitution, extortion, and murder, especially of African American inmates.

Sex in Prison

Most inmates serve their sentences in same-sex institutions with no heterosexual outlet, producing a number of problems for inmates and prison staff. The normal need for sexual release by inmates leads to various adaptations. Some inmates engage in homosexual activity; others resort to prostitution, engaging in homosexual sex while maintaining a heterosexual identity.

LINK A few states (e.g., California, Mississippi, and New York) allow conjugal visits with spouses, as discussed in Chapter 13.

Many male inmates experience intense sexual harassment included unwanted and offensive sexual advances, which frequently produce an acute fear of sexual assault. Within prison, sexual harassment involves both physical incidents, such as kissing, touching, or fondling, and verbal harassment, including statements that feminize an inmate, sexual propositions, and sexual extortion.[81]

Male Inmates

Sexual harassment of inmates is relatively common in men's prisons. Daniel Lockwood's study of two New York prisons found nearly 30 percent of inmates reported being sexually targeted, 33 percent propositioned, and 7 percent touched or grabbed.[82] When Peter Nacci and Thomas Kane looked at inmate sex

Headline Crime The Aryan Brotherhood

On July 28, 2006, a jury in Santa Ana, California, convicted four leaders of the Aryan Brotherhood prison gang—Barry Mills, Tyler Bingham, Edgar Hevle, and Christopher Gibson—under the Racketeer Influenced and Corrupt Organization law on charges of using murder and intimidation to protect their drug trafficking activities in prison. The four had been charged with 32 murders and attempted murders involving members of the Aryan Brotherhood over nearly three decades. Mills and Bingham were found guilty of ordering the murder of African American inmates from their cells at the super-max prison at Pelican Bay.

Over the years, most of the gang's violence was directed at African American inmates. The bloodiest attack occurred on August 28, 1997, when gang members at the Lewisburg federal prison armed themselves with shivs (homemade knife-like weapons) and stabbed six African American inmates, killing two.

While originally formed to strike at African American inmates belonging to the Black Guerrilla Family, the Aryan Brotherhood soon evolved into a full-fledged criminal enterprise involving extortion, drug trafficking, and the sale of "punks" (inmates forced into prostitution). To be accepted into the gang, a recruit had to "make his bones," which often required killing another inmate. Although only a few hundred inmates belong to the gang, its impact has been far-reaching. For example, the Aryan Brotherhood accounts for most of the drug distribution that occurs in state and federal prisons.

The conviction of the four gang leaders suggests an attempt by prison officials to break up one of the most violent prison gangs and to restore a greater control over prison gangs, even in super-max prisons.

Sources: Greg Risling, "Jury Convicts 4 White-Supremacists," *Washington Post* (July 28, 2006), available at http://www.washingtonpost.com/wp-dyn/content/article/2006/07/28/AR2006072801079.html, accessed June 1, 2007; Associated Press, "4 Aryan Brotherhood Leaders Convicted," *CBSNews* (July 28, 2006), available at http://www.cbsnews.com/stories/2006/07/28/national/main1847262.shtml, accessed August 1, 2007; David Grann, "The Brand: How the Aryan Brotherhood Became the Most Murderous Prison Gang in America," *New Yorker* 80:156-171 (2004).

VOICE OF THE COURT *Gideon v. Wainwright*

One of the most important landmark decisions affecting the criminal justice process was handed down by the U.S. Supreme Court in 1963. The *Gideon v. Wainwright* case established the right of a defendant to be represented by counsel in state criminal proceedings. Clarence Gideon, a 51-year-old petty thief, was arrested in Panama City, Florida, on June 4, 1961, for breaking into and entering a poolroom and stealing beer, soft drinks, and coins from a cigarette machine. At his trial, Gideon stood before Judge Robert McCrary and stated that he was unable to proceed because he did not have the assistance of an attorney. He requested that the court appoint counsel to represent him and insisted, "The United States Supreme Court says I am entitled to be represented by counsel."

The court refused to appoint an attorney to represent Gideon. Acting as his own counsel, Gideon was unable to present his defense in the manner required by law. He was found guilty by the jury and sentenced to five years in the Florida State Prison. Gideon subsequently submitted a handwritten petition to the Florida Supreme Court, which was denied without a hearing; he then filed a petition with the U.S. Supreme Court. His petition was granted on June 4, 1962, and his case was argued before the Court by an appointed attorney, Abe Fortas, who was later appointed as a justice on the Supreme Court. On March 18, 1963, Justice Hugo Black delivered the opinion of the Court:

> Put to trial before a jury, Gideon conducted his defense about as well as could be expected from a layman. He made an opening statement to the jury, cross-examined the State's witnesses, presented witnesses in his own defense, declined to testify himself, and made a short argument "emphasizing his innocence to the charge contained in the information filed in this case. . . ." Reason and reflection require us to recognize that in our adversary system of criminal justice, any person hauled into court, who is too poor to hire a lawyer, cannot be assured a fair trial unless counsel is provided for him. This seems to us to be an obvious truth. Governments, both state and federal, quite properly spend vast sums of money to establish machinery to try defendants accused of crime. Lawyers to prosecute are everywhere deemed essential to protect the public's interest in an orderly society. Similarly, there are few defendants charged with crime, few indeed, who fail to hire the best lawyers they can get to prepare and present their defenses. That government hires lawyers to prosecute and defendants who have the money to hire lawyers to defend are the strongest indications of the widespread belief that lawyers in criminal courts are necessities, not luxuries. The right of one charged with crime to counsel may not be deemed fundamental and essential to fair trials in some countries, but it is in ours. From the very beginning our state and national constitutions and laws have laid great emphasis on procedural and substantive safeguards designed to assure fair trials before impartial tribunals in which every defendant stands equal before the law. This noble ideal cannot be realized if the poor man charged with crime has to face his accusers without a lawyer to assist him.

The U.S. Supreme Court reversed the judgment and sent the case back to the Supreme Court of Florida for retrial.

Source: Gideon v. Wainwright, 372 U.S. 335 (1963).

End of Chapter Material: Wraps up each chapter with a complete review, including a case study conclusion, chapter recap, critical thinking questions, a glossary of key terms, and a bibliography for further reading on chapter subjects.

WRAP UP

THINKING ABOUT CRIME AND JUSTICE: CONCLUSION

The tragic events at Virginia Tech brought to light the poor communication and coordination between the mental health and criminal justice systems. For privacy reasons, information about Seung-Hui Cho's mental health status was not linked to criminal justice databases, so law enforcement had little reason to investigate his bizarre behaviors prior to the shootings. Just as the 9/11 Commission found that poor communication between intelligence and law enforcement agencies contributed to the ability of Al Qaeda members to plot and execute the September 11, 2001, attacks, so a barrier today exists between the privacy expectations of mentally ill individuals and public safety. It is likely that the national debate that arose after these shootings will lead to policy changes in this area to help better protect the U.S. public.

Chapter Spotlight

- All societies experience crime. To combat crime, societies create criminal justice systems that define the rules that govern social interactions, enforce the standards of conduct necessary to protect their citizens, and determine the punishments for violating those rules so as to establish social order.
- The significant increase in serious crime through the early 1990s and the recent international terrorist attacks have caused the U.S. federal government to define crime—and specifically terrorism—as a major problem. In response, Congress has passed a number of pieces of legislation aimed at reducing crime levels.
- Democratic and open societies deal with crime differently than authoritarian and closed societies. In a democratic society, the criminal justice system is accountable to the public and operates within a legal framework established by legislators who are elected by the people.
- The criminal justice system is an interrelated set of subsystems that operate horizontally and vertically across federal, state, and local levels. The laws and jurisdictions of each subsystem sometimes overlap, which sometimes leads to conflicts among the various agencies.
- The criminal justice process is composed of well-defined legal rules that specify the procedures through which suspects enter and are treated by the criminal justice system.
- Perceptions of the seriousness of the crime problem and the best ways to solve it are influenced by popular culture, and particularly by violent television programs that may increase the aggressive behavior of their viewers.

Putting It All Together

1. Is the amount of crime in society normal, or have we simply come to accept the level of crime that we have as being normal?
2. What are the advantages and disadvantages of a democratic and open system of criminal justice as compared to systems of criminal justice in countries with authoritarian governments, such as Burma, China, Iran, North Korea, and the Sudan?
3. Why are perceptions of crime and justice so important to our thinking about crime and to the daily operation of the criminal justice system?
4. The crime rate has decreased significantly in the past decade, and only a very small percentage of the public now views crime as the most serious problem in the United States. Why do you believe the crime rate has dropped? Do you feel safer today than you did in the immediate past?
5. Violent television programs may increase the aggression of some viewers. Should television programs be censored for violence? Is it important for the government to regulate what viewers may watch on television? Do television networks have a First Amendment right (freedom of speech) to present whatever material they choose?

Key Terms

criminal justice process The procedures that occur in the criminal justice system, from a citizen's initial contact with police to his or her potential arrest, charging, booking, prosecution, conviction, sentencing, and incarceration or placement on probation.

criminal justice system A complex set of interrelated subsystems composed of three major components—police, courts, and corrections—that operate at the federal, state, and local levels.

norms Rules and expectations by which a society guides the behavior of its members.

Notes

1. Solomon Herbert, "King Litigation May Spur Healing or Unrest: Rodney King's Civil Lawsuit against the Los Angeles Police Department," *Black Enterprise*, September, 1992, available at http://www.findarticles.com/p/articles/mi_m1365/is_n2_v23/ai_12539601, accessed June 8, 2007.
2. Rodney King, "Quotation of the Day," *New York Times*, May 2, 1992, p. 6A.
3. Émile Durkheim, *The Rules of Sociological Method*, translated by Sarah Solovay and John Mueller (New York: Free Press, 1893/1938), p. 68.
4. David Wallechinsky, "*Parade*'s Annual List of the World's 10 Worst Dictators," *Parade Magazine*, February 11, 2007, p. 7.
5. Émile Durkheim, *The Division of Labor and Society* (New York: Free Press, 1893/1964); Durkheim, note 3.
6. Kai Erikson, *Wayward Puritans* (New York: John Wiley & Sons, 1966).
7. Erich Goode, *Drugs in American Society*, 7th ed. (New York: McGraw-Hill, 2007); Charles Faupel, Alan Horowitz, and Greg Weaver, *The Sociology of American Drug Use* (New York: McGraw, 2003).
8. George Gallup, Jr., *The Gallup Poll* (Princeton, NJ: Gallup Poll, September 1989).
9. Darrell Gilliard and Allen Beck, *Prisoners in 1993* (Washington, DC: U.S. Department of Justice, 1994), p. 1.
10. George Gallup, Jr., *The Gallup Poll* (Princeton, NJ: Gallup Poll, September 1993); Gary Burden, "U.S. Drug Policy May Be in for a New Look," *Law Enforcement News* 15:5 (1993).
11. *The Violent Crime Control and Law Enforcement Act of 1994* (HR 3355) (Washington, DC: U.S. Government Printing Office, 1994).
12. Kristen Hughes, *Justice Expenditures and Employment in the United States, 2003* (Washington, DC: U.S. Department of Justice, 2006).
13. Federal Bureau of Investigation, *Crime in the United States, 2006* (Washington, DC: U.S. Government Printing Office, 2007); U.S. Congress, HR 1572; *The Omnibus Crime Control and Safe Streets Act* (P.L. 90-351) (Washington, DC: U.S. Printing Office, 1968); The Disaster Center, available at http://www.disastercenter.com/crime/ederal, accessed May 4, 2007; *The Comprehensive Crime Control Bill of 1984* (P.L. 98-473) (Washington, DC: U.S. Government Printing Office, 1984); *The Crime Control Act of 1990* (P.L. 101-647) (Washington, DC: U.S. Government Printing Office, 1990); *The Violent Crime Control and Law Enforcement Act of 1994* (HR 3355), note 11; *USA Patriot Act of 2001* (HR 3162 RDS) (Washington, DC: U.S. Government Printing Office, 2001).

RESOURCES

Instructor Resources

This text is the core of an integrated teaching and learning system, which includes ancillaries developed by criminal justice educators.

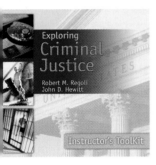

Instructor's Toolkit CD-ROM
ISBN-13: 978-0-7637-5924-7

Preparing for class is easy with the resources found on this CD-ROM, including:

- **Lecture Outlines** offer complete, ready-to-use lesson plans that outline all of the topics covered in the text. All outlines can be adapted to your unique course.

- **PowerPoint Presentations** provide ready-to-use presentations that are both educational and engaging. Slides can be modified and tailored to meet your needs.

- **Electronic Test Bank** contains multiple-choice and scenario-based questions, allowing you to create custom-made classroom tests and quizzes quickly by selecting, editing, organizing, and printing a test along with an answer key that includes page references to the text.

- **Image Bank** supplies images from the text that can be used in multimedia lectures and in handouts.

The resources found on the Instructor's ToolKit CD-ROM are formatted for online course management systems and can be customized to fit your course needs.

Technology Resources

Essential components to the teaching and learning system include additional resources available on this companion website. Students are able to grasp key concepts about the criminal justice system and explore topics in more depth. Instructors can easily download the Lecture Outlines, PowerPoint Presentations, and Electronic Test Bank.

www.jbpub.com/ExploringCJ

Make full use of today's teaching and learning technology with www.jbpub.com/ExploringCJ. This site has been specifically designed to complement *Exploring Criminal Justice*. Some of the resources available include:

- Practice Quizzes
- Sample Legal Documents
- Key Term Explorer
- Web Links

Crime is something that affects all of us in some way. Crime can determine where we live, where we send our children to school, when and where we go out, how much we pay for insurance, and whether our tax dollars are used to build new roads or new prisons. The agencies and entities that deal with crime are complex and far-reaching, and the system they form must continually adapt to changes in the nature of crime. Explaining that system is no easy task, but that is exactly what Bob Regoli and John Hewitt have accomplished in *Exploring Criminal Justice.*

The field of criminal justice is not for the faint-hearted. It can be exciting and challenging, precisely because of its complexity and the importance of crime in our daily lives. *Exploring Criminal Justice* provides a crucial understanding of the criminal justice system and its key components. The text begins by laying a strong foundation—examining the criminal justice system as it exists today, as well as its history and important aspects of criminal law, and presenting the philosophical bases that underlie the criminal justice system. The rest of the book is appropriately devoted to understanding the pillars of the criminal justice system: police, courts, and corrections. The final section—Special Issues—delves into terrorism and cybercrime, two enormously important topics that rarely receive the level of attention provided in this book.

Many years have passed since I first met Bob Regoli and John Hewitt. Both are well-respected scholars who have devoted their careers to teaching and studying criminal justice. In this book, they have pulled together their many years of experience and the vast knowledge they have about crime and the system designed to address it to create a well-written text that provides a thorough review of the field and practices of criminal justice. Professors Regoli and Hewitt skillfully discuss technical and complex topics in understandable language. They also encourage exploration and learning, presenting all aspects of the criminal justice system, and they do it in a way that is both interesting and educational.

In some instances, the reader may be led to disagree with the authors of this book and argue about apparent inadequacies in the system and the failures of assumptions and practices. That is to be expected; after all, the criminal justice system is often controversial. More importantly, the reader will undoubtedly be challenged by the ideas presented in this text to overcome politics and ideology and work toward a new vision of the criminal justice system—one that will seek to be less reactive and more prescriptive in solving problems; one that abandons isolation and restriction in favor of collaboration. For the serious student of the criminal justice system, this book will help guide the way.

Edward J. Latessa
Professor and Division Head
Division of Criminal Justice
University of Cincinnati

ACKNOWLEDGMENTS

Thank you, Edward Latessa of the University of Cincinnati, for writing the foreword to *Exploring Criminal Justice*. Your comments will serve to inspire the critical study of the criminal justice process by undergraduates, students who pursue graduate work in the discipline, and those who choose to become practitioners and work in the field, hands-on, with offenders and their victims.

The people with whom we have worked at Jones and Bartlett have been spectacular! Jones and Bartlett assembled a terrific team to produce our book, and we are honored and proud to be counted among the company's authors. We are especially thankful to Stephanie Boucher, who displayed the courage to originally sign the text. Following her departure, we found our new acquisitions editor, Jeremy Spiegel, to be not only a most wonderful and gracious person, but also someone whose knowledge of the publishing business from top to bottom is unparalleled. His guidance, direction, critical assessment, and encouragement made it possible for us to "keep on keeping on." We will forever be grateful for what he has taught us. The marketing manger for the text, Wendy Thayer, has done an extraordinary job of getting our message out to the criminal justice community. Words alone cannot describe how greatly appreciative we are of her tireless efforts on the project. Our developmental editor, Robyn Schafer, has provided good, strong consistent insights regarding what must be done to make the book better and more competitive in every way. She is absolutely the best developmental editor we have worked with on the many books we have written. Her commitment to the project and to us as authors went well beyond anything that we originally expected. We were blessed to have Robyn on our team! And, of course, the guiding light of the project, Kim Brophy, was the best coach of any team we have played on.

Many colleagues, some of whom are personal friends and others of whom we have not yet met, unselfishly shared information, insights, and criticisms with us and helped us grow and formulate our ideas about crime and justice. Our appreciation to these people extends well beyond the appreciation we express here, but we do thank you. Each of you will always hold a special place in our hearts: Geoff Alpert, Alex Alverez, Rosalie Arndt, Patrick Bacon, Sarah Bacon, Rachel Bandy, Gregg Barak, Allison Bayless, Kevin Beaver, Janet Behrens, Joanne Belknap, Ingrid Bennett, Richard Bennett, Mark Berg, Dennis Blewitt, Bob Bohm, Lisa Campione, Todd Clear, Shannon Coffey, Mark Colvin, Herb Covey, Sarah Corcoran, John Crank, Bob Culbertson, Frank Cullen, Lois DeFleur, Walter DeKeseredy, Matt DeLisi, Brendan Dooley, Delbert Elliott, John Fuller, Sarah Getman, Michelle Goetz, Lindsey Grall, Richard Grossenbacher, Ed Grosskopf, Mark Hamm, Kraig Hays, Laura Hettinger, Eric Hickey, Andy Hochstetler, Lou Holscher, Charles Hou, James Houston, Peter Iadicola, Eric Jensen, Christopher Kierkus, Brian Kingshott, Beverly Kingston, Georgia Kinkade, Blair Lachman, Paul Lasley, Richard Lawrence, Stephanie Lichtenauer, Andy Miracle, Tina Miracle, Marilyn McDowell, Jean McGloin, Gloria Mendoza, Bill Miller, Tiare Moorman, Katie Murphy, Hal Pepinsky, Mark Pogrebin, Joycelyn Pollock, Hillary Potter, Eric Primm, Beverly Quist, Michael Radelet, Ronald Reed, Tom Reed, Adam Regoli, Andrea Regoli, George Rivera, Geoff Rivers, Richard Rogers, Clarence Romig, Joe Sanborn, Jeff Schrink, Eric Schwartz, Marty Schwartz, Madison Serdinak, Rajshree Shrestha, James F. Short, Jr., Chad Simcox, Victor Strieb, Terry Thornberry, Larry Travis, Beverly Quist, Jay Watterworth, Laura VanderDrift, Amanda VanHoose, Michael Vaughn, Regina Verna, Jerry Vito, Lisa Hutchinson Wallace, Jules Wanderer, Tom Winfree, Ralph Weisheit, and John Wright.

There is no doubt that we unintentionally omitted some very important people in our lives from the list. Please know that this was not our intention, and accept our heart-felt apology for the omission. Other colleagues selected to review the book improved the text in innumerable ways. Each went beyond what was asked and provided us with insights that we considered seriously and in most instances incorporated into *Exploring Criminal Justice*. There is no way to put into words how much your guidance along the way has meant to us.

Writing does not take place in a vacuum, and authors are not unaffected by—nor do they fail to affect—those closest to them. Our children Adam, Andrea, Eben, and Sara; our grandchildren Zoë, Henry, and Hugh; and especially our wives, Debbie and Avis, have sacrificed and shared with us in various ways along the road of this most incredible journey. They stood alongside us in the long and difficult process of writing this book. Their encouragement, support, tolerance, patience, and humor are appreciated much more than they ever will know.

Reviewers

Alissa Ackerman
John Jay College of Criminal Justice

Sarah Bacon
Florida State University

Deborah Barrett
American InterContinental University

Kevin Borgeson
Salem State College

Raymond E. Foster
California State University, Fullerton

Lisa Therese Fowler
Keiser University

Rebekah Garrett
American InterContinental University

M. Arif Ghayur
Iowa Wesleyan College

Michael Hogan
Colorado State University

Lisa Hutchinson
University of Arkansas at Little Rock

James Jabbour
American InterContinental University

Stacy Mallicoat
California State University, Fullerton

Vincent A. Sainato
John Jay College of Criminal Justice

Fred Sams
Canyon College

Special thanks to the following people for their contributions to the text:

Lee Ayers
Southern Oregon University

Peter J. Conis
Iowa State University

Amy Harrell
Nash Community College

ABOUT THE AUTHORS

Robert M. Regoli

In addition to his extensive experience researching and studying police and corrections officers, Robert M. Regoli also has been a crime victim, secret delinquent, criminal complainant and witness, jury member, and legal consultant. Born in Antioch, California, he earned his B.S. in psychology, M.A. in police science and administration, and Ph.D. in sociology at Washington State University. Over the course of his career, Dr. Regoli has taught at Indiana State University, Texas Christian University, and, currently, the University of Colorado at Boulder.

Dr. Regoli has held several positions in the Academy of Criminal Justice Sciences, including being the President and named a Fellow and was former Executive Editor of *The Social Science Journal*. He is the recipient of two William J. Fulbright awards and is a member of Phi Beta Kappa. In addition to being a consultant to local law enforcement agencies, the Colorado Department of Corrections, Indiana Department of Corrections, National Institute of Corrections, and National Institute of Law Enforcement and Criminal Justice, Dr. Regoli has also served as a member of the Indiana Juvenile Justice Task Force and the Task Force for the Children's Constitutional Rights. With over three decades of experience writing extensively about criminal justice issues, Dr. Regoli has amassed more than 100 professional journal publications and numerous books.

John D. Hewitt

Currently a professor of criminal justice at Grand Valley State University, John D. Hewitt was born in Carmel, California, while his father was stationed at the Presidio during World War II. Dr. Hewitt completed his undergraduate work at Western Washington State College and his Ph.D. at Washington State University. He has taught for more than 30 years at small and large state colleges and universities as well as in small liberal arts colleges in the Midwest and West.

During his career, Dr. Hewitt has worked in the Indiana Department of Corrections and was a member of the Board of Directors of the Delaware County Youth Services Bureau and Bethel Place for Boys. He has testified as an expert witness in Arizona, Indiana, and Michigan in cases dealing with the death penalty, drug trafficking, judicial sentencing, and youth gangs. The author of many publications, Dr. Hewitt has written extensively about issues of crime, criminal justice, and delinquency. He co-authored *Delinquency in Society* and *The Impact of Sentencing Reform* as well as numerous articles on issues ranging from judicial sentencing, rule enforcement by correctional officers, victim–offender relationships in homicide cases, and work stress among police executives.

Although more than 1.5 million people are currently confined in state and federal prisons in the United States, crime in this country has actually been declining for the past decade—that is, at least until recently, when there has been a small increase in offenses committed. Whether this upward trend will continue and what the future holds are hotly debated topics among criminologists. Nevertheless, some of the most recent data produced by the Federal Bureau of Investigation suggest that both reported crimes and arrests are at their lowest levels since the early 1970s, and national surveys of victims report similar declines in victimization rates. The number of people under sentence of death and the number of actual executions each year have fallen significantly since they reached their peak in 2000. The U.S. Supreme Court heard arguments for only half the number of cases in 2006 than it did in 1987.

Do these trends mean that policymakers have discovered how to control (or possibly even eliminate) crime? The answer is emphatically *no*. Criminologists are not sure about why crime has declined or, conversely, why there was a small increase in the amount of serious violent crime reported to police in 2006. Certainly, expenditures on the criminal justice system have not declined; some states spend more on their prison systems than they do on public education. The criminal courts are clogged with cases, and local communities continue to build ever more jails and detention centers. Since the terrorist attacks of September 11, 2001, the criminal justice system has also faced a plethora of new challenges—for law enforcement, in the detection and response to terrorist activity; for the courts, in the detention and prosecution of alleged terrorists; and for corrections, in the incarceration of those convicted of terrorism.

Many students of criminal justice are curious about how crime is changing and how the criminal justice system is responding to those changes. Some have already decided to major in criminal justice, whereas others are just beginning to explore what this major has to offer them. We wrote *Exploring Criminal Justice* to be a useful guide for all students interested in learning more about the U.S. criminal justice system.

More than 60 years of collective teaching experience has taught us that students come to courses with a variety of backgrounds and expectations. We believe this truism is especially valid for students taking their first course in criminal justice. Some students have family members who work in the system as police officers, attorneys, counselors, or probation, parole, or correctional officers. Other students have already had first-hand experience in the system, either as employees, volunteers, or perhaps even crime victims or criminal offenders. Students also come into this course with knowledge of the criminal justice system based on television dramas, documentaries, news programs, films, music, and other media. The common ground that these students share is an interest in learning more about the criminal justice system and its operation.

A good textbook is more than just a compendium of facts and figures, flow charts, and case studies. Recognizing that fact, *Exploring Criminal Justice* engages students in an exploration of the intricate details of the criminal justice process. It breathes life into the criminal justice system by examining its many components and investigating how they operate in relationship to one another as well as considering the people who work in the system and what they do.

In responding to the needs of professors and their students, we have provided a text that is comprehensive, current, accurate, and engaging. The issues, research, policies, programs, and court cases are presented and discussed in a balanced manner intended to foster discussion and debate in addition to learning and greater understanding. We hope that both professors and students will benefit from this text as they explore our presentation of the criminal justice system and its processes.

Robert M. Regoli
University of Colorado at Boulder

John D. Hewitt
Grand Valley State University

Crime and Criminal Justice

The study of criminal justice must first begin with the social and historical context of both crime and justice. This section examines the nature of criminal justice in society (past and present), extending beyond the components of the criminal justice system (namely, the police, courts, and corrections) to consider its foundations.

Chapter 1 looks at how society defines crime and provides a brief overview of both the criminal justice system and the criminal justice process. The origins of this system are the focus of Chapter 2, which presents the development of criminal law through the ages and the ways that law defines crime and criminal responsibility. Focusing on crime, its criminals, and its victims, Chapter 3 discusses the nature, extent, and measurement of crime and victimization. Chapter 4 places these concepts in a theoretical framework, outlining key criminological theories and the various policies which are designed to prevent or reduce crime, emerging from these schools of thought. This chapter also examines why some theories are more easily translated into policy than others and explores why some theories that present plausible explanations of criminality do not provide practical guidelines for action.

How society defines, measures, and explains crime and criminal victimization provides the necessary framework to understand the organization and operation of the U.S. criminal justice system.

OBJECTIVES

- Define crime and explain how criminal behaviors change over time.
- Describe how effective crime control legislation has been at reducing crime.
- Know the three primary components of the criminal justice system.
- Identify the principal functions of the police, courts, and corrections.
- Grasp the steps in the criminal justice process.
- Be aware of the influence of public perceptions on crime control policy.

Overview of Criminal Justice

CHAPTER

1

On April 16, 2007, Seung-Hui Cho perpetrated the deadliest mass killing in modern U.S. history when he murdered 32 people and attempted to murder 29 others before killing himself at the campus of Virginia Polytechnic Institute and State University (Virginia Tech) in Blacksburg, Virginia.

In the wake of the massacre, a disturbing picture emerged of Cho, a senior English major at the university. He was plagued by severe psychiatric problems and had been in several incidents involving worrisome antisocial behavior, including stalking female students and submitting essays in a creative writing class that were so violent and disturbing that a professor referred Cho for mental health counseling. Indeed, in 2005 Cho had been diagnosed with mental illness and was declared a danger to himself and others, but follow-up was not completed. Despite his troubled history, Cho had no previous criminal record and was able to legally obtain the firearms and ammunition he used in the shootings.

The American public got a glimpse of the extent of Cho's illness through a multimedia package containing video, photographs, and a manifesto that Cho sent to NBC during a two-hour lull between the shootings. These items revealed feelings similar to what Cho had previously expressed in his writing assignments—namely, intense hatred of other students, especially females and those with money, and feelings of persecution. Acquaintances recall that Cho was an extreme loner who often refused to talk with others.

The shootings at Virginia Tech initiated national debates about privacy, mental health treatment and public safety, gun control, university preparedness for disasters, criminal justice response to emergency events, and the relationship between the news media and the public. The massacre also showcased the amazing grace and dignity of the victims of the shooting, the Virginia Tech community, and the American public when exposed to these horrific events.

- Which factors prevented the criminal justice system from intervening in Cho's life before the shootings?
- In what ways were the Virginia Tech shootings similar to the terrorist attacks of September 11, 2001, as they relate to criminal justice?

Source: http://www.cnn.com/SPECIALS/2007/virginiatech.shootings/, accessed April 25, 2007.

Introduction

As horrific as Cho's conduct was, it is important to recognize that it is not only disturbed killers like Cho who engage in violent behavior and cause widespread fear, social unrest, and conflict. Sometimes criminal justice authorities themselves are to blame, especially when they act in ways that undermine their authority with the public or participate in behaviors that endanger the lives of innocent citizens.

Such was the case of Rodney King, who was captured on camera in one of the most horrifying, racially charged incidents of excessive police force ever exposed. Across the nation, millions of Americans were shocked by the fuzzy video images of four officers from the Los Angeles Police Department (three white and one Latino) beating an African American man—Rodney Glen King—with their batons. The officers believed that King was extremely dangerous, possi-

bly under the influence of drugs, and thus felt the force they used to subdue him was justified. Even so, the officers were indicted on charges of "assault by force likely to produce great bodily injury" and "filing false reports." In 1992, a jury consisting of ten whites, one Latino, and one Asian acquitted the officers. The verdict triggered massive rioting in Los Angeles, which ultimately left hundreds of buildings damaged or destroyed and dozens of people dead.[1] During this turbulent time, King appeared on television, pleading for peace: "People, I just want to say, you know, can we all get along?"[2]

Throughout history, there have always been people—whether killers like Cho, the four Los Angeles police officers, or ordinary citizens—who cannot "get along." The criminal justice system defines the rules that govern social interaction, enforces the standards of conduct necessary to protect individuals and the community, and establishes the pun-

In the United States, the government has declared a war on drugs. It imposes severe penalties for possession of marijuana and hashish (cannabis), which U.S. drug laws classify as Schedule I drugs—placing them in the same category as heroin, cocaine, and LSD. Although the legal penalties for possession of these drugs in each state vary greatly, federal law mandates a prison sentence and a hefty fine for someone who distributes 1000 pounds or more of marijuana. People who are caught with even very small amounts of these drugs may face the forfeiture of their cars, homes, or other possessions.

In the Netherlands, the government views these drugs quite differently. Beginning in the mid-1970s, the Netherlands quietly decriminalized the personal use of marijuana and hashish. The Dutch designed their approach to limit the negative and stigmatizing effects of drug use on individual users. They did so by drawing a clear distinction between "hard" drugs, such as opiates, and "soft" drugs, such as cannabis, and gave law enforcement priority over controlling the production, importation, and trafficking of hard drugs. Dutch law enforcement also decided to ignore the sale of small amounts of cannabis for personal use. Dutch officials believed that if cannabis was decriminalized, it would separate the soft and hard drug markets. This would reduce the likelihood that marijuana users would come into contact with heroin users, and reduce the likelihood that young people experimenting with marijuana would become involved with more dangerous and addictive drugs. The Dutch drug policies were also aimed at normalizing the drug problem. That is, the Dutch admitted that extensive cannabis use had gained a firm foothold in society, as was the case with alcohol and tobacco, and that it was far more realistic to try to reduce the personal and social harms associated with drug use through education and treatment programs.

In Amsterdam, there are more than 400 coffee shops where people can legally buy and smoke marijuana and hashish. These shops provide a relaxed but controlled environment where people can use the drugs, thus reinforcing tolerance of soft drug use, while condemning the sale or use of harder drugs. No other drugs may be sold or used in these shops, and the shops may not advertise or sell cannabis to people younger than age 16.

Although there is minimal support for decriminalizing marijuana in the United States, groups such as the National Organization for the Reform of Marijuana Laws and American Civil Liberties Union do support such legislation. Some states have changed their laws to reduce sentences for possession of small amounts of marijuana.

In contrast to the views espoused by supporters of the legalization of marijuana, the U.S. Drug Enforcement Agency and the American Academy of Pediatrics see marijuana as a "gateway" drug and insist that its decriminalization would lead to greater use of harder drugs and erode public morals. Indeed, many Americans believe that decriminalizing cannabis would lead to a dramatic increase in drug use. However, studies in the Netherlands suggest that this perception is not accurate. In a household survey of people age 12 and older in Amsterdam, Dutch sociologist Peter Cohen found that the number of people smoking marijuana was similar to the rate of reported use in the United States. The Dutch drug laws have been in place for more than 20 years, and any negative impact on Dutch society has been negligible.

Sources: "Drug War Facts," available at http://www.drugwarfacts.org, accessed May 30, 2007; "Response to the American Academy of Pediatrics Report on the Legalization of Marijuana," *Pediatrics* 116:1256–1257 (2005); Manja Abraham and Hendrien Kaal, *Licit and Illicit Drug Use in the Netherlands* (Amsterdam: Mets and Schilt, 2002); Rudolph Gerber, *Legalizing Marijuana* (New York: Praeger, 2004); "Marijuana: A Continuing Concern for Pediatricians," *Pediatrics* 104:982–985 (1999).

WEED MENU

INDOOR BIO	NLG	
GLORIA	17.-	1 GRAM
CRYSTAL	16.-	1 GRAM
WHITE WIDOW	16.-	1 GRAM
DUTCH DELIGHT	15.-	1 GRAM
AFGHAAN SKUNK	15.-	1 GRAM
MILLENIUM WEED	15.-	1 GRAM
K 2	14.-	1 GRAM
JUICY FRUIT	14.-	1 GRAM
ORANGE BUD	14.-	1 GRAM
SUPER SKUNK	11,5	1...

OUTDOOR BIO		
PURPLE POWER	10.-	1 GRAM

FOREIGN		
JAMAICA	14.-	1 GRAM
SWAZI	7,5	1 GRAM

WWW.COFFEESHOP-ABRAXAS.COM

crime rate in cities such as New York, Seattle, and San Francisco, among others, has been attributed to different factors, including improvements in the U.S. economy, the imposition of longer prison sentences and longer amounts of time actually served, better policing patrolling strategies, and legalized abortion.[15]

> **LINK** Crime rates in the United States have fluctuated greatly over the past few decades and are tracked by the Federal Bureau of Investigation's (FBI) Uniform Crime Report, presented in Chapter 3.

In contrast, some crime-related legislation passed by Congress has not been effective. The 1968 Omnibus Crime Control Bill and the Comprehensive Crime Control Act of 1984, for example, had only a negligible effect on the crime rate.[16] In fact, crime rates continued to soar during the decades of the 1970s and early 1980s. Throughout most of this period, the public did not perceive crime as a serious problem, however. It was not until the early 1990s that public perceptions regarding the seriousness of crime changed—and when they did change, those perceptions changed *very* quickly **FIGURE 1-2**. By August 1994, shortly before Congress voted on the Violent Crime Control and Law Enforcement Act, 37 percent of Americans indicated that crime was the most important problem facing the country; today, only about 2 percent believe this to be true.[17]

Crime Control in a Democratic Society

It is difficult for a heterogeneous society to agree on the nature of the crime problem or on ways to con-trol it. Which behaviors should be considered crimes? Which limitations (if any) should be imposed on law enforcement officers when they are enforcing the law? How severely should criminals convicted of violent offenses be published? Opinions differ, yet maintenance of order in society is clearly a primary function of government. The criminal justice system in a democratic nation is accountable to the public when members of that system maintain order and ensure due process.

The Study of Criminal Justice

The academic study of criminal justice can be traced back to the early 1900s, when students studied criminal sociology, criminal anthropology, and criminalistics as different disciplines. By the 1920s and 1930s, courses on criminology, juvenile delinquency, and penology (the study of corrections) were increasingly offered within sociology departments. During the same period, efforts to professionalize the police led to the creation of police science departments at many universities. Many of these programs, however, were little more than training programs intended to teach police specific skills, such as weapons firing and collection of evidence.[18]

As crime and disorder became increasingly of concern to the public and policy makers in the 1960s, more attention was focused on the study of crime and its control. In 1968, the Law Enforcement Assistance was established along with the Law Enforcement Education Program to provide funding

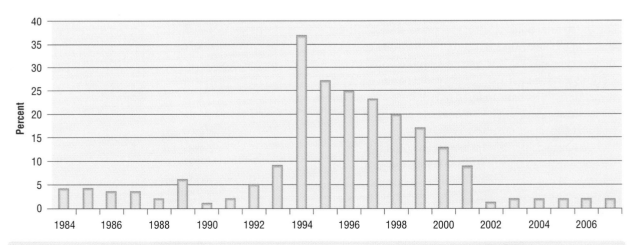

FIGURE 1-2 Percentage of Americans Who Believe Crime Is the Most Important Problem Facing the United States

Source: Ann Pastore and Kathleen Maguire, *Sourcebook of Criminal Justice Statistics, 2003,* 31st ed. (Washington, DC: U.S Department of Justice, 2006), available at http://www.albany.edu/sourcebook/pdf/t212006.pdf, Table 2.1, accessed May 23, 2007.

Governments in the United Kingdom and other European countries are cracking down with on-the-spot fines of $75 for people who spit their chewing gum onto the street. This "crime" costs taxpayers millions of dollars annually in clean-up expenses for sidewalks covered with wads of chewing gum. Ireland, in an effort to clean up the streets of Dublin, now fines gum litterbugs $160. In Great Britain, approximately 80 percent of all major streets have gum spots, and the government spends about $15 million dollars each year to remove wads of gum from its sidewalks.

Just about every major city in Europe has grappled with this sticky situation, and an entire industry has evolved around gum removal. The Dutch-developed GumBusters machine uses steam and a nontoxic solvent to remove gum at sites such as Barcelona's Las Ramblas pedestrian mall and Amsterdam's Schiphol airport. Launched in 2000, GumBusters International earns about $10 million annually from cleaning walkways covered with the sticky, black remains of abandoned wads of chewing gum in Europe, the United States, Australia, and Japan. The simplest and least expensive solution to the problem, of course, is for gum chewers to put their gum in a piece of paper and toss it into a trash bin.

Sources: Cesar Soriano, "Europe Tries to Eradicate Gum Crime," *USA Today*, July 25, 2006, p. 7A; Russell Leadbetter and Jonathan Rennie, "Is Bubble about to Burst for City's Careless Gum Users?" *Evening Times*, Glasgow, UK, January 26, 2007.

for in-service criminal justice personnel and students aspiring to careers in criminal justice.[19] Enrollment in criminal justice courses skyrocketed, and criminal justice departments soon split off from sociology and political science programs. With a greater focus on the interrelationships between the police, courts, and corrections, criminal justice emerged as a unique academic discipline.[20]

Both criminal justice and criminology established their own national professional organizations, and both disciplines now have well-defined graduate and undergraduate programs as well as academic journals and federally funded research efforts that support these fields. While the field of criminal justice continues to draw from a variety of disciplines, including sociology, psychology, public administration, economics, and law, it has clearly become a distinct subject of its own. Its study includes not only the criminal justice system itself, but also issues of criminal behavior, professional ethics, and the history and philosophy of crime and crime control.

The Structure of the Criminal Justice System

The United States does not have a single, monolithic criminal justice system that is centralized and controlled by the federal government. Instead, the U.S. criminal justice system is a loosely coupled combination of three major components—law enforcement, courts, and corrections—that operate across federal, state, and local levels `TABLE 1-1`.

- *Federal Level.* The federal criminal code defines federal crimes. Dozens of federal government law enforcement agencies enforce laws, such as the FBI and the Drug Enforcement Administration. The federal government also has a system of courts, including District Courts, Courts of Appeals, and the U.S. Supreme Court, as well as a system of corrections, including the Federal Bureau of Prisons and federal probation and parole agencies.

- *State Level.* Each state has a criminal code that defines state crimes and provides statutes setting punishments for offenders. Every state also has its own system of law enforcement, courts, and corrections for both adult and juvenile offenders.

- *Local Level.* Counties and cities have sheriff's departments and municipal police agencies, city lock-ups and county jails, community corrections programs, and city and county criminal courts, Justice of the Peace courts, and town courts.

Each agency has its own unique operating system, though the jurisdictions and activities of these organizations may overlap. For example, the Virginia Tech shootings were investigated by federal, state, local, and campus law enforcement agencies.[21] Because

TABLE 1-1

Criminal Justice System Agencies

	Federal	State	Local
Law enforcement	Federal Bureau of Investigation Drug Enforcement Administration Secret Service U.S. Marshals	State police Highway patrol State bureau of narcotics State fish and game agency	Municipal police County police Town constables
Courts	U.S. Supreme Court U.S. Courts of Appeal U.S. District Courts Federal Magistrates Courts	State supreme court State court of appeals	Criminal court City or town court Justice of the Peace court Traffic Juvenile
Corrections	Federal Bureau of Prisons Federal probation Federal parole	State department of corrections State parole	County jail City lock-up County probation Community corrections

of this overlap, "turf wars" often occur, such that the various components of the criminal justice agencies may not cooperate smoothly. Conflicts may arise from differences in opinion about how crime should be controlled and how justice should be achieved. Because one of the most important functions of law enforcement is to investigate crime and apprehend criminals, officers often see defense attorneys as working against them when they obtain acquittals for defendants. Similarly, police may feel that judges contribute to the crime problem when they hand out reduced sentences to chronic offenders. Sometimes, federal and local prosecutors may engage in battles when, after investigating criminal activity and preparing cases for prosecution, each party wants to control the investigation and receive recognition for any success.

Tensions also exist between the courts and corrections personnel. Although state prison administrators may believe they are best suited to establish institutional rules and policies, federal judges sometimes intervene on behalf of inmates, declaring single institutions or even entire correctional systems to be in violation of inmates' constitutional rights. Also, in an effort to get tough on crime, local judges may sentence increasing numbers of offenders to prison—only to aggravate the problem of prison overcrowding. Even within a prison, the demands and interests of correctional officers (e.g., security) may conflict with those of treatment staff.

More typically, the different parts of the criminal justice system work very well together. Prosecutors cooperate with police to investigate crimes, such as drug trafficking, and may assist of-

ficers when evaluating evidence to ensure that the evidence will later be admissible in court. Judges review and sign search and arrest warrants brought to them by police. At trial, law enforcement officers are frequently witnesses for the prosecution. When offenders are released from prison and on parole, they are supervised by parole officers, who work closely with law enforcement to monitor their activities.

Police

In the United States, nearly 18,000 local, state, and federal policing agencies collectively employ about 1 million people, nearly 90 percent of whom work for state and local agencies.[22] Most Americans are aware of their local police and sheriff's personnel because they see them almost daily, patrolling streets, managing crowds, and making traffic stops. The most important goals of police are to enforce the law and to maintain order. To achieve these goals, police officers actively engage in crime prevention, investigate reported crimes, participate in community-based programs such as the Police Athletic League, and arrest criminal suspects. Officers also respond to domestic disturbance calls, settle disputes, calm down loud parties, and remove drunks and transients from city streets.

State and local law enforcement officers also provide citizens with valuable services and regulate traffic. Over time, the service function performed by these personnel has changed dramatically. Early in the twentieth century, for instance, police provided shelter for the homeless. Today, however, the service activities of officers include opening locked car doors, searching for lost children, providing citizens with directions, assisting the elderly, and more. When officers are performing traffic duties, for example, you will often see them directing vehicles at concerts and sporting events, investigating accidents, enforcing speed limits and other traffic laws (such as running a red light and failure to wear a seat belt), and arresting motorists who are driving under the influence of alcohol or drugs.

LINK Police perform a variety of functions, including patrol, criminal investigation, public service, and traffic enforcement. All of these are critical to the police mission of protecting citizens. The link between these functions and order maintenance and crime control is discussed in Chapter 6.

There are 65 federal law enforcement agencies that employ more than 105,000 persons; these agencies include the U.S. Secret Service, the Capitol Police, U.S. Customs, the Mint Police, and the Bureau of Indian Affairs.[23] Unlike their state and local counterparts, federal officers do not perform service or traffic functions; they only investigate violations of federal law.

Courts

The United States has a dual system of courts, composed of parallel court systems at the federal and state levels. Every state (plus the District of Columbia and all U.S. territories such as Guam and the Virgin Islands) has its own court system. Each state court interprets and applies state laws, whereas the federal court applies federal laws. These systems operate largely independently. Occasionally, however, cases at the state level that involve constitutional issues are appealed to the federal courts. Nearly all decisions decided by the U.S. Supreme Court originated from cases originally prosecuted at the state level.

Both federal and state court systems are organized into three tiers: lower courts, intermediate appellate courts, and courts of last resort, also known as supreme courts. The lower courts are further divided into courts of limited jurisdiction and general trial courts, or courts of general jurisdiction. Courts of limited jurisdiction handle the majority of criminal cases, dealing with infractions of city ordinances (e.g., dog leash laws) and misdemeanors (e.g., shoplifting). More than 3000 general trial courts operate in the United States, plus 94 U.S. District Courts that hear felony cases.[24]

LINK U.S. courts are the core of the criminal justice process and are divided into a dual system, with local, state, and federal courts of different levels. Chapter 9 examines the history and structure of the U.S. court system.

The lower courts are the first to hear a case. The process begins with a preliminary hearing, which is an early hearing to review the charges against the defendant, set the defendant's bail, present witnesses, and determine probable cause. If it is determined that the suspect probably committed the crime, bail is set depending on the nature of the crime, and counsel is assigned to suspects who cannot afford an attorney. Guilty pleas are accepted from defendants who decide to forfeit their right to a trial. If the defendant pleads "not guilty," a trial is held to determine the guilt or innocence of the alleged offender. If an offender is found guilty, he or she is then sentenced by the court.

Intermediate appellate courts hear appeals of cases brought to them from the lower courts. They do not retry cases, but rather review transcripts from cases and hear testimony on issues concerning violations of legal procedure, such as the admission of illegally obtained evidence, which may form a basis for overturning or modifying a lower court's decision. In 2002, for instance, Andrea Yates was sentenced to life in prison for murdering her five children.[25] The jury rejected the insanity defense, concluding that Yates knew right from wrong at the time she killed her children. In 2005, the case was appealed to the Texas First Court

Andrea Yates killed her five children by drowning them in the family bathtub. At her first trial, she was convicted of first-degree murder; at her second trial, her conviction was overturned on appeal. Yates was found not guilty by reason of insanity.

Should Andrea Yates have been saved from a death sentence?

of Appeals, which reversed the conviction because an expert witness for the state, Dr. Park Dietz, had presented false testimony when he said that Yates might have been influenced by an episode of the *Law and Order* television program, though no such episode ever aired. Given that one or more jurors might have been influenced by this false testimony, a new trial was ordered. At this second trial, Yates was found not guilty by reason of insanity and was sentenced to a state-run, maximum-security mental hospital.

The U.S. Supreme Court has jurisdiction over all cases involving a federal question or constitutional issue. It reviews federal district court decisions as well as decisions appealed from state courts focusing on issues of federal law. The U.S. Supreme Court does not have jurisdiction over cases involving state law or violations of a state's constitution. In these instances, each state's own Supreme Court is the final arbiter of justice.

Corrections

Federal, state, and local correctional systems are responsible for the custody, punishment, and rehabilitation of convicted persons. In 2006, more than 2 million individuals were incarcerated in local, state, and federal corrections facilities (i.e., jails, halfway houses, correctional and detention centers, reformatories, and prisons) in the United States. In addition, more than 4 million offenders were on probation and an additional 1 million offenders were on parole.[26]

The Federal Bureau of Prisons operates more than 175 prisons, including detention centers, medical centers, prison camps, and penitentiaries. An additional 1300 state-run correctional facilities house persons convicted of state-level crimes. Nearly 400,000 people work in state correctional systems, and more than 35,000 individuals work for the federal correctional system.[27]

Both federal and state correctional systems classify inmates based on various factors, such as the seriousness of the offense committed, treatment needs, and perceived dangerousness. Once an inmate is classified, he or she is assigned to a suitable facility or program. Offenders who are convicted of felonies are confined in prisons; those who are convicted of misdemeanors are detained in local jails or minimum-security corrections facilities. Correctional institutions are categorized as super-maximum, maximum-, medium-, minimum-, or low-security facilities.

The primary functions of correctional institutions are to provide offenders with treatment, pun-

LINK Camp Delta in Guantanamo Bay in Cuba is a special prison where suspected terrorists are detained. It is operated by the U.S. military rather than the Federal Bureau of Prisons or the states. More information and the unique features of Camp Delta are discussed in Chapter 13.

ish them for their wrongdoings, and shield society from any harm they might otherwise cause. Sometimes these institutions offer counseling, job training, and education to aid in the rehabilitation of offenders. Community corrections, including probation and parole services, focus on reintegrating offenders into society through supervision and participation in counseling that works to resolve job, family, education, and drug- or alcohol-related problems. What pulls all of these separate parts of the system together is the criminal justice process.

The Criminal Justice Process

The structure of the criminal justice system is different from the criminal justice process.[28] The structure refers to the institutions, agencies, and personnel who enforce and apply the criminal law. The criminal justice process is a complex process that includes the stages discussed in this section.

Police

Initial Contact

For most people, their initial contact with the criminal justice system begins with the police. Usually, it entails an officer observing a crime in progress, a victim or a witness reporting a crime, or an ongoing investigation providing law enforcement officials with enough evidence to take action.

Criminal Investigation

Once the police determine that a crime has been committed, they will gather evidence and may identify a suspect. Occasionally a suspect is apprehended at the crime scene, but most often he or she is identified later, through information obtained from victims and witnesses, physical evidence (i.e., blood or hair samples, fingerprints, tire marks), or a prior criminal record.

Arrest and Booking

If police believe that a suspect committed a crime, they arrest him or her. Once an arrest is made, authorities will book the individual, recording the name of the person arrested, the place and time of the arrest, the reason for the arrest, and the name of the arresting authority. At booking, the suspects also are fingerprinted, photographed, and placed in holding cells, where they await further interrogation.

> **LINK** Arresting a criminal suspect is a complex process. When a suspect is arrested, the police officer must not violate the suspect's constitutional rights. If he or she does, either intentionally or by accident, the case may be dismissed and the suspect set free. Chapter 7 discusses the procedures that police are required to follow when making an arrest.

Courts

Charging

After making an arrest, police turn over the information they have gathered about the crime to the prosecutor, who decides which charges, if any, will be filed with the court. The prosecutor may decide either to dismiss the case, leading to the suspect's release, or to proceed with the case.

Initial Appearance, Preliminary Hearing, or Arraignment

If the case proceeds, the defendant next makes his or her initial appearance in court, where the charges are read, bail is set, and the defendant is informed of his or her rights. If the defendant is charged with a misdemeanor, he or she may enter a plea. If this plea is "guilty," the judge may impose a sentence immediately.

If the defendant is charged with a felony, he or she may choose not to enter a plea at the initial appearance. Instead, a judge may schedule a preliminary hearing to determine probable cause (i.e., to determine that there is sufficient evidence that a crime was committed and that the accused person likely committed it).

In cases where a defendant has been indicted by a grand jury and probable cause has been established through the grand jury investigation, the defendant's first appearance in court is at an arraignment, where the trial date is set.

Bail and Detention

Following the initial appearance, a defendant will typically post bail (a sum of money that the arrested person pays to guarantee that he or she will appear at future hearings). As an alternative to posting bail, most jurisdictions allow "good risk" defendants to make a personal promise to appear in court, called release on recognizance (ROR).

Defendants who cannot post bail or who do not qualify for ROR will be transferred to the city or county jail, where they are likely to remain until their arraignment date. Some defendants are not eligible for bail because they are viewed by the court as posing a serious threat to the community (including victims or witnesses who may testify against the defendant) or because they are likely to run away. These defendants are held in preventive detention.

Plea Bargaining

Very few cases actually go to trial. Instead, most are resolved through the process of plea bargaining—a negotiation between the prosecutor and the defense attorney to arrive at a mutually satisfactory disposition of the case without going to trial.

Trial

Defendants who choose to go to trial are guaranteed the right to a trial by jury, although they may request a bench trial, in which the judge alone determines their guilt or innocence. In either situation, the trial concludes with a verdict of not guilty (acquittal), guilty (conviction), or undecided (hung jury). When a trial ends in a hung jury, the prosecutor may refile the charges and prosecute the defendant again.

Sentencing and Appeals

Following a guilty plea or a verdict, a sentencing hearing is set. The judge decides the appropriate sentence by considering both characteristics of the offense and characteristics of the offender that might increase or decrease the severity of the sentence (known as aggravating or mitigating factors) as well as other relevant materials. Often, the sentence the judge imposes is mandated by law.

If defendants believe they were unfairly tried, convicted, or sentenced, they may appeal their verdicts or sentences to an appellate court. This court reviews the lower court's transcripts solely for procedural errors, such as admission of illegally obtained evidence. If an appellate court determines significant errors were made at trial, it may overturn the conviction and order a new trial.

> **LINK** Judges often examine presentence investigation reports prepared by the probation office, which include information about the crime, the offender's background, and the offender's prior criminal record. They also consider victim impact statements—oral statements by the family of a victim that explain the nature and extent of the crime's impact on the victim and/or his or her family and friends. These and other relevant considerations are examined further in Chapter 12.

In 2006, John Mark Karr voluntarily confessed to police that he had killed, drugged, and had sex with 10-year-old JonBenet Ramsey. Karr was arrested and charged with criminal offenses related to the murder. Before Karr's first scheduled appearance in a courtroom, however, Boulder (Colorado) District Attorney Mary Lacy dropped the charges against Karr after DNA tests failed to tie him to the crime, despite his own statements of involvement. This case demonstrated that when a prosecutor thinks there is insufficient evidence to proceed with a prosecution, he or she will dismiss the charges. In Karr's case, Lacy concluded that she would not be able to establish that Karr had committed the crime, even though he repeatedly insisted that he did.

Source: Tom Kenworthy, "Ramsey Suspect's DNA Not a Match," *USA Today,* August 29, 2006, pp. 1A, 3A.

Corrections

Probation

A convicted offender may be placed on probation, a sentencing option typically involving a suspended prison sentence and supervision in the community. The conditions of probation might include paying a fine, participating in psychological counseling, taking part in a drug or alcohol treatment program, obtaining a job, or regularly reporting to a probation officer. If the offender violates any of these conditions, the court may revoke his or her probation and return the probationer to prison.

Incarceration and Rehabilitation

If the court decides that probation is not an appropriate sentence, the offender may be incarcerated in jail (for misdemeanor convictions) or placed in prison (for longer-term imprisonment).

Release and Parole

Few offenders serve their full prison sentence. Some are released early as a result of earning "good time reductions" in their sentences; others receive parole, a type of conditional release. If released early, the offender is supervised in the community by a parole officer and must follow a set of clearly articulated conditions. If any of the conditions are violated, the offender may be returned to prison.

LINK Not all prisons provide inmates with identical programs, as noted in Chapter 14. The greatest disparities are found in the differences in programming in men's and women's correctional institutions.

Perceptions of the Criminal Justice System

A major goal of the criminal justice system is to achieve justice. To do so, the system must balance both the rights of the victim and the rights of the accused. One way of understanding justice is to recognize that it entails the fair and impartial treatment of all persons when applying the law. In reality, of course, people are not treated equally by the criminal justice system.

When considering the meaning of justice, several questions immediately arise with respect to the behavior of the criminal justice system and its agents:

- When should police officers use force? How and what type of force should they use?
- When should convicted offenders be eligible for probation and returned to the community?
- Should persons who are convicted of comparable crimes receive similar sentences?
- How should the courts use their limited resources to balance the needs of victims, their families, and the public against the rights of the accused and convicted?

If citizens believe that law enforcement officers abuse their power, that courts are insensitive, and that prisons discriminate against racial and ethnic minorities, then they will be less likely to cooperate and support criminal justice officials. Conversely, if the public thinks that the criminal justice system puts the rights of criminals above the rights of victims, then the system is once again jeopardized. Developing a positive public perception of the criminal justice system is necessary for it to function effectively and efficiently across all three components of law enforcement, courts, and corrections.

Police

Since the 1960s, public opinion polls have generally reported that the majority of the U.S. public has a favorable view of the police.[29] Police today are better trained and educated, more conscientious, and less corrupt than they once were. Although there are also more police today in the United States than at any time in history, law enforcement personnel continue to depend on public support and participation to ensure their effectiveness. The public assists police in many ways, especially by providing them with information to help them solve crimes and make arrests.

> **LINK** Despite dramatic improvements in police ethics, police corruption continues to persist in the twenty-first century. It is a problem that has haunted police operations in the United States for more than a century and is discussed in greater detail in Chapter 8.

Crime-Reporting Habits

Citizens usually report crime only if they believe that
- The offense is serious enough to warrant official intervention.
- Police will respond to the call.
- Police will provide a useful service at the crime scene.

Unfortunately, citizens sometimes do question police effectiveness, and crime victims often do not report criminal victimizations. Only about 47 percent of serious violent crimes and 40 percent of serious crimes are reported to police. The following reasons have been cited to explain this low rate:
- A fear of retaliation
- A decision to handle the matter informally
- A belief that the crime is minor
- A feeling that police will not be able to do anything
- A desire not to bother police[30]

> **LINK** It is difficult to measure the total amount of crime. Chapter 3 discusses a variety of ways criminologists have developed to provide a credible gauge of crime in the United States.

Few cases illustrated racial differences in criminal justice opinion more clearly than the case of O. J. Simpson. Most African Americans believed that Simpson was not guilty of murder, but most whites believed that he was.

? What explains this difference in perception?

Perceptions Based on Race

Researchers have discovered that African Americans and whites have different attitudes about the police. Whites are generally more satisfied with police than African Americans, who over the years have become increasingly mistrustful of the police, are fearful of becoming victims of excessive force, and believe that officers are prejudiced. This finding has been documented for more than 40 years in dozens of studies and polls and has been publicized by media across the United States. For example, studies evaluating public satisfaction with police in 12 cities found that 90 percent of whites and 76 percent of African Americans were generally satisfied with their experience with police. This disparity may have a spiraling effect on police–citizen relations. African Americans who feel more alienated and hostile toward police may, in turn, be less likely to cooperate with them; as a result, police may develop negative perceptions of African Americans.[31] Yet, curiously, African Americans are more likely than whites to report to police both serious violent and property crimes **FIGURE 1-3** .[32]

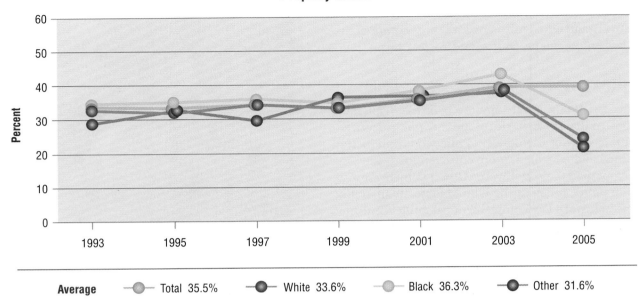

Property Crime

Average   Total 35.5%   White 33.6%   Black 36.3%   Other 31.6%

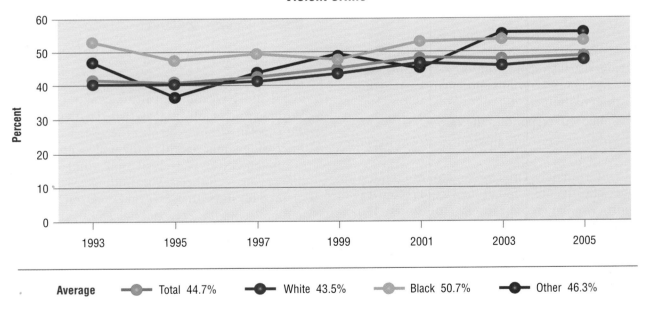

Violent Crime

Average   Total 44.7%   White 43.5%   Black 50.7%   Other 46.3%

FIGURE 1-3 Percentage of Serious Crime Reported to Police by Race of Victim

Source: Shannon Catalano, *Criminal Victimization, 2005* (Washington, DC: U.S. Department of Justice, 2006).

Courts

Most Americans have a favorable view of the U.S. court system: In 2006, 80 percent of all Americans reported having confidence in the courts and the criminal justice system.[33] This confidence is critical to the effective operation of the courts, not only because judicial officials depend on public support for reelection, but also because public participation is necessary for forming citizen juries and providing witness testimony.

Because most judges and prosecutors are elected officials, they need to maintain the public's confidence to win reelection. In addition, jury duty requires people to take time away from their jobs and family, disrupts their normal routines, and offers little compensation. Citizens who are selected for jury duty are paid a small daily or hourly fee, although the amount of money they receive is less than half the federal 2007 minimum wage of $5.85. (The Federal minimum wage will increase $0.70 each summer until 2009, when all minimum wage jobs will pay no less than $7.25 per hour.)[34] If people see jury duty as more of an inconvenience than a civic responsibility, they are more likely to try to find ways to avoid serving.

With the exception of expert witnesses (whose expenses may be paid by the criminal defendant or the court), witnesses receive no remuneration for their time or testimony. Some witnesses testify in court because they believe they can make a contribution; others testify only because they have been ordered to appear by a subpoena. For many people, being questioned on the witness stand produces anxiety and is an unpleasant experience. Like jurors, witnesses must arrange to appear at trial, and they may be frustrated and angry if trials are canceled or rescheduled. Negative experiences in the court may lead to a reduced willingness to cooperate in the future.

Corrections

Prisons and jails need public support for funding to build, staff, and operate corrections facilities. Finding an acceptable location for a new prison is not always easy, however. Many people object to prisons being built in or near their communities, owing to fears that inmates may escape or commit new crimes after release. Although most Americans favor building new prisons, reformatories, halfway houses, and detention centers, few want them near where they live. There are some exceptions: Some communities have lobbied their legislatures to build new facilities in their towns so as to provide jobs and bolster the local economy.[35]

Criminal Justice in Popular Culture

Popular culture affects public perceptions of the criminal justice system as well as personal beliefs about acceptable or appropriate thoughts and feelings about groups, interpersonal relationships, and behaviors.[36] Through the repetition of certain themes and content, popular culture may distort viewers' beliefs about the real world or confirm previously held beliefs.

Television often provides exaggerated images of crime and justice. One report found that violence was portrayed on more than 80 percent of television programs. Murder alone accounted for nearly one-fourth of all television crimes; in reality, murder accounts for only 1 percent of all violent crime and only 0.2 percent of all serious crimes reported to the police.[37] Furthermore, viewers rarely face the sort of violence that occurs on television. Likewise, the high rates of arrest and conviction of criminals on television are fictional.

Television news and documentaries may affect public perceptions of crime and justice as well. These programs emphasize dramatic crimes, such as murder, forcible rape, and gang violence. Additionally, crime has become a staple of the nightly news, and even mundane crimes committed by public figures (such as shoplifting) are widely publicized by the media.

Americans are fascinated by extreme offenders such as serial killer Jeffrey Dahmer.

What does this fascination suggest about America's appreciation of both deviance and justice?

Criminal Justice and Popular Culture

Film, the Internet, music, television, and video games are powerful forces in shaping the values of contemporary American culture. The first films to focus on crime and criminal justice were released in the 1930s, when movies filled with gangsters and crooks became popular. These films followed a pattern established in radio programming that presented the listening public with an underlying moral lesson: Villains do not win and heroes never lose. They also stressed the importance of good over evil, truth over lies, and civilization over anarchy. Many films in the 1960s and 1970s, such as *Bonnie and Clyde* and *The Godfather,* also emphasized the theme of good versus evil, despite their glorification of outlaws and mobsters.

A second theme of crime films, which emerged after the turbulent 1960s, reflected a conservative public backlash against crime and what was perceived to be an ineffective criminal justice system. In the 1970s, a number of vigilante movies were released, including *Dirty Harry, Sudden Impact,* and the *Death Wish* series. These films blamed crime on liberal permissiveness and hypocrisy, and showed that the crime problem could be solved only by resorting to violence.

More recent crime films have focused on injustices within the criminal justice system, illustrating problems caused by racial and ethnic tensions, gang violence, and illegal drugs. *Boyz 'N the Hood, American History X,* and *Traffic,* for example, presented rather bleak images of society's ability to solve pervasive social problems.

On television, police have dominated images of crime and its control for several decades, starting with programs in the 1950s such as *Dragnet* and *Highway Patrol* and more recently with shows such as *Law and Order, The Shield,* and *CSI.* Television also presents a mythology about crime: The offender's guilt is clear, and police are portrayed as keepers of the peace and protectors of the public. Of more significance is the fact that television shows the law being applied equally to all citizens—black and white, female and male, poor or wealthy—illustrating the notion that no citizen is above the law.

Criminal courts have also been portrayed on television through both older and more contemporary shows (such as *Perry Mason, Law and Order, The Practice, Ally McBeal,* and *Boston Legal*). These programs provide a brief but distorted portrayal of defense attorneys, prosecutors, and judges. Unlike in real life, nearly all the criminal cases presented on television go to trial and few are plea bargained.

Viewers have the opportunity to see the "realities" of the criminal justice system through "infotainment," or programs that blend news with entertainment. Programs such as *Cops, Court TV,* and *24* show real or recreated events in which police officers arrest and interact with victims and witnesses. The Courtroom Television Network, a 24-hour cable channel, also presents live coverage of criminal trials. It was launched in 1991 to inform viewers about the inside workings of the criminal justice system.

Music has also been one of the most pervasive and accessible elements of popular culture. Folk music and the blues have always contained images of outlaws, gunfights, and prison, whereas country music has usually emphasized people languishing in prison, the hangovers and pains from drinking, fighting, lost love, and occasional challenges to unjust intrusions of authority. Some contemporary rock and rap songs also contain images of crime, violence, and the injustice of the criminal justice system. Songs such as Bob Marley's "I Shot the Sheriff" and Lenny Kravitz's "Bank Robber Man," and those reflecting acceptance of drug use, such as Eminem's "My Fault" and J. J. Cale's "Cocaine" present challenges to conventional morality and law.

While critics complain about the glorification of crime and extensive violence in film, music, television, and video games, the target of many individuals is the content of "gangsta rap" music that discusses forcible rape, murder, violent robbery, and the sexual exploitation of women. Defenders of gangsta rap argue that these songs and their messages reflect the economic and social frustrations of the urban underclass and provide a window into urban culture. At the same time, artists such as Eminem, Tupac, 50 Cent, Bone Thugs-N-Harmony, and Biggie have been criticized for degrading women and glorifying drugs and violence. Critics complain that these artists encourage adolescents to take unnecessary risks, consume drugs, and commit violent crime, thereby putting unnecessary strain on law enforcement agencies that are already operating with sparse and meager resources.

Sources: Ray Surette, *Media, Crime, and Criminal Justice,* 3rd ed. (Belmont, CA: Wadsworth, 2006); Harry Benshoff and Sean Griffin, *America on Film* (Malden, MA: Blackwell, 2004).

Public fear of crime and the perceived likelihood of victimization are clearly related to television viewing habits: The more television people watch, the more likely they are to be afraid of becoming a crime victim.[38] In addition, the more intensely news programs focus on local crime, the more likely residents are to be fearful. Other research has shown that watching television for many hours increases public acceptance of police violence.[39] Some studies point to the possibility that frequent viewing of violence on television may actually cause crime by increasing the aggressive behavior of viewers. The more violence to which a person is exposed, the more likely he or she is to perceive aggressive behavior as acceptable. However, exposure to violent content may not directly cause violent behavior; instead, it may be that people who are predisposed to act violently are more likely to be stimulated to commit violence following exposure to violent television.[40] Alternatively, perhaps exposure to violence leads to the individual becoming desensitized to violence and, therefore, more inclined to commit aggressive acts.[41]

WRAP UP

The tragic events at Virginia Tech brought to light the poor communication and coordination between the mental health and criminal justice systems. For privacy reasons, information about Seung-Hui Cho's mental health status was not linked to criminal justice databases, so law enforcement had little reason to investigate his bizarre behaviors prior to the shootings. Just as the 9/11 Commission found that poor communication between intelligence and law enforcement agencies contributed to the ability of Al Qaeda members to plot and execute the September 11, 2001, attacks, so a barrier today exists between the privacy expectations of mentally ill individuals and public safety. It is likely that the national debate that arose after these shootings will lead to policy changes in this area to help better protect the U.S. public.

Chapter Spotlight

- All societies experience crime. To combat crime, societies create criminal justice systems that define the rules that govern social interactions, enforce the standards of conduct necessary to protect their citizens, and determine the punishments for violating those rules so as to establish social order.

- The significant increase in serious crime through the early 1990s and the recent international terrorist attacks have caused the U.S. federal government to define crime—and specifically terrorism—as a major problem. In response, Congress has passed a number of pieces of legislation aimed at reducing crime levels.

- Democratic and open societies deal with crime differently than authoritarian and closed societies. In a democratic society, the criminal justice system is accountable to the public and operates within a legal framework established by legislators who are elected by the people.

- The criminal justice system is an interrelated set of subsystems that operate horizontally and vertically across federal, state, and local levels. The laws and jurisdictions of each subsystem sometimes overlap, which sometimes leads to conflicts among the various agencies.

- The criminal justice process is composed of well-defined legal rules that specify the procedures through which suspects enter and are treated by the criminal justice system.

- Perceptions of the seriousness of the crime problem and the best ways to solve it are influenced by popular culture, and particularly by violent television programs that may increase the aggressive behavior of their viewers.

Putting It All Together

1. Is the amount of crime in society normal, or have we simply come to accept the level of crime that we have as being normal?

2. What are the advantages and disadvantages of a democratic and open system of criminal justice as compared to systems of criminal justice in countries with authoritarian governments, such as Burma, China, Iran, North Korea, and the Sudan?

3. Why are perceptions of crime and justice so important to our thinking about crime and to the daily operation of the criminal justice system?

4. The crime rate has decreased significantly in the past decade, and only a very small percentage of the public now views crime as the most serious problem in the United States. Why do you believe the crime rate has dropped? Do you feel safer today than you did in the immediate past?

5. Violent television programs may increase the aggression of some viewers. Should television programs be censored for violence? Is it important for the government to regulate what viewers may watch on television? Do television networks have a First Amendment right (freedom of speech) to present whatever material they choose?

criminal justice process The procedures that occur in the criminal justice system, from a citizen's initial contact with police to his or her potential arrest, charging, booking, prosecution, conviction, sentencing, and incarceration or placement on probation.

criminal justice system A complex set of interrelated subsystems composed of three major components—police, courts, and corrections—that operate at the federal, state, and local levels.

norms Rules and expectations by which a society guides the behavior of its members.

Notes

1. Solomon Herbert, "King Litigation May Spur Healing or Unrest: Rodney King's Civil Lawsuit against the Los Angeles Police Department," *Black Enterprise,* September, 1992, available at http://www.findarticles.com/p/articles/mi_m1365/is_n2_v23/ai_12539601, accessed June 8, 2007.
2. Rodney King, "Quotation of the Day," *New York Times,* May 2, 1992, p. 6A.
3. Émile Durkheim, *The Rules of Sociological Method,* translated by Sarah Solovay and John Mueller (New York: Free Press, 1893/1938), p. 68.
4. David Wallechinsky, "*Parade*'s Annual List of the World's 10 Worst Dictators," *Parade Magazine,* February 11, 2007, p. 7.
5. Émile Durkheim, *The Division of Labor and Society* (New York: Free Press, 1893/1964); Durkheim, note 3.
6. Kai Erikson, *Wayward Puritans* (New York: John Wiley & Sons, 1966).
7. Erich Goode, *Drugs in American Society,* 7th ed. (New York: McGraw-Hill, 2007); Charles Faupel, Alan Horowitz, and Greg Weaver, *The Sociology of American Drug Use* (New York: McGraw, 2003).
8. George Gallup, Jr., *The Gallup Poll* (Princeton, NJ: Gallup Poll, September 1989).
9. Darrell Gilliard and Allen Beck, *Prisoners in 1993* (Washington, DC: U.S. Department of Justice, 1994), p. 1.
10. George Gallup, Jr., *The Gallup Poll* (Princeton, NJ: Gallup Poll, September 1993); Gary Burden, "U.S. Drug Policy May Be in for a New Look," *Law Enforcement News* 15:5 (1993).
11. *The Violent Crime Control and Law Enforcement Act of 1994* (HR 3355) (Washington, DC: U.S. Government Printing Office, 1994).
12. Kristen Hughes, *Justice Expenditures and Employment in the United States, 2003* (Washington, DC: U.S. Department of Justice, 2006).
13. Federal Bureau of Investigation, *Crime in the United States, 2006* (Washington, DC: U.S. Government Printing Office, 2007); U.S. Congress, HR 1572; *The Omnibus Crime Control and Safe Streets Act* (P.L. 90-351) (Washington, DC: U.S. Printing Office, 1968); The Disaster Center, available at http://www.disastercenter.com/crime/ederal, accessed May 4, 2007; *The Comprehensive Crime Control Bill of 1984* (P.L. 98-473) (Washington, DC: U.S. Government Printing Office, 1984); *The Crime Control Act of 1990* (P.L. 101-647) (Washington, DC: U.S. Government Printing Office, 1990); *The Violent Crime Control and Law Enforcement Act of 1994* (HR 3355), note 11; *USA Patriot Act of 2001* (HR 3162 RDS) (Washington, DC: U.S. Government Printing Office, 2001).

14. *The Crime Control Act of 1990,* note 13; *The Violent Crime Control and Law Enforcement Act of 1994,* note 11; *USA Patriot Act of 2001* (HR 3162 RDS) (Washington, DC: U.S. Government Printing Office, 2001); *USA Patriot Improvement Act of 2005* (HR 199) (Washington, DC: U.S. Department of Justice, 2006).

15. Steven Levitt and Stephen Dubner, *Freakonomics,* revised ed. (New York: HarperCollins, 2006).

16. *The Omnibus Crime Control and Safe Streets Act,* note 13; *The Comprehensive Crime Control Bill of 1984,* note 13; *The Crime Control Act of 1990,* note 13.

17. Ann Pastore and Kathleen Maguire, *Sourcebook of Criminal Justice Statistics, 2003,* 31st ed. (Washington, DC: U.S Department of Justice, 2006), available at http://www. albany.edu/sourcebook/pdf/t212006.pdf, accessed July 14, 2007; Brian Reaves, *Federal Law Enforcement Agencies, 2004* (Washington, DC: U.S. Department of Justice, 2006).

18. James Finckenauer, "The Quest for Quality in Criminal Justice Education," *Justice Quarterly* 22:415 (2005).

19. President's Commission on Law Enforcement and Administration of Justice, *The Challenge of Crime in a Free Society* (Washington, DC: U.S. Government Printing Office, 1967).

20. Elizabeth Monk-Tuner, Ruth Triplett, and Green Kim, "Criminology/Criminal Justice Representation in the Discipline of Sociology: Changes between 1992 and 2002," *Journal of Criminal Justice Education* 17:323–335 (2006).

21. David Maraniss, "That Was the Desk I Chose to Die Under," *Washington Post,* April 19, 2007, p. A01, available at http://www.washingtonpost.com/wp-dyn/content/article/ 2007/04/18/AR2007041802824_pf.html, accessed June 28, 2007; Federal Bureau of Investigation, "School Shooting: Role of FBI at Virginia Tech," available at http://www.fbi. gov/page2/april07/shootings041607.htm, accessed June 8, 2007.

22. Pastore and Maguire, note 17.

23. U.S. Courts, The Federal Judiciary, availalable at http:// www.uscourts.gov/, accessed May 23, 2007.

24. U.S. Department of Justice, Bureau of Justice Statistics, Law Enforcement Statistics, available at http://www.ojp.usdoj.gov/ bjs/lawenf.htm, accessed May 23, 2007.

25. Laura Parker, "Yates Rejects State's Plea Offer," *USA Today,* February, 28, 2006, p. 3A.

26. U.S. Department of Justice, Bureau of Justice Statistics, "The Number of Adults in the Correctional Population Has Been Increasing," available at http://www.ojp.usdoj.gov/bjs/glance/ corr2.htm, accessed July 23, 2007.

27. Federal Bureau of Prisons, available at http://www.bop.gov/, accessed June 20, 2007.

28. The discussion is derived from Yale Kasimar, Wayne Lafave, Jerold Isreal, and Nancy King, *Modern Criminal Procedure,* 11th ed. (Eagen, MN: Thomson/West, 2005).

29. Catherine Gallagher, Edward Maguire, Stephen Mastrofski, and Michael Reisig, *The Public Image of the Police* (Alexandria, VA: International Association of Chiefs of Police, 2001).

30. Shannan Catalano, *Criminal Victimizations, 2005* (Washington, DC: U.S. Department of Justice, 2006).

31. Jake Horowitz, *Making Every Encounter Count* (Washington, DC: National Institute of Justice, 2007); Steven Smith, Greg Steadman, Todd Mintor, and Meg Townsend, *Criminal Victimization and Perceptions of Community Safety in 12 Cities, 1998* (Washington, DC: Bureau of Justice Statistics, 1999).

32. Catalano, note 30.

33. U.S. Courts, "Jury Pay," available at http://www.uscourts.gov/newsroom/confidence.html, accessed June 4, 2007.

34. "Americans Have Confidence in Courts, Justice System," available at http://www.uscourts.gov/jury/jurypay.html, accessed May 3, 2007; "Federal Minimum Wage Increase for 2007," available at http://www.laborlawcenter.com/federal-minimum-wage.asp, accessed May 3, 2007; Jesse J. Holland, "Federal Minimum Wage to Rise by 70 Cents," Associated Press, July 23, 2007, available at http://www.forbes.com/feeds/ap/2007/07/23/ap3942117.html, accessed July 24, 2007.

35. Sharon Dunn, "Progress on Ault Prison Is Dragging," *Greeley Tribune,* September 10, 2006, available at http://www.greeleytrib.com/article/20060910/NEWS/109100090/-1/rss02, accessed May 27, 2007; Kevin Dayton, "Prison Keeps Impoverished Town Alive," *Honolulu Advertiser,* October 3, 2005, available at http://the.honoluluadvertiser.com/article/2005/Oct/03/ln/FP510030313.html, accessed May 19, 2007; Ryan King, Marc Mauer, and Tracy Huling, Big Prisons, Small Towns (Washington, DC: Sentencing Project, 2003), available at http://www.soros.org/initiatives/justice/articlespublications/publications/bigprisons20030201/bigprisons.pdf, accessed June 3, 2007.

36. Ray Surette, *Media, Crime, and Criminal Justice,* 3rd ed. (Belmont, CA: Wadsworth, 2006); Danlo Yanich, "Kids, Crime, and Local Television," *Crime & Delinquency* 51:103–122 (2005); Kathryn Greene and Marina Krcmar, "Predicting Exposure to and Liking of Media Violence," *Communication Studies* 56:71–93 (2005); Jeffrey Johnson, Patricia Cohen, Elizabeth Smailes, Stephanie Kasen, and Judith Brook, "Television Viewing and Aggressive Behavior During Adolescence and Adulthood," *Science* 295:2468–2471 (2002); Daniel Anderson, Aletha Huston, Kelly Schmitt, Deborah Linebarger, and John Wright, *Viewing and Adolescent Behavior* (Malden, MA: Blackwell, 2001); Franklin Gilliam, Jr., and Shanto Iyengar, "Prime Suspects: The Influence of Local Television News on the Viewing Public," *American Journal of Political Science* 44:560–573, 2000.

37. Federal Bureau of Investigation, note 13.

38. Barry Glassner, *The Culture of Fear* (New York: Basic Books, 2000); Sarah Eschholz, Ted Chiricos, and Marc Gertz, "Television and the Fear of Crime," *Social Problems* 50:395–415 (1993).

39. Kenneth Dowler, "Media Consumption and Public Attitudes toward Crime and Justice," *Journal of Criminal Justice and Popular Culture* 10:109–126 (2003).

40. Peter Klopher, Shamett Bakshi, Richard Hockey, Jeffrey Johnson, Patricia Cohen, Elizabeth Smailes, Stephanie Kasen, and Judith Brook, "Kids, TV Viewing, and Aggressive Behavior," *Science* 297:49–50 (2002); Craig Anderson and Brad Bushman, "Psychology: The Effects of Media Violence on Society," *Science* 295:2377–2379 (2002).

41. Craig Anderson, Douglas Gentile, and Katherine Buckley, *Violent Video Game Effects on Children and Adolescents* (New York: Oxford, 2007).

JBPUB.COM/**ExploringCJ**

Interactives

Key Term Explorer

Web Links

OBJECTIVES

■ Understand the origins and development of criminal law.

■ Grasp the relationship between civil and criminal law, and describe how the law distinguishes different levels of seriousness.

■ Outline the functions of criminal law, both within the criminal justice system and within society at large.

■ Grasp the essential elements of a crime, including *actus reus* and *mens rea*.

■ Know the meaning and uses of the various justifications, excuses, and exemptions that may bar legal liability.

■ Highlight the Constitutional amendments that deal with due process, the rights of the accused, and the applicability of these principles.

Criminal Law

CHAPTER

2

On December 22, 1984, on a New York subway, Bernard Goetz shot and injured four unarmed teenage boys who he thought were trying to rob him. The case caused a national firestorm. Goetz, nicknamed the "Subway Vigilante," was viewed as a folk hero for standing up to criminals in a city and an era plagued with crime. Despite the racially charged undertones of the case (Goetz was white and his would-be attackers were African Americans), Goetz received support from vigilante and civil rights groups and was even encouraged by media commentators to run for mayor.

Despite his popularity, Goetz was indicted for attempted murder, assault, and weapons charges. He openly and unapologetically confessed to the shooting, claiming that his actions were taken in self-defense. The young men later admitted that they had intended to rob Goetz, who was convicted of felony possession of an illegal weapon and served eight months of a one-year sentence. Eleven years later, Goetz was sued by one of the teens who was injured and paralyzed in the shooting, and Goetz was found liable for $43 million in damages. Two of the youths were later imprisoned for other violent crimes including robbery, assault, and rape.

- Why did the jurors acquit Goetz of attempted murder and assault charges despite his confession?
- Why should Goetz have to pay damages to the youth who had threatened and intended to rob him?

Sources: People v. Goetz, 73 N.Y. 2d 751 (1988).

Introduction

Throughout history, the creation and evolution of law have been instrumental in promoting and regulating social behavior. Aristotle, for example, believed that law is the essence of social order: Good social order can be built only on good law; bad law can also produce social order, but such order may not be desirable.[1] Law, however, is not inherently good or bad, nor has it always accomplished its goals. Law is good to the extent that it is used or adhered to lawfully. If those individuals who are responsible for administering law fail to operate according to the accepted rules, law may become oppressive and a tool of manipulation.

Laws are formalized rules that prescribe or limit actions. Criminal law is one category of law, which consists of substantive criminal law and procedural criminal law. Substantive criminal law identifies behaviors considered harmful to society, labels those behaviors as crimes, and specifies their punishments. Procedural criminal law specifies how crimes are to be investigated and prosecuted. Together, substantive criminal law and procedural criminal law form the foundation of the U.S. system of criminal justice.

Exploring criminal law helps us to answer the following questions:

- Which behaviors are crimes?
- Under what circumstances do those behaviors constitute crimes?
- Which punishments may be imposed on people who commit crimes?
- How should those punishments be administered?
- Which rules govern the police in the investigation, arrest, and processing of people accused of crimes?
- Which rules control the behavior of the state in criminal prosecutions?

The Origins of Criminal Law

Each year, the U.S. Congress and state legislatures create new laws and, sometimes, abolish old ones. In the process, they identify additional behaviors as crimes, establish appropriate punishments, and clarify rules of criminal procedure.

Most members of the public believe that laws are created through debate and compromise. The common perception is that various interest groups, struggling to identify wrongs and ways to deal with them, influence legislatures, which then hammer out new laws in response to their pleas. In reality, this is not usually the case. The creation of criminal law today, for the most part, consists of the expansion or revision of existing laws, often reflecting changes in society. For example, the introduction of new technology, such as computers, DVDs, and MP3 players, created opportunities for such acts as tampering with computer files, pirating movies, and downloading copyrighted music. Because existing laws did not specifically prohibit these acts, new laws were created to deal with them. The moral, philosophical, and legal foundations of contemporary law, and hence the bases for contemporary legal debate, have evolved over many centuries.

Early Codification of Law

The Code of Hammurabi, which dates back to the eighteenth century B.C., is one of the earliest legal codes in Western culture. The laws contained in this code were attributed to divine guidance by the gods and made no distinction between secular and religious interests in controlling behavior. In addition, the notion that law should be fair in both substance and process was stressed, a concern clearly found in our own Bill of Rights. Of course, what the Babylonians considered fair may not be viewed as fair today; for example, the Code of Hammurabi considered the social position of the victim in determining the appropriate punishment for a crime.[2]

Mosaic law, or the Law of Moses (about 1250 B.C.), was also considered divine in origin and reflected both an interest in governing the religious life of the people and a desire to regulate social behavior (i.e., admonitions to not kill, steal, commit adultery, bear false witness, or work on the Sabbath). Present-day law in the United States retains many of the same regulations. Indeed, until recently, most states had so-called "blue laws," which prohibited businesses from operating on Sunday, the Christian Sabbath.

Roman law was secular. It began with the Law of the Twelve Tables (about 450 B.C.), which codified the duties, rights, and expectations of citizens. After the fall of Rome in the fifth century, the Justinian Code—a comprehensive body of civil law to regulate behavior—was created. The Justinian Code defined civil and criminal wrongs and established the first legal defense of insanity.

The Romans, like the Greeks before them, also made a distinction between positive law and natural law. Positive law consisted of those legal codes governing citizens and foreigners, and formalized by the state and based on reason. Natural law, which was viewed as being created by a

Criminal law is derived from ancient texts that were heavily influenced by religion.

? Why does the United States struggle with the issue of separation of church and state?

higher power or deity, reflected those binding rules and principles that guide behavior. Roman law allowed positive law and natural law to coexist and viewed neither as ultimately superior.

Although many people view natural law as ultimately superior to positive law, criminologist Hermann Mannheim suggests that natural law has been open to a variety of interpretations:

> While its underlying idea is the longing of mankind for an absolute yardstick to measure the goodness or badness of human actions and the law of the State and to define their relations to religion and morality, the final lesson is that no such yardstick can be found.[3]

Common Law and the Concept of *Stare Decisis*

Although the early legal codes laid a foundation for formalizing principles and customs into law, it was the emergence of English common law that held the greatest significance for the development of criminal law in the United States. During the reign of Henry II (1154–1189), who attempted to impose Norman values on the conquered Anglo-Saxons, a centralized system of courts was established, with judges being appointed by the king to represent the interests of the Crown. These judges traveled to the countryside, where they encountered diverse regional customs that often conflicted both with each other and with the king's new laws. The tradition of <u>common law</u> allowed these judges to determine which behaviors constituted crimes and what appropriate punishment should be imposed when they were violated, thus establishing a body of law common to the entire nation.

One of the most important concepts operating in common law was the doctrine of precedent, or <u>*stare decisis*</u> (literally, "to stand by the decisions"). This doctrine allows courts to interpret and apply law based on previous court decisions. According to *stare decisis*, judges were required to decide new cases in a manner consistent with principles established in prior cases. To the extent that a new case was substantially similar to a previous one, the judge was required to interpret the law in the same way and follow the precedent. Judges were not supposed to create laws, but they could study past legal decisions, discover the principles embodied in them, and apply those principles to new situations.[4]

Common law, with its reliance on precedent, continued to evolve with little centralized planning or deliberation about what the law should contain. Consequently, the common law of England existed as an unsystematic compilation and recording of thousands of cases over the years.[5] Although the signing of the Magna Carta by King John in 1215 established the first set of statutory laws (formal written enactments of a governing or legislative body), it was not until the sixteenth century that the English Parliament began enacting legislation, thereby shifting the country's legal system from common law to codified statutory law.

Contemporary Sources of Criminal Law

Criminal law in the United States has largely grown out of English common law that was brought over to America during the colonial period. Although much of this common law eventually became part of the nation's legal codes, many Americans viewed it as antiquated and inadequate to the task of maintaining order in their new nation. Many feared that if judges were the only source for determining which actions were criminal, citizens would be left ignorant of their duties and responsibilities and, consequently, would be subject to judicial arbitrariness. Instead, early colonists looked to a codified system of law to provide greater uniformity, standardization, and predictability. As a result, the states and the federal government began to formalize law by developing statutes and by drawing

upon a number of other sources—case law, administrative rules, and the constitutions of the various states and the federal government.

Statutes

Criminal law is contained in written codes called <u>statutes</u>. According to the balance of powers established in the Constitution, the law-making function resides in the legislative branch rather than the judicial branch of government. Congress and state legislatures are responsible for enacting statutes that define crimes (substantive laws) and specify the applicable penalties for their violation as well as law governing legal procedures (procedural laws).

Case Law

<u>Case law</u> is a continuation of the common-law tradition in which judicial decision making in individual cases involves interpreting existing law, looking at relevant precedent decisions, and making judgments about the legitimacy of the law. Because gaps will inevitably exist between what a legislative body intends when it passes a law and what actually happens when that law is enforced, the practice of case law allows the courts to interpret the law as they apply it.[6]

For example, suppose a state legislature creates a statute defining assault with a dangerous weapon. The statute states that the definition of a dangerous weapon includes, but is not limited to, rifles, handguns, shotguns, and knives. In a current criminal trial, the defendant is charged with assault with a dangerous weapon, but the weapon used was a baseball bat—a device not specifically identified in the state statute. The prosecution searches through previous criminal cases tried in the state and finds a case in which the trial court held that a baseball bat was, indeed, a dangerous weapon. The court can then cite this precedent in applying the statute to the current case.

Administrative Rules

The rules, orders, decisions, and regulations established by state and federal administrative agencies are another source of law. The Federal Trade Commission, Internal Revenue Service, Food and Drug Administration, and Environmental Protection Agency, for example, have all established a multitude of rules and regulations that have the full force and effect of law. These agencies investigate and impose criminal sanctions for such violations as securities fraud, the willful failure to pay income tax, the intentional sale of contaminated food, and the dumping of toxic wastes.

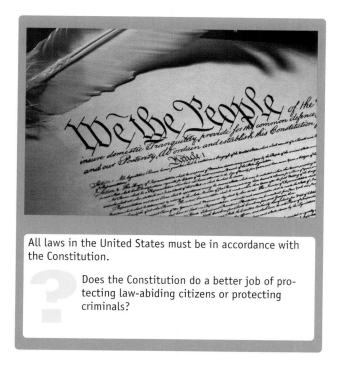

All laws in the United States must be in accordance with the Constitution.

? Does the Constitution do a better job of protecting law-abiding citizens or protecting criminals?

Constitutions

The U.S. Constitution and each of the 50 state constitutions are the final arbiters of substantive and procedural law. A law enacted by a state legislature may be found to be in violation of either that state's constitution or the U.S. Constitution. Federal laws, regulations, or administrative acts may be judged only against the U.S. Constitution. In addition, the Bill of Rights, which was added to the U.S. Constitution in 1791, includes protections afforded to citizens in criminal prosecutions (such as the right to counsel, prohibitions against illegal search and seizure, and the right to due process), reflecting the framers' fear of a strong centralized government. Although most state constitutions include similar protections, the U.S. Supreme Court, in its interpretations in various decisions over the years, has required all the states to ensure that defendants in state prosecutions be granted the specific protections enumerated in the federal Constitution.

Conceptualizing Crime

The official definition of crime is an intentional act or omission in violation of criminal law, committed without defense or excuse, and sanctioned (i.e., punishable) by the state. In this way, crime is a legal construct, since the law narrowly defines the specific elements of the forbidden act and the conditions under which they occur. For example, intentionally taking of the life of another person may or may not constitute a crime. Although it would be a crime for a person to intentionally kill his or her spouse to collect life insurance, it would not be a criminal act for a police officer to intentionally kill an armed suspect in self-defense.

Crime is also a social construct; social thinking and interaction play an important role in determining which acts are defined as criminal. For example, before the U.S. Supreme Court's 1973 decision in *Roe v. Wade,* the law in most states, which conformed to the thinking of many citizens, defined abortion as a crime.[7] As social and legal views about the right of women to control their own reproductive systems shifted, the Court, in *Roe,* held that laws restricting abortion were unconstitutional. While many people agreed with the decision in *Roe,* some of those supporters—including current Supreme Court Justice Ruth Bader Ginsburg—disagreed with the unusual reasoning the Court used to arrive at its ruling.

Most of our current drug laws also reflect shifts in thinking about which kinds of drugs (both prescription and nonprescription) should be controlled and how violations of drug laws should be punished. Addictive dosages of opiates, including heroin, morphine, and powdered opium, were widely marketed in the United States in the nineteenth and early twentieth centuries. These "medicines" were sold to parents to quiet infants who were sickly, colicky, or teething.[8] In 2005, cold medicines containing pseudoephedrine (a key ingredient in making methamphetamine) were taken off the shelves at retailers such as Target and Wal-Mart and sold only in limited quantities by their pharmacies. Today, at least 30 states and the federal government have legislation regulating the sale of pseudoephedrine, including requirements for the collection of personal information from purchasers, maintenance of logbooks, and daily sales limits and 30-day purchase limits.

The modern definition of crime is an act (armed robbery) or a failure to act (not paying income tax). At various times in history, a condition of being or status was included in definitions of crime. For example, during the seventeenth century, Massachusetts Bay Colony made it a crime to be a Quaker. Until 1962, in California it was illegal to "be addicted to the use of narcotics" (the statute was eventually declared unconstitutional by the U.S. Supreme Court).[9]

The Relationship between Civil Law and Criminal Law

Civil law is similar to criminal law in that it is designed to control behavior and protect the interests of people and property. However, unlike criminal law, civil law is private in nature. That is, violations of civil law are settled between the parties involved, whereas violations of criminal law are considered offenses against the state. In a criminal case, the prosecutor brings charges against the defendant. In a civil case, the plaintiff (the injured party) is responsible for bringing suit against the defendant (the party that inflicted the injury); if the defendant loses the case, he or she is required to pay damages to the plaintiff. The burden of proof required to decide a case also differs between civil and criminal law. In a civil case, the jury need be convinced only by a preponderance of the evidence, which simply means that the evidence presented by the plaintiff must outweigh the evidence presented by the defendant. Criminal cases, by contrast, require proof beyond a reasonable doubt.

LINK As Chapter 11 explains, proof beyond a reasonable doubt is required to convict, because a person is considered innocent until proven guilty.

Civil legal claims may arise in connection with criminal actions, and victims or their families may choose to pursue a civil suit following the outcome of a criminal trial. Because the required degrees of proof and the parties involved may differ between the two types of cases, an acquittal in a criminal trial does not preclude awarding damages to the victim in a civil suit. For example, in 1995, O. J. Simpson was charged with the murder of his estranged wife, Nicole, and her friend, Ron Goldman. Simpson was ultimately acquitted of the criminal charge, but two years later he was found liable for their deaths by a jury in a civil trial.

Seriousness of the Crime

Generally speaking, acts that are defined as crimes are considered more serious violations of norms (rules that regulate behavior) than are noncriminal acts. Nevertheless, perceptions of the seriousness of certain crimes may vary between different times, cultures, and societies. According to public opinion polls, most Americans agree that violent crimes are more serious than property crimes, but there are gradations—most people see a parent's assault on a child as more serious than a husband's assault on his wife,

and selling heroin is generally considered to be a more serious crime than selling marijuana.[10] In the United States, people who engage in sexual relations before marriage may be breaking the law in some states, though there is little chance of prosecution. In China, however, they may be charged with prostitution (for the female) and rape (for the male).[11]

Mala in Se Crimes versus Mala Prohibita Crimes

In the early development of criminal law, all crimes were considered wrong for one of two reasons: They were considered inherently wrong or evil (*mala in se*) or they were wrong merely because they were prohibited by a criminal statute (*mala prohibita*). Only nine common-law crimes were classified as *mala in se* offenses:

- Murder
- Manslaughter
- Rape
- Sodomy
- Robbery
- Larceny
- Arson
- Burglary
- Mayhem

These offenses were also the first group of crimes to be referred to as felonies. The *mala prohibita* crimes, by comparison, were considered less serious and consequently were classified as misdemeanors.

The significant historical distinction between these two categories of crimes reflects perceptions of the degree of public harm they present. Because *mala in se* crimes were believed to be inherently evil and to pose a major threat to the social order, it was understandable that they would be sanctioned by the law and more severely punished. *Mala prohibita* crimes, such as public drunkenness, loitering, prostitution, and gambling, did not carry the same broad moral condemnation. **TABLE 2-1** presents a brief list of examples of *mala in se* and *mala prohibita* crimes today.

The basic distinction between these two groups of crimes persists in our own present-day criminal law. The offenses classified as *mala in se* crimes have largely remained the same, but the number of *mala prohibita* crimes has greatly expanded. For example, statutes have been enacted to prohibit driving under the influence of alcohol or drugs, copyright infringement, and the manufacture, distribution, and possession of illegal drugs. Statutes have been created to

TABLE 2-1

Examples of Contemporary *Mala In Se* and *Mala Prohibita* Crimes

Mala In Se	Mala Prohibita
Murder	Prostitution
Rape	Gambling
Robbery	Vagrancy (loitering)
Larceny	Panhandling
Arson	Fraud
Burglary	Public intoxication
Aggravated assault	Public nudity
Incest	Trespass
	Possession of drug paraphernalia
	Copyright infringement
	Illegal possession of weapon
	Disorderly conduct

control cybercrime, including theft of information, creation of computer viruses to cause mischief or damage data, copying software, downloading of copyright-protected music or movies, and identity theft.

LINK The potential loss from cybercrime runs into the millions of dollars, which is why they are increasingly becoming a focus of both federal and state prosecutions. Cybercrime is explored in detail in Chapter 17.

Felonies, Misdemeanors, and Infractions

U.S. criminal law distinguishes between felonies, misdemeanors, and infractions and assigns punishments accordingly.

The most serious crimes, called <u>felonies</u>, result in a more severe punishment. In most states, felonies carry maximum sentences of death or imprisonment for a term greater than one year in a state prison and typically carry higher fines than misdemeanors. A felony conviction also may result in the loss of certain rights, such as the loss of a person's right to vote, hold public office, carry a gun, or be licensed in certain professions.

Crimes classified as <u>misdemeanors</u> carry less severe punishments than are meted out for felonies. Typically, the maximum incarceration sentence is one year or less in a local jail and a smaller fine than would be incurred in a felony.

The third category of crimes, called <u>infractions</u>, is composed of petty offenses. These involve violations of city or county ordinances and include such offenses as illegal parking, jaywalking, cruising, or violations of noise ordinances. Infractions are gen-

erally not punishable by incarceration; rather, fines or community service may be imposed.

Curiously, in a number of states, certain crimes may be charged by a prosecutor as either a felony or a misdemeanor, with the charging decision depending on the suspect's prior record or the specific details of the case. In California, these cases are called "wobblers," because the offense may "wobble" in either direction, and the same crime can lead to significantly different punishments. For example, a "wobbler" offense could end up being a misdemeanor that is punished by a probation sentence or the "third strike" that sends a person to prison for a term of 25 years to life.

The distinction between these categories of crimes is not always clear, especially with regard to felonies and misdemeanors. What may be a felony in one state may be a misdemeanor in another state. For example, in Texas, the possession of less than two ounces of marijuana is a misdemeanor punishable by a fine of up to $2000 and the possibility of six months in jail; in North Dakota, possession of one ounce or more of marijuana is a felony punishable by up to five years in prison and a $5000 fine.

One distinction between felonies and misdemeanors involves the authority of law enforcement officers to make arrests. When an officer has reasonable grounds to believe that a felony has been committed, even when he or she did not directly observe the act, the officer may arrest a suspect. By contrast, many states have an in-presence requirement, for misdemeanors, meaning that an arrest may not be made unless the criminal act was committed in the presence of the officer. However, if the victim of a misdemeanor files a formal complaint and the court issues an arrest warrant, the officer may then arrest the suspect.

Functions of Criminal Law

Criminologists often hold differing views of the functions of law in society, but they agree that in complex, diverse societies such as the United States, law becomes increasingly necessary to regulate behavior. Three of the most commonly identified functions of criminal law are:

- Defining serious forms of socially unacceptable behavior
- Controlling behavior and maintaining social order
- Regulating punishment

Cybercrime presents some of law enforcement's greatest challenges today. In August 2003, a 14-year-old boy released a computer worm that infected computers across the globe. Known as the RPCSdbot worm, this variant of the highly publicized Blaster worm attacked a Windows vulnerability that gave the juvenile control over the infected computers. He directed them to attack Microsoft's website, causing the site to shut down for several hours. After an investigation by the Northwest Cyber Crime Task Force, which included local law enforcement, the Federal Bureau of Investigation, and the U.S. Secret Service, the youth was arrested. He pleaded guilty to the federal charge of intentionally causing damage and attempting to cause damage to protected computers and was sentenced to three years of probation, 300 hours of community service, and probation department monitoring of his computer use.

In a different case, a task force known as Operation Firewall combined the efforts of the Secret Service, the U.S. Attorney's Office, the Computer Crime and Intellectual Property Section of the Criminal Division of the Department of Justice, and a number of local law enforcement agencies in a year-long investigation of the international Internet crime organization called Shadowcrew. At its conclusion, the government arrested 21 hackers, most of whom pleaded guilty to federal charges involving trafficking in stolen credit and bank card numbers and identity information, purchasing equipment used to encode counterfeit credit cards with stolen numbers, credit card fraud, and gift card vending (selling counterfeit merchandise gift cards from retail stores).

Andrew Mantovani, age 23, was a co-founder of the criminal organization that operated the Shadowcrew .com website, through which the enterprise conducted much of its business. Mantovani admitted using techniques such as phishing and spamming to illegally obtain credit and bank card information, which was then used to make purchases online. The illegally obtained goods were sent to a "drop" or mailing address specifically set up to receive the stolen goods. In addition, Mantovani admitted that he illegally acquired approximately 18 million e-mail accounts with associated user names, passwords, dates of birth, and other personal identifying information.

In yet another case, in May 2005, Operation D-Elite—a task force that included members of federal, state, and local law enforcement agencies— announced the first criminal enforcement action targeting individuals who had committed copyright infringement via peer-to-peer (P2P) file-sharing networks. The P2P network known as the Elite Torrents attracted more than 133,000 members and allegedly facilitated the illegal distribution of more than 17,800 files, including movies and software, which were downloaded more than 2 million times. Authorities estimate that Internet pirates cost U.S. corporations hundreds of billions of dollars in lost revenue every year from the illegal sale of copyrighted goods.

You can learn much more about how federal law enforcement is attacking the problem of cybercrime by visiting a website managed by the U.S. Department of Justice. Public information about federal statutes and law enforcement efforts can be found at www.cybercrime.gov.

Source: U.S. Department of Justice, Computer Crime and Property Section of the Criminal Division of the U.S. Department of Justice, Washington, DC, 2005.

Defining Socially Unacceptable Behavior

The law helps to establish the boundaries for social interactions. When people overstep the social boundaries, the law holds them accountable.[12] This does not mean, however, that all members of society agree that a particular behavior is objectionable or intolerable. It simply means that a significant number of people—or a number of significant people—define some act as falling outside the bounds of acceptable behavior, decide that this behavior should be punished by law, and are able to get lawmakers to implement their concern.[13]

Ideally, as public perceptions about inappropriate behavior change, so does the law. Unfortunately, the law often lags behind public opinion. For example, only recently have laws prohibiting fornication (sexual intercourse between unmarried persons) and seduction (the action of a man to entice a woman to engage in intercourse through persuasion or promise) been removed from the statute books in many states. In 2003, in *Lawrence v. Texas*, the Supreme Court held, in a 6 to 3 decision, that the Texas law prohibiting sodomy was unconstitutional.[14] This decision effectively invalidated similar laws in other states that were designed to criminalize homosexual activity between consenting adults in private.[15]

Controlling Behavior and Maintaining Social Order

Criminal law goes beyond only stating what will not be tolerated; through its enforcement, it also controls objectionable behaviors. Not all undesirable behaviors are defined as crimes, of course. Many are informally controlled by what sociologist Thorsten Sellin referred to as conduct norms, or norms that are specific to localized groups and that may or may not be consistent with crime norms (those found in the criminal law).[16] Conduct norms reflect the values, expectations, and behaviors of groups in everyday life. As such, they exert powerful control over the behaviors of members of the group. For example, many Mormon communities have only a few members who smoke, drink alcohol, or engage in extramarital affairs because the communities' conduct norms strongly prohibit such behavior.

The law also functions to maintain the larger social order by settling disputes between individuals and mediating confrontations when the conduct norms of various groups come into conflict. In some countries, the law is used not only to maintain order, but also to ensure that the general population poses no threat to the absolute control by the government.

Regulating the Punishment of Behavior

The law also specifies the punishments that should be imposed for criminal offenses, which helps prevent arbitrary or excessive punishments by the state—for example, imposing the death penalty for shoplifting. The law also regulates punishments to prevent unauthorized persons from imposing their own punishments. If people do not believe that the severity and certainty of the punishment mandated for a person accused of a particularly heinous crime are sufficient, they may be tempted to take the law into their own hands. Historically, this attitude of-

ten led to vigilante justice, although more often such instances have been carried out by individuals.

Elements of a Crime

As a legal definition, crime also includes what is known as the corpus (literally, "body of the crime"), which refers to the facts, or foundation, of the crime that must be established in a court of law. These include *actus reus* (criminal act), *mens rea* (criminal intent), and the concurrence of these two concepts.

Actus Reus

In his novel *1984*, George Orwell described a society in which both thoughts and acts were restrained and regulated by the Think Pol, or thought police.[17] Through constant surveillance, the Think Pol were able to monitor any expression of prohibited thoughts. U.S. law, however, generally limits criminal responsibility to *actus reus*—an actual act, the planning or attempt to act in violation of the law, or the specific omission to act when the law requires action. The written or oral expression of certain thoughts, such as making threats or intimidating remarks to a witness, may also be viewed as *actus reus* and, therefore, may be prohibited by criminal law.

If a person does not fully complete a criminal act or does not directly participate in the act, he or she may have still committed a crime. The law defines such partial acts as inchoate crimes (pronounced "in-KO-ate")—acts with the potential for harm, though they have not yet produced that harm. An inchoate crime may take one of the following forms:

- An *attempt* to act, which involves an intent to commit a crime and the taking of a substantial step toward its completion.
- A *conspiracy* to act, in which two or more people agree to commit a criminal act. In some states, at least one co-conspirator must commit an overt act toward accomplishing the crime for the act to qualify as a conspiracy.
- A *solicitation* of another to act, which includes asking, enticing, encouraging, or hiring another person to engage in a criminal act.
- Being an *accomplice* to an act, which does not require that the individual participate directly in the crime. Someone may serve as an accomplice by giving the perpetrator assistance or encouragement before or after the crime.

Both the Cuban Constitution and the Criminal Code of Cuba function to suppress activities deemed contrary to the "goals of the socialist state" or the "people's decision to build socialism and communism." Such activities include certain behaviors, speech, and even the printing of books, pamphlets, or magazines that would not constitute criminal acts in more open, democratic societies. For example, the "dangerousness provision" in Cuban law allows authorities to detain, for an undetermined length of time, persons believed to have a "special proclivity . . . to commit crimes, demonstrated by conduct that is observed to be in manifest contradiction with the norms of socialist morality." Indicators of dangerousness include "antisocial conduct" or behavior that "perturbs the order of the community." If a person is determined to be dangerous, he or she may be confined "until the dangerousness disappears." Anyone who threatens or offends the dignity or decorum of a public authority may be punished by up to a year in prison.

The Constitution and Criminal Code recognize certain due-process rights of defendants in criminal cases, but there are clear limitations on these rights:

- Trials may be conducted behind closed doors, in which case the defendant may be accompanied only by his or her lawyer.

- The police and other authorities may conduct warrantless arrests of anyone accused of a crime against state security.

- Suspects may be detained for a week without review of the legality of the detention. *Habeas corpus* petitions (a judicial order to bring a person immediately before the court to determine the legality of his or her detention) were eliminated from the Cuban Constitution in 1975, although they are still recognized under the Criminal Code.

- Coerced confessions are prohibited by the Cuban Constitution, and judges are required to inform defendants of their right to remain silent at trial.

- While defendants are guaranteed the right to an attorney under the Constitution, Cuban law prohibits an independent bar association, and private law firms were banned in 1973. All lawyers who did not work directly for the government were required to join collective law firms.

Recent estimates are that Cuba's prisons may hold 100,000 inmates, including more than 350 political prisoners.

Sources: "Cuba's Repressive Machinery: Human Rights Forty Years After the Revolution," Human Rights Watch, available at http://www.hrw.org/reports/1999/cuba/index.htm, accessed May 11, 2007; *Amnesty International Annual Report: Cuba, 1999,* available at http://www.amnesty.org/ailib/aireport/ar99/amr25.htm, accessed May 11, 2007.

Mens Rea

According to an old Latin maxim, an act does not make a person guilty unless the mind is guilty. In other words, a defendant is not criminally liable for conduct unless *mens rea* (criminal intent) was present at the time of the act. For a crime to exist, the person must intend for his or her action to have a particular consequence that is a violation of the law. The mere fact that a person engages in conduct in violation of law is not sufficient to prove criminal liability; rather, the defendant must also intend to commit the crime. As former Supreme Court Justice Oliver Wendell Holmes once noted, "Even a dog distinguishes between being stumbled over and being kicked."

Different degrees of criminal intent exist, and there are even some exceptions to the requirement that intent be present. In an attempt to create greater legal uniformity between the states, the American Law Institute wrote a Model Penal Code in 1962. It identifies levels of criminal responsibility, or culpability, reflecting differing degrees of intent to act: The person must have "acted (1) purposely, (2) knowingly, (3) recklessly, or (4) negligently, as the law may require, with respect to each material element of the offense."[18]

- *Purposely* means to act with conscious deliberation, planning, or anticipation to engage in some conduct that will result in specific harm.

- A person acts *knowingly* when he or she is aware that the conduct is prohibited or will produce a forbidden result.
- Acting *recklessly* involves conscious disregard of a known risk, although there is no conscious intent to cause the harm (such as speeding and unintentionally causing an automobile accident).
- *Negligent* conduct creates a risk of harm when an individual is unaware, but should have been aware. In other words, to be negligent, a person must engage in conduct that a reasonable person would not engage in, or an individual must fail to act (an omission) in the manner in which a reasonable person would act under the same or similar circumstances.

Common law has historically distinguished between general intent and specific intent. *General intent* requires the willful commission of a criminal act or an omission of a legal duty to act; in other words, the person intends to violate criminal law. However, the prosecution need not prove that the defendant intended the precise harm that resulted. For example, if the defendant intended to harm but not kill a victim, but the victim's death resulted as an unanticipated consequence of that harm, the offender may not be found guilty of murder because a lower level of culpability—specifically, recklessness—was present.

Specific intent, by contrast, involves the intent to engage in a precise act prohibited by law, such as assault with intent to rape. Specific intent is present when circumstances demonstrate that "the offender must have subjectively desired the prohibited result"; general intent exists when the circumstances show "the prohibited result may reasonably be expected to follow from the offender's voluntary act, irrespective of any subjective desire to have accomplished this result."[19] Because specific intent carries a higher level of culpability than general intent, the seriousness of the offense—and consequently the punishment—is greater.

Some criminal laws do not require any particular criminal intent; that is, the person's state of mind (*mens rea*) at the time of the act is not relevant. In these strict liability laws, there is liability without culpability. Strict liability laws provide for criminal liability without requiring the presence of either general or specific intent; in other words, a person may be held criminally responsible even though he or she had no intent to produce the harm. For example, bartenders have been held criminally liable for the intoxication of patrons and hosts of parties have been held criminally liable for the intoxication of their guests who are later involved in fatal accidents. The fact that neither the bartender nor the host had any intention to cause the intoxication or the subsequent accident is neither a required element of proof nor a valid defense. Penalties for strict liability violations typically involve fines rather than jail time.

Concurrence of *Actus Reus* and *Mens Rea*

For an act to be considered criminal, both the act (*actus reus*) prohibited by criminal law and the intent (*mens rea*) prohibited by the criminal law must be present before the crime is completed. It is not sufficient for an act to be defined as a crime if the person has only the guilty mind but commits no act. Nor is it sufficient for a person to have acted without criminal intent, with the exception noted earlier for strict liability offenses.

Concurrence may exist even if the act and the intent do not coincide as the offender intended. Suppose Jim aimed a gun at Terry and shot with the intent to kill him, but missed, hitting and killing John instead. Jim is still liable for murder under the doctrine of transferred intent. The intent to kill, in other words, is transferred from Jim to John. If the bullet missed both Terry and John but instead hit an electrical transformer and caused a fire, Jim would not be responsible for the crime of arson, since he did not intend to commit this specific act, though he may still be held responsible for reckless behavior.

Defenses and Responsibility

Society and criminal law have long recognized that certain actions may be justified or excused, such that the offender does not bear legal liability for the act. Sometimes these justifications and excuses, which are called defenses, are based on the mental state of the person at the time the act was committed. At other times, circumstances beyond the individual's control may come into play that may negate criminal liability. Both justifications and excuses are affirmative defenses; that is, the defendant must prove that his or her act was justified or excused.

John Hinckley, Jr.'s shooting of President Ronald Reagan in 1981 was seen by millions of people as they watched the television news, and yet Hinckley's defense of not guilty by reason of insanity prevented him from being convicted for the crime. In this case, the defendant did not deny engaging in the action:

Hinckley did shoot Reagan. Nevertheless, his defense of insanity successfully allowed him to avoid being held criminally responsible for the assault.

Justifications

Justifications are based on a defendant admitting responsibility but arguing that, under the circumstances, he or she did what was right.

Self-Defense

Defendants who raise the claim of self-defense as a justification for avoiding criminal responsibility argue that they acted in a lawful manner to defend themselves, others, or their property, or to prevent a crime. Most states permit a person to use as much force as is reasonably necessary for such protection. The individual must also have an honest and reasonable belief that he or she is in immediate danger from unlawful use of force by another person. The degree of force used in one's self-defense must be limited to a reasonable response to the threat: A person should meet force only with like force. Thus a person who is attacked by an unarmed assailant should not respond with a weapon.

According to the Model Penal Code, deadly force may be used only in response to a belief that there is imminent threat of death, serious bodily harm, kidnapping, or rape. It may *not* be used if the defendant provoked the offender to use force. Some jurisdictions also require that when there is a safe escape route from a house, a person must retreat instead of using deadly force. Thus, if a person has an opportunity to retreat safely from the person posing the threat, deadly force would not be justified as self-defense. This retreat rule has several exceptions, such as cases of battered woman syndrome (see "Focus on Criminal Justice").

In the past, deadly force generally has not been accepted as a response intended solely to protect property, because owners could have taken steps to protect their property and prevent the criminal act from occurring. For example, it is illegal to set a deadly trap or device for intruders in a home or business, even if the home or business has previously been burglarized. Because our society has long believed that a human life is more valuable than property, the use of deadly force to protect property was viewed as unreasonable. In 2005 and 2006, however, at least 15 states expanded the right to use deadly force in self-defense and defense of property. These new laws—referred to as "stand your ground" laws by their supporters and "shoot first" laws by their opponents—permit people to use

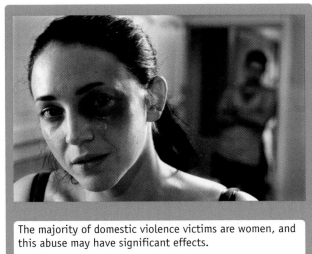

The majority of domestic violence victims are women, and this abuse may have significant effects.

 Is the use of deadly force by a battered woman justifiable? Should battered women be held to the same legal standards as other people?

deadly force against intruders who have illegally and forcefully entered their homes. For example, Florida law states that citizens no longer need to prove that they feared for their safety before they responded with deadly force, only that the intruder illegally entered their home or vehicle. Moreover, according to the law, a person "has no duty to retreat and has the right to stand his or her ground and meet force with force, including deadly force." The law also prohibits the arrest, detention, or prosecution of such a person and disallows the victim of such a shooting from bringing a civil suit against the shooter.[20]

Necessity

Necessity, as a defense, represents the dilemma of choosing between two evils. A person may violate the law out of necessity when he or she believes that the act, which is a violation of law, is required to avoid a greater evil. According to the Model Penal Code, conduct that a person "believes to be necessary to avoid a harm or evil to himself or to another is justifiable, provided that the harm or evil sought to be avoided by such conduct is greater than that sought to be prevented by the law defining the offense charged."[21] For example, breaking into a mountain cabin to secure shelter or food during a snowstorm or into a home to use the telephone to report an emergency may establish the defense of necessity and thereby negate the crime of breaking and entering. In either case, the individual must intend to avoid a greater harm than the crime charged to justify the act.

The Battered Woman Syndrome and Deadly Force

Nancy Seaman, a 52-year-old elementary school teacher, may have suffered from *battered woman syndrome* after enduring years of alleged physical abuse by her husband, Robert. At her trial, Nancy admitted to killing her husband with a hatchet but claimed it was an act of self-defense initiated during one of his attacks soon after she asked him for a divorce. She testified that when she told Robert she wanted a divorce, he became furious, cut her with a knife, chased her into the garage, forced her to the ground, and repeatedly kicked her. According to Nancy, she grabbed the closest object she could find—a hatchet—and drove it into her husband's skull. She then stabbed and beat him to ensure he was dead.

Prosecutors told another story. They claimed the act was premeditated and that Nancy purchased the ax and took great care to conceal the crime scene. Surveillance video from a hardware store showed Nancy stealing a hatchet identical to the one used in the murder. Two days later, she returned it using the receipt from the purchase of the hatchet used in the murder, perhaps in an attempt to erase the purchase from her credit card record. Prosecutors said that Nancy slammed the hatchet into Robert's skull more than a dozen times, dragged his body into the garage, and then stabbed him 21 times, severing his jugular vein and voice box. The next morning, she stopped to purchase a tarp, bottles of bleach, and latex gloves to clean up the mess.

Other testimony also appeared to contradict Nancy's account. When police first arrived at the couple's home, Nancy claimed Robert's death was an accident. A co-worker said he had overheard Nancy talking with another teacher at school about poisoning her husband.

After five hours of deliberation, the jury found Nancy guilty of first-degree murder, rejecting her claim of self-defense. She was subsequently sentenced to life in prison.

According to feminist criminologist Cynthia Gillespie, laws regulating deadly force have been created by men based on a code of "manly" behavior that expects a person to be fearless and confront an attacker directly. Such laws do not consider the woman's assessment that she cannot escape further injury as long as the abuser is alive.

Victims of battered woman syndrome are often unable to leave their abusers, even when circumstances appear to permit their escape. Over time, a battered woman may lose all hope of controlling her husband's or boyfriend's violence. Many such women succumb to learned helplessness: They become emotionally dependent on their abusive partners and learn to be passive as a result of beatings when they tried to assert themselves. In addition, some women do not leave because no safe refuge exists where an enraged partner cannot find them or their children. Abusive husbands often threaten to harm or take custody of the children if the woman leaves. Even with the increase in numbers of shelters for abused women, shelters must turn away more women than they serve.

Sources: Mike Martindale, "Seaman Gets Life in Prison," available at http://www.detnews.com/2005/metro/0501/25/B01-68930.htm, accessed June 28, 2007; "Teacher Claims Self-Defense in Husband's Ax-Murder," *Court TV,* March 30, 2005, available at http://www.courttv.com/trials/seaman/background_ctv.html, accessed June 28, 2007; Cynthia Gillespie, *Justifiable Homicide: Battered Women, Self-Defense, and the Law* (Columbus, OH: Ohio State University Press, 1989); Walker quoted in *State v. Kelly,* 478 A.2d 364 (N.J. 1984).

Consent

The defense of consent arises when a defendant claims the victim consented to the act. Certain common-law offenses, such as theft and rape, require a clear demonstration that the victim did not give consent. For example, if the owner of an automobile voluntarily consented to a neighbor taking her car, then the neighbor has not committed motor vehicle theft.

During the summer of 2003, professional basketball star Kobe Bryant was charged with raping a 19-year-old female hotel employee while he was staying at a Colorado resort. Bryant admitted he had sex with the woman but claimed that she had consented. Shortly after the charges were filed, Bryant stated, "Nothing that happened June 30 was against the will of the woman who now falsely accuses me." The rape charge was eventually dismissed. Afterward, Bryant stated, "Although I truly believe this encounter between us was consensual, I recognize that she did not and does not view this incident the same way I did. After months of reviewing [evidence submitted at] discovery, listen-

Kobe Bryant and his accuser had starkly different assessments of their sexual relationship, which resulted in a civil settlement.

? Why do Americans often respond emotionally to charges of sexual assault as opposed to other serious crimes?

ing to her attorney, and even her testimony in person, I now understand how she feels that she did not consent to this encounter." The woman then filed a civil law suit against Bryant; the suit was settled out of court, and terms of the settlement were not released.[22]

Many cases of consent reflect different interpretations of the situation, or simply come down to hearsay. Other claims of consent present additional dilemmas for the court, such as doctor-assisted suicide or the "mercy killing" of a terminally ill spouse who has been experiencing great pain. In 2006, the U.S. Supreme Court, in *Gonzales v. Oregon,* held that doctors who prescribed lethal doses of drugs to terminally ill patients under the Oregon Death with Dignity Act (sometimes known as the assisted-suicide act) were not acting in a criminal fashion because the patients had consented to the act.[23]

Excuses

Excuses are based on a defendant admitting that what he or she did was wrong but arguing that, under the circumstances, he or she was not responsible for the criminal act.

Insanity

Probably no other legal defense has resulted in more public scrutiny and debate than the insanity defense. In reality, insanity pleas are very rare. The insanity defense is raised in less than 1 percent of all criminal cases, and only in 25 percent of those cases is the person actually found not guilty because of insanity.[24] Even so, many people are concerned when a clearly dangerous person avoids incarceration and punishment after being found legally insane at the time the crime was committed. It is important to recognize that people who are released from criminal charges owing to insanity do not go free, but instead are sent to mental hospitals until they are considered sane. Only then are they released back into the community.

The insanity defense is based on a legal concept, rather than a medical or psychiatric definition of insanity. Legally, "insanity" refers to a person's state of mind at the time he or she committed the crime charged, though actual legal definitions of insanity have been—and continue to be—rather vague. In the past, concepts such as madness, irresistible impulse, states of unsound mind or weak-mindedness, and mental illness, disease, defect, or disorder have all been used to inform the law.[25]

It should not be surprising that the legal notion of insanity has varied greatly over time and across cultures. According to the Justinian Code (535 C.E.), "There are those who are not to be held accountable, such as a madman and a child, who are not capable of wrongful intention."[26] In 1723, an English judge established the "wild beast test" of insanity: "In order to avail himself of the defense of insanity, a man must be totally deprived of his understanding and memory so as not to know what he is doing, no more than an infant, a brute, or a wild beast."[27]

The M'Naghten Rule The M'Naghten rule, which is also known as the "right from wrong" test, is based on an English case that was decided in 1843. Until recently, it was the most widely accepted standard of insanity in the United States. Daniel M'Naghten, a Scottish woodcutter, believed that the English Prime Minister, Sir Robert Peel II, was persecuting him. In an attempt to assassinate Peel, M'Naghten mistakenly shot and killed Peel's assistant. At the trial, the court instructed the jury that:

[To] establish a defense on the ground of insanity, it must be clearly proved that, at the time of the committing of the act, the party accused was labouring under such a defect of reason from disease of the mind as not to know the nature and quality of the act he was doing; or, if he did know it, that he did not know he was doing what was wrong.[28]

M'Naghten was tried and found not guilty by reason of insanity.

Under the M'Naghten rule, the defendant is presumed to be sane and must prove that he or she suffered from a "disease of the mind" and, therefore, lacked a sufficient degree of reason to distinguish between right and wrong. This test of insanity has been criticized on several grounds:

- "Disease of the mind" is not clearly defined.
- Too much stress is placed on the requirement of knowing.
- It is unclear how a person must know that an act is wrong.
- Some people may be insane but still able to distinguish right from wrong.

Subsequent rules have sought to overcome these weaknesses in the M'Naghten rule.

Irresistible Impulse Test   In 1897, the U.S. federal courts and a number of the states added the irresistible impulse test to supplement the M'Naghten rule. According to this test, defendants may be found not guilty by reason of insanity if they can prove that a mental disease caused loss of self-control over their conduct. This test arose from an 1886 Alabama Supreme Court decision in *Parsons v. State,* which held that it may be possible for a person to know that the action was wrong but nevertheless be so overcome by emotion that he or she temporarily lost self-control or the ability to reason to a degree sufficient to prevent the act.[29] In revising the M'Naghten rule, the irresistible impulse test allowed defendants to raise the insanity defense and plead that, although they knew that what they were doing was wrong, they were unable to control their behavior.

Durham Rule   The Durham rule, which states that "an accused [person] is not criminally responsible if his unlawful act was the product of mental disease or mental defect," was formulated in *Durham v. United States* in 1954.[30] Judge David Bazelon, the presiding judge in the case, established new case law in his rejection of the M'Naghten rule, arguing that insanity is actually a product of many personality factors. According to the Durham rule, a mental condition may be either a disease (a condition capable of improving or deteriorating) or a defect (a condition not considered capable of improving or deteriorating). Further, the Durham rule states that a defect could be congenital, the result of injury, or the residual effect of either physical or mental disease. Under the Durham test, the prosecutor must prove beyond a reasonable doubt that the defendant was not acting as a result of mental illness, but the jury determines whether the act was a *product* of such disease or defect.

The Substantial Capacity Test   The Durham rule, like its predecessors, was soon criticized. Specifically, critics argued that it provided no useful definition of "mental disease or defect." In 1962, the American Law Institute offered a new test for insanity in its Model Penal Code. Known as the substantial capacity test or Model Penal Code Test, it includes the following provisions:

1. A person is not responsible for criminal conduct if, due to mental disease or defect, he or she lacks the substantial capacity to appreciate the criminality (wrongfulness) of his or her conduct or to conform to the requirements of law.

2. The terms "mental disease or defect" do not include an abnormality manifested only by repeated criminal or antisocial conduct.[31]

The substantial capacity test is broader than the M'Naghten rule because it substitutes the notion of "appreciate" for "know," thereby eliminating the M'Naghten requirement that a person be able to fully distinguish right from wrong. In other words, a defendant may know the difference between right and wrong yet not be able to appreciate the significance of that difference. The substantial capacity test absolves from criminal responsibility a person who knows what he or she is doing, but is driven to act by delusions, fears, or compulsions.[32] Like the Durham rule, the substantial capacity test places the burden of proof beyond a reasonable doubt on the prosecutor.

In 1972, in *United States v. Brawner,* the federal courts rejected the Durham rule and adopted a modified version of the substantial capacity test.[33] By 1982, it was being used in 24 states, the District of Columbia, and the federal courts.

Insanity Defense Reform Act of 1984   Until 1981, the substantial capacity test dominated federal and state practice. Matters changed after March 30, 1981, when

State	Insanity Defense Rule	Location of Burden of Proof	Allows GBMI and GBI Verdicts
Utah	Insanity defense abolished		Yes
Vermont	Model Penal Code	Defendant	No
Virginia	M'Naghten rule; irresistible impulse test	Defendant	No
Washington	M'Naghten rule	Defendant	No
West Virginia	Model Penal Code	State	No
Wisconsin	Model Penal Code	Defendant	No
Wyoming	Model Penal Code	Defendant	No

Source: The Defense of Insanity: Standards and Procedures, *State Court Organization, 1998* (Washington, DC: U.S. Department of Justice Statistics, 2000).

killed two of her three young children by beating them in the head with rocks.

- In 2004, Kenneth Pierott was convicted of murder and sentenced to 60 years in a Texas prison after the jury deliberated only four hours before rejecting the defense claim of insanity. Pierott had smothered his girlfriend's six-year-old child and stuffed him in an oven. Only eight years earlier, Pierott had been found not guilty by reason of insanity after beating his sister to death with a dumbbell, crushing her skull and displacing her eyeballs. He was sent to a state mental hospital, where he spent four months and was then released.[36]

Intoxication

The defense of intoxication is based on the claim that the defender had diminished control over himself or herself owing to the influence of alcohol, narcotics, or drugs and, therefore, lacked criminal intent. According to the Model Penal Code, the defense of intoxication should not be used unless it negates an element in the crime, such as criminal intent. The courts recognize a difference between involuntary intoxication and voluntary intoxication. Involuntary intoxication that results from mistake, deceit of others, or duress (for example, if a drug was unknowingly put in the person's drink, or if liquor was forcibly poured down the person's throat) will excuse the defendant from responsibility for criminal action that resulted from the intoxication. For example, Louisiana law states that "Where the production of the intoxicated or drugged condition has been involuntary, and the circumstances indicate this condition is the direct cause of the commission of the crime, the offender is exempt from criminal responsibility."[37]

Voluntary intoxication is generally not a defense, but it may be presented in an effort to mitigate the seriousness of the crime. A person charged with committing premeditated murder while voluntarily intoxicated, for example, may be able to have the charge reduced to the less serious charge of homicide. In 2005 in Lawrence, Kansas, Jason Dillon, who was charged with murdering his girlfriend's three-year-old daughter, cited "voluntary intoxication" as a factor in his crime. Dillon claimed that he had consumed 16 beers the night before he babysat the young girl and was incapable of acting intentionally. Dillon did not dispute that he struck the girl on the head more than a dozen times after she refused to help pick up laundry and told him she didn't want him to be her daddy anymore. As a result of a plea bargain, Dillon was convicted of second-degree murder rather than first-degree murder as initially charged; he was sentenced to a reduced term of 16½ years in prison.[38]

Entrapment

The defense of <u>entrapment</u> is an excuse for criminal actions based on the claim that the defendant was encouraged or enticed by agents of the state to engage in an act that he or she would not have committed otherwise. The courts have generally held that it is permissible for law enforcement agents to solicit information from informants, use undercover officers, and even place electronic monitoring devices on informants or officers to record conversations regarding criminal behavior. It is not, however, considered legitimate for police to encourage or coerce individuals to commit crimes when they had no previous predisposition to commit such acts. Government agents may not "originate a criminal design, implant in an innocent person's mind the disposition to commit a criminal act, and then induce commission of the crime so that the Government may prosecute."[39]

For an entrapment defense to be valid, two related elements must be present:

1. Government inducement of the crime
2. The defendant's lack of predisposition to engage in the criminal conduct[40]

Predisposition is generally considered to be the more important of these two elements. Thus entrapment may not have occurred if the government induces a person who is predisposed to commit such an act.

In recent years, many state and federal law enforcement agencies have conducted "sting" operations that are designed to trap people who are engaged in crime or predisposed to commit crimes. Such operations often involve law enforcement agents posing as prostitutes, drug buyers, buyers of stolen auto parts, and people attempting to bribe government officials. Do such activities create an illegal inducement to commit crime? In *Sherman v. United States*, the U.S. Supreme Court held that "to determine whether entrapment has been established, a line must be drawn between the unwary innocent and the trap for the unwary criminal."[41] Entrapment occurs when government activity in the criminal enterprise crosses this line.

The line suggested by the Supreme Court, however, is often ambiguous. The use of deceit by the police to create a circumstance in which a person then commits a crime does not necessarily constitute entrapment. In *United States v. Russell*, the Court held that

> [T]here are circumstances when the use of deceit is the only practicable law enforcement technique available. It is only when the government's deception actually implants the criminal design in the mind of the defendant that the defense of entrapment comes into play.[42]

In this case, the defendant, who had been convicted of unlawfully manufacturing and selling methamphetamine (speed) to an undercover agent, was viewed by the Court as "an active participant in an illegal drug manufacturing enterprise which had begun before the government agent appeared on the scene, and continued after the agent had left the scene." According to the Court, the defendant was not an unwary innocent but rather an unwary criminal.

Duress

The defense of duress presents the claim that the defendant is actually a victim, rather than a criminal. For example, if someone holds a gun to a person's head, threatening to shoot unless he or she steals money, the resulting theft would be considered an action under duress, and the thief should not be held criminally responsible for complying with the demand to steal. The Model Penal Code's provision on duress states that

> [It] is an affirmative defense that the actor engaged in the conduct charged to constitute an offense because he was coerced to do so by the use of, or a threat to use, unlawful force against his person or the person of another, which a person of reasonable firmness in his situation would have been unable to resist.[43]

This defense is *not* applicable to people who intentionally, recklessly, or negligently place themselves in situations in which it is probable that they will be subject to duress. For example, a person who, in the course of escaping from prison, commits a kidnapping to avoid being caught cannot claim duress as a defense against the charge of kidnapping.

Mistake

Everyone has probably heard the expression, "Ignorance of the law is no excuse." But what does it mean? Although we may be familiar with many laws, must we be aware of all the laws? Must we know exactly what they prohibit and under what circumstances? When the federal and state criminal codes were first developed, they were fairly limited in scope; the relative homogeneity of the population meant that most laws were generally understood. More recently, given the increase in language and cultural differences among the general population as well as the vast proliferation of laws, it is fair to assume that many people today are ignorant of many laws.

Ignorance of what the law requires or prohibits generally does not excuse a person from committing a crime, but, under some circumstances, ignorance has been accepted as a defense. A federal court of appeals held in 1989 that "Under the proper circumstances . . . a good faith misunderstanding of the law may negate willfulness."[44] Mistake, as a criminal defense, takes two forms: mistake of law and mistake of fact.

Mistake of law occurs when the defendant does not know a law exists; only in rare cases is it a legitimate defense. Such a case might exist when a new law is passed but not published so as to give the public adequate notice of it. For example, in 1957 the Supreme Court, in *Lambert v. California*, reversed the conviction of a petitioner who had claimed ig-

VOICE OF THE COURT — *Jacobson v. United States*

On September 24, 1987, Keith Jacobson was indicted for violating a provision of the Child Protection Act of 1984, which criminalizes the knowing receipt through the mail of a "visual depiction [that] involves the use of a minor engaging in sexually explicit conduct." Jacobson was found guilty after a jury trial.

In the same month that the Child Protection Act became law, postal inspectors found Jacobson's name on a mailing list of a California bookstore from which he had ordered two magazines containing photographs of nude preteen and teenage boys. Over the next two and a half years, the government attempted—through the repeated efforts of two government agencies, five fictitious organizations, and a bogus pen pal—to explore Jacobson's willingness to break the new law by ordering sexually explicit photographs of children through the mail. One of the bogus organizations created by the government, the Heartland Institute for a New Tomorrow (HINT), sent Jacobson materials proclaiming that sexual freedom and freedom of choice should be protected against government restrictions. Eventually, Jacobson ordered a pornographic magazine depicting young boys engaged in various sexual activities. He was arrested after a controlled delivery of a photocopy of the magazine.

After his conviction, Jacobson appealed to the U.S. Supreme Court, claiming that he did not know the magazines would depict minors and that he was a victim of police entrapment. The Court, in a 5 to 4 decision, overturned the earlier conviction. It held that the government had failed to demonstrate that Jacobson was predisposed to commit a criminal act. In the verdict, the court said that Jacobson's responses to the many communications prior to ordering the last magazine were

At most indicative of certain personal inclinations, including a predisposition to view photographs of preteen sex. . . . Even so, petitioner's responses hardly support an inference that he would commit the crime of receiving child pornography through the mails. Furthermore, a person's inclinations and fantasies . . . are his own and beyond the reach of government . . .

More importantly, the Court argued that any indication that Jacobson was ready and willing to receive child pornography through the mail came only after two and a half years of the government's attempts to convince him that he had or should have the right to receive such material. According to the Court, "Rational jurors could not say beyond a reasonable doubt that the petitioner possessed the requisite predisposition to the Government's investigation and that it existed independent of the Government's many and varied approaches to petitioner. . . ." Furthermore, the ruling declared that the government "played on the weaknesses of an innocent party and beguiled him into committing crimes which he otherwise would not have attempted"—which is the definition of entrapment.

Source: Jacobson v. United States, 503 U.S. 540 (1992).

---

norance of the law.[45] The Los Angeles Municipal Code at the time contained a provision requiring any person who had previously been convicted of a felony to register with the police within five days of entering the city. Lambert, who was unaware of the law's existence, failed to register. Upon his conviction, he was fined $250 and placed on probation for three years. In reversing Lambert's conviction, Justice William O. Douglas stated:

Engrained in our concept of due process is the requirement of notice. Notice is sometimes essential so that the citizen has the chance to defend charges . . . Notice is required in a myriad of situations where a penalty or forfeiture might be suffered for mere failure to act.[46]

Mistake of fact occurs when a person unknowingly violates the law because he or she believes some fact to be true when it is not. In other words, had the facts been as a defendant believed them to be, the defendant's action would not have been a crime. For example, a woman who is charged with the crime of bigamy may have believed that her divorce was final before she remarried when, in fact, it was not. Mistake of fact is often raised as a defense by people who are charged with selling alcohol to a minor or with committing statutory rape. In such cases, defendants may have been led to believe that the minor was older than he or she claimed because the claim appeared consistent with the minor's appearance.

Exemptions

In some situations, a defendant may raise the defense that he or she is legally exempt from criminal responsibility. Unlike the defenses discussed earlier, legal exemptions are not based on the question of the defendant's mental capacity or culpability for committing the crime. Rather, they are seen as concessions to the defendant for the greater good of the public welfare.[47]

Double Jeopardy

The Fifth Amendment to the Constitution states that "no person shall be subject for the same offense to be twice put in jeopardy of life or limb." This protection against double jeopardy is not intended to provide protection for guilty defendants, but rather to prevent the state from repeatedly prosecuting a person for the same charge until a conviction is finally achieved.[48] Jeopardy in a bench trial (a case tried before a judge rather than a jury) attaches (i.e., becomes activated) when the first witness is sworn in. In jury trials, some jurisdictions consider a defendant in jeopardy once the jury is selected, though a few define it at the point of indictment, when criminal charges are filed.

Double jeopardy does not apply when a case is ruled a mistrial on the motion of the defense or when a jury is unable to agree on a verdict and the judge declares a mistrial. In both circumstances, the prosecutor may retry the case. Also, if upon conviction a defendant appeals to a higher court and has the conviction reversed, he or she may be retried on the original charge.

Some exceptions to the double jeopardy rule also exist. For example, some crimes are violations of both state and federal law. In 1959, in *Bartkus v. Illinois,* the Supreme Court held that people may be prosecuted for the same criminal acts by both a state and the federal government because people are considered citizens of both a state and the United States.[49] This was the basis for the federal case brought against the Los Angeles police officers in the Rodney King incident after they had been acquitted in the state court.

Statute of Limitations

Under common law, there was no limit to the amount of time that could pass between a criminal act and the state's prosecution of that crime. More recently, however, the states and the federal government have enacted statutes of limitations establishing the maximum time allowed between the act and its prosecution by the state for most crimes. Thus a defendant may raise the defense that the statute of limitations for the crime has expired, which requires a dismissal of the charges.

Statutes of limitations vary by jurisdiction and are generally longer for more serious offenses. For instance, murder has no statute of limitations, whereas in many states burglary carries a five-year limitation, and misdemeanors have a two-year limitation period in most jurisdictions. Although murder has no limits on when charges may be filed, some states limit the filing of charges for attempted murder to a period as short as three years. The statute of limitations may, however, be interrupted if the defendant leaves the state. For example, if a person who is charged with assault leaves the state for a period of two years, an additional two years would be added to the statutory limit of five years. In Missouri, criminal prosecutions for sexual offenses involving persons 18 years of age or younger must be initiated within 10 years after the victim reaches the age of 18.[50] In Maine, by contrast, the statute of limitations for both criminal and civil claims of child sexual abuse have been abolished.[51]

Age

On March 8, 2000, six-year-old Kayla Rolland was shot in the neck in her first-grade classroom with a .32-caliber pistol and died a half hour later. Her killer, Dedrick Owens, was also six years old. He had gotten into a quarrel with Kayla on the playground the day before. Dedrick had found the loaded pistol in his home and brought it to school tucked in his pants. After shooting Kayla, Dedrick ran into a nearby bathroom and tossed the gun into a trashcan. Because of his age, the court determined that Dedrick could not be held criminally responsible for Kayla's death.[52]

Although not considered either a justification or an excuse for a criminal act, a person's age may establish a defense against criminal prosecution. Under early English common law, children younger than 7 years of age were considered incapable of forming criminal intent and, therefore, could not be convicted of crimes. Children between the ages of 7 and 14 were considered to have limited criminal responsibility, and chil-

According to public opinion and the inmate code, sex offenders are the most reviled of all criminals.

? Do *mala in se* crimes violate natural law and, if so, should they be punishable by death? Should crimes like child molestation and rape carry harsh punishments?

First-grader Kayla Rolland was shot and killed by her classmate, Dedrick Owens, after a playground dispute.

? Is justice achieved when criminal offenders like Owens are not prosecuted because of their age? Do children lack the capacity to understand the wrongfulness of their actions?

dren older than age 14 were presumed to have the capacity to form criminal intent and could be criminally prosecuted. With the creation of the juvenile court system in the United States at the end of the nineteenth century, most youths between ages 7 and 18 who were charged with crimes were processed through the more informal proceedings of that court. The minimum age for transfer to adult court today varies by state but is commonly set at age 14, 15, or 16. Consequently, a child of 12 who breaks into a neighbor's home and steals a television set, for example, is younger than the minimum age for prosecution in criminal court and, therefore, may not be charged with a crime. Nevertheless, he or she may be petitioned by the juvenile court for having committed a delinquent act.

LINK Chapter 16 discusses the juvenile justice system and the inclusion of juveniles in the adult criminal justice system.

Due Process and the Rights of the Accused

Due process, which is established in procedural criminal law, ensures the constitutional guarantees of a fair application of the rules and procedures in criminal proceedings, from the investigation of crimes to an individual's arrest, prosecution, and punishment. Unfortunately, there is not always agreement over the concept of due process, its specific applications,

Headline Crime | A Three-Year Limitation on Attempted Murder

On the night of June 22, 1977, only a few days into their cross-country bicycle trip along the Bicentennial Trail, two undergraduates at Yale University pitched their tent near a river at Cline Falls State Park in central Oregon. Not long after falling asleep, Terri Jentz heard the cries of her roommate and woke up to a crushing pain on her chest. A man had driven over their tent with his truck. The man got out, hatchet in hand, and wildly attacked the two young women. Miraculously, both women survived the attack, although Terri's roommate lay unconscious with a severe wound to the back of her head, which later caused permanent loss of sight. Terri's right lung had collapsed, her collarbone was fractured, her left forearm was broken, and the right side of her rib cage was crushed; she also received several hatchet blows to her head. The vicious attack was briefly

investigated by the Oregon State Police, with some minor support from the Redmond Police Department and the Deschutes County Sheriff's Office, but no suspect was apprehended.

Fifteen years after the attack, Terri returned to investigate the crime herself. Her investigation, which was aided by two dedicated Oregon victim advocates, Bob and Dee Dee Kouns, eventually led to the discovery of the man authorities believed to be the attacker. Because of Oregon's three-year statute of limitations on attempted murder, however, he was immune from prosecution for the crime under state law. Only after involvement in another criminal incident did he finally serve time behind bars for an unrelated crime—five years for kidnapping, unlawful use of a dangerous weapon, and coercion.

It was not until 1997 that Oregon passed Bill 614, which changed the

state's statute of limitations for attempted murder. Under this legislation, "A prosecution for aggravated murder, attempted aggravated murder, murder or attempted murder or manslaughter may be commenced at any time after the attempt to kill." Unfortunately for Terri and her roommate, the legislation was not retroactive.

Source: Terri Jentz, *Strange Piece of Paradise* (New York: Farrar, Straus and Giroux, 2006); personal interview with Terri Jentz, April 10, 2007.

---

or even who is eligible to claim the rights associated with the guarantees of due process.

LINK Currently, there is intense debate over whether foreign citizens (and even U.S. citizens) who are charged with terrorist acts should be given all of the rights typically accorded people charged with crimes in the United States. Chapter 17 explores the debate over the due process rights of those accused of terrorism.

When a person is arrested, the immediate concern is typically focused on whether he or she committed the crime. How does a court of law make this determination? The police might threaten or coerce the suspect to extract a confession, and some people might confess to crimes they did not commit to avoid further mistreatment. Evidence might also be presented to establish the individual's guilt even though that evidence was obtained by devious or unethical means (for example, searching a person's private property without a search warrant). The accused might be held in jail without bail and denied access

to an attorney while the government builds a convincing case. Although convictions might be obtained in such instances, such procedures would offend the public's sense of fairness related to the criminal process.

The principles of procedural fairness in criminal cases are designed to reduce the likelihood of erroneous convictions. Criminal procedures that produce convictions of large numbers of innocent defendants would be patently unfair. The evolution of procedural safeguards against unfair prosecution is based on a relative assessment of the interests at stake in a criminal trial. According to law professor Thomas Grey, "While it is important as a matter of public policy (or even of abstract justice) to punish the guilty, it is a very great and concrete injustice to punish the innocent."[53] In U.S. criminal law, procedural safeguards have been established in the Fourth, Fifth, Sixth, Eighth, and Fourteenth Amendments to the U.S. Constitution to prevent that problem from occurring.

The Bill of Rights

The first 10 amendments, known as the <u>Bill of Rights</u>, were added to the Constitution on December 15, 1791—only three years after the Constitution had been ratified by the states. The framers of the Constitution had intended it to provide citizens with protections against a possible future dictatorship by establishing a clear separation of powers between the three branches of government (executive, legislative, and judicial). All too soon, they realized that the individual rights of citizens were not adequately protected against possible intrusions and violations by the newly formed federal government. To correct this deficiency, they added a series of amendments to the Constitution. Four of these amendments enumerate the rights of citizens in criminal proceedings.

The Fourth Amendment

The Fourth Amendment protects citizens against unreasonable governmental invasion of their privacy. As it has come to be interpreted by the courts, this amendment means that agents of the government may not arbitrarily or indiscriminately stop and search people on the street or in their vehicles, search their homes or other property, or confiscate materials without legal justification. Such justification must be based on sufficient probable cause to convince a judicial magistrate to issue a search warrant specifically describing who or what is to be searched and what is to be seized. Any evidence seized as a result of searches in violation of the Fourth Amendment cannot be used in a subsequent criminal prosecution.

The Fifth Amendment

The Fifth Amendment contains four separate procedural protections:

1. A person may not face criminal prosecution unless the government has first issued an indictment stating the charges against the person.
2. No person may be tried twice for the same offense (double jeopardy).
3. The government may not compel a defendant to testify against himself or herself (this provision includes protection against self-incrimination during questioning and the right to refuse to testify during a criminal trial).
4. No person may be deprived of due process, which means that people should be treated fairly by the government in criminal prosecutions.

The Sixth Amendment

The Sixth Amendment was designed to ensure a fair trial for defendants. Toward this end, it established six specific rights:

1. Speedy and public trial
2. Trial by an impartial jury (which has been interpreted by the courts to mean a jury of one's peers)
3. Notification of the nature and cause of the charges
4. Opportunity to confront witnesses called by the prosecution
5. Ability to obtain witnesses on the defendant's own behalf
6. Assistance of an attorney in presenting the defendant's defense

The Eighth Amendment

The Eighth Amendment simply states, "Excessive bail shall not be required, nor excessive fines imposed, nor cruel and unusual punishments inflicted." Although this amendment does not guarantee a defendant the constitutional right to be released on bail while awaiting trial, it does prohibit the imposition of excessive bail.

The Fourteenth Amendment

For nearly 80 years after the adoption of the Bill of Rights, the federal government and the various states interpreted the rights enumerated in these amendments to apply only to cases involving disputes between citizens and the federal government: The protections did not extend to citizens prosecuted by the states. (Actually, many state constitutions included these same rights, but if they were violated, the federal courts were not empowered to intervene.)

States' rights advocates believe in a strong separation of federal and state powers and the right of states to interpret and enforce particular laws. They base their arguments on the Tenth Amendment to the Constitution, which states that "The powers not delegated to the United States by the Constitution, nor prohibited by it to the states, are reserved to the states respectively or to the people." In the past, the states frequently cited this principle to nullify federal laws through their courts, such as when state courts refused to obey mandates from the U.S. Supreme Court. Not surprisingly, federal and state interests came into conflict over a variety of issues, such as South Carolina's attempt to nullify the Tariff

First Amendment

Congress shall make no law respecting an establishment of religion, or prohibiting the free exercise thereof; or abridging the freedom of speech, or of the press; or the right of the people peaceably to assemble, and to petition the government for a redress of grievances.

Second Amendment

A well-regulated militia, being necessary to the security of a free state, the right of the people to keep and bear arms, shall not be infringed.

Third Amendment

No soldier shall, in time of peace, be quartered in any house, without the consent of the owner, nor in time of war, but in a manner to be prescribed by law.

Fourth Amendment

The right of the people to be secure in their persons, houses, papers, and effects, against unreasonable searches and seizures, shall not be violated, and no warrants shall issue, but upon probable cause, supported by oath or affirmation, and particularly describing the place to be searched, and the persons or things to be seized.

Fifth Amendment

No person shall be held to answer for a capital, or otherwise infamous crime, unless on a presentment or indictment of a grand jury, except in cases arising in the land or naval forces, or in the militia, when in actual service in time of war or public danger; nor shall any person be subject for the same offense to be twice put in jeopardy of life or limb; nor shall be compelled in any criminal case to be a witness against himself, nor be deprived of life, liberty, or property, without due process of law; nor shall private property be taken for public use, without just compensation.

Sixth Amendment

In all criminal prosecutions, the accused shall enjoy the right to a speedy and public trial, by an impartial jury of the State and district wherein the crime shall have been committed, which district shall have been previously ascertained by law, and to be informed of the nature and cause of the accusation; to be confronted with the witnesses against him; to have compulsory process for obtaining witnesses in his favor, and to have the assistance of counsel for his defense.

Seventh Amendment

In suits at common law, where the value in controversy shall exceed twenty dollars, the right of trial by jury shall be preserved, and no fact tried by a jury shall be otherwise re-examined in any court of the United States, than according to the rules of the common law.

Eighth Amendment

Excessive bail shall not be required, nor excessive fines imposed, nor cruel and unusual punishments inflicted.

Ninth Amendment

The enumeration in the Constitution, of certain rights, shall not be construed to deny or disparage others retained by the people.

Tenth Amendment

The powers not delegated to the United States by the Constitution, nor prohibited by it to the States, are reserved to the States respectively, or to the people.

---

of 1832, which had been passed by Congress; the refusal of the governors of Connecticut, Massachusetts, and Rhode Island to place their militia under federal command during the War of 1812; and, of course, slavery.[54]

Although the Thirteenth Amendment, which was ratified on December 18, 1865, at the end of the Civil War, abolished slavery, debate over the application of the Bill of Rights continued. Finally, on July 28, 1868, the Fourteenth Amendment to the Constitution was

After lengthy debate and several drafts in wording, the Fourteenth Amendment to the U.S. Constitution passed Congress on June 13, 1866, and was sent to the states for ratification. When the last of the necessary 28 states voted to ratify the amendment, Secretary of State William Henry Seward proclaimed the Fourteenth Amendment part of the Constitution of the United States on July 28, 1868. It states:

All persons born or naturalized in the United States, and subject to the jurisdiction thereof, are citizens of the United States and of the State wherein they reside. No State shall make or enforce any law which shall abridge the privileges or immunities of citizens of the United States; nor shall any State deprive any person of life, liberty, or property, without due process of law; nor deny to any person within its jurisdiction the equal protection of the laws.

ratified. It was eventually interpreted to mean that the Bill of Rights did, indeed, apply to all citizens and that the states must ensure these rights.

Early court interpretations of the Fourteenth Amendment emphasized that its fundamental principle was "an impartial equality of rights"[55] and that its "plain and manifest intention was to make all the citizens of the United States equal before the law."[56] These initial decisions did not interpret the amendment to necessarily apply the Bill of Rights to the states. For example, in 1884, in *Hurtado v. California*, the Supreme Court held that the Fifth Amendment's guarantee of a grand jury indictment in criminal proceedings applied only to federal trials, not those conducted by the state.[57]

It was not until the early decades of the twentieth century that the due process clause of the Fourteenth Amendment, which guaranteed that no state shall "deprive any person of life, liberty, or property, without due process of the law," began to specifically incorporate the Bill of Rights. This move ultimately made the rights described in these amendments applicable to the states.

Incorporation of the Bill of Rights

The process of incorporation of the Bill of Rights occurred only gradually and reflected a major split on the Supreme Court. In 1947, Associate Justice Hugo Black strongly called for total incorporation, arguing that the framers of the Fourteenth Amendment originally intended the Bill of Rights to place limits on state action.[58] At the time, Black's

position was in the minority on the Court. The majority opinion, led by Justice Felix Frankfurter, held that, although the concept of due process incorporated fundamental values—one of which was fairness—it was left to judges to objectively and dispassionately discover and apply these values to any petitioner's claim of injustice. Due process expressed local values arising from different historical and practical considerations.

Indeed, the suggestion that the Fourteenth Amendment incorporates the first eight Amendments as such is not unambiguously urged. Even the boldest innovator would shrink from suggesting to more than half the states that they may no longer initiate prosecutions without indictment by grand jury, or that thereafter all the states of the Union must furnish a jury of twelve for every case involving a claim above twenty dollars. There is suggested merely a selective incorporation of the first eight Amendments into the Fourteenth Amendment. Some are in and some are out . . .[59]

Therefore, according to Frankfurter, the due process clause only selectively incorporated those provisions necessary to fundamental fairness. In a series of cases, the fundamental values protecting the First Amendment freedoms of speech, religion, and assembly were held to be binding on the states, but the Fifth Amendment's protection against double jeopardy was not.[60]

In 1953, when Earl Warren was appointed Chief Justice of the Supreme Court, a liberal majority was formed on the Court. It rapidly expanded the application of the due process clause to the states. Over the next two decades, the Warren Court handed down numerous decisions establishing individual and civil rights, and clearly moved the Court from its fundamental fairness position to one of absolute compliance.

During the 1960s, Chief Justice Earl Warren presided over a great expansion of due process rights for those accused of crimes.

 Were the decisions of the Warren Court beneficial for American society? How might crime control and due process enthusiasts debate this question?

LINK The application of these due process guarantees are discussed in several chapters of this book. Chapter 7 describes the application of due process in police decisions to arrest, search, and interrogate suspects. Chapter 10 examines the prosecution of suspects and considers due process issues in the determination of probable cause, charging decisions, notification of defendant rights, and entering of pleas. Due process issues are also raised in Chapter 11 in the discussion of a defendant's right to a speedy, public, and fair trial. Chapter 12 examines the application of due process to sentencing. Guarantees of due process also apply to people who are incarcerated in state and federal correctional facilities; these issues are examined in a discussion of inmates' rights in Chapter 14. Chapter 16 explores the decision of whether juveniles arrested for crimes are tried in the juvenile or adult criminal justice systems and considers how their constitutionally protected due-process rights differ from those of adults. Finally, Chapter 17 examines due process issues arising from the war on terror.

WRAP UP

Even though Goetz admitted to shooting the four youths in this case, he was acquitted of the most serious charges because, at times in the criminal justice system, there is a distance between the letter of the law and the spirit of the law. The "victims" in this case admitted that they planned to attack Goetz, and several of them had juvenile criminal records and went on to commit other serious violent crimes. The attackers were unsympathetic, and the defendant, who had previously been robbed and assaulted on the streets of New York, was acting in self-defense in a city in a time period in which crime was rampant. Given these facts, citizens and jurors alike held a certain degree of admiration for Goetz's willingness to take action before he was victimized. These considerations were likely in the minds of the jurors who chose to acquit Goetz of the most serious charges.

Chapter Spotlight

- Laws are formalized rules that reflect a body of principles prescribing or limiting people's actions. The laws collectively known as criminal law are generally divided into substantive law and procedural law. Together, they provide the framework for the criminal justice system.

- Most criminal law in the United States has its origins in English common law, although some basic precepts date back to early notions of natural and positive law found in Mosaic, Roman, and Greek codes. Probably the most important contribution from common law was *stare decisis* (the doctrine of precedent).

- Crimes have generally been conceptually divided between those considered to be *mala in se* (inherently wrong or evil acts) and those considered to be *mala prohibita* (acts that are wrong because they are prohibited by a criminal statute). Criminal codes further distinguish crimes as felonies, misdemeanors, and infractions.

- Criminal law functions in a variety of ways: It defines seriously socially unacceptable behavior, it provides for the control of behavior and maintenance of social order, and it regulates the punishment of behavior.

- For an act to be defined as a crime, a number of elements must be present: *actus reus* (criminal act), *mens rea* (criminal intent), and the concurrence of these two concepts.

- In certain circumstances, an individual might engage in an act defined as a crime yet not be held criminally responsible for that action. These circumstances involve legal justifications and excuses, or defenses that negate a person's criminal responsibility.

- Although the insanity defense is successfully raised in less than 1 percent of all criminal cases, it is still very controversial. The federal government and many of the states have revised their insanity statutes in recent years, and several have developed "guilty, but mentally ill" statutes to supplement other insanity defenses.

- Procedural criminal law establishes protections for individuals against unfair prosecution. These safeguards are found in the Fourth, Fifth, Sixth, and Eighth Amendments contained in the Bill of Rights.

- The constitutional protections found in the Bill of Rights were initially interpreted to apply only in federal prosecutions. It was not until the adoption of the Bill of Rights that they began to be applied to the states as well.

Putting It All Together

1. Is it possible for natural law and positive law to coexist? Explain your answer.

2. Discuss the major functions of law in society. What alternatives to law might fulfill the same functions?

3. Identify four distinctions between criminal law and civil law. Do you think that it is reasonable for a crime victim to be able to file a civil suit against an offender? Why or why not?

4. Should the insanity defense be allowed? Should all states adopt "guilty, but mentally ill" or "guilty but insane" statutes? Why or why not?

5. Are the guarantees of due process for people accused of crimes reasonable? Do they make it more difficult to deal with the crime problem? Why are they so important to protect?

actus reus Guilty act; a required material element of a crime.

Bill of Rights First 10 amendments to the U.S. Constitution.

case law Law that emerges when a court modifies a law in its application in a particular case.

civil law A body of private law that settles disputes between two or more parties to a dispute.

common law Case decisions by judges in England that established a body of law common to the entire nation.

corpus The body of the crime; the material elements of the crime that must be established in a court of law.

crime An intentional act or omission to act, neither justified nor excused, that is in violation of criminal law and punished by the state.

double jeopardy Trying a person for the same crime more than once; it is prohibited by the Fifth Amendment.

Durham rule An insanity test that determines whether a defendant's act was a product of a mental disease or defect.

entrapment The claim that a defendant was encouraged or enticed by agents of the state to engage in a criminal act.

excuses Claims based on a defendant admitting that what he or she did was wrong but arguing that, under the circumstances, he or she was not responsible for the criminal act.

felony A serious crime, such as robbery or embezzlement, that is punishable by a prison term of more than one year or by death.

guilty, but mentally ill (GBMI) A substitute for traditional insanity defenses, which allows the jury to find the defendant guilty and requires psychiatric treatment during confinement. Also called guilty but insane (GBI).

Hurtado v. California Supreme Court decision that the Fifth Amendment guarantee of a grand jury indictment applied only to federal—not state—trials, and that not all constitutional amendments were applicable to the states.

incorporation The legal interpretation by the Supreme Court in which the Fourteenth Amendment applied the Bill of Rights to the states.

infraction A violation of a city or county ordinance, such as cruising or noise violations.

irresistible impulse test An insanity test that determines whether a defendant, as a result of a mental disease, temporarily lost self-control or the ability to reason sufficiently to prevent the crime.

justification Defense wherein a defendant admits responsibility but argues that, under the circumstances, what he or she did was right.

laws Formalized rules that prescribe or limit actions.

mala in se Behaviors, such as murder or rape, that are considered inherently wrong or evil.

mala prohibita Behaviors, such as prostitution and gambling, that are considered wrong because they have been prohibited by criminal statutes, rather than because they are evil in themselves.

mens rea Guilty mind, or having criminal intent; a required material element of a crime.

misdemeanor A crime that is less serious than a felony, such as petty theft or possession of a small amount of marijuana, and that is punishable by less than one year in prison.

M'Naghten rule Insanity defense claim that because of a defect of reason from a disease of the mind, the defendant was unable to distinguish right from wrong.

procedural criminal law A body of law that specifies how crimes are to be investigated and prosecuted.

self-defense Claim that a defendant acted in a lawful manner to defend himself or herself, others, or property, or to prevent a crime.

stare decisis Literally, "to stand by the decision"; a policy of the courts to interpret and apply law according to precedents set in earlier cases.

statute Legislation contained in written legal codes.

statute of limitations The maximum time period that can pass between a criminal act and its prosecution.

strict liability laws Laws that provide for criminal liability without requiring either general or specific intent.

substantial capacity test An insanity test that determines whether the defendant lacked sufficient capacity to appreciate the wrongfulness of his or her conduct.

substantive criminal law A body of law that identifies behaviors harmful to society and specifies their punishments.

Notes

1. Aristotle, *The Politics*, Book VII, trans. Carnes Lord (Chicago: University of Chicago Press, 1994), p. 4.
2. Charles Thomas and Donna Bishop, *Criminal Law: Understanding Basic Principles* (Newbury Park, CA: Sage, 1987), p. 14.
3. Hermann Mannheim, *Comparative Criminology* (Boston: Houghton Mifflin, 1967), p. 47.
4. Thomas and Bishop, note 2, p. 25.
5. Thomas and Bishop, note 2, pp. 25–26.
6. P. S. Atiyah and Robert Summers, *Form and Substance in Anglo-American Law: A Comparative Study in Legal Reasoning, Legal Theory and Legal Institutions* (Oxford, UK: Oxford University Press, 1987), p. 97.
7. *Roe v. Wade,* 410 U.S. 113 (1973).
8. Jack Douglas and Frances Waksler, *Sociology of Deviance* (Boston: Little, Brown, 1982).
9. *Robinson v. California,* 370 U.S. 660 (1962).
10. U.S. Department of Justice, *Report to the Nation on Crime and Justice,* 2nd ed. (Washington, DC: Bureau of Justice Statistics, 1988), p. 16.
11. John D. Hewitt, "Gardeners Shape a New Future for Delinquents," *China Reconstructs,* August 1987, pp. 29–31.
12. Kai Erikson, *Wayward Puritans: A Study in the Sociology of Deviance* (New York: John Wiley & Sons, 1966).
13. Herbert Blumer, *Symbolic Interactionism* (Englewood Cliffs, NJ: Prentice Hall, 1969); Howard Becker, *Outsiders: Studies in the Sociology of Deviance* (New York: Free Press, 1963).
14. *Lawrence v. Texas,* 539 U.S. 558 (2003).
15. *Lawrence,* note 14.
16. Thorsten Sellin, *Culture Conflict and Crime* (New York: Social Science Research Council, 1938).
17. George Orwell, *1984* (New York: Penguin, 1981).
18. American Law Institute, *Model Penal Code, Proposed Official Draft,* Section 2.02 (1) (Philadelphia: American Law Institute, 1962).
19. *State v. Daniels,* 109 So.2d 896 (1959).
20. Adam Liptak, "15 States Expand Right to Shoot in Self-Defense," *New York Times,* August 7, 2006, available at http://www.nytimes.com/2006/08/07/us/07shoot.html, accessed June 28, 2007.

21. American Law Institute, note 18, Section 3.02. 1.
22. Gary Tuchman and Brian Cabell, "Kobe Bryant Charged with Sexual Assault," *CNN.com,* July 19, 2003, available at http://www.cnn.com/2003/LAW/07/19/kobe.bryant/index.html, accessed August 6, 2007; "Kobe Bryant Sexual Assault Case," *CourtTV* News, available at http://www.courttv.com/trials/bryant/071804_ctv.html, accessed August 6, 2007; "Judge Dismisses Kobe Bryant Rape Case," available at http://www.courttv.com/trials/bryant/090104_ctv.html, accessed June 28, 2007; Patrick O'Driscoll, "Kobe Bryant, Accuser Settle Her Civil Lawsuit," *USA Today,* March 2, 2005, available at http://www.usatoday.com/sports/basketball/nba/2005-03-02-bryant-settles_x.htm, accessed June 28, 2007.
23. *Gonzales v. Oregon,* No. 04-623 (2006).
24. Henry Steadman and Jeraldine Braff, "Defendants Not Guilty by Reason of Insanity," pp. 200–237 in John Mohanan and Henry Steadman (Eds.), *Mentally Disordered Offenders* (New York: Plenum, 1983); John P. Martin, "The Insanity Defense: A Closer Look," *Washington Post,* February 27, 1998, available at http://www.washingtonpost.com/wp-srv/local/longterm/aron/qa227.htm, accessed June 28, 2007; "Texas Lawmakers Studying State's Insanity Defense," *Associated Press,* January 4, 2005, available at http://www.demaction.org/dia/organizations/ncadp/news.jsp?key=1080&t=, accessed June 28, 2007.
25. Rita Simon and David Aaronson, *The Insanity Defense* (New York: Praeger, 1988).
26. Simon and Aaronson, note 25, p. 10.
27. Rita Simon, *The Jury and the Defense of Insanity* (Boston: Little, Brown, 1967), p. 17.
28. *M'Naghten's Case,* 8 Eng. Rep. 718 (H.I. 1843).
29. *Parsons v. State,* 81 Alabama 577 (1886).
30. *Durham v. United States,* 214 F.2d 862 (D.C. Cir. 1954).
31. American Law Institute, note 18, Section 4.01.
32. Irving Kaufman, "The Insanity Plea on Trial," *New York Times Magazine,* August 8, 1982, p. 18.
33. *United States v. Brawner,* 471 F.2d 969 (1972).
34. U.S. Congress, *The Comprehensive Crime Control Act of 1984,* P.L. 98-473 (Washington, DC: U.S. Congress, 1984).
35. John Klofas and Janette Yandrasits, "'Guilty But Mentally Ill' and the Jury Trial: A Case Study," *Criminal Law Bulletin* 24:425 (1988).
36. "Yates Not Guilty by Reason of Insanity," *CNN.com,* July 27, 2006, available at http://www.cnn.com/2006/LAW/07/26/yates.verdict/, accessed June 28, 2007; Patty Reinert, "Supreme Court Hears Argument on Insanity Defense," *Houston Chronicle,* June 14, 2006, available at http://www.chron.com/disp/story.mpl/nation/3805879.html, accessed June 28, 2007; Pam Easton, "Texas Jurors Convict Man in Boy's Death," *Associated Press,* May 12, 2005, available at http://news.ewoss.com/articles/D8A1CCVO1.aspx, accessed June 28, 2007.
37. *Louisiana Revised Statutes,* 14:15.
38. Eric Weslander, "Dillon Plans 'Intoxication Defense,'" *Lawrence Journal-World,* November 2, 2005, available at http://www2.ljworld.com/news/2005/nov/02/dillon_plans_intoxication_defense/?city_local, accessed June 28, 2007; Eric Weslander, "Child's Killer Expresses Regret as He's Given 16-Year Term," *Lawrence Journal-World,* March 4, 2006, available at http://www2.ljworld.com/news/2006/mar/04/childs_killer_expresses_regret_hes_given_16year_te/, accessed June 28, 2007.
39. *Jacobson v. United States,* 503 U.S. 540, 548 (1992).
40. *Mathews v. United States,* 485 U.S. 58, 63 (1988).
41. *Sherman v. United States,* 356 U.S. 369 (1958).
42. *United States v. Russell,* 411 U.S. 423 (1973).
43. American Law Institute, note 18, Section 2.09.1.
44. *United States v. Cheek,* 498 U.S. 192 (1994).
45. *Lambert v. California,* 365 U.S. 225 (1957).
46. *Lambert v. California,* note 45.
47. David Jones, *Crime and Criminal Responsibility* (Chicago: Nelson-Hall, 1978), pp. 67–68.
48. Jones, note 47, p. 68.
49. *Bartkus v. Illinois,* 359 U.S. 121 (1959).
50. Mo. Rev. Stat. § 556.037.
51. Me. Rev. Stat. Ann. 14 § 752-C.
52. Christy McDonald and Chris Pavelich, "First-Grader Shot Dead at School," March 8, 2000, available at http://www.ABC-NEWS.com/2000/3/8, accessed June 28, 2007.
53. Thomas Grey, "Procedural Fairness and Substantive Rights," in J. Roland Pennock and John Chapman (Eds.), *Due Process* (New York: New York University Press, 1977), p. 185.
54. William Nelson, *The Fourteenth Amendment: From Political Principle to Judicial Doctrine* (Cambridge, MA: Harvard University Press, 1988), pp. 27–29.
55. *State v. Hairston and Williams,* 63 N.C. 439 (1869).
56. *State v. Gibson,* 36 Ind. 393 (1871).
57. *Hurtado v. California,* 110 U.S. 516 (1884).
58. *Adamson v. California,* 332 U.S. 46 (1947).
59. *Adamson v. California,* note 58.
60. *Palko v. Connecticut,* 302 U.S. 319 (1937).

JBPUB.COM/**ExploringCJ**

Interactives
Key Term Explorer
Web Links

OBJECTIVES

■ Understand the strengths and weaknesses of the Uniform Crime Reports (UCR) and National Crime Victimization Survey (NCVS).

■ Explain the relationship between UCR and NCVS data.

■ Outline how self-report studies add to our understanding of criminality.

■ Know what the "dark figure of crime" is and how it affects estimates of criminality.

■ Describe the prevalence and incidence of crime in the United States.

■ Identify the differences and similarities among juvenile, white-collar, and senior citizen offenders.

■ Understand the nature and extent of the criminal victimization experienced by infants, teenagers, college students, senior citizens, and intimate partners.

■ Explain how crime victims are a part of the criminal justice system.

Crime, Offenders, and Victims

CHAPTER

3

When college students are asked to estimate the number of crimes they have committed in the past year (generally acts such as underage drinking, vandalism, or shoplifting) and create a ratio of their criminal activity to their arrests, the results may be surprising—ratios like 150:0 and 500:1 are not uncommon. Prison interviews show that many offenders have committed nearly 10,000 crimes each year with only a minimum number of arrests. This discrepancy should be cause for alarm, as demonstrated by the case of Coral Eugene Watts.

Watts was arrested in May 1982 in Houston, Texas, for attempting to murder two women after breaking into their homes. Watts was suspected to be responsible for 12 additional murders in Texas and Michigan, but authorities believed that they did not have enough evidence to convict him of those crimes. Texas prosecutors offered Watts immunity from prosecution for the 12 murders in exchange for a guilty plea to the charge of burglary with intent to murder and a 60-year prison sentence. Watts accepted the deal.

After Watts served 22 years in prison, he became eligible for conditional release, even though he openly admitted that he would kill upon release and confessed to committing between 40 and 80 murders in the past. Michigan authorities quickly mobilized an investigation before Watts could be released. In December 2004, Watts was convicted of murder and sentenced to life imprisonment, thereby averting the release of a serial killer.

- What do the common ratios of criminal activity to arrests suggest about crime statistics?

- What are the strengths and weaknesses of using arrests and self-reports to calculate the prevalence of crime?

Source: Rachael Bell, "Coral Eugene Watts: The Sunday Morning Slasher," available at http://www.crimelibrary.com/serial_killers/predators/coral_watts/, accessed March 5, 2007.

Introduction

Historically, crime has always been difficult to measure. Years ago, the economist Sir Josiah Stamp complained that crime statistics "come in the first instance from the village watchman, who just puts down what he damn pleases."[1] Even today, criminologists agree that public information about crime is not very accurate, in part because crime is both context- and time-specific. That is, behavior is evaluated differently depending on where and when it occurs. For example, in 1992, chewing gum was illegal in Singapore. This ban has since been relaxed but not entirely removed. Singaporeans today may purchase chewing gum only in a pharmacy and must submit their names and ID card numbers when buying it.[2] Another problem with crime data is that some people commit crimes relatively frequently but are never caught, whereas others may be arrested when committing their first offense. Thus arrest records do not always reflect a person's actual involvement in crime. To ease these problems, criminologists have developed multiple yardsticks that, when taken together, provide a respectable approximation of the extent and nature of criminality.

LINK The sticky crime of gum litterbugs in Europe is explored in "Around the Globe" in Chapter 1.

Two main measures of crime are distinguished: official crime statistics, which are based on the aggregate records of offenders and offenses processed by police, courts, and corrections agencies; and unofficial crime statistics, which are produced by people and agencies outside the criminal justice system. The majority of criminal statistics come from three sources: the URC, which is produced by the Federal Bureau of Investigation (FBI); the NCVS, which is produced by the U.S. Bureau of Justice Statistics; and an unofficial measure—self-report studies.

Uniform Crime Reports

One of the earliest national measures of crime was the Uniform Crime Reports (UCR).[3] Since its inception in 1929, the UCR has collected statistics from local and state law enforcement agencies on Part I offenses, also known as Crime Index offenses, which include the violent crimes of murder and non-negligent manslaughter, forcible rape, robbery, aggravated assault, and the property crimes of burglary, larceny, motor vehicle theft, and arson (the last of which was added in 1979) **TABLE 3-1**. In addition to crimes known to the police, the UCR provides information on number of arrests and characteristics of person arrested including the suspect's sex, race, and age. Today, the UCR represents a nationwide, cooperative effort involving roughly 17,000 law enforcement agencies (about 95 percent of all U.S. policing agencies) that voluntarily report data on crime to the FBI. Participation in the UCR program is not mandatory, however, so some state and local law enforcement agencies do not supply data. These omissions make it difficult for criminologists to assemble comprehensive crime data for the United States. In 2005, for instance, law enforcement agencies participating in the program represented approximately 296 million U.S. residents, or about 94 percent of the population. Thus crimes committed by 6 percent of the U.S. population (about 18 million people) were not included in UCR data.[4] The data that law enforcement agencies willingly report to the FBI are published annually in a report titled *Crime in the United States*.

Problems with UCR Data

While UCR data certainly represent an improvement over the village watchmen's haphazard guesses, there are still several criticisms about the data's accuracy:

The UCR reports only crimes known to the police. Because a high percentage of crime victims do not report their experiences to law enforcement agencies participating in the UCR program, the data that are reported underestimate both the incidence (number of crimes committed) and the prevalence (number of offenders) of crime. Victims tend to not report crime for a variety of reasons, including that they consider the crime to be a private or personal matter, they do not think it is important enough, or they fear reprisal. Because most crime is not reported to the police, there is a large gap between the actual number of crimes committed and the number of crimes reported to the police. The dark figure of crime is the term used by criminologists to describe the amount of unreported or undiscovered crime, which calls into question the reliability of UCR data. For instance, the UCR does not report drug offenses or information on federal crimes, including insurance fraud and tax evasion, as "known" crimes.[5]

The UCR reports on only the most serious crime incident. The information reported in the UCR is based

---

TABLE 3-1

UCR Serious Criminal Offenses

Beginning in 2004, the UCR no longer reports a Crime Index; rather, it simply provides data on the number of people arrested and crimes known to the law enforcement agencies for the following eight categories of serious violent and property crimes.

Serious Violent Crimes

Murder and nonnegligent manslaughter—the willful killing of one person by another
Forcible rape—the carnal knowledge of a female forcibly and against her will
Robbery—the taking or attempting to take anything of value from the care, custody, or control of a person or persons by force or threat of force or violence and/or by putting the victim in fear
Aggravated assault—the unlawful attack by one person upon another for the purpose of inflicting severe or aggravated bodily injury

Serious Property Crimes

Burglary—the unlawful entry into a structure to commit a felony or theft
Larceny–theft—the unlawful taking, carrying, leading, or riding away of property from the possession or constructive possession of others
Motor vehicle theft—the theft or attempted theft of a motor vehicle
Arson—any willful or malicious burning or attempt to burn, with or without intent to defraud, a dwelling house, public building, motor vehicle or aircraft, or the personal property of another

Source: Federal Bureau of Investigation, *Crime in the United States, 2006* (Washington, DC: U.S. Department of Justice, 2007).

---

on the hierarchy rule: For a single crime incident in which multiple offenses were committed, only the most serious offense is reported. (Arson is an exception; it is always reported to the FBI.) Thus, if an offender robs and murders a victim, only the murder is reported. The hierarchy rule also affects international crime rate comparisons because many other countries include each crime in a multiple-offenses incident in their statistics.

The UCR does not collect all relevant data. The UCR collects crime details about the victim, the offender, and the circumstance only for homicide cases. The types of weapons used are gathered only for murder, robbery, and aggravated assault. Weapons used in forcible rape are not reported, and data for rapes include only female victims. (Forcible rape of a male victim is recorded as aggravated assault.)

The UCR reveals more about police behavior than it does about criminality. Some law enforcement agencies falsify the reports they submit to the FBI. Once a citizen reports a crime, police must make an official record for the crime to be counted in the UCR. Sometimes, however, law enforcement officers do not complete a crime report. For example, in Atlanta, crimes that were reported to police were not recorded for a number of years to help the city land the 1996 Olympic Games and boost tourism. In 2002, researchers discovered more than 22,000 missing police reports that never were submitted to the FBI. Those reports included more than 4000 violent offenses that were committed but never counted.[6] The National Center for Policy Analysis also has discovered that police agencies in Boca Raton (Florida), New York, and Philadelphia have systematically underreported or downgraded crimes in their cities. Both individual officers and police departments may take these steps in response to the extreme pressure they face on a daily basis to demonstrate that they are doing their job. The findings from the National Center for Policy Analysis lend credence to research conducted by Donald Black, who discovered that police filed a crime report in only 64 percent of crimes where no suspect was present, even though a complainant had reported the crime to law enforcement.[7]

In spite of these criticisms, the UCR continues to be widely used and remains the primary source of national estimates of the nature and extent of criminality in the United States. Although there are certainly flaws in the UCR data, the data are still stable enough to provide trend data on changes over the years. Ultimately, the mass of research accumulated over many years provides basic support regarding the validity of the UCR.[8]

Reforming the UCR

Efforts are continuously underway to improve the reliability and validity of official statistics. For instance, recognizing the need for more detailed crime statistics, law enforcement called for a thorough evaluative analysis that would modernize the UCR program. In 1982, these studies led to the creation and implementation of the National Incident-Based Reporting System (NIBRS).[9]

The NIBRS is currently a component of the UCR program and is expected to eventually replace it. The NIBRS differs from the UCR in several significant ways TABLE 3-2 . For example, the NIBRS collects data on each single incident and arrest. When a crime becomes known to the police, information is then gathered for the following categories: the crime incident, the victim, the nature of the property, and the characteristics of the arrested suspect. A total of 53 data elements are recorded for crimes in 22 categories TABLE 3-3 .

Data produced by NIBRS are of significant benefit to local agencies. When they have access to such comprehensive crime data, local police agencies may be more effective in making arguments for acquiring and then effectively allocating the resources needed to respond to crime.[10] Although only about 36 percent (about 6500) of all U.S. law enforcement agencies currently report data to the NIBRS, it is expected that over the next few years the number of participating agencies will increase dramatically.[11]

In addition to the creation of the NIBRS, another important change to the UCR occurred in 2004 when the FBI discontinued use of the Crime Index. The original purpose of the Crime Index was to indicate whether overall serious crime was increasing or decreasing in the United States. Today, only the number of serious violent and serious property crimes known to and reported to the police is included in *Crime in the United States.*[12]

National Crime Victimization Survey

Victimization surveys, which ask victims of crime about their experiences, were first developed in the late 1960s, partly in response to the inability of the UCR to provide accurate estimates of the dark figure of crime. Like the NIBRS, victim surveys gather specific information about such crime characteristics as when and where the crime occurred, whether a weapon was used, and whether there was any known relationship between victim and offender.[13]

The first nationwide victimization survey was conducted by the National Opinion Research Center in 1967, which contacted 10,000 households. Interviewers asked a knowledgeable person in each household a few short screening questions that were used to determine whether a member of the household had been a crime victim during the preceding year. Surprisingly, survey results showed that the victimization rate reported for Crime Index offenses was greater than twice the rate reported in the UCR.[14]

TABLE 3-2

Differences Between UCR and NIBRS Data

	UCR	NIBRS
Offenses reported	Part I offenses (8 crimes)	Group A offenses (22 crimes)
Rape	Female victims only	Male and female victims
Attempted versus completed offenses	Does not differentiate	Does differentiate
Multiple-offenses crime incidents	Hierarchy rule: reports only the most serious offense	All offenses are reported
Weapons	Recorded only for cases of murder, robbery, and aggravated assault	All weapons data are recorded
Crime categories	Crimes against persons (e.g., murder, rape, and aggravated assault) Crimes against property (e.g., robbery, burglary, and larceny—theft)	Crimes against persons Crimes against property Crimes against society (e.g., drug or narcotic offenses)

TABLE 3-3

NIBRS Serious Criminal Offenses

The following offense categories, known as *Group A Offenses,* are those for which extensive crime data are collected in the NIBRS:

1. Arson
2. Assault offenses: aggravated assault, simple assault, intimidation
3. Bribery
4. Burglary/breaking and entering
5. Counterfeiting/forgery
6. Destruction/damage/vandalism of property
7. Drug/narcotic offenses: drug/narcotic violations, drug equipment violations
8. Embezzlement
9. Extortion/blackmail
10. Fraud offenses: false pretenses/swindle/confidence game, credit card/automatic teller machine fraud, impersonation, welfare fraud, wire fraud
11. Gambling offenses: betting/wagering, operating/promoting/assisting gambling, gambling equipment violations, sports tampering
12. Homicide offenses: murder and non-negligent manslaughter, negligent manslaughter, justifiable homicide
13. Kidnapping/abduction
14. Larceny/theft offenses: pocket picking, purse snatching, shoplifting, theft from building, theft from coin-operated machine or device, theft from motor vehicle, theft of motor vehicle parts or accessories, all other larceny
15. Motor vehicle theft
16. Pornography/obscene material
17. Prostitution offenses: prostitution, assisting or promoting prostitution
18. Robbery
19. Sex offenses, forcible: forcible rape, forcible sodomy, sexual assault with an object, forcible fondling
20. Sex offenses, nonforcible: incest, statutory rape
21. Stolen property offenses (receiving, etc.)
22. Weapon law violations

Source: Federal Bureau of Investigation, *Developments in the National Incident-Based Reporting System* (Washington, DC: U.S. Department of Justice, 2004).

In 1972, the U.S. Bureau of Justice Statistics began the National Crime Survey, which was renamed the National Crime Victimization Survey (NCVS) in 1990; it is the most comprehensive and systematic survey of victims in the United States.[15] The NCVS produces data on both personal and household crimes:

Personal Crimes
- Aggravated Assault
- Robbery
- Rape
- Sexual assault

Household Crimes
- Burglary
- Motor vehicle theft
- Theft

These seven offenses constitute the crimes of interest, so selected because victims are likely to report them to police and victims are usually able to recall them when questioned about them.

Approximately 134,000 people, representing about 77,000 houses, are interviewed each year for the NCVS. If the member of the household who was a victim was younger than age 12 at the time of the

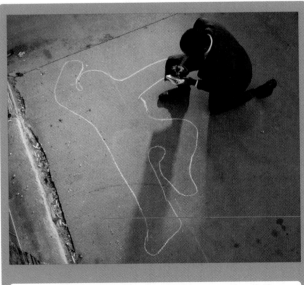

Murder is not measured by the NCVS because the victim obviously cannot be interviewed.

 Should the NCVS researchers interview family members of murder victims, a group known as "co-victims of homicide"? How could the NCVS further its understanding of homicide?

crime, an older member of the household is asked to provide the information. A few screening questions are asked of each person interviewed. These screening questions are used to determine whether the interviewee was a victim of one or more of the crimes of interest; if so, additional questions are then asked.

In addition to providing better estimates on the dark figure of crime, NCVS data perform the following functions:

- The data help criminologists to better understand why so many victims do not report crime incidents to police.
- The data demonstrate that variations in crime reporting depend greatly on the type of offense, crime situation factors, the characteristics of the victim (e.g., his or her race, sex, and social class), and the nature of the victim–offender relationship.
- The data allow criminologists to establish theoretical explanations for how crime often results from social interactions between victims and offenders.[16]

Problems with NCVS Data

As with the UCR data, there are several criticisms about the NCVS:

The NCVS is limited in scope. Obviously, the small number of crimes of interest is a problem because they represent only a tiny fraction of all crimes committed. The majority of crimes committed in the United States involve alcohol and illegal drugs, and the many robberies, burglaries, and crimes committed against commercial establishments such as bars, businesses, and factories are not included in the NCVS.[17] By excluding these and other crimes such as computer hacking, insider trading, and public order crimes (e.g., driving under the influence of alcohol or drugs, illegal gambling, and prostitution), the NCVS provides data on only a small subset of all crime incidents. Also, because murder victims cannot be interviewed, the most serious of all crimes cannot be included in the survey.

Interview data may be unreliable. Because respondents do not have to meet legal or evidentiary standards to report crimes committed against them, NCVS data may overreport crimes that law enforcement would have considered unfounded and excluded from UCR data.

Additionally, because the NCVS is based on the answers that people give to questions regarding past and sometimes troublesome events, their responses are vulnerable to several types of biases:

- *Memory errors*—difficulty recalling details about the event.
- *Telescoping*—difficulty remembering the time of the crime, because it may feel as though the event occurred more recently than it did given that the incident remains vivid in the interviewee's memory.
- *Errors of deception*—difficulty reporting events that are embarrassing, unpleasant, or self-incriminating. (Some people also fabricate crime incidents.)
- *Sampling error*—difficulty including populations outside of ordinary households and resolving discrepancies between sample estimates of behavior and the actual amount of behavior. For instance, because the sampling unit in the NCVS is households, homeless people, who are at greater risk of victimization, are excluded from the sample.[18]

In the 1990s, changes were made to the NCVS to increase the likelihood of respondents recalling events accurately. These changes included a variety of questions and cues designed to more accurately elicit the victims' memories of crime incidents. In addition, interviewers were asked more explicit questions about sexual victimizations. For instance, interviewees are now asked: "Have you been forced or coerced to engage in unwanted sexual activity by (1) someone you didn't know before, (2) a casual acquaintance, or (3) someone you know well?"

The first results from the redesigned survey were released in 1992. One of the most significant findings was that victims appeared to recall and report more types of crime incidents than respondents had in the past. This trend was particularly apparent for aggravated assault, rape, and simple assault.

Acclaim for the NCVS

Because NCVS data tend to match official crime reports, the survey helps to provide an understanding about the true magnitude of crime with greater confidence, validity, and reliability. When researchers Janet Lauritsen and Robin Schaum recently compared UCR and NCVS data for robbery, burglary, and aggravated assault in Chicago, Los Angeles, and New York from 1980 to 1998, they found that for burglary and robbery, UCR crime rates were generally similar to NCVS estimates over the study period. Police and victim survey data were more likely to show discrepancies in levels and trends related to aggravated assault. Lauritsen and Schaum also found that even when UCR and NCVS

data were different, the differences were not statistically significant.[19] In other words, the UCR and NCVS tell the same story about the extent of these three serious crimes. Indeed, for more than 30 years, criminologists have found that UCR and victimization data generally report similar results regarding the incidence of criminality in the United States.[20]

Self-Report Surveys

Self-report surveys are an unofficial source of crime data. They provide criminologists with a method for collecting data without having to rely on government resources. In self-report surveys, criminologists ask respondents to identify their own criminality during a specific time period, such as during the prior year. Self-report surveys have generally focused on juvenile crime because youths are more easily surveyed in schools, courts, detention centers, and correctional facilities and are more likely than adults to report their illegal behaviors. For more than 60 years, criminologists have consistently found that 85 to 90 percent of persons report having committed criminal behavior that could have led to their arrest had they been caught.[21]

The most comprehensive self-reported survey so far developed is the National Youth Survey (NYS).[22] The NYS interviews a random sample of approximately 1700 youths who were between the ages of 11 and 17 when they were first interviewed. These youths were originally drawn from more than 100 communities around the country and are representative of the socioeconomic, race, and ethnicity of youths in the United States. The original sample has now been reporting on their criminal activity for more than 30 years.

Problems with Self-Report Surveys

While valuable sources of information, self-report surveys do have some problems:

Self-report survey data are not always reliable. When people are asked to tell strangers about their illegal acts, they may lie about their criminal involvement. In addition, many people forget, misunderstand, or distort their participation in crime. Typically, the most active criminals do not participate in self-report surveys because they are not likely to reveal themselves or their activities to strangers.[23]

Self-report studies often exclude serious chronic offenders. Because many of these surveys sample college student populations, it is not surprising that only a small amount of serious crime is detected. In fact, when Stephen Cernkovich and his colleagues compared the self-reported behavior of incarcerated youths and nonincarcerated youths, they discovered that self-report studies often focus on less serious, occasional offenders. These researchers observed significant differences in the offending patterns of the two groups, leading them to conclude that "institutionalized youth are not only more delinquent than the 'average kid' in the general population, but also considerably more delinquent than the most delinquent youth identified in the typical self-report."[24]

These concerns have inspired criminologists to develop methods to validate the findings from self-report studies. Findings from studies that employed such validity checks have provided general support of the self-report method. In a comprehensive review of the reliability and validity of self-reports, Michael Hindelang and his colleagues concluded that the difficulties of self-report instruments are surmountable and that the self-report method is not fundamentally flawed.[25] Thus, like the UCR and the NCVS, self-report studies provide criminologists with a variety of data for use in making generalizations about the nature and extent of crime in the United States.

Self-report studies typically discover trivial events. Studies often find an abundance of respondents who report stealing a small sum of money, using fake identification, occasionally smoking marijuana, and having premarital sex prior to age 18 (a crime in some states). These crimes do not help criminologists to better understand or construct policies to address the problem of crime.[26] Unfortunately, some criminologists who rely on self-report studies may lump together children who commit innocuous offenses with adjudicated delinquency and conclude, for instance, that there is no relationship between socioeconomic status and crime.

Acclaim for Self-Report Surveys

Self-report surveys provide criminologists with much information about crime. For example, they have established that more than 90 percent of all juveniles have committed at least one criminal act.[27] Similarly, these surveys have added to our awareness of the real extent of the dark figure of crime, pinning it down to somewhere between 4 and 10 times greater than the crime rate reported in the UCR. Finally, self-report research provides clear evidence of race, ethnic, and gender bias in the official processing of suspects.[28]

Crime Statistics for the United States

There is no perfect measure of crime. Each method of unearthing crime data has both strengths and weaknesses. The best single source of data for estimating serious violent and property crimes is the UCR, which describes 1-year, 5-year, and 10-year trends. Reasonable estimates of less serious crimes can be derived from victimization surveys and self-report studies, although self-report studies are preferable for gauging drug offenses. All of these measures provide some useful information about crime in the United States. When the data from the various sources are merged, they provide criminologists with a much better understanding of the nature and extent of crime.

There are now more than 300 million people living in the United States, and the population is projected to increase in the coming decades FIGURE 3-1. In this country, a serious violent crime is committed every 22 seconds, and a serious property crime is committed every 3 seconds FIGURE 3-2.[29] In 2005, more than 1 million serious violent crimes and more than 10 million serious property crimes were reported to law enforcement agencies.[30] These figures represent a very conservative estimate of the amount of crime committed, however, because most crime is not reported to police.

Based on estimates derived from crime victim studies, from 1993 to 2003, on average, only 44 percent of serious violent crimes and 35 percent of

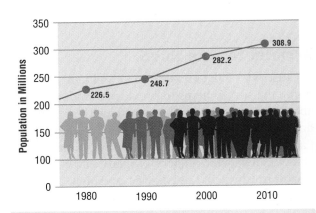

FIGURE 3-1 U.S. Population, 1980–2010

Source: U.S. Census Bureau, *Statistical Abstract of the United States, 2007* (Washington, DC: U.S. Census Bureau, 2006), available at http://www.census.gov/compendia/statab/, accessed June 28, 2007.

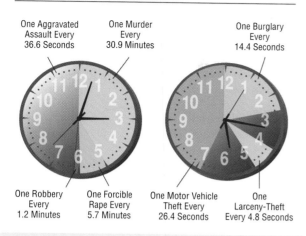

One Violent Crime Every 22.2 Seconds

One Property Crime Every 3.2 Seconds

One Aggravated Assault Every 36.6 Seconds

One Murder Every 30.9 Minutes

One Burglary Every 14.4 Seconds

One Robbery Every 1.2 Minutes

One Forcible Rape Every 5.7 Minutes

One Motor Vehicle Theft Every 26.4 Seconds

One Larceny-Theft Every 4.8 Seconds

FIGURE 3-2 2006 UCR Crime Clock

Source: Federal Bureau of Investigation, *Crime in the United States, 2006* (Washington, DC: U.S. Department of Justice, 2007).

serious property crimes were reported to police. In addition, NCVS data published in 2007 indicate that Americans age 12 and older reported being the victims of more than 16 million property and violent victimizations. These criminal victimizations included more than 14 million property crimes and 3.3 million violent crimes, and nearly 200,000 personal thefts.[31] Yet, between 1994 and 2005, the overall U.S. crime rate dropped 34 percent, as did rates of violent and property crime **FIGURE 3-3** .

Several explanations for the lower crime rate in recent decades have been suggested, including the controversial idea that legalized abortion may have played a significant role (see "Focus on Criminal Justice").[32] Other explanations for the dramatic decrease in crime during the past decade include changes in the following areas:

- *The economy.* The lower crime rate may be tied to changes in the U.S. economy. During a recession, fewer youth-initiated crimes may occur because unemployed parents are more likely to stay home and supervise their children. In times of economic expansion, the economy might provide people with legitimate opportunities to earn money, making crime a less desirable option.

- *Prisons.* Policies designed to incarcerate a greater number of offenders for longer periods of time tend to reduce crime rates. For example, the U.S. incarceration rate increased from 313 inmates per 100,000 population in 1985 to 737 inmates per 100,000 population in 2005—a 135 percent increase over this 20-year span. The crime rate dropped by more than 25 percent during this same period.[33]

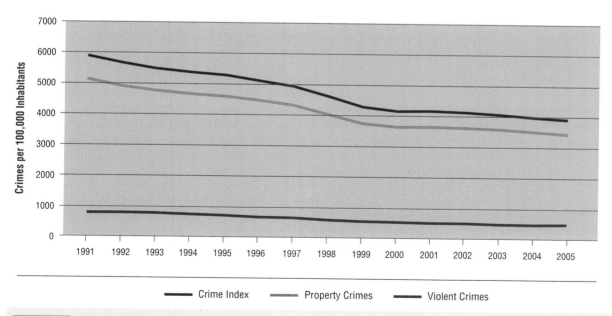

FIGURE 3-3 U.S. Crime Rate Index

Source: The Disaster Center, *U.S. Crime Rates, 1960–2005* (Washington, DC: Federal Emergency Disaster, 2006), available at http://www.disastercenter.com/crime/uscrime.htm, accessed May 24, 2007.

- *Policing.* Strengthened policing efforts may have led to a reduction in crime. Law enforcement has implemented more effective crime-controlling strategies, such as community policing, than in the past and assigned a larger number of officers to the streets to fight crime.

- *Age.* Changes in crime rates are closely related to changes in the age distribution of the population. Males between the ages of 20 and 39 are the most likely to commit crimes. When young males account for a smaller portion of the total population, it would seem to follow that there will be less crime.[34]

Despite these correlated trends, criminologists do not know with certainty why the crime rate has

LINK Community policing, which is discussed in Chapter 6, brings together the police and the public in the fight against crime.

fallen over recent decades, although it ticked upward in both 2006 and 2007. Most likely, the decline in crime observed in recent years is the result of several of these factors being entangled in complex and as yet unknown ways.

Criminal Offenders

Data from the UCR, NCVS, and self-report studies have provided criminologists with an abundance of information about criminal offenders. For example, these sources show that 60 percent of all people who are arrested are between the ages of 19 and 39, even though this age group accounts for only 28 percent of the total U.S. population. With regard to race, about 13 percent of the U.S. population is African American, yet African Americans are arrested for 39 percent of serious violent crimes and 29 percent of serious property crimes. In terms of sex, about 49 percent of the U.S. population is male, but men are arrested for 82 percent of serious violent crimes and 68 percent of serious property crimes. Pulling the data together, the persons who are most likely to be arrested for both serious violent and serious property crimes are African American males between the ages of 19 and 39.[35]

Offenders by Age

Age and crime are closely related. The age–crime curve FIGURE 3-4 shows that crime rates increase during preadolescence, peak in late adolescence, and then

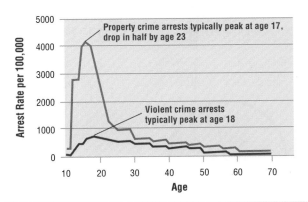

FIGURE 3-4 Age–Crime Curve

Source: Federal Bureau of Investigation, *Crime in the United States, 2006* (Washington, DC: U.S. Department of Justice, 2007).

decline steadily thereafter.[36] The high point of the curve is different for violent offenses and property offenses. Serious violent crime arrests peak at age 18 and then decline, whereas property crime arrests hit their highest point at age 16 and then decrease.

Juveniles (persons younger than age 18) constitute about 26 percent (about 80 million people) of the U.S. population. In 2005, members of this age group accounted for 15 percent of arrests for serious violent crimes and 28 percent of arrests for serious property crimes. Between 1996 and 2005, the number of serious violent crimes committed by juveniles decreased by 25 percent. In the same period, arrests of juveniles for serious property crimes dropped sharply, decreasing by 42 percent.[37]

As the age–crime curve shows, older persons typically commit fewer crimes than younger people, a trend known as the aging-out phenomenon. The aging-out phenomenon has its roots in the fact that reductions in strength, energy, and mobility with age make it physically more difficult to commit crime as a person grows older. Crime rates also decrease with age for the following reasons:

- Personality changes
- Increased awareness of the cost of crime
- Decreased importance of peer influences
- Lower testosterone levels linked with decreases in male aggression

Some people, however, do not age out of crime. These chronic offenders often started breaking the law at a very young age and continue to commit crime throughout the course of their lives. Unfortunately, many chronic juvenile offenders will likely become adult criminals who cannot be rehabilitated.[38]

The Criminal Unborn

There are many explanations for the dramatic decrease in crime in the late 1990s, but no explanation is more controversial than that offered by economists John Donohue III and Steven Levitt. They attribute the decrease in crime to the U.S. Supreme Court's 1973 *Roe v. Wade* decision, which legalized abortion.

Donohue and Levitt provide evidence linking the legalization of abortion to recent crime reductions, demonstrating that a steep rise in abortions after 1973, along with declining crime rates, has meant that many persons who might have been prone to criminal activity in the 1990s were never born. Given that the legalization of abortion reduces the general population, it also means a reduction in the number of people who reach the age at which they are most prone to commit crimes. Also, the populations that are more likely to have abortions (e.g., teenagers, unmarried women, the poor, and African Americans) are also more likely to have children who are at risk for committing crimes later in life. Finally, there is a reduction in the number of children born to women with unwanted pregnancies, who may be less likely to be good parents or to harm their fetus during pregnancy by drinking alcohol and taking drugs, thereby increasing the child's likelihood of future criminality.

The following evidence appears to support Donohue and Levitt's claim:

- The significant decline in crime in the United States occurred during the same years in which *Roe v. Wade* would have had its greatest impact on reducing the number of persons who would have reached the peak of their criminal activity.

- The five states that first experienced a drop in crime legalized abortion in 1970, fully three years before the *Roe v. Wade* decision.

- The largest declines in crime since 1985 have occurred in states with high abortion rates from 1973 to 1976, even when the analysis takes into account differences in incarceration rates, racial composition, and income.

Donohue and Levitt conclude that, if abortion had not been legalized, current crime rates would be 10 to 20 percent higher in the United States. They estimate that legalized abortion may account for as much as 50 percent of the recent drop in crime. According to Donohue and Levitt, legalized abortion has saved Americans more than $30 billion annually.

Sources: Steven Levitt and Stephen Dubner, *Freakonomics,* revised ed. (New York: Harper, 2006); John Donohue III and Steven Levitt, "The Impact of Legalized Abortion on Crime," *Quarterly Journal of Economics* 116:379–420 (2001).

Offenders by Socioeconomic Status

When most Americans think of crime, they think of street crime, which includes acts of personal violence and crimes against property. These perceptions of crime are reinforced by the news media, whose stories typically stress street crime and magnify people's fears about their personal safety and belongings. Although the data collected via the UCR and NCVS emphasize street crime, white-collar crime also causes serious harm to society. Examples of white-collar crime include the following:

- ATM fraud
- Cellular phone fraud
- Computer fraud
- Counterfeiting
- Credit card fraud
- Embezzlement
- Forgery
- Identity theft
- Illegal dumping of toxic waste
- Insider stock market trading
- Telemarketing fraud
- Welfare fraud

The origin of the phrase "white-collar crime" can be traced to sociologist Edwin Sutherland, who defined it as "a crime committed by a person of respectability and high social status in the course of his or her occupation."[39]

LINK In differential association theory, as outlined in Chapter 4, Sutherland explained how deviants such as white-collar criminals acquire the motivation and technical knowledge for criminal activity through verbal and nonverbal communications with other offenders.[40]

In 2006 in Lima, Ohio, a bank robber might have gotten away with her crime if she hadn't left an important clue left behind—her book bag. At just 15 years old, this girl is one of the youngest bank robbers in the country.

The robbery was typical in most every way of a robbery committed by an adult. Bank employees reported that two masked men walked into the bank at 10 A.M., one with a gun demanding money. They walked out with $5000 in stolen cash and sped away in a getaway car to Chicago, where they spent it on drugs and shopping.

Authorities later determined that one of the masked robbers was actually a teenage girl, a freshman at a local high school who confessed to participating in the bank robbery. The girl was not sentenced to a juvenile prison; a magistrate decided that she should receive mental health care.

Source: Sharon Coolidge, "Bank Heists Linked to Teens Catch FBI's Eye," *USA Today,* August 14, 2006, p. 3A, available at http://www.usa today.com/news/nation/2006-08-14-young-bank-robbers_x.htm, accessed June 19, 2007.

Over the years, innumerable instances of white-collar crime have occurred.[41] The monetary cost of white-collar crime is unknown because very little information on its scope is available. The best source of data is the NIBRS, which suggests that more than 6 million people in the United States are the victims of white-collar crime annually. According to the FBI, the annual monetary cost of white-collar crime in the United States exceeds $300 billion.[42]

Ken Lay's white-collar crimes led to the collapse of energy giant Enron.

? Questions: Why are most people more tolerant of white-collar crime than of street crime? Should white-collar criminals receive harsher punishments than street criminals because their offenses can affect many more people?

Crime Victims

When a crime is committed, there usually is a victim. <u>Victimology</u> is the study of the characteristics of crime victims and the reasons why certain people are more likely than others to become victims of crime. The field of victimology covers a wide range of disciplines, including sociology, psychology, criminal justice, and law.

Crime victims play an important role in the efficient operation of the criminal justice system. Although police depend on victims to report crimes and act as complainants, and prosecutors, defense attorneys, judges, and juries rely on victims to participate in trials as witnesses, interest in studying crime victimization remained limited for many years. That situation changed in 1967, when the President's Commission on Law Enforcement and Administration of Justice, in a special task force report, declared that crime victims had been neglected.[43] Since then, more than 3000 state and federal pieces of legislation have been passed to aid crime victims, and many states have added victims' rights amendments to their state constitutions FIGURE 3-5 .

Children

Children of all ages are vulnerable to becoming victims of crime. Often the offenders in these crimes are family members, which is especially the case for infants. Some juveniles are also the victims of abuse and neglect at the hands of their caregivers. Research has

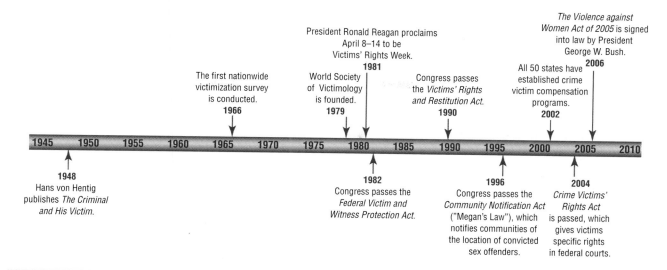

The first nationwide victimization survey is conducted.
1966

President Ronald Reagan proclaims April 8–14 to be Victims' Rights Week.
1981

World Society of Victimology is founded.
1979

Congress passes the *Victims' Rights and Restitution Act.*
1990

The Violence against Women Act of 2005 is signed into law by President George W. Bush.
2006

All 50 states have established crime victim compensation programs.
2002

1948
Hans von Hentig publishes *The Criminal and His Victim.*

1982
Congress passes the *Federal Victim and Witness Protection Act.*

1996
Congress passes the *Community Notification Act* ("Megan's Law"), which notifies communities of the location of convicted sex offenders.

2004
Crime Victims' Rights Act is passed, which gives victims specific rights in federal courts.

FIGURE 3-5 Victimology Timeline

Source: Office for Victims of Crime, available at http://www.ojp.usdoj.gov/ovc/, accessed May 14, 2007.

shown that child victimization and abuse are associated with problem behaviors later in life, such as teen pregnancy, alcohol and drug abuse, and criminality.[44] Understanding childhood victimization can lead to a better understanding of criminality in general.

One of the leading causes of death for children ages 1 to 11 is homicide. When young children are murdered, they are usually killed by family members (61 percent by their parents); by contrast, older children who are the victims of homicide are usually murdered by acquaintances (58 percent).[45]

From 1993 to 2002, children between the ages of 12 and 17 were about three times more likely than adults to be victims of nonfatal violent crimes. Children in this age group were also twice as likely as adults to be victims of robbery or aggravated assault, three times as likely to be victims of rape or sexual assault, and roughly three times as likely to be victims of simple assault. In addition, the nonfatal victimization rate for male youths is about 50 percent greater than the corresponding rate for female youths. Over the 10-year period from 1993 to 2002, urban youth ages 12 to 17 had a significantly higher nonfatal violent victimization rate than did suburban or rural youth.[46]

One of the most common forms of child victimization is <u>child maltreatment</u>, which is an act or omission (i.e., failure to act) by a parent or other caregiver that results in harm or serious risk of harm to a child. It can take one of several forms:

- Physical abuse
- Sexual abuse
- Neglect
- Emotional abuse

Each year, approximately 3 million cases of child maltreatment are referred to child protective service agencies in the United States. Almost 6 million children

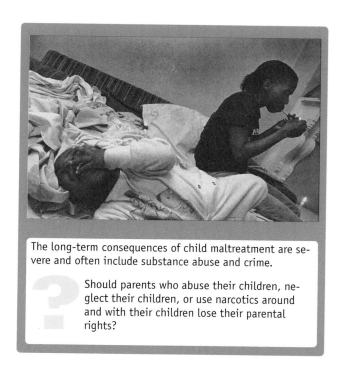

The long-term consequences of child maltreatment are severe and often include substance abuse and crime.

? Should parents who abuse their children, neglect their children, or use narcotics around and with their children lose their parental rights?

are included in these referrals. Of these cases, about one-third are confirmed, which means that roughly 1 million children are victims of maltreatment each year. Moreover, experts estimate that the actual number of incidents of abuse and neglect are three times greater than the number reported to authorities.[47]

The rate of maltreatment victimization is inversely related to age, with the highest rate occurring among the youngest children FIGURE 3-6 , and with girls being slightly more likely than boys to be victims. The vast majority of perpetrators are parents (80 percent), including birth parents, adoptive parents, and step-parents FIGURE 3-7 .

Child maltreatment is very serious and produces dire consequences:

- Eighty percent of young adults who had been victims of child maltreatment experienced one or more psychiatric disorders by age 21, including depression, anxiety, eating disorders, and post-traumatic stress disorder.
- Maltreated children are 25 percent more likely to experience teen pregnancy than their peers who were not maltreated.
- Children who are victims of maltreatment are 59 percent more likely to be arrested as juveniles, 28 percent more likely to be arrested as adults, and 30 percent more likely to commit violent crimes.
- Children who have been sexually abused are nearly three times more likely to develop alcohol abuse and four times more likely to develop drug addictions than their counterparts who were not sexually abused.
- Nearly 67 percent of all people in treatment for drug abuse report being maltreated as children.[48]

Senior Citizens

There are approximately 36 million senior citizens (age 65 and older) living in the United States today, accounting for more than 12 percent of the total population.[49] NCVS data show that persons age 65 or older generally experience victimizations at much lower rates than younger groups of people FIGURE 3-8 .

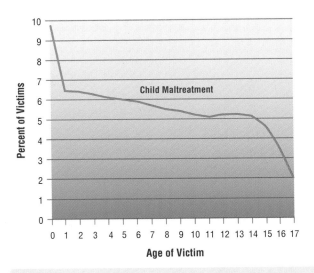

FIGURE 3-6 The Relationship between Child Maltreatment and Age

Source: Howard Snyder and Melissa Sickmund, *Juvenile Offenders and Victims: 2006 National Report* (Washington, DC: U.S. Department of Justice, 2006).

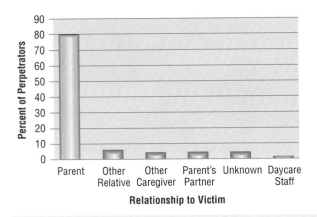

FIGURE 3-7 Perpetrators of Child Maltreatment

Source: Howard Snyder and Melissa Sickmund, *Juvenile Offenders and Victims: 2006 National Report* (Washington, DC: U.S. Department of Justice, 2006).

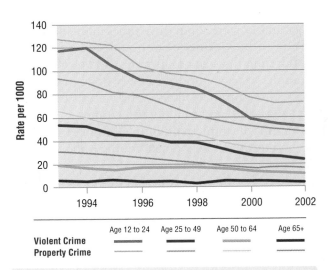

FIGURE 3-8 Rate of Violent and Property Crime Victimization per 1000 Persons, by Age of Victim

Source: Patsy Klaus, *Crimes against Persons Age 65 or Older, 1993–2002* (Washington, DC: U.S. Department of Justice, 2005).

Seniors experience nonfatal violent crime at 5 percent that of young persons (only 4 victims per 1000 persons 65 and older). Households headed by seniors experienced property crimes at a rate about 25 percent of that for households headed by persons younger than age 25. Although persons age 65 or older experienced lower victimization rates, when they are victimized they are most often the victims of property crimes; such offenses accounted for about 90 percent of victimizations among people older than age 50. Compared to younger victims of personal crimes, seniors are disproportionately victimized by thefts of their purses and wallets, an act that accounts for 20 percent of personal crimes against seniors.

While it is uncommon, elderly persons may be offenders—like J. L. Hunter Rountree who was convicted of bank robbery at age 92 and sentenced to 12 years in prison—or, more often, victims.

? Should elderly offenders receive lighter punishments? Should offenders who target elderly victims receive harsher punishments?

Although seniors are less likely to be victims of violence, when victimized they are equally likely to face offenders with weapons, are more likely to offer no resistance, and are equally likely to receive serious injuries as are younger persons. However, compared to younger victims, persons age 65 or older are more likely to report violence, purse snatching, and pocket picking to the police.

Victims of Intimate Partner Violence

One characteristic shared by many victims of crime is that they run the risk of <u>intimate partner violence (IPV)</u>—violent victimization by intimates, including current or former spouses, boyfriends, girlfriends, or romantic partners. IPV includes violent acts such as murder, rape, sexual assault, robbery, aggravated assault, and simple assault.

Data released in 2007 indicated that less than 1 percent of households experienced IPV. While this number appears to be low, another way to frame this issue is to recognize that IPV takes place in 1 in 320 U.S. households each year.[50] Recent estimates of IPV indicate that approximately 1 million violent crimes are committed against intimates each year. Women are five times more likely to be victims of violence by intimates than are men; each year on average, more than 570,000 women are injured by intimates.[51] African American, young, divorced or separated women who are earning lower incomes, living in rental housing, and living in an urban area are most likely to be victimized. Men who are young, African American, divorced or separated, or living in rented housing also have significantly higher rates of IPV than other men.[52]

IPV is more prevalent and more severe in disadvantaged neighborhoods and occurs more often in households facing economic distress. About one-third of all IPV against women and about half of all IPV against men occurred in the victim's home. Nearly two-thirds of both male and female victims of IPV are physically attacked, with the remaining third being victims of threats or attempted violence. Overall, 65 percent of IPV against women and 68 percent of IPV against men involves a simple assault, the least serious form of IPV FIGURE 3-9 .[53]

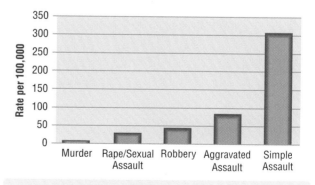

FIGURE 3-9 Rate of Intimate Partner Violence, by Type of Crime
Source: Callie Rennison and Sarah Welchans, *Intimate Partner Violence* (Washington, DC: U.S. Department of Justice, 2000).

Intimate Partner Homicide

Most victims of intimate partner homicide are killed by spouses, although this proportion has declined in recent years. Boyfriends or girlfriends are now as likely as spouses to murder their partners, while the number of intimate partner homicides by ex-spouses have remained steady and infrequent FIGURE 3-10. The decline in IPV-related homicide numbers has been most notable for males, although the number of female intimate homicide victimizations has also dropped substantially FIGURE 3-11.

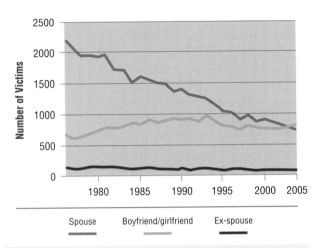

FIGURE 3-10 Homicides of Intimates by Relationship of the Victim to the Offender

Source: James Alan Fox and Marianne Zawitz, *Homicide Trends in the U.S.: Intimate Homicide,* available at http://www.ojp.usdoj.gov/bjs/homicide/intimates.htm, accessed May 28, 2007.

Among minorities of both sexes, a greater percentage of IPV victims are killed by boyfriends or girlfriends than by current spouses, and firearms are the weapons most commonly used in the deaths of victims of both sexes. Males are more likely than females to be killed by knives, whereas females are more likely than males to be killed by blunt objects or blows delivered by the hands or feet.[54]

Intimate Partner Rape

Almost 18 million women and 3 million men in the United States have been raped. It is estimated that in a single year, more than 300,000 women and almost 93,000 men are raped. One in every six women has been raped at some time in her life, and younger women are significantly more likely to report being raped than older women. More than half of all female victims and nearly 70 percent of all male victims were raped before age 18. Women who are raped before age 18 are also twice as likely to report being raped as adults. Females are nearly twice as likely as males to be victims of gang rape (8.3 percent versus 4.6 percent, respectively), whereas 78 percent of female rape victims and 83 percent of male rape victims are raped by only one person.[55] Despite commonly held beliefs, the majority of rapists are not strangers FIGURE 3-12.[56]

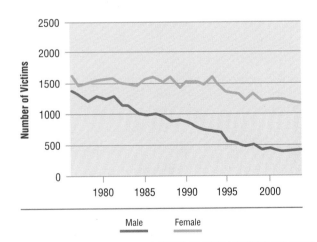

FIGURE 3-11 Homicides of Intimates by Gender of Victim

Source: James Alan Fox and Marianne Zawitz, *Homicide Trends in the U.S.: Intimate Homicide,* available at http://www.ojp.usdoj.gov/bjs/homicide/intimates.htm, accessed June 28, 2007.

For same-sex partners, the proportion of intimate partner homicides committed by males is much greater than the percentage committed by females.

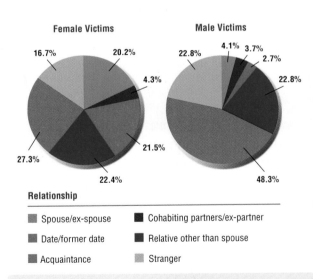

FIGURE 3-12 Rape by Relationship of Victim to Offender
Note: Percentages by victim gender exceed 100% because some victims were raped by more than one person.

Source: Patricia Tjaden and Nancy Thoennes, *Extent, Nature, and Consequences of Rape Victimization: Findings from the National Violence against Women Survey* (Washington, DC: National Institute of Justice, 2006).

Even though thousands of hate crimes are reported annually, some hate crimes are so atrocious that they capture the attention of a nation. One such incident took place on the night of September 27, 2003, in Linden, Texas.

On that night, Billy Ray Johnson, a 42-year-old, mentally disabled African American, was brought to a "pasture party" where more than a dozen young white partygoers were having a bonfire. They got Johnson drunk, humiliated him, and jeered him with racial epithets. When Johnson started to get angry, one of the men, Colt Amox, punched him, knocking Johnson to the ground. For nearly an hour, Johnson lay unconscious bleeding from the head as the group debated what to do. Eventually, they loaded Johnson into the back of a pickup truck and drove two miles down a rural back road (rather than one mile to the nearest hospital). Eventually, they dropped Johnson onto a pile of stinging fire ants near a mound of rotten rubber at a used tire dumpsite. Several hours later, some of the assailants notified law enforcement.

The beating left Johnson severely injured with irreversible brain damage. Today he lives in a nursing home where he drools and soils himself. His speech has been severely impaired, and he has trouble swallowing food and walking unassisted.

Four men were arrested for the assault: Colt Amox, age 20; Dallas Stone, age 18; Cory Hicks, age 24; and Wes Owens, age 19. Amox defended his actions by claiming that Johnson aggressively charged toward him. The all-white jury acquitted Amox and Hicks of serious felony charges and instead handed down lesser convictions, with a recommended sentence of probation. The men were fined and sentenced to both imprisonment and probation, though none served more than 60 days in jail.

Several groups, including the National Association for the Advancement of Colored People (NAACP), have since looked into the case. The Southern Poverty Law Center filed a civil suit against the four men, alleging that the defendants were liable for assault and negligence and seeking compensatory damages to help pay for Johnson's care. On April 22, 2007, the

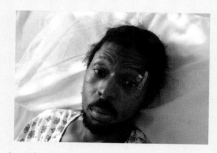

jury found Hicks and Amox responsible for Johnson's injuries (Stone and Owens had previously reached confidential settlements), and a jury awarded Johnson $9 million in damages.

Sources: "Billy Ray Johnson Trial Set for April 17, 2007," Southern Poverty Law Center, available at http://www.splcenter.org/news/item.jsp?aid=246, accessed March 31, 2007; "Why I'm Angry—Billy Ray Johnson," *Angry Black Woman,* available at http://theangryblackwoman.wordpress.com/2007/03/07/why-im-angry-billy-ray-johnson/, accessed March 31, 2007; Pamela Colloff, " The Beating of Billy Ray Johnson," *Texas Monthly,* February 2007, available at http://www.texasmonthly.com/2007-02-01/index.php, accessed March 29, 2007; Andre Coe, "'Good Ole Boys': Weapons of Black Destruction," available at http://www.blackpressusa.com/news/Article.asp?SID=3&Title=National+News&NewsID=4552, accessed March 31, 2007; Ed Lavandera, "Billy Ray Johnson," *Paula Zahn Now,* aired March 8, 2007, available at http://transcripts.cnn.com/TRANSCRIPTS/0703/08/pzn.01.html, accessed March 31, 2007; "Texas Jury Awards $9 Million to Beating Victim," available at http://www.cnn.com/2007/LAW/04/22/texas.beating.ap/index.html, accessed April 22, 2007.

Victims of Hate Crime

Crimes of hatred and prejudice are a sad fact of U.S. history. The FBI has been investigating these offenses as far back as the early 1920s, when it opened their first case against the Ku Klux Klan. The term "hate crime" did not enter the national vocabulary until the 1980s, when emerging hate groups such as Skinheads launched a wave of bias-related crime. Today, a hate crime (also known as a bias crime) is defined as a crime in which an offender chooses a victim based on a specific characteristic and evidence is provided that hate or personal disapproval of this characteristic prompted the offender to commit the crime.

Following passage of the Hate Crime Statistics Act of 1990, the FBI has gathered and published hate crime statistics every year since 1992 FIGURE 3-13 .[57] Some of these statistics are summarized here:

The Primary Motivation for a Hate Crime
- The victim's race (51 percent)
- The victim's sexual orientation (18 percent)
- The victim's ethnicity (17 percent)
- The victim's religion (13 percent)
- The victim's disability (1 percent)

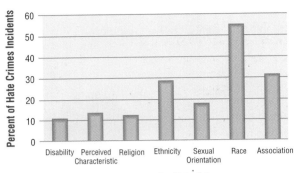

FIGURE 3-13 Motivations for Hate Crimes

Source: Caroline Wolf Harlow, *Hate Crime Reported by Victims and Police* (Washington, DC: U.S. Department of Justice, 2005).

Few images embody the idea of hate crime like the activities of the Ku Klux Klan.

Should belonging to a hate group be a crime? Should hate crimes be given special treatment within the criminal justice system?

The Most Frequently Committed Hate Crimes

- Verbal threats or intimidation (40 percent)
- Simple assault (23 percent)
- Vandalism (18 percent)
- Aggravated assault (14 percent)

The Most Likely Victims of Hate Crimes

- Males (65 percent)
- Whites (67 percent)
- Persons 30–49 years old (35 percent)

NCVS estimates suggest that only about 44 percent of hate crimes are reported to police, or roughly the same percentage of crimes without a hate component that are reported to police. In half of the reported victimizations, police took a report; in about one-third of cases, they questioned witnesses or suspects; and in about 25 percent of cases, they made an arrest. Victims who reported hate crimes to police had several motivations, including punishing the offender and preventing future crimes. Other victims reported hate crimes so they could receive professional help or alert police about their situation, hoping to improve police surveillance in particular areas. In contrast, many who did not report hate crimes decided to handle the matter on their own. Approximately 25 percent of hate crime victims did not report the incident to police because they considered the crime to be minor or were not certain that the offender intended harm.[58]

Victims and Criminal Justice

The victim's relationship to the criminal justice system has changed considerably from what it was in the earlier part of the twentieth century. Crime victims remained largely invisible until the middle of the twentieth century. Then, as the public became more attuned to the plight of crime victims, they became increasingly included in decisions made about criminal offenders. As public awareness of crime victims increased, significant changes were made in the way the criminal justice system and its agents responded to victims.

Interaction with Police

Sometimes police and victims may come into conflict over the pace and conduct of the investigation of the crime. Police often move cautiously early in the investigation process, especially when they are not certain whether a crime has been committed because they lack evidence to make legally binding decisions. For their part, victims may complain that the police are not responding quickly enough. Victims also may be uncomfortable with or dislike the process of questioning by police, when officers try to determine whether victims are telling the truth.

When police make an arrest, in addition to reading the suspect his or her rights, in some states officers must provide the following information to the victim to alleviate or intensify a victim's sense of personal safety:

- Names, badge numbers, and contact information for the officer and detective handling the case
- Case ID number
- Whether a suspect has been arrested
- Whether a suspect is in jail or out on bail

Victim impact statements convey the emotion inherent to criminal sentencing.

? Does the emotional toll on victims deserve a place in criminal sentencing?

Interaction with Prosecutors

After an arrest has been made and charges are being prepared, victims typically come into contact with the prosecutor. The prosecutor has the following responsibilities related to victims of crime:

- Keep crime victims informed of the status of the case
- Articulate victims' interests to the court on matters such as bail, continuances, negotiated pleas, dismissed cases, dropped charges, sentences, and restitution arrangements
- Protect victims from harassment, threats, injuries, intimidation, and retaliation
- Resolve cases as quickly as possible
- Help victims avoid needless waste of their time and money by advising them of court appearances and scheduling changes
- Expedite the return of recovered property[59]

Some of these responsibilities may be met through victim assistance programs. Most such programs provide crisis intervention, follow-up counseling, and assistance to victims throughout the criminal justice process. These programs' staff members may also provide victims with counseling, emergency shelter, food, and transportation; inform victims of court dates; and help them appear in court if necessary.

A significant challenge facing prosecutors is protecting victims who may be witnesses for the prosecution from threats and intimidation. If victim testimony is not available, the prosecutor must often drop the charges against the accused. It is difficult to estimate the extent of this problem because much victim intimidation goes unreported. Nevertheless, one report issued in 2006 estimated that as many as 36 percent of victims are intimidated in some way after they report a crime.[60] Victims may be intimidated in a variety of ways, including direct confrontation by the accused or a member of his or her family, vandalizing of their property, or, occasionally, assault.

As is the case with police and victims, conflicts may sometimes arise between victims and prosecutors. Victims may get upset when a case is dropped, charges are reduced, or a plea is negotiated. Additionally, victims are often frustrated because they have very little input into the sentencing decision. Victims are only one of several parties that may sway the outcome—prosecutors, defense attorneys, defendants, and probation officers may also exercise some influence over the sentence. The best chance victims have for affecting the sentencing decision is through a victim impact statement, in which they explain to the judge the emotional, financial, and physical harm they have suffered as a result of the crime.

Interaction with Parole Boards

Most incarcerated offenders will eventually be returned to the community. Perhaps not surprisingly, the decision to grant early release to an inmate often becomes a point of conflict between victims and parole boards. Many parole boards appear to give little consideration to victims' statements, considering them alongside input from prosecutors, judges, prison officials, and offenders. Victims' rights groups are trying to change this situation and to increase the role played by the victim in the parole hearing. These groups would like victims to be notified of parole board hearings, to be able to speak at the hearings, and to be notified if the convict is released.

WRAP UP

The disparity between a person's actual criminal activity and a person's history of arrest merely reinforces the dark figure of crime, the difference between official measures and the true amount of crime. When using arrests as an indicator, it is clear that the criminal justice system fails to measure most criminal activity. Official measures such as the Uniform Crime Reports include data on crimes known to the police, number of arrests, and persons arrested, but they do not count crimes that occur without official notice.

By contrast, self-reports of criminal activity are able to quantify the actual amount of crime that a person commits but have limited accuracy because people may be ignorant, dishonest, or forgetful about their crimes. Additionally, it is unlikely that people who commit serious crimes such as rape and robbery would voluntarily tell others for fear that they would be arrested. As in the case of Coral Watts, even though an arrest record may only hint at a person's actual criminal activity, self-reports may also prove to be unreliable. Ultimately, only the offender may know exactly how many offenses he or she has committed—regardless of what his or her criminal records indicate.

Nevertheless, arrest statistics have been shown to be a valid and reliable measure of crime, because official measures and victimization reports overlap as indicators of crime. Arrests are also a reliable indicator of differences in levels of criminality, such as those that exist between students and prisoners. Whereas students commit generally benign crimes at low rates, prisoners commit an array of crimes, including serious violent offenses, at significantly higher rates. The fact that the average student has zero arrests and the average prisoner has more than 10 arrests reflects not only their differences in criminal activity, but also their differences in criminality.

Chapter Spotlight

- The Uniform Crime Reports (UCR) includes data on crimes known to the police, number of arrests, and persons arrested.
- The National Crime Victimization Survey (NCVS)—and victimization surveys in general—ask people directly whether they have been victims of a crime during the past year.
- Self-report studies ask people directly about which crimes they have committed during the past year.
- The "dark figure of crime" represents the gap between the actual amount of crime committed and the number of crimes reported to police.
- For more than 30 years, UCR and NCVS data have generally told the same story about the incidence and prevalence of crime.
- The primary motivation for committing a hate crime is the victim's race. The most frequently committed hate crimes are verbal threats and intimidation.
- Crime victims today play a much more visible role in the criminal justice system than they did in the past.

Putting It All Together

1. Of what value are UCR, NCVS, and self-report surveys? What are their strengths and weaknesses?

2. If you could use only one source of crime data on which to base your decision making, would you choose the UCR, NCVS, or self-report surveys?

3. Why are senior citizens less likely to commit crimes than younger people?

4. What are the arguments in support of and opposing hate crime legislation?

5. Why do crime victims so often feel ignored by the criminal justice system? What would you propose as a solution to this problem?

age–crime curve A curve showing that crime rates increase during preadolescence, peak in late adolescence, and steadily decline thereafter.

aging-out phenomenon The decline of participation in crime after the teenage years.

child maltreatment The physical, sexual, or emotional abuse or neglect of children.

crime index A statistical indicator consisting of eight offenses that is used to gauge the amount of crime reported to the police. It was discontinued in 2004.

crimes of interest The seven offenses in the National Crime Victimization Survey that people are asked whether they have been a victim of during the past year.

dark figure of crime A term used by criminologists to describe the amount of unreported or undiscovered crime; it calls into question the reliability of UCR data.

hate crime A crime in which an offender targets a victim based on a specific characteristic (i.e., ethnicity, race, or religion), and evidence is provided that hate or personal disapproval of this characteristic prompted the offender to commit the crime.

hierarchy rule A rule dictating that only the most serious crime in a multiple-offenses incident will be recorded in the Uniform Crime Reports.

intimate partner violence (IPV) Violence in intimate relationships, including that committed by current or former spouses, boyfriends, or girlfriends

National Crime Victimization Survey (NCVS) An annual survey of criminal victimization in the United States conducted by the U.S. Bureau of Justice Statistics.

National Youth Survey (NYS) A comprehensive, nationwide self-report study of 1700 youths who reported their illegal behaviors each year for more than 30 years.

official crime statistics Statistics based on the aggregate records of offenders and offenses processed by police, courts, corrections agencies, and the U.S. Department of Justice.

self-report surveys Surveys that ask offenders to self-report their criminal activity during a specific time period.

Uniform Crime Report (UCR) An annual publication from the Federal Bureau of Investigation that presents data on crimes reported to the police, number of arrests, and number of persons arrested.

unofficial crime statistics Crime statistics produced by people and agencies outside the criminal justice system, such as college professors and private organizations.

victimization survey A method of producing crime data in which people are asked about their experiences as crime victims.

victimology The study of the characteristics of crime victims and the reasons why certain people are more likely than others to become victims of crime.

Notes

1. Sir Josiah Stamp, *Some Economic Matters in Modern Life* (London: King and Sons, 1929), pp. 258–259.
2. R. W. Apple, Jr., "Asian Journey; Snacker's Paradise: Devouring Singapore's Endless Supper," *New York Times*, September 10, 2003, available at http://query.nytimes.com/gst/fullpage .html?sec=travel&res=9C01E0DF153BF933A2575AC0A965 9C8B63&fta=y, accessed May 17, 2007.
3. The discussion of the UCR is from Federal Bureau of Investigation, *Crime in the United States, 2006* (Washington, DC: U.S. Department of Justice, 2007).
4. Federal Bureau of Investigation, note 3, available at http:// www.fbi.gov/ucr/05cius/, accessed August 6, 2007.
5. Adolphe Quetelet, *Research on the Propensity for Crime at Different Ages* (Cincinnati: Anderson, 1831/1984); Adolphe

Quetelet, *Treatise on Man and the Development of His Faculties* (Edinburgh, Scotland: S.W. and R. Chambers, 1842); Marc Mauer, *New Incarceration Figures* (Washington, DC: Sentencing Project, 2006), available at http://www.sentencing project.org/Admin/Documents/publications/inc_ newfigures.pdf, accessed May 28, 2007.

6. Samuel Walker and Charles Katz, *The Police in America*, 4th ed. (New York: McGraw-Hill, 2002).

7. National Center for Policy Analysis, *Does Punishment Deter?* (Dallas: National Center for Policy Analysis, 1998); Donald Black, "The Production of Crime Rates," *American Sociological Review* 35:733–748 (1970).

8. Walter Gove, Michael Hughes, and Michael Geerken, "Are Uniform Crime Reports a Valid Indicator of the Index Crimes? An Affirmative Answer with Minor Qualifications," *Criminology* 23:451–501 (1985); Gary LaFree and Kriss Drass, "The Effect of Changes in Intraracial Income Inequality and Educational Attainment on Changes in Arrest Rates for African Americans and Whites, 1957 to 1990," *American Sociological Review* 61:614–634 (1996).

9. Federal Bureau of Investigation, *Developments in the National Incident-Based Reporting System* (Washington, DC: U.S. Department of Justice, 2004).

10. Federal Bureau of Investigation, *Crime Reporting in the Age of Technology* (Washington, DC: U.S. Department of Justice, 1999), available at http://www.ojp.usdoj.gov/bjs/pub/pdf/ v4no1nib.pdf, accessed June, 2007.

11. Federal Bureau of Investigation, notes 3 and 9.

12. Federal Bureau of Investigation, note 3.

13. Patsy Klaus, *Crime and the Nation's Households, 2002* (Washington, DC: U.S. Department of Justice, 2004).

14. Philip Ennis, *Criminal Victimization in the United States* (Washington, DC: U.S. Government Printing Office, 1967).

15. Robert O'Brien, *Crime and Victimization Data* (Beverly Hills, CA: Sage, 1985); Shannon Catalano, *Criminal Victimization, 2003* (Washington, DC: U.S. Department of Justice, 2004); David Cantor and James Lynch, "Self-Report Measures of Crime and Criminal Victimization," in U.S. Department of Justice, *Criminal Justice 2000*, Volume 4 (Washington, DC: National Institute of Justice, 2000), pp. 85–138.

16. L. Edward Wells and Joseph Rankin, "Juvenile Victimization," *Journal of Research in Crime and Delinquency* 32:287–307 (1995).

17. Anthony Walsh and Lee Ellis, *Criminology* (Thousand Oaks, CA: Sage, 2007); Albert Biderman and James Lynch, *Understanding Crime Incidence Statistics* (New York: Springer-Verlag, 1991).

18. Wesley Skogan, *Issues in the Measurement of Victimization* (Washington, DC: U.S. Department of Justice, 1981).

19. Janet Lauritsen and Robin Schaum, *Crime and Victimization in the Three Largest Metropolitan Areas, 1980–1998* (Washington, DC: U. S. Department of Justice, 2005).

20. Alfred Blumstein, Jacqueline Cohen, and Richard Rosenfeld, "Trend and Deviation in Crime Rates: A Comparison of UCR and NCS Data for Burglary and Robbery," *Criminology* 29:237–264 (1991); Michael Hindelang, "The Uniform Crime Reports Revisited," *Journal of Criminal Justice* 2:1–17 (1974); David MacDowall and Colin Loftin, "Comparing the UCR and NCS Over Time," *Criminology* 30:125–132 (1992); Steven Messner, "The 'Dark Figure' and Composite Indexes of Crime: Some Empirical Explorations of Alternative Data Sources," *Journal of Criminal Justice* 12:435–444 (1984).

21. Austin Porterfield, *Youth in Trouble* (Austin, TX: Leo Potishman Foundation, 1946); James Wallerstein and J. C. Wyle, "Our Law-Abiding Lawbreakers," *Federal Probation* 25:107–112 (1947); James F. Short, Jr., "A Report on the Incidence of Criminal Behavior, Arrests and Convictions in Selected Groups," *Research Studies of the State College of Washington* 22:110–118 (1954); James F. Short, Jr., and F. Ivan Nye, "Extent of Unrecorded Juvenile Delinquency," *Journal of Criminal Law, Criminology, and Police Science* 49:296–302 (1958); Maynard Erickson and LaMar Empey, "Court Records, Undetected Delinquency and Decision-Making," *Journal of Criminal Law, Criminology, and Police Science* 54:456–469 (1963); Jay Williams and Martin Gold, "From Delinquent Behavior to Official Delinquency," *Social Problems* 20:209–229 (1972).

22. Suzanne Ageton and Delbert Elliott, *The Incidence of Delinquent Behavior in a National Probability Sample* (Boulder, CO: Behavioral Research Institute, 1978); Delbert Elliott and Suzanne Ageton, "Reconciling Race and Class Differences in Self-Reported and Official Estimates of Delinquency," *American Sociological Review* 45:95–110 (1980).

23. Steven Levitt and Stephen Dubner, *Freakonomics,* revised ed. (New York: HarperCollins, 2006).

24. Stephen Cernkovich, Peggy Giordano, and Meredith Pugh, "Chronic Offenders," *Journal of Criminal Law and Criminology* 76:705–732 (1985).

25. Michael Hindelang, Travis Hirschi, and Joseph Weis, *Measuring Delinquency* (Beverly Hills, CA: Sage, 1981), p. 114.

26. Walsh and Ellis, note 17.

27. Wallerstein and Wyle, note 21; Short, note 21; Short and Nye, note 21; Erickson and Empey, note 21; Williams and Gold, note 21.

28. William Chambliss and Richard Nagasawa, "On the Validity of Official Statistics," *Journal of Research in Crime & Delinquency* 6:71–77 (1969); Leroy Gould, "Who Defines Delinquency?" *Social Problems* 16:325–336 (1969); Michael

Leiber, "Comparison of Juvenile Court Outcomes for Native Americans, African Americans, and Whites," *Justice Quarterly* 11:257–279 (1994).

29. Federal Bureau of Investigation, note 3.
30. Federal Bureau of Investigation, note 3.
31. Patsy Klaus, *Crime and the Nation's Households, 2005* (Washington, DC: U.S. Department of Justice, 2007)
32. Levitt and Dubner, note 23.
33. Paige Harrison and Allen Beck, *Prisoners in 2005* (Washington, DC: U.S. Department of Justice, 2006); The Disaster Center, *United States Crime Rates, 1960–2005,* availale at http://www.disastercenter.com/crime/uscrime.htm, accessed July 21, 2007; Christopher Mumola and Allen Beck, *Prisoners in 1996* (Washington, DC: U.S. Department of Justice, 1997).
34. Gordon Witkin, "The Crime Bust," *U.S. News & World Report,* May 25, 1998, pp. 28–40.
35. Federal Bureau of Investigation, note 3.
36. Adolphe Quetelet, *Research on the Propensity for Crime at Different Ages,* note 4; Daniel Nagin, David Farrington, and Terrie Moffitt, "Life-Course Trajectories of Different Types of Offenders," *Criminology* 33:111–139 (1995); David Farrington, "Age and Crime," pages 189–250 in Michael Tonry and Norval Morris (Eds.), *Crime and Justice,* Volume 7 (Chicago: University of Chicago Press, 1983); Travis Hirschi and Michael Gottfredson, "Age and the Explanation of Crime," *American Journal of Sociology* 89:552–584 (1983).
37. Federal Bureau of Investigation, note 3.
38. Matt DeLisi, *Criminal Careers in Society* (Thousand Oaks, CA: Sage, 2005).
39. Edwin Sutherland, *White Collar Crime* (New York: Dryden, 1949), p. 9.
40. Edwin Sutherland, *Principles of Criminology,* 4th ed. (Philadelphia: Lippincott, 1947).
41. CNNMoney.com, "Stewart Convicted on All Charges," available at http://money.cnn.com/2004/03/05/news/companies/martha_verdict/, accessed March 24, 2007; Brooke Masters, "WorldCom's Ebbers Convicted," *Washington Post,* March 16, 2005, p. A01; Gregg Farrell, "Trial Judge Vacates Conviction of Late Enron Founder Lay," *USA Today,* October 18, 2006, p. 3A; Roben Farzad, "Jail Term for Two Top of Adelphia," *New York Times,* June 21, 2005, p. C1; Andrew Ross Sorkin and Jennifer Bayot, "Ex-Tyco Officers Get 8 to 25 Years," *New York Times,* September 20, 2005, p. A1; Susan Haigh, "Cendant Official Must Pay Back $3.27 Billion," *Washington Post,* August 4, 2005, p. D03.
42. Cynthia Barnett, *The Measurement of White-Collar Crime Using Uniform Crime Reporting (UCR) Data* (Washington, DC: U.S. Department of Justice, 2003), available at http://www.fbi.gov/ucr/whitecollarforweb.pdf, accessed May 24, 2007.
43. President's Commission on Law Enforcement and Administration of Justice, *Task Force on Assessment* (Washington, DC: U.S. Government Printing Office, 1967), p. 80.
44. Howard Snyder and Melissa Sickmund, *Juvenile Offenders and Victims: 2006 National Report* (Washington, DC: U.S. Department of Justice, 2006).
45. Snyder and Sickmund, note 44.
46. Shannan Catalano, *Criminal Victimizations, 2005* (Washington, DC: U.S. Department of Justice, 2006); Snyder and Sickmund, note 44.
47. Childhelp—Prevention and Treatment of Child Abuse, "National Child Abuse Statistics," available at http://www.childhelp.org/resources/learning-center/statistics, accessed March 27, 2007; National Center for Injury Prevention and Control, *Child Maltreatment: Fact Sheet, 2006* (Atlanta: Centers for Disease Control and Prevention, 2006); Lisa Jones, "Child Maltreatment Trends in the 1990s," *Child Maltreatment* 11:107–120 (2006).
48. Childhelp, note 47.
49. Patsy Klaus, *Crimes against Persons Age 65 or Older, 1993–2002* (Washington, DC: U.S. Department of Justice, 2005).
50. Klaus, note 31.
51. National Center for Injury Prevention and Control, *Intimate Partner Violence: Fact Sheet, 2006* (Atlanta: Centers for Disease Control and Prevention, 2006).
52. Callie Rennison and Sarah Welchans, *Intimate Partner Violence* (Washington, DC: U.S. Department of Justice, 2000).
53. Michael Benson and Greer Fox, *When Violence Hits Home: How Economics and Neighborhood Play a Role* (Washington, DC: U.S. Department of Justice, 2004).
54. Leonard Paulozzi, Linda Saltzman, Martie Thompson, and Patricia Holmgreen, *Surveillance for Homicide among Intimate Partners—United States, 1981–1998* (Atlanta: Centers for Disease Control and Prevention, 2001).
55. Patricia Tjaden and Nancy Thoennes, *Extent, Nature, and Consequences of Rape Victimization: Findings from the National Violence against Women Survey* (Washington, DC: National Institute of Justice, 2006).
56. Tjaden and Thoennes, note 55.
57. Federal Bureau of Investigation, note 3.
58. Caroline Wolf Harlow, *Hate Crime Reported by Victims and Police* (Washington, DC: U.S. Department of Justice, 2005).
59. National Center for Victims of Crime, available at http://www.ncvc.org/, accessed June 22, 2007; President's Task Force, *Victims of Crime: Final Report* (Washington, DC: U.S. Government Printing Office, 1982).
60. Kelly Dedel Johnson, *Witness Intimidation* (Madison: University of Wisconsin–Madison Law School's Frank J. Remington Center, 2006).

JBPUB.COM/**ExploringCJ**

Interactives

Key Term Explorer

Web Links

- Distinguish between choice theories and trait theories.
- Describe the similarities and differences among biological, psychological, and sociological theories.
- Understand critical theories such as labeling and conflict theory.
- Know the principal ideas behind developmental theories.
- Identify the assumptions of theoretical perspectives on crime and their social policy applications.

Crime Theory and Social Policy

CHAPTER 4

Andrei Chikatilo has the dubious distinction of being the most prolific serial killer in history. From 1978 to 1990, Chikatilo murdered 53 women and children in his native Russia—crimes for which he was executed in 1994. In addition to the sheer volume of violence, Chikatilo was infamous for the sadism of his crimes, which included sexual assault, torture, mutilation, and cannibalism.

The details of Chikatilo's life show unparalleled depravity. His childhood spanned both the violence of World War II and the extreme poverty and famine of Stalinist Russia. His brother was kidnapped from his family's home and purportedly cannibalized. Frequently beaten during his childhood, Chikatilo suffered severe head trauma that may have caused injuries to the frontal lobes of his brain, the region of the brain that handles executive functions such as self-control. In addition to classic psychopathic symptoms, Chikatilo had a personality described as dissociative and paranoid.

- How would you explain Chikatilo's crimes from the biological, psychological, and sociological perspectives?

- If he had not been executed, would Chikatilo have made a good candidate for rehabilitation efforts?

Source: Stephen Giannangelo, *The Psychopathology of Serial Murder: A Theory of Violence* (Westport, CT: Praeger, 1996).

Introduction

For more than 200 years, criminologists have studied the causes of crime and constructed many theories, or integrated sets of ideas, that explain when and why people commit crime. Theories are important because ideas have consequences. In other words, different theories suggest preventing crime in different ways and often, therefore, lead to different policy recommendations (see the "Wrap Up" section for a summary of theories and policy applications).

Most criminological theories fall into one of three categories:

- Choice theories contend that people have free will, are rational and intelligent, and make informed decisions to commit crimes based on whether they believe they will benefit from doing so.

- Trait theories blame crime on biological and psychological factors over which individuals have little—if any—control, such as low intelligence and personality disorders.

- Sociological theories attribute crime to a variety of social factors external to the individual,

Susan Smith gained international scorn in 1994 when she drove her car into a lake, purposely drowning her two children.

What theory of criminology could explain this behavior?

focusing on how the environment in which the person lives affects his or her behavior.

These theories present different explanations of crime based on different beliefs about crime and criminal offenders. Because these theories suggest preventing crime in different ways, they often lead to different policy recommendations. For example, criminologists who blame crime on faulty brain chemistry might suggest drug therapy for offenders, whereas those who attribute crime to economic inequality would more likely recommend policies to provide equal access to legitimate opportunities. To better understand these theories, it is first important to examine two basic questions about punishment.

Why Are Offenders Punished?

The purpose of punishment is to control and change behavior. Since biblical times, it has been argued that punishment deters crime. Modern-day scholars such as James Q. Wilson and Ernest van den Haag have likewise suggested that greater use of more severe, certain, and swift punishment will deter criminality.[1] If people choose to commit crimes, then punishment should deter criminality. As such, punishment should have four goals:

1. *General deterrence.* Punishment deters would-be offenders from committing crime.
2. *Specific deterrence.* Offenders are less likely to repeat their crimes after receiving punishment.
3. *Incapacitation.* Criminals who are removed from society cannot commit crime.
4. *Retribution.* Offenders deserve to be punished for the damage they caused.

The legal system in the United States embraces all of these purposes. Recent crime control policies, however, appear to place greater emphasis on retribution.

Does Punishment Deter Crime?

The answer to this question is not simple. Sometimes punishment does deter unwanted behavior; at other times, it does not. Punishment is most effective in the following circumstances:

• There is a short time lapse between the offense and the punishment.
• Offenders are told the reason they are being punished.
• Punishment is delivered consistently.
• There is a good, strong relationship between the offender and the punisher.[2]

Unfortunately, punishment is not carried out effectively in the modern U.S. criminal justice system for several reasons. First, it is not executed swiftly. The time lag between the crime and the offender's

arrest, prosecution, conviction, and sentencing is usually many months, sometimes years. Second, the likelihood (or certainty) of being arrested is very low. Most crime is not detected; when it is observed by others, only a small percentage of crime incidents are reported to the police, even fewer offenders are arrested, still fewer are convicted, and only a small number are incarcerated.

The current system of criminal justice in the United States, while imperfect, is the result of a complex evolution of policy based on varying theories of criminality. Each type of theory—choice theories, trait theories, and sociological theories—has offered practical suggestions for improving the effectiveness of punishment based on its unique understanding about the origin of criminality.

In 2002, 15-year-old Charles Bishop committed suicide by crashing his stolen plane into a building in Tampa, Florida. Bishop left a suicide note expressing support for Osama bin Laden and the terrorist attacks of September 11, 2001.

 What does this event suggest about the rationality of people who commit crime? Do rational people smash into buildings with airplanes?

Choice Theories

Choice theories from the classical and neoclassical schools of criminology, popular in the eighteenth and nineteenth centuries, believe that criminals are intelligent people who choose to commit crime. Choice theories state that the decision to commit (or refrain from) crime is an exercise of free will based on the offender's efforts to maximize pleasure and minimize pain.

LINK When people commit crime and are caught, they face punishment. Chapter 12 discusses the various punishments they may receive.

Classical School

The founding father of the classical school was Cesare Beccaria, who developed his ideas about crime and punishment at a time when European systems of criminal justice displayed callous indifference toward human rights. People were held accountable and punished for crimes against religion (including blasphemy and witchcraft) as well as for crimes against the state. Offenders were often punished without explanation, and anyone could be taken to jail for any variety of reasons.[3]

These seemingly capricious conditions inspired Beccaria to write an essay titled *On Crimes and Punishments,* in which he laid out the framework for a new system of justice—a system emphasizing consistency, rationality, and humanity.[4] This influential essay became the cornerstone of the 1791 criminal code of France and laid the groundwork for many of the important ideas in the U.S. Constitution, including protection from self-incrimination, the right to counsel, the right to confront one's accusers, and the right to a speedy trial by a jury of one's peers.

Another prominent figure within the classical school was an eighteenth-century British lawyer and philosopher, Jeremy Bentham. His greatest contribution toward reforming criminal justice was his promotion of the principle of utility, which prescribed "the greatest good for the greatest number."[5] Bentham judged whether behavior is immoral or moral by assessing its effect on the happiness of the community. According to him, law needed to maximize the pleasure and minimize the pain of the largest number of people in society—that is, "the greatest good for the greatest number."[6] Bentham also proposed the hedonism doctrine, in which he argued that people seek to maximize pleasure and minimize pain as their main goal in life and that actions taken to further all other goals are simply means to achieving pleasure and avoiding pain. Therefore, according to Bentham, crime is rational because criminals act purposely to acquire what they desire.

"Hedonistic calculus" was Bentham's term for the process in which people logically weigh the benefits of their behavior against its possible costs. Through this process, Bentham argued, people use their free will to determine their actions and will pursue crime if they decide that committing the crime will bring more pleasure than pain. When people commit crime, they understand that they are seeking to increase their pleasure illegally and that their actions may be wrong. Therefore, it is society's responsibility to punish them. The best punishment is one that produces more pain for the criminal than any pleasure that he or she may have derived from committing the criminal act.[7]

Bentham's ideas radically transformed the nineteenth-century English penal code and reduced the number of capital crimes from 222 to 3—murder, treason, and piracy.[8] However, despite its good intentions, the classical school ultimately failed, owing to its own rigidity. Beccaria's and Bentham's theories ultimately did not explain *why* people committed crime—only that they did. Classical theory held that people were equally responsible for their behavior; those who committed similar crimes received comparable punishments, regardless of the reason why crime was committed. In other words, the classical school focused solely on the criminal act, not on the actor. Of course, in reality, people are quite different. People who are insane, incompetent, or still children may not be as responsible for their behavior or as criminally culpable as normative adults. The idea that people are different from one another led to the formation of the neoclassical school.

LINK Many criminologists believe that juveniles are not as competent as adults. This is one reason why in 2005 the U.S. Supreme Court, in *Roper v. Simmons* (discussed in Chapter 16), ruled that persons who commit their crimes as juveniles cannot receive the death penalty.

Neoclassical School

The neoclassical school built on the works of the classical school by focusing on the role of the criminal justice system in preventing crime. While the founders of the neoclassical school were sympathetic to what the classical school had hoped to achieve, they also recognized that crime may be influenced by factors that are beyond the offender's control.

Reality television programs show both the rationality and the irrationality of criminal behavior.

Could classical theorists Beccaria and Bentham ever have envisioned the poor decision making of today's criminals? Do the decisions made by many criminals undermine rational choice theory?

Mitigating circumstances, such as age or mental illness, affect the choices that people make and influence a person's ability to form criminal intent (*mens rea*). The introduction of mitigating circumstances at criminal trials led to the notion of individual justice—the idea that criminal law must take into consideration the significant differences among people and the unique circumstances under which their crimes occur. Individual justice led to a number of important developments in the operation of criminal justice systems, including the introduction of the insanity defense and the inclusion of expert witnesses. Most importantly, individual justice became the foundation for a new explanation of crime, one that blamed criminality on individual characteristics that were in place before the act was committed. This new way of thinking about crime, which was called scientific determinism, relied on science to explain crime and became the centerpiece of the positive school of criminology (discussed later in this chapter).

The Get-Tough Movement

During the 1970s, criminologists began to ask new questions about the effectiveness of rehabilitation. A flurry of evaluation studies of rehabilitation programs concluded that, at best, some treatment plans work, sometimes, for some offenders, in certain settings.[9] This unconvincing endorsement of the rehabilitation model led many scholars to recommend that criminals be punished to prevent them from committing more crimes.

This new way of thinking launched the "get-tough" movement in criminal justice. Supporters of this movement argued that policies needed to increase prison terms, reinstitute corporal punishment, and apply the death penalty more frequently.[10] Another component of the "get-tough" movement included making incarceration a more painful experience. As a result, inmates wore striped uniforms in some jurisdictions and "chain gangs" reemerged in several states.

One proponent of this movement, Sheriff Joe Arpaio of Maricopa County in Phoenix, Arizona,[11] required that the inmates adhere to the following policies:

- Inmates must wear pink underwear and striped uniforms.
- Coffee, movies, pornographic magazines, smoking, and unrestricted TV are banned.
- Inmates are fed only twice per day; their meals will not include additional salt and pepper.
- Inmate chain gangs clean streets, paint over graffiti, and bury the indigent in the county cemetery.

Rational Choice Theory

Rational choice theory, a theory formulated by Ronald Clarke and Derek Cornish, explores the reasoning process of criminals and suggests that offenders are rational people who make calculated choices before they commit a criminal act.[12] According to this theory, offenders collect, process, and evaluate information about the crime; they weigh the costs and benefits of the crime before they make the decision to commit it. Offenders then use the same calculated process to decide where to commit crimes, who to target, and how to execute the crime.

Are Criminals Rational? Some research reports that burglars, drug dealers, prostitutes, rapists, robbers, serial killers, and street criminals do calculate the risks of getting caught. It has been discovered that gang members make rational decisions when deciding who to target, which business deals to make, and how to recruit new members. Other studies, however, report the opposite finding—that offenders make irrational decisions.[13]

Are Crimes Rational? Some predatory crimes, such as robbery, may be easily understood from a logical perspective, and even violence can be rationalized in

Calculated Crime

People commit crime because the likelihood of getting caught is very low. Morgan Reynolds has found that offenders are arrested for only about one out of every 1000 crimes they actually commit. Reynolds went on to estimate the cost of committing a crime, by calculating the actual punishment received by multiplying the probability of arrest, prosecution, conviction, and incarceration by the median time served for a crime.

For example, for every 100 burglaries, only 50 incidents are reported to police, and only 7 of those burglars are ultimately arrested. Of those arrested, 90 percent will be prosecuted (6.2 percent of all burglaries). Two-thirds of those prosecutions will result in felony convictions (4.2 percent of all burglaries). Of those defendants convicted, only 1.9 will be incarcerated; the remaining 2.3 will receive some combination of probation, fines, or jail time. Therefore, the overall probability of serving time in prison for burglary is only 1.9 percent.

To complete Reynolds's calculation, the median time served for burglary must be multiplied by this probability of incarceration. The median prison term per act of burglary committed in the United States is only 9.4 days. Therefore, according to Reynolds's calculations, a rational, risk-neutral criminal should find burglary profitable so long as what is stolen is worth more than 17.86 days in prison.

Reynolds also calculated the following costs of committing crimes:

Crime	Rational Cost
Murder	1.8 years
Rape	60 days
Robbery	23 days
Arson	6.7 days
Aggravated assault	6.4 days
Larceny/theft	3.8 days
Motor vehicle theft	1.5 days

As the expected punishment for serious crimes decreases, the rate of serious crime increases. Today, the expected punishment for serious crimes is about one-third of what it was in 1950, and the rate at which serious crimes are committed has increased by nearly five times.

Reynolds's findings have significant implications if the goal of punishment is deterrence. For the threat of punishment to prevent crime, there must be a reasonable chance that offenders will be caught, arrested, prosecuted, convicted, and incarcerated. If would-be offenders perceive the likelihood of detection as being low, they may more readily commit crimes.

Source: Morgan Reynolds, *Crime and Punishment in America* (Dallas: National Center for Policy Analysis, 2000).

circumstances where offenders believe it will produce desired rewards. For instance, when two people participate in a barroom fight, the reward may be an enhanced reputation; spouses may assault each other in an attempt to win arguments; and employees have murdered their co-workers to put an end to harassment and bullying. Thus, it is not uncommon for some people to use violence to get what they want.[14]

Routine Activities Theory

Expanding on rational choice theory, Lawrence Cohen and Marcus Felson formulated <u>routine activities theory</u>, which argues that for crime to occur, three elements must converge:

1. A motivated offender

2. A suitable target

3. The absence of protective measures (such as people)

In other words, crime increases when there are motivated offenders who are focusing on vulnerable targets (e.g., keys in the car ignition), with few people (e.g., bystanders) available to protect them.[15]

Two objections have been raised to routine activities theory. First, it does not identify those factors that would motivate someone to commit a crime in the first place; the theory incorrectly assumes that people are inclined to commit crimes when opportunities arise (some people are and others are not). Second, the theory overlooks factors that cause the criminalization of some behavior (i.e., smoking marijuana) and the legalization of equally harmful behavior (i.e., drinking alcohol).[16]

Milwaukee has recently experienced a spate of mob violence and serious violent crime that is overwhelming the resources available to the local police. In a three-month period in 2006, police arrested more than 14,000 people, confiscated more than 700 guns, and removed nearly 40 pounds of cocaine from the city's streets. They also witnessed the following horrific acts:

- An 11-year-old girl informed police that she went to the home of a 16-year-old girl, where she was directed to perform oral sex on three teenage boys, have vaginal sex with a 40-year-old man, and engage in sexual acts with 15 other males, sometimes with more than one at a time.

- More than a dozen people chased a man through the streets and beat him to death with shovel handles, rakes, and tree limbs.
- A mentally ill man died after being beaten and robbed by a mob.
- A 14-year-old boy was in a coma for two weeks after he was kicked, punched, and hit on the head with a piece of lumber after he exchanged words with a girl.
- Four brothers were beaten by a group armed with bats, bottles, sticks, and socks stuffed with canned food.

While city leaders see these acts as inhumane and an embarrassment to the city, they do not know how to prevent them from taking place. Is it possible there is no remedy for senseless

violence? Are the offenders and their crimes rational? Which policies should be put in place to put a stop to these seemingly irrational acts of violence?

Sources: John Diedrich and Mary Zahn, "We Have a Societal Crisis," *Milwaukee Journal Sentinel,* September 6, 2006, available at http://www. jsonline.com/story/index.aspx?id=492945, accessed March 30, 2007 ; John Bacon, "Milwaukee Facing 'Societal Crisis'?", *USA Today,* September 10, 2006, p. 2A; "Girl, 11, Put through Sex Ordeal by Mob," *Sydney Morning Herald,* September 8, 2006, p. 1A.

Lifestyle Theory

Michael Hindelang, Michael Gottfredson, and James Garofalo have proposed lifestyle theory, which is closely related to routine activities theory, to explain why some people are more likely than others to be crime victims.[17] Hindelang and his colleagues suggest that people become crime victims because of the situations in which they put themselves (for example, spending time alone on dimly lit streets at night with no bystanders). Lifestyle theory proposes that the more time that people spend away from home, the greater their risk of being victimized owing to their increased visibility and accessibility. Additionally, when they are out, their cars and homes may be left unattended and are thus more likely to be victimized.

LINK Chapter 3 reported crime victimization rates that are higher for some categories of people. Does lifestyle theory explain these differences? Which other aspects of individuals' lifestyles may contribute to varying levels of victimization?

Social Policy Applications: Choice Theories

When crime rates soared in the 1960s and 1970s, Americans became increasingly concerned about whether it was possible to rehabilitate criminals. As a consequence, they looked for new ways to fight crime. Crime control legislation modeled in the neoclassical school of thought and its efforts were devised to increase the certainty, severity, and swiftness of punishment. These measures included:

- Cell phone tracking surveillance
- Three-strikes sentencing and truth-in-sentencing laws
- Hiring more police officers

LINK Chapter 12 explores habitual offender (or three-strikes) laws, which require mandatory imprisonment for offenders convicted of their third serious criminal offense, and truth-in-sentencing laws, which require offenders to serve a substantial portion of their original sentence before they become eligible for release from incarceration.

Many other, more simple and direct strategies for reducing crime were implemented to reduce the opportunities for situational crimes:

- Making it physically more difficult to commit crimes (e.g., locks on steering wheels)
- Increasing the perceived risk of crime (e.g., electronic merchandise tags in department stores)
- Reducing anticipated rewards of crime (e.g., requiring a personal identification number to operate a car stereo)[18]

Taken together, these strategies have helped to lower the overall crime rate as well as the crime rates for serious violent and property crimes.

Trait Theories

Trait theories offer a different way of thinking about crime and criminals. Unlike choice theories, which assume that people are basically alike except for the choices they make, trait theories are rooted in the biology, psychology, or behavior of individuals and argue that offenders commit crimes because of particular traits, characteristics, deficits, or psychopathologies they possess.

Biological Theories

The popularity of theories from the classical school declined in the late nineteenth century, in part because these theories had only a negligible impact on crime, and in part because they were considered too simplistic. During this time, the positive school of criminology emerged, founded on the work of sociologist August Comte. Comte proposed that human behavior was caused by forces over which the individual had no control. He suggested that societies pass through a series of stages, beginning with primitive understandings of the social order, and culminating with a more rational, scientific awareness known as positivism. Thus, scholars who adhered to Comte's views called themselves positivists.[19]

Positivists sought out scientific explanations of crime, and their early theories suggested that criminals possessed biological traits or defects that caused them to commit crime FIGURE 4-1 . Early biological theories of crime were crude and deterministic, typifying criminals as subhuman and implicating dubious social policies such as eugenics. As biological theories evolved, however, they became more interdisciplinary, focusing their attention on personal traits, an individual's environment, and the interaction between them.

Atavism

In 1876, an Italian psychiatrist named Cesare Lombroso published *The Criminal Man,* the first book devoted exclusively to the causes of criminality. Lombroso argued that not all people had evolved to the same degree and that many criminals were evolutionary throwbacks (or atavists). These individuals, Lombroso suggested, could be distinguished by certain features that closely resembled our apelike ancestors in traits, abilities, and dispositions. These stigmata (distinctive physical features) included the following characteristics:

- Asymmetrical face
- Enormous jaw
- Large or protruding ears
- Receding chin

Lombroso proposed that, through no fault of their own, criminals were unable to obey the complex rules and regulations of modern society and should be placed in restrictive institutions such as prisons.[20]

Lombroso's findings were challenged in 1913 by Charles Goring, who compared the physical measurements of 3000 English convicts with the physical measurement of a sample of university students, but found no evidence of any physical disparities between the two groups.[21] Goring's conclusion remained unchallenged for nearly 40 years, until Earnest Hooton discovered that Goring had ignored data that refuted his argument and supported Lombroso's theory. Once Hooton reexamined Goring's data, he concluded that criminals were inferior to civilians in nearly all of their bodily measurements.[22]

More than 125 years have passed since Lombroso made his original claims. Recently, researchers have discovered that people with asymmetrical extremities (i.e., ears, fingers, or feet of different sizes or shapes) tend to react aggressively when annoyed or provoked. These slight physical imperfections may be derived from adverse prenatal conditions such as the mother smoking during a pregnancy, which could cause poor impulse control and subsequently lead to an increase in the likelihood of criminal behavior.[23]

Body Type

In 1949, William Sheldon proposed another theory based on the relationship between body type and criminality. His so-called somatotype theory posited that persons with particular body types are likely to be inclined toward certain behaviors. Sheldon identified three body types with associated characteristics:

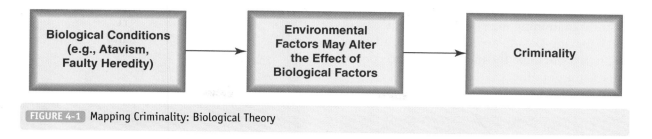

FIGURE 4-1 Mapping Criminality: Biological Theory

- *Endomorphs*—rounder, relaxed, comfortable, extroverted "softies"
- *Mesomorphs*—muscular, active, assertive, lust for power
- *Ectomorphs*—lean, thin, introverted, overly sensitive, love privacy

To evaluate the validity of his theory, Sheldon measured the bodies of 200 incarcerated juvenile offenders and 4000 male college students on different dimensions. He found that the young criminals were more likely to be mesomorphs and much less likely to be ectomorphs. No significant differences were detected among the groups for endomorphs.[24] Sheldon's research has since been widely replicated and, in general, criminologists have found support for his thesis.[25]

Some criminologists unequivocally reject the idea of a relationship between body type and criminality, yet evidence continues to mount suggesting that there may be a relationship between the two. For example, when neurobiologist Adrian Raine and his colleagues studied the effects of body size on delinquency among 1130 children, they found that large body size in young children correlated with increased aggression later in life. Large children tended to be more fearless and stimulation seeking, and the effects of body size on delinquency remained after controlling for temperament.[26]

Intelligence

Intelligence is among the most thoroughly studied correlates of crime. In 1575, the Spanish physician Juan Huarte defined "intelligence" as the ability to learn, exercise judgment, and be imaginative. Since then, scientists have developed many different ways to measure intelligence. The first usable intelligence test, which is the basis for the contemporary intelligence quotient (IQ) test, was developed late in the nineteenth century by Alfred Binet and Theodore Simon.[27]

Until recently, the possibility of a relationship between intelligence and crime was a taboo topic among criminologists because of possible race and class issues. When the relationship was finally stud-

ied, however, criminologists generally found a connection between IQ and crime. Studies have reported an IQ difference between offenders and nonoffenders ranging from 9 to 14 points.[28] Travis Hirschi and Michael Hindelang, for example, have concluded that IQ is a better predictor of involvement in crime than either race or social class and suggest that the average IQ of delinquents is about eight points lower than the average IQ of nondelinquents.[29] Other criminologists agree. For instance, in one study it was found that IQ predicted delinquency even when controlling for good, strong correlates of delinquency, such as social class, race, and academic motivation.[30]

Intelligence has also been found to be correlated with adult crime. After evaluating the intelligence-crime link among 261 sex offenders and 150 nonsexual violent offenders, Jean-Pierre Guay and his colleagues reported that sex offenders had significantly impaired cognitive abilities compared to other criminals in areas such as vocabulary, comprehension, arithmetic, mental math computations, object assembly, letter-number sequencing, and perception.[31]

Low Resting Heart Rate

Some modern criminologists propose that people may be biologically predisposed to crime because they suffer from chronic nervous system underarousal, a condition that manifests itself as a low resting heart rate. These individuals seek out stimulating events, such as thrill-seeking activities like bungee jumping, skydiving, or crime. Some research also points to the fact that known psychopaths show slower heart rate responses to fear-provoking incidents than other people. In addition, a low resting heart rate is also more common among chronic offenders than among one-time offenders, violent criminals as compared to nonviolent criminals, and prisoners as compared to nonprisoners.

The finding of an association between low resting heart rate and antisocial behavior has been replicated in many different studies in different countries, including Canada, England, Germany, Mauritius,

New Zealand, and the United States.[32] In an unrelated study examining many possible predictors of violent behavior, David Farrington also found a low resting heart rate to be the most robust and consistent predictor of crime.[33]

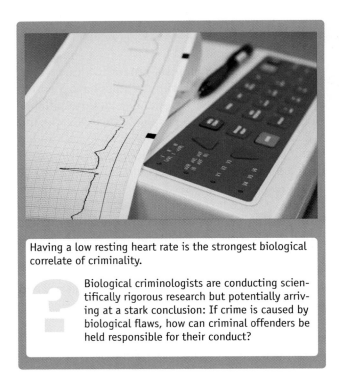

Having a low resting heart rate is the strongest biological correlate of criminality.

? Biological criminologists are conducting scientifically rigorous research but potentially arriving at a stark conclusion: If crime is caused by biological flaws, how can criminal offenders be held responsible for their conduct?

Brain Structure and Process

The brains of different individuals are both structured differently and process information differently. There exists strong evidence suggesting that criminality is associated with particular differences in brain structure that affect individuals' ability to exercise self-control and to respond to changes in their environment. In particular, people with low self-control appear to be more likely to commit crimes than persons with high self-control.[34]

The brains of some people produce more or fewer chemicals than are needed; as a result, their brains process information differently than the brains of people with normal levels of the same chemicals. A brain that produces too little serotonin, for example, may cause a behavioral condition called attention deficit/hyperactivity disorder (ADHD) that causes impulsivity, aggression, and violent offending.[35] People with ADHD often make careless mistakes, appear not to listen when spoken to, are forgetful, answer a question before the speaker has finished, and interrupt the activities of others at inappropriate times.

Attention deficit/hyperactivity disorder is a prevalent and costly behavioral disorder. Most children with ADHD lead normal lives and safely manage their medical condition. For some, however, it is the forerunner of a lifetime of serious criminal behavior.

? How can the same condition lead to such diverse outcomes? Are powerful medications safe treatments, or are they as damaging as ADHD itself?

Furthermore, research has continued to find a positive correlation between ADHD and criminality.[36]

Genetics

While genes do not cause crime, some people might inherit traits or tendencies that make them more predisposed toward committing crime than other people. Genetic structure does influence hormones and enzymes that interact with environmental factors to lead to aggressive behavior. To discover the possible effects of genes on behavior, criminologists often study twins and adoptees.

Twin Studies Two types of twins are distinguished: monozygotic twins (identical twins), who come from one fertilized egg and therefore have identical DNA, and dizygotic twins (fraternal twins), who come from two separate eggs that are fertilized at the same time. Fraternal twins are no more alike genetically than nontwin siblings. Given this fact, criminologists examining the link with genetics argue that if there is a genetic factor in crime, then monozygotic twins should be more alike in criminality than are dizygotic twins. Many twin studies have been completed, and they generally show that there is more concordance (in which both twins are similar with regard to criminal behavior) among monozygotic twins than among dizygotic twins. From these findings, criminologists have reached a simple conclusion: Heredity matters.[37]

Adoption Studies   Adoption studies offer another way to evaluate the role of genetics on crime. These studies try to separate the effects of heredity from the effects of the environment. After all, many adopted children have little or no contact with their biological parents. Thus, if the children resemble their biological parents in behavior, this finding would provide support for the argument that heredity affects behavior.

In one of the most frequently cited studies of adoptees, Sarnoff Mednick and Karl Christiansen concluded that the criminality of the child was more closely related to the criminality of the biological parents rather than the adoptive parents.[38] Since this pioneering research, many other studies of adoptees have been conducted and have generally reached a similar conclusion: Heredity matters.[39]

Social Policy Applications: Biological Theories

Policies derived from biological theories include recommendations that society invest more money in prenatal and postnatal care for women, closer monitoring of young children during their most crucial developmental years, paid maternal leave, and nutritional programs for pregnant women, newborns, and young children. Although biologists do not believe that any of these programs are a cure for internal deficiencies, they may nevertheless help to prevent future criminality.

In addition, some biologists recommend that offenders receive pharmacological treatments, such as medication for those diagnosed with biological disorders such as ADHD. The idea behind prescribing the drugs to offenders—and particularly to young children—is that early intervention in their lives may help to promote factors that insulate them from crime, minimize or erase the risk factors that contribute to crime, and equalize the life chances for them to develop into healthy, prosocial adults.

Finally, some biologists believe that certain individuals pose a serious risk to public safety because their internal deficiencies cannot be controlled with drug therapy and, therefore, must be incarcerated as a last resort. If offenders cannot control their biological predisposition to commit crimes on their own, public safety mandates that when the cause of the behavior is known, it must be neutralized.

Psychological Theories

Trait theorists believe that in addition to biological factors, psychological conditions may cause crimi-

Monozygotic (identical) twins share the same DNA. Studies of such twins have shed light on the genetic underpinnings of behavior. While identical twins have the same DNA, however, all people share more than 99 percent of the same DNA.

 Does the fact that all human variation is explained by less than 1 percent differences in genetics mean that genes are not responsible for human behavior?

nal behavior. Many criminals grew up in dysfunctional families and lived in a state of conflict with their parents, neighbors, peers, classmates, and teachers. The conflict these individuals experienced throughout their lives is a red flag to criminologists, who believe that these offenders may have mental deficiencies that cause them to commit crime.

Psychoanalytic Theory

In 1910, psychologist Sigmund Freud published *The Interpretation of Dreams*, followed in 1923 by *The Ego and the Id*. Freud's thinking about mental processes provided a radically new approach to the study of antisocial behavior known as <u>psychoanalytic theory</u>, which suggests that unconscious mental processes developed in early childhood control the personality.

Freud theorized that the personality consists of three parts: the id, the ego, and the superego.[40]

- In the beginning of life, the child possesses an <u>id</u>, which consists of blind, unreasoning, instinctual desires and motives. The id cannot differentiate between fantasy and reality. It is antisocial and knows no rules, boundaries, or limitations. If the id is left unchecked, it will destroy a person.

- From the id grows the <u>ego</u>, which represents the problem-solving dimension of the personality. The ego is able to deal with reality, differentiating it from fantasy. From the ego the child learns to delay gratification, because he or she figures out that acting on impulse will simply get him or her into trouble.

- Next comes the <u>superego</u>, which develops from the ego. The superego represents the moral code, norms, and values the young person has acquired. It is responsible for causing feelings of guilt and shame in the child and is closely aligned with the person's conscience.

When people are mentally healthy, the id, ego, and superego work together. When these three parts of the personality are in conflict, however, it is likely that the person will become maladjusted and possibly commit crimes. Although Freud did not write specifically about crime, his theory influenced many others who did. They generally suggested that criminals have either under- or overdeveloped superegos `FIGURE 4-2` .[41]

Behavioral Theory

<u>Behavioral theory</u> proposes that behavior is a product of interactions people have with others throughout their lifetime `FIGURE 4-3`. This theory was developed by B. F. Skinner, who suggested that children learn conformity and deviance from the punishments and reinforcements they receive in response to their behavior. Skinner also believed that the environment shapes behavior such that children are more likely to repeat behavior that was rewarded and to desist from committing behavior that was punished.[42]

Social Learning Theory

Behaviorists used Skinner's theory as a springboard to construct new theories. For example, Albert Bandura suggested that learning and experiences combine with values and expectations to determine behavior. In his <u>social learning theory</u>, he stated that children learn by modeling or imitating the behav-

Social learning theory examines how children learn criminal behavior from their parents.

? Why is it controversial to examine how crime runs in families?

ior of others.[43] For instance, children who see their parents fighting, watch violent television programs, or play violent video games are more likely than other children who do not have similar experiences to be aggressive.

Researchers have shown that violent video games are linked to antisocial behavior.

? Why is American culture so tolerant of violent video games? Why do children prefer games with violent content?

Research testing Bandura's ideas has generally found support for them. Indeed, of the more than 3500 published studies on the topic, only 18 have

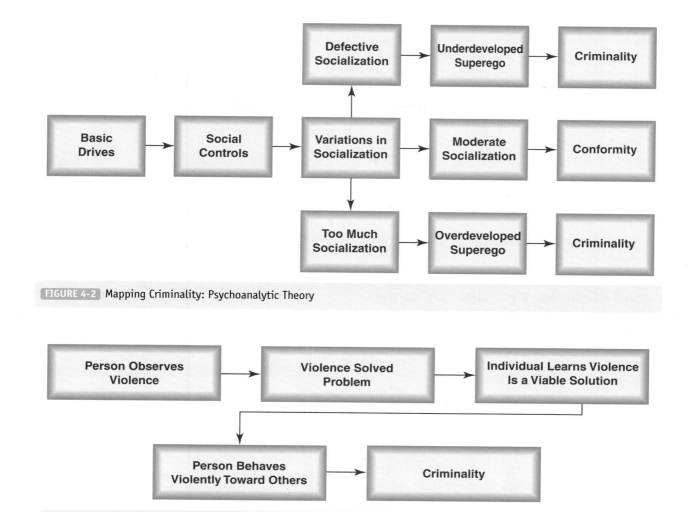

FIGURE 4-2 Mapping Criminality: Psychoanalytic Theory

FIGURE 4-3 Mapping Criminality: Behavioral Theory

not reported a positive association between media exposure and violent behavior.[44] Seeing violent entertainment affects children in one of three ways:

1. Children see violence as an effective way to settle conflicts.

2. Children become emotionally desensitized toward violence in real life.

3. Entertainment violence feeds a perception that the world is a violent place and increases fear of criminal victimization.[45]

Social Policy Applications: Psychological Theories

Freud and his protégés suggest that crime is a symptom of deep-seated psychological problems in which an individual's instinctual drives are not balanced and controlled. As a consequence, offenders require counseling to help them understand how destruc-

tive thinking has adversely affected their lives and has caused mental or emotional disturbance. Behavioral theorists, such as Skinner and Bandura, take a different view, blaming crime on individuals' interactions with their environments. Policies derived from behavioral theory, therefore, emphasize teaching people different ways of responding to their environment through techniques such as behavior modification.

Two behavior modification strategies that are widely used in correctional settings are aversion therapy and operant conditioning.

- In aversion therapy, the person learns to connect unwanted behavior with punishment. Alcohol offenders, for instance, receive treatment where they are required to ingest a drug that causes nausea or vomiting if they drink alcohol. It is thought they will connect drinking with this unpleasant experience and stop drinking to avoid the ill effect.

- Operant conditioning uses rewards to reinforce wanted behavior and punishments to extinguish unwanted behavior. One example of operant conditioning used in federal penitentiaries today is the token economy, whereby inmates earn tokens in exchange for good behavior, which they can then exchange for items from the commissary or for privileges such as watching television. Conversely, inmates who participate in unwanted behavior have their tokens taken from them.

Sociological Theories

In the 1920s, criminologists began to look beyond individual-focused theories about crime. From this work, sociological theories began to emerge. These theories suggested that the causes of crime were located outside the offender. Rather than blaming criminality on some biological or psychological flaw of the person, sociologists suggested that crime might be caused by social factors found in people's environments, including neighborhoods, schools, and family.

Cultural Deviance Theory

The first sociological explanation of crime was published in the 1920s by Clifford Shaw and Henry McKay. Their work focused on the role of the neighborhood in which the offender lived. In their cultural deviance theory, Shaw and McKay suggested that crime is the product of social and economic factors located within a neighborhood. In other words, crime is a function of how neighborhoods are structured.[46]

In more affluent neighborhoods, Shaw and McKay discovered, crime rates were low. These neighborhoods, they proposed, provided consistency in values and norms and met the needs of the children who were closely supervised by parents. In low-income neighborhoods, where crime rates were higher, conflicting values and norms were in place. In such socially disorganized neighborhoods, children did not receive the support or supervision they needed to encourage them to obey the law.

Shaw and McKay demonstrated that immigrant families living in areas with high crime rates had not committed criminal offenses in their countries of origin. It was only after they moved to the inner city in the United States that members of these families started to participate in crime. The criminal involvement of immigrants typically did not persist after

Decades of sociological research indicate that neighborhoods exert powerful effects on crime.

? Is it possible that the neighborhood-crime link exists because a core of criminally inclined people live in the worst neighborhoods?

these individuals moved away from the inner city. Moreover, crime rates in the inner city remained stable and high, regardless of the race or ethnicity of the people living there, and these rates continued to be high even after most of the original residents left. Shaw and McKay concluded that it was neighborhoods themselves that were responsible for crime and that criminal values were passed from one generation to the next through a process of cultural transmission FIGURE 4-4.

Differential Association Theory

Expanding on the idea of cultural transmission, Edwin Sutherland constructed differential association theory to explain both individual criminality and group crime by identifying those conditions that need to be present for crime to take place (and that must be absent when there is no crime). This theory suggests that criminal behavior is learned in interactions with friends, family, and other intimates through verbal and nonverbal communication.[47] These interactions teach the techniques of committing crime and the specific direction of motives, drives, rationalizations, and attitudes. According to this theory, the longer, earlier, more intensely, and more often someone is exposed to attitudes about criminality (either positive or negative), the more likely it is that he or she will be influenced in a particular direction.

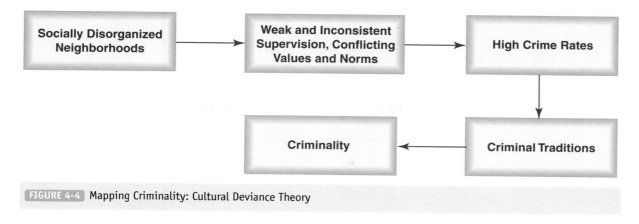

FIGURE 4-4 Mapping Criminality: Cultural Deviance Theory

Strain Theory

Developed by Robert Merton, strain theory faults American culture for teaching all of its members to strive for economic success (the American dream) while restricting access to legitimate means to achieve success for some people. This theory blames crime on a lack of integration between cultural goals (what people are told they should want) and institutionalized means (what the social structure allows them to achieve). When goals and ways to achieve those goals are not in line, Merton argued, social norms break down, creating a condition called anomie in which people feel alienated and uncertain about society's expectations and are less able to control their behavior.[48]

In essence, strain theory suggests that crime is a normal response to social conditions that limit the opportunities for some members of society to obtain the economic success that all members try to achieve. For example, a cultural goal in the United States is acquisition of money. The socially approved ways to acquire money are through training, education, career advancement, and hard work. These means are aided by getting a good education, receiving job training, and pursuing career advancement. It is easy to see that some people have a much shorter path to success under this rubric than others who are born in less advantaged circumstances **FIGURE 4-5** .

The playing field is clearly not equal, and some members of society have more access to these means to achieve success than other individuals. Their increased opportunities are often attributable to the following factors:

- Ascribed qualities (gender and race)
- Resources (wealthy parents and good connections)
- Environmental advantages (growing up in a good neighborhood)

- Instilled values (education, good work ethic, and the pursuit of worthwhile goals)

For many other people, access to legitimate means to achieve success is blocked. These barriers create problems because even those people without these resources desire wealth and status.

Merton believed that strain between means and goals is always present. He identified five ways that people adapt to such frustration **TABLE 4-1** :

TABLE 4-1

Merton's Modes of Adaptation

Modes of Adaptation	Cultural Goals	Institutionalized Means
Conformity	Accept	Accept
Innovation	Accept	Reject
Ritualism	Reject	Accept
Retreat	Reject	Reject
Rebellion	Reject prevailing goals and means and substitute new ones	

Source: Adapted from Robert Merton, *Social Theory and Social Structure,* revised ed. (New York: Macmillan, 1968).

1. *Conformity*—buying into the system, accepting both cultural goals and the means approved to achieve those goals

2. *Innovation*—deviating from socially acceptable ways to achieve cultural goals

3. *Ritualism*—abandoning accepted cultural goals and accepting the status quo

4. *Retreat*—withdrawing from society altogether

5. *Rebellion*—refusing to accept socially accepted goals or ways to achieve those goals

Strain theorists suggest that people are inherently good and participate in crime only out of ne-

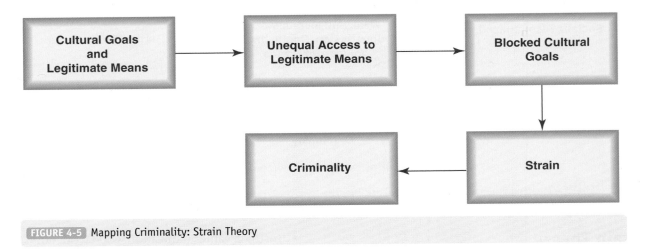

FIGURE 4-5 Mapping Criminality: Strain Theory

cessity. Thus, if society found a way to eliminate the conditions that produce strain, it could eliminate crime.

Social Control Theory

Social control theory claims that people are, by their very nature, amoral (without morals) and will break the law unless obstacles are thrown in their paths **FIGURE 4-6** . Crime, therefore, is the expected behavior, so social control theorists seek to explain why people do *not* commit crime. The originator of this theory, Travis Hirschi, believed that crime is something all people will engage in if there are no controls on their behavior.[49] Controls are attitudes that are implanted quite effectively in most people but less so in some other individuals who have a weak social bond or connection to society. The stronger a person's bond to society, the less likely he or she will commit crime.

Self-Control Theory

More than two decades after he introduced social control theory, Hirschi reevaluated his position and joined with his colleague, Michael Gottfredson, in proposing self-control theory. This theory suggested that people are self-gratifying and seek pleasure; they commit crime because they are unable to regulate their behavior owing to low self-control.[50] Hirschi and Gottfredson proposed that some people are more impulsive, insensitive, and short-sighted; these risk

FIGURE 4-6 Mapping Criminality: Social Control Theory

takers have a low tolerance for frustration, making them more likely to engage in criminal behavior. In contrast, people with high self-control are less likely to commit crime. According to Hirschi and Gottfredson's theory, the amount of self-control someone exhibits is a product of early childhood rearing; post-childhood experiences have little effect on self-control. Parents who monitor the behavior of their children, supervise them closely, recognize unacceptable behavior, and administer punishment are, therefore, more likely to have children who have the self-control necessary to resist easy gratification and the desire to commit crime **FIGURE 4-7** .

- Deviants and nondeviants are more similar than they are different.
- Whether people are labeled deviant depends on how people react to their behavior, rather than on the behavior itself.
- Behavior is neither moral nor immoral; it becomes one or the other depending on people's reaction to it.[53]

In addition, Becker described the process of becoming deviant, which may also be applied to the process of becoming a criminal. The first step is the commission

FIGURE 4-7 Mapping Criminality: Self-Control Theory

Labeling Theory

Labeling theory examines the role of societal reactions in shaping behavior; in other words, it focuses on why some people and behaviors are considered criminal while others are not. Frank Tannenbaum was one of the first criminologists to explore these issues. He believed that the only real difference between criminals and noncriminals is that the former had been caught and labeled "criminal."[51] Edwin Lemert added two new ideas to Tannenbaum's work: primary deviance and secondary deviance.[52] Primary deviance is the original behavior that leads to the application of the deviant label; secondary deviance occurs once someone has internalized the deviant label and uses it as a means of defense, attack, or adjustment to the problems caused by the label **FIGURE 4-8** .

Labeling theory reached the height of its popularity in the 1960s with the work of Howard Becker. Becker made the following arguments:

of a deviant act; it is followed by getting caught or being accused of the act. Once caught, the spotlight is placed on the offender, giving him or her new status with a label (i.e., "liar," "drug user," or "thief"). After the person is labeled, he or she is presumed by others to be more likely to commit other deviant behaviors. The negative label or stigma is, therefore, generalized to the whole person, such that someone who is accused of one type of deviance (i.e., cheating) is expected to commit other types of deviance (i.e., stealing).

When the label is successfully applied, being a deviant becomes the person's master status (what others think about him or her when they first meet). The final step in the process is for the deviant to join an organized group (e.g., delinquent or criminal gang) in which members have learned to rationalize their deviant activities so they may continue to commit crime without experiencing any feelings of remorse, guilt, or shame.

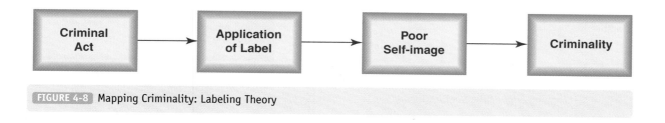

FIGURE 4-8 Mapping Criminality: Labeling Theory

A proposal recently introduced by English Policing Minister Hazel Blears has reignited an important debate that first appeared in the 1960s. Blears has recommended that 125,000 English children with incarcerated fathers be tracked or monitored. Research shows that children with incarcerated family members are more likely to commit crimes.

The program Blears has proposed would target these "at-risk" children and provide assistance to their families throughout the child's adolescence. One goal of the program is to teach parenting skills; to help do so, it would incorporate social work visits, after-school activities, and other measures designed to promote pro-social behavior.

Critics complain that the program may further stigmatize the very children who it is designed to help. The argument critics raise is based on labeling theory. If the system labels the children of incarcerated fathers as potential delinquents, it may create a self-fulfilling prophecy. If experts continuously tell a child that he or she is likely to grow up to be a criminal, they run the risk that the child might eventually get the (wrong) message.

The proposed intervention faces another hurdle. The central lesson of life-course theory is that young people must be targeted by the correct type of intervention if they are to stay out of trouble. If it is true that children follow different pathways to crime, then "one size does not fit all" and no single prevention will always be best for all children, all of the time, in all situations. For some adolescents (particularly ones with neuropsychological deficits), the cycle of misbehavior may begin very early in their lives. For others, their participation in crime may start much later and represent nothing more than "run-of-the-mill" adolescent rebellion. The latter children, who may have fathers in prison, likely do not need to be tracked and targeted. Yet because of the inflexibility of the proposed program, they will be, because children are treated as equals regardless of their personal needs.

It is too early to say whether this program might be beneficial or whether it will stigmatize children. If life-course theory is correct, the program will help some children and harm others.

Source: "Youth Crime Plans 'Could Misfire,'" *BBC News Online,* available at http://news.bbc.co.uk/nolpda/ukfs_news/hi/newsid_3568000/3568492.stm, accessed May 14, 2007.

Conflict Theory

Conflict theory views society in terms of inequalities in power and influence. The most influential conflict theory was constructed by Karl Marx and Friedrich Engels. They stated that in industrialized societies, the economic interests of those who own the means of production (the bourgeoisie) and those who sell their labor (the proletariat) are incompatible. The ensuing class conflict produces conditions ripe for criminality.[54] In addition, the bourgeoisie exert control over all aspects of social life, including the production of ideas, which means they control the creation of the criminal law. Their beliefs form the basis for both law and its enforcement, which become important tools to protect their economic interests. Thus Marx and Engels believed that crime is the product of a disheartened working class. Criminologist Willem Bonger later constructed a theory of crime based on a Marxist analysis, suggesting that crime results from people who are trying to get ahead and thinking only of their own personal needs FIGURE 4-9 .[55]

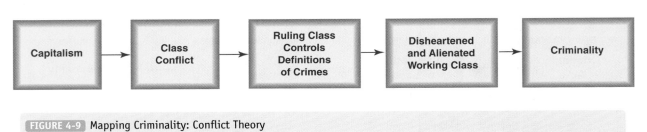

FIGURE 4-9 Mapping Criminality: Conflict Theory

Life-Course Theory

Life-course theory (also called developmental theory) has recently gained popularity for its emphasis on treating behavior as constantly changing, evolving from demands, opportunities, interests, and events that people experience as they grow older. Central to this theory is an examination of offenders' early childhood and the way in which these experiences influence the onset of their participation in crime at later stages in life over the course of their criminal career. At the core of life-course theory is the idea that human development does not end either in childhood or during adolescence, but rather is a continuous process that stretches throughout the entire life cycle. Therefore, life-course theorists look beyond what happened in the lives of young people immediately before they got into trouble and instead assess what has been going on in their lives during all the years preceding that criminal activity.[56]

Associating with criminal peers is among the strongest indicators of crime.

? Why are youths so susceptible to the negative influences of peers? Why does this effect decline with age?

Social Policy Applications: Sociological Theories

Sociological theories point to a variety of external factors (such as experiencing strain while trying to achieve cultural goals, having weak social bonds, or maintaining poor self-control) as the root causes of crime. Crime prevention policies based on these theories are designed to affect the relationship between offenders and society.

Policies based on cultural deviance theory seek to change the landscape of a neighborhood so as to make it easier to mainstream people—that is, to bring them into the larger society. One successful program is the Chicago Area Project (CAP), started in 1931, which mobilizes residents living in neighborhoods with high crime rates by focusing on direct service, advocacy, and community involvement. Community residents work with CAP officials to keep children out of trouble, help children when they do get into trouble, and keep the neighborhood clean.[57]

Strain theorists advocate policies that seek to reduce crime by creating new opportunities for disadvantaged or underprivileged people. Opportunities for offenders to "go straight" and be successful in the legitimate world can reduce the need to commit crime. Several far-reaching prevention programs based on strain theory were implemented in the 1960s. The best known of these initiatives is Project Head Start, which is a comprehensive child development program serving the needs of children from birth to age 5. More than 1 million children are enrolled in Head Start programs today across the United States.[58]

Policies drawn from social control theory aim to strengthen the bond between children and their parents, other adults, schools, and the community by involving young people in pro-social activities. The Police Athletic League is an example of this kind of effort. It offers youth a positive experience with police and provides at-risk children with guidance, discipline, and the inculcation of values from adults who serve as mentors, along with educational support, increased awareness of career options, and assistance for setting pro-social goals.[59]

Programs guided by self-control theory touch a child's life at a very young age, such as early childhood intervention programs that assist single mothers with child care. Successful programs include a parent-training curriculum that aims to strengthen parents' monitoring and disciplinary skills and build confidence in their parenting abilities.[60]

Labeling theory suggests the best strategies for reducing crime are to ignore minor acts of deviance, react informally by diverting people from the formal criminal justice system, and bring the offender, victim, and community together to restore justice. According to this theory, formal intervention should be a last resort, and diversion programs should be used whenever possible.[61]

Conflict theory has had a negligible impact on crime policy. For the most part, this theory is too radical for state and federal governments to implement, because it calls for sweeping changes to the social and economic organization of society (such as eliminating inherited wealth). Nevertheless, conflict theory has led to useful discussions about the consequences of structural inequalities and efforts to eliminate overt discrimination within the criminal and juvenile justice systems.

LINK Despite attempts to eliminate discrimination within the criminal justice system, it often persists in both arrest decisions (Chapter 8) and sentencing (Chapter 12).

Life-course theory has many policy applications, most of which focus on prevention programs for at-risk people of various ages. For younger children, programs might focus on strengthening family ties and engaging in effective communication. For youths in high school, it might be more important to focus on peer pressure, gang involvement, and other issues related to the peer groups of at-risk children. Later intervention strategies should focus on making effective transitions to the job market and avoiding dysfunctional personal relationships.

Crime Theory, Causes of Crime, and Social Policy

Theory	Cause of Crime	Social Policy
Choice Theories	People are rational and intelligent, and they weigh the costs and benefits of behavior before they act; they have free will.	Three-strikes sentencing; truth in sentencing; cell phone tracking surveillance; hiring more police officers.
Trait Theories		
Biological theory	Biological flaw or defect within the individual.	Isolation; sterilization.
Psychoanalytic theory	Defect in personality.	Treatment, counseling, drug therapy.
Behavioral theory	Behavior is a product of interactions with other people throughout one's lifetime.	Aversion therapy; operant conditioning.
Social learning theory	Behavior is learned through imitation and the person imagining the consequences if he or she committed the criminal act.	More negative reinforcement; more positive reinforcement.
Sociological Theories		
Cultural deviance theory	Disorganized neighborhoods.	Community empowerment such as through the Chicago Area Project.
Strain theory	Some people are blocked from achieving cultural goals through legitimate means.	Increase opportunities to become successful through legitimate avenues.
Social control theory	Offenders are poorly connected to society as a consequence of faulty socialization.	Improve childrearing programs and develop other programs such as the Police Athletic League to strengthen the bond between wayward youth and the community.
Self-control theory	Low self-control, which stems from faulty parenting.	Programs that teach fathers and mothers parenting skills and appropriate techniques for handling the difficult situations they will face.
Labeling theory	Societal reactions to behavior are responsible for crime.	Diversion programs; reintegration of the offender into the community; restorative justice.
Conflict theory	Power differentials among members of society; competition.	Programs that equalize power relations among all of society's members.
Life-course theory	Crime is caused by cumulative factors that vary from childhood to early adulthood.	Age-appropriate interventions to interrupt the cycle of crime.

WRAP UP

According to the biological perspective, the source of Chikatilo's crimes may have been the severe head trauma he experienced as a result of frequent beatings throughout his childhood. Damage to the brain caused a lack of self-control and self-restraint, which might explain the feverish, insatiable nature of Chikatilo's crimes.

From a psychological perspective, Chikatilo's crimes are the outcome of his antisocial personality traits, which facilitate his involvement in crime and complicate his ability to function in society. Psychiatric diagnoses showed that Chikatilo suffered from overlapping personality disorders, including psychopathy. A hallmark of the psychopathic personality is the inability to empathize with or feel an emotional connection to other people. Instead, psychopaths are cruel and without conscience. This personality profile makes it easier to victimize other people—or, in Chikatilo's case, to kill and eat them.

The sociological perspective views Chikatilo's crimes as the outcome of an abusive and neglectful childhood in which he had little or no exposure to conventional role models and caregivers. Without emotional support and material resources, crime became an attractive way to satisfy his desires and deal with glaring and unfair social inequality.

As studies of rehabilitation programs have shown, treatment works only some of the time, for specific offenders, in certain settings. Given the extremity of biological, psychological, and sociological factors in this case, it is doubtful that any rehabilitative effort would have helped rehabilitate Chikatilo.

Chapter Spotlight

- Classical and neoclassical theories suggest that criminals are rational and intelligent and choose to commit crime. To prevent crime, they suggest, the pain of punishment must be greater than the pleasure the offender would receive from committing the crime.

- Biological theories see criminals as being inherently different from noncriminals. Social policies derived from these theories focus on the separation and imprisonment of offenders.

- Psychological theories such as psychoanalytic theory and behavioral theory encourage the use of counseling and behavior modification as means to prevent crime.

- Sociological theories contend that crime is the product of external forces—that is, forces outside the offender. According to these theories, crime can be reduced by improving the community, eliminating social and structural obstacles to achievement, strengthening the bond of people to society, and improving parenting practices.

Putting It All Together

1. How are theory and public policy related?

2. Are people rational when they engage in crime? What evidence supports your position?

3. Some theorists believe that crime is learned in the same way any other behavior is learned. What are the implications of this idea for parenting and for the community's responsibility for crime?

4. Labeling theory argues that crime is a social construct—that is, something evaluated negatively by others. If this is true, is society justified in assigning labels to criminals in an effort to control crime?

5. What are the inherent flaws of conflict theories of crime? Are state legislatures likely to adopt their basic ideas? If not, of what value are such theories?

Key Terms

anomie A social condition where the norms of society have broken down and cannot control the behavior of its members.

atavists Individuals who are throwbacks to an earlier, more primitive stage of human development, and more closely resemble their apelike ancestors in traits, abilities, and dispositions.

aversion therapy Therapy in which people are taught to connect unwanted behavior with punishment.

behavioral theory Theory that views behavior as a product of interactions people have with others throughout their lifetime.

bourgeoisie People who own the means of production.

choice theories Theories that assume that people have free will, are rational and intelligent, and make informed decisions to commit crimes based on whether they believe they will benefit from doing so.

classical school A school of thought that holds that criminals are rational, intelligent people who have free will and the ability to make choices.

conflict theory Theory that blames crime on inequalities in power.

criminal career The progression of criminality over time or over the life-course.

cultural deviance theory Theory that proposes that crime is the product of social and economic factors located within a neighborhood.

cultural transmission The process through which criminal values are transmitted from one generation to the next.

differential association theory Theory that explains the process by which a person becomes involved in criminality.

dizygotic twins Twins who do not share the same set of genes (fraternal twins).

ego Component of the personality that represents problem-solving dimensions.

id Component of the personality that is present at birth, and consists of blind, unreasoning, instinctual desires and motives.

individual justice Concept that criminal law must reflect differences among people and their circumstances.

labeling theory Theory that examines the role of societal reactions in shaping a person's behavior.

life-course theory Theory that explains the change in the progression of criminality over time.

lifestyle theory Theory that proposes that the way people live their lives can place them in settings with a higher or lower risk of criminal victimization.

master status The status bestowed on an individual and perceived by others as a first impression.

mitigating circumstances Factors such as age or mental illness that influence the choices people make and affect a person's ability to form criminal intent.

monozygotic twins Twins who share the same set of genes (identical twins).

neoclassical school A school of thought that argues that there are real differences among people that must be taken into consideration when administering punishment.

operant conditioning Treatment in which rewards are used to reinforce desired behavior and punishments are used to curtail undesired behavior.

positive school A school of thought that blames criminality on factors that are present before a crime is actually committed.

primary deviance The behavior that originally leads to the application of the "deviant" label.

proletariat People who sell their labor to the bourgeoisie.

psychoanalytic theory Theory that unconscious mental processes developed in early childhood control the personality.

rational choice theory Theory in which criminals are rational people who make calculated choices regarding their actions before they act.

routine activities theory Theory that examines the crime target or whatever it is the offender wants to take control of, whether it is a house to break into, a bottle of beer, or illegal music to download from the Internet.

secondary deviance Acts of deviance that occur after someone has internalized the "deviant" label and uses it as a means of defense, attack, or adjustment to the problems caused by the label.

self-control theory Theory in which people seek pleasure, are self-gratifying, and commit crimes owing to their low self-control.

social bond A measure of how strongly people are connected to society.

social control theory Theory that holds that people are amoral and will break the law unless obstacles are thrown in their path.

social learning theory Theory that suggests that children learn by modeling and imitating others.

sociological theories Theories that attribute crime to a variety of social factors external to the individual, focusing on how the environment in which the person lives affects his or her behavior.

somatotype theory Theory that suggests that individuals with particular body types are likely to be inclined toward certain behaviors.

stigmata Distinctive physical features.

strain theory Theory that proposes that a lack of integration between cultural goals and institutionalized means causes crime.

superego Component of the personality that develops from the ego and comprises the moral code, norms, and values the person has acquired.

theories Integrated sets of ideas that explain when and why people commit crime.

token economy A system used in penitentiaries of handing out and taking away rewards that can be exchanged for privileges such as watching television.

trait theories Theories that argue that offenders commit crimes because of traits, characteristics, deficits, or psychopathologies they possess.

Notes

1. James Q. Wilson, *Thinking about Crime* (New York: Basic Books, 1985); Ernest van den Haag and John P. Conrad, *The Death Penalty* (New York: Plenum, 1983).
2. Ross Parke and Richard Walters, "Some Factors Determining the Efficacy of Punishment for Inducing Response Inhibition," *Monographs of the Society for Research in Child Development* 32:109 (1967); Ross Parke, "Effectiveness of Punishment as an Interaction of Intensity, Timing, Agent Nurturance and Cognitive Structuring," *Child Development* 40:213–235 (1969).
3. Leon Radzinowicz, *Ideology and Crime* (New York: Columbia University Press, 1966); "Why Did Josh Kill?" *CBS News,* June 12, 2000, available at http://www.cbsnews.com/stories/1999/10/07/48hours/main65411.shtml, accessed May 18, 2007.
4. Cesare Beccaria, *On Crimes and Punishments* (Indianapolis: Bobbs-Merrill, 1764/1963).
5. Jeremy Bentham, *A Fragment on Government and an Introduction to the Principles of Morals and Legislation* (Oxford, UK: Basil Blackwell, 1789/1948).
6. Bentham, note 5, p. 151.
7. Bentham, note 5, p. 151.
8. Bentham, note 5.
9. Robert Martinson, "What Works? Questions and Answers about Prison Reform," *The Public Interest* 35:22–54 (1974); Douglas Lipton, Robert Martinson, and Judith Wilks, *The Effectiveness of Correctional Treatment* (New York: Praeger, 1975); William Bailey, "Correctional Outcome: An Evaluation of 100 Reports," *Journal of Criminal Law, Criminology, and Police Science* 57:153–160 (1966); Hans Eysenck, "The Effects of Psychotherapy," *International Journal of Psychiatry* 1:99–144 (1965); Rachel Pergament, "Susan Smith: Child Murderer or Victim," available at http://www.crimelibrary.com/notorious_murders/famous/smith/susan_3.html, accessed March 24, 2007.
10. Graeme Newman, *The Punishment Response* (Philadelphia: Lippincott, 1985); van den Haag and Conrad, note 1.
11. Maricopa County Sheriff's Office, available at http://www.mcso.org/index.php?a=Home, accessed May 8, 2007.
12. Ronald Clarke and Derek Cornish, "Modeling Offender's Decisions: A Framework for Research and Policy," in Michael

Tonry and Norval Morris (Eds.), *Crime and Justice,* Volume 6 (Chicago: University of Chicago Press, 1985), pp. 145–167.

13. Lisa Maher, "Hidden in the Light," *Journal of Drug Issues* 26:143–73 (1996); Neal Shover, *Great Pretenders* (Boulder, CO: Westview Press, 1996); John Petraitis, Brian Flay, and Todd Miller, "Reviewing Theories of Adolescent Substance Use," *Psychological Bulletin* 117:67–86 (1995); Paul Cromwell, James Olson, and D'Aunn Avary, *Breaking and Entering* (Beverly Hills, CA: Sage, 1991); Richard Wright and Scott Decker, *Burglars on the Job* (Boston: Northeastern University Press, 1994); Eric Hickey, *Serial Murderers and Their Victims,* 4th ed. (Belmont, CA: Thomson, 2005); Janet Warren, Roland Reboussin, Robert Hazlewood, Andrea Cummings, Natalie Gibbs, and Susan Trumbetta, "Crime Scene and Distant Correlates of Serial Rape," *Journal of Quantitative Criminology* 14:231–245 (1998).

14. James Tedeschi and Richard Felson, *Violence, Aggression and Coercive Actions* (Washington, DC: American Psychological Association, 1994).

15. Lawrence Cohen and Marcus Felson, "Social Change and Crime Rate Trends: A Routine Activity Approach," *American Sociological Review* 44:588–608 (1979); Marcus Felson, *Crime and Everyday Life,* 4th ed. (Thousand Oaks, CA: Sage, 2007).

16. Ronald Clarke and Derek Cornish, "Modeling Offender's Decisions," in Michael Tonry and Norval Morris (Eds.), *Crime and Justice: An Annual Review of Research,* Volume 7 (Chicago: University of Chicago Press, 1985), pp. 145–167.

17. Michael Hindelang, Michael Gottfredson, and James Garofalo, *Victims of Personal Crime* (Cambridge, MA: Ballinger, 1978).

18. Anthony Walsh and Lee Ellis, *Criminology* (Thousand Oaks, CA: Sage, 2007).

19. Harriet Martineau, *The Positive Philosophy of August Comte* (Whitefish, MT: Kessinger, 1855/2003).

20. Cesare Lombroso, *The Criminal Man* (Milan, Italy: Hoepli, 1876); Marvin Wolfgang, "Pioneers in Criminology: Cesare Lombroso," *Journal of Criminal Law, Criminology, and Police Science* 52:361–369 (1961).

21. Charles Goring, *The English Convict* (London: His Majesty's Stationary Office, 1913).

22. Earnest Hooton, *The American Criminal* (Westport, CT: Greenwood Press, 1939/1969).

23. Zeynep Benderlioglu, Paul Sciulli, and Randy Nelson, "Fluctuating Asymmetry Predicts Human Reactive Aggression," *American Journal of Human Biology* 16:458–469 (2004).

24. William Sheldon, *Varieties of Delinquent Youth* (New York: Harper & Row, 1949).

25. Sheldon Glueck and Eleanor Glueck, *Physique and Delinquency* (New York: Harper & Row, 1956); Juan Cortes and Florence Gatti, *Delinquency and Crime* (New York: Seminar Press, 1972).

26. Adrian Raine, Chandra Reynolds, Peter Venables, Sarnoff Mednick, and David Farrington, "Fearlessness, Stimulation-Seeking, and Large Body Size at Age 3 Years as Early Predispositions to Child Aggression at Age 11 Years," *Archives of General Psychiatry* 55:745–751 (1998); Patricia Brennan, Adrian Raine, Fini Schulsinger, Lis Kirkegaard-Sorrensen, Joachim Knop, Barry Hutchings, Rabin Rosenberg, and Sarnoff Mednick, "Psychophysiological Protective Factors for Male Subjects at High Risk for Criminal Behavior," *American Journal of Psychiatry* 154:853–855 (1997); Adrian Raine, P. H. Venables, and Sarnoff Mednick, "Low-Resting Heart Rate at Age 3 Years Predisposes to Aggression at Age 11 Years," *Journal of the American Academy of Child and Adolescent Psychiatry* 36:1457–1464 (1997); Christopher Patrick, Bruce Cuthbert, and Peter Lang, "Is 'Fear Image Processing' Defective in Psychopaths?" *Crime Times* 1:6–7 (1995).

27. Alfred Binet and Theodore Simon, *A Method of Measuring the Development of Intelligence of Young Children,* 3rd ed. (Kila, MT: Kessinger, 2007).

28. Lee Ellis and Anthony Walsh, "Crime, Delinquency and Intelligence," in H. Nyborg (Ed.), *The Scientific Study of General Intelligence* (Amsterdam: Pergamon, 2003), pp. 343–365.

29. Travis Hirschi and Michael Hindelang, "Intelligence and Delinquency," *American Sociological Review* 42:571–586 (1977).

30. Donald Lynam, Terrie Moffitt, and Magda Stouthamer-Loeber, "Explaining the Relation between IQ and Delinquency: Class, Race, Test Motivation, School Failure, or Self-Control?", *Journal of Abnormal Psychology* 102:187–196 (1993).

31. Jean-Pierre Guay, Marc Ouimet, and Jean Proulx, "On Intelligence and Crime: A Comparison of Incarcerated Sex Offenders and Serious Non-sexual Violent Criminals," *International Journal of Law and Psychiatry* 28:405–417 (2005).

32. Adrian Raine, "Annotation: The Role of Prefrontal Deficits, Low Autonomic Arousal, and Early Health Factors in the Development of Antisocial and Aggressive Behavior in Children," *Journal of Child Psychology and Psychiatry* 43:417–434 (2002).

33. David Farrington, "The Relationship between Low Resting Heart Rate and Violence," in Adrian Raine, Patricia Brennan, David Farrington, and Sarnoff Mednick (Eds.), *Biosocial Bases of Violence* (New York: Plenum, 1997).

34. Michael Gottfredson and Travis Hirschi, *A General Theory of Crime* (Stanford, CA: Stanford University Press, 1990).

35. Diana Fishbein, *Biobehavioral Perspectives in Criminology* (Belmont, CA: Wadsworth, 2001); Terrie Moffitt et al., "Whole Blood Serotonin Relates to Violence in an Epidemiological Study," *Biological Psychiatry* 43:446–457 (1998).

36. Michael Reiff, Sherill Tippins, and Anothony Letourveau, *ADHD* (Elk Grove Village, IL: American Academy of Pediatrics, 2004); Travis Pratt et al., "The Relationship of ADHD to Crime and Delinquency," *International Journal of Police Science and Management* 4:344–360 (2002).

37. C. Robert Cloninger, Theodore Reich, and Samuel Guze, "The Multifactorial Model of Disease Transmission: II. Sex Differences in the Familial Transmission of Sociopathy (Antisocial Personality)," *British Journal of Psychiatry* 127:11–22 (1975); C. Robert Cloninger and Sauel Guze, "The Multifactorial Model of Disease Transmission: Familial Relationships between Sociopathy and Hysteria (Briquet's Syndrome)," *British Journal of Psychiatry* 127:23–32 (1975); David Rowe and David Farrington, "The Familial Transmission of Criminal Convictions," *Criminology* 35:177–201 (1997), p. 199; David Farrington, Darrick Jolliffe, Rolf Loeber, Magda Stouthamer-Loeber, and Larry Kalb, "The Concentration of Offenders in Families and Family Criminality in the Prediction of Boys' Delinquency," *Journal of Adolescence* 24:579–596 (2001); Johannes Lange, *Crime as Destiny* (London: Allen & Unwin, 1929).

38. Sarnoff Mednick and Karl Christiansen, *Biosocial Basis of Criminal Behavior* (New York: Gardner Press, 1977).

39. David Rowe and D. Wayne Osgood, "Heredity and Sociological Theories of Delinquency: A Reconsideration," *American Sociological Review* 49:526–540 (1984); David Rowe, "Genetic and Environmental Components of Antisocial Behavior: A Study of 265 Twin Pairs," *Criminology* 24:513–532 (1986); David Rowe, *Biology and Crime* (Los Angeles: Roxbury, 2002); David Rowe and Bill Gulley, "Sibling Effects on Substance Abuse and Delinquency," *Criminology* 30:217–223 (1992); Barry Hutchings and Sarnoff Mednick, "Criminality in Adoptees and Their Adoptive and Biological Parents," pp. 127–143 in Sarnoff Mednick and Karl Christiansen (Eds.), *Biosocial Basis of Criminal Behavior* (New York: Gardner Press, 1977), Sarnoff Mednick, William Gabrielli, and Barry Hutchings, "Genetic Factors in the Etiology of Criminal Behavior," in Eugene McLaughlin, John Muncie, and Gordon Hughes (Eds.), *Criminological Perspectives,* 2nd ed. (Thousand Oaks, CA: Sage, 2003), pp. 67–80; Michael Bohman, C. Robert Cloninger, Soren Siguardson, and Anne-Liss von Knorring, "Predisposition to Petty Criminalistics in Swedish Adoptees," *Archives of*

General Psychiatry 39:1233–1241 (1982); Raymond Crowe, "The Adopted Offspring of Women Criminal Offenders," *Archives of General Psychiatry* 27:600–603 (1972).

40. Sigmund Freud, *The Interpretation of Dreams* (New York: Avon, 1910/1980); Sigmund Freud, *The Ego and the Id* (New York: W.W. Norton, 1923/1962).

41. Franz Alexander and William Healy, *Roots of Crime* (New York: Knopf, 1935); August Aichhorn, *Wayward Youth* (New York: Viking Press, 1936); Fritz Redl and David Wineman, *Children Who Hate* (New York: Free Press, 1951).

42. B.F. Skinner, *The Behavior of Organisms* (New York: Appleton, 1938); B. F. Skinner, "Are Theories of Learning Necessary?", *Psychological Review* 57:211–220 (1950); B. F. Skinner, *Science and Human Behavior* (New York: Macmillan, 1953); C. Ray Jeffery, "Criminal Behavior and Learning Theory," *Journal of Criminal Law, Criminology, and Police Science* 56:294–300 (1965); Robert Burgess and Ronald Akers, "A Differential Association-Reinforcement Theory of Criminal Behavior," *Social Problems* 14:128–147 (1966).

43. Albert Bandura, *Social Learning Theory* (Englewood Cliffs, NJ: Prentice Hall, 1977).

44. Diane Eicher, "TV Tempest That Never Dies," *Denver Post,* June 10, 2001, pp. 1K, 12–13K.

45. Craig Anderson, Douglas Gentile, and Katherine Buckley, *Violent Video Game Effects on Children and Adolescents* (New York: Oxford University Press, 2007); Ann Oldenburg, "TV, Films Blamed for Child Violence," *USA Today,* July 26, 2000, p. 9D.

46. Clifford Shaw and Henry McKay, *Juvenile Delinquency in Urban Areas* (Chicago: University of Chicago Press, 1942); Clifford Shaw and Henry McKay, *Juvenile Delinquency in Urban Areas,* revised ed. (Chicago: University of Chicago Press, 1969).

47. Edwin Sutherland, *Principles of Criminology,* 4th ed. (Philadelphia: Lippincott, 1947).

48. Robert Merton, "Social Structure and Anomie," *American Sociological Review* 3:672–682 (1938); Robert Merton, *Social Theory and Social Structure,* revised ed. (New York: Macmillan, 1968).

49. Travis Hirschi, *Causes of Delinquency* (Berkeley, CA: University of California Press, 1969).

50. Gottfredson and Hirschi, note 34.

51. Frank Tannenbaum, *Crime and the Community* (New York: Columbia University Press, 1938).

52. Edwin Lemert, *Social Pathology* (New York: McGraw-Hill, 1951).

53. Howard Becker, *Outsiders* (New York: Free Press, 1963).

54. Karl Marx and Friedrich Engels, *Capital* (New York: International Publishers, 1867/1967).

55. Willem Bonger, *Criminality and Economic Conditions* (New York: Agathon Press, 1916/1967).

56. David Farrington and Henry Pontell, *Developmental and Life Course Theories of Offending* (Englewood Cliffs, NJ: Prentice-Hall, 2007); Terence Thornberry, Marvin Krohn, Alan Lizotte, Carolyn Smith, and Kimberly Tobin, *Gangs and Delinquency in Developmental Perspective* (New York: Cambridge University Press, 2003).

57. Chicago Area Project, available at http://www.chicagoarea project.org/, accessed May 18, 2007.

58. Office of Head Start (Washington, DC: U.S. Department of Health and Human Services, 2007), available at http://www.acf.hhs.gov/programs/hsb/, accessed May 18. 2007.

59. National Association of Police Athletic League Activities, available at http://www.nationalpal.org/, accessed May 19, 2007.

60. Carolyn Webster-Stratton, *The Incredible Years Training Series* (Washington, DC: Office of Juvenile Justice and Delinquency Prevention, 2000).

61. Edwin Schur, *Radical Nonintervention* (Englewood Cliffs, NJ: Prentice-Hall, 1973).

JBPUB.COM/ExploringCJ

JBPUB.COM/ExploringCJ

Interactives
Key Term Explorer
Web Links

CHAPTER 4 Crime Theory and Social Policy   **111**

OBJECTIVES

- Describe the origins of policing in the United States, including the various systems that formed the foundation of modern policing.

- Outline policing reforms that shaped U.S. police systems from the 1700s to today.

- Explain the purpose and goals of modern models of policing, particularly community policing.

- Grasp the guiding principles of police operations and organization.

- Identify the functions of local, state, federal, and international policing agencies.

- Describe the role of private security police, its members, and its limitations.

Police History and Systems

CHAPTER
5

New York City experienced rampant crime during the 1980s, followed by great reductions in the crime rate during the tenure of Mayor Rudolph Giuliani in the 1990s. One reason for the change in direction was that Giuliani, a former federal prosecutor, advocated the use of aggressive policing that strictly enforced the law and combated public disorder.

Compelling evidence suggests that the tougher policing methods resulted in safer streets in New York. During Giuliani's tenure, public disorder crimes—such as graffiti, turnstile jumping, and open-air drug sales—declined significantly. More serious crimes also decreased: For example, gun homicides declined by 75 percent. Law enforcement officers became more efficient, as evidenced by a 25 percent reduction in officer use of firearms, 67 percent reduction in shootings per officer, and 150 percent increase in arrests.

These improvements were not lost on New Yorkers. At the end of Giuliani's time in office, nearly 85 percent of the city's residents held favorable views of police.

- Is it reasonable for politicians to take credit for crime reductions?
- How much credit should Mayor Giuliani receive for the trends in New York, given that crime was also declining all over the United States?
- Is aggressive crime control the only effective way to reduce crime?

Sources: Heather Mac Donald, *Are Cops Racist? How the War Against Police Harms Black Americans* (Chicago: Ivan R. Dee, 2003); William Bratton, *The Turnaround: How America's Top Cop Reversed the Crime Epidemic* (New York: Random House, 1998); Bruce Smith, *Police Systems in the United States,* revised edition (New York: Harper and Row, 1960).

Introduction

Police are a product of their history. To fully understand the structure and functions of the police, it is first necessary to examine them in their historical context. Studying police history provides valuable insights into modern police agencies and procedures. The knowledge of how and why changes in policing occurred can help guide police in the future as they encounter new demands and challenges. Likewise, familiarity with police history informs modern police problems, such as the use of excessive force and corruption, and provides assistance in grappling with these issues.[1]

In addition, by knowing about the problems of the past, police can avoid repeating these mistakes in the future. Unfortunately, not all police tactics used in the past have been effective. In the 1840s, for instance, the police controlled the civilians on their beats (assigned areas) with their nightsticks. One notorious New York City police officer, Alexander Williams, was nicknamed "Clubber" because he regularly assaulted citizens. Hailed as a hero among his police peers, Williams ascended rapidly through the police ranks and became a police inspector.[2] In time, however, his brutal method of controlling the public proved to be ineffective. Beating alleged offenders had a negligible impact on the crime rate but greatly influenced police–citizen relations, which deteriorated to the point of police resentment. As a result, future generations of police received extensive training on more humane methods for maintaining order.

LINK Police brutality has plagued generations of police administrators, and notable cases like those involving Rodney King and Amadou Diallo have made headlines around the globe. The nature and extent of police use of force are discussed in more detail in Chapter 8.

English Heritage

The English police system developed over several centuries. Over that span, numerous models of policing were implemented: the kin police system, the frankpledge police system, the parish–constable police system, and the uniformed police system.

Police systems of the Political Era were characterized by corruption and brutality.

? How can single acts of police brutality damage the reputation of an entire police force? Will the uneasiness that some citizens feel about the police ever fully go away?

Kin Police System

Prior to the Norman Conquest of 1066, the responsibility for law enforcement was placed in the hands of ordinary citizens. Under the <u>kin police system</u>, there were no formal police officers. Instead, each individual was responsible for helping his or her neighbors, following the adage, "I am my brother's keeper." Over time, this model was slowly replaced with a more formalized, community police system known as the frankpledge system.[3]

Frankpledge Police System

In the <u>frankpledge police system</u>, citizens shared the responsibility of policing.[4] This system was founded on principles of self-policing. Adult males (older than age 12) were organized into small groups of 10 people (called tithings). A <u>constable</u>, appointed by the local nobleman, ruled over 10 tithings (called a hundred). Ten hundreds were grouped into shires (similar to today's counties) that were supervised by a shire reeve (or sheriff), who was appointed by the king.

The men in each tithing were required to report criminal behavior by any of the other nine members of their tithing to the constable. If they did not, and their negligence was discovered, all members of the tithing received a heavy fine. The frankpledge sys-

tem operated effectively from 1066 to 1234. Its demise can be traced to inadequate supervision by the king and his appointees.

Parish–Constable Police System

In 1285, the Statute of Winchester established the first official English police force and created the <u>parish–constable police system</u>, which defined English policing for the next 500 years.[5] Under this model, one man from each parish (or county) served a one-year term as constable on a rotating basis. This model had several other key components:

- *Watch and ward system.* Constables had the authority to draft any male citizens into positions as night watchmen. These guards protected the town gates and arrested law violators.
- *Hue and cry system.* When a watchman confronted more resistance than he could handle, he delivered a loud call for help (the "hue and cry"). Upon hearing the call, the men of the town were required by law to stop what they were doing and lend assistance. Anyone who did not join in this effort could be arrested for aiding and abetting the criminal.
- *Weapons ordinance.* Semiannual inspections ensured that all male town residents owned and maintained a short, broad-bladed saber to protect themselves.
- *Curfew.* At a set time determined by the constable, the city's gates were locked to keep out wanderers and other insalubrious characters from entering the township.

The parish–constable system eventually failed because it depended on unpaid watchmen, most of whom had no interest, skills, or training in police work. It was common practice for citizens who could afford to do so to pay others to fulfill their guarding duties. Because of the nature of police work, coupled with its very low pay, the profession was able to attract only the poor, illiterate, and unskilled. The parish–constable system eventually collapsed in the late 1700s.

Uniformed Police System

Near the end of the 1700s, London had become a metropolitan, industrial city that was booming as a result of the Industrial Revolution. While the Industrial Revolution brought much good to the city, including more factories and marketplaces, this industrial development came at a steep price: a breakdown in social order, which led to a precipitous increase in theft and vandalism. The upper and middle classes who were concerned about these issues eventually sought more police protection.[6] Short-term solutions for controlling the growing civil unrest and crime rates included calling out the cavalry, appointing more law-abiding citizens as constables, and using the army to quell riots. Unfortunately, these tactics proved to be only temporary solutions to a long-term problem.[7]

Another proposal sought to replace the parish–constable system with a model of policing that adopted a centralized police force whose mission would be crime prevention.[8] However, citizens and politicians worried that a centralized police force would look too much like a standing army, giving the government too much control over citizens. After much debate, the British Parliament approved the Metropolitan Police Act in 1829, which was drafted by England's home secretary, Sir Robert Peel II.[9] The

The modern police force was created in 1829 with the Metropolitan Police Act drafted by Sir Robert Peel, II.

 Which social forces necessitated the creation of modern police forces? What does the need for a formal police force suggest about the nature of crime and its control?

Act created the first large-scale, uniformed, organized, paid, civil police force in London. The officers wore unique uniforms that included three-quarter-length white pants, royal blue coats, and a top hat. The officers were armed with truncheons—the equivalent of today's police batons.

Sir Robert Peel, II is considered by many to be the founder of modern policing. He contributed several important principles to the field of policing:

1. The mission of police is to prevent crime.

2. Police and the public must work together to preserve the interests of the community; police must have the support and willing cooperation of the public to effectively perform their duties.

3. Police should use physical force only as a last resort; the public's respect of the police diminishes proportionally to the amount of force police use.

4. Police gain the respect of the public by impartial application of law.

5. Police should only enforce the law and never act toward civilians as if they are judge, jury, and executioner.

6. The test of police efficiency is the absence of crime and disorder.[10]

Peel further organized the police into a uniformed, salaried, full-time police force in which officers were assigned to specific jurisdictions and were expected to become familiar with the individuals living in their neighborhood. Curiously, these officers were not well received. Some civilians viewed them as an occupying army, and there were constant battles between police and the public. As time passed, however, these officers earned the respect of the public and became very popular with civilians. Eventually all cities in England adopted a similar system.

American Policing

When the colonists arrived in America, they brought the parish–constable system along with them to monitor the many widely scattered villages that evolved into America's first towns and cities.[11] In the beginning, this system worked well—at the time, "major crimes" included working on the Sabbath, failure to pen animals, and cursing in public. As towns became more populated and their economies prospered, however, crime became a more serious issue. With this change, policing obligations became more time-consuming and less attractive. Like their English counterparts, many Americans found ways to evade

The First "Bobbies" in London

The English Parliament created the first salaried, bureaucratic police unit in 1829 with the passage of the Metropolitan Police Act. The London Metropolitan Police force was responsible for maintaining order and preventing and detecting crime and was structured like a military unit with ranks and a formal chain of command. London was divided into 17 districts with more than 3000 officers. The force was headed by Home Secretary Sir Robert Peel II.

Job requirements for becoming a constable (or "bobbie," as they were called in honor of Sir Robert) were substantial: In addition to being physically fit and literate, constables had to have good moral character and manners. They were to be smartly dressed in a blue swallow-tailed coat, a leather strap, and a black top hat. Constables carried rattles, with which they could summon help, but did not carry firearms. A traditional rattle was made of wood, with one or two blades that were held in a frame and a ratchet turned to make the blades "snap," making a very loud noise. The constables were trained to be authoritative, to be fair, and to exercise self-control; any use of force was to be measured, limited, and minimal. To help ensure accountability, constables wore a personal identification number on their collar where it could easily be seen. Gradually, whistles replaced rattles, helmets replaced top hats, and—in 1994—London police gained authorization to carry firearms.

In 1830, 80,000 men initially applied to be constables. Only 2800 officers were hired, and only 562 were still in the force four years later. The high rate of attrition was caused by the demanding schedules and low pay accorded to the officers. Despite their complaints, Peel refused to change the requirements or the salary, a policy that created serious problems in attracting and retaining good constables to the world's first organized police force.

Sources: Clive Emsley, *Crime and Society in England,* 1750–1900, 3rd ed. (Upper Saddle River, NJ: Longman, 2005); Haia Shpayer-Makov, *The Making of a Policeman* (Hampshire, UK: Ashgate, 2002); Wilbur Miller, *Cops and Bobbies,* 2nd ed. (Columbus, OH: Ohio State University Press, 1999); Philip John Stead, *The Police of Britain* (New York: Macmillan, 1985).

their policing obligation, leading cities to pass ordinances that imposed fines on individuals who abandoned their policing responsibility, although the ordinances and fines proved to be ineffective.[12]

1700s: Origins of Organized Policing in America

Since fines were not sufficient to coax citizens into fulfilling their police duties, the responsibility of law enforcement shifted, as it had in England, from *all*

male citizens to only those men who could not afford to hire others, with similar results. City managers quickly realized that public ownership of policing did not work and that what was needed was a salaried, full-time police force.[13]

Philadelphia was the first city to implement a solution to the growing problem of crime control. In 1749, it passed two pieces of legislation: one law permitting constables to hire as many guards as they needed and a second law that established a tax to pay them. Other cities soon followed Philadelphia's lead.

The first organized police in the United States were slave patrols, which originated in Georgia in 1757 and later spread to other southern states. The purpose of slave patrols was to maintain white supremacy by breaking up meetings where slaves were gathering and possibly planning a revolt. These police often had the support of local townships and would break into homes of slaves suspected of keeping arms, whip those slaves who confronted them, and apprehend any slaves who were suspected of running away or committing a crime. Slave patrols were also responsible for monitoring suspicious activities, confiscating contraband (such as liquor and weapons), and enforcing the law that no slave could receive an education.

Anywhere they operated, slave patrols had considerable authority, which they exercised at their own discretion. If a patroller seriously harmed a slave (thus lowering the slave's economic value), however, he could be held liable and required to pay restitution to the slave's owner.

All white citizens were eligible for slave patrol duty, because it was usually considered a community effort to keep the slaves in line. Initially, patrollers were residents of the county who worked without pay. As the years passed, many counties began to pay patrollers a minimal fee (one dollar per night). The patrollers on duty were also allowed to benefit from any reward granted for the return of an escaped slave. If a patroller was unavailable for duty, he could hire a replacement. If the replacement did not show up for work, however, the patroller would be fined.

Eventually, the slave patrols gave way to city police officers following the end of the Civil War. With the Emancipation Proclamation and newfound freedom of African Americans in 1865, slave patrols came to an end. Unfortunately, the social unrest and bigotry toward African Americans persisted, and the founders and patrollers formed a new organization—the Ku Klux Klan.

Sources: Sally Hadden, *Slave Patrols: Law and Violence in Virginia and the Carolinas* (Cambridge, MA: Harvard University Press, 2001); Kendall Clark, *Patrols and Privilege,* March 8, 2002, available at http://monkeyfist.com/articles/813, accessed June 25, 2007; Sally Hadden, "Slave Patrols," May 14, 2003, available at http://www.georgia encyclopedia.org/nge/Article.jsp?id=h-900, accessed June 25, 2007.

Unfortunately, city leaders soon realized that this new approach to policing was not effective, because cities had difficulty finding capable men for the job. Police work had become increasingly dangerous, and the pay was still too low. As a consequence, some officers sought to increase their paltry salaries by accepting bribes from gambling houses and prostitution rings.[14]

Problems for police intensified between 1750 and 1800 as a result of political and social unrest sparked by severe economic depression. Riots (precursors to the American Revolution) were common, and crime flourished. In response to the problem of growing crime, vigilante groups—bands of citizens who took the law into their own hands to suppress crime and protect themselves against robberies and gangs of outlaws—formed in the rural South.[15] In addition, slave patrols emerged and gained power. These patrols consisted of small, organized groups who worked to control the slave population and halt outbreaks of slave revolts (see "Focus on Criminal Justice" above).[16]

On the Western frontier, vigilante groups also emerged despite the presence of U.S. marshals and elected sheriffs who were hired to protect citizens in the growing towns. While the slave patrols were eliminated after the Civil War, vigilante groups continued to operate into the late nineteenth century, although their activities generally led to greater demands for more formal policing.[17]

1800s: Growth, Brutality, and Corruption

The U.S. population grew rapidly in the nineteenth century, especially in urban areas. For example, the population of New York City skyrocketed from only 33,000 in 1790 to 150,000 by 1830.[18] U.S. cities attracted large numbers of migrants from rural areas and immigrants from foreign countries, whose presence caused difficulties for city governments. While they welcomed those aspects of growth that promoted business, government officials feared some unexpected consequences caused by increasing number of strangers who spoke foreign languages and had different customs and religious practices. They also were concerned about the increasing numbers of poor people languishing on city streets (both adults and children), which merely worsened the crime problem. All of these forces came together to set the stage for new police reforms.

As they had in the past, Americans turned to England in search of a model of effective policing. Police reformers were particularly attracted to one of the guiding principles of the English system: the responsibility of government to provide for the social well-being of its citizens. With this new philosophy in hand, cities demanded that police departments with paid, full-time, uniformed officers be established. By 1860, these departments had become a fixture in the largest U.S. cities. These new police assumed a diverse set of duties—from lighting gas street lamps to monitoring elections, from apprehending criminals to providing lodging for the poor. Over time, many of these duties were transferred to other government agencies, allowing the police to concentrate on crime prevention.

However, all was not well. As was the case with earlier generations of police, the "new" police received a small salary. In turn, like their predecessors, some officers supplemented their incomes by accepting bribes to overlook illegal activities.

As the nineteenth century came to a close, the newly created police organizations were plagued with problems. Thus the early years of the twentieth century were characterized by a series of reform efforts aimed at changing the way police did their jobs.

LINK In 1892, Reverend Charles Parkhurst delivered a public sermon criticizing the New York City police department. His charges captured the attention of business and civic leaders, who pressured the New York Chamber of Commerce to launch a full investigation. This was the first review committee to examine police corruption, though, as discussed in Chapter 8, it was certainly not the last.

Early 1900s: Development of Organizations and Technology

An early police reformer in America, Theodore Roosevelt, served as Police Commissioner for New York City before becoming the twenty-sixth President of the United States. When Roosevelt became commissioner, the New York Police Department (NYPD) was one of the most corrupt police agencies in the

country. Bringing his iron will to the office, Roosevelt immediately changed how the department was run. He started by establishing new disciplinary rules, requiring officers to arm themselves with 32-caliber pistols and insisting that officers take annual physical exams. He appointed 1600 new officers based on their physical and mental qualifications, rather than their political affiliations, and created opportunities within the department for women and racial and ethnic minorities. Roosevelt also created the city's first bicycle squad to patrol the city's streets, established the NYPD's first police meritorious service medals, and installed telephones in station houses. These changes paved the way for others to further professionalize policing in the United States.[19]

Many of Roosevelt's ideas were built upon by a young, emerging police leader, August Vollmer, who served as Chief of Police in Berkeley, California, from 1905 to 1932. Expanding on the innovations introduced by Roosevelt, Vollmer was responsible for bringing more change to the profession than any other single individual:

- Vollmer installed the first basic records system in the United States.
- He conducted the first scientific investigation of a crime, utilizing the analysis of blood, fibers, and soil.
- He established a unique police school based on law and evidence procedures.
- He organized the first motorcycle and automobile patrols.
- He established the first School of Criminology at the University of California at Berkeley.
- He required officers to attain a college degree and began using intelligence testing to recruit police officers.
- He introduced the use of the first lie detector instrument and established one of the first fingerprint systems.

Vollmer believed that police departments must become more efficient to protect the public. In its totality, Vollmer's reform agenda was based on several guiding principles:

- *Police work must be defined as a profession.* Police should be public servants and interact with citizens politely and without bias.
- *Political influence must be eliminated.* For too long, police departments had been tools of the political party in power. To enforce laws fairly, the police must be free to choose the best course of

August Vollmer is widely considered the founder of the professional American police department.

 What changes marked the shift from brawn to brains in U.S. policing? Has this been a good shift? Are more professional departments more effective at reducing crime rates?

action for the situation they confront, rather than what is best for a particular political party or politician.

- *Police administrators must be proven leaders.* The position of police chief is a very difficult one and should be filled by an individual with experience managing large organizations.

- *Standards for becoming a police officer must be raised.* This includes having a minimum education requirement and tests measuring health, intelligence, and moral character.

- *Specialized units must be developed.* Police departments are confronted with highly varied tasks, ranging from criminal investigations to traffic control to patrol units to juvenile units. Effective police agencies develop specialized units to meet the most pressing demands of the community.[20]

During the same era, advances in technology increased the proficiency of police in innumerable ways. For example, the patrol car was introduced. The first police patrol cars hit the streets prior to World War I; by the end of the 1920s, patrol vehicles were being used by nearly all police departments in the United States. Automobiles allowed officers to cover larger geographical areas more quickly, although they also had the unanticipated effect of isolating police from the public. With the addition of telephones and two-way radios, citizen reports informed police and response time quickened dramatically.[21]

A downside of the changes and improvements inspired by the reform agenda was that the public came to expect much more from the police than they had in the past. This included not only faster response times to calls for service, but also citizens now expected more arrests and less overall crime. These high expectations led to difficult times for police in the 1960s when police were unable to deliver on what citizens believed they had promised.

Mid-1900s: Responses to Increasing Crime Rates

The 1960s was a period of ruthless and intense conflict between the police and the public, particularly in terms of clashes between police and both civil rights demonstrators and antiwar activists. Police responded to the new social conditions by implementing some of the reforms they had made during the era of professionalism, such as sending specialized riot units to suppress incidents of public disorder and having police chiefs address citizen concerns and media.[22] However, when put into action, these and other reforms did not calm the complaints of an increasingly disgruntled public. Crime rates continued to soar, and fear of crime increased. Racial and ethnic minorities loudly protested perceived police mistreatment and discrimination, and protesters challenged the legitimacy of the police. Additionally, the national media publicized riots and police responses to them, even as a struggling economy forced local governments to slash police budgets.[23] Ultimately, both the police and the public discovered that the changes made during the reform era did not reduce crime and, in fact, crime rates jumped considerably FIGURE 5-1 .

The federal government turned to legislation in an effort to improve the ability of police to respond to crime. In 1965, the Law Enforcement Assistance

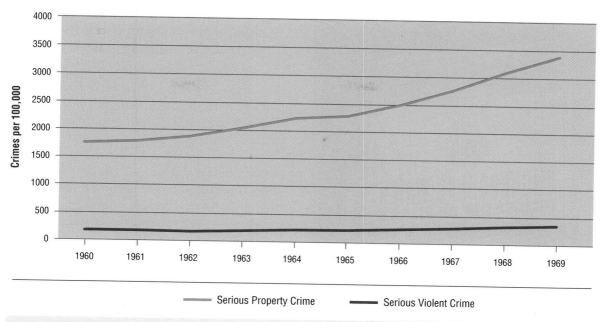

FIGURE 5-1 U.S. Crime Rates

Source: Federal Bureau of Investigation, *Crime in the United States, 1969* (Washington, DC: U.S. Department of Justice, 1970).

Act created the Law Enforcement Assistance Administration, which provided funding to local police agencies to help them identify the causes of urban disorders, train personnel, provide students loans for law enforcement recruits, and distribute grants to departments for crime-fighting equipment.[24]

The judicial branch also affected policing practices and policies in the 1960s. During this era, the U.S. Supreme Court handed down several important rulings that dramatically changed police practices:

- *Mapp v. Ohio* (1961) prohibited the police from conducting unreasonable searches and seizures and excluded illegally obtained evidence from federal criminal trials.[25]

- *Miranda v. Arizona* (1966) established that suspects in police custody have the right to refuse to answer police questions.[26]

- *Katz v. United States* (1967) required the police to obtain a court order to listen to conversations in which the parties have a reasonable right to expect privacy.[27]

LINK Police are limited by law in terms of how they may handle many different situations they face, from conducting high-speed chases to interrogating a suspect. These laws were developed from real-world cases and court rulings, as presented in Chapter 7.

The Supreme Court, in essence, stated that the police must enforce the law and maintain order within the due process requirements of the Constitution. These rulings produced concern for police, who feared that their normal—but now illegal—practices might result in their cases being dismissed in court. Police administrators responded by increasing the amount of pre-service training that recruits received (from 300 hours in 1967 to more than 1000 hours in 1987) and by having police academies focus less on firearms training and more on providing information on constitutional law, public relations, and strategies for managing terrorism, conducting sobriety tests, identifying suicide bombers, and defensive driving.[28]

Late 1990s: The Quiet Revolution

The reforms implemented in the 1960s and 1970s achieved only modest success. Perhaps the greatest change to police strategies came from recommendations made by federal commissions charged with studying police problems. While previous commission reports had focused solely on how the police could be more efficient, the new reports tried to strike a balance between efficiency and effectiveness. Whereas prior recommendations encouraged police to respond to calls quickly, new recommendations also emphasized the importance of taking time to talk with and listen to victims, witnesses, and other members of crime-plagued communities. These two ideas—efficiency and effectiveness—triggered the community policing movement.

Community policing attempts to strengthen the bond between the police and the public they serve, with a focus on solving problems and community empowerment rather than strict enforcement.

? Should police be social services providers? What does the future of community policing mean for law enforcement?

- Neighborhood Watch programs
- Mini- and storefront-police stations
- Police-sponsored athletic leagues
- Citizen auxiliary police[31]

The implementation of these programs seeks to create improved mutual understanding between the police and the public, uniting them in their efforts to prevent crime. To the extent police are able to become familiar with the people living in their jurisdiction, befriend them, and come to understand their daily concerns, they will become more effective crime fighters who enjoy greater public support. A closer relationship between residents and the police helps to make the neighborhood a safer place for all. Of course, these programs must be customized to the particular neighborhood, because residents of some neighborhoods may be more suspicious of police or demanding of police services.

The roots of community policing can be traced to an essay written in 1979 by criminologist Herman Goldstein, which argued that police officers must look not just at crime incidents but at the connection between crime incidents and the underlying causes of crime.[32] Goldstein believed that traditional policing efforts often fail because they approach crime as though each incident is an isolated and self-contained event. For police to be effective crime fighters, they must notice how crime incidents relate to one another and develop a more in-depth understanding of those factors that tend to be highly correlated with criminality.

In 1982, George Kelling and James Q. Wilson expanded Goldstein's work.[33] They argued that the changes in the ways traditional policing is practiced, such as improvements in radio communications, will not reduce serious crime. Instead, police need to eliminate conditions in neighborhoods (such as graffiti, drug dealing, and gambling) that produce fear and lead to neighborhood decay. This idea is called broken windows theory, based on Kelling and Wilson's popular metaphor for neighborhood signs of deterioration: Once a window is broken and is not repaired, other windows will be broken. Similarly, when a "social window" is broken and it is not repaired (e.g., roadside litter), other social windows will be broken (e.g., vandalism). The broken windows theory argues that small signs of public disorder set in motion a downward spiral of deterioration, neighborhood decline, and increasing crime.[34]

The purpose of community policing is to prevent these social windows from being broken in the

Community policing includes the understanding that police cannot control crime alone and need help from citizens to prevent crime. At the core of community policing is good, strong police–community relations. They are important because they "reduce tensions, develop mutual trust, promote the free exchange of information, and acquaint officers with the culture and lifestyle of those being policed."[29] Rural police officers have worked toward these ends for many years—protecting the community while sharing its values and fully participating in its activities.[30] When the police and the public are connected through their everyday activities, they are more likely to cooperate and be less suspicious of each other than when they are strangers.

The community policing model requires police to become involved in a variety of activities:

Community Policing in Japan

The "new" innovation of community policing in the United States has actually long been the foundation of police work in Japan. Japanese policing is organized around a system of mini-precinct houses called *kobans*. Each of the approximately 6700 *kobans* in Japan has between 2 and 12 officers per shift. The goal of *kobans* is to bring the police as close as possible to the community they serve. These sites serve as the focal points of community police activities and as community safety centers for local residents.

Officers who are assigned to a particular *koban* make regular house calls and determine whether anything specific requires attention. They also stand guard at the *koban* to watch for suspicious activity, prevent crime by their presence, detect fires, control traffic if necessary, and give directions. *Koban* officers do very little in-depth police work, although they may patrol the neighborhood regularly—either on foot or by bicycle—to maintain a visible presence. The aim is to encourage citizens to cooperate with the police, enhance crime-prevention measures, and maintain good public relations.

The *koban* system is part of Japan's frontline against crime and a crucial component to preserve the safety of Japan's neighborhoods.

Sources: Narayan Chand Thakur, "Overview of Japanese Police," *National Police Agency,* April 6, 2005; Ordway Burden, "In Japan, the Ultimate in Community Policing," *Law Enforcement News,* 15:5 (1993).

---

first place by paying attention to the overall quality of life in a neighborhood, not just the serious crime. Both physical incivilities, such as trash and graffiti, and social incivilities, such as gamblers in alleys and drunks in public areas, greatly diminish the quality of life in an area. Through face-to-face communications with a neighborhood's residents, police officers first identify the sources of incivility and then work with the residents to eliminate them.

The community policing approach to crime control has been implemented in different ways, such as by increasing patrols and expanding face-to-face interactions with citizens. In cities such as Houston and New York, police have enjoyed some success in mobilizing the public to fight crime by pressuring business, employers, schools, and government agencies to assume partial responsibility for community crime problems.

2000s: Intelligence-Led Policing

The terrorist attacks on the United States on September 11, 2001, ushered in a form of policing never before experienced by the American public.[35] The Homeland Security Act of 2002 established the U.S. Department of Homeland Security, which leads a unified, national effort in securing the country by preventing and deterring terrorist attacks. This agency has granted large amounts of financial assistance to local police departments to form special intelligence units, which serve as the foundation of the modern intelligence-led policing system. Intelligence-led policing includes the following features:

- Police intelligence units that identify security threats from terrorists groups, extremists, and gangs
- Federal guidelines for police conduct
- Advances in police computing and network systems

Increasingly, police agencies are requiring college degrees for entry-level positions.

What does this trend suggest about the nature of policing and the changes it has undergone in recent decades? Is it more important for police to be "book smart" or "street smart"?

LINK Protecting Americans from terrorist attacks is a principal concern of all law enforcement agencies. Guidelines for police to follow are set forth in the Homeland Security Act of 2002, which is discussed in Chapter 17.

Intelligence-led policing is a new model of policing that reflects present-day realities. This crime-fighting strategy is driven by computer databases, intelligence gathering, and analysis. Whereas in the past only big-city police departments had the resources to maintain intelligence units to target drug smugglers and organized crime, today law enforcement agencies of all sizes are developing these capabilities.

In addition, these agencies are working together to increase communication and information sharing. Regional hubs (called fusion centers) assist networking efforts by pooling information from multiple jurisdictions. Fusion centers are one of the hottest new developments in policing and are rapidly expanding to every state, despite their potential problems. It is unclear, for instance, exactly which functions the fusion centers will perform. Another potential problem is the considerable variability among states regarding their guidelines and procedures.

Additional concerns have arisen regarding this new style of policing. Intelligence databases are not easily accessible across all departments, and there are not yet enough intelligence analysts or well-trained support staff available to fill all possible positions. Critics also worry about the use of corrupt police tactics, possible civil rights violations, and the potential for compromising citizens' privacy.

LINK The war on terror has led to the development of several controversial intelligence databases. Critics see these databases as violations of privacy, but supporters believe they are a necessary tool to thwart terrorist attacks on U.S. soil. These conflicting opinions and the common ground that critics and advocates share are discussed in Chapter 17.

Whether intelligence-led policing is good or bad, it represents a dramatic transformation in the practice and operation of law enforcement in the United States. Supporters of this new police system believe that by arming law enforcement with the latest information on crime and criminal activity, it will be easier to protect Americans from another terrorist attack.

Police Systems

The police system in the United States consists of all of the local, state, and federal agencies that enforce laws. While many different law enforcement agencies exist, there is little uniformity among them. Nevertheless, three general principles guide their operations and distinguish them from police systems in many other countries:

1. *Police have limited authority.* Police must follow specific rules and regulations to protect individual liberties.

2. *There is local control of police.* Some countries in Asia, Europe, the Middle East, and South America have centralized national police forces. By contrast, in the United States it is the responsibility of cities and counties to provide individuals with police protection. There also is "home rule" in the United States, which gives cities and counties the right of self-government. As a consequence, they may establish specific laws, such as curfews and speed limits.

3. *Agencies are decentralized and fragmented.* Instead of having a single, national police force, the United States supports roughly 18,000 separate law enforcement agencies that are loosely coordinated, with much duplication among them.[36]

There are three levels of public law enforcement in the United States: local, state, and federal. The efforts of these organizations are supplemented by international police and private police. Each police force, regardless of its level, has a jurisdiction, which is the territory or body of law it controls. The more than 18,000 police agencies employ more than 1 million individuals and have a total annual operating budget of nearly $89 billion FIGURE 5-2.[37] Nearly

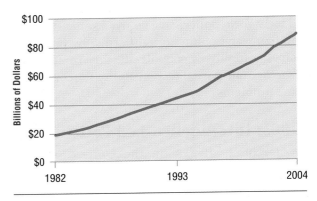

Percent Increase from 1982–2004: 367%

FIGURE 5-2 Direct Expenditures for Police

Source: Bureau of Justice Statistics, *Expenditure and Employment Statistics, 1982–2004* (Washington, DC: U.S. Department of Justice, 2007), available at http://www.ojp.usdoj.gov/bjs/glance/exptyp.htm, accessed June 26, 2007.

700,000 persons employed by law enforcement are sworn officers, men and women who are empowered to arrest suspects, serve warrants, carry weapons, and use force.[38] The overall police–population ratio is approximately 2.5 sworn officers per 1000 persons, although the actual ratio varies widely across different cities.[39] Washington, D.C., for instance, has 6.6 officers per 1000 population, whereas the police–population ratio in San Francisco is 1.6. Research has not found any statistically significant relationship between the police–population ratio and the crime rate.[40]

Local Police

There are more than 17,000 local law enforcement agencies in the United States; this number includes city, county, and some special-jurisdiction agencies, such as campus police, park rangers, and transit police.[41] Communities range greatly in their size and complexity, as do their police departments. Some local police agencies are very small and homogeneous, whereas others are extremely large and diverse. The nation's largest police department is the NYPD, which has almost 38,000 uniformed officers.[42] Most local police departments are small, employing fewer than 25 sworn officers and serving fewer than 10,000 residents.[43] Approximately 67 percent of local police work for municipalities (cities), while much of the remainder are employed by county sheriff departments.

Regardless of the size of the department, local police officers have similar duties and responsibilities—for example, controlling traffic, patrolling streets, and investigating crimes. Officers in many departments also handle animal control, operate search-and-rescue missions, provide emergency medical care, and control crowds. In big-city departments, special units may deal with counterterrorism and community problems such as drunk driving, missing children, victim assistance, and gang violence. Local police also assist in meeting the needs of special populations, including persons with human immunodeficiency virus (HIV)/acquired immune deficiency syndrome (AIDS), the homeless, victims of domestic violence, and abused or neglected children.

One difference between past and present local police departments is their racial, ethnic, and gender composition. Today, racial and ethnic minorities account for 24 percent of full-time sworn officers in local departments, up from 15 percent in 1987; women represent 12 percent of officers, up from 8 percent in 1987.[44] Nevertheless, in many departments in large cities, African Americans and Latinos are still underrepresented relative to their share of the general population TABLE 5-1 .

In city police departments, the police chief is usually appointed by the city council or mayor. A city police department's jurisdiction is limited by statute to the geographic boundaries of the city. By contrast, all states are divided into districts called *counties* (called *parishes* in Louisiana and *boroughs* in Alaska). The chief law enforcement officer of a county is the sheriff, who is an elected official, except in Rhode Island (where the sheriff is appointed by the governor) and Hawaii (where the sheriff is appointed by the Department of Health). The sheriff's department has the following duties:

- Investigate crimes
- Operate jails
- Process court orders
- Provide security for county courts
- Collect county fees and property taxes[45]

TABLE 5-1

Ratio of Minority Officers in 25 Large City Police Departments to Minority Residents

City	Officer–Resident Ratio	
	African American	Latino
1. New York City	0.50	0.66
2. Los Angeles	1.21	0.71
3. Chicago	0.70	0.49
4. Houston	0.77	0.48
5. Philadelphia	0.80	0.66
6. Phoenix	0.40	0.35
7. San Diego	0.61	0.63
8. Dallas	0.56	0.38
9. San Antonio	0.70	0.71
10. Las Vegas	0.48	0.33
11. Detroit	0.77	0.60
12. San Jose (CA)	1.60	0.75
13. Honolulu	0.67	0.39
14. San Francisco	1.24	0.96
15. Indianapolis	0.69	0.18
16. Jacksonville	0.67	0.29
17. Columbus	0.59	0.12
18. Austin	1.07	0.54
19. Baltimore	0.60	0.94
20. Memphis	0.72	0.63
21. Charlotte	0.64	0.23
22. Milwaukee	0.57	0.80
23. Boston	0.95	0.42
24. Washington (DC)	1.11	0.63
25. Nashville	0.72	0.94

Source: Bureau of Justice Statistics, *Police Departments in Large Cities, 1990–2000* (Washington, DC: U.S. Department of Justice, 2002), p. 11.

Often the duties of a sheriff are more demanding than those of city police. The sheriff's jurisdiction, for instance, may pose special obstacles because of its large geographical size and often small population compared to that of municipalities. A sheriff may need to drive 100 or more miles to respond to a citizen's call for assistance. This requirement poses special problems for sheriffs, because they are more likely than city police officers to ride alone and backup units may not be readily available in an emergency. Given the people they serve, sheriffs are also more likely to confront armed citizens than city police: Per capita gun ownership is higher in rural areas than in either suburban or urban areas.[46]

Los Angeles County, which has 10 million residents, has the largest sheriff's department in the United States, employing more than 8500 officers.[47] Most of the nation's 3000 sheriff's departments are small, with two-thirds employing fewer than 25 sworn officers and 71 percent serving fewer than 50,000 residents.[48]

The U.S. highway system is the main jurisdiction of state police departments.

? Why do local, state, and federal police have so much overlapping jurisdiction in the same geographic area? Is this the most efficient way to police?

State Police

There are 49 state police agencies in the United States (Hawaii does not have a state police agency).[49] State police agencies have statewide authority to conduct criminal investigations, enforce traffic laws, and respond to calls for service. Normally they perform functions outside of the county sheriff's jurisdiction, such as enforcing traffic laws on state highways and interstate expressways. They also protect state capital buildings and the governor, train officers for local jurisdictions that are too small to operate their own training facilities, and provide local police access to state crime laboratories as needed.

State police agencies originated with the Texas Rangers, which was founded by volunteers in 1835 to protect the Texas border from Mexican outlaws.[50] The Pennsylvania Constabulary, formed in 1905 to suppress the growing conflict between business and labor, is recognized as the first *formal,* full-service, nonvolunteer state police agency.[51]

State police departments may go by a variety of names, including State Police, State Highway Patrol, Highway Patrol, State Patrol, Department of Public Safety, or State Troopers.[52] States may also extend police and investigative powers to other agencies, such as Alcohol Beverage Control, Department of Criminal Investigation, or State Bureau of Investigation.

Federal Police

Although federal police powers originate in the Constitution, the primary responsibility for crime control resides at the state level; state agencies, in turn, delegate some powers to individual cities or counties. By contrast, federal police agencies enforce federal laws. Typically their work includes controlling immigration, investigating counterfeiting, policing airports, and protecting the President and other members of federal institutions. Federal police may also investigate crimes that are not local to just one state—for example, kidnapping, narcotics trafficking, and Internet and mail fraud. In addition, they enforce the law in federal buildings and national parks.

In the United States, roughly 65 federal police agencies employ more than 100,000 employees who are authorized to make arrests and carry firearms **TABLE 5-2** .[53] The total annual budget of these agencies is approximately $4 billion.[54] Some of the most widely recognized federal police agencies are the U.S. Marshals Service, U.S. Secret Service, U.S Postal

TABLE 5-2

Federal Agencies Employing 500 or More Officers

Agency	Number of Officers
U.S. Customs Service and Border Protection	27,705
Federal Bureau of Prisons	15,214
Federal Bureau of Investigation	12,242
Immigration and Naturalization Service	10,399
U.S. Secret Service	4769
Drug Enforcement Administration	4400
U.S. Federal Probation Office	4126
U.S. Marshals Service	3135
U.S. Postal Inspection Service	2976
Internal Revenue Service	2777
Veterans Health Administration	2423
Bureau of Alcohol, Tobacco, and Firearms	2375
National Park Service	2148
U.S. Capitol Police	1535
Bureau of Diplomatic Security	825
U.S. Fish and Wildlife Service	708
USDA Forest Service	600

Source: Bureau of Justice Statistics, *Federal Law Enforcement Officers,*
2004 (Washington, DC: U.S. Department of Justice, 2006).

Inspection Service, and the Federal Bureau of Investigation.

U.S. Marshals Service

The oldest federal police agency is the U.S. Marshals Service, which was established by the Judiciary Act of 1789. Marshals occupy a unique position in law enforcement: They are the enforcement arm of the federal courts, are involved in every federal policing program, and have the broadest authority and jurisdiction of all federal officers. The President or U.S. Attorney General holds the power to appoint one U.S. marshal for each of the 94 federal district offices. These marshals are assisted by 4800 deputy marshals.

The duties of marshals and their deputies vary, but include the following responsibilities:

- Protect federal judicial officials, including judges, attorneys, and jurors
- Arrest persons who commit federal crimes
- Arrest fugitives
- Operate the Witness Security Program
- Operate the Justice Prisoner and Alien Transportation Program
- Operate the Asset Forfeiture Program
- Provide prison services to approximately 53,000 inmates in 1300 federal prisons each day[55]

U.S. Secret Service

Congress established the U.S. Secret Service as a bureau of the Department of Treasury in 1865.[56] Today this agency has roughly 5000 employees, with field officers throughout the United States and liaison offices in several international cities, including Bangkok, London, Paris, and Rome. The Secret Service was originally created to combat a growing threat to the U.S. economy—counterfeit currency. Today, the Secret Service protects the President and Vice President, their families, and heads of state, as well as investigating violations of law relating to counterfeiting, identity theft, computer fraud, and cybercrimes. The more than 2100 special agents in the Secret Service are rotated throughout their careers between investigative and protective assignments.

U.S. Postal Inspection Service

Benjamin Franklin founded the U.S. Postal Inspection Service in 1775 to protect the U.S. Post Office's employees, customers, and mail.[57] Today, the U.S. Postal Inspection Service employs approximately 2000 postal inspectors and an additional 1100 postal police officers. The agency's mission is to investigate crimes that may adversely affect U.S. mail, the postal system, or postal employees. Such crimes include theft of mail, use of the mail system to commit fraud, delivery of illegal pornography through the mail, and identity theft.

Postal inspectors are federal police with the authority to serve warrants and subpoenas, make arrests without warrants for postal-related felonies, carry firearms, and seize property as provided by law. The Postal Inspection Service also operates several forensic crime laboratories that analyze evidence for criminal trials.

Federal Bureau of Investigation

Established in 1908, the Federal Bureau of Investigation (FBI) is the principal investigative arm of the U.S. Department of Justice.[58] The FBI has a threefold mission:

- To defend the United States against terrorist and foreign intelligence threats
- To uphold and enforce the criminal laws of the United States
- To provide leadership and law enforcement assistance to federal, state, municipal, and international agencies

Secret Service officers are active in the investigation and seizure of counterfeit notes, coins, bonds, and food stamps. Counterfeiting is a more serious problem than most people realize. During World War II, for example, Adolph Hitler used this "paper weapon" to undermine his enemies. Relying on slave labor in concentration camps, Hitler counterfeited the paper money of the countries he planned to invade. He used this counterfeit money first to buy war materials and then to flood the local economies, causing economic panic. During the war, Hitler also manufactured about half a billion dollars in counterfeit notes, fueling and funding his machinery of war.

Source: John Cooley, "The False-Money Weapon," *Christian Science Monitor,* January 15, 2002, available at http://www.globalpolicy.org/nations/launder/regions/2002/0115euro.htm, accessed June 22, 2007.

This mission is performed by the agency's more than 30,000 employees who are special agents and support professionals, including intelligence analysts, language specialists, scientists, and information technology specialists. Although the FBI has the authority and responsibility to investigate specific crimes assigned to it, it primarily focuses on counterterrorism, cybercrime, white-collar crime, organized crime, major thefts, and violent crime. The FBI is also authorized to provide other law enforcement agencies with support, including fingerprint identification, laboratory examinations, and police training.

International Police: Interpol

While there is no official international police force, the International Criminal Police Organization (Interpol) facilitates international police cooperation and supports organizations in preventing and combating international crimes. Interpol was founded in Austria in 1923 and now includes 186 member countries. This organization has the following responsibilities:

- *Secure global police communication services.* Interpol provides police around the world with a common platform through which they can share crucial information about criminals and criminality.

Interpol is an international criminal police organization that has combated international crimes since 1914.

? As the fields of criminal justice and intelligence have merged in the war on terror, what role should Interpol play in counterterrorism efforts?

- *Organize data services and databases for police.* Interpol ensures that police worldwide have access to the information and services they need to prevent and investigate crimes, such as names, fingerprints, and DNA profiles of offenders in addition to information about stolen property such as passports, vehicles, and works of art.
- *Operate police support services.* Interpol provides emergency support and operational activities related to fugitives, public safety and terrorism, drugs and organized crime, human trafficking, and financial and high-tech crimes.

Worldwide, Interpol receives more than 1 million messages each year. In 2005, this agency assisted member countries in making more than 3000 arrests.[59]

Private Security Police

In addition to public police, private security police assist local policing efforts. Private police are not sworn officers, so their arrest powers are significantly restricted. Private security forces provide police services to individuals, small businesses, and large corporations, including amusement parks, healthcare facilities, hotel and resort complexes, industrial plants, museums, office buildings, professional sport teams, restaurants, schools, and shopping malls. Some of the most common duties carried out by these police include:

- Installing and servicing burglar alarms
- Transporting valuable commodities
- Patrolling buildings or parks

- Providing protection at schools
- Monitoring public transportation systems

The first private police agency in the United States was formed by Allan Pinkerton in 1850 to protect the national railway system and quell labor disputes.[60] Today, private security in the United States is a $12 billion annual industry, employing approximately 2 million people in roughly 90,000 private security agencies.[61] There are roughly three private security police officers for every one public sworn officer in the United States.[62] One of the largest private security employers, the Sears Roebuck Company, has nearly 6000 security officers; by comparison, the Denver Police Department has only 1500 sworn officers.[63]

Many part-time private security officers are public police officers, who moonlight to earn extra money. Although some police departments realize that police salaries are low and encourage their officers to find additional work, they worry that such moonlighting may lead to serious problems such as conflicts of interest, increased levels of stress, and sleep deprivation. Recognizing these potential hazards, some public police departments have placed restrictions on the amount and type of work that off-duty police officers can perform.

Private security officers do not have the same legal authority as public police. Generally they are not permitted to carry concealed weapons and, unless deputized, have no greater or fewer arrest powers than ordinary citizens. States vary in just how much authority private police are given in the areas of search and seizure. The principal difference between private police and the general public is that private police exercise their right to make a citizen's arrest, whereas ordinary citizens usually do not.

Unfortunately, private police are often poorly trained. Indeed, many states have no licensing or training requirements and do not conduct background investigations on private police officers. Because these officers are under no obligation to protect the rights of the public and take no oath that serves as a guide for their behavior, some critics worry about the growth of the private policing industry.[64]

Private Policing at Macy's

Macy's is one of the largest department stores in the United States. It is also one of the largest employers of private security police, spending an average of $28 million annually on security. Store detectives apprehend and interrogate more than 12,000 customers every year at the 105 Macy's stores across the country. Shoplifting is the most common reason for detaining customers, and most people who are apprehended for this offense (95 percent) confess to the crime and pay a penalty to the store. The exact penalty a shoplifter faces varies from state to state, but generally ranges from one to five times the value of the stolen item. Even so, Macy's loses about $15 million to shoplifters each year.

In Macy's Manhattan store alone, the company employs about 100 private police, including both uniformed security guards and plain-clothed store detectives. Four German shepherd dogs, 300 cameras, and a closed-circuit television center enforce store security measures. If police observe a customer shoplifting, they take that individual to a room and interrogate him or her about the incident. If a shopper asks for legal representation, Macy's contacts the New York City police.

This system has drawn some criticism from police administrators, lawyers, and shoppers, who contend that customers may be deprived of their basic legal rights (such as the right to have legal representation present during questioning) or coerced into confessions. Additionally, they complain that the security officers conducting the interrogation may be poorly trained and not adequately supervised.

Source: Andrea Elliott, "In Stores, Private Handcuffs for Sticky Fingers," *New York Times,* June 17, 2003, available at http: query.nytimes.com/gst/fullpage.html?sec=technology&res=9903EFD71238F934A25755C0A9659C8B63, accessed June 24, 2007.

WRAP UP

In the case of New York City, Mayor Rudy Giuliani and NYPD Commissioner William Bratton deserve the credit for the city's dramatic reductions in crime. Their zero-tolerance approach to policing is based on the idea that police must eliminate conditions in neighborhoods (such as rowdy teenagers, public drunks, and panhandlers) that generate fear and ultimately lead to neighborhood decay if police hope to successfully prevent crime. One reason why crime began declining around the United States during this period was that police departments paid attention to what was happening in New York City and tailored their policing methods accordingly. In fact, William Bratton is currently the Commissioner of the Los Angeles Police Department; Los Angeles hopes to experience similar results of increased quality of life and reduced crime from Bratton's zero-tolerance initiatives. Widespread arrests and detention of active criminal offenders are means that have consistently been proven to reduce crime. While aggressive crime control is not the only way to reduce crime, it is certainly one of the most effective and tangible strategies.

Chapter Spotlight

- The U.S. police system is based on English models, which evolved from volunteer citizens who served as night watchmen.

- American police systems have gone through several distinct phases, including periods marked by corruption, growth, organization, and reform.

- In community policing, police officers work in partnership with neighborhood residents to prevent and respond to crime.

- Intelligence-led policing is the newest model for law enforcement. It entails using information centers or hubs to coordinate intelligence reports from national, state, and local agencies for more effective crime fighting.

- Law enforcement agencies in the United States work at the local, state, federal, and international levels. In addition to these public agencies, private security police serve private-sector clients.

- There are more than 60 federal police agencies in the United States, each with specific rights, duties, responsibilities, and obligations.

- Private policing agencies are increasing in popularity. Critics worry that private police officers may be poorly trained and thus may not respect the constitutional rights of citizens.

Putting It All Together

1. The United States does not have a centralized national police force, but instead has many separate police agencies. Is this the best approach for controlling crime?

2. The introduction of new technology changed policing in the early part of the twentieth century. What are some of the new technologies being used in the twenty-first century that are once again transforming the U.S. system of policing?

3. Reforms were undertaken during the early part of the twentieth century to address the influence of politics on policing. To what extent were these reforms successful? How does politics influence policing today?

4. How do more recent developments in community policing compare to earlier models of policing? Are present-day models more effective at controlling crime?

5. Police in private security agencies may not be well trained or always have the best interests of the community in mind. Knowing that, should private police be permitted to take citizens into custody and detain them?

broken windows theory A theory that proposes that small signs of public disorder set in motion a downward spiral of deterioration, neighborhood decline, and increasing crime.

community policing A policing model that was popular in the 1990s, in which police and citizens unite to fight crime.

constable An elected law enforcement officer in a small town without a police force.

frankpledge police system An English policing system that spanned the eleventh through thirteenth centuries, in which every male older than age 12 assumed responsibility for fighting crime.

fusion centers A police intelligence operation in which regional hubs pool information from multiple jurisdictions.

intelligence-led policing A crime-fighting strategy driven by computer databases, intelligence gathering, and analysis.

Interpol An international criminal police organization that facilitates international police cooperation.

jurisdiction The territory over which a law enforcement agency has authority.

kin police system An English policing system used between 400 C.E. and 500 C.E. in which each male citizen assumed responsibility for protecting his neighbor.

Metropolitan Police Act The 1829 act that established the London Metropolitan Police force, which was the first salaried, uniformed police agency.

parish–constable police system A police system that operated in England between 1285 and 1829, in which constables and watchmen were appointed to prevent crime.

police–population ratio The number of sworn officers per 1000 citizens.

private security police Individuals who are employed by citizens and businesses to provide security.

sheriff The principal law enforcement officer in a county.

slave patrols Small, organized groups who controlled the slave population and outbreaks of slave revolts in pre-Civil War United States.

state police agencies Law enforcement agencies that protect the interests of the state.

sworn officers Officers who are empowered to arrest suspects, serve warrants, carry weapons, and use force.

Notes

1. Eric Monkkonen, *Policing in Urban America, 1860–1920* (Cambridge, UK: Cambridge University Press, 1981); Wilbur Miller, *Cops and Bobbies*, 2nd ed. (Columbus, OH: Ohio State University Press, 1999).

2. Lincoln Steffens, *The Autobiography of Lincoln Steffens* (New York: Harcourt, Brace, and Co., 1931); James Lardner and Thomas Reppetto, *Behind the NYPD's Blue Wall of Silence* (New York: Henry Holt, 2000).

3. Charles Reith, *A New Study of Police History* (Edinburgh, UK: Oliver and Boyd, 1956); Samuel Walker, *A Critical History of Police Reform* (Lexington, MA: Lexington Books, 1977); Samuel Walker and Charles Katz, *The Police in America*, 6th ed. (New York: McGraw-Hill, 2008).

4. The discussion of English policing history is derived from David Johnson, *American Law Enforcement* (St. Louis: Forum Press, 1981); James Richardson, *The New York Police* (New

York: Oxford University Press, 1970); Herbert Johnson, *History of Criminal Justice* (Cincinnati: Anderson, 1988).

5. Clive Emsley, *The English Police* (New York: Palgrave Macmillan, 1961).
6. Miller, note 1.
7. James F. Richardson. *The New York Police* (New York: Oxford University Press, 1970), p. 10.
8. Craig Uchida, "The Development of the American Police," in Roger Dunham and Geoffrey Alpert (Eds.), *Critical Issues in Policing,* 5th ed. (Prospect Heights, IL: Waveland Press, 2005), pp. 20–30.
9. Uchida, note 8.
10. Thomas Reppetto, *The Blue Parade* (New York: Free Press, 1978).
11. Uchida, note 8.
12. Andrew Harris, *Policing the City* (Columbus, OH: Ohio State University Press, 2004); Robert Fogelson, *Big-City Police* (Cambridge, MA: Harvard University Press, 1977); Herman Goldstein, *Problem-Oriented Policing* (New York: McGraw-Hill, 1990); Herbert Johnson, *History of Criminal Justice* (Cincinnati: Anderson, 1988); Carl Klockars, *The Idea of Police* (Beverly Hills, CA: Sage Publications, 1985); Roger Lane, *Policing the City* (Cambridge, MA: Harvard University Press, 1967); Walker and Katz, note 3.
13. Johnson, note 12; Lane, note 12; Walker and Katz, note 3.
14. Harris, note 12; Fogelson, note 12; Johnson, note 12.
15. Johnson, note 12; Lane, note 12; Walker and Katz, note 3.
16. Sally Hadden, *Slave Patrols* (Cambridge, MA: Harvard University Press, 2001).
17. Frank Richard Prassel, *The Western Peace Officer* (Norman, OK: University of Oklahoma Press, 1972).
18. Uchida, note 8.
19. The New York City Police Museum, "Leadership of the City of New York Police Department, 1845–1901," available at http://www.nycpolicemuseum.org/html/tour/leadr1845.htm, accessed July 25, 2007; H. W. Brands, *T.R.: The Last Romantic* (New York: Basic Books, 1998); H. Paul Jeffers, *Commissioner Roosevelt* (New York: John Wiley, 1996); Jay Stuart Berman, *Police Administration and Progressive Reform: Theodore Roosevelt as Police Commissioner of New York* (Westport, CT: Greenwood Press, 1987).
20. August Vollmer, *The Police and Modern Society* (Berkeley: University of California Press, 1936); Gene Carte and Elaine

Carte, *Police Reform in the United States* (Berkeley: University of California Press, 1975); Walker and Katz, note 3.
21. Walker and Katz, note 3.
22. Uchida, note 8; Walker and Katz, note 3.
23. Fogelson, note 12; Johnson, note 12.
24. Fogelson, note 12; Johnson, note 12.
25. *Mapp v. Ohio,* 367 U.S. 643 (1961).
26. *Miranda v. Arizona,* 384 U.S. 436 (1966).
27. *Katz v. United States,* 389 U.S. 347 (1967).
28. Matthew Hickman and Brian Reaves, *Local Police Departments, 2003* (Washington, DC: U.S. Department of Justice, 2006).
29. Herman Goldstein, "The New Policing," paper presented at the National Institute of Justice Conference on Community Policing, 1993, p. 5.
30. Ralph Weisheit, L. Edward Wells, and David Falcone, "Community Policing in Small Towns and Rural America," *Crime & Delinquency* 40:549–567 (1994).
31. Jerome Skolnick and David Bayley, *Community Policing* (Washington, DC: U.S. Department of Justice, 1988).
32. Herman Goldstein, "Improving Policing: A Problem-Oriented Approach," *Crime & Delinquency* 25:236–258 (1979).
33. George Kelling and James Q. Wilson, "Broken Windows: The Police and Neighborhood Safety," *Atlantic Monthly* 249:29–38 (1982).
34. Samuel Walker, *The Police in America,* 2nd ed. (New York: McGraw-Hill, 1992), p. 178.
35. David Kaplan, "Spies Among Us," *U.S. News & World Report* 8:40–49 (2006).
36. Walker and Katz, note 3.
37. Brian Reaves, *Census of State and Local Law Enforcement Agencies, 2004* (Washington, DC: U.S. Department of Justice, 2007); Brian Reaves, *Federal Law Enforcement Officers, 2004* (Washington, DC: U.S. Department of Justice, 2006); Brian Reaves and Andrew Goldberg, *Campus Law Enforcement Agencies, 1995* (Washington, DC: U.S. Department of Justice, 1996).
38. Reaves, *Census of State and Local Law Enforcement Agencies, 2004,* note 37.
39. Hickman and Reaves, note 28.
40. Hickman and Reaves, note 28.
41. Brian Reeves and Matthew Hickman, *Police Departments in Large Cities, 1990–2000* (Washington, DC: U.S. Department of Justice, 2002).

42. New York City Police Department, available at http://www.nyc.gov/html/nypd/html/misc/pdfaq2.html, accessed June 26, 2007.
43. Hickman and Reaves, note 28.
44. Hickman and Reaves, note 28.
45. Matthew Hickman and Brian Reaves, *Sheriffs' Offices, 2003* (Washington, DC: U.S. Department of Justice, 2006); National Sheriffs' Association, available at http://www.sheriffs.org/, accessed June 25, 2007.
46. Reaves, *Census of State and Local Law Enforcement Agencies, 2004*, note 37.
47. "More Gun Ownership Statistics," April 23, 2007, available at http://www.halfsigma.com/2007/04/more_gun_owners.html, accessed June 27, 2007; Lee Brown, "The Role of Sheriff," in Alvin Cohn (Ed.), *The Future of Policing* (Beverly Hills, CA: Sage Publications, 1978), pp. 227–247; Dana Brammer, *A Study of the Office of the Sheriff in the United States* (Jackson, MS: University of Mississippi Press, 1968).
48. Hickman and Reaves, note 45; U.S. Census Bureau, "Los Angeles County, California," available at http://quickfacts.census.gov/qfd/states/06/06037.html, accessed June 26, 2007.
49. Hickman and Reaves, note 45.
50. Reaves, *Census of State and Local Law Enforcement Agencies, 2004*, note 37.
51. Robert Utley, *Lone Star Justice* (New York: Oxford University Press, 2002); Charles Robinson, *The Men Who Wear the Star* (New York: Modern Library, 2001).
52. "The Pennsylvania State Police," *New York Times*, February 28, 1910, available at http://select.nytimes.com/mem/archive/pdf?res=FB0E10FC3F5417738DDDA10A94DA405B808DF1D3, accessed June 26, 2007.
53. Hickman and Reaves, note 45; Tom O'Connor, "Police Structure and Organization: A State-by-State Guide," available at http://faculty.ncwc.edu/TOCONNOR/polstruct.htm, accessed June 27, 2007.
54. Brian Reaves, *Federal Law Enforcement Officers, 2004*, note 37, Kristen Hughes, *Justice Expenditure and Employment in the United States, 2003* (Washington, DC: U.S. Department of Justice, 2006).
55. United States Marshals Service, available at http://www.usdoj.gov/marshals/, accessed June 25, 2007; James Cheno-with, *Down Darkness Wide* (Frederick, MD: Publish America, 2004); Robin Langley Sommer, *A History of U.S. Marshals* (Philadelphia: Running Press, 1993).
56. United States Secret Service, available at http://www.ustreas.gov/usss/, accessed June 25, 2007; Philip Melanson and Peter Stevens, *Secret Service* (New York: Carroll and Graf, 2005).
57. U.S. Postal Inspection Service, available at http://www.usps.com/postalinspectors/, accessed June 25, 2007; John Makris, *The Silent Investigators* (Boston: E.P. Dutton, 1959).
58. Federal Bureau of Investigation, available at http://www.fbi.gov/, accessed August 24, 2007.
59. Interpol, available at http://www.interpol.int/, accessed August 24, 2007; Michael Fooner, *Interpol* (New York: Plenum Press, 1989), p. 177; Jean Bashfield, *Interpol* (Milwaukee: World Almanac Library, 2004); Charles Hanley, "Long Arm of Interpol Widens Its Reach," *Arizona Daily Sun*, May 30, 1993, p. 24.
60. Frank Morn, *The Eye That Never Sleeps* (Bloomington, IN: Indiana University Press, 1982).
61. U.S. Department of Justice, *National Policy Summit: Building Private Security/Public Policing Partnerships to Prevent and Respond to Terrorism and Public Disorder* (Washington, DC: International Association of Chiefs of Police, 2004).
62. Elizabeth E. Joh, "The Paradox of Private Policing," *Journal of Criminal Law and Criminology* 95:49–131 (2004).
63. Brian Frost, "The Privatization and Civilianization of Policing," in Charles Friel et al. (Eds.), *Criminal Justice 2000: Boundary Changes in Criminal Justice* (Washington, DC: National Institute of Justice, 2000), pp. 19–79; Hickman and Reaves, note 28, p. 2.
64. Karen Hess and Henry Wrobleski, *Introduction to Private Security* (Belmont, CA: Wadsworth, 2005); Brian Johnson, *Principles of Security Management* (Englewood Cliffs, NJ: Prentice-Hall, 2004); Charles Nemeth, *Private Security and the Law*, 3rd ed. (St. Louis: Butterworth-Heinemann, 2003); Robert Fischer, *Introduction to Security*, 7th ed. (St. Louis: Butterworth-Heinemann, 2003); William Cunningham, John Strauchs, and Clifford Van Meter, *Private Security Trends, 1970–2000: The Hallcrest Report II* (Woburn, MA: Butterworth, 1990).

POLICE LINE — DO NOT

- Identify the formal structure of police agencies, including their organization, rules, and limitations.

- Comprehend the informal aspects of police organization, such as the "working personality" and styles of operation.

- Understand police operations and functions.

- Know the different tasks executed by patrol officers and the methods by which they complete their work.

- Grasp the process of criminal investigations and the duties of police in these investigations.

- Describe the principle role of police in traffic enforcement and their important service functions.

Police Organization and Operations

CHAPTER 6

OSS

On December 28, 2006, in Santa Ana, California, police officers were conducting surveillance on a career criminal—Oscar Gabriel Gallegos, who was wanted for the attempted murder of two Long Beach police officers. An illegal alien with prior deportations to Mexico, Gallegos had an extensive criminal history that included arrests for assault, weapons violations, making terrorist threats, and narcotics possession. After being pulled over for a routine traffic stop, Gallegos jumped out of his vehicle and shot the two law enforcement officers. Luckily, both officers survived. Six days later, when the surveillance team contacted Gallegos in a strip mall, the fugitive opened fire. The officers returned fire and killed him.

• Which aspects of police organization led to the use of lethal force in this case?

• What does this incident show about police subculture?

Source: Charles Montaldo, "Police Kill Cop-Shooting Suspect," December 28, 2006, available at http://crime.about.com/b/a/257244.htm?nl=1, accessed August 5, 2007.

Introduction

What do police do? Are police trying to prevent crime or only respond to crime? Is it the responsibility of police to reduce fear of crime among civilians? Should police be concerned with whether civilians are satisfied with the services they provide?

These difficult questions sit at the core of policing in the United States. In American society, police occupy two unique roles: They fight crime and they maintain order. As crime fighters, police practice law enforcement. When police are maintaining order, they provide civilians with services and keep the peace.[1]

The police roles of crime fighters and order maintainers are carried out within a police agency. Inside of this agency is a formal and informal structure that shapes the roles of all department members, including the rules that they must follow, their specific duties, and their responsibilities to civilians as police officers. The formal structure of large departments resembles a military-style structure FIGURE 6-1 in that it establishes relationships among department members and clarifies the responsibilities of each position.

Police work also involves an underlying, informal structure, which represents the unofficial relations that exist among officers. These relations affect police operations and the ways in which police perform their duties of patrol, criminal investigations, traffic enforcement, and community services. It is important for every police agency to strike a balance between the official rules and the informal structure when trying to achieve department goals and successfully carry out these functions.

Formal Structure of Police Organizations

A police department is a bureaucracy, a model of organization with strict rules, close supervision, and reliance on authority. This organizational model, which was developed by nineteenth-century German sociologist Max Weber, includes the following components:

• Chain of command
• Delegation of authority
• Specialization
• Rules and regulations
• Limited rewards
• Competency[2]

While this model works well in theory, the inflexibility of this structure makes it difficult for real-world bureaucracies to adapt to external change. As such, it is difficult for police agencies to respond quickly to changes in the patterns of crime in a community. This rigidity often means that patrol officers are seldom given a voice in shaping department pol-

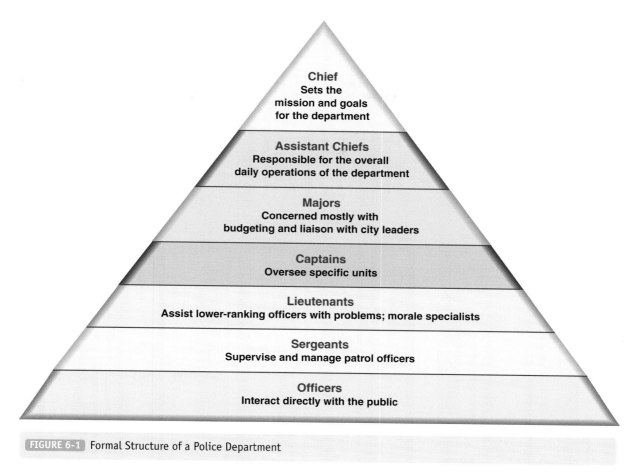

FIGURE 6-1 Formal Structure of a Police Department

icy within the organization. In police departments, this may have a negative effect on morale and the manner in which police officers do their work.[3] In addition, the rigid "chain of command" does not always work the way it is intended to, and communication often breaks down. As a consequence, police departments or officers may not receive the information they need to perform their jobs effectively. Another flaw in the bureaucratic model is that officers become isolated from the members of the public whom they serve. On a police force, this isolation may breed contempt, which negatively affects the officer-citizen relationships that are essential to solving crimes. Despite the problems associated with bureaucracies, the model has been widely adopted by both legitimate and illegitimate enterprises throughout the world for more than 150 years.

Chain of Command

A chain of command, or hierarchy of authority, identifies who communicates with and gives orders to whom. In police departments, this chain establishes the working relationships among the different ranks.

The purpose of the chain is to make the line of authority clear and precise. Sergeants, for example, know they have less authority than lieutenants but more authority than patrol officers. The complexity of a department's chain is related to its size (e.g., number of employees). In a small department, the bureaucratic structure is less complex than it is in large agencies. Regardless of the department's size, however, all police departments have an organization chart **FIGURE 6-2** which identifies the flow of information within the department, making it clear who is responsible for specific tasks and operations.

The chain of command is determined by specific criteria, and law enforcement agencies use a variety of screening methods such as personal interviews, medical exams, drug tests, and psychological evaluations when hiring new officers and determining their rank.[4] Most departments also conduct criminal records checks, background investigations, and driving record checks. In addition, nearly all local law enforcement agencies have a minimum education requirement for new officer recruits **FIGURE 6-3** .[5] Most also require an average of 800 to 1500 hours of training to cover

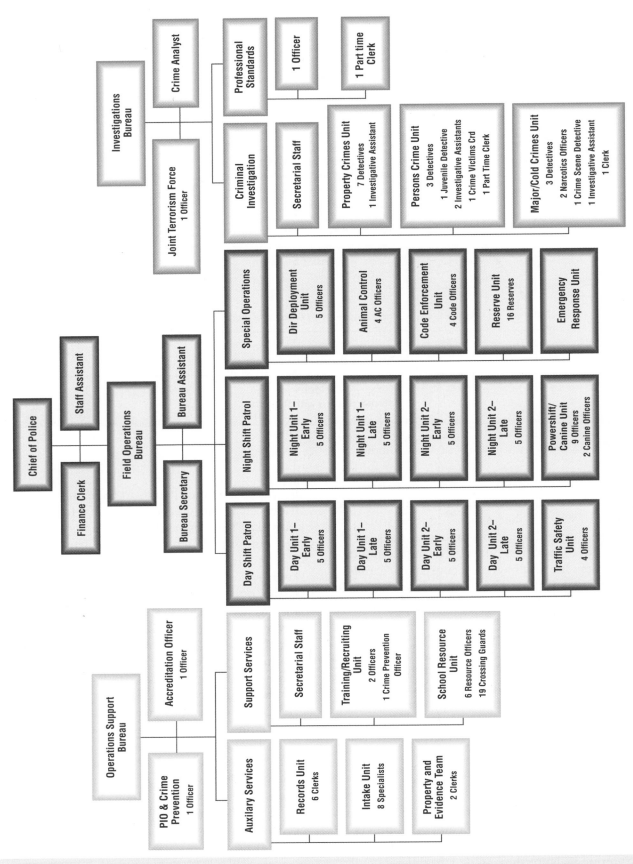

Why Drug Dealers Live at Home

Drug dealing gangs are similar to police agencies in that both have a well-identified organizational structure. This fact was discovered by economist Sudhir Venkatesh who went deep inside one of Chicago's poorest neighborhoods to learn why so many people wanted to be a part of crack gangs. He discovered that, in fact, most crack dealers did not make much money. Venkatesh found that crack gangs are actually complex, structured hierarchies that include a board of directors, treasurer, and clearly defined roles for those persons charged with ensuring gang safety, money transport, and drug delivery.

In the gang that Venkatesh studied, the lowest members of the chain of command—the rank-and-file members—paid the gang $200 to $300 per month for gang protection and the possibility of "career" advancement. These dues, plus revenue from drug sales and extortion taxes (paid by local businesses to be left in peace), meant the gang's monthly income averaged $32,000. Like every business, the gang also had expenses. These expenses, which averaged $14,000 per month, covered the cost of drugs, board of director salaries, weapons, and miscellaneous expenditures such as bribes, legal fees, community service events, funeral costs, and death benefits for murder victims' families.

Given these costs, Venkatesh calculated average annual gang member salaries based on level in the chain of command:

Board of directors	$500,000
Gang leaders	$100,000
Officers	$8400
Foot soldiers	$4000
Rank-and-file members	$0

There is a stark contrast between the annual income of the street dealers and that of the bosses. There is also a dramatic difference in lifestyle: Dealers face much higher risks of arrest, injury, and death. Venkatesh found that members of this gang were, on average, six times more likely to be arrested than a boss and had a one in four chance of being killed, making drug dealing one of the most dangerous jobs in the United States. Despite the bleak outlook, many young men vied for the position of foot soldiers in the gang in hopes of working up the chain of command and becoming wealthy, even though common drug dealers did not make enough money to live on their own. Instead, dealers lived at home with their parents until they either dropped out of the gang, were incarcerated, or were killed.

Sources: Steven Levitt and Stephen Dubner, *Freakonomics,* revised edition (New York: William Morrow, 2006).

all state and federal mandates and agency requirements FIGURE 6-4 .[6]

Police salaries also correspond with the chain of command. After completing the necessary training, an entry-level patrol officer can expect, on average, to earn approximately $30,000 annually, depending on the size of the jurisdiction and department, plus special pay and bonuses. At the top of the chain, starting local police chiefs earn an average of $50,800 per year.[7]

Delegation of Authority

The practice of passing decision-making responsibilities through the chain of command is called delegation of authority. Police chiefs will delegate some authority to assistant chiefs, who in turn pass some authority to captains, and so forth down the line. Authority is delegated because chiefs cannot monitor every sit-uation or make every decision on their own. In efficient organizations, department leaders must share the management of the various responsibilities.

Specialization

No one person has the time or the skills to complete all of the responsibilities of running a police department. Big-city police agencies, for example, often specialize or concentrate their efforts on specific activities. Specialization requires an agency to focus its resources on a narrow area of knowledge, skill, or activity. It typically involves a law enforcement agency adapting itself to perform some particular function, such as forming a special unit on domestic violence, gang suppression, or homicide investigations. Specialization is similar to the department implementing a division of labor that outlines and assigns tasks to officers, such as patrolling a particular

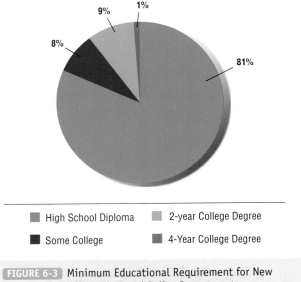

FIGURE 6-3 Minimum Educational Requirement for New Officers in Local Police Departments

Source: Matthew Hickman and Brian Reaves, *Local Police Departments, 2003* (Washington, DC: U.S. Department of Justice, 2006), p. 9.

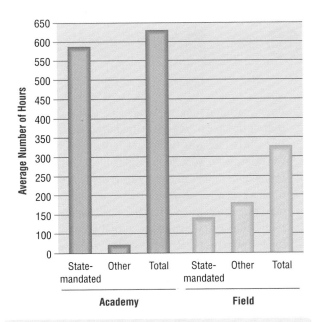

FIGURE 6-4 Training Requirements for New Recruits

Source: Matthew Hickman and Brian Reaves, *Local Police Departments, 2003* (Washington, DC: U.S. Department of Justice, 2006), p. 8.

neighborhood, controlling traffic at an event, or updating the press about an ongoing investigation. Usually both specialization and division of labor come about from trial and error; during this process, it may be discovered that certain groups of individuals are better at performing particular assignments than others. By specializing their work forces, police departments are typically able to increase their productivity.

Rules and Regulations

Police organizations and internal operations are governed by a detailed set of rules. A rule, in policing, is a proscription about behavior (i.e., do this, don't do that). Department rules govern behavior and specific courses of action to achieve particular goals.

For instance, in some departments the rules manual states that when officers are on duty they must conduct themselves as follows:

- Be neat and clean in appearance and wear standard uniforms.
- Avoid cigarettes, alcohol, and vulgar or profane language.
- Do not engage in political or religious arguments.
- Be obedient and loyal to the department at all times.

These and other rules help police to maintain order by portraying a positive image to the public.

The formal rules in police departments are clear, widely understood, and intended to be fairly applied. The reason for having clear-cut rules and regulations is they reduce ambiguity, decrease internal conflicts, and increase the likelihood that work will be completed satisfactorily and on time. The responsibilities and authority for each role in the police agency are plainly spelled out in the department manual. During a criminal investigation, for instance, patrol officers—who usually are the first members of the department to reach the crime scene—complete the preliminary inquiry. Later, detectives arrive and interrogate suspects. Both the patrol officers and the detective on the scene know when and where the responsibilities of one party end and the responsibilities of the other party begin.

Rules also make clear which behaviors are unacceptable. A significant amount of research has examined occurrences of improper police behavior, such as being drunk on the job, pilfering supplies from the department, and accepting bribes and sexual favors from alleged offenders.[8]

Limited Rewards

Most police officers begin and end their careers as patrol officers. Promotion opportunities in police agencies are rare for several reasons:

Body Art on the Squad

Tattoos are becoming increasingly popular among young adults, particularly among police personnel. Today, an increasing number of police agencies across the United States are putting in place stricter policies regarding

just how much body art is acceptable for their officers. Departments are concerned that tattooed officers do not present a professional image to the public and that the tattoos themselves do not comply with the grooming standards of the profession.

Police agencies across the country differ in terms of what they regard as an excessive tattoo. Some agencies disqualify applicants with any body art; other departments allow tattoos that may cover 25 percent or more of an arm or a leg. For example, in Baltimore, no tattoos are permitted; in Los Angeles, police must cover tattoos with skin-colored patches or clothing; in San Diego, officers must hide any markings that cover 30 percent or more of exposed body parts; and in Houston, police must wear clothing to cover tattoos year-round.

In Hartford, Connecticut, police officers challenged the department's rules on tattoos. Police argued that their tattoos were protected under the First Amendment, specifically their right to free speech. A ruling in 2006 by a U.S. appeals court disagreed, stating that police officer tattoos do not enjoy First Amendment protection and can be subject to department rules.

Source: Matt Reed, "Tattoos: Official Blots on Reputations?" *USA Today,* July 23, 2007, p. 3A, available at http://www.usatoday.com/news/nation/2007-07-22-tattoos_N.htm, accessed July 30, 2007.

- Civil service regulations mandate that officers serve for a specific number of years in a particular rank before they become eligible for promotion.
- Promotional exams are given at irregular intervals because department promotions depend more on the financial well-being or health of a city than on the needs of the police agency.
- Promotions are based on a formal testing process that usually consists of an oral interview and written exams that may favor applicants with more privileged educational backgrounds.

The nature of the selection process may also restrict opportunities of employment and promotion for women and racial and ethnic minority members. As a result, members of these groups are likely to be underrepresented on the police force. In nearly all instances, hiring decisions are made by a selection committee that consists almost entirely of white males, so the criteria used to select officers may unintentionally favor members of this group.[9] For example, physical agility tests may present obstacles for females, and personal interviews may place whites at an advantage, because people are generally more attracted to persons like themselves.[10] These kinds of inherent biases in the police selection and promotion process effectively keep the numbers of women and minorities on the police force proportionally low.[11]

Competency

In bureaucracies, personnel should be hired and promoted based on their knowledge, skills, and behavior to perform the job. Collectively, these capabilities are called competency. Competency is made appar-

Law enforcement is one of the most physically demanding and challenging occupations.

 Should different standards be used for male and female officers regarding physical tests for policing jobs? How might gender influence the ability of a police officer to perform the duties of his or her job?

ent by evidence that reflects the desired abilities or skills, such as qualifications, test scores on promotional exams, and field performance. Ascribed attributes, such as gender, ethnicity, and race, should not influence hiring, promotion, or retention decisions in a bureaucracy—but sometimes they do.

LINK Detective Kathleen Burke of the New York Police Department (NYPD) busted through the "glass ceiling." Throughout her illustrious career, she faced a series of obstacles in her daily struggle to perform her job and to be promoted. Some of the barriers she confronted are discussed in Chapter 8.

Informal Structure of Police Organizations

The foundation of the informal structure of police departments is a <u>police subculture</u>, whose beliefs, values, and patterns of behavior separate officers from the public and police administrators. Subcultures are not unique to police work; they are found in both legitimate and illegal lines of work ranging from lawyers and physicians to criminal gangs and auto thieves. Police subculture has a strict code of conduct that teaches police officers to adhere to the following conduct:

- Take care of your partner(s).
- Never back down.
- Do not interfere in another officer's sector or work area.
- Do not snitch on another officer.
- Never trust new officers until they have been "checked out."
- Protect yourself.
- Do not obviously ingratiate yourself to supervisors.[12]

LINK Informal subcultures also develop in prisons (see Chapter 14) in which prisoners operate according to a unique code of conduct not terribly different than the code of the police subculture.

Like many other subcultures, police subculture enforces a "code of silence," ensuring that what goes on "behind closed doors" stays private. A 1950s study by William Westley discovered that police often believe that the public is their enemy.[13] This perception pushes police to turn to fellow officers for support, a tendency that is strengthened by officers' strong commitment to secrecy. To shield themselves from

Decades of research suggest that police view themselves as the guardians of society.

 Does the adoption of a secretive and cynical police personality make officers more effective police?

outsiders, police officers may go to great lengths to protect one another, including covering up improper behavior and lying to supervisors.

The idea that there is a strong, unified, and influential police subculture has been challenged over the past several decades. Some criminologists contend that while police have slightly different attitudes and beliefs than the public, these differences are negligible.[14] For instance, Eugene Paoline, who studied big-city police agencies, concluded that the notion of a police subculture has been highly overrated. Paoline also concluded that many different groups of officers exist in police departments, some of which share the same values and some of which do not. Moreover, whether officers share similar beliefs may have more to do with their particular shift and assigned neighborhood than with their ethnicity, race, or sex.[15]

Working Personality

Criminologist Jerome Skolnick hypothesized that police develop a working personality to deal with the danger and authority inherent in their professional position. Slowly, through their interactions with the public and the police administration, officers become more authoritarian and suspicious than they were before they entered the police academy. Police also learn to carefully protect their authoritative position, which often means establishing a disdain for the rights of criminals and a high suspicion of the stereotypical criminal—poor, young, minority males to the point where "every hostile glance directed at the passing patrolman is read as a sign of possible guilt."[16]

Expanding on this work, retired NYPD police officer and criminologist Arthur Niederhoffer reported on the attitude of police cynicism, the belief that police think people are selfish and motivated by evil.[17] Niederhoffer argued that police are cynical about both the outside world and the world inside the police department, thinking department promotions are based on favoritism instead of ability, and that the public is more likely to interfere with police work than to help officers solve crimes. Police cynicism increases during the first 7 to 10 years that an officer is on the job and declines thereafter, although it never totally disappears.

Researchers have found that the relationship between police work and shifts in personality is complex and may be influenced by a wide range of factors. For example, the law, police bureaucracy, compe-

tency, and morality may all strongly affect how police perceive their jobs and the decisions they make.[18]

Operational Styles

Once officers leave the training academy and begin work on the street, they will likely develop an operational style—that is, a way they interact with their fellow officers and the public. Political scientist James Q. Wilson and criminologist John Broderick have constructed separate typologies to represent common operational styles of policing.

Wilson found that police departments develop one of three operational styles that affect the behavior of the department's officers when they are reacting to misdemeanor crimes and noncrime incidents:

- *Legalistic departments* adopt a zero-tolerance approach to serious crime. In these departments, administrators believe arrest deters crime. For minor infractions, police may not always arrest the perpetrators, but they will almost always use the threat of arrest to maintain order.

- *Watchmen departments* resolve disputes and community problems informally before they resort to making an arrest. They believe arrests exacerbate an already tenuous relationship between police and the public.

- *Service departments* emphasize helping the public and are not overly concerned with enforcing the law for minor violations. Rather than making arrests, officers are more likely to refer offenders to neighborhood treatment agencies for guidance and assistance. Taking formal action against someone who has committed a minor crime is a last resort.[19]

Focusing on individual officers rather than entire departments, John Broderick proposed that officers have one of four personality types:

- *Enforcers* are more concerned with enforcement of law than with the protection of individual rights.

- *Idealists* highly value individual rights.

- *Realists* are concerned with neither individual rights nor enforcement of laws.[20]

- *Optimists* value individual rights and are service-oriented.

A corresponding set of attitudes accompanies these personality types. That is, both the enforcer and the idealist embrace feelings of resentment. The realist has a "to hell with it" attitude, while the optimist shows little resentment and a strong commitment to the profession.[21]

The operational styles of departments and officers derive from a city's political culture, climate, financial resources, and organization.[22] However, it is important to note that the operational styles identified by both Wilson and Broderick may not exist in all departments. In her study of the Dallas Police Department, for example, Ellen Hochstedler did not find evidence of either the department or its officers having developed a particular type of operational style; instead, she concluded, police work is too unpredictable and complicated for either the organization or its officers to utilize one specific style consistently.[23]

Police Operations

Police operations refer to the services that police agencies provide and the methods they use to deliver these services. In the words of August Vollmer, former Chief of Police in Berkeley, California:

The citizen expects police officers to have the wisdom of Solomon, the courage of David, the strength of Samson, the patience of Job, the leadership of Moses, the kindness of the good Samaritan, the strategical training of Alexander, the faith of Daniel, the diplomacy of Lincoln, the tolerance of the Carpenter of Nazareth, and finally, an intimate knowledge of every branch of the natural, biological, and social sciences. If he had all of these, he might be a good policeman![24]

In many ways, the public expects more from the police than it is possible for officers to deliver.[25] Police agencies do their best to meet the demands placed on them. They are available 24 hours a day, 7 days a week, 365 days a year. They provide citizens with a 9-1-1 telephone number to call during emergencies and use automobiles with flashing lights and sirens to offer rapid response to citizen complaints.

While police perform a wide variety of tasks, there are four main components of police operations: patrol, criminal investigations, traffic enforcement, and community services.

Patrol

Despite common misconceptions, police actually spend only one-fourth of their time, on average, as

law enforcers FIGURE 6-5 .[26] Instead, police spend the majority of their time on patrol—that is, moving through assigned areas by foot or vehicle to enforce regulations, manage traffic, control crowds, prevent crime, and arrest violators. Criminologist Geoffrey Alpert and his colleagues have summarized the accomplishments of patrol:

- Deterring crime
- Allowing police to apprehend criminals committing crimes in progress
- Providing opportunities for police to assist citizens who find themselves in dangerous situations
- Helping police to maintain order
- Keeping motor vehicle and pedestrian traffic flowing[27]

Patrol officers are the most visible component of all criminal justice personnel, in that they have the greatest amount of contact with the public. On their assigned beat (area), patrol officers become the eyes and ears of the department. Other divisions of a police department are necessary only to the extent that patrol officers are not 100 percent effective in carrying out their duties, particularly in terms of preventing crime.[28] Therefore, a large percentage of the police budget is allocated to patrol. In big-city departments, patrol may account for as much as 60 percent of the

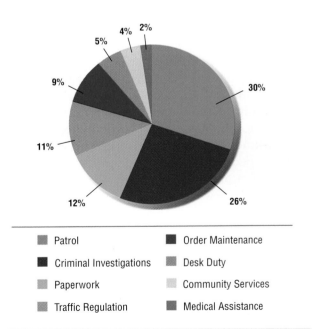

■ Patrol	■ Order Maintenance
■ Criminal Investigations	■ Desk Duty
■ Paperwork	■ Community Services
■ Traffic Regulation	■ Medical Assistance

FIGURE 6-5 Police Responsibilities

Source: Adapted from Jack Green and Carl Klockars, "What Police Do," in Carl Klockars and Stephen Mastrofski (Eds.), *Thinking about Police,* 2nd ed. (New York: McGraw-Hill, 1991), pp. 273–284.

As time passes, methods of policing evolve to keep pace. In the twenty-first century, police are finding inventive new technologies that help them track down and catch criminals quickly and effectively.

In crowded cities such as New York, police departments have figured out a way to avoid the traffic-filled streets that often assist their suspects in escaping—helicopters. The newest additions to the New York Police Department are Bell 412 helicopters that can hover over a crime scene and use telephoto lenses to get up close and personal with evidence, suspects, and victims. These helicopters are equipped with infrared sensors to assist in apprehension of criminals at night. These state-of-the-art choppers provide a bird's-eye view of the city and can help chase teenagers out of cemeteries at night and give traffic advice to local citizens. Since September 11, 2001, these helicopters have also been outfitted with the latest technological devices to help fight terrorism, such as improved cameras and chemical and radiation detection sensors. They are used to peep into train yards, high-profile landmarks, and shipping terminals in an effort to ferret out criminals.

Another development in police patrol is the use of Segways®—two-wheeled stand-up scooters that have been used successfully in airport terminals and on city sidewalks. These electric scooters have a top speed of 13 miles per hour and give off no carbon emissions. They are operated simply by leaning in the desired direction, allowing police to easily maneuver through tight spaces and move quickly through pedestrian traffic.

Both helicopters and Segways are used by police as they try to catch

potential criminals and aid victims in need, reducing response time and helping officers do their jobs more efficiently.

Sources: Roger Yu, "Segways Are Airports' Latest Arrivals," *USA Today*, July 10, 2006, p. 3B, available at http://www.usatoday.com/tech/products/2006-07-09-segway-airports_x.htm, accessed July 30, 2007; Andrew Jacobs, "Instead of Walking a Beat, Flying One," *New York Times*, July 2, 2006, available at http://www.nytimes.com/2006/07/02/nyregion/02aviation.html?ex=1309492800&en=3c44ecd6cc60ae56&ei=5090&partner=rssuserland&emc=rss, accessed June 28, 2007.

---

total personnel and budget. In small departments, virtually every officer participates in patrol.[29]

Historically, police patrolled their beats on foot. With the introduction of the automobile, this method of patrol was supplanted in the 1930s by motor patrol. Beginning in the 1950s, police administrators insisted on relying more heavily on the automobile because officers in cars could cover areas more quickly than when on foot, and officers were less accessible to the public, thus making it less likely they would be tempted by corrupting influences. In addition to foot and automobile patrols, police today use a variety of other methods of patrolling their beats, including bicycles, horses, motorcycles, snowmobiles, and watercrafts.

Foot Patrol

The first formal police patrols in the United States were on foot, a practice carried over from England. Currently, however, only 6 percent of all U.S. patrol officers work their beats on foot. During the reform era of the early twentieth century, foot patrol was stopped for two reasons: (1) response times were slow and (2) officers were too isolated from the department, which permitted the emergence of corruption and differential law enforcement.

Foot patrol actually has several advantages, such as reducing the public's fear of crime, improving community relations, and increasing morale and job satisfaction among police.[30] Foot patrol also provides police with the opportunity to produce "new" information that may help them solve future crimes. Additionally, foot patrol allows officers to establish long-term, face-to-face relationships with citizens; this rapport can later help police when they try to solve crimes, because officers may be able to call upon citizens for assistance.

Foot patrol is regaining its popularity today as departments work to develop closer connections with the public and work with citizens to prevent crime.

Research evaluating the effectiveness of foot patrol shows that this approach generally produces positive results. Two cities in particular—Flint, Michigan, and Newark, New Jersey—have recently implemented large foot patrol programs.[31] In Newark, the results suggest that foot patrol may have only a small effect on the crime rate and may not be a cost-effective approach. In Flint, the foot patrol program has produced more positive results:

- Decreased crime
- Fewer calls to police for assistance
- Increased public satisfaction with police
- Reduced fear of victimization[32]

In fact, at a time when Flint had one of the highest unemployment rates in the country, on three separate occasions its residents voted to continue and expand the foot patrol program and pay the bill through increased taxation.[33] While foot patrol is more likely to reduce citizens' fear of crime, its widespread implementation is too costly for most cities.

Automobile Patrol

The use of police patrol cars first occurred in the heart of the U.S. automobile industry—in Detroit in 1929. As the geographic area to be patrolled expanded, more and more departments implemented automobile patrol. Since the 1930s, police have predominantly patrolled their beats in automobiles. Today motor vehicles are the most widely used method of patrol for several reasons:

- Police can respond to calls more quickly.
- Police can patrol a larger physical area, even patrolling more than one beat.
- Cars protect officers from inclement weather.
- Cars provide police with a shield from bullets and thrown objects.
- Officers in cars may be fully equipped with a radio, first-aid kit, report forms, weapons, and other necessary tools and supplies.
- Cars can be used to confine and transport criminals.

Today, patrol vehicles have been enhanced by technology including the global positioning system (GPS), a satellite-based radio navigation system developed and operated by the U.S. Department of Defense. GPS makes it make it easier for police to track suspects and determine the whereabouts of undercover surveillance officers during emergency situations and search and rescue missions.[34]

The use of automobile patrol revolutionized U.S. policing.

? Which new technological changes will the twenty-first century bring to policing? Should the public be wary of any of these changes?

Automobile patrols may be either one- or two-officer patrols. Police chiefs favor one-officer cars because they are more cost-effective, but unions argue that two-officer vehicles are safer. Research has supported the position of police administrators, finding that officers in one-person vehicles were assaulted less frequently and were less likely to be involved in resisting-arrest incidents than officers in two-person units.[35] Officers in one-person cars also are more productive, make more arrests, write more criminal incident reports, and are more likely to avoid potentially volatile situations when patrolling.[36]

Automobile patrol carries some costs as well. Some critics blame the deteriorating relations between the police and the public on the increased use of automobile patrol, which isolates officers from the public. Additionally, this separation may cause many citizens—especially members of minority groups—to feel alienated from the police.[37]

Police Patrol Strategies

It is common for a police agency to assign more than 60 percent of its personnel to patrol work.[38] Most officers are assigned a particular geographic area, called a beat. An entire collection of beats in a particular geographic area forms a precinct. In small departments, usually one precinct serves as the department's headquarters or station house for the entire agency. In contrast, in New York, where the police department employs more than 38,000 officers who serve 8 million people, there are nearly 80 precincts. Departments adopt one of two patrolling strategies: preventive patrol or directed patrol.

Preventive Patrol

Preventive patrol was introduced in the 1950s by Orlando W. Wilson, then superintendent of the Chicago Police Department. Wilson believed that if the police established an omnipresence in a neighborhood by driving conspicuously marked cars randomly through the city's streets and gave special attention to hot spots of crime (i.e., specific areas with high crime rates), they would deter criminal activity and alleviate the public's fears.[39]

Wilson's idea dominated police patrol operations for two decades, until 1972 when the idea was scientifically tested by George Kelling and his associates. Using a well-developed experimental design, Kelling and his colleagues conducted the Kansas City Preventive Patrol Experiment, in which they gathered crime data from 15 patrol beats, each of which they had assigned to one of three levels of patrol:

1. *Reactive beats*: Police did not patrol and only responded to citizen calls for service.
2. *Proactive beats*: Police regularly patrolled in vehicles at a higher rate than usual (two or three cars per beat).
3. *Control beats*: Police patrolled at regular rates (one car per beat).

They discovered that increasing or decreasing patrol activity in an area had no measurable impact on crime rates, citizens' fear of crime, public attitudes toward police effectiveness, police response time, or the number of traffic accidents.[40] Follow-up studies in Houston, New York, San Diego, and Syracuse produced similar results. Criminologists concluded that preventive patrol made about as much sense as fire fighters driving their trucks around city streets looking for houses in flames.[41] Taken collectively, these findings forced police administrators to conclude that "random patrol produced random results" and to reevaluate police operations.[42]

Directed Patrol

The Kansas City Preventive Patrol Experiment changed the way police administrators viewed the effectiveness of patrol. Initially, they responded to the study's findings by developing alternative methods of patrol. One strategy they introduced was directed patrol, in which police identify hot spots of crime. Another strategy involved the application of geographic information systems (GIS)—that is, systems for capturing, storing, analyzing, and managing data and associated attributes that are spatially referenced to the Earth. With this technology, police are able to see a visual map of the times, offenses, and places where crime most frequently occurs. This kind of crime mapping helps police identify the locations and days and times of major sources of community problems, such as crack houses, prostitution rings, illegal gambling, and gang hangouts. Armed with this knowledge, police dispatchers know the best times to send officers to patrol particular areas. In these areas, police departments saturate the neighborhoods with officers, making them highly visible, and establish decoy units to catch potential offenders, conduct sting operations, and assign special units to track offenders.[43]

Advocates of directed patrol contend that crime decreases when departments aggressively enforce the law by being vigilant and intrusive, adopting a "zero tolerance" stance, and maximizing interrogations and street stops of citizens.[44] Research supports these claims. In an analysis of 171 cities, Robert Sampson and Jacqueline Cohen found that aggressive policing reduced both the incidence (the number of offenses committed) and the prevalence (the number of people committing the crime) of robbery.[45] Other studies have reported that directed patrol has substantially reduced crime rates when it targeted a specific crime in a particular location, such as firearm crimes in areas with high rates of violent crime.[46]

In reality, directed patrol strategies may not be the panacea they once were thought to be. Critics suggest that directed patrol tactics do not reduce crime, but merely displace it from one neighborhood to another. Research testing their concern has shown that this is not true, however. In addition to having a specific deterrent effect (apprehending particular offenders), directed patrol appears to have a general deterrent effect on criminal activity, sending a strong message to would-be offenders.[47]

A second criticism of directed patrol is that it strains relations with innocent citizens who live in neighborhoods targeted by police.[48] While some residents do resent the increased police presence, many others welcome the officers and believe that the increased patrols add to their sense of safety and security.[49]

Whether a strategy of directed patrol is effective depends in part on response time (how long it takes for an officer to arrive at the scene) and reporting time (how much time passes between when a crime was committed and when the police are called). Reducing response time has been found to only slightly increase the likelihood of arrest for serious crimes, because most crimes are reported to police only after the offender has left the scene. If a crime is not reported within 60 seconds of being committed, police generally cannot respond quickly enough to apprehend the suspect.[50] For involvement crimes—crimes in which an offender directly confronts the victim (such as a sexual assault or mugging)—a fast response time has a greater effect.

To help reduce reporting time, the 9-1-1 telephone dispatch system was developed in Alabama in 1973. Unfortunately, it has not had a large effect in decreasing reporting time, only making a difference of about 10 seconds. In some instances, the 9-1-1 system has actually increased reporting time because some citizens—particularly elders—may delay contacting authorities for fear of using the system in an improper situation and angering the police.[51]

Criminal Investigations

During a criminal investigation, the police try to uncover evidence that will assist prosecutors in proving beyond a reasonable doubt whether a suspect committed an offense. First, however, the police must determine whether a crime has been committed, who committed it, and how the crime occurred.

When responding to a call, the first officer on the scene conducts a preliminary investigation that includes possibly arresting a suspect, assisting any victims, securing the crime scene, collecting physical evidence, and writing an initial report. If the crime cannot be solved immediately, a detective is assigned to the case for a follow-up investigation.[52]

The detective division includes investigative officers and, depending on the size of the department, may have forensic laboratories or specialized units that focus on specific types of crimes (i.e., homicide, narcotics). Because crime labs are expensive to oper-

ate and maintain, only federal and state governments and big-city police departments have forensic labs; smaller agencies typically send their forensic evidence to state-run or regional crime centers for analysis.

Within the police department, detectives have higher status and enjoy more prestige and autonomy than patrol officers. The job of the detective is more specialized and encompasses the following responsibilities:

- Interviewing suspects, witnesses, and informants
- Discussing the case with patrol officers, their peers, and supervisors
- Searching crime scenes for physical evidence
- Attending autopsies
- Reviewing state and federal computer databases for clues

In spite of how the popular media and fiction portray crime investigation, detective work is often tedious, routine, mundane, and boring. Detectives spend most of their day writing reports and examining computer files, and they solve only about 3 percent of all crimes they investigate.[53] Even in those cases, detectives rarely discover the key evidence to solve a case, but instead rely on data derived from the preliminary investigation or witness testimony.[54] The majority of cases are, in fact, solved by the interrogation of suspects and witnesses or information provided by informants (often insiders within criminal gangs).

Detectives also solve crimes by examining a suspect's public and private records, which may include:

- Fingerprint records (The Federal Bureau of Investigation (FBI) maintains records of people who have committed felonies, applied for a federal security clearance, or served in the U.S. armed forces as well as some fingerprints of misdemeanor offenders.)
- Records of criminal arrests and convictions
- Photographs or mug shots of persons arrested
- Motor vehicle records
- Credit card and bank statements
- Hotel registration records
- Credit reports
- Computer files

Detectives may also gather forensic evidence, which includes fingerprints, DNA analysis, bloodstains, footprints, tire tracks, and the presence of narcotics. This evidence is sent to crime laboratories, where it is analyzed by scientific experts called criminalists or forensic scientists.[55]

Can Police Patrol Tactics Go Too Far?

When police implement a policy of aggressive patrol, their objective is to directly confront crime. This strategy, however, raises constitutional concerns: In a democratic society, restrictions are placed on police action.

Consider the tactics used in Lawrence, Massachusetts: After police spent months fighting drug dealers in one hot spot of crime with little success, they implemented a new tactic. They closed off four blocks with roadblocks to keep drug customers out and set up a checkpoint at the only intersection left open. Police handed out yellow passes to residents of the neighborhood, which permitted them to come and go freely. All other vehicles were stopped, and the occupants were handed cards warning them of the neighborhood surveillance. Police also recorded the license plate numbers of nonresident vehicles and sent letters to the vehicles' owners, informing them that their cars were stopped at a police checkpoint. The American Civil Liberties Union criticized and challenged the legality of this operation, charging that the police had imposed martial law.

Police disagreed. They argued that aggressive police action was a neighborhood initiative, and most of the residents saw the tactic as what needed to be done to stop drugs and crime. The aggressive approach used by police was ultimately effective: A drug-infested, crime-ridden neighborhood became a safe, quiet community in less than two weeks.

By sealing off a neighborhood and handing out passes to its residents, do the police increase divisions among citizens? Should neighborhoods become barricaded and isolated islands within the larger community? Although this particular neighborhood solved its immediate drug problem, will the solution merely shift the problem to somewhere else?

Sources: John Larrabee, "Mass. Town's Roadblock to Crime," *USA Today*, December 23, 1992, p. 3A; Lawrence Sherman, "Police Crackdowns," *Crime and Justice* 12:1–48 (1990); Mark Kleiman, *Heroin Crackdowns in Two Massachusetts Cities* (Cambridge, MA: Harvard University, 1989); Mark Kleiman, *Crackdowns: The Effects of Intensive Enforcement on Retail Heroin Dealing* (Cambridge, MA: Harvard University, 1988); Mark Kleiman, *Street-Level Drug Enforcement* (Washington, DC: U.S. Department of Justice, 1988).

Forensic evidence has assisted in crime solving for more than 2000 years. Its earliest recorded use occurred in the legend of "Eureka," when Archimedes (287–212 B.C.) used the principles of water displacement to examine the density and buoyancy of a gold crown to prove it was a fake. The first application of fingerprints in crime solving took place in the seventh century, when a print taken from a merchant's bill established the identity of a customer who had failed to pay. The scientific analysis of forensic evidence has improved considerably over the centuries. Today it is routine for criminalists to gather and examine evidence for trial, as when forensic evidence was used to solve the mystery of a presidential scandal through the analysis of semen taken from a dress worn by Monica Lewinsky.[56] As science advances, technology is greatly enhancing the field of forensic science (also called criminalistics) through developments such as DNA analysis, examination of fiber and gunshot residue, and use of gas chromatography-mass spectrometry for analysis of chemicals and other materials left at crime scenes.

Traffic Enforcement

Traffic enforcement encompasses all traffic safety functions, including law enforcement, accident investigation, impoundment of abandoned or stolen vehicles, and roadside sobriety checkpoints. Each year, more than 1 million traffic accidents occur in the United States. More than three times as many Americans are killed in traffic accidents each year than die as the result of criminal activity, and the dollar value of property damaged from traffic accidents is much higher than the value of property stolen or damaged by criminal acts. The traffic situation in the United States gets worse every year as more and more vehicles are registered and placed in operation.[57] Yet, even though approximately 50,000 traffic fatalities have occurred every year since 1980, traffic enforcement receives a low priority in nearly all police departments and does not seem to be much of a concern to the public FIGURE 6-6 .

Citizens do not express much fear about traffic accidents, nor do they demand much from the police

DNA and Criminal Offenders

Deoxyribonucleic acid (DNA) is the fundamental building block of a person's genetic makeup. Because a strand of DNA in one part of the body is exactly the same as any other DNA strand anywhere else in the body, DNA can prove extremely valuable for crime solving, in that it creates a "genetic fingerprint." With DNA evidence, it becomes possible to convict—or to acquit—someone of murder, forcible rape, child molestation, and other serious crimes by examining evidence at the crime scene, such as a cigarette butt (saliva), pubic hair, or a used condom (semen).

DNA testing also enables members of law enforcement to trace suspects who live in different cities and states. The Combined DNA Index System (CODIS) is a database containing the DNA of convicted offenders; law enforcement officials use CODIS records to match suspects with crimes just as they would with fingerprint records. As with fingerprints, the effective use of DNA may require the collection and analysis of elimination samples. That is, it is often necessary to use elimination samples to determine whether the evidence comes from the suspect or from someone else. For instance, in a case of forcible rape, police may need to collect and analyze DNA evidence from the woman's most recent sexual partners to eliminate them as criminal suspects.

DNA may be found on evidence that is many years old, but several factors will affect the usability of DNA discovered at a crime scene. Heat, moisture, bacteria, sunlight, and mold can all render DNA unsuitable for assessment, for example. Also, DNA testing cannot tell police when the suspect was at the crime scene or for how long. Finally, DNA evidence can be easily contaminated.

The first DNA-based exoneration of a convicted individual occurred in 1989. Since then, there have been 203 exonerations; 15 of these people had served time on death row. The average length of time served in prison for exonerees is approximately 12 years. Of the exonerees, 122 have been African Americans, 57 whites, 19 Latinos, and 1 Asian American; the race or ethnicity of 5 exonerees is unknown. The actual offenders have been identified in 76 of the DNA-exonerated cases.

Sources: The Innocence Project, "Facts on Post-Conviction DNA," available at http://www.innocenceproject.org/Content/351.php, accessed June 27, 2007; David Lazer, *DNA and the Criminal Justice System: The Technology of Justice* (Cambridge, MA: MIT Press, 2004); Ross Gardner, *Practical Crime Scene Processing and Investigation.* (New York: CRC Press, 2005); Barry Scheck and Peter Neufeld, *The Innocence Project* (New York: Benjamin N. Cardozo Law School, 2006), available at http://www.innocenceproject.org, accessed June 27, 2007; National Institute of Justice, *What Every Law Enforcement Officer Should Know about DNA Evidence* (Washington, DC: Department of Justice, 1999).

regarding enforcement of traffic laws. When police departments face financial strain, they often take funds away from traffic enforcement, with little public outcry. At the same time, traffic enforcement is the only police function that produces revenues (through fines for violations) and essentially pays for itself. Traffic police also provide intangible benefits by functioning as reserve officers and provide support in emergencies.

Enforcing traffic laws consumes a large amount of time and resources but can be an effective tool for reducing criminal activity, capturing fugitives, and recovering stolen property. Routine traffic enforcement stops have often led to significant arrests and apprehensions for other offenses, such as when an officer from the Oklahoma Highway Patrol stopped Timothy McVeigh for speeding after his bombing of the Alfred P. Murrah Federal Building in Oklahoma City.

Community Services

As society changes, so do the services provided by police. The service function has been an important part of police work for more than 100 years. Typical services include:

- Rendering first aid
- Rescuing animals
- Giving tourists information

Using Computer Technology in Traffic Enforcement

Now in development is a new device that will allow an officer who stops a motorist to touch a button and instantly have access to all available pertinent information about the vehicle. As the scanner behind the vehicle's grill reads the bar-coded license plate, the mobile digital terminal will display complete registration information. It will also check the National Crime Information Center, which will warn the officer if there is a warrant on the plate. By touching a second button, the officer can ensure that the terminal identifies the patrol unit's location through a satellite locator system and sends a signal to the police dispatcher, who will then know the exact location of the officer.

After advising the violator that he or she is being cited for an offense, the officer can position the driver's

license on a screen, with bar-coded information about the motorist on the license then being displayed. If no "red flag" appears, touching a pre-coded button will activate a mini-printer that generates the proper citation to be given to the offender. If the officer has reason to be suspicious of either the motorist or the vehicle, the motorist will be instructed to put his or her finger into a small box and touch a screen surface inside; this device will scan the state's automated fingerprint identification system file to confirm the driver's identity.

Source: Katie Hafner, "Wanted by the Police: A Good Interface," *New York Times,* November 11, 2004, p. 4, available at http://www.nytimes.com/2004/11/11/technology/circuits/11cops.html?ex=1257829200&en=0f1603bf6b60fb9d&ei=5090, accessed July 30, 2007.

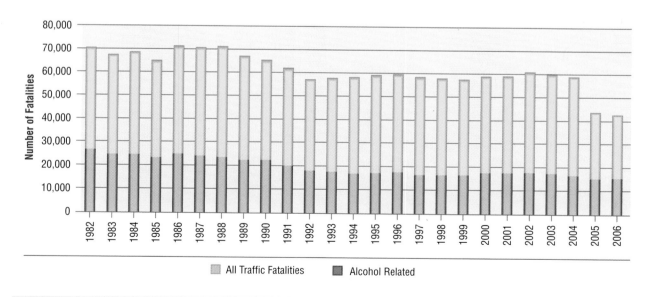

FIGURE 6-6 Motor Vehicle Fatalities

Source: U.S. Bureau of the Census, *Statistical Abstract of the United States, 2006,* 125th edition (Washington, DC: U.S. Department of Commerce, 2006); Transportation Board of the National Academies, *2006 Estimates of Motor Vehicle Traffic Crash Fatalities and People Injured,* available at http://www.trb.org/news/blurb_detail.asp?id=7748, accessed June 28, 2007.

- Providing roadside assistance
- Finding lost pets
- Checking door locks on vacationers' homes
- Opening doors for people who are locked out of their vehicles

Before 1900, police provided lodging and food for the needy as an informal way of controlling the poor. This practice was strongly criticized by social welfare reformers, who argued that it perpetuated poverty. One critic was Theodore Roosevelt, who believed this policy contributed to crime. He thought crime was class based and suggested that the first step in eliminating the "dangerous class" should be to end police lodging. Roosevelt also argued that police had better things to do than to dispense charity, such as patrolling the streets and apprehending crim-

LINK Roosevelt's thinking laid the groundwork for the early stirrings of the police professionalization movement (discussed in Chapter 5), which began in the twentieth century. Central to that movement was a shift in society's definition of the police role from one of controlling dangerous people to one of controlling crime.

inals. He also wanted police to distance themselves from the urban poor. In the 1890s, Roosevelt was instrumental in bringing an end to the practice of police lodging in New York City.[58]

Police today deliver services to citizens as part of their overall crime-fighting strategy. These services generate goodwill with members of the public, who, in turn, may be more motivated to help police solve crimes by providing information. James Q. Wilson, however, disagrees with this position. He suggests police should not provide services that help only a few individuals, but instead should let private industry assume this responsibility.[59] Wilson believes that providing these kinds of services drains police departments of scarce resources and is not cost-effective for the agency.

Interestingly, even proponents of community policing find merit in Wilson's position. Criminologist Herman Goldstein, for example, says that for community policing to be successful, police must redefine and streamline their role.[60]

It is difficult to predict how a reduction of police services might affect community relations. Police agencies across the country are currently trying to strike a balance between the two positions: They want to streamline the services they offer, while offering the services the public wants.

Despite the excitement occasionally associated with law enforcement, police departments are primarily community service agencies.

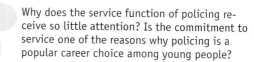

Why does the service function of policing receive so little attention? Is the commitment to service one of the reasons why policing is a popular career choice among young people?

WRAP UP

Law enforcement everywhere is galvanized whenever a police officer is attacked. After the two Long Beach officers were shot, an intensive manhunt was undertaken to find their assailant. After officers received information indicating that Gallegos was the perpetrator, this manhunt was transformed into an intense surveillance operation. When Gallegos attempted to use force again, officers believed they were left with few options except to shoot. Tactically, this decision made sense. However, the use of lethal force against a potential "cop killer" was also in line with the solidarity that binds police officers in a strong, coherent subculture. This incident shows that when felons use violence against police, the response is often swift—and harsh—from the law enforcement community.

Chapter Spotlight

- Police departments have both formal and informal structures that guide the activities of their members.

- The formal structure is bureaucratic, stressing adherence to established rules and regulations, and characterized by a chain of command, clear delegation of authority, division of labor, and an emphasis on competency with regard to hiring and promotion.

- The bureaucratic model has several shortcomings. In particular, its limited rewards create obstacles for females and minorities when they seek to be hired and to be promoted.

- The informal police structure is organized around a set of beliefs, values, and behavior patterns that set officers apart from the rest of society. This police subculture produces group solidarity and reinforces the development of a unique working personality among officers, which may breed cynicism.

- The law enforcement function of policing consists of three major activities: patrol, crime investigation, and traffic enforcement.

- In addition to enforcing laws, police provide the public with community services, including rendering first aid, rescuing animals, and giving tourists information.

Putting It All Together

1. Traditional hiring and promotion practices in police departments have been criticized because they limit opportunities for racial and ethnic minorities and for women. Should quota systems be used to create diversity in police departments?

2. What kinds of problems are associated with the existence of the police subculture? Can anything be done to reduce them?

3. Aggressive patrol tactics have reduced crime rates in certain hot spots. However, some citizens complain that these tactics cause police to target neighborhoods with higher concentrations of people living in poverty and racial and ethnic minorities. How should departments balance the priorities of effective crime control with the potential biasing effects of such policies?

4. Police officers complain that many citizens are unpleasant and resentful when they are stopped for traffic violations. Why do people react this way toward officers? How might police make these encounters less confrontational?

5. Given the large number of traffic fatalities that occur each year, should a larger percentage of the police budget be allocated to traffic law enforcement and taken from the crime-fighting and service functions?

beat An assigned area of police patrol.

bureaucracy A model of organization in which strict and precise rules are used as a way of effectively achieving organizational goals.

chain of command A hierarchical system of authority that prescribes who communicates with (and give orders to) whom.

competency A list of factors that reflect abilities or skills, including qualifications, test scores on promotional exams, and field performance.

crime mapping Computerized mapping by address of crime occurrences which helps police identify the locations and days and times of major sources of community problems.

criminal investigation The process of searching for evidence to assist in solving a crime.

criminalists Scientists who work in crime laboratories and examine forensic evidence, which includes fingerprints, DNA analysis, bloodstains, footprints, tire tracks, and the presence of narcotics.

delegation of authority Decision making made through a chain of command in a bureaucracy.

detective division A police division consisting of investigative officers, and possibly a forensic laboratory or specialized unit that focuses on specific types of crime (i.e., homicide, narcotics).

directed patrol A patrol technique in which officers are given specific instructions on how to use their patrol time.

division of labor A system of assigning duties for the routine jobs completed in bureaucracies.

forensic evidence Physical evidence found at a crime scene, including such things as fingerprints, DNA analysis, bloodstains, footprints, tire tracks, and the presence of narcotics.

geographic information systems (GIS) A system for capturing, storing, analyzing, and managing data and associated attributes that are spatially referenced to the Earth.

global positioning system (GPS) A satellite-based radio navigation system.

hot spots of crime Locations characterized by high rates of crime.

involvement crimes Crimes in which the offender directly confronts the victim, such as an armed mugging.

Kansas City Preventive Patrol Experiment A study done to assess how allocating patrol at different levels of enforcement affected the crime rate and perceptions of public safety.

operational style The way in which police officers interact with fellow officers and the public.

patrol Police responsibility to move through assigned areas by foot or vehicle to enforce laws, regulate traffic, control crowds, prevent crime, and arrest violators.

police cynicism Belief of police that people are selfish and motivated by evil.

police operations Services that police agencies provide and the methods they use to deliver these services.

police subculture Beliefs, values, and patterns of behavior that separate officers from police administrators and the public.

precinct The entire collection of police beats in a specific geographic area.

preventive patrol A crime control strategy based on the idea that crime is deterred by the mere presence of police.

reporting time The time lag between when a crime is committed and when the police are called.

response time The time it takes for police to respond to a call.

service function Role of police to assist citizens with noncriminal matters, such as providing emergency medical assistance.

specialization The practice of dividing work among employees for it to be completed more effectively and efficiently.

traffic enforcement Police duties related to highway and traffic safety and accident investigations.

working personality A term that distinguishes an officer's off-the-job persona from his or her on-the-job behavior.

1. Jack Green and Carl Klockars, "What Police Do," in Carl Klockars and Stephen Mastrofski (Eds.), *Police: Contemporary Readings,* 2nd ed. (New York: McGraw-Hill, 1991), pp. 273–284.

2. Max Weber, *The Theory of Social and Economic Organizations,* translated by Talcott Parsons (New York: Free Press, 1947).

3. John P. Crank, *Understanding Police Culture* (Cincinnati: Anderson Publishing, 2003); Jerome Skolnick, *Justice Without Trial,* 3rd ed. (New York: Macmillan, 1994).

4. Matthew Hickman and Brian Reaves, *Local Police Departments, 2003* (Washington, DC: U.S. Department of Justice, 2006).

5. Hickman and Reaves, note 4.

6. Hickman and Reaves, note 4.

7. Hickman and Reaves, note 4.

8. Thomas Barker and David Carter, *Police Deviance,* 3rd ed. (Cincinnati: Anderson Publishing, 1993); Victor Kappeler, Richard Sluder, and Geoffrey Alpert, *Forces of Deviance,* 2nd ed. (Prospect Heights, IL: Waveland Press, 1998); Victor Kappeler, *Police Civil Liability,* 2nd ed. (Prospect Heights, IL: Waveland Press, 2006); Victor Kappeler, *Critical Issues in Police Civil Liability,* 4th ed. (Prospect Heights, IL: Waveland Press, 2006).

9. Hickman and Reaves, note 4; William Helreich, *The Things They Say Behind Your Back* (New Brunswick, NJ: Transaction Books, 1997); Geoffrey Alpert, Roger Dunham, and Meghan Stroshine, *Policing: Continuity and Change* (Long Grove, IL: Waveland University Press, 2006); James Conser, Gregory Russell, Rebecca Paynich, and Terry Gingerich, *Law Enforcement in the United States,* 2nd ed. (Sudbury, MA: Jones and Bartlett Publishers, 2005); Samuel Walker and Charles Katz, *The Police in America,* 6th ed. (New York: McGraw-Hill, 2008); Larry Gaines and Victor Kappeler, *Policing in America* (Cincinnati: Lexis/Nexis, 2005).

10. Alpert et al., note 9; Conser et al., note 9; Walker and Katz, note 9; Gaines and Kappeler, note 9.

11. Alpert et al., note 9; Conser et al., note 9; Walker and Katz, note 9; Gaines and Kappeler, note 9.

12. Elizabeth Reuss-Ianni, *Two Cultures of Policing* (New Brunswick, NJ: Transaction Books, 1982).

13. William Westley, *Violence and the Police* (Cambridge, MA: MIT Press, 1971).

14. Steve Herbert, "Police Subculture Revisited," *Criminology* 36:343–369 (1998).

15. Eugene Paoline, *Rethinking Police Subculture* (New York: LFB: Scholarly Publishing, 2001).

16. Skolnick, note 3.

17. Arthur Niederhoffer, *Behind the Shield* (New York: Doubleday, 1967).

18. Kraig Hays, Robert M. Regoli, and John D. Hewitt, "Police Chiefs, Anomie, and Leadership," *Police Quarterly* 10:3–22 (2007); Ann Parker, *Differential Use of Discretionary Powers,* Ph.D. dissertation, University of South Australia, 2005; Robert M. Regoli, Robert Culbertson, and John P. Crank, "Using Composite Measures in Police Cynicism Research," *Journal of Quantitative Criminology* 7:41–58 (1991); Robert M. Regoli, Robert Culbertson, John P. Crank, and James Powell, "Career Stage and Cynicism among Police Chiefs," *Justice Quarterly* 7:701–722 (1990); Robert M. Regoli, John P. Crank, and Robert Culbertson, "Police Cynicism, Job Satisfaction, and Work Relations of Police Chiefs," *Sociological Focus* 22:161–172 (1989).

19. James Q. Wilson, *Varieties of Police Behavior* (Cambridge, MA: Harvard University Press, 1968).

20. John Broderick, *Police in a Time of Change,* 2nd ed. (Prospect Heights, IL: Waveland Press, 1986).

21. Broderick, note 20.

22. Gad Barzical, *Communities and Law* (Ann Arbor, MI: University of Michigan Press, 2003); William Lyons, *The Politics of Community Policing* (Ann Arbor, MI: University of Michigan, 1999); Charles Bonjean, *Community Politics* (New York: Free Press, 1971).

23. Ellen Hochstedler, "Testing Types: A Review and Test of Police Types," *Journal of Criminal Justice* 9:451–466 (1981).

24. Read Bain, "The Policeman on the Beat," *Science Monthly* 48:5 (1939).

25. Carl Klockars, *The Idea of Police* (Beverly Hills, CA: Sage Publications, 1985), pp. 15–16.

26. Elaine Cumming, Ian Cumming, and Laura Edell, "Policeman as Philosopher, Friend, and Guide," *Social Problems* 12:276–286 (1965); Wilson, note 19; Dorothy Guyot, *Policing as Though People Mattered* (Philadelphia: Temple University Press, 1991); Thomas Bercal, "Calls for Police Assistance," *American Behavioral Scientist* 13:681–691 (1970); Stephen Mastrofski, "The Police and Non-Crime Services," in Gordon Whitaker and Charles Phillips (Eds.), *Evaluating Performance of Criminal Justice Agencies* (Beverly Hills, CA: Sage Publications, 1983), pp. 33–62; Albert Reiss, Jr., *The Police and the Public* (New

Haven, CT: Yale University Press, 1971); Green and Klockars, note 1; Herman Goldstein, *Policing a Free Society* (Cambridge, MA: Ballinger Publishing, 1977), p. 35.

27. Alpert et al., note 9.
28. Orlando W. Wilson and Roy McLaren, *Police Administration*, 4th ed. (New York: McGraw-Hill, 1977); James Fyfe, Jack Green, William Walsh, and Orlando W. Wilson, *Police Administration*, 5th ed. (New York: McGraw-Hill, 1996).
29. Hickman and Reaves, note 4.
30. Lawrence Sherman, "Police Crackdowns," *NIJ Reports* (Washington, DC: National Institute of Justice, 1990).
31. The Police Foundation, *The Newark Foot Patrol Experiment* (Washington, DC: The Police Foundation, 1981); Robert Trojanowicz, *An Evaluation of the Neighborhood Foot Patrol Program in Flint, Michigan* (East Lansing, MI: Michigan State University Press, 1982); Robert Trojanowicz and Dennis Banas, *The Impact of Foot Patrol on Black and White Perceptions of Policing* (East Lansing, MI: Michigan State University Press, 1985).
32. The Police Foundation, note 31; Trojanowicz, note 31; Trojanowicz and Banas, note 31.
33. The Police Foundation, note 31; Trojanowicz, note 31; Trojanowicz and Banas, note 31.
34. Donna Rogers, "GPS Gains a Stronger Position," *Law Enforcement Technology*: 64–68 (June 2001).
35. William Spelman and Dale Brown, *Calling the Police* (Darby, PA: Diane Publishing Company, 1984).
36. Spelman and Brown, note 35.
37. Laure Weber Brooks, "Police Discretionary Behavior," in Roger Dunham and Geoffrey Alpert (Eds.), *Critical Issues in Policing*, 5th ed. (Long Grove, IL: Waveland Press, 2005), pp. 89–105.
38. Hickman and Reaves, note 4.
39. Orlando W. Wilson, *Police Administration*, 3rd ed. (New York: McGraw-Hill, 1977).
40. George Kelling, Tony Pate, Duane Dieckman, and Charles Brown, *The Kansas City Preventive Patrol Experiment* (Washington, DC: The Police Foundation, 1974).
41. Carl Klockars, *Thinking About Police* (New York: McGraw-Hill, 1983).
42. Henry Wroblewski and Karen Hess, *Introduction to Law Enforcement and Criminal Justice*, 8th ed. (Belmont, CA: Thomson/Wadsworth, 2006).
43. Herman Goldstein, *Problem-Oriented Policing* (Philadelphia: Temple University Press, 1990); Anthony Braga, *Problem-Oriented Policing and Crime Prevention* (Monsey, NY: Willow Tree Press, 2002).
44. James Q. Wilson and Barbara Boland, "The Effect of Police on Crime," *Law and Society Review* 12:367–387 (1978).
45. Robert Sampson and Jacqueline Cohen, "Deterrent Effects of Police on Crime," *Law and Society Review* 22:163–190 (1988).
46. Edmund McGarrell, Steven Chermak, Alexander Weiss, and Jeremy Wilson, "Reducing Firearms Violence Through Directed Patrol," *Criminology and Public Policy* 1:119–148 (2001); Michael Scott, *The Benefits and Consequences of Police Crackdowns* (Washington, DC: U.S. Department of Justice, 2003).
47. Scott, note 46.
48. Robert M. Regoli, "Directed Patrol, Hot Spots of Crime, and Community Satisfaction," unpublished manuscript, 2008.
49. Regoli, note 48; Lawrence Sherman, "Police Crackdowns," *NIJ Reports* (Washington, DC: National Institute of Justice, 1990).
50. Spelman and Brown, note 35.
51. Spelman and Brown, note 35.
52. James Poland, "Detectives," in William Bailey (Ed.), *The Encyclopedia of Police Science* (New York: Garland Publishing, 1995), p. 142.
53. Goldstein, note 26; Goldstein, note 43; Herman Goldstein, "The New Policing," paper presented at the National Institute of Justice Conference on Community Policing, 1993.
54. Keith Wingate, "The O.J. Simpson Trial: Seeing the Elephant," *Hastings Women's Law Journal* 19:121–133 (1996).
55. Michael Baden and Marion Roach, *Dead Reckoning: The New Science of Catching Killers* (New York: Simon & Schuster, 2001); Stuart Kind and Michael Overman, *Science Against Crime* (New York: Doubleday and Company, 1972); Joe Nickell and John Fischer, *Crime Science: Methods of Forensic Detection* (Lexington, KY: University Press of Kentucky, 1999).
56. Edward Conners, *Convicted by Juries, Exonerated by Science* (Darby, PA: Diane Publishing Company, 1996); David Eggers, Lola Vollen, and Scott Turow, *Surviving Justice* (San Francisco: McSweeny's Books, 2005); David Lazer, *DNA and the Criminal Justice System* (Cambridge, MA: MIT Press, 2004); Saundra Westervelt and John Humphrey, *Wrongly Convicted* (New Brunswick, NJ: Rutgers University Press, 2001); Edmund Morris, *The Rise of Theodore Roosevelt* (New York: Random House, 2001).
57. John Moffart and Philip Salzberg, *The FBI Law Enforcement Bulletin* 68:18–20 (1999).
58. Eric Monkkonen, *Police in Urban America, 1860–1920* (Cambridge, UK: Cambridge University Press, 1981), p. 107.
59. Wilson, note 19, p. 5.
60. Goldstein, note 53, p. 5.

Interactives

Key Term Explorer

Web Links

OBJECTIVES

- Understand what constitutes a legal search and seizure.
- Know the exclusionary rule, its significance, and its importance to the U.S. criminal justice system.
- Identify the exceptions that govern warrantless searches and seizures.
- Outline the process of arrest, including obtaining an arrest warrant.
- Explain the *Miranda* warning and the controversy that surrounds it.
- Describe the rules that police must follow during booking, lineup, interrogation, and confession.

Police and the Law

CHAPTER

7

Cases involving the abduction, rape, and murder of children are among the most horrifying to pass through the U.S. criminal justice system. They are precisely the cases that can least afford being weakened by legal technicalities. Yet, in the case of John Couey—the convicted kidnapper, rapist, and murderer of nine-year-old Jessica Lunsford—police misconduct ruined crucial evidence in the case, making a taped confession inadmissible. While being questioned by police detectives, Couey confessed to the crime and directed police to Jessica's body, which was buried in his lawn. But Couey also told police that he wanted to consult with an attorney—an opportunity he was denied. Because his right to counsel was violated, a judge ruled that the initial confession could not be admitted as evidence. The following year, a second confession by Couey was also ruled inadmissible, after police again continued questioning Couey after he requested a lawyer.

In court, Couey pleaded not guilty to charges of first-degree murder, sexual battery, kidnapping, and burglary. Even without the confessions, strong physical evidence linked Couey to the murder, and he was convicted of kidnapping, rape, and murder. On August 24, 2007, John Couey was sentenced to death for his crimes.

- Which constitutional issues are raised by this thrown-out confession?

- Why should apparently guilty defendants be protected by legal technicalities?

Source: Associated Press, "Statements in Lunsford Girl's Killing Thrown Out," January 8, 2007, available at http://www.msnbc.msn.com/id/16526588/, accessed March 3, 2007; CNN.com, "Covey Jury Wants Death for Florida Girl's Killer," http://www.cnn.com/2007/Law/03/14/Covey.sentence/, accessed September 19, 2007.

Introduction

Police perform a wide variety of functions, one of which is criminal investigation. During criminal investigations, police frequently arrest suspects and gather evidence for prosecutors to use. However, in a democratic society where individual liberty is highly valued, police are held accountable for their actions and behavior. Formal limitations are placed on the rights of police to search, seize, arrest, and interrogate citizens. Criminal procedure law prescribes how the government enforces criminal law and protects citizens from overzealous police, prosecutors, and judges.

Police in the United States have been granted many powers, including the right to stop, frisk, question, and detain citizens. In accordance with the law, they may also arrest individuals, use force against them, and search and seize a person and his or her property. At the same time, police must respect the rights of the accused. An officer who violates the constitutional rights of a suspect may be reprimanded by his or her superiors, be reassigned to a less desirable position, be suspended without pay, and, in extreme cases, be criminally prosecuted. Sometimes, police must choose between adhering to the Constitution and allowing criminal activity to occur.

Because all persons are considered innocent until proven guilty in the United States, even criminal suspects have important, inviolate constitutional rights, granted by the following constitutional amendments:

- Fourth Amendment: A person cannot be searched or have his or her property seized, except in ways that are consistent with the law.

- Fifth Amendment: No person can be forced to say anything that would help convict him or her of a crime (self-incrimination).

- Sixth Amendment: A person accused of a crime may have counsel appointed for his or her defense.

- Fourteenth Amendment: States may not violate the aforementioned rights.

LINK These amendments (discussed in detail in Chapter 10) guide and restrict police behavior during searches and seizures, arrests, bookings, lineups, interrogations, and confessions to protect the rights of the accused.

Search and Seizure

Before the police may initiate official action against a citizen, they must establish probable cause—that is, they must uncover enough evidence for a reasonable person to conclude a crime has been committed by a particular individual. Once probable cause has been established, police may search for additional evidence and seize (take into police custody) objects relating to the crime. Additionally, people may be legally seized when they are held in police custody at a crime scene and are not free to leave.[1]

To comply with the Fourth Amendment prohibition against unreasonable searches and seizures, police must receive voluntary consent for search and seizure from the individual or property owner or an authorized third party.[2] If permission is not granted or police do not request it, they must obtain a search warrant, a written order that grants permission for search and seizure, outlines the specific location of a certain property or person(s) relating to a crime, and requires police to account for the results to the judicial officer.

To obtain a warrant, police must present an affidavit for search warrant (a document that outlines the evidence against the suspect and the circumstances of the crime) to a judicial officer. This officer then evaluates all available information when deciding whether a search warrant should be issued. According to the totality-of-the-circumstances rule, the judicial officer makes a judgment based on all information outlined in the affidavit and determines whether there is a good probability that what the police are looking for will be found in its proposed location.

LINK Search warrants are a vital component of the criminal justice system, ensuring that the rights of the accused are protected while giving law enforcement and criminal prosecutors the tools they need to make sure guilty parties are brought to justice. Because of the crucial nature of search warrants, reviewing and signing warrants is one of the key responsibilities of judges, as discussed in Chapter 9.

Exclusionary Rule

The exclusionary rule prohibits the introduction of illegally obtained evidence, including confessions, into a criminal trial to protect the constitutional rights of citizens by restricting police powers. The U.S. Supreme Court established the exclusionary rule in 1914, in its ruling in *Weeks v. United States*.[3] In this case, the defendant, Freemont Weeks, was accused of using the U.S. mail system for illegal purposes and was arrested without a warrant. After his arrest, police searched his residence—again without a warrant—and found incriminating papers, which they confiscated. The papers were given to a U.S. Marshal, who—again without a warrant—searched Weeks's home a second time, finding and seizing additional incriminating documents. At Weeks's trial, his lawyer petitioned the court to have the materials returned. The request was denied, and Weeks was convicted.

On appeal to the Supreme Court, Weeks's conviction was reversed on the grounds that the procedures utilized by the police were unconstitutional under the Fourth Amendment. According to the Court, "If letters and private documents can be seized and used as evidence against a citizen accused of an offense, the protection of the Fourth Amendment, declaring his right to be secure against such searches and seizures, is of no value."

The ruling in *Weeks* had a monumental effect in changing the rules governing the admissibility of evidence, but it was limited to actions taken by federal officers in federal cases. Even then, evidence that had been illegally seized by state officers remained admissible in federal courts. Furthermore, the Supreme Court said nothing about what evidence could be used in state cases or how that evidence might be secured. This ruling led to the silver platter doctrine, so called because it allowed officers in one jurisdiction to hand over "on a silver platter" evidence that had been illegally obtained by officers in another jurisdiction to use in court.

Between 1920 and 1960, in a series of court decisions, the silver platter doctrine was gradually eroded. In 1920, for example, in *Silverthorne Lumber Company v. United States*, the U.S. Supreme Court created the "fruits of the poisonous tree" doctrine, which excluded from trial any evidence the police had secured through illegal practices, such as a forced confession.[4] In 1956, in *Rea v. United States*, the Court decided that federal officers could no longer provide state officers with evidence they had illegally seized for use in state courts.[5] Finally, in 1960, in *Elkins v. United States*, the Court announced that state officers could not provide federal officers with evidence they had seized illegally to be used in federal courts.[6]

VOICE OF THE COURT — *Illinois v. Gates*

In *Illinois v. Gates* (1983), the U.S. Supreme Court considered whether probable cause could be based solely on hearsay evidence. The Court ruled that it depends on the "totality of the circumstances" surrounding the case. Prior to *Gates,* probable cause for a search warrant could not be based solely on hearsay information. Under this restriction, information received by the police from an informant had to be corroborated if it was to be used to establish probable cause and thereby secure an affidavit for a search warrant. In *Illinois v. Gates,* however, the Supreme Court held that a search warrant could be issued on the basis of hearsay evidence.

In this case, the Bloomingdale, Illinois, Police Department received a handwritten letter in the mail on May 3, 1978, that read:

> This letter is to inform you that you have a couple in your town that strictly makes their living on selling drugs. They are Sue and Lance Gates; they live on Greenway, off Bloomingdale Road in the condominiums. Most of their buys are done in Florida. Sue, his wife, drives their car to Florida, where she leaves it to be loaded up with drugs, then Lance flies down and drives it back. Sue flys [sic] back after she drops the car off in Florida. May 3 she is driving down there again and Lance will be flying down in a few days to drive it back. At the time Lance drives the car back he has the trunk loaded with over $100,000.00 in drugs. Presently they have over $100,000.00 worth of drugs in their basement. They brag about the fact that they never have to work, and make their en-

tire living on pushers. I guarantee that if you watch them carefully you will make a big catch. They are friends with some big drug dealers, who visit their house often.

Bloomington police acted on the tip. They discovered that Lance Gates had made plane reservations for a flight and made arrangements with an agent of the Drug Enforcement Administration to observe Gates. Lance Gates took the flight, stayed overnight in a motel room registered in Sue Gates's name, and left the following morning in a car also registered to Sue Gates. On the basis of this information and the anonymous tip about the illegal drug activity, officers secured a search warrant for the Gates's residence and automobile. When Lance Gates arrived at his home, the police were waiting. When the police searched both the home and the car, they found marijuana and other contraband. The couple was arrested and charged with violating drug laws.

Before trial, the suspects moved to suppress the evidence seized during the search on the grounds that the search warrant was based on hearsay information. The lower court and the Illinois Supreme Court agreed that the evidence should be suppressed. The state of Illinois then appealed the case to the U.S. Supreme Court. After examining the totality of the circumstances, the Court reversed the lower court's decision and ruled the affidavit for a search warrant had established probable cause.

Source: Illinois v. Gates, 462 U.S. 213 (1983).

Even though the ruling in *Elkins* ended the silver platter doctrine, it was not until 1961, in <u>Mapp v. Ohio</u>, that the Court took definitive action and extended the exclusionary rule, making evidence that was obtained illegally inadmissible in both state and federal trials.[7] The exclusionary rule has had a profound effect on police practices. Fearing that cases would be lost due to technicalities, police administrators reacted by increasing the quantity and quality of training officers receive. As a result of these efforts, relatively few criminal suspects now go free because of mistakes in evidence collection. The ex-

clusionary rule also prompted Americans to question more fully the nature of police actions and the ways in which these actions may relate to violations of their constitutional rights.

While the exclusionary rule means that no one— not even the government and its agents—is above the law, critics fear that it may cause the public to lose respect for the entire criminal justice system because it sometimes means that guilty people go free because of "technicalities." Critics also believe that the exclusionary rule further clogs an already congested criminal court system, because attorneys may

VOICE OF THE COURT — *Mapp v. Ohio*

On May 23, 1957, three police officers arrived at the residence of Dollree Mapp, having received information that a suspect in a recent bombing would be found at her home. The police were also told by an informant that Mapp was concealing a large amount of gambling paraphernalia. The officers knocked on Mapp's door and demanded entrance, but Mapp, after telephoning her attorney, refused to admit them without a search warrant. Officers advised their headquarters of the situation and undertook surveillance of the home.

Three hours later, the officers again sought entrance into Mapp's house. When Mapp did not come to the door immediately, the police forcibly opened the door and entered. Mapp demanded to see the search warrant, and an officer held up a piece of paper that he claimed to be a warrant. Mapp grabbed the paper and placed it in her bosom. A struggle ensued in which the officers recovered the piece of paper and handcuffed Mapp, claiming she had resisted their official rescue of the "warrant" from her person. The police forcibly took Mapp upstairs

to her bedroom, where they thoroughly searched Mapp's home and personal belongings.

Police found neither the bombing suspect nor the gambling equipment, but they did find some pornographic materials, which were illegal under Ohio law at that time. Police arrested Mapp and charged her with possessing "lewd and lascivious books, pictures, and photographs." At trial, the prosecution did not produce a search warrant, nor was the failure to produce one ever explained or accounted for. The Court determined that there was "considerable doubt as to whether there ever was any warrant for the search of [Mapp's] home." In spite of the doubt, Mapp was convicted.

On appeal, the U.S. Supreme Court held that police violated Mapp's rights under the Fourth Amendment. Perhaps most importantly, the Supreme Court's ruling extended the application of the exclusionary rule to state criminal cases.

Source: Mapp v. Ohio, 367 U.S. 643 (1961).

---

file motions to suppress evidence that they falsely claim was illegally obtained, thereby obstructing the criminal judicial process TABLE 7-1 .

The Supreme Court agreed that the exclusionary rule must have some exceptions to avoid these potential concerns. First, in 1965, the Court held that the *Mapp* ruling would not be applied retroactively; in other words, this decision could not be used to throw out previously decided cases.[8] Then, in 1984, the Court ruled in *Nix v. Williams* that evidence that was obtained illegally but could have eventually been discovered by lawful means is admissible in court—an exception known as the inevitable discovery rule.[9] That same year, the Court established the good faith exception in *United States v. Leon*, which allows evidence collected illegally to be admitted at trial if the police had good reason to believe that their actions were legal.[10]

In 1987, the Court further expanded the good faith exception to include state searches and seizures in its ruling in *Illinois v. Krull*.[11] Later, in *Arizona v. Evans*, the Court expanded the exception to apply to evidence seized during an arrest. Specifically, in this

case, the evidence resulted from an inaccurate computer record indicating there was an outstanding arrest warrant for the suspect in question.[12] The exception applied regardless of whether police or court personnel were responsible for the erroneous record's continued presence in the police computer. Given the proliferation of computers in society and criminal justice, *Arizona v. Evans* is sometimes referred to as the "computer errors exception" to the exclusionary rule.[13]

Finally, in 2006 in a 5–4 decision in *Hudson v. Michigan*, the Supreme Court ruled against the knock-and-announce rule, which had required police to announce their presence and wait about 20 seconds before entering a home, arguing that this provision could mean the destruction of evidence and the potential escape of dangerous criminals.[14] In this case, the Court held that drugs or other evidence seized during "no-knock" searches (in which police did not knock and announce their presence) could be admitted in trial. This ruling reversed a near-century-old precedent established in *Weeks*, which says evidence found during search without a warrant cannot be used. The Court also ruled that officers may

TABLE 7-1

Exceptions to the Search Warrant Requirement

When police want to execute a search of a person or property, a search warrant is usually required. The U.S. Supreme Court has recognized that sometimes it is impossible for police to secure a search warrant at all or prior to their search. Exceptions to the search warrant requirement, along with the case in which the exception arose, are presented here in the order that they appear in this chapter.

Exception	Cases
School searches	*New Jersey v. T.L.O.* (1985)
Incident to arrest	*Chimel v. California* (1969)
Stop and frisk	*Terry v. Ohio* (1968)
Motor vehicle searches	*Carroll v. United States* (1925)
Plain view doctrine	*Harris v. United States* (1968)
Open field searches	*Hester v. United States* (1924)
Consent searches	*Schneckloth v. Bustamonte* (1973)
Abandoned property	*California v. Greenwood* (1988)
Border searches	*United States v. Martinez-Fuerte* (1976)
Electronic surveillance	*Olmstead v. United States* (1928)

In 2006, the Supreme Court held in *Hudson v. Michigan* that no-knock searches of homes were constitutional.

? Would requiring police to knock and announce themselves give an advantage to criminals? Should police powers sometimes trump certain civil liberties?

conduct legal search and seizures without either consent or a search warrant under limited circumstances.

School Searches

In 1985 in <u>*New Jersey v. T.L.O.*</u>, the U.S. Supreme Court ruled that school officials can conduct warrantless searches of students at school on the basis of reasonable suspicion (that which would lead a reasonable person to believe a crime has been or is about to be committed).[15] In this case, a 14-year-old high school student was caught smoking cigarettes in a school lavatory, in violation of school policy. When the student was taken to the principal's office, she denied that she had been smoking and informed the principal that she did not smoke at all. The principal opened the student's purse without her consent and found a pack of cigarettes and rolling papers of the type often used to make marijuana cigarettes (commonly called joints). The principal proceeded to search the purse thoroughly and found marijuana, a pipe, plastic bags, a fairly substantial amount of money, an index card with a list of students' names who owed this student money, and two letters that implicated the student in marijuana dealing. The principal reported the student to the police, who filed delinquency charges against the student in juvenile court. The judge in the case denied the student's motion to suppress the evidence found in her purse due to claims of unreasonable search and seizure, instead ruling that school officials had a right to search students' property in cases of reasonable suspicion.

"One Arm's Length" Rule

When the police make an arrest, they are permitted, without a warrant, to search the suspect as well as the immediate area (colloquially referred to as "one arm's length") that he or she occupies, albeit to a limited extent. This rule seeks to place limits on warrantless searches while preventing the suspect from using a weapon or destroying evidence.

The "one arm's length" rule was established in 1969 in <u>*Chimel v. California*</u>.[16] In this case, three police officers arrived at the home of Ted Chimel in Santa Ana, California, with a warrant for his arrest in connection with the burglary of a coin shop. Police knocked on the door, identified themselves to Chimel's wife, and asked if they could come inside. She let the officers in, where they waited for Chimel to return from work. When Chimel entered the house, one of the officers handed him the arrest warrant and asked if he could look around. Chimel refused, but the officer insisted that they search the premises without permission or a search warrant on the basis of a lawful arrest. The police seized numerous items in their search that included some coins that were admitted as evidence in Chimel's trial and led to his conviction.

On appeal, the U.S. Supreme Court reversed Chimel's conviction, ruling that the police had no constitutional justification to search his residence. Justice Potter Stewart explained that, when an arrest is made, it is "reasonable for the arresting officer to search the person arrested in order to remove any weapons that the latter might seek to use in order to resist or affect his escape" to protect the safety of the officer. Additionally, the Court ruled, it is "entirely reasonable for the arresting officer to search for and seize any evidence on the arrestee's person in order to prevent its concealment or destruction," noting that a gun on the table in front of a suspect can be just as dangerous as one in the suspect's clothing. Therefore, Justice Potter held:

> There is ample justification; therefore, for a search of the arrestee's person and the area "within his immediate control." . . . There is no comparable justification, however, for routinely searching rooms other than that in which the arrest occurs. . . . Such searches, in the absence of well-recognized exceptions, may be made only under the authority of a search warrant.

Stop and Frisk Rule

In 1968 in _Terry v. Ohio_, the Supreme Court considered the issue of what constitutes a legal stop of a civilian by a police officer. The Court held that a stop must be based on reasonable suspicion by the officer that the individual may have been engaged in criminal activity. Reasonable suspicion is a legal standard used in the United States that suggests an individual is about to commit a crime based on specific facts and inferences. Reasonable suspicion is evaluated using the criteria of what a "reasonable person" would be doing or how a "reasonable officer" would respond to what he or she observed. In contrast, to make an arrest the officer must have probable cause to believe that the suspect committed a criminal offense. Probable cause exists when known facts and circumstances justify a reasonably prudent officer to conclude that a crime has been or is being committed.[17]

If an officer observes behavior that leads him or her to conclude that criminal activity may be in progress and that the suspect is armed and dangerous, the officer may stop and frisk and question a suspect after identifying himself or herself as a police officer.[18] If, in

Terry v. Ohio provides the constitutional basis for law enforcement to use profiling.

 Is profiling a genuine concern or problem in the United States? Are there certain situations in which profiling clearly is or is not appropriate?

doing so, an officer has reasonable fear for his or her own safety or that of others, the Supreme Court has ruled that the officer may conduct a "carefully limited search of the outer clothing of such persons in an attempt to discover weapons . . . Such a search is a reasonable search under the Fourth Amendment, and any weapons seized may properly be introduced in evidence against the person from whom they were taken."

The stop-and-frisk rule does not apply in all cases, however. Before frisking a suspect, police must first establish "reasonable articulable suspicion" to justify an investigative search and seizure. These suspicions should be established by considering the totality of the circumstances.[19] For example, airport narcotics agents may identify drug couriers by using profiles based on the following characteristics:

- The suspect traveled from a city which is a common source of drugs.
- The suspect carried little or no baggage.
- The suspect appeared nervous.
- The suspect purchased his or her airline ticket with cash.
- The suspect made one or more telephone calls at the airport.[20]

Even when the police lack reasonable suspicion, they may engage a citizen in conversation and ask questions, as long as the citizen feels free to leave.[21]

The _Terry_ ruling gave police a way to stop people against their will and conduct a quick search, but the stop-and-frisk rule also has specific limitations, which the Court outlined in a series of cases.

Kolender v. Lawson (1983)

The Supreme Court ruled that a California statute that required an individual to present "credible and reliable" identification in the context of an investigative stop permitted police too much discretion and was a violation of the Fourth Amendment.

The defendant in this case, Edward Lawson, was a 36-year-old African American with dreadlocks who frequently took walks in white neighborhoods, sometimes late at night.[22] During a two-year period, police stopped Lawson approximately 15 times, justifying the stops based on a California law that required citizens to produce "credible and reliable" identification if asked to do so by the police. Refusal to provide such identification was deemed a crime.

Lawson appealed his conviction on the charge. A federal appeals court agreed with Lawson's position and overturned his conviction. Attorneys for California appealed the decision to the Supreme Court, which affirmed the ruling of the federal appellate court and remanded the case back to the lower courts for further consideration.

California v. Hodari D. (1991)

The Supreme Court ruled that a police officer's confiscation of materials discarded by a fleeing suspect prior to a warrantless arrest does not violate Fourth Amendment search and seizure standards. In this case, Hodari D. (whose last name is abbreviated for anonymity because he was a minor at the time of the offense) was apprehended by police in an alleyway after he and his friends ran away from a police patrol car. While pursuing Hodari on foot, a police officer saw him throw a small object into the street. The officer was suspicious. When he finally caught, tackled, and handcuffed Hodari, he retrieved the item, which turned out to be crack cocaine. Hodari was also found to be carrying $130 in cash and a pager.[23]

Hodari's attorney argued to suppress the evidence, holding that the officer did not have the "reasonable suspicion" required by *Terry* to justify stopping the youth, so that his retrieval of the cocaine constituted an illegal seizure. The U.S. Supreme Court ruled that the police chase did not constitute a seizure but only a "show of force," which is not limited by the Constitution. It also ruled that Hodari established probable cause by discarding the cocaine prior to the seizure.

Minnesota v. Dickerson (1993)

The Supreme Court issued a ruling that helped to clarify and limit the lawful scope of a *Terry* frisk, and refused to extend to *Terry* a "plain feel" test. The Court held that if a police officer lawfully pats down a suspect's outer clothing and feels an object whose contour or mass makes its identity immediately apparent, there has been no invasion of the suspect's privacy beyond that already authorized by the officer's search for weapons.[24]

In this case, a Minneapolis police officer stopped Timothy Dickerson, who was acting suspiciously as he exited a known crack house. A protective pat-down frisk of Dickerson revealed no weapons, but the officer felt a small lump inside the front pocket of Dickerson's nylon jacket. The officer continued his search but then returned to Dickerson's pocket and squeezed, slid, and manipulated the object. Eventually, he determined that the item was a rock of crack cocaine and seized it.

Dickerson was found guilty of possession of a controlled substance. When the U.S. Supreme Court reviewed the case, it held that police may seize non-threatening contraband detected through the sense of touch during a protective pat-down search of the sort permitted by *Terry*. However, the Court also ruled that the officer in this case violated the lawful bounds marked by *Terry* because the officer did not recognize the object as contraband until after a further search—one not authorized by *Terry*. Under *Terry*, the officer was entitled to place his hands on Dickerson's jacket and to feel the lump in the pocket. His continued exploration of the pocket after determining that the lump was not a weapon, however, was unrelated to the sole justification for the search in the first place under *Terry*. Therefore, this further search was unconstitutional, making the seizure of the cocaine invalid. The U.S. Supreme Court upheld the state appellate court's reversal of the conviction.

Maryland v. Wilson (1997)

The Supreme Court held that officers may order both passengers and the driver out of a vehicle during a stop to deny any of these parties access to any possible concealed weapons. In this case, after stopping a speeding car, a Maryland state trooper ordered a nervous passenger (Wilson) out of the car.[25] When he exited the vehicle, a large quantity of cocaine fell to the ground. Wilson was arrested and charged with possession of cocaine with intent to distribute.

The Baltimore County Circuit Court granted Wilson's motion to suppress the evidence, deciding that the trooper's demand of Wilson to vacate the car constituted an unreasonable seizure under the Fourth Amendment. The Maryland Court of Special Appeals affirmed the decision. The case was subsequently appealed to the U.S. Supreme Court, which reversed the decision. The Court held that after lawfully stopping a speeding vehicle, an officer may order both its driver and any passengers to step out without showing suspicion or probable cause regarding the passenger's activities. Although the Court noted that this policy may burden the liberties of passengers, officers must be permitted such authority because the overriding government interest in officer safety must be protected. The Court did not address whether police could then search such individuals.

Motor Vehicle Searches

Police may conduct a warrantless search of a vehicle (defined by the Supreme Court as an automobile, mobile home, or water vessel) as long as officers have established probable cause to believe it is involved in illegal activity.[26] There are two key exceptions to this general rule:

1. Police cannot randomly stop motorists to search for evidence of illegal activity.

2. Automobiles parked on private property cannot be searched without a warrant.[27]

The landmark decision governing automobile searches was handed down in 1925, during Prohibition, in *Carroll v. United States*.[28] George Carroll, a well-known bootlegger, was stopped by the police, who searched his car and found alcohol. Carroll was arrested and convicted of transporting liquor for sale, in violation of federal law and the Eighteenth Amendment. Because the liquor had been seized by federal agents acting without a search warrant, Carroll appealed the decision to the Supreme Court, which upheld his conviction. The Court's ruling established the <u>Carroll doctrine</u>, which permits the warrantless search of vehicles whenever the police have a reasonable basis for assuming illegal activities are taking place.

The decision about how exhaustive such a vehicle search may be evolved through a series of related cases:

- *United States v. Chadwick* (1977): The Court ruled that a warrantless search of a footlocker in the trunk of an automobile was not allowed because, in the Court's opinion, the search violated the defendant's rights under the Fourth Amendment.[29]
- *United States v. Ross* (1982): The Court reversed its ruling in *United States v. Chadwick*, saying that police may search "every part of a vehicle that might contain the object of the search" if they have probable cause.[30]
- *California v. Acevedo* (1991): The Court clarified its ruling in *United States v. Ross*, stating that po-

The Carroll Doctrine asserts that police may search vehicles without warrants as long as they have probable cause.

Is probable cause entirely subjective? Can police do whatever they want as long as their actions have the basic appearance of reasonableness?

lice "may search an automobile and *any container* found within it when they have probable cause to believe contraband or evidence will be found."[31]

- *Illinois v. Caballes* (2005): The Court ruled that police may use a drug-sniffing dog around the outside of a vehicle during a routine traffic stop, even when police have no grounds to suspect illegal activity.[32] Once the dog indicates the likelihood of drugs, the officer then has probable cause to search further.

Additionally, the Supreme Court ruled in *Delaware v. Prouse* (1978) that police cannot randomly stop vehicles.[33] An exception to this rule is roadside sobriety checkpoints, which the Court decided are constitutional in *Michigan Department of State Police v. Sitz* (1990).[34] The Court ruled that these checkpoints are legal because they represent only a minor intrusion that is easily counterbalanced by the public interest.

Plain View Doctrine

The <u>plain view doctrine</u>, which was established in the 1968 case of *Harris v. United States*, states that when police inadvertently discover evidence (such as weapons, contraband, or stolen property) in a place where they have a legal right to be, they have a right to seize it.[35] However, the Court also ruled that police may not move an object to gain a better view of things that might otherwise remain hidden without probable cause.

In a series of cases, the U.S. Supreme Court attempted to further clarify its position on searches:

- *Coolidge v. New Hampshire* (1971): The Court held that evidence found by inadvertent discovery may be seized under the plain view exception to the warrant requirement for searches and seizures.[36]
- *Arizona v. Hicks* (1987): The Court ruled that police could not move stereo speakers even a few inches to view the serial numbers (necessary so that police could verify that they were stolen property).[37]
- *Horton v. California* (1990): The Court ruled that inadvertent discovery is not a necessary condition for the application of the plain view exception to seizures.[38]

Open Field Searches

Open field searches are becoming increasingly popular and have played an important role in the government's war on drugs. The issue raised in open field searches is whether police should be allowed to fly over or enter private land without a search warrant and seize illegal substances, such as marijuana plants. The Supreme Court originally ruled that open field searches were legal in *Hester v. United States,* a decision later affirmed in both *Oliver v. United States* and *Florida v. Riley:*

- *Hester v. United States* (1924): The Court held that the Fourth Amendment did not protect "open fields" and, therefore, that police searches in such areas as pastures, wooded areas, open water, and vacant lots need not comply with the requirements of warrants and probable cause.[39]
- *Oliver v. United States* (1984): This case involved the warrantless search of a Kentucky farm for marijuana plants. The police received a tip that Ray Oliver was growing marijuana. Upon going to his residence, they found that Oliver had posted a "No Trespassing" sign on a locked gate. The police found a path around the gate that led into the fields, where they found marijuana plants. Oliver was arrested and convicted of growing marijuana, and the Supreme Court upheld his conviction. According to the Court, police officers who trespass upon posted and fenced private land do not violate the Fourth Amendment, because the protections of the Amendment do not include or extend to open fields.[40]
- *Florida v. Riley* (1989): In this case the police were flying a helicopter at an altitude of 400 feet and observed marijuana plants growing inside a greenhouse located in a residential backyard. The Florida Supreme Court ruled that police had the right to seize the property, and the U.S. Supreme Court subsequently affirmed this decision.[41]

The warrantless search of open fields is permissible and serves as a police tool in the war on drugs.

 Do people have a constitutional right to privacy? Do exceptions to illegal searches and seizures infringe on privacy rights?

Consent Searches

The police may conduct a warrantless search if consent is given voluntarily, known as a <u>consent search</u>. Furthermore, the police have no constitutional obligation to inform the suspect that he or she may deny permission to search.[42] However, when voluntary consent is obtained through deception, any evidence police seize during the search will not be admissible in court. For example, the U.S. Supreme Court ruled in *Bumper v. North Carolina* that officers who entered the home of an elderly woman after misinforming her that they possessed a search warrant did so unlawfully and could not submit the evidence they found relating to a rape case.[43] Under just what circumstances consent is present was considered in a number of cases decided by the Supreme Court:

Schneckloth v. Bustamonte (1973)

After pulling Robert Bustamonte over for a traffic violation, a police officer searched Bustamonte's car, with his expressed consent, and seized a stolen check. Bustamonte was subsequently charged with possessing a stolen check with intent to defraud.[44] The Supreme Court upheld his conviction on an appeal, after evaluating the totality of the circumstances—which included both characteristics of the accused (such as his age and level of education) and details of the interrogation (length of detention and repeated and prolonged nature of the questioning). The Court concluded that, under the circumstances, the officer was within his rights to conduct the search after voluntary consent had been given.

Illinois v. Rodriguez (1990)

The Supreme Court has held that voluntary searches may be conducted with third-party consent under certain conditions.[45]

Gail Fischer told the police that Edward Rodriguez had beaten her in his apartment earlier in the day. Police asked Fischer to accompany them to Rodriguez's apartment, and she consented. The police had neither an arrest warrant nor a search warrant. When they arrived at the apartment, Fischer unlocked the door and gave the police permission to enter. In the living room, they observed drug paraphernalia and containers filled with cocaine. They proceeded to the bedroom, where they found Rodriguez asleep. Police arrested him and seized the drugs and related paraphernalia. The Court ruled that the police could reasonably have believed that Fischer had the authority to grant consent and permit police entry, so the search was valid.

Florida v. Bostick (1991)

As a result of this case, police may board buses, trains, and planes and ask passengers to consent to being searched, without a warrant or probable cause to ask for consent.[46]

Two armed officers, wearing their badges and insignia, boarded a bus bound for Atlanta during a stopover in Fort Lauderdale. Once on board, and without reason for suspicion, the officers randomly selected Terrance Bostick and asked to inspect his ticket and identification. Bostick complied, and the officers found nothing unusual about his papers. Then they told Bostick they were narcotics officers who were in search of illegal drugs. They asked Bostick if they could search his luggage. Police found one pound of cocaine, and Bostick was arrested and charged with drug trafficking.

At Bostick's trial, his attorney moved to suppress the cocaine as evidence on the grounds that it had been obtained by an illegal seizure, thus violating Bostick's Fourth Amendment rights. The trial court denied the motion, and Bostick was convicted. The case was appealed to the U.S. Supreme Court, which

held that in evaluating the "whole picture" (totality of the circumstances), Bostick had not been "seized" at the time he voluntarily consented to the search and, therefore, the evidence had not been illegally obtained.

Georgia v. Randolph (2006)

The Supreme Court ruled that police may not enter a home to conduct a search if one resident gives permission but the other does not.[47]

Janet Randolph called police to report a domestic disturbance and asked the police to come to the home that she shared with her husband, Scott Randolph. Police asked if they could search the home. Scott objected, but Janet consented. She then led police to their bedroom, where cocaine was found. Police used the cocaine to charge Scott with drug possession. While the Supreme Court had previously ruled that cohabitants were allowed to give consent to search, in *Randolph* it established that this decision did not apply if another cohabitant was present and objected, except when there is evidence of abuse or other circumstances that may require immediate entry by police.[48]

State v. Schwartz (2006)

The Court ruled that police cannot search a home when they have only the consent of a juvenile and the owner of the property is absent. While it is legal for a third party with shared ownership to consent to a search, in general a juvenile does not have authority equal to that of a parent or guardian regarding their shared property. The Court also stated that a person younger than age 18 does not have the capacity or authority to relinquish his or her parents' privacy rights.[49]

Abandoned Property

The Supreme Court has consistently held that police can conduct warrantless searches of abandoned property, which is property that is intentionally left behind or placed in a situation in which others may reasonably take it into their possession. For example, in *California v. Greenwood*, the Court ruled that garbage disposed of in a public place where others might have access to it is considered abandoned property.[50] In this case, a police officer received information that Billy Greenwood was selling illegal drugs. The officer acted on the tip by conducting surveillance on Greenwood's home, during which she observed behavior consistent with drug trafficking. The officer then asked the neighborhood's regular trash collector to pick up the plastic garbage bags that Greenwood had left on the curb in front of his house and give the bags to her without mixing their contents with the garbage from the other houses. The officer searched through the garbage and found items suggesting narcotics use. She subsequently applied for and received a warrant to search Greenwood's residence, where she found quantities of cocaine and hashish.

At trial, the lower court dismissed the charges, holding that the warrantless search of garbage violated the defendant's right to privacy under the Fourth Amendment. The Supreme Court reversed this decision, however, and the charges were reinstated.[51] In short, the Court held that because Greenwood left his garbage in a public place, he had no reasonable expectation of privacy.

Border Searches

Individuals entering the United States have been subject to warrantless searches for many years. Such searches are designed to seize illegal drugs and control illegal immigration. The Supreme Court has historically placed limits on border searches by prohibiting the police from stopping automobiles for the sole purpose of seizing illegal aliens.[52] In recent decades, however, increased public concern over the number of illegal immigrants entering the United States has influenced some Court rulings. For instance, in *United States v. Martinez-Fuerte* (1976), the Court held that police did not need probable cause to stop cars and to question passengers at fixed checkpoints.[53]

Electronic Surveillance

In 1928, in *Olmstead v. United States,* the U.S. Supreme Court ruled that electronic eavesdropping did not constitute a violation of an individual's constitutional rights. The *Olmstead* decision was severely criticized by civil libertarians, and Congress responded by passing the Federal Communications Act in 1934. This Act declared that no person who is not authorized by the sender shall intercept any communication, or divulge or publish the existence, contents, substance, purport, effect, or meaning of such intercepted communication to any person.[54]

Further restrictions were placed on wiretapping in 1967. In the case of *Katz v. United States,* federal agents believed that the defendant, Charles Katz, was using a pay telephone to transmit gambling information.[55] The agents placed a listening device outside a public telephone booth and secured information sufficient to convict Katz. The Supreme Court later overturned Katz's conviction on the grounds that the

Since the terrorist attacks of September 11, 2001, border enforcement has shifted from drug interdiction to national security and counterterrorism efforts.

 Should law enforcement renew its commitment to drug searches and border enforcement? Why haven't international terrorists made more attempts to enter the United States from Canada and Mexico?

Fourth Amendment protects the privacy of people, regardless of the place of their offense. The ruling in *Katz,* however, did not prohibit electronic surveillance.

In 1968, Congress authorized the passage of the Omnibus Crime Control and Safe Streets Act, which included a provision on electronic surveillance. The Act established that the federal use of wiretaps and other electronic eavesdropping equipment is legal, as long as officers have previously obtained a warrant.

Kyllo v. United States (2001) tested the legality of this Act.[56] Police believed that Danny Kyllo was growing marijuana in his home and used a thermal imaging device to scan the area. The scan revealed that Kyllo's garage roof and a side wall were hotter than normal, which police suspected was due to the presence of the high-intensity lighting required to grow marijuana indoors. Police requested and received a warrant to search Kyllo's home. During this search, they found marijuana and arrested Kyllo on federal drug charges. Kyllo unsuccessfully moved to suppress the evidence and eventually appealed his conviction to the U.S. Supreme Court. The Court ultimately ruled that such surveillance by the government—even without physical intrusion—qualifies as an unreasonable search under the Fourth Amendment and requires a warrant.

Since the September 11, 2001, terrorist attacks in the United States and the terrorist bombings that followed in subsequent years in Madrid and London, countries across Europe have begun granting their governments more power to eavesdrop electronically on conversations than exists in the United States. For example, in December 2005, as part of the European Union's antiterrorism program, the U.K. Parliament passed legislation requiring telecommunication companies to retain phone data and Internet logs for a minimum of six months.

Following the July 7, 2005, subway bombing in London, Italy passed a terrorism law that stated if an attack is feared, only approval from a prosecutor—not a judge—is required for a wiretap. Since this law was passed, the number of authorized wiretaps in Italy has tripled, up from 32,000 in 2001 to 106,000 in 2006. This increased focus on wiretapping has also resulted in two major accomplishments for the Italian government. The first occasion involved the capture of one of the men wanted in the London bombing. After the suspect left England, Italian authorities tracked his cell phone, recorded his conversations, and located him in an apartment in Rome. In a second incident, Italian authorities arrested an Egyptian man who had recruited suicide bombers for Iraq and was a suspect in the Madrid train bombings of March 11, 2004. After weeks of listening to phone calls from his Milan apartment, police made the arrest.

In the Netherlands, the government has passed similar sweeping measures that have lowered the threshold for wiretapping and surveillance. Dutch public attitudes regarding wiretapping changed dramatically following the 2004 murder of filmmaker Theo van Gogh by a Muslim extremist who claimed that a film van Gogh made insulted Islam.

Likewise, France passed new anticrime legislation in 2004 that made wiretapping easier to implement. Prosecutors can now apply for wiretaps when investigations are still in a preliminary phase, rather than having to wait for an investigating magistrate to assume control of the case.

The evidence from other countries thus far suggests that in the future electronic surveillance will become more frequent than it is today. As terrorism threats and attacks spread to more countries, nations around the world will take whatever steps they believe are necessary to protect themselves—even if that requires some sacrifice of civil liberties.

Source: Victor L. Simpson, "Wiretapping on the Increase in Europe," *USA Today,* April 9, 2006, available at http://www.usatoday.com/tech/news/2006-04-09-wiretapping-europe_x.htm, accessed June 7, 2007.

Arrest

Once the police have established probable cause or directly observed a crime, they may make an arrest. An <u>arrest</u> occurs when the police physically take a suspect into custody on the grounds that they believe the suspect has committed a criminal offense.

Arrest Warrant

When the police have established probable cause to arrest someone, but the suspect is not in police custody, they apply to a judicial officer for an <u>arrest warrant</u>. An arrest warrant is a written court order instructing the police to arrest a specific person for a specific crime. It will be granted if the police can demonstrate probable cause. A faulty arrest warrant, however, can create issues of legal liability.

Police Liability

When applying for an arrest warrant, the police must be certain that the affidavit establishes probable cause. If it does not, the officer may be held civilly liable for damages if an arrest is made without probable cause. If a plaintiff files charges against a police officer, the plaintiff has the burden to prove the officer deprived him or her of a constitutional right. Police officers who are so charged usually rely on a good faith defense, which stems from the U.S. Supreme Court's decision in *United States v. Leon.*[57] Under the good faith defense, the accused officer must prove the following:

- The officer believed in good faith that his or her actions were lawful.
- The officer believed the legality of his or her conduct was reasonable when measured against some objective standard.

An emerging issue related to electronic surveillance in the United States (and throughout the industrialized world) is whether the government may track suspects using cell phones without probable cause. The question that U.S. courts have been asked to answer is whether police may monitor individuals through their cell phones without having to provide evidence of criminal activity.

On several occasions, U.S. courts have received requests from the U.S. Department of Justice (DOJ) for permission to track suspects without showing probable cause. Except in one instance, judges have denied the DOJ's motions, stating that the government did not have the authority to track cell phone locations without a warrant. In the lone exception, which is also the most recent case dealing with this issue, Judge Gabriel Gorenstein of the U.S. District Court, Southern District of New York, ruled in the government's favor, stating that the USA Patriot Act and the 1986 Electronic Communications Privacy Act permit warrantless cell phone tracking. According to Judge Gorenstein, a cell phone user who chooses to voluntarily transmit a signal to a cell phone provider assumes the risk that the provider might reveal cell-site information to police.

On the one hand, proponents of cell phone tracking view it as an effective tool for law enforcement, in that this approach saves police time when searching for suspects—that is, it allows police to rely on cell phone technology to determine a suspect's general whereabouts. On the other hand, critics contend that the practice of real-time tracking of cell phones constitutes a privacy violation, which conflicts with an individual's rights under the Fourth Amendment.

The issue of whether police may use cell phones to track criminal suspects has yet to be resolved. At some point, the government will undoubtedly present evidence in a criminal trial obtained through cell phone

tracking without having shown probable cause, at which point the defendant will challenge the admissibility of that evidence. When this happens, such a case may reach the Supreme Court. It may also place increased pressure on Congress to revisit and redefine current wiretapping statutes.

Sources: Teresa Baldas, "Feds' Cell Phone Tracking Divides the Courts," *The National Law Journal,* January 19, 2006, pp. 1–3; Matt Richtell, "Live Tracking of Mobile Phones Prompts Court Fights on Privacy," *New York Times,* December 10, 2005, pp. A1–A2; Daniel Wise, "U.S. Wind Bid to Collect Cell Tower Location Data," *New York Law Journal,* December 23, 2005, available at http://www.law.com/jsp/article.jsp?id=1135245911048, accessed July 6, 2007.

In more extreme cases, police officers may be held criminally liable, which also requires the plaintiff to prove that the officer willfully violated his or her constitutional rights.[58] Police officers have only limited immunity from civil and criminal prosecution, so they need to be well informed about citizens' constitutional rights and court decisions interpreting them.[59]

> **LINK** Should police have immunity from prosecution? Consider the case of *Scott v. Harris* (2007), discussed in Chapter 8, in which an officer rammed a speeding vehicle to end a high-speed chase and left the teenage driver a quadriplegic.

Judicial Liability

Unlike police officers, who have limited immunity from civil and criminal prosecution, judicial officers have absolute immunity because they must rely on the information given to them—and that information can sometimes be faulty.[60] If a judge or magistrate signs a search warrant when no probable cause exists, the trial judge will in all likelihood determine the arrest was invalid and dismiss the case. The defendant cannot file a civil suit against the judge who originally issued the search warrant.

Warrantless Arrests

Most arrests that police make are without a warrant. Indeed, they often make decisions to arrest on the scene during their interactions with suspects. Police may make a warrantless arrest in the following circumstances:

- Police observe a felony in progress.
- Police have knowledge that a felony has been committed and have probable cause to believe that the crime was committed by a particular suspect.
- The law of the particular jurisdiction permits police to arrest without a warrant.[61]

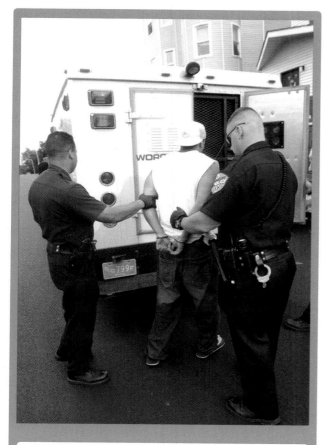

Police officers can make warrantless arrests in several situations.

 How has the U.S. Supreme Court limited or expanded the ability of the police to make warrantless arrests? Are criminals the only people who should be concerned about warrantless exceptions to arrest?

- *Steagald v. United States* (1981): The Court ruled that if a criminal suspect is in the home of another person, the police must obtain both an arrest warrant and a search warrant or third-party consent before entering the home to make an arrest.[64]
- *Welsh v. Wisconsin* (1984): The Court ruled that having hearsay information in a driving while intoxicated case was not sufficient to establish an emergency situation.[65]
- *Minnesota v. Olson* (1990): The Court held that police acted properly when they entered the home of a fleeing armed robber and executed an arrest.[66]
- *Brigham City, Utah v. Stuart* (2006): The Court held that police may enter a home without a warrant when they have an objectively reasonable basis for believing that an occupant is seriously injured or imminently threatened with such injury.[67]

Miranda Warning

When suspects are arrested, with or without a warrant, they must be informed of their rights. To do so, police read the <u>*Miranda* warning</u> to all suspects who are taken into custody.

In 1966, in its ruling in <u>*Miranda v. Arizona*</u>, the U.S. Supreme Court ordered the police to create an environment that would produce only voluntary confessions. The result of this case—the *Miranda* warning—is well known:

> You have the right to remain silent; anything you say can be used against you in a court of law. You have the right to the presence of an attorney; if you cannot afford an attorney, one will be appointed for you prior to any questioning.[68]

The accused must be informed of these constitutional rights before his or her arrest.

This decision came as a result of a case involving an assault against Patricia Ann Weir, age 18. Weir was walking to a bus in downtown Phoenix, Arizona, in 1963, when she was accosted by a man who shoved her into his car, tied her hands and ankles, drove her to a location somewhere outside the city, and raped her.[69] Weir was then driven to a street near her home and let out of the car. Immediately afterward, Weir called the police to report the crime.

Shortly after the incident, police arrested Ernesto Miranda, who voluntarily agreed to speak with police about the incident and willingly participated in a line-up in which Weir identified him as the man who raped her. Miranda, age 23, was an eighth-grade dropout who

In general, police do not need to delay an arrest because they lack a warrant.[62] If warrants were always required before an arrest was made, many suspects would escape while such a warrant was being secured. However, officers may not make a warrantless arrest for a misdemeanor offense unless it is committed in their presence, a limitation known as the <u>in-presence requirement</u>.

One exception to the warrantless arrest occurs when the police want to enter a suspect's home. The Supreme Court ruled in 1980, in *Payton v. New York*, that an arrest warrant is a prerequisite to valid entry into a home in all nonemergency situations.[63] An emergency is determined by the seriousness of the offense, as has been detailed in a series of Supreme Court rulings:

Career criminal Ernesto Miranda became an unlikely pillar of jurisprudence in *Miranda v. Arizona* (1966).

? Are the effects of *Miranda* laws positive? Who would benefit if *Miranda* was overturned?

In spite of these violations, and over Moore's objection, the judge ruled that Miranda's written confession could be admitted into evidence. Miranda was found guilty and sentenced to 20 to 30 years in prison for kidnapping and rape. His case was appealed to the Arizona Supreme Court, which upheld the decision of the lower court. The case was then appealed to the U.S. Supreme Court, with Miranda's new attorney, John Flynn, asking the Court to decide whether the confession of "a poorly educated, mentally abnormal, indigent defendant" who was not informed of his right to counsel should be admitted, despite a "specific objection based on the absence of counsel." In short, Flynn stated to the Court that Miranda had been manipulated into confessing as a result of his poor education and lack of information about his rights.

On June 13, 1966, the Supreme Court announced its decision. In a 5-to-4 vote, Chief Justice Earl Warren, writing for the majority, stated that Miranda's Fifth Amendment right to protection from self-incrimination and his Sixth Amendment right to counsel had been violated. As a result of this decision, police are required to give a person who is taken into custody a *Miranda* warning if they wish to question the person and use the answers as evidence at trial.

Although the Supreme Court overturned Miranda's conviction, it did not nullify the indictment that had been brought against him, and he was not released after the ruling. In 1967, Miranda was retried and found guilty of raping and kidnapping Weir after his common-law wife, Twila Hoffman, testified that Miranda had confessed to her that he had committed the crimes.[71] Miranda was sentenced to 11 years in prison and was subsequently paroled after serving one-third of his sentence. (In a twist of fate, four years after being paroled, Miranda was stabbed to death in a bar fight; when his assailant was arrested, police officers followed the law and read him his *Miranda* rights.[72])

While some experts hailed the *Miranda* decision as long overdue, others criticized it. Justice John Marshall Harlan dissented from the Supreme Court majority position and denounced the decision, calling it a "dangerous experimentation" at a time when the "high crime rate is of growing concern."[73] The reaction of the police to the *Miranda* outcome was also generally negative. Most officers believed the ruling would interfere with their efforts to protect society. Their sentiment largely stemmed from the erroneous belief that a large percent of major crime convictions resulted from confessions. Across the country, crim-

had a police record dating back nearly a decade. He had been diagnosed by psychiatrists as being seriously ill mentally and having pronounced sexual fantasies. Police asked Miranda to write a confession, which he did, stating his guilt, describing the crime, acknowledging that his confession was voluntary, and agreeing that he had full knowledge of his legal rights. Police then charged Miranda with kidnapping and rape.

At trial, Miranda's court-appointed attorney, Alvin Moore, questioned the officers at length about their interrogation. The officers admitted that during the two-hour interrogation, neither of them had advised Miranda of his right to have counsel present during questioning (a right that had been established two years earlier in *Escobedo v. Illinois*).[70] Additionally, during interrogation and prior to the line-up, two officers misled Miranda when they told him Weir had already identified him as her attacker.

What Are Your *Miranda* Rights?

Most Americans are aware of the *Miranda* warning. Nevertheless, few Americans fully understand the warning, its significance, and its relevance to them.

The *Miranda* warning does not apply under all circumstances. There are a few key exceptions:

- Police may ask questions at a crime scene.
- Police may question citizens for fact-finding purposes.
- When police stop someone, they may ask that person questions.
- Police may use statements that are made voluntarily by a suspect as evidence in court.

These exceptions to the *Miranda* warning are very important. They encompass situations in which police might acquire evidence from suspects who are under no legal obligation to speak to them. In some cases, the exceptions to the *Miranda* warning make it possible for police to establish probable cause to arrest a suspect based on the answers he or she offers at a crime scene.

Nevertheless, police must read the *Miranda* warning before making an arrest, informing the suspect of his or her right to remain silent and speak with an attorney to avoid self-incrimination.

Other than providing your name, you have no legal obligation to disclose additional information to police.

Source: Miranda v. Arizona, 384 U.S. 436 (1966).

---

inal justice personnel believed that if suspects were informed of their rights, they would not confess.

The Supreme Court's decision, however, asserted that police interrogation procedures may not intimidate the suspect. The Court defined custodial interrogation as any "questioning initiated by law enforcement officers after a person has been taken into custody" and established procedural safeguards for its conduct:

Prior to any questioning, the person must be warned that he has the right to remain silent, that any statement he does make may be used as evidence against him, and that he has a right to the presence of an attorney, either retained or appointed. The defendant may waive effectuation of these rights, provided the waiver is made voluntarily, knowingly, and intelligently. If, however, he indicates in any manner and at any stage of the process that he wishes to consult with an attorney before speaking, there can be no questioning. Likewise, if the individual is alone and indicates in any manner that he does not wish to be interrogated, the police may not question him. The mere fact that he may have answered some questions on his own does not deprive him of the right to refrain from answering any further inquiries until he has consulted with an attorney and thereafter consents to be questioned.[74]

This statement formed the basis for the *Miranda* warning, but several subsequent cases have more clearly defined its parameters.

Oregon v. Hass (1975)

In some circumstances, the police do not need to give the *Miranda* warning to a suspect. In the case of *Oregon v. Hass,* police arrested Hass and read him his *Miranda* warning. Hass acknowledged receiving it, but as he sat in the back seat of a police car, an officer initiated a conversation in which Hass made self-incriminating statements.[75] The Court allowed the officer to give testimony about the content of those statements for the explicit purpose of casting doubt on the credibility of Hass, who had provided different testimony on the witness stand. According to the Court, the shield provided by *Miranda* is not to be perverted to a license to testify inconsistently, or even perjuriously, free from the risk of confrontation with prior inconsistent utterances. The ruling in *Hass* was confirmed in *Oregon v. Mathiason* (1977), when the Court decided that *Miranda* warnings are not required if suspects voluntarily submit to questioning.[76]

Brewer v. Williams (1977)

Ten-year-old Pamela Powers was abducted, raped, and strangled on Christmas Eve, 1968, in Des Moines, Iowa.[77] Two days later, Robert Williams surrendered to the police in Davenport, Iowa. Police booked Williams and read the required *Miranda* warning.

The Davenport police telephoned their counterparts in Des Moines to inform them that Williams had surrendered and was in custody. Williams' attorney was at the Des Moines police headquarters when Williams called and spoke with him on the telephone. In the presence of the Des Moines chief of police and a detective, the attorney advised Williams that Des Moines police officers would drive to Davenport to pick him up, that the officers would not interrogate him, and that he should not talk to the officers about Pamela Powers until after meeting with him (the attorney).

While transporting Williams from Davenport to the Des Moines police headquarters, a detective appealed to Williams' religious inclinations. The detective noted that Williams was the only person who knew where the body was buried, stated that they would be driving right by the area, and asked if they could stop and locate the body because "the parents of this little girl should be entitled to a Christian burial." Soon after the detective finished the speech, Williams directed the officers to where he had buried the body.

Williams was indicted for first-degree murder. After his conviction on the charge, the U.S. Supreme Court overturned the case on three grounds:

1. The police questioning was improper because Williams's counsel was not present.
2. The officers used psychological coercion.
3. The detectives were advised not to interrogate Williams without his attorney present.

Edwards v. Arizona (1981)

On January 19, 1976, Robert Edwards was arrested and charged with robbery, burglary, and first-degree murder. At the police station, he was informed of his *Miranda* rights. Edwards indicated that he was willing to talk with the police, however, so he was questioned until he said he wanted his attorney. The next day, detectives came to the jail and said they wanted to talk with Edwards, informed him of his *Miranda* rights again, and subsequently obtained a confession.

At trial, his attorney's motion to suppress the confession was denied and Edwards was convicted. The case was eventually appealed to the U.S. Supreme Court, which held that when an accused has invoked his or her right to have counsel present during custodial interrogation, a valid waiver of that right cannot be established by showing only that the suspect responded to police-initiated interrogation after being again advised of his or her rights. If at any point

a suspect requests the presence of counsel, the police cannot resume questioning until counsel has been made available unless the accused initiates further communication, which did not occur in this instance.[78]

During questioning, a suspect's request for an attorney must be stated clearly. Officers may continue questioning until a suspect clearly makes a request for an attorney. For example, in *Davis v. United States*, the suspect, Robert Davis, stated to investigators, "Maybe I should talk to a lawyer." At that point, the investigators asked Davis outright whether he wanted a lawyer. Davis replied, "No, I'm not asking for a lawyer." The court ruled that Davis had suffered no violation of his rights, because police are not required to seek clarification of ambiguous statements.[79]

New York v. Quarles (1984)

In this case, the U.S. Supreme Court established a public safety exception to *Miranda*. In *New York v. Quarles*, police officers pursued a rape suspect after receiving notification of the crime by the alleged victim. They found the suspect in a supermarket, frisked him, and found an empty shoulder holster. When asked, the man told officers where he had discarded a gun. Only after they retrieved the gun did the officers read Quarles his *Miranda* rights and proceed to question him about the details of the weapon. The Court ruled that under such circumstances, "The need for answers to questions posing a threat to the public safety outweighs the need for the prophylactic rule protecting the Fifth Amendment's privilege against self-incrimination."[80]

Arizona v. Roberson (1988)

Are self-incriminating statements admissible if a suspect makes them while being interrogated a second time about an unrelated offense? Ronald Roberson was arrested at the scene of a burglary and, after being advised of his *Miranda* rights, requested counsel. Three days later, while Roberson was in custody and still had not spoken with an attorney, a second police officer, who was unaware that Roberson had previously requested assistance of counsel, advised Roberson of his *Miranda* rights and asked him about a second burglary. During that round of questioning, Roberson made incriminating statements about both crimes.

At the trial, citing the ruling in *Edwards v. Arizona*, the judge suppressed the statements, noting that once a defendant has invoked his or her right to counsel, the suspect may not be interrogated a second time unless counsel has been made available or unless the suspect initiates the conversation. The judge determined that even though the second questioning regarded a separate, unrelated crime, both sets of confessions should be suppressed because no counsel was present at the time of Roberson's statements.

The state of Arizona appealed the case to the U.S. Supreme Court, which affirmed the trial court's decision, reasoning that failure on the part of the police to honor Roberson's initial request could not be justified because a police officer was unaware that Roberson had requested assistance of counsel.[81] The *Roberson* decision was later affirmed in *Minnick v. Mississippi* (1990), in which the Supreme Court held that interrogation must cease if no counsel is present, even after the defendant has had an opportunity to consult with an attorney.[82]

Commonwealth v. Santiago (2002)

In *Commonwealth v. Santiago,* the Supreme Judicial Court of Massachusetts ruled that the admissibility of a statement under the public utterance exception to the hearsay rule regarding testimony quoting persons not in court is based on two criteria:

1. An exciting event took place.
2. The statement was spontaneously made in reaction to the event.

In this case, the mother of an alleged rape victim testified that her daughter had told her the alleged rapist put his finger into the victim's vagina, but did not have intercourse with her.[82] At trial, the prosecutor offered the mother's hearsay statement as a spontaneous utterance. The boyfriend was convicted of indecent assault and battery on a child but appealed the verdict. The Appeals Court reversed his conviction, finding that the trial court made an error when it admitted the mother's statement as a spontaneous utterance since sufficient time had passed between the crime and the boyfriend's arrest.

The case was then appealed to the Supreme Judicial Court of Massachusetts, which affirmed the conviction ruling that, in this instance, there was a strong connection between the exciting event (e.g., the arrest) and the underlying event to which the event related (e.g., the crime).

Additional Rulings

During the past several years, a series of Supreme Court rulings have further defined the limits of the original *Miranda* ruling:

- *Harris v. New York* (1971): Evidence produced in violation of *Miranda* can be used to cast doubt on the credibility of a defendant's testimony.[84]

- *Michigan v. Tucker* (1974): The prosecution can question a witness whose name had been divulged by defendants who were being interrogated by the police but who had not yet been advised of their *Miranda* rights.[85]
- *Rhode Island v. Innis* (1980): An interrogation includes any actions or words that the police "should have known were reasonably likely to elicit an incriminating response."[86]
- *Berkemer v. McCarty* (1984): Police do not need to give the *Miranda* warning before questioning motorists detained as part of routine traffic stops because such stops do not exert significant pressure to impair "free exercise of [one's] privilege against self-incrimination to require that [the stopped motorist] be warned of his constitutional rights."[87]
- *Moran v. Burbine* (1986): Police do not need to tell suspects of attempts by their attorneys to contact them because the Constitution does not require police to supply a suspect with information to help him or her make a decision regarding the execution of his or her legal rights.[88]
- *Illinois v. Perkins* (1990): *Miranda* warnings do not apply to prison inmates.[89]
- *Arizona v. Fulminante* (1991): Conversations between suspects and undercover agents do not evoke the concerns underlying *Miranda*.[90]
- *Dickerson v. United States* (2000): The *Miranda* warning is a constitutional issue and may not be overruled by an act of Congress.[91]

Booking

Once arrested, a suspect is taken to a police station and booked. Booking is the process of officially recording the name of the person arrested, the place and time of the arrest, the reason for the arrest, and the name of the arresting authority. During booking, a suspect is photographed and fingerprinted, and samples of his or her handwriting, voice, and blood may be taken. Taking such samples, many have argued, is self-incrimination, from which citizens are protected by the Fifth Amendment. The U.S. Supreme Court considered this issue in 1966, in *Schmerber v. California*, and decided that such actions do not amount to testimonial compulsion, which is what the Fifth Amendment protects against; accordingly such samples are admissible.[92]

Once booked, a suspect is permitted to make one telephone call. He or she is then assigned to a holding cell pending transportation to court or jail.

Line-Up

After booking, a suspect may be required to participate in a line-up.[93] A line-up is a pretrial identification procedure in which several people are shown to a victim or a witness of a crime, who is then asked whether any person in the line-up committed the crime. Suspects have a constitutional right to have counsel present during a line-up.[94]

Although police cannot suggest to a victim that one of the people in the line-up is the suspect, they can ask each person in the line-up to speak words that were allegedly spoken at the crime scene.[95] If the police suggest to the victim or witness that one of the people in the line-up is the suspect, then police will be in violation of the suspect's right to due process as guaranteed by the Fifth and Fourteenth Amendments. Thus, in a line-up, the police cannot do or say anything that might lead or encourage the victim or witness to identify the suspect.

Interrogation and Confession

Interrogation is the stage in the pretrial process where police ask suspects questions to obtain information the individual might not otherwise willingly disclose. It generally occurs before the suspect is placed in a line-up and may continue well after a line-up is conducted. A confession is a voluntary declaration to another person by someone who has committed a crime in which he or she admits to involvement in the offense.

Interrogations may be very emotionally charged, and police may become upset and angry. In the past, police sometimes used physical force to extract a confession. Today this is not allowed, and if police do use force to obtain a confession, the charges against the suspect will be dismissed, as in *State v. Jenkins* (1991) in which police kicked Jenkins in the groin, stomach, and back and threatened to kill him if he did not confess.[96]

In 1884, in *Hopt v. Utah*, the Supreme Court established the parameters for an involuntary confession—specifically, a confession cannot be precipitated by a promise, threat, fear, or torture.[97] However, the holding of *Hopt* was restricted to the inadmissibility of involuntary confessions at federal (not state) trials. Twelve years later, in *Wilson v. United States*, the Supreme Court ruled that it is important to broadly review the circumstances surrounding a confession before deciding whether it was voluntarily given.[98]

The photo line-up is a staple of crime dramas and a way to use eyewitness identification.

? Given that psychologists report that eyewitness identification is often unreliable, should the line-up remain a part of standard protocol?

As with the *Hopt* decision, the ruling in *Wilson* applied only to federal cases.

In 1936, in *Brown v. Mississippi*, the Supreme Court first ruled on the inadmissibility of coerced confessions at the state level.[99] In this case, three African American men—Yank Ellington, Ed Brown, and Henry Shields—were arrested for the murder of a white man, Raymond Stewart. The only evidence against the three suspects was their confessions, which were used to convict them and sustain death sentences. The defendants appealed their case to the Supreme Court, where their convictions were reversed. The Court ruled that the suspects' confessions had been extorted only after the suspects had endured brutal treatment, including whipping, and therefore the confessions were regarded as inadmissible coerced confessions. This ruling raised several important questions about the nature and extent of confessions.

What Is a Coerced Confession?

In *Chambers v. Florida* (1940), four males were convicted of murder on the basis of confessions obtained several days after their arrests.[100] The Supreme Court ruled that such a long delay, coupled with constant interrogation, constituted a form of psychological coercion that, in all likelihood, produced involuntary confessions. Since *Chambers,* the Court has consistently ruled that coercion can be either mental or physical.[101] Police cannot, for instance, deliberately make suspects uncomfortable during an interrogation by not letting them sit, rest, or use a toilet.

In a series of three cases, the Supreme Court further detailed what amount of pressure constitutes coercion:

- *Spano v. New York* (1959): Police may not continue an interrogation after the defendant expresses fatigue.[102]
- *Rogers v. Richmond* (1961): Police may not threaten to bring in the defendant's family for questioning; the Court determined that the confession in such a case was inadmissible.[103]
- *Lynumn v. Illinois* (1963): Police may not threaten to take away a defendant's children (and place them in foster care).[104]

In general, the courts determine the admissibility of a confession by examining the totality of the circumstances surrounding the arrest and interrogation.

Can Unnecessary Delay Constitute Coercion?

In 1943, in *McNabb v. United States*, the Supreme Court ruled there can be no unnecessary delay between arrest and arraignment.[105] In *McNabb*, a confession extorted after almost 36 hours of continuous interrogation was judged to be "inherently coercive" and, therefore, was not allowed to be presented in court. In 1957, in *Mallory v. United States*, the Court broadened the ruling in *McNabb* when it reversed the conviction of a rapist who confessed after a delay of 18 hours.[106]

In 1968, in the Omnibus Crime Control and Safe Streets Act, Congress abolished the *McNabb-Mallory* rule and created a new guideline: The admissibility of a confession will be based on the voluntary nature of the confession, and delay in arraignment will be one factor that will be considered in determining whether the confession was "voluntary."[107]

Do Suspects Have a Right to Counsel During Interrogation?

In the 1963 case of <u>*Gideon v. Wainwright*</u>, the Supreme Court handed down a far-reaching decision when it ruled that any person who is charged with a felony has the right to counsel.[108] The Court expanded the ruling in *Argersinger v. Hamlin* (1972), when it decided that accused persons also have a right to counsel in misdemeanor cases in which they face the possibility of incarceration.[109]

In the *Gideon* ruling, the Court did not explicitly state at what point in the criminal justice process a suspect was entitled to an attorney. That issue was decided in 1964 in <u>*Escobedo v. Illinois*</u>, when the Court ruled that suspects accused of a felony may have an attorney present during police interrogation.[110]

VOICE OF THE COURT
Escobedo v. Illinois

On the night of January 19, 1960, Danny Escobedo's brother-in-law, Manuel Valtierra, was fatally shot. Escobedo was arrested the next morning without a warrant and interrogated by the police. He made no statement and was subsequently released.

On January 30, Benedict DiGerlando, who was then in police custody, told the police that Escobedo had fired the fatal shots. Escobedo was arrested later that day. On the way to the police station, one of the arresting officers told Escobedo that DiGerlando had named him as the killer. Police also told Escobedo that they had a pretty tight case and that he might as well confess. Escobedo replied, "I am sorry, but I would like to have advice from my lawyer."

When Escobedo's attorney, Warren Wolfson, arrived at the police station, police would not let him talk with Escobedo as they continued their interrogation. Throughout the interrogation, Escobedo repeatedly asked to consult with his attorney. Each time, his request was denied. Escobedo eventually confessed, and he was convicted of murder.

The case was later reviewed by the Supreme Court. In a 6-to-3 decision, the Court vacated Escobedo's conviction on the grounds that the accused must be permitted to consult with his or her lawyer before or during interrogation.

Source: Escobedo v. Illinois, 378 U.S. 478 (1964).

WRAP UP

All criminal defendants, even those guilty of heinous crimes such as the rape or murder of a child, are entitled to their constitutional rights, including the protection against self-incrimination and access to counsel. Fortunately for the interests of justice, other evidence proved Couey's guilt beyond the voided confession. The adversarial system of justice that exists in the United States centers on procedure and obeying the letter of the law. Even when victims' emotions want to run roughshod over the law to achieve justice against serious criminals, it is the duty of the courts to ensure that procedural justice is administered. For this reason, lawful police behavior is essential for securing just convictions.

Chapter Spotlight

- In democratic societies, the police are expected to investigate crimes according to established procedural laws governing what they may (and may not) do to produce evidence. These laws are designed to protect citizens from abuses of their constitutional rights.

- If the police believe a suspect committed a crime, but the suspect is not yet in custody, they must demonstrate probable cause in their application for an arrest warrant from a judicial officer.

- If the police have probable cause to believe that a particular search will produce the evidence they are seeking, they may apply to a judicial officer for a search warrant, which identifies who or what is to be searched and seized.

- The exclusionary rule excludes from a criminal trial any evidence or confessions that have been unlawfully obtained. This rule prohibits the police from conducting illegal searches and seizures, although the U.S. Supreme Court has established a number of exceptions.

- Once a suspect is arrested, he or she is booked at the police station. The suspect is also informed of the basic *Miranda* rights regarding the right to remain silent and the right to consult with an attorney.

- Police are constrained by law during interrogations so that any statements or confessions made by a suspect are not produced through coercion or other illegal means.

- A confession may be deemed inadmissible at a trial if the suspect was not informed of his or her *Miranda* rights.

Putting It All Together

1. Should police be allowed to stop citizens randomly and search for evidence?

2. Are warrantless searches necessary? Are there other situations in which a warrantless search should be permitted?

3. When should third parties be permitted to give police consent to search the residence of another person? What should the relationship be between the person who gives consent and the suspect?

4. Does the *Miranda* warning impede crime prevention? Why or why not?

5. How does the release of guilty offenders because of legal "technicalities" affect your view of the criminal justice system?

abandoned property Property that is intentionally left behind or placed in a situation in which others may reasonably take the item into their possession.

affidavit for search warrant A document that outlines the evidence against the suspect and the circumstances of the crime.

arrest Police action of physically taking a suspect into custody on the grounds that there is probable cause that he or she committed a criminal offense.

arrest warrant A written court order instructing the police to arrest a specific person for a specific crime.

booking The process of officially recording the name of the person arrested, the place and time of the arrest, the reason for the arrest, and the name of the arresting authority.

Brown v. Mississippi Ruling that established that involuntary confessions are inadmissible in state criminal prosecutions.

Carroll doctrine Doctrine that permits the warrantless search of vehicles whenever police have a reasonable basis for believing illegal activities are taking place.

Chimel v. California Ruling that established the "one arm's-length" rule, which allows police without a warrant to search suspects and, to a limited extent, the immediate area they occupy.

confession A voluntary declaration to another person by someone who has committed a crime in which the suspect admits to involvement in the offense.

consent search A legal, warrantless search conducted after a person gives expressed consent to police.

criminal procedure law A body of law that prescribes how the government enforces criminal law and protects citizens from overzealous police, prosecutors, and judges.

custody Assumed legal control of a person or object.

Escobedo v. Illinois Ruling that held that suspects accused of a felony may have an attorney present during interrogation.

exclusionary rule The rule of law prohibiting the introduction of illegally obtained evidence or confessions into a trial.

Gideon v. Wainwright Ruling that determined that every person who is charged with a felony has the right to appointed counsel.

good faith exception An exception to the requirement for a warrant for search and seizure; it allows evidence collected in violation of the suspect's privacy rights under the Fourth Amendment to be admitted at trial if the police had good reason to believe their actions were legal.

Hopt v. Utah Ruling that established guidelines for involuntary confessions.

inevitable discovery rule Rule that if illegally obtained evidence would have eventually been discovered by lawful means, it is admissible regardless of how it was originally discovered.

in-presence requirement A requirement that police may not make a warrantless arrest for a misdemeanor offense unless the offense is committed in their presence.

interrogation A method police use during an interview with a suspect to obtain information that the suspect might not otherwise disclose.

involuntary confession A confession precipitated by a promise, threat, fear, torture, or other external factor such as mental illness.

knock-and-announce rule Rule that requires police to announce their presence and wait about 20 seconds before entering a home.

line-up A pretrial identification procedure in which several people are shown to a victim or witness of a crime, who is then asked if any of those individuals committed the crime.

Mapp v. Ohio Ruling that expanded the exclusionary rule to state courts.

Miranda v. Arizona Ruling that established that criminal suspects must be informed of their right to consult with an attorney and their right against self-incrimination prior to questioning by police.

***Miranda* warning** A warning required by law to be recited at the time of arrest, informing suspects of their constitutional right to remain silent and have an attorney present during questioning.

New Jersey v. T.L.O. Ruling that established that school officials can conduct warrantless searches of students at school on the basis of reasonable suspicion.

plain view doctrine Standard that provides when police discover evidence in a place where police have a legal right to be, they have a right to seize that evidence.

probable cause A set of facts and circumstances that would lead a reasonable person to believe that a crime had been committed and that the accused committed it.

search warrant A written order instructing police to examine a specific location for a certain property or persons relating to a crime, to seize the property or persons if found, and to account for the results to the judicial officer who issued the warrant.

silver platter doctrine Doctrine that permitted officers in one jurisdiction to hand over "on a silver platter" evidence that had been illegally obtained to officers in another jurisdiction to use in court.

stop-and-frisk rule Rule that police may stop, question, and frisk individuals who look suspicious.

Terry v. Ohio Ruling that determined if police observe behavior that leads them to conclude criminal activity may be in progress and the suspect is armed and dangerous, they may stop and frisk and question a suspect after identifying themselves as police officers.

totality-of-the-circumstances rule Rule that requires a judge to evaluate all available information when deciding whether to issue a search warrant.

United States v. Leon Ruling that established the good faith exception to the exclusionary rule, under which evidence that is produced in good faith and later discovered to be obtained illegally may still be admissible in court.

warrantless arrest Arrest without a warrant when an officer has probable cause to believe that a crime has been or is being committed.

Weeks v. United States Ruling that established the exclusionary rule in federal cases.

Notes

1. *Payton v. New York*, 445 U.S. 573 (1980).
2. William Cohen, *Constitutional Law*, 12th ed. (Belmont, CA: Foundation Press, 2005).
3. *Weeks v. United States*, 232 U.S. 383 (1914).
4. *Silverthorne Lumber Co. v. United States*, 251 U.S. 385 (1920).
5. *Rea v. United States*, 350 U.S. 214 (1956).
6. *Elkins v. United States*, 364 U.S. 206 (1960).
7. *Mapp v. Ohio*, 367 U.S. 643 (1961).
8. *Linkletter v. Walker*, 381 U.S. 618 (1965).
9. *Nix v. Williams*, 467 U.S. 431 (1984).
10. *United States v. Leon*, 468 U.S. 897 (1984), also see the companion case, *Massachusetts v. Sheppard*, 468 U.S. 981 (1984).
11. *Illinois v. Krull*, 480 U.S. 340 (1987).
12. *Arizona v. Evans*, 514 U.S. 1 (1995).
13. *Arizona v. Evans*, note 12.
14. *Hudson v. Michigan*, 547 U.S. 1096 (2006).
15. *New Jersey v. T.L.O.*, 469 U.S. 325 (1985).
16. *Chimel v. California*, 395 U.S. 752 (1969).
17. *Draper v. United States*, 358 U.S. 307 (1959).
18. *Terry v. Ohio*, 392 U.S. 1 (1968).
19. *United States v. Sokolow*, 490 U.S. 1 (1989).
20. Morgan Cloud, "Search and Seizure by the Numbers," *Boston University Law Review* 65:843–921 (1985); Gerald Robin, "Inquisitive Cops, Investigative Stops, and Drug Courier Hops," *Journal of Contemporary Criminal Justice* 9:41–49 (1993).
21. *Florida v. Royer*, 460 U.S. 491 (1983).
22. *Kolender v. Lawson*, 461 U.S. 352 (1983).
23. *California v. Hodari D.*, 499 U.S. 621 (1991).
24. *Minnesota v. Dickerson*, 113 S. Ct. 2130 (1993).
25. *Maryland v. Wilson*, 519 U.S. 408 (1997).
26. *California v. Carney*, 471 U.S. 386 (1985); *United States v. Villamonte-Marquez*, 462 U.S. 579 (1983); *Brinegar v. United States*, 338 U.S. 160 (1949).
27. *Delaware v. Prouse*, 440 U.S. 648 (1979); *Coolidge v. New Hampshire*, 403 U.S. 443 (1971).
28. *Carroll v. United States*, 267 U.S. 132 (1925).
29. *United States v. Chadwick*, 433 U.S. 1 (1977).

30. *United States v. Ross*, 456 U.S. 798 (1982).
31. *California v. Acevedo*, 500 U.S. 565 (1991).
32. *Illinois v. Caballes*, 543 U.S. 405 (2005).
33. *Delaware v. Prouse*, 440 U.S. 648 (1978).
34. *Michigan Department of State Police v. Sitz*, 496 U.S. 444 (1990).
35. *Harris v. United States*, 390 U.S. 234 (1968).
36. *Coolidge v. New Hampshire*, note 27.
37. *Arizona v. Hicks*, 480 U.S. 321 (1987); *United States v. Irizarry*, 673 F.2d 554 (1st Cir. 1982).
38. *Horton v. California*, 496 U.S. 128 (1990).
39. *Hester v. United States*, 265 U.S. 57 (1924).
40. *Oliver v. United States*, 466 U.S. 170 (1984).
41. *Florida v. Riley*, 488 U.S. 445 (1989).
42. *Schneckloth v. Bustamonte*, 412 U.S. 218 (1973).
43. *Bumper v. North Carolina*, 391 U.S. 543 (1968).
44. *Schneckloth v. Bustamonte*, note 42.
45. *Illinois v. Rodriguez*, 497 U.S. 177 (1990).
46. *Florida v. Bostick*, 501 U.S. 429 (1991).
47. *Georgia v. Randolph*, 547 U.S. 103 (2006).
48. *Illinois v. Rodriguez*, note 46; *United States v. Matlock*, 415 U.S. 164 (1974).
49. *State v. Schwartz*, 2006 MT 120 (2006).
50. *California v. Greenwood*, 486 U.S. 35 (1988).
51. *California v. Greenwood*, note 50.
52. *Almeida-Sanchez v. United States*, 413 U.S. 266 (1973); *United States v. Brignoni-Ponce*, 422 U.S. 873 (1975); *United States v. Ortiz*, 422 U.S. 891 (1975).
53. *United States v. Martinez-Fuerte*, 428 U.S. 543 (1976).
54. *Olmstead v. United States*, 277 U.S. 438 (1928).
55. *Katz v. United States*, 389 U.S. 347 (1967).
56. *Kyllo v. United States*, 533 U.S. 27 (2001).
57. *United States v. Leon*, note 10.
58. *Screws v. United States*, 325 U.S. 91 (1945).
59. *Malley v. Briggs*, 475 U.S. 335 (1986).
60. *Heimbach v. Village of Lyons*, 597 F.2d 344 (2d Cir. 1979); *Stump v. Sparkman*, 435 U.S. 349 (1978); *Pierson v. Ray*, 386 U.S. 547 (1967).
61. Cohen, note 2.
62. *United States v. Watson*, 423 U.S. 411 (1976); *United States v. Santana*, 427 U.S. 38 (1976).
63. *Payton v. New York*, note 1.
64. *Steagald v. United States*, 451 U.S. 204 (1981).
65. *Welsh v. Wisconsin*, 466 U.S. 740 (1984).
66. *Minnesota v. Olson*, 495 U.S. 91 (1990).
67. *Brigham City, Utah v. Stuart*, 547 U.S. 1067 (2006).
68. *Miranda v. Arizona*, 384 U.S. 436 (1966).
69. *State v. Miranda*, 401 P.2nd 721 (1965).
70. *Escobedo v. Illinois*, 378 U.S. 478 (1964).
71. David Davies, "Second Rape Conviction for Miranda," *The Arizona Republic*, February 25, 1967, pp. 1A, 4A; Liva Baker, Miranda: *Crime, Law and Politics* (New York: Atheneum, 1983).
72. Baker, note 71.
73. Fred Graham, "High Court Puts New Curb on Powers of the Police to Interrogate Suspects," *New York Times*, June 14, 1966, p. A1.
74. *Miranda v. Arizona*, note 68.
75. *Oregon v. Hass*, 420 U.S. 714 (1975).
76. *Oregon v. Mathiason*, 429 U.S. 492 (1977).
77. *Brewer v. Williams*, 430 U.S. 387 (1977).
78. *Edwards v. Arizona*, 451 U.S. 477 (1981).
79. *Davis v. United States*, 114 S. Ct. 2350 (1994).
80. *New York v. Quarles*, 467 U.S. 649 (1984).
81. *Arizona v. Roberson*, 486 U.S. 675 (1988).
82. *Minnick v. Mississippi*, 498 U.S. 146 (1990).
83. *Commonwealth v. Santiago*, 432 Mass. 620 (2002).
84. *Harris v. New York*, 401 U.S. 222 (1971).
85. *Michigan v. Tucker*, 417 U.S. 433 (1974).
86. *Rhode Island v. Innis*, 446 U.S. 291 (1980); *Miranda v. Arizona*, note 68.
87. *Berkemer v. McCarty*, 468 U.S. 420 (1984).
88. *Moran v. Burbine*, 475 U.S. 412 (1986).
89. *Illinois v. Perkins*, 496 U.S. 292 (1990).
90. *Arizona v. Fulminante*, 499 U.S. 279 (1991).
91. *Dickerson v. United States*, 530 U.S. 428 (2000).
92. *Schmerber v. California*, 384 U.S. 757 (1966); *United States v. Dionisio*, 410 U.S. 1 (1973); *United States v. Mara*, 410 U.S. 19 (1973); *Gilbert v. California*, 388 U.S. 263 (1967); *United States v. Wade*, 388 U.S. 218 (1967); *United States v. Euge*, 444 U.S. 707 (1980).
93. *United States v. Wade*, note 92.
94. *Kirby v. Illinois*, 406 U.S. 682 (1972).
95. *Simmons v. United States*, 390 U.S. 377 (1968); *Neil v. Biggers*, 409 U.S. 188 (1972).
96. "Beating Nullifies Conviction," *Seattle Times*, July 6, 1991, pp. A1, 12.
97. *Hopt v. Utah*, 110 U.S. 574 (1884).
98. *Wilson v. United States*, 162 U.S. 613 (1896).
99. *Brown v. Mississippi*, 297 U.S. 278 (1936).
100. *Chambers v. Florida*, 309 U.S. 227 (1940).
101. *Blackburn v. Alabama*, 361 U.S. 199 (1960).
102. *Spano v. New York*, 360 U.S. 315 (1959).
103. *Rogers v. Richmond*, 365 U.S. 534 (1961).
104. *Lynumn v. Illinois*, 372 U.S. 528 (1963).
105. *McNabb v. United States*, 318 U.S. 332 (1943).
106. *Mallory v. United States*, 354 U.S. 449 (1957).
107. Leonard Levy, Kenneth Karst, and Dennis Mahoney, *Encyclopedia of the American Constitution*, 2nd ed. (New York: Macmillan Reference Books, 2000).
108. *Gideon v. Wainwright*, 372 U.S. 335 (1963).
109. *Argersinger v. Hamlin*, 407 U.S. 25 (1972).
110. *Escobedo v. Illinois*, note 70.

JBPUB.COM/ExploringCJ

Interactives

Key Term Explorer

Web Links

RISH PLAZA
0TH ANNIVERSARY

EY OFM CAP
OF BOSTON
6 2005

- Understand the purpose and consequences of police discretion.
- Know the legal and extralegal factors that influence police decisions to make an arrest.
- Explain the history and extent of police corruption.
- Outline research on police brutality and excessive force.
- Understand federal and state laws and policies regulating high-speed chases.
- Be familiar with the factors responsible for police stress, and know how stress affects the overall performance of law enforcement agencies.
- Describe the history of women and minorities in policing and the unique problems that these groups face.

Critical Issues in Policing

CHAPTER 8

Although the vast majority of police–citizen interactions in the United States are civil and peaceful, some are lethal. Criminals murder 79 police officers annually, on average, and officers kill nearly 400 suspected felons. One in six murders of police officers directly results in the justifiable killing of the officer's murderer.

These statistics are not uniformly distributed across social groups. Between 1976 and 1998, young African American males (who constitute 1 percent of the U.S. population) accounted for 14 percent of felons who were killed by police and 21 percent of felons who murdered a police officer. Young white males (7 percent of the U.S. population) accounted for 15 percent of felons who were killed by police and 20 percent of felons who murdered a police officer.

When police officers are killed, there are often severe legal consequences. The first death sentence in New York City since 1954 was pronounced for Ronnell Wilson, who was convicted of the execution-style murder of two New York City police officers.

- In what ways might legal and extralegal factors explain racial differences in lethal police–citizen interactions?

- How do prior police contacts affect subsequent police perceptions of suspects?

Sources: Jodi Brown and Patrick Langan, *Policing and Homicide, 1976–1998: Justifiable Homicide by Police, Police Officers Murdered by Felons* (Washington, DC: U.S. Department of Justice, 2001); Jose Martinez, "Cop Killer Gets Federal Death Sentence," available at http://www.nydailynews.com/01-31-2007/news/story/493459p-415661c.html, accessed July 1, 2007.

Introduction

Police officers are members of one of the most dangerous professions in the United States.[1] Every day police confront violent or angry people, and occasionally they may be assaulted or killed in the line of duty **FIGURE 8-1** **FIGURE 8-2** . On every shift, officers must make difficult decisions about how to deal with offenders and exercise discretion in stressful situations such as high-speed chases and cases that require the use of deadly force. Sometimes they cross a line and apply unnecessary or inappropriate force against citizens. In addition to the issues of stress, deadly force, and brutality, police departments constantly face challenges in the form of corruption.

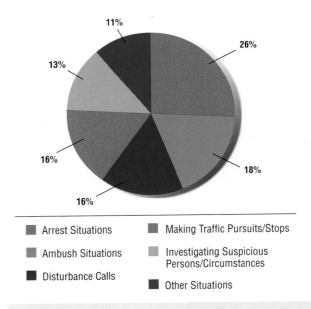

- ■ Arrest Situations
- ■ Ambush Situations
- ■ Disturbance Calls
- ■ Making Traffic Pursuits/Stops
- ■ Investigating Suspicious Persons/Circumstances
- ■ Other Situations

FIGURE 8-1 Percentage of Police Officers Feloniously Killed or Assaulted in Different Situations, 2005

Source: Federal Bureau of Investigation, *Law Enforcement Officers Killed and Assaulted, 2005* (Washington, DC: U.S. Department of Justice, 2006).

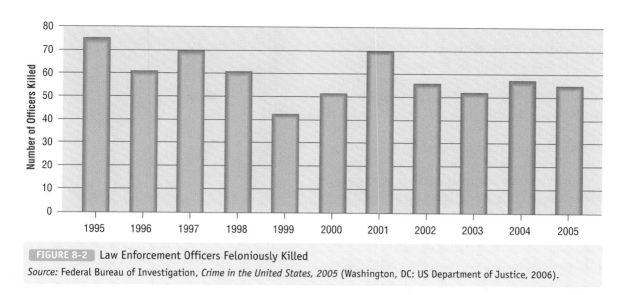

FIGURE 8-2 Law Enforcement Officers Feloniously Killed

Source: Federal Bureau of Investigation, *Crime in the United States, 2005* (Washington, DC: US Department of Justice, 2006).

Police Discretion

The heart of policing is discretion. For police officers, discretion typically involves answering three questions:

1. Should I intervene?
2. What should I do?
3. How should I do it?

These are difficult questions to answer, particularly for an officer who may need to make a critical, split-second decision about how the law should be applied.

Formally, police discretion is the authority of officers to choose one course of action over another. For example, officers use discretion in deciding whether to stop and question two youths walking down a sidewalk, assisting a motorist stalled at the side of the road, releasing a criminal suspect, or searching a vehicle for illegal drugs. Discretion is so widely used by police because it is not possible to have rules that would cover every specific situation. A policy of full law enforcement, in which officers respond formally to all suspicious behavior, is impractical for several reasons:

- Most violations are minor and do not require full enforcement.
- The criminal justice system has insufficient resources to react formally to all violations of law TABLE 8-1 .[2]
- Full enforcement would mean that the majority of officers' time would be spent completing paperwork and testifying in court, not policing the streets.
- Even well-defined legal statutes are sometimes vague and open to interpretation.

TABLE 8-1	
Average Ratio of Police Officers per 1000 Residents	
Population Served	Number of Officers per 1000 Residents
All sizes	2.5
250,000 or more	2.5
100,000–249,999	1.9
50,000–99,999	1.8
25,000–49,999	1.8
10,000–24,999	2.0
2500–9999	2.2
1000–2499	2.6

Source: Matthew Hickman and Brian Reaves, *Local Police Departments, 2003* (Washington, DC: U.S. Department of Justice, 2006).

- Full enforcement would create an extraordinary strain between the police and the public, reducing citizen cooperation and increasing crime.

Because full enforcement is not a realistic approach, agencies end up practicing selective law enforcement, in which police under-enforce some laws and over-enforce other laws. While on its face selective law enforcement is the only practical approach for policing, it also brings its own challenges. Selective enforcement has historically proven to be problematic for the following reasons:

- It is inherently unfair that police are allowed to respond differently to similar situations.
- Under this kind of system, officers may abuse their power, targeting specific individuals or populations.
- Selective enforcement may lead to favoritism and corruption, with those empowered to choose being able to help their friends, take bribes, and threaten parties from whom they desire favors.

Despite these concerns, discretion is a necessary component of the criminal justice system. Selective enforcement may certainly be warranted for minor offenses. For instance, in some circumstances, a warning may be equally effective at preventing future violations without draining the government's legal resources.

In particular, discretionary authority is used by police when making decisions to arrest suspects. Only about 13 percent of encounters between police and suspects result in arrests.[3] In an ideal world, police might use only legal criteria to make their arrest decisions. In reality, studies have shown that extralegal factors such as race, gender, or socioeconomic class influence police decisions as well **TABLE 8-2** .

The presence of physical evidence of a crime significantly increases the likelihood of arrest.

? Why do police sometimes decline to make an arrest when there is obvious evidence of a crime (such as in drunk driving, domestic violence, or other assault cases)? Do police too often play the part of prosecutor and decline to make arrests they think will not result in convictions?

TABLE 8-2

Factors Influencing the Decision to Arrest

Legal Factors

Seriousness of the Offense: People who commit more serious offenses are more likely to be arrested.

Prior Arrest Record: Police are more likely to arrest persons with a prior arrest record.

Presence of Evidence: When police have strong evidence, they are more likely to arrest the suspect.

Suspicious Behavior: Police are more likely to arrest people who engage in behavior that is out of place in the specific circumstances (e.g., wearing a heavy coat in hot and humid weather).

Extralegal Factors

Race and Ethnicity: Research has produced mixed results regarding the relationship between race and arrest.

Attitude and Disrespect: Suspects who are disrespectful of police are more likely to be arrested.

Socioeconomic Class: Police generally treat poor and wealthy persons similarly for comparable offenses.

Gender: Research on the relationship between gender and arrest has been mixed.

Characteristics of Police Officers: Younger, less educated, and African American police officers are the most likely to arrest suspects.

Organization of Police Departments: The social organization of the police agency has only a small effect on the arrest decision.

Legal Factors

Seriousness of the Offense

People who engage in more serious crimes are much more likely to face arrest than are those who engage in more minor offenses.[4] A suspect is also more likely to be arrested if he or she possesses a weapon.[5] In addition, crimes that are perceived by police as sophisticated, premeditated, or malicious are more likely to result in arrest.[6]

Prior Arrest Record

Police are more likely to arrest persons who have previously been arrested. For example, juveniles with five or more previous arrests account for more than 66 percent of juvenile arrests, whereas first-time offenders account for only 7 percent of those taken into custody.[7] This factor is more likely to become an issue when the decision about whether to formally process a suspect is made at the police station rather than on the street, although studies have shown that police consider having a prior record as confirmation of the suspect's involvement in criminality.[8]

Presence of Evidence

Police have sufficient evidence to link a suspect to a crime in approximately 75 percent of police–citizen contacts, and nearly 20 percent of these cases result in an arrest. In contrast, when no situational evidence is available, only 0.5 percent of cases result in arrest.[9] Suspects are significantly more likely to be arrested when evidence is present, for example, when an officer hears a suspect confess, hears claims from others regarding the suspect's involvement in the crime, observes physical evidence, or sees the suspect commit the act.[10]

Suspicious Behavior

Merely acting suspiciously does not provide a legal justification for an arrest. Nevertheless, a police officer's decision to stop and possibly arrest a suspect often begins when the officer observes someone engaging in suspicious behavior, such as wearing a long coat while shopping in a department store during the heat of the summer or driving a car very slowly in an area known for drug sales.

Extralegal Factors

Extralegal factors are elements of a police–citizen encounter or characteristics of a suspect or of the officer that have nothing to do with the actual crime, yet nonetheless influence the decision-making process. Factors such as race, ethnicity, sex, socioeconomic class, and demeanor may all affect an officer's perception of a suspect. Ultimately, however, the decision to arrest tends to be the result of behavioral cues such as the person's appearance, the location, or the time the suspect is observed.

Race

Extensive research on the relationship between race and police discretion has produced inconsistent findings. Most research shows that police discretion is affected by the race and ethnicity of a suspect when offenses are comparable.[11]

Proportionally, police arrest more African Americans than whites **TABLE 8-3**.[12] A variety of explanations have been suggested for this disparity:

- Law enforcement agencies receive a disproportionate number of calls for assistance from African American neighborhoods and, therefore, assign more vehicles to patrol those neighborhoods, resulting in more opportunities to observe minorities engaging in crimes.[13]

- Police stop and question African Americans at higher rates, and record these encounters, which increases the likelihood that arrests will be made.[14]

- Police perceive African Americans as being more likely to engage in serious crimes than whites.[15]

- African Americans commit a disproportionate amount of serious criminal behavior.[16]

At least one study found that police treated minorities more leniently than they treated whites, but the majority of African Americans continue to believe that they are personally harassed by the police, that

TABLE 8-3

Proportionate Race–Arrest Ratio for Whites and African Americans, 2006

Offense Charged	Proportionate White–African American Arrest Ratio
Murder and non-negligent manslaughter	1:8
Forcible rape	1:4
Robbery	1:10
Aggravated assault	1:4
Burglary	1:3
Larceny-theft	1:3
Motor vehicle theft	1:4
Arson	1:2
Violent crime	**1:5**
Property crime	**1:3**
Total crime	**1:4**

To calculate the proportionate white–African American arrest ratio, the number of arrests of African Americans was multiplied by 7.7 (there are 7.7 times more whites than African Americans in the U.S. in 2006). That number was then divided by the *actual* number of white persons arrested for the same crime. For example, to read this table, go to the column labeled "White–African American Arrest Ratio" and see the proportionate ratio corresponding to forcible rape. The ratio is 1:4. In 2006, one white person was arrested for this crime, while the proportionate number of African Americans arrested for forcible rape was 4.

Sources: Federal Bureau of Investigation, *Crime in the United States, 2006* (Washington, DC: U.S. Department of Justice, 2007) Table, 43; U.S. Census Bureau, *Statistical Abstract of the United States, 2007,* 126th ed. (Washington, DC: U.S. Census Bureau, 2007).

police surveillance is discriminatory, and that clear racial differences exist in terms of whom police officers watch and stop.[17]

Joan McCord and her colleagues believe that the vast amount of research on police–citizen encounters supports claims of racial bias by police, although the findings are somewhat inconsistent. They explain away the inconsistencies by contending that police practices vary by location (e.g., city versus rural departments), and by noting that policies and administrations change over time. Evidence of police suspiciousness of minorities, coupled with alleged discriminatory practices and beliefs by police, frequently produces hostile feelings among African Americans toward police. As a consequence, African Americans are more likely to interact with police in a more antagonistic or disrespectful manner than whites do, which may in turn produce a greater likelihood of arrest.[18]

LINK Racial and ethnic minorities may also experience discrimination in sentencing. This concern and the extent of racial disparities in sentencing are examined in Chapter 12.

African American police officers act more harshly toward African American citizens than white officers do.

? Is this research surprising? How can this phenomenon be explained?

Sex

Research has shown that police officers are more suspicious of males than of females. In fact, one study found that more than 76 percent of police officers agreed with the statement that "If two or more males are together, they are probably committing a delinquent act."[19] Similarly, other studies have reported that police are less suspicious of women than they are of men.[20] Males typically commit much more serious crimes and commit crime in general more frequently than do females, and men are significantly more likely to be arrested (accounting for approximately 76 percent of all arrests) than women.[21]

Some studies report that women are treated more leniently in the criminal justice system than men, although other studies have failed to confirm this finding.[22] For example, researchers have found evidence to support the following assertions:

- Police generally treat female suspects more leniently, but they are more likely to arrest females than males for sex offenses.[23]
- Police treat females with greater compassion, even when the case is serious.[24]
- Although females who commit serious felonies are less likely to be arrested than men, they are

more likely than males to be arrested for less serious crimes.[25]
- Law enforcement officers adopt a more paternalistic and punitive attitude toward young females in an attempt to deter them from engaging in further inappropriate sex-role violations.[26]
- Females who violate middle-class expectations of traditional female roles do not receive more lenient treatment by police.[27]

Socioeconomic Class

Most research has suggested that the police treat people similarly for comparable offenses regardless of the suspects' social standing, although the seriousness of offending varies between classes.[28] Researchers report that suspects police encounter in lower-class neighborhoods are more likely to be arrested than persons whom law enforcement stop in middle- or upper-class areas.[29] This difference may, in part, reflect two facts: (1) lower-income persons are more likely to be repeat offenders and (2) persons from lower-class neighborhoods account for a larger proportion of petty offenses that generally result in high arrest rates.[30] In addition, police allocation of resources (patrolling activities) is likely influenced by neighborhood-level social class. While individual officers may respond to suspicious behavior consistently across classes, police may be more likely to observe suspicious behavior in neighborhoods characterized by lower socioeconomic status, simply because they tend to have a greater presence in those communities in the first place because of the larger number of calls from the public reporting crimes.

Social class also plays a role in police arrests of juveniles. That is, juveniles from middle- and upper-class families are often treated more leniently (perhaps because their families have more resources to help minimize their involvement with the juvenile justice system), whereas parents of lower-class youths more frequently look to the police and probation officers to help them control their children.[31]

Demeanor

While race, sex, and socioeconomic class may all play some role in an officer's decision to make an arrest, the attitude and demeanor of the suspect almost always affect the decision-making process.[32] Studies show that police decisions to arrest suspects are frequently based on character cues found in police–citizen interactions. A person's demeanor is a primary predictor of arrest decisions in 50 to 60 percent of the cases.[33] Specifically, an arrest is a more

Adam "Pacman" Jones has had more than 10 police contacts since entering the NFL and was suspended for the 2007-08 season due to pending charges.

Why are certain people more likely to draw police attention than others? Which personality traits increase a person's likelihood of being arrested?

curring in only 4 percent of observed encounters. Such a tendency toward disrespectful behavior by police varies widely based on the age of the citizen: Police are three times more likely to disrespect children (ages of 6 to 12) than seniors (more than 60 years old).[38] Studies have also shown that race is a factor in terms of police disrespect: White children are more likely to be disrespected by the police than are minority youth among nonresistant suspects.[39]

Additional Extralegal Factors

Characteristics of the police officers themselves may also affect the arrest decision:[40]

- Younger officers are more likely to arrest suspects than are older officers.[41]
- College-educated officers are less likely to make arrests than officers with no college education.[42]
- African American officers generally adopt a more aggressive patrol style and make proportionally more arrests than white police, especially among African American citizens.[43]
- Female and male police arrest suspects at about the same rate.[44]

In addition to these personal characteristics, aspects of the social organization within an officer's police department may affect the arrest decisions that he or she makes. For example, James Q. Wilson found that three factors of agency organization influenced the way officers treated suspects:[45]

- Department organization
- Strength of connections to the local community
- Formal and informal organizational norms

Wilson found that in bureaucratized agencies characterized by direct supervision of officers, police were expected to apply a strict interpretation of department rules when dealing with suspects. In contrast, police officers in more fraternal agencies without systematic rules for guiding decision making used personal judgments to make arrest decisions, which were then affected by individual and situational differences.

Wilson's study suggests that a combination of centralized management and close supervision creates situations in which officers in the field are more likely to follow department policy. Other studies have demonstrated that departments with greater bureaucratic control are also more likely to have policies emphasizing counsel and release dispositions, which tends to result in higher rates of counseling and releasing of suspects. By contrast, in departments characterized by low bureaucratic control, an emphasis on following department policies has little effect on

likely outcome for citizens who are disrespectful of the police.[34] People who are hostile are nearly three times more likely to be arrested than are those who are not hostile.[35]

Research has found that while police are initially trying to establish a relationship with a suspect, they may interpret the suspect's demeanor as evidence of acceptance or rejection of the attempt to build trust. Failure to display an appropriate demeanor (i.e., deference to authority, contriteness, politeness) is viewed by officers as a violation of that trust and, therefore, is more likely to lead to an arrest.[36] Noncompliance or verbal resistance in front of other officers further increases the likelihood of arrest. Indeed, suspects who are hostile toward an officer when other police are present are four times more likely to be arrested than nonhostile suspects.[37]

Police may also treat citizens with disrespect, though this kind of unprovoked behavior is rare, oc-

Cop Moonlights as Prostitute

A New Zealand policewoman who was moonlighting as a prostitute was censured by the Auckland Police Department. Prostitution is legal in New Zealand, and police officers are permitted to take approved second jobs, but the Auckland police department has decided that prostitution was unauthorized for officers, even when they are working undercover. The officer had moonlighted as a prostitute only for a short time in Auckland to make some extra money before her concealed activity was uncovered, though neither her name nor her rank have been made public.

The officer has been allowed to keep her day job as a police officer but was told that she would have to give up her job as a prostitute. When asked about the officer's activity in the sex trade industry, New Zealand Police Minister Annette King said it would be inappropriate for her to comment because the matter was an internal police employment issue.

Source: "Lady Cop Goes Undercover . . . Um . . . *Really* Undercover," *FoxNews.com,* July 20, 2006, available at http://www.foxnews.com/story/0,2933,204774,00.html, accessed August 30, 2007.

---

disposition rates.[46] However, in a recent study, Robin Engel and her colleagues examined the effect of close supervision on arrest decisions and found that management styles of police supervisors had little or no impact on the decision to arrest.[47]

Regulating Police Discretion

Police discretion is a double-edged sword: It means that justice is not evenly applied to all members of society, so some citizens may be denied due process of law or given preferential treatment.[48] As a result, police administrators create safeguards through written rules and technology to help regulate police discretion.

Written rules are the most widely used method for controlling discretion. Such rules provide police officers with guidelines regarding what action they may take in specific situations. Nearly every municipal and county law enforcement agency today has specific regulations for controlling the following issues:

- When force may be used and to what extent
- How and when to participate in high-speed chases
- How to handle special populations (e.g., juveniles, mentally ill persons, and the homeless)
- How many hours per week police may work
- Which types of employment police may accept outside of their regular shift hours[49]

In addition to written rules, police administrators may rely on technological developments to track officers while they are on duty. For example, the Automatic Vehicle Locator (AVL) system uses a Global Positioning System (GPS) device to monitor patrol cars. With an AVL system, a police dispatcher can pinpoint the longitude, latitude, ground speed, and course direction of every patrol vehicle in operation at a given time; the dispatcher can also route the vehicle to a particular location if necessary. The AVL system provides the dispatcher with a real-life snapshot of where police vehicles are so that he or she can advise citizens as to when an officer will arrive. This system also reduces police response time because the dispatcher can direct the closest patrol vehicle in the area to the scene. With this approach, administrators are able to more closely monitor officer activities.[50]

Police Corruption

As with any position of power, there is always the potential for corruption within police departments—that is, the misuse of authority by officers for the benefit of themselves or others. There are innumerable ways for an officer to become involved in corruption: Some seek out opportunities for economic gain, some are tempted as they observe other officers engage in corrupt activities, and some find themselves becoming corrupt as a result of bad decisions involving deals made with criminals. Some corrupt activities are benign, whereas others are much more serious.

Several criminologists, including Julian Roebuck and Thomas Barker, have developed typologies of police corruption that group such actions into conceptual categories, in increasing order of seriousness.[51] These include:

AROUND THE GLOBE

Noble Cause Corruption

Noble cause corruption is a type of corruption that some police and civilians believe is justified because it may serve the greater public good. For example, they believe police should be permitted to beat a confession out of a known murderer or to fabricate evidence against him or her so that the offender will be convicted and be put behind bars. Noble cause corruption is at the heart of a recent case in Victoria, Australia, where police detectives assaulted a group of suspected armed robbers. Their attack on the suspects, unbeknownst to the officers, was recorded by the Office of Police Integrity. The video showed the detectives beating suspects to obtain a confession.

When this video was exposed to the public, many civilians stated that the detectives did nothing wrong. In fact, a poll found that 80 percent of respondents said they did not care if squad detectives assaulted suspects and that the crooks deserved what they received.

The victims of noble cause corruption are often the most vulnerable individuals in society, particularly those who are least likely or able to complain. Police may justify their behavior toward these individuals as a means of "cleaning up the community" and making the streets safer for law-abiding citizens or point to the pressures put upon them to solve crimes and lock up criminals.

Some of the most common forms of this type of corruption include the following:
- Perjury
- Planting evidence
- Denial of basic rights
- Assaults and pressure to induce confessions
- Posing as a solicitor to advise suspects to cooperate with police
- Tampering with electronic interception

Many police who are frustrated with the criminal justice system's lax treatment of offenders see noble cause corruption as one way of "evening the score." They also believe they are doing what civilians want them to do, and the results of the public opinion poll suggest they may be right.

Source: Gary Hughes, "Police Ostracise Clean-Up Officers," July 19, 2007, available at http://www.theaustralian.news.com.au/story/0,25197,22096719-5006785,00.html, accessed July 25, 2007.

- *Corruption of authority.* The most common form of corruption occurs when an officer accepts a small gratuity for services, such as a free meal for being at a restaurant while in uniform.[52]
- *Kickbacks.* An officer may receive goods or services for referring business to individuals or companies.
- *Opportunistic theft.* Officers may take advantage of situations they are in—for example, stealing from intoxicated citizens.[53]
- *Shakedowns.* An officer may extort money from a citizen with a threat to enforce a law if the officer is not paid, or an officer may offer to accept a bribe in return for ignoring an offense.[54]
- *Protection of illegal activities.* Officers may systematically accept bribes for protecting ongoing criminal activity, thereby allowing individuals and businesses to commit crimes such as those committed in drug operations.[55]
- *"Fixing" charges.* Police sometimes undermine criminal investigations or proceedings, such as "fixing" a traffic ticket by failing to show up to testify in court against the defendant.
- *Direct criminal activities.* Some police commit crimes against persons or property, such as using their patrol vehicles to transport drugs for dealers.[56]
- *Internal payoffs.* Officers may barter, buy, and sell favors to other officers, such as falsifying the score of promotional exams.[57]
- *"Flaking" or "pudding."* An officer may place a firearm at a crime scene to give the impression that a suspect who was shot and killed by police was armed so as to justify the shooting.[58]

Department Corruption

In addition to the corruption of individual officers, entire departments may be corrupt. Corrupt departments range from those in which there are only a few dishonest officers to those characterized by pervasive organized corruption.

When police supervisors are asked about corruption, many will admit that a few corrupt officers accept bribes and sometimes commit crimes. Few will admit to the existence of small groups of corrupt officers who work together in a manner similar to a criminal gang. On the agency level, criminal activity may be widespread but unorganized (i.e., officers regularly take advantage of situations without coordination or discussion among other officers), or it may be organized into a complex system of corruption replete with payoffs, theft, and extortion (the use of or implicit threat of the use of violence or other criminal means to cause harm to person, reputation, or property as a means to obtain property from someone else with his or her consent).

Investigating Police Corruption

If widespread corruption is discovered, city managers or mayors may form a commission to investigate the breadth and depth of the illegal activities. The most important commissions to investigate police corruption in the United States to date have been the Chicago Crime Commission, the Wickersham Commission, the Knapp Commission, and the Mollen Commission.

Started in 1919 to combat organized crime, the *Chicago Crime Commission* was the first of these watchdog groups to be formed, and it continues to operate today. The purpose of the Commission is to keep a watchful eye on organized crime throughout Chicago. Today, the Commission's efforts focus on monitoring the city's criminal justice system primarily for carelessness, corruption, and leniency. The Commission is also a strong proponent of a more efficient criminal justice system in Chicago and promotes deterrence through severe punishment.[59]

In 1929, President Herbert Hoover appointed George Wickersham to head the *Wickersham Commission,* which had been charged with identifying the causes of crime, recommending social policies for preventing it, and examining the failure by federal, state, and local police to enforce Prohibition, which had been established by the Eighteenth Amendment in 1920.[60] In their report, the members of the Commission documented innumerable instances of police participating in bribery, entrapment, coercion of witnesses, fabrication of evidence, and illegal wiretapping. Curiously, the Commission recommended to President Hoover that Prohibition should not be repealed. Its recommendations on this front were ignored, however, and in 1933 Congress passed the Twenty-First Amendment repealing

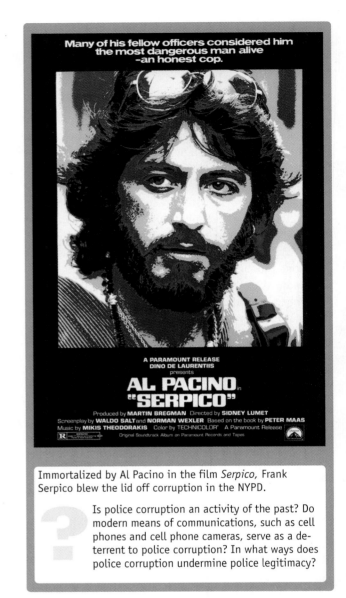

Immortalized by Al Pacino in the film *Serpico,* Frank Serpico blew the lid off corruption in the NYPD.

? Is police corruption an activity of the past? Do modern means of communications, such as cell phones and cell phone cameras, serve as a deterrent to police corruption? In what ways does police corruption undermine police legitimacy?

Prohibition, with state ratification conventions quickly endorsing the amendment.

The *Knapp Commission* was formed in April 1970 by Mayor John Lindsay to investigate police corruption inside the New York Police Department (NYPD). Its roots can be traced to the publicity generated by the public revelations of police corruption made by Patrolman Frank Serpico and Sergeant David Durk. Following an exhaustive review of hundreds of documents and countless interviews with officers and supervisors, the Knapp Commission issued its final report in 1973. It identified two types of corrupt police officers: *grass-eaters* (those who accept bribes when offered) and *meat-eaters* (those who aggressively misuse their power for personal gain). It was the conclusion of the Knapp Commission that the majority of corrupt police officers were in the former category.[61]

In 1992, Mayor David Dinkins appointed former judge Milton Mollen to head a commission to once again investigate corruption in the NYPD. The *Mollen Commission* issued its final report in 1994, concluding that the corruption it uncovered in the NYPD was different from what the Knapp Commission had found just two decades earlier. Corruption in the 1970s was largely a matter of accommodation: criminals and police officers giving and taking bribes, and buying and selling protection. In other words, corruption was consensual. By the 1990s, however, corruption had become characterized by brutality, theft, abuse of authority, and active police criminality. Corruption within the NYPD was not only widespread, but well organized and allowed to persist by Internal Affairs investigators and high-level police officials who turned a blind eye to its presence. The Mollen Commission charged that virtually all of the corruption it unearthed involved groups ("crews") of officers who protected and assisted one another's criminal activities. On average, each of these crews consisted of 8 to 12 officers, who operated with set rules and used a group name. They worked in flexible networks, planning and coordinating their criminal raids with the help of intelligence, communications, and special equipment from their departments.[62]

Reasons for Police Corruption

Criminologists offer many explanations for police corruption, including:

- *Limited accountability.* Police are often poorly supervised and are not held accountable for many of their actions.
- *Officer secrecy.* The police subculture isolates officers from the public, creating a "blue wall of silence" that prevents police from "snitching" on one another or discussing police business with outsiders.

LINK Chapter 6 explores the strict code of police conduct that reinforces secrecy, emphasizes taking care of one another, and has significant consequences for police–citizen relations.

- *Managerial secrecy.* Supervisors are not exempt from the police subculture and often support the "code of silence." As a consequence, they may hesitate to investigate charges of corruption due to group loyalty.

Studies have shown that certain characteristics help explain the predictability of police corruption:

- *Pre-employment history.* Officers whose life histories include records of arrest, traffic violations, and failure in other jobs are more likely than others to become involved in corruption.
- *Education.* Officers who hold associate or higher degrees are less likely to be terminated due to criminal involvement.
- *Training.* Officers who do well in the Police Academy's recruit training program are less likely than marginal recruits to eventually be terminated due to corruption.
- *Diversity.* Agencies with more racial and ethnic diversity among officers tend to have less corruption.[63]

The message to police administrators is clear: To minimize corruption, police agencies must hire officers with clean histories and strong educations. Once hired, officers must be well trained and closely supervised to make certain that minor problems with the department's internal disciplinary system do not escalate into career-ending misconduct.[64] Police supervisors must admit when corruption exists and confront the problem. Furthermore, they must recognize that corruption often begins at the top and drifts downward through the ranks, so police managers must lead by example. Sincere and candid administrators establish the parameters for what is considered acceptable behavior, which strongly affects the recruitment and promotion processes.

Police Brutality

Police brutality is the unlawful use of force, language, and application of the law. Generally speaking, it refers to excessive force and all "unnecessary force" used by police. Such use of force is a crime.

The use of excessive use of force by police officers is an unfortunate but constant aspect of policing history.[65] As early as 1931, an investigative commission found the widespread, systematic use of coerced confessions with force, violence, and psychological threats as well as many incidents of excessive force during street encounters with suspects.[66] Even today, the Human Rights Watch Organization estimates that thousands of incidents of police use of excessive force occur each year, only a fraction of which are reported and even fewer of which are formally investigated.[67]

One night in New Orleans' famed French Quarter, a 64-year-old retired school teacher named Robert Davis was out for a walk when he encountered a police officer on horseback. What exactly happened next is unclear. According to Davis's attorney, a second officer approached and made some rude remarks to Davis, who responded by saying, "I think that was unprofessional." As Davis turned to walk across the street, he claims that one of the officers struck him from behind.

What is known is that after this brief encounter, officers hit Davis at least four times on the head and dragged him to the ground. One officer kneed Davis and punched him twice.

The entire incident was caught on camera by a television news crew covering the aftermath of Hurricane Katrina. The video ended with Robert Davis lying on a sidewalk with his head and shirt soaked in blood.

Police charged him with public intoxication, resisting arrest, battery on a police officer, and public intimidation. Besides the concerns about police brutality, this violent incident also raised civil rights issues since Davis is African American, and the three officers are white.

Two officers involved in the attack, Robert Evangelist and Lance Schilling, were fired. Officer Evangelist was charged with false imprisonment and second-degree battery. The third officer, Stewart Smith, was charged with simple battery and, if convicted, will face up to 6 months in jail and a $500 fine. Smith was suspended for 120 days but remains on the force today.

Davis pleaded "not guilty" to all the charges which were later dropped. On June 11, 2007, Officer Schilling was found dead from a self-inflicted gunshot only one month before his trial was to begin.

Sources: "New Orleans Officers Plead Not Guilty," *CNN.com,* October 11, 2005, available at

http://www.cnn.com/2005/LAW/10/10/taped.beatings/index.html, accessed June 7, 2007; Mary Foster, "New Orleans Officers Indicted in Beating," *Sacbee,* March 30, 2006, available at http://www.sacbee.com/24hour/special_reports/katrina/story/3244298p-12001341c.html, accessed May 29, 2007; "Man in Court after Police Beating," *CBS News,* October 12, 2005, available at http://www.cbsnews.com/stories/2005/10/12/national/main937449.shtml, accessed June 26, 2007; "Victim of Police Beating Says He Was Sober," *Associated Press,* October 10, 2005, available at http://www.msnbc.msn.com/id/9645260/, accessed June 26, 2007; Cyndi Nguyen, "A Former NOPD Officer Accused in a Videotaped Beating Takes His Own Life," available at http://abc26.trb.com/news/wgno_071207suicide,0,2503818.story?coll=wgno-news-1, accessed June 28, 2007.

LINK In one of the most infamous cases of police brutality, the 1991 assault on Rodney King (discussed in Chapter 1), the LAPD officers charged in the case were eventually acquitted, and the streets of Los Angeles filled with angry rioters in response.

A commission formed after the 1991 assault of Rodney King by members of the Los Angeles Police Department (LAPD) found that 5 percent of all officers accounted for more than 20 percent of allegations of excessive force, and 28 percent of officers agreed that prejudice may have led to the use of excessive force in these situations.[68] This finding is supported by further research indicating that victims of excessive force are typically younger, lower-class, minority, and male.[69] Additionally, victims of excessive force tend to be under the influence of alcohol or drugs.[70] Other victims of police brutality are suspected of committing a violent crime.[71] The officers charged with using excessive force in such incidents are usually less-experienced males.[72]

These findings led the commission to recommend that specific steps be taken to identify "violence-prone" officers before they act out.[73] One way to reduce police brutality is by creating a more balanced approach to address citizen complaints of excessive use of force by bringing such complaints to independent review boards. As long as police continue to investigate themselves, suspicions of undisclosed corruption and brutality will inevitably persist. Means suggested to remedy this situation include more effective disciplinary procedures, refined police selection criteria, more thorough police training on appropriate use of force, and instruction on alternative methods to maintain control when a suspect is resisting arrest.[74]

SWAT Teams and "No-Knock" Raids

Special Weapons Attack Teams (SWAT teams) were originally formed in 1966 by Los Angeles Police Chief Daryl Gates. At that time, SWAT teams were used mostly for entering barricades and resolving hostage situations. They were not regularly used to capture drug couriers and dealers until the escalation of the "war on drugs" in the 1980s.

Prior to entering a residence, SWAT teams are typically required to knock on the door, announce themselves, and present a search warrant to the occupants. However, in the past few decades many exceptions have been made to this rule, so that teams may often forcibly enter a residence now without making their presence known. Such "no-knock" raids are becoming more common, in part because more drug search warrants permit such entries to occur, with the goal of preventing evidence from being destroyed or flushed down the toilet.

Unfortunately, no-knock raids are prone to mistakes. Some have even turned tragic when officers mistakenly used lethal force to defend themselves when the situation did not necessitate such conduct. Consider these examples:

- Brooklyn, New York: A SWAT team stormed into the apartment of Martin Goldberg, 84, a decorated World War II veteran, and his wife Leona, 82. Both were ordered to the floor. Leona Goldberg was hospitalized after the raid for an irregular heartbeat. The police later reported that

they had the wrong address and meant to enter an apartment on the opposite side of the street.

- Sunrise, Florida: Anthony Diotaiuto, a part-time community college student, had 10 bullets fired into his chest, head, torso, and limbs when police raided his house looking for drugs, weapons, and drug paraphernalia. The police found less than one ounce of marijuana in his home, but they claimed Diotaiuto had a gun in his possession and was pointing it at the invading officers. Although the gun was later found near Diotaiuto's body, he had a conceal-carry permit for it. Neighbors speculate that because there was no announcement by the officers before entering, Diotaiuto sought to use the gun to defend himself against what he believed to be intruders.

Such potential tragedies of mistaken intrusions, humiliation, and unnecessary violence have increased in recent years as no-knock raids have become more common. Although SWAT teams are undoubtedly necessary in highly volatile situations, the number of no-knock raids that either occur at the wrong residence or end in death continue to provoke public debate over this tactic.

Sources: Peter Kraska and Louis Cubellis, "Militarizing Mayberry and Beyond," *Justice Quarterly* 14:607–629 (2006); Radley Balko, *Overkill* (Washington, DC: Cato Institute, 2006); Randley Balko, "Drug War Police Tactics Endanger Innocent," *foxnews.com,* July 21, 2006, available at http://www.foxnews.com/story/0,2933,205040,00.html, accessed May 29, 2007.

Deadly Force

When police find themselves in dangerous and volatile situations, they must act quickly. An officer does not have time to call a supervisor and ask what he or she should do. When an explosive situation presents itself along with the possibility that the officer may be prosecuted if he or she makes the wrong decision, the officer is in a particularly precarious position.[75] If the officer chooses to use too little force, the officer may endanger both his or her life and the lives of other officers and innocent bystanders. When too much force is used, suspects may be killed or seriously injured, and the officer may face public con-

demnation, discipline by the department, and possibly prosecution.[76]

The most severe action an officer can take against a citizen is deadly force. The standards regulating deadly force have changed considerably over the years. The colonial approach to deadly force mirrored its English predecessor: the fleeing felon doctrine, which stated that if an individual suspected of committing a felony fled, a police officer was permitted to use deadly force to stop the suspect.[77] However, in 1985, the U.S. Supreme Court ruled in *Tennessee v. Garner* that the fleeing felon doctrine was unconstitutional. In this case, police shot 15-year-old Edward Garner in the back as he ran from a house. The Court stated

VOICE OF THE COURT
Tennessee v. Garner

In 1974, two Memphis police officers were dispatched to answer a call from a woman about a prowler. When they arrived on the scene, they saw a woman standing outside on her porch gesturing toward the adjacent house. She told the officers that she heard glass shattering and that someone was breaking in next door. One of the officers went behind the house, while the other officer radioed for backup. While looking outside, one of the officers heard a door slam and saw someone running across the backyard. With the aid of a flashlight, the officer reported that he saw no signs of a weapon and was reasonably sure that the suspect was unarmed. The officer yelled at the man—Edward Garner—to halt. Instead, Garner attempted to climb over a fence. When he did, the officer shot him. Garner died from the gunshot wounds. Police later found that Garner had stolen 10 dollars and a purse.

At that time, Tennessee statute instructed police to shoot to kill fleeing felons rather than risk their possible escape. The U.S. Supreme Court ruled that the fleeing felon law was unconstitutional, arguing that deadly force is a seizure and that seizures must conform to the reasonableness requirement of the Fourth Amendment. Because Garner had posed no immediate threat to the officer or to others, the legal force used to apprehend him did not justify the resulting harm.

Source: Tennessee v. Garner, 471 U.S. 1 (1985).

---

that "When the suspect poses no immediate threat to the officer and no threat to others, the harm resulting from failing to apprehend him does not justify the use of deadly force to do so."[78]

A new standard for the use of deadly force, the <u>defense of life standard</u>, essentially says that officers may use deadly force only in defense of their own lives or another's life. For police, the impact of the *Garner* ruling has been profound. The decision in *Garner* opened the door for all use of force by police to be looked at from the "reasonableness standard." In response to the *Garner* outcome, police departments across the nation have quietly expanded the Supreme Court's ruling by implementing a <u>preservation of life policy</u>, which mandates that officers use every other means possible to maintain order before turning to deadly force.

Police in Springdale, Arkansas, for example, recently introduced a graduated use of force scale instructing officers to take the following steps when confronting a dangerous suspect:

1. Identify himself or herself as a police officer
2. Give the suspect a verbal command to terminate his or her activities
3. Use (in order) soft hand restraints, chemical spray or stun gun, physical restraints, or a baton
4. Rely on deadly force as a last resort[79]

The policy used in Springdale parallels many of the newer standards being implemented in police agencies across the United States.

Even with policies like these which seek to reduce the use of deadly force, between 1976 and 2004 police justifiably killed more than 10,000 citizens, an average of 371 people each year FIGURE 8-3 .[80] If an officer contributes to the unnecessary death of a citizen because of his or her reckless behavior, the officer may be held criminally liable and be prosecuted. Research has shown that the most likely victim of deadly force is an unarmed, African American male between the ages of 17 and 30, who is out at night in a public location, with some connection to an armed robbery.[81] In fact, racial and ethnic minorities are killed by police in disproportionate numbers.[82]

Research on deadly force has also uncovered the following relationships:

- Use of deadly force corresponds with neighborhood crime rates.
- African American officers are more likely than White police to use deadly force.
- Male officers are more likely to use deadly force than female officers.[83]

In reality, very few police–citizen contacts end with the use of lethal force. Of the more than 45.3 million police–citizen contacts that occur annually, only about 1.5 percent of citizens report police use of force. Nearly all force used against citizens is nonlethal. The most frequently used forms of force by police include these measures:

- Pushing or grabbing
- Kicking or hitting

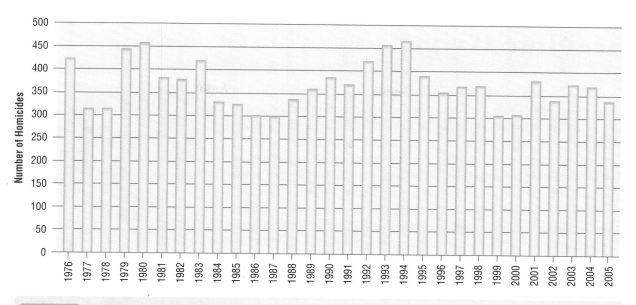

FIGURE 8-3 Justifiable Homicides by Police

Source: Federal Bureau of Investigation, *Crime in the United States, 2006* (Washington, DC: U.S. Department of Justice, 2007).

- Pointing a gun
- Threatening to use force

Such nonlethal force is used on more than 650,000 people per year.[84]

High Speed Chases

LINK The automobile is the police officer's most deadly weapon. During a high-speed chase, it becomes a 3000-pound mass of steel accelerating through crowded neighborhoods and residential streets at speeds sometimes exceeding 100 miles per hour. The introduction of patrol cars and the important role they play in policing strategies are discussed in Chapter 5.

Nearly as many citizens are killed as a result of high-speed chases as from police shootings; there are roughly 350 deaths from the former cause each year.[85] Fleeing suspects and innocent bystanders are not the only ones at risk during high-speed pursuits; sometimes a police officer is killed or seriously injured during a chase.[86] A high-speed chase becomes dangerous very quickly. In 50 percent of such pursuits, a collision is likely to occur within the first two minutes, and 70 percent of all high-speed chase collisions take place within the first six minutes.

To decrease the danger associated with high-speed chases, officers in some departments today are trained in defensive driving tactics. Nevertheless, the most effective method for reducing fatalities is for the officer to terminate the chase. In a study of

146 jailed suspects who had been involved as drivers in high-speed chases, more than 70 percent of them said they would have slowed down if police had stopped chasing them.[87]

Training alone will not prevent high-speed chases; department policy is equally important. Most departments are formalizing procedures and enforcing written policies regarding when police may participate in a pursuit of a fleeing suspect **TABLE 8-4**. Ninety-four percent of local police departments, including all of those serving 25,000 or more residents, have a written policy governing high-speed chases. Sixty-one percent of departments, employing 82 percent of all officers in the United States, have a restrictive pursuit driving policy—one that restricts pursuits according to specific criteria such as type of offense or maximum speed. Twenty-five percent of departments, employing about 13 percent of police officers in the United States, have a judgmental pursuit policy—one that leaves pursuit decisions to the officer's discretion. Only 6 percent of departments, employing 3 percent of all officers nationwide, have a policy that discourages high-speed chases.[88]

There are no federal guidelines regulating police chases. The decision to chase is initially made by the department and executed by the officer using the following criteria:

- Severity of the offending infraction
- Speed of travel
- Number of pedestrians and vehicles on the street
- Weather conditions

Headline Crime — Officer Killed in Traffic Pursuit

On March 17, 2005, about 2:50 P.M., Thomas Catchings, a patrol officer with the Jackson (Mississippi) Police Department, was shot and mortally wounded during a traffic pursuit. The 40-year-old officer, who had nine years of law enforcement experience, was patrolling on his motorcycle when he was flagged down by a man who told the officer that just minutes before he had been the victim of a carjacking. The man provided the officer with a description of his vehicle.

Shortly thereafter, Catchings spotted the stolen vehicle and began to pursue it in a high-speed chase. The driver of the stolen car lost control and wrecked the vehicle in a ditch. The officer pulled up to the scene, parked his motorcycle, and walked toward the driver's side of the vehicle. The 18-year-old suspect climbed out of the car from the driver's-side window and fell onto the ground. He then stood up, pulled out a .38-caliber semiautomatic handgun, and shot the officer once in the stomach. Catchings returned fire, hitting the suspect three times and fatally wounding him. The officer was taken to a local hospital where he died of his wound.

Sources: Jackson (MS) Police Department, "Officer Thomas Catchings," available at http://www.nleomf.org/cgi-bin/boardposting.cgi?id=11463, accessed September 19, 2007; The Officer Down Memorial Page, "Patrolman Thomas Drumane Catchings," available at

http://www.odmp.org/officer.php?oid/17627, accessed September 19, 2007.

TABLE 8-4

Circumstances in which Officers May Engage in a High-Speed Chase

	Suspect's Offense			
	Traffic Violation	Misdemeanor	Nonviolent Felony	Violent Felony
Police Agency				
Colorado Springs	Yes	Yes	Yes	Yes
Detroit	No	No	Yes	Yes
Los Angeles	No	Yes	Yes	Yes
New Orleans	No	No	No	No
Oakland	No	No	Yes	Yes
San Francisco	No	No	No	Yes
San Jose	No	No	Yes	Yes
Philadelphia	No	No	No	Yes
Phoenix	No	No	No	Yes

Source: Gabe Cabrera, *Police Pursuit Policies in Other Jurisdictions* (San Francisco: San Francisco City Government, 2004), available at http://www.sfgov.org/site/bdsupvrs_page.asp?id=24020, accessed September 15, 2007.

- Whether the suspect is known and could be apprehended at a later time
- Whether the benefits of apprehending the suspect outweigh the risks of endangering officers, the public, and the suspect[89]

Occasionally, high-speed chases end in death or serious injury, and police may be held accountable.

Because of risks to public safety, many police departments have developed more conservative policies regarding high-speed chases.

 Are there any situations in which motorists might be justified in evading the police and initiating a high-speed chase?

The courts have awarded third parties (e.g., passengers) injured in high-speed chases a monetary settlement. For example, in *Travis v. City of Mesquite (TX)* (1992), the court determined that the officer did not calculate the risk involved in the chase and was liable for damages.[90] However, the U.S. Supreme Court has ruled that police officers and departments cannot be held liable when suspects are injured in high-speed chases as long as they had no intention of physically harming the suspect or worsening the suspect's potential criminal charges. Bystanders, by contrast, may file lawsuits for damages against the officer and the department if it can be shown that the officer did not drive responsibly.[91]

Police Stress

Stress frequently interferes with police officers' ability to perform their jobs to the best of their ability. <u>Stress</u> is an upsetting condition that occurs in response to adverse external influences and is capable of affecting an individual's physical health. Stress often leads to an increased heart rate, a rise in blood pressure, muscular tension, irritability, and depression.[92]

Officers experience stress for a variety of reasons **TABLE 8-5**. In addition to individual characteristics, the most common sources of police stress are difficult decisions, conflict with supervisors, frustration with the courts, and criticism from the public.

During the course of performing their duties, officers regularly experience role conflict and role ambiguity. They are expected to maintain order and provide citizens with services while enforcing the law. Oftentimes they find themselves having to be a counselor, law enforcer, public servant, and social worker, all at the same time. In these situations officers are supposed to follow strict policies and procedures, yet the situations themselves often are ambiguous and "black-and-white" decisions are at best impractical solutions. Volatile situations force officers into a difficult position: They may need to make split-second decisions for their own safety and the safety of others, without knowing whether their decisions will be supported by their supervisiors.[93]

It is not unusual for police to believe they are not supported by their supervisors and their department. In fact, the most common source of stress for officers comes from supervisors who may either overwhelm or under-support officers, providing them with too much paperwork and not enough structure. Other supervisors may apply discipline and enforce

TABLE 8-5

Leading Causes of Police Stress

1. Encountering child abuse
2. Harming/killing an innocent person
3. Disagreement with department policies
4. Harming/killing a police officer
5. Domestic violence
6. Hate groups/terrorists
7. Poor supervisory support
8. Riot control
9. Public disrespect
10. Shift work

Source: Adapted from Dennis J. Stevens, "Police Officer Stress," *Law and Order* 47(9):77–81 (1999).

rules inconsistently, adding to officers' uncertainty.[94] The courts may also be seen as unsupportive, issuing rulings that are viewed as too lenient on offenders and too restrictive on procedural issues (such as rules governing the admissibility of evidence at criminal trials). As a consequence, police may view the courts as making their job more dangerous than it already is and be resentful of the courts' actions.

LINK The courts have set limits on what police can do. Chapter 7 provides a thorough analysis of the procedural restrictions and expansion of police powers determined by the judicial branch.

Officers also frequently complain that they are treated unfairly by the media and the public. Police may think reporters distort the truth to meet publication deadlines, do not understand the complexity of the cases they are reporting on, or simply fail to report the facts. Police may also believe that the public does not support them, instead preferring to challenge what police do and how they perform their jobs.[95] Citizens may submit complaints to the mayor's office, police chiefs, and newspapers criticizing speed traps, slow response times, busy 9-1-1 numbers, or—even worse—police discrimination and brutality. These actions reinforce a feeling among police that they are "damned if they do, and damned if they don't." This belief further alienates police from the public, builds solidarity among police, and contributes to police stress.

In addition to these factors, stress levels are strongly affected by the officer's individual personality and background characteristics such as amount of experience, level of education, and assigned duties. For example, being assigned for a long period of time to a neighborhood with a high crime rate will likely produce more stress for an officer than if he

Sacramento v. Lewis and Scott v. Harris

In *Sacramento v. Lewis* (1988) and *Scott v. Harris* (2007), the question before the Supreme Court was whether the Fourteenth Amendment's substantive due process protection was violated when police caused the death of a suspect by deliberate or reckless indifference during a chase. The Court has answered this question with a near-unanimous "no." The facts of the cases are presented below.

On May 22, 1990, county sheriff James Smith, along with officer Murray Stapp, responded to a call regarding a motorcycle being driven at a high speed by Brian Willard and carrying passenger Philip Lewis. In an attempt to stop the motorcycle, Stapp turned on his emergency lights, yelled for the motorcycle to stop, and pulled his car next to Smith's vehicle in an attempt to pen the motorcycle. This tactic failed, as Willard maneuvered between the two cars and sped off. Smith immediately switched on his siren and began the high-speed chase. The chase ended when the motorcycle tipped over; Smith slammed on his brakes, but his car skidded into Lewis, causing his death.

Representatives of Lewis's estate sued, alleging that Lewis had been deprived his Fourteenth Amendment right to life. Following conflicting decisions by lower courts, the Supreme Court held that a police officer does not violate substantive due process by causing death through deliberate or reckless indifference in a high-speed automobile chase aimed at apprehending a suspect.

More recently, in *Scott v. Harris,* the Court ruled that a police officer should not be held responsible for damages caused when he or she rammed a speeding vehicle in an attempt to stop it. In March 2001, during a high-speed chase, Deputy Timothy Scott bumped Harris's car; the car left the road and crashed, leaving the teenage driver a quadriplegic.

Harris sued for damages, claiming that the use of deadly force was a violation of the Fourth Amendment. Scott disagreed and claimed he had qualified immunity from this suit because the law at the time was not sufficiently clear to put Scott on notice that his actions were unlawful. The lower courts sided with Harris; the case was appealed to the 11th Circuit, which also held that Scott did not have qualified immunity because (1) a jury could find that his actions were unreasonable under the Fourth Amendment and (2) the law was clear at the time of the incident. However, in an 8–1 vote, the U.S. Supreme Court reversed the rulings from the lower courts, deciding that a police officer's attempt to terminate a high-speed chase by forcing the vehicle off the road was reasonable and, therefore, did not violate the Fourth Amendment.

Sources: Sacramento v. Lewis, 523 U.S. 833 (1988); *Scott v. Harris,* 127 S. Ct. 1769 (2007).

---

or she is assigned to patrol a low-crime neighborhood. Additionally, officers with more education and training tend to handle stress better than other officers.[96]

When officers experience stress, it can produce emotional, psychological, and physical problems. Studies have shown that officer stress may lead to a variety of extremely negative consequences **TABLE 8-6** .[97]

Critics of these studies complain that the studies are based on small samples that cannot be generalized and that the causal order between stress and these outcomes is difficult to establish; in other words, these destructive consequences may, in fact, be precursors to stress.[98] Additionally, critics of these studies suggest that police may simply do a poor job of managing their stress, such that the maladaptive coping strategies contribute to increased stress levels and negative outcomes.

Women and Minorities in Policing

Women have worked in policing for more than 100 years, but they did not receive full police powers until early in the twentieth century.[99] Lola Baldwin, the first female police officer in the United States, was hired by the Portland (Oregon) Police Department in 1905 to shelter women and children from the unruly crowds and seedy characters that would be roaming the streets when the city anticipated a large influx of people due to a large event (the Lewis and Clark Exposition).[100] The first regularly commissioned police woman was Alice Stebbins-Wells, who was hired in 1910 by the LAPD. By 1925, women

Policing is among the most stressful occupations.

? Do the stressful situations that police face result in depression and other maladaptive behaviors? Should police officers earn higher salaries because of their sacrifice in the name of public service?

TABLE 8-6
Consequences of Stress for Police Officers

Professional Consequences
Absenteeism
Burnout
Corruption
Excessive force

Physical Consequences
Alcoholism
Cancer
Heart disease
Suicide
Fatigue
Lack of exercise

Psychological Consequences
Depression
Post-traumatic stress disorder
Loss of interest in hobbies and activities

Relationship Consequences
Child abuse/neglect
Divorce
Spousal abuse

Source: Adapted from Kent Kerley, "The Costs of Protecting and Serving," in Heith Copes (Ed.), *Policing and Stress* (Upper Saddle River, NJ: Prentice-Hall, 2005), pp. 73–86.

were employed in more than 145 police departments across the United States.[101]

During the next 40 years, the hiring of female officers stalled and the status of women who were working in policing changed very little from what it was at the turn of the twentieth century: working with children, caring for prisoners, and performing secretarial duties. In 1940, only 141 of the 417 largest cities employed any females. Then, in 1967, the President's Commission on Law Enforcement and Administration of Justice released a ground-breaking report that stated women should perform the same duties in policing as men.[102] As a result of this recommendation, women were hired by police forces throughout the country, opening new opportunities for women in the profession.[103]

Since the 1960s, police departments around the country have made great strides in recruiting larger numbers of women and minorities. Today, more than 11 percent of all police officers nationwide are fe-

males (more than 50,000 officers), and they account for nearly 13 percent of all sworn officers in large agencies (those with more than 100 sworn officers).[104] Even though more women are being hired, they still tend to hit the "glass ceiling": Only a few female officers advance beyond the rank of patrol officer over the course of their careers.

Studies have shown that female police are equally as effective as their male counterparts: They consistently perform as well as men, generally use the same techniques to gain and keep control, and are no more likely than male officers to display or use a weapon.[105] In response to domestic violence incidents, female officers have been found to respond more effectively than their male counterparts.[106] Yet women in the force face several unique problems, such as trying to balance pregnancy with regular work assignments (see "Focus on Criminal Justice").

Women are not the only ones entering the police force in great numbers; there are also more African Americans and Latinos in policing today than at any other time in history. The first African American officers were hired in 1861 in Washington, D.C., and the first Latino officer was hired in 1896 in New York City. Since then, their numbers have climbed significantly. By 1900, approximately 3 percent of all U.S.

The 1993 Family and Medical Leave Act (FMLA) established federal minimum leave requirements but it does not cover the full range of issues facing women in policing when they become pregnant. According to the National Center for Women and Policing, pregnant officers still face several crucial concerns:

- *Duties.* Some pregnant officers want to shift their assignments to lighter duty and are unable to do so, while others are forced to move to lighter-duty assignments against their will.
- *Firearms.* Firearms training and range qualification present potential harm to the fetus from exposure to lead poisoning and noise from firearms firing.
- *Suitable uniforms.* Police officers in uniformed divisions face the problem of finding maternity uniforms.
- *Length of maternity leave.* Currently, female officers receive only 12 weeks of parental leave, but some require more before they are ready to return to their duties.

Source: National Center for Women and Policing, "Workplace Issues," available at http://www.womenandpolicing.org/workplace4%7E pregnancy.asp, accessed June 30, 2007.

---

police officers were African Americans, most of whom worked in northern metropolitan areas. The percentage of African American officers held steady until after World War II. Since then, there has been a steady increase in African Americans' numbers on police forces relative to their share of the U.S. population.[107] In 2004, approximately 12 percent of all law enforcement officers nationwide were African Americans (totaling 54,000 officers), representing an increase of more than 1500 officers since 2000, and the number of African American officers is now roughly equivalent to their share of the U.S. population.[108] In contrast, Latinos account for about 14 percent of the U.S. population but only 9 percent of police officers, although they—like African Americans and women—are making gains in the profession FIGURE 8-4 .[109]

In part, the increase in minority officers is a result of lawsuits filed by African Americans and Latinos charging police agencies with racial and ethnic discrimination regarding the entrance requirements and promotion examinations. The U.S. Supreme Court decisions handed down in *Griggs v. Duke Power Company* (1971), and *Albemarle Paper Co. v. Moody* (1975), for example, supported defendants' claims of discriminatory practices, and today law enforcement agencies are deemed to be in violation of federal law

if their hiring practices are discovered to be unfair.[110] The decisions in these two cases make it clear that police agencies must be able to demonstrate that their entrance requirements for hiring and promotion are job related, bias free, fairly administered, and properly graded.

LINK Even when the selection process is fair and unbiased, there are still only limited rewards available in police organizations, as discussed in Chapter 6.

The Equal Employment Opportunity Act of 1972 laid the groundwork for the establishment of affirmative action programs and quota systems for hiring and promotion of police officers. In 1987, in *United States v. Paradise,* the U.S. Supreme Court questioned the use of promotion quotas by the Alabama State Police and required the state to promote one African American officer for each white officer promoted until 25 percent of the top ranks were occupied by African Americans.[111] Critics of this ruling contend that quotas that tie employment decisions to race or ethnicity violate the Civil Rights Act of 1964. They also believe that lowering standards to achieve a targeted quota creates resentment among employees and jeopardizes the ability of law enforcement to serve and protect

It often takes a legal challenge for a law enforcement agency to change its recruiting and promotion practices. In a most memorable case of sexual discrimination, now-retired New York City Police Department (NYPD) detective Kathleen Burke struggled against sexism every day of her career. Burke eventually filed a lawsuit against the department, claiming that she was not promoted because of her gender and that the NYPD instead promoted her male colleagues despite their drunkenness and incompetence.

Burke was subject to a great deal of harassment within the department. Even before she officially became a member of the force, at the pre-selection physical she was forced to listen to "wolf calls" and crude remarks about her body. Burke's immediate supervisor told her that a mother was not fit to be a police officer, but

rather should be at home changing diapers and baking bread. The same supervisor told Burke, "It will be a cold day in hell before I will promote a woman over a man."

Confronting sexism became routine for Burke, who describes what it was like for a woman to work in a testosterone-fueled environment where she was forced to continuously avoid advances from her male colleagues:

If a woman rebuffed a male cop's advances, she was labeled a lesbian. If she accepted his advances, she was elevated to the status of a whore. If she had a drink with another cop, she was a drunk. If she declined, she was stuck up.

Burke won her lawsuit, was promoted to the highest level for a de-

tective, and had an exceptional 23-year career with the NYPD. During her extraordinary career, Burke earned the department's Medal of Honor, the highest commendation for heroism that an officer may receive.

Sources: Kathy Burke and Neal Hirschfeld, *Detective* (New York: Scribner, 2006); Adam Nichols, "They Told Me to Go Change Diapers," *New York Daily News,* October 1, 2005, pp. 1–2, available at http://www.nydailynews.com/news/localv-pfriendly/story/351514p-2999827c/html, accessed June 28, 2006.

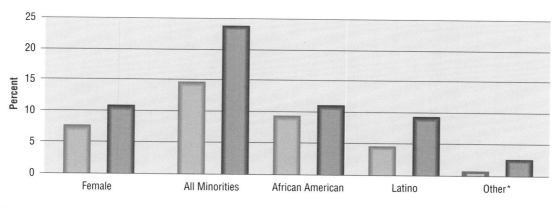

*Includes Asians, Pacific Islanders, American Indians, and Alaska Natives.

■ 1997   ■ 2003

FIGURE 8-4 Female and Minority Local Police Officers

Source: Matthew Hickman and Brian Reaves, *Local Police Departments, 2003* (Washington, DC: U.S. Department of Justice, 2006).

citizens. Conversely, proponents of quotas see these measures as an obligatory remedy for past wrongs. They also contend there is no evidence to show that department standards are lowered when an affirmative action plan is in place.

For police to do their jobs effectively and safely, they must be able to communicate with the people they protect and serve. In particular, they need to immediately understand the complaint of a victim or the information being provided by a witness to a crime. A recent U.S. Census report noted that approximately 20 percent of U.S. residents speak a language other than English at home and approximately 10 percent qualify as limited English proficient. For example, 26 percent of all Spanish speakers, 30 percent of all Chinese speakers, and 28.2 percent of all Vietnamese speakers report that they speak English "not well" or "not at all."[112] For police officers, the task of communicating and building trust with these residents can be immense and presents an enormous challenge for law enforcement agencies committed to developing community policing in neighborhoods through their city.

Communicating in a Multilingual Society

Police agencies are working to develop effective strategies for communicating with suspects, victims, and witnesses who speak little or no English. Additionally, Title VI of the Civil Rights Act of 1964 requires that all police departments that receive federal assistance take reasonable steps to ensure that their services are accessible to *all* citizens—not just to those who speak English. Failure to do so leaves the department vulnerable to lawsuits charging discrimination on the basis of national origin.

A few of the approaches police agencies are taking to overcome linguistic barriers include:

- *Using translators.* Electronic handheld translators, secured telephonic interpretation contracts, and rosters of volunteer interpreters may assist police in communicating with non-English speakers.
- *Language training.* Officers are receiving training in the languages spoken in their communities.
- *Communication tips.* Officers are taught tactics for effective communication with non-English speakers (such as asking children for assistance in translation, because they are often more likely than their parents to understand English).

- *Recruiting immigrant students.* Departments are turning to local high schools and colleges for assistance and are creating new cadet programs to assist in communicating with Asian and Hispanic populations.
- *Combining resources.* Some departments are joining together to share bilingual (or multilingual) officers and translators to reduce the financial burden and overcome language barriers.

While these strategies hold promise, more effort is needed. Regardless of which strategy a department settles on, its ultimate goal must be to enhance communication with people with limited English proficiency; to develop strong, trusting relationships with these residents; and to ensure that officers are able to protect and serve all citizens effectively.

Source: U.S. Department of Justice, "Department of Justice Sends Limited English Proficiency Guidance to the *Federal Register* That Will Help Reduce Language Barriers," April 12, 2002, available at http://www.usdoj.gov/opa/pr/2002/April/02_crt_218.htm, accessed June 30, 2007.

WRAP UP

Legal factors such as the seriousness of the offense seriousness or a prior criminal record influence police behavior in contacting and arresting suspects. At the same time, extralegal factors are also considered by police. These factors—such as race, gender, socioeconomic class, and demeanor—are elements of an encounter that, while not directly related to the crime, may still influence police decisions.

Every type of data source, including official estimates, indicates that minorities (especially African Americans) commit disproportionately more serious crime than whites. Primarily for this reason, police are often suspicious of minority youth, which may lead them to stop African American and Latino juveniles more often than whites and make records of these encounters. Later, when police stop and question these youths again, their prior contact with the police may become a basis for more severe treatment. Police officers' suspiciousness of minorities, coupled with discriminatory practices and beliefs on the part of police, generates feelings of hostility among African Americans. In turn, African Americans are more likely than whites to interact with police in a more antagonistic or disrespectful manner, which may lead to higher instances of arrest.

Chapter Spotlight

- The exercise of police discretion (decision of police to choose between alternative courses of action) is central to officers carrying out their duties. This decision is influenced by several important factors, including legal considerations and extralegal characteristics of the suspect.

- A major problem in law enforcement is the issue of police corruption—that is, the misuse of authority by an officer in a manner designed to obtain some sort of personal gain. Corruption among officers ranges from receiving or demanding minor items from businesses during the course of their duties to extorting cash from suspects, accepting bribes, and engaging in perjury or premeditated theft.

- Police are among the few public servants authorized to use force, but sometimes their use of force is excessive. Police brutality refers to instances when officers use unlawful, unnecessary, or extreme force with suspects.

- The automobile is a law enforcement officer's most deadly weapon. More than 350 citizens are killed annually as a result of high-speed chases—more than are killed from police shootings.

- Stress is inherent in police work. Sources of police stress include rotating shift assignments, fear and danger, limited opportunities for career growth and development, and inadequate rewards.

- The number of women and minorities in policing is increasing every year, but these groups are underrepresented at the supervisory ranks and face unique challenges in police work.

Putting It All Together

1. Should police have limits on their discretion? How can officers exercise discretion and still be fair in their treatment of suspects?

2. Is it possible to regulate police corruption? What might be an effective policy for controlling police corruption?

3. Under what conditions should police use force? At what point does force become excessive? At what point is deadly force reasonable?

4. Should police participate in high-speed chases? Is the benefit of potentially capturing an offender worth the potential costs of such a chase?

5. Should female and male officers be expected to perform the same duties? Does more attention need to be devoted to the recruitment and promotion of racial and ethnic minorities in policing?

Key Terms

corruption Misuse of authority by officers for the benefit of themselves or others.

defense of life standard Policy mandating that officers may use deadly force only in defense of their own lives or another's life.

extortion The use of or implicit threat of the use of violence or other criminal means to cause harm to a person, reputation, or property as a means to obtain property from someone else with his or her consent.

fleeing felon doctrine Law (prior to 1985) stating that an officer could use deadly force to stop a felony suspect from fleeing.

full law enforcement Law enforcement technique in which officers respond formally to all suspicious behavior.

police brutality The unlawful use of force, language, and application of the law.

police discretion Authority of police to choose between alternative courses of action.

preservation of life policy Policy mandating that police use every other means possible to maintain order before turning to deadly force.

selective law enforcement Law enforcement technique in which officers under-enforce some laws and over-enforce others.

stress A condition that occurs in response to adverse external influences and is capable of affecting an individual's physical health.

Tennessee v. Garner U.S. Supreme Court ruling that eliminated the "shoot a fleeing felon" policy and replaced it with a defense of life standard.

Notes

1. Francis Cullen, Bruce Link, Lawrence Travis, and Terrence Lemming, "Paradox in Policing," *Journal of Police Science and Administration* 11:457–462 (1983); Federal Bureau of Investigation, *Uniform Crime Reports, 2005* (Washington, DC: U.S. Department of Justice, 2006); William Westley, "Violence and the Police," *American Journal of Sociology* 59:34–41 (1953).
2. Matthew Hickman and Brian Reaves, *Local Police Departments, 2003* (Washington, DC: U.S. Department of Justice, 2006).
3. Stephanie Myers, *Police Encounters with Juvenile Suspects: Explaining the Use of Authority and Provision of Support* (Washington, DC: National Institute of Justice, 1999).
4. Robert Terry, "Discrimination in the Handling of Juvenile Offenders by Social-Control Agencies," *Journal of Research in Crime & Delinquency* 4:218–230 (1967); Donald Black and Albert Reiss, "Police Control of Juveniles," *American Sociological Review* 35:63–77 (1970).
5. Myers, note 3.
6. Nathan Goldman, *The Differential Selection of Juvenile Offenders for Court Appearance* (New York: National Council on Crime and Delinquency, 1963).
7. Terry, note 4.
8. Ronald Farrell and Lynn Swigert, *Law and Society Review* 12:437–453 (1978).
9. Black and Reiss, note 4; Irving Piliavin and Scott Briar, "Police Encounters with Juveniles," *American Journal of Sociology* 70:206–214 (1964).
10. Kenneth Novak, James Frank, Brad Smith, and Robin Engel, "Revisiting the Decision to Arrest: Comparing Beat and Community Officers," *Crime & Delinquency* 48:70–98 (2002).

11. Black and Reiss, note 4.
12. Federal Bureau of Investigation, *Crime in the United States, 2006* (Washington, DC: U.S. Department of Justice, 2007).
13. Samuel Walker, Cassia Spohn, and Miriam DeLone, *The Color of Justice,* 4th ed. (Belmont, CA: Wadsworth, 2006).
14. John Boydstun, *San Diego Field Interrogation* (Washington, DC: The Police Foundation, 1975); Robert Sampson, "Effects of Socioeconomic Context on Official Reaction to Juvenile Delinquency," *American Sociological Review* 51:876–885 (1986).
15. Black and Reiss, note 4.
16. Federal Bureau of Investigation, note 12.
17. Sandra Lee Browning, Francis Cullen, Liqun Cao, Renee Kopache, and Thomas Stevenson, "Race and Getting Hassled by the Police," *Police Studies* 17:1–11 (1994); Matt DeLisi and Robert M. Regoli, "Race, Conventional Crime, and Criminal Justice," *Journal of Criminal Justice* 27:549–558 (1999).
18. Carl Werthman and Irving Piliavin, "Gang Members and the Police," in David Bordua (Ed.), *The Police: Six Sociological Essays* (New York: John Wiley & Sons, 1967), pp. 56–98.
19. Terrence Allen, "Taking a Juvenile into Custody: Situational Factors that Influence Police Officers' Decisions," *Journal of Sociology and Social Welfare* 32:121–129 (2005).
20. Michael Smith, Matthew Makarios, and Geoffrey Alpert, "Differential Suspicion," *Justice Quarterly* 23:271–295 (2006); Geoffrey Alpert, Roger Dunham, Meghan Stroshine, Katherine Bennett, and John MacDonald, *Police Officers' Decision Making and Discretion* (Washington, DC: National Institute of Justice, 2004).
21. Federal Bureau of Investigation, note 12.
22. Novak et al., note 10; Joan McCord, Cathy Widom, and Nancy Crowell, *Juvenile Crime/Juvenile Justice* (Washington, DC: National Academy Press, 2001), p. 245.
23. Thomas Monahan, "Police Dispositions of Juvenile Offenders," *Phylon* 31:91–107 (1970).
24. Delbert Elliott and Harwin Voss, *Delinquency and Dropout* (Lexington, MA: Lexington Books, 1974).
25. Gail Armstrong, "Females under the Law: Protected but Unequal," *Crime & Delinquency* 23:109–120 (1977); Meda Chesney-Lind, "Judicial Paternalism and the Female Status Offender," *Crime & Delinquency* 23:121–130 (1970); Ruth Horowitz and Ann Pottieger, "Gender Bias in Juvenile Justice Handling of Seriously Crime-Involved Youth," *Journal of Research in Crime and Delinquency* 28:75–100 (1991).
26. Christy Visher, "Gender, Police Arrest Decisions, and Notions of Chivalry," *Criminology* 21:5–28 (1983).
27. Visher, note 26, pp. 22–23.
28. Paul Tracy, *Decision Making in Juvenile Justice* (New York: Praeger, 2002); Christopher Uggen, "Class, Gender, and Arrest," *Criminology* 38:835–862 (2000); Robert Sampson, "Effects of Socioeconomic Context of Official Reaction to Juvenile Delinquency," *American Sociological Review* 51:876–885 (1986).
29. Douglas Smith and Christy Visher, "Street-Level Justice," *Social Problems* 29:167–177 (1981); Stephen Mastrofski, Robert Worden, and Jeffrey Snipes, "Law Enforcement in a Time of Community Policing," *Criminology* 33:539–563 (1995); Ronet Bachman, "Victims' Perceptions of Initial Police Responses to Robbery and Aggravated Assault," *Journal of Quantitative Criminology* 12:363–390 (1996); David Huizinga and Delbert Elliott, "Juvenile Offenders," *Crime & Delinquency* 33:206–223 (1987).
30. George Bodine, "Factors Related to Police Dispositions of Juvenile Offenders," paper presented at the annual meeting of the American Sociological Association, 1964.
31. Sampson, note 28.
32. Robert Brown, "Black, White, and Unequal: Examining Situational Determinants of Arrest Decisions from Police-Suspect Encounters," *Criminal Justice Studies* 18:51–68 (2005).
33. Werthman and Piliavin, note 18.
34. Richard Lundman, "Demeanor and Arrest: Additional Evidence from Previously Unpublished Data," *Journal of Research in Crime and Delinquency* 33:306–323 (1996); Robert Worden and Robin Shepard, "Demeanor, Crime, and Police Behavior: A Reexamination of the Police Services Study Data," *Criminology* 34:83–105 (1996); Robert Worden, Robin Shepard, and Stephen Mastrofski, "On the

Meaning and Measurement of Suspects' Demeanor toward the Police: A Comment on 'Demeanor and Arrest,'" *Journal of Research in Crime and Delinquency* 33:324–332 (1996).

35. Novak et al., note 10.

36. Aaron Cicourel, *The Social Organization of Juvenile Justice* (New York: John Wiley & Sons, 1976).

37. Robin Engel, James Sobol, and Robert Worden, "Further Exploration of the Demeanor Hypothesis: The Interaction Effects of Suspects' Characteristics and Demeanor on Police Behavior," *Justice Quarterly* 17:249 (2000).

38. Stephen Mastrofski, Michael Reisig, and John McCluskey, "Police Disrespect Toward the Public: An Encounter-Based Analysis," *Criminology* 40:519–551 (2002); Michael Reisig, John McCluskey, Stephen Mastrofski, and William Terrill, "Suspect Disrespect Toward the Police," *Justice Quarterly* 21:241–268 (2004).

39. Mastrofski et al., note 38, p. 534.

40. Theodore Ferdinand, "Police Attitudes and Police Organization," *Police Studies* 3:46–60 (1980).

41. Lonn Lanza-Kaduce and Richard Greenleaf, "Age and Race Deference Reversals," *Journal of Research in Crime and Delinquency* 37:221–236 (2000); Charles Crawford and Ronald Burns, "Resisting Arrest," *Police Practice and Research* 3:105–117 (2002).

42. Douglas Smith and Jody Klein, "Police Control of Interpersonal Disputes," *Social Problems* 31:468–481 (1984).

43. Robert Friedrich, "Police Use of Force," *Annals of the American Academy of Political and Social Science,* 452:82–97 (1980).

44. Bodine, note 30.

45. James Q. Wilson, *Varieties of Police Behavior* (Cambridge, MA: Harvard University Press, 1968).

46. Richard Sundeen, "Police Professionalization and Community Attachments and Diversion of Juveniles from the Justice System," *Criminology* 11:570–580 (1974).

47. Robin Engel, "The Effects of Supervisory Styles on Patrol Officer Behavior," *Police Quarterly* 3:283 (2000).

48. Samuel Walker and Charles Katz, *The Police in America,* 6th ed. (New York: McGraw-Hill, 2008).

49. Hickman and Reaves, note 2; Westley Jennings and Edward Hudak, "Police Response to Persons with Mental Illness," in Roger Dunham and Geoffrey Alpert (Eds.), *Critical Issues in Policing,* 5th ed. (Long Grove, IL: Waveland Press, 2005), pp. 115–128.

50. "Automatic Vehicle Locator," November 7, 2001, available at http://whatis.techtarget.com/definition/0,,sid9_gci523967, 00.html, accessed May 29, 2007; Peter Manning, "Information Technologies and the Police," in Michael Tonry and Norval Morris (Eds.), *Modern Policing* (Chicago: University of Chicago Press, 1993), pp. 349–399.

51. David Carter, "Drug-Related Corruption of Police Officers," *Journal of Criminal Justice* 18:85–98 (1990); Ellwyn Stoddard, "Organizational Norms and Police Discretion," *Criminology* 17:159–171 (1979); Julian Roebuck and Thomas Barker, "A Typology of Police Corruption," *Social Problems* 21:423–437 (1974).

52. William DeLeon-Granados and William Wells, "Do You Want Extra Coverage with Those Fries?" *Police Quarterly* 1:71–85 (1998).

53. James Spradley, *You Owe Yourself a Drunk* (Prospect Heights, IL: Waveland Press, 1970).

54. Samuel Walker and Dawn Irlbeck, "Driving While Female: A National Problem in Police Misconduct" (Omaha, NE: University of Nebraska at Omaha, 2002), available at http://www.pennyharrington.com/drivingfemale.htm, accessed May 29, 2007.

55. Geoffrey Alpert, Roger Dunham, and Meghan Stroshine, *Policing* (Long Grove, IL: Waveland Press, 2006).

56. Milton Mollen, *Commission to Investigate Allegations of Police Corruption and the Anti-corruption Procedures of the Police Department: Final Report* (New York: Mollen Commission, 1994).

57. W. Doherty, "Ex-sergeant Says He Aided Bid to Sell Exam," *Boston Globe,* February 26, 1987, p. 61.

58. Maurice Punch, *Conduct Unbecoming* (London: Tavistock, 1985).

59. Chicago Crime Commission, available at http://www.chicagocrimecommission.org, accessed May 29, 2007.

60. George Wickersham, *Enforcement of Prohibition* (Washington, DC: U.S. Government Printing Office, 1931).

61. Knapp Commission, *The Knapp Commission Report on Police Corruption* (New York: George Braziller, 1973).

62. Mollen, note 56, p. 2.

63. James Fyfe and Robert Kane, *Bad Cops: A Study of Career-Ending Misconduct among New York City Police* (Washington, DC: U.S. Department of Justice, 2006).

64. Fyfe and Kane, note 63.

65. Lincoln Steffens, *The Autobiography of Lincoln Steffens* (New York: Harcourt, Brace and Co., 1931).

66. Wickersham, note 60.

67. Allyson Collins, *Shielded from Justice* (New York: Human Rights Watch, 1998).

68. Independent Commission on the Los Angeles Police Department, *Report of the Independent Commission on the Los Angeles Police Department* (Los Angeles: Independent Commission of the Los Angeles Police Department, 1991).

69. William Terrill and Stephen Mastrofski, "Situational and Officer-Based Determinants of Police Coercion," *Justice Quarterly* 19:215–248 (2002); Joel Garner and Christopher Maxwell, *Understanding the Use of Force by and Against the Police in Six Jurisdictions* (Washington, DC: U.S. Department of Justice, 2002).

70. Alpert et al., note 55.

71. Terrill and Mastrofski, note 69; Garner and Maxwell, note 69.

72. Terrill and Mastrofski, note 69; National Center for Women and Policing, "Men, Women, and Police Excessive Force: A Tale of Two Genders," April 2002 (Los Angeles: National Center for Women and Policing, 2002), available at http://www.womenandpolicing.org/PDF/2002_Excessive_Force.pdf, accessed June 4, 2007.

73. Independent Commission on the Los Angeles Police Department, note 68; Victor Kappeler, *Critical Issues in Police Civil Liability*, 4th ed. (Long Grove, IL: Longman Press, 2006).

74. Kappeler, note 73.

75. Federal Bureau of Investigation, "Homicide Trends in the U.S.," *Supplementary Homicide Reports, 1976–2004* (Washington, DC: Bureau of Justice Statistics, 2006), available at http://www.ojp.usdoj.gov/bjs/homicide/tables/justifytab.htm, accessed May 27, 2007.

76. William Geller and Hans Toch, *Police Violence* (New Haven, CT: Yale University Press, 2005); William Geller and Kevin Karales, *Shootings of and by Chicago Police* (Chicago: Law Enforcement Study Group, 1981), p. 56.

77. Eric Weslander, "Child's Killer Expresses Regret as He's Given 16-Year Term," March 4, 2006, available at http://www2.ljworld.com/news/2006/mar/04/childs_killer_expresses_regret_hes_given_16year_te/, accessed August 25, 2007.

78. *Tennessee v. Garner,* 471 U.S. 1 (1985).

79. Steve Caraway, "Agencies Set Own Deadly Force Policy," *The Morning News* (Springdale, AR), March 9, 2006, p. 6A.

80. Kenneth Matulia, *A Balance of Forces* (Gaithersburg, MD: International Association of Chiefs of Police, 1982).

81. James Fyfe, "Police Use of Deadly Force," *Justice Quarterly* 5:165–205 (1988); William Geller, *Deadly Force* (Washington, DC: Police Executive Research Forum, 1992); Catherine Milton, Jeanne Halleck, James Lardner, and Garry Albrecht, *Police Use of Deadly Force* (Washington, DC: Police Foundation, 1977).

82. Jodi Brown and Patrick Langan, *Policing and Homicide, 1976–98: Justifiable Homicide by Police, Police Officers Murdered by Felons* (Washington, DC: U.S. Department of Justice, 2001).

83. Fyfe, note 81; Geller and Karales, note 76; Sean Grennan, "Findings on the Role of Officer Gender in Violent Encounters with Citizens," *Journal of Police Science and Administration* 15:78–85 (1987).

84. Matthew Durose, Erica Schmitt, and Patrick Langan, *Contacts between Police and the Public* (Washington, DC: U.S. Department of Justice, 2005).

85. Mark Sherman, "High Speed Chase Reaches Supreme Court," *USA Today,* available at http://www.usatoday.com/news/washington/judicial/2007-02-24-police-chase-case_x.htm, accessed June 29, 2007.

86. Alpert et al., note 55; Geoffrey Alpert and Lorie Fridell, *Police Vehicles and Firearms* (Prospect Heights, IL: Waveland Press, 1992).

87. John Hill, "High-Speed Police Pursuits," *FBI Law Enforcement Bulletin*, July 2002, pp. 1–5, available at http://www.findarticles.com/p/articles/mi_m2194/is_7_71/ai_89973554, accessed May 29, 2007.

88. Gabe Cabrera, "Police Pursuit Policies in Other Jurisdictions," April 12, 2004 (San Francisco: City and County of San Francisco, 2004), available at http://www.sfgov.org/site/bdsupvrs_page.asp?id=24020, accessed May 29, 2007.

89. Cabrera, note 88.

90. *Travis v. City of Mesquite*, 830 S.W.2d 94 (1992).

91. *Sacramento v. Lewis*, 523 U.S. 833 (1998); *Indianapolis v. Garman*, 49S00-0602-CV-55 (2006); Kevin Corcoran, "Bystanders Can Sue Police over Chases," *The Indianapolis Star*, June 15, 2006, available at http://www.indystar.com/apps/pbcs.dll/article?AID=/20060615/NEWS01/606150492&SearchID=73248239963237, accessed May 26, 2007; Kappeler, note 73; *Scott v. Harris*, 127 S. Ct. 1769 (2007).

92. Kent Kerley, "The Costs of Protecting and Serving," in Heith Copes (Ed.), *Policing and Stress* (Upper Saddle River, NJ: Prentice Hall, 2005), pp. 73–86.

93. Vivian Lord, *Changes in Social Support Sources of New Law Enforcement Officers* (doctoral dissertation, North Carolina State University, 1992).

94. John Crank and Michael Caldero, "The Production of Occupational Stress in Medium-Sized Police Agencies," *Journal of Criminal Justice* 19:339–349 (1991).

95. Cara Donlon-Cotton, "How Dangerous Is That Reporter?" *Law and Order* 55:20–22 (2007); Kenneth Dowler and Valerie Zawilski, "Public Perceptions of Police Misconduct and Discrimination: Examining the Impact of Media Consumption," *Journal of Criminal Justice: An International Journal* 35:193–203 (2007).

96. Kerley, note 92.

97. Dennis J. Stevens, "Police Officer Stress," *Law and Order* 47(9):77–81 (1999).

98. Victor Kappeler and Gary Potter, *The Mythology of the Criminal Justice System*, 4th ed. (Prospect Heights, IL: Waveland Press, 2004).

99. Alissa Pollitz Worden, "The Attitudes of Women and Men in Policing," *Criminology* 31:203–242 (1993).

100. Gloria Myers, *A Municipal Mother* (Corvallis, OR: Oregon State University Press, 1995).

101. Penny Harrington, "History of Women in Policing," available at http://www.pennyharrington.com/herstory.htm, accessed May 29, 2007.

102. President's Commission on Law Enforcement and Administration of Justice, *The Challenge of Crime in a Free Society* (Washington, DC: U.S. Government Printing Office, 1967).

103. President's Commission, note 102.

104. National Center for Women and Policing, "Equality Denied: The Status of Women in Policing, 2001," available at http://www.womenandpolicing.org/PDF/2002_Status_Report.pdf, accessed June 30, 2007.

105. Alissa Worden, "The Attitudes of Women and Men in Policing," *Criminology* 31:203–242 (1993); Joyce Sichel, *Women on Patrol* (Washington, DC: U.S. Government Printing Office, 1978).

106. Robert Homant and Daniel Kennedy, "Police Perceptions of Spouse Abuse: A Comparison of Male and Female Officers," *Journal of Criminal Justice* 13:29–47 (1985).

107. Roger Able, *The Black Shields* (Bloomington, IN: AuthorHouse, 2006); National Black Police Association, available at http://www.blackpolice.org/, accessed June 29, 2007; Jack Kuykendall and David Burns, "The Black Officer," *Journal of Contemporary Criminal Justice* 1:103–113 (1980).

108. Hickman and Reaves, note 2.

109. Hickman and Reaves, note 2.

110. *Griggs v. Duke Power Company*, 401 U.S. 424 (1971); *Albemarle Paper Co. v. Moody*, 422 U.S. 405 (1975).

111. *United States v. Paradise*, 480 U.S. 149 (1987).

112. U.S. Department of Justice, "Department of Justice Sends Limited English Proficiency Guidance to the *Federal Register* That Will Help Reduce Language Barriers," available at http://www.usdoj.gov/opa/pr/2002/April/02_crt_218.htm, accessed June 29, 2007.

Courts

The criminal courts are the core of the criminal justice process. People look to the courts as the place where justice is done. However, public perceptions of justice often differ. For some, justice is measured by the severity of the punishment imposed; for others, justice may be measured by how well a defendant's constitutional rights have been protected. For still others, justice is achieved only if all defendants are treated in an equal fashion by the courts.

Once suspects have been arrested and booked by the police, they are brought before the court to be informed of the charge(s) they face. Over the next few days, weeks, months, or sometimes years, each defendant's guilt or innocence will be determined. The movement of individual cases through the court system can vary greatly depending on whether the defendant is charged with an infraction, misdemeanor, or felony, and whether the case is prosecuted as a state or federal offense. Although the vast majority of cases are resolved through the process of plea bargaining, a significant minority go to trial. If defendants are acquitted, they are released. If defendants plead guilty or are found guilty as a result of a trial, they face sentencing by the court.

Part 3 of this book explores the history and structure of the U.S. court system as well as the major participants in the court and the process of prosecution, trial, and sentencing. Chapter 9 traces the history of the courts, illuminates the contemporary structure of the state and federal court systems, and identifies the primary court participants, including the judge, prosecutor, and defense attorney. Chapter 10 focuses on the process of prosecuting people accused of criminal offenses, including the defendant's initial appearance in court, grand jury investigations and indictments, the prosecutor's discretion in charging decisions, preliminary hearings, the entering of pleas, and plea bargaining. Chapter 11 examines the trial, including pretrial motions, jury selection, the trial itself, and jury decision making. Chapter 12 discusses the sentencing of offenders and examines the goals of sentencing, the types of sentences available to the courts, the determination of sentences, disparities, and discrimination in sentencing, and issues surrounding capital punishment.

THE·TRUE·ADMINI

Criminal Courts History and Structure

CHAPTER

9

The investigative process is begun by the police and continued by the prosecutor. Given the need for consistency throughout this process, effective coordination between the police and the courts is needed to present the best evidence to secure a conviction. When the police and prosecutor do not communicate well, the results can be disastrous.

In December 2006, Ronald Dominique was indicted on nine counts of first-degree murder and rape of several male victims in Louisiana. In 1993, a man told police that Dominique had tied him up and raped him at gunpoint, but an officer chose not to make an arrest. In 1996, another man went to the same officer with an almost identical account. This time, Dominique was arrested, charged with aggravated rape, and jailed. The prosecutor later reduced the charge to forcible rape, and three months later he dropped the charge. Eight months later, on July 14, 1997, the first of Dominique's 23 alleged victims was found dead. Today, Dominique is awaiting trial on 23 counts of murder and other charges.

- Why would officers be reluctant to further investigate such serious charges in the first place?

- Why might prosecutors reject filing charges even though two alleged victims reported nearly identical accounts of armed rape?

Source: Charles Montaldo, "Serial Killer Ronald Dominique," available at http://crime.about.com/od/serial/a/dominique.htm, accessed March 5, 2007.

Introduction

The modern dual system of courts incorporates both federal and state or local courts. This system is the product of many years of gradual development. Outside this formally established structure, however, personal relationships between key court participants can guide court proceedings and procedures. This chapter examines the history of the criminal courts; their structure; issues of jurisdiction; the roles of judges, prosecutors, and defense attorneys; and the informal workgroups that emerge through their dealings.

History of Criminal Courts

The modern court structure is constantly evolving, and its history reflects the impact of conquest, population growth, industrialization and urbanization, economic change, political struggle, the rise of various interest groups, and changes in beliefs about what is just and fair.

Emergence of Criminal Courts

The origins of the contemporary criminal courts can be traced back through their colonial predecessors— the Anglo-Saxon and English court systems. Contributions from outside this lineage were minor, although Louisiana's legal system reflects a strong French influence, owing to the substantial early French settlement there.

Anglo-Saxon Courts

The earliest records of legal procedures for criminal matters in Anglo-Saxon England are found in proclamations (known as dooms) issued by King Aethelbert of Kent in 601–604 C.E. that prohibited theft and provided for a variety of punishments for "violations against the king's peace" (any violation of the king's interests).[1] Anglo-Saxon courts, in an attempt to move away from blood feuds (the long-running cycle of violent retaliations, typically between families or clans), used a variety of oaths and ordeals to determine an individual's truth or guilt. The compurgatory oath required that the accused swear an oath of innocence: If the defendant's testimony was supported by statements of a sufficient number of others (known as oath helpers, who were often relatives of the accused), the defendant would be acquitted and released. However, if the testimony was not convincing, the accused would face either trial by ordeal or trial by battle. The absence of burns or scars from an ordeal or simple survival in battle was an indication of innocence.

Rise of the English Jury System

In 1066, the Normans invaded England. As king of England, the Norman leader William the Conqueror worked to consolidate his power and unify the country by sending representatives to the local shire (similar to a county) courts. Within 100 years after the Norman Conquest, a system of mandatory jury service had been established in England in which people were selected to sit on presenting juries (similar to today's grand juries) that heard the initial testimony in criminal cases. If the presenting jury decided there was merit to a case, the accused was ordered to stand trial.

Common law courts were established in England during the late twelfth century. They were called common law courts because their judgments of what acts constituted crimes and what punishments should be imposed became part of the law common to the entire country.[2]

One of the most significant innovations was the development of the doctrine of *stare decisis*, or precedence. Early in the thirteenth century, people involved in the courts as judges or prosecutors were able to refer to Year Books, or records of court proceedings for prior years. Previous decisions, and the bases for those decisions, were then cited as precedents for the current case.

The groundwork for the Bill of Rights and subsequent procedural protections contained in the U.S. Constitution was laid in 1215, when the English barons forced King John to sign the Magna Carta. In essence, the Magna Carta specified that there must be sufficient evidence to try the accused, that the accused had a right to be tried by a jury of his or her peers, and that the trial should not be unnecessarily delayed or biased by the social position of the accused.[3]

In the late fifteenth century, the Star Chamber was established. Within a few decades of its creation, the Star Chamber trial methods eventually included secrecy and torture. Frequently, the accused were not allowed to know the charges against them, the identity of witnesses, or the nature of the evidence condemning them. The Star Chamber was eventually abolished in 1660, leading to an increase in judicial

responsibility for the common law courts.[4] Countries following the traditions of English common law and courts have rarely supported courts that operate in secret like the Star Chamber. Nevertheless, one rather secret federal court currently operates in the United States: the Foreign Intelligence Surveillance Court (see Focus on Criminal Justice).

History of the American Court System

English common law and the English court system were the primary role models for the beginnings of the American court system.

Colonial Courts

Colonial courts performed a variety of functions, ranging from legislative and executive activities, such as the determination of tax assessments, to more traditional activities associated with the judicial branch.[5] These courts were relatively simple, with most of the judicial personnel being local influential citizens who were appointed to their positions by the colonial governor. Justice of the Peace courts were established at the local county level; they were typically administered by a person with some degree of status or recognition within the community rather than someone with formal legal training.

Lawyers tended to be drawn from the propertied classes, but their wealth and status as merchants or landowners did not confer a similar high status on law as a profession. Frequently, lawyers were treated with contempt and hostility—attitudes generally prompted by resentment of their wealth and the negative experiences less affluent people had had with lawyers in England.[6]

Equal Protection of the Law

The English use of bail, grand juries, inquests into suspicious deaths, jury trials, and appeals of verdicts were included in many early colonial legal codes. For example, the early Massachusetts Bay Colony legal codes included the Body of Liberties, written in 1648, which later served as the model for the Bill of Rights. The Body of Liberties established a number of provisions to ensure due process for the accused throughout a criminal prosecution and outlined liberties guaranteed to members of the colony and to foreigners:

- Prohibition of search or seizure without warrant
- Right to reasonable bail
- Right to a reasonably speedy determination of guilt

The Foreign Intelligence Surveillance Court

The Foreign Intelligence Surveillance Court was created in 1978 by the Foreign Intelligence Surveillance Act (FISA). This court consists of 11 district court judges from seven of the U.S. judicial circuits, each of whom serves a term of seven years. No fewer than three of the judges are required to reside within 20 miles of the District of Columbia.

The court meets every two weeks to secretly review applications by the Department of Justice and grant orders approving electronic surveillance, including wiretaps, telephone record traces, and physical searches anywhere within the United States. As noted in Chapter 7, normally search warrants and requests for wiretaps from local law enforcement need the approval of a judge after those personnel have demonstrated there is probable cause a crime has been committed or is being planned. The warrant must be specific and, in case of a search warrant, the target of the search must be notified. Warrantless searches are very narrowly circumscribed. The FISA court operates differently from local courts, however. Because it involves issues of foreign intelligence, law enforcement agents need simply identify the target as a foreign power or its agent, whether the person is a U.S. citizen or not. Between 1978 and 2004, the court approved nearly 19,000 warrants and rejected only five. All documents pertaining to the cases brought to the Foreign Intelligence Surveillance Court are classified, and targeted individuals are unable to challenge any evidence used in the requests for warrants.

The passage of the USA Patriot Act in 2001 and subsequent investigations of suspected terrorist activities, both within the United States and abroad, have led to questions about the FISA court and its operation. In 2006, controversy arose when some of the secret activities of the court became publicly known. Sometime after the September 11, 2001, terrorist attacks, President George W. Bush issued a secret order to permit the National Security Agency to monitor emails and telephone conversations between persons in the United States and suspected terrorists overseas. Critics believed that this warrantless surveillance may have been used to obtain additional wiretap warrants from the FISA court. While some members of Congress demanded an investigation into the warrantless wiretaps, the Bush administration defended it as being quite constitutional. It is important to note that because the FISA court operates in secrecy, both congressional and media interest in its operations are necessarily limited.

On August 17, 2006, U.S. District Judge Anna Taylor for the 6th Circuit ruled that the eavesdropping program was unconstitutional, stating that it violates citizens' rights to free speech and privacy. Less than two months later, however, a unanimous ruling by a three-judge panel of the 6th Circuit ruled that the Bush administration could continue the warrantless surveillance program while the administration appeals Judge Taylor's decision.

Sources: Carol Leonnig, "Secret Court's Judges Were Warned about NSA Spy Data," *Washington Post*, February 9, 2006, p. A1; Elizabeth Bazan, *The Foreign Intelligence Surveillance Act: An Overview of the Statutory Framework and Recent Judicial Decisions* (Washington, DC: Library of Congress, 2004); *The Foreign Intelligence Surveillance Act*, P.L. 95-511, Title I, § 103, 50 U.S.C. § 1803 (1978); Dan Eggen and Dafina Linzer, "Judge Rules Against Wiretaps," *Washington Post*, August 18, 2006, p. A01; David Ashenfelter, "Eavesdropping Will Continue During Appeal: Ruling Against Spying Is on Hold," *Detroit Free Press*, October 5, 2006.

- Right to hire counsel to represent the accused
- Right to be informed of charge(s)
- Right to confront one's accusers
- Protection against double jeopardy
- Right to trial by jury
- Right to challenge jurors (to exclude those who might be biased)
- Right to compel the appearance of witnesses for the defense
- Prohibition of cruel and unusual punishment
- Right to appeal verdicts

The Body of Liberties also contained the seeds of the concept of equal protection of the law: the idea that all citizens—including women, children, and servants—should have equal access to the courts.

The American Revolution and Creation of State Courts

When the American Revolution took place, the royal colonial courts were closed down and then reestablished as state courts by the new state assemblies. Although the basic structure of the courts remained essentially the same, the new state courts were more decentralized than the colonial courts, and judges were either elected or appointed by the state legislature or governor. With the growing emphasis on popular democracy and responsiveness to the local community by the middle of the nineteenth century, each new state entering the Union required the popular election of all or most of its judges, although only white males were allowed to vote.[7]

Specialized local courts, such as small claims courts and family courts, were eventually created in the larger cities to handle the growing number of cases as the U.S. population expanded. Most state courts were assigned general trial jurisdiction over both criminal and civil matters, and each state created at least one court of appeals. As the states developed their individual constitutions, outlining both the structure and the process of governmental operations, most also included in their constitutions a section protecting many of the rights of citizens accused of crimes that had been stipulated in the earlier colonial laws.[8]

The Bill of Rights

To protect themselves from the possibility of having their rights trampled by the new federal government, members of Congress added a Bill of Rights in 1791 as the first 10 amendments to the U.S. Constitution. Because the states had already implemented many of these safeguards, Congress intended the Bill of Rights to protect citizens from abuses by the federal government alone; abuses of citizens by the states were to be regulated by state protections.

Creation of the Federal Court System

The Judiciary Act of 1789, which was created by the First Congress, laid out the basic structure of the federal court system. Four federal circuit courts were established, along with 13 federal district courts. The circuit courts were assigned general trial jurisdiction and the responsibility to hear appeals from the lower district courts. Separate federal circuit courts of appeal were created in 1891. Two decades later, the trial functions of the circuit and district courts were combined under the aegis of U.S. District Courts. Today, there are 94 District Courts, 12 regional U.S. Courts of Appeal, and a single U.S. Court of Appeals for the Federal Circuit.

The U.S. Constitution states that the ultimate judicial power of the government is to be placed in a supreme court. In addition, it gives Congress the right to create more federal courts and provides for the lifetime tenure of federal judges, thus protecting them from political pressures and possible capricious

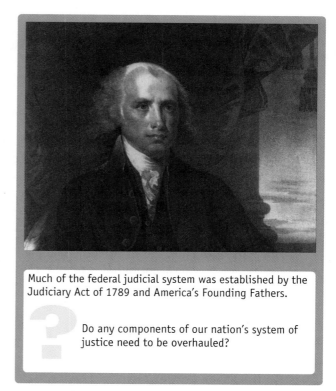

Much of the federal judicial system was established by the Judiciary Act of 1789 and America's Founding Fathers.

? Do any components of our nation's system of justice need to be overhauled?

attempts by Congress to remove judges who made unpopular decisions.

In 1803, in _Marbury v. Madison_, the U.S. Supreme Court established the principle of judicial review—the power of the Supreme Court to review and determine the constitutionality of acts of Congress and orders by the executive branch.[9] In 1810, the Supreme Court extended this principle to acts by the states in _Fletcher v. Peck_.[10] During the 35 years that John Marshall was Chief Justice, the Court handed down decisions overturning more than a dozen state laws, ruling that they were in conflict with the Constitution.[11]

Frontier Justice

Much of the nineteenth-century history of the United States was dominated by westward expansion. Pioneers pushed the frontier back from the Allegheny Mountains to the Pacific Ocean. This expansion created a variety of problems, not least of which was how to provide justice for citizens living in rural and sparsely populated communities along the frontier.[12] Law and order, as it developed, was administered by local and federal marshals, sheriffs, and circuit-riding judges.

The scarcity of trained judges and lawyers in the frontier meant that judges had to ride by coach or horseback on a traveling circuit to provide judicial services to rural areas. Those accused of crimes were held in local jails until the circuit-riding judge arrived. Court was often held in the front room of a general

Doing Justice in Dodge City

During the late nineteenth century, Dodge City, Kansas, was served by circuit-riding district judges, local Justice of the Peace courts, and municipal or police courts. On occasion, the Justice of the Peace and police courts provided a degree of entertainment in the form of comedy or drama for the local citizens. The Dodge City newspaper described the procedures for a case of disturbing the peace:

"The Marshal will preserve strict order," said the Judge. "Any person caught throwing turnips, cigar stumps, beets, or old quids of tobacco at this Court will be immediately arraigned before this bar of Justice."

The policeman looked savagely at the mob in attendance, hitched his ivory handle a little to the left, and adjusted his moustache. "Trot out the wicked and unfortunate, and let the cotillion commence," said his Honor.

"Again," said the judge, as he rested his alabaster brow on his left paw, "do you appear within this sacred realm, of which I, and only I, am high muck-i-muck. You have disturbed the quiet of our lovely village. Why, instead of letting the demon of passion fever your brain into this fray, did you not shake hands and call it all a mistake? Then the lion and the lamb would have lain down together and white-robed peace would have fanned you with her silvery wings and elevated your thoughts to the good and pure by her smiles of approbation; but no, you went to chawing and clawing and pulling hair. It is $10.00 and costs, Mr. Martin."

Source: Reprinted with permission from C. Robert Haywood, *Cowtown Lawyers: Dodge City and Its Attorneys, 1876–1886*, Norman: University of Oklahoma Press, pp. 33–34.

store or before a jury gathered in the local saloon. Regardless of the location, when the traveling judge arrived, the local residents gathered for the occasion.

As the frontier became more settled, court was held on a more regular basis in local communities. Courtroom trials—and especially criminal trials—also often functioned as a source of local entertainment. In these trials, issues of guilt and innocence often were secondary in importance to the oratory ability of the participants. Justice was generally served, but often with little regard for the technicalities of law and judicial procedure. Despite these shortcomings, by the end of the nineteenth century, the basic dual court system was largely in place in the United States.

Jurisdiction of the Criminal Courts

At the heart of the U.S. court structure is the concept of judicial jurisdiction—the power or authority a court has to hear a case or consider a particular legal motion. Rules of jurisdiction have been developed over the years in both federal and state constitutions and statutes to specify the kinds of cases a court may (or must) hear.

Jurisdictional questions may concern geographical boundaries, a significant characteristic of the case, general subject matter of the case, or some characteristic of the parties involved. For example, where a crime was committed or where the suspect was arrested may determine which court has jurisdiction to hear the case. Whether the criminal charge is a felony or a misdemeanor, or a state or federal offense, will also determine its jurisdiction. In addition, the jurisdiction of some courts may be limited to juvenile or criminal cases, criminal or civil cases, or special matters dealing with tax or copyright laws.

There are four main divisions of jurisdiction: original, appellate, exclusive, and concurrent. A court that has original jurisdiction is the first court with the authority to hear a particular case and render a verdict; it is sometimes known as the court of first instance. Courts with appellate jurisdiction have the power to review and reverse the judicial action(s) of a lower court. Courts with exclusive jurisdiction in a case are the only ones with statutory authority to hear the case. Finally, two or more courts may have concurrent jurisdiction, in which both are authorized to hear and rule in a particular case. For example, crimes that violate both state and federal laws could be tried in either state or federal court. The most basic jurisdictional division in the U.S. court system, however, is between state courts and federal courts—the basis of the dual system of courts.

Dual System of Courts

The Constitution provided for the establishment of a federal judicial system, even as the states developed their own court structures, thereby creating a dual system of courts FIGURE 9-1 . Today, there are 50 independent state court systems as well as separate court systems in the District of Columbia and the Commonwealth of Puerto Rico and territorial courts in the Virgin Islands, American Samoa, Guam, and the Northern Mariana Islands. For the most part, these systems have very similar structures and procedures. Each provides for general trial courts, appellate courts, and some sort of supreme court.

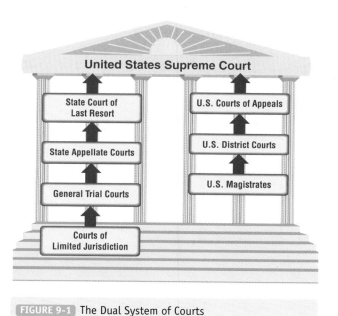

FIGURE 9-1 The Dual System of Courts

Each state court system administers and interprets its own state's laws, whereas the federal system deals with federal laws and violations. When a case that has been tried in a state court is appealed, it is appealed through the state appeals court system. In contrast, cases that were originally tried in the federal courts are appealed through the federal appellate courts.

In issues of the constitutionality of a particular law or procedure, state courts must consider both the state constitution and the federal Constitution, whereas federal courts are generally limited to issues stemming from the federal Constitution. It should also be noted that decisions made by a U.S. Court of Appeals are binding only on its specific jurisdiction, not on the entire country.

State Court Systems

Although the general structures of the court systems of the 50 states are very similar, some differences between them are apparent. Because each state has the sovereign authority to establish its own criminal code and organize its judiciary as it sees fit, the variations in the complexity and the names of the specific courts make generalizations difficult. For example, Illinois has a very simplified court structure that features only three tiers: 22 circuit courts with general jurisdiction (the authority to hear both criminal and civil cases), 5 intermediate district appellate courts, and a supreme court FIGURE 9-2 . By comparison, Texas has a much more complex structure FIGURE 9-3 . Its system includes nearly 2200 courts of limited jurisdiction (divided into county courts, probate courts, county courts at law, municipal courts, and Justice of the Peace courts), 410 district courts, 10 criminal district courts with general jurisdiction, 14 intermediate courts of appeals, a supreme court to hear civil cases, and a last-resort Court of Criminal Appeals.

No single state court system is representative of all the other state systems. However, regardless of the actual structure and complexity of each state court system, all make distinctions between the courts according to their basic jurisdiction: limited, general, or appellate.

Limited Jurisdiction

Courts of limited jurisdiction, often referred to as lower or inferior courts, are the most numerous of the various state courts and handle much heavier caseloads than do general trial courts. In 2005, more than 37 million cases were filed in state courts. Of these, over 60 percent were initially filed in courts of limited jurisdiction.[13] The vast majority of cases handled by limited jurisdiction courts deal with traffic offenses. Cases involving minor misdemeanors, civil and domestic disputes, juvenile offenses, and local ordinance violations account for most of the remaining filings. These courts, which are limited to hearing specific kinds of cases, bear a wide variety of titles:

- Justice of the Peace courts
- City courts
- Juvenile courts
- County courts
- Police courts
- Mayor courts
- Traffic courts

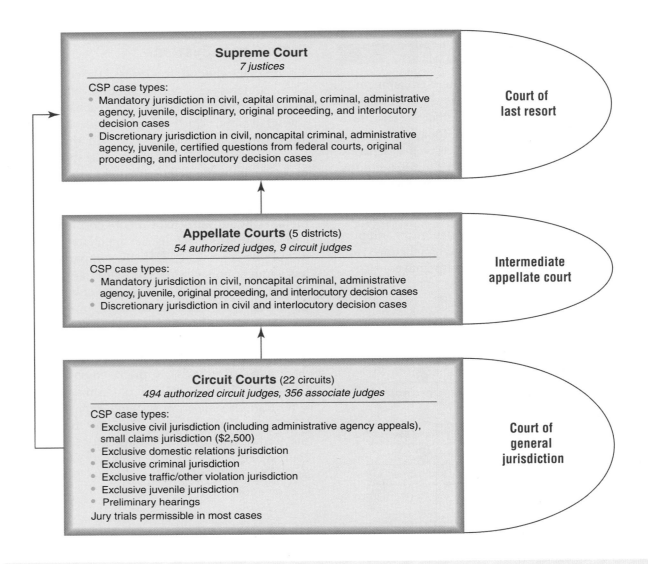

Supreme Court
7 justices

CSP case types:
- Mandatory jurisdiction in civil, capital criminal, criminal, administrative agency, juvenile, disciplinary, original proceeding, and interlocutory decision cases
- Discretionary jurisdiction in civil, noncapital criminal, administrative agency, juvenile, certified questions from federal courts, original proceeding, and interlocutory decision cases

Court of last resort

Appellate Courts (5 districts)
54 authorized judges, 9 circuit judges

CSP case types:
- Mandatory jurisdiction in civil, noncapital criminal, administrative agency, juvenile, original proceeding, and interlocutory decision cases
- Discretionary jurisdiction in civil and interlocutory decision cases

Intermediate appellate court

Circuit Courts (22 circuits)
494 authorized circuit judges, 356 associate judges

CSP case types:
- Exclusive civil jurisdiction (including administrative agency appeals), small claims jurisdiction ($2,500)
- Exclusive domestic relations jurisdiction
- Exclusive criminal jurisdiction
- Exclusive traffic/other violation jurisdiction
- Exclusive juvenile jurisdiction
- Preliminary hearings

Jury trials permissible in most cases

Court of general jurisdiction

FIGURE 9-2 Illinois Court Structure

Source: Adapted with permission from *The State Court Caseload Statistics, 2004* (Williamsburg, VA: National Center for State Courts, 2005), p. 21.

In many jurisdictions, the lower courts handle the defendant's initial appearance in a criminal case—that is, the reading of charges, the setting of bail in felony cases, and the determination of the need for a court-appointed attorney. If the offense is a misdemeanor or violation of a local ordinance, the case may be dealt with immediately by the defendant entering a plea of guilty, by the presiding judge making a determination of guilt, or by the judge dismissing the case. In felony cases, if the defendant's initial appearance is in a lower court, the case is then transferred to a general trial court. Typically, lower courts are not considered courts of record. That is, most do not create transcripts of hearings because of the time and expense involved in processing the large number of cases. Consequently, if a defendant wishes to

appeal a decision from the lower court, the absence of a transcript requires that the case be reheard in a general trial court.

General Jurisdiction

General trial courts, or <u>courts of general jurisdiction</u>, go under many different names but are most commonly known as superior courts, circuit courts, district courts, or courts of common pleas. These trial courts have the power to hear virtually any criminal or civil case as a court of first instance. They are often formally divided into criminal, civil, probate, juvenile, and domestic or family courts. In criminal cases, such courts typically hear felony cases. In 2005, more than 6.4 million criminal cases were filed in general trial courts.[14]

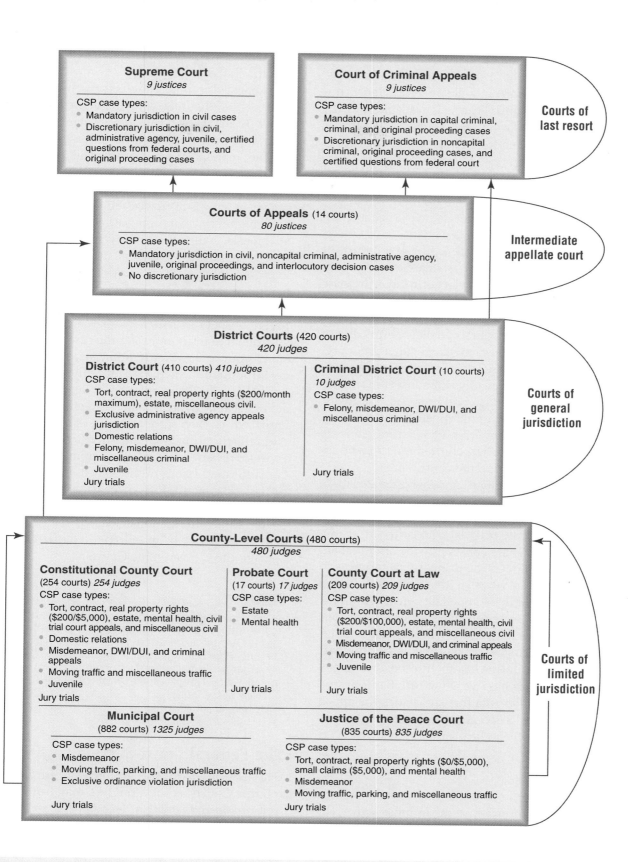

Supreme Court
9 justices

CSP case types:
- Mandatory jurisdiction in civil cases
- Discretionary jurisdiction in civil, administrative agency, juvenile, certified questions from federal courts, and original proceeding cases

Court of Criminal Appeals
9 justices

CSP case types:
- Mandatory jurisdiction in capital criminal, criminal, and original proceeding cases
- Discretionary jurisdiction in noncapital criminal, original proceeding cases, and certified questions from federal court

Courts of last resort

Courts of Appeals (14 courts)
80 justices

CSP case types:
- Mandatory jurisdiction in civil, noncapital criminal, administrative agency, juvenile, original proceedings, and interlocutory decision cases
- No discretionary jurisdiction

Intermediate appellate court

District Courts (420 courts)
420 judges

District Court (410 courts) *410 judges*
CSP case types:
- Tort, contract, real property rights ($200/month maximum), estate, miscellaneous civil.
- Exclusive administrative agency appeals jurisdiction
- Domestic relations
- Felony, misdemeanor, DWI/DUI, and miscellaneous criminal
- Juvenile

Jury trials

Criminal District Court (10 courts)
10 judges
CSP case types:
- Felony, misdemeanor, DWI/DUI, and miscellaneous criminal

Jury trials

Courts of general jurisdiction

County-Level Courts (480 courts)
480 judges

Constitutional County Court
(254 courts) *254 judges*
CSP case types:
- Tort, contract, real property rights ($200/$5,000), estate, mental health, civil trial court appeals, and miscellaneous civil
- Domestic relations
- Misdemeanor, DWI/DUI, and criminal appeals
- Moving traffic and miscellaneous traffic
- Juvenile

Jury trials

Probate Court
(17 courts) *17 judges*
CSP case types:
- Estate
- Mental health

Jury trials

County Court at Law
(209 courts) *209 judges*
CSP case types:
- Tort, contract, real property rights ($200/$100,000), estate, mental health, civil trial court appeals, and miscellaneous civil
- Misdemeanor, DWI/DUI, and criminal appeals
- Moving traffic and miscellaneous traffic
- Juvenile

Jury trials

Municipal Court
(882 courts) *1325 judges*

CSP case types:
- Misdemeanor
- Moving traffic, parking, and miscellaneous traffic
- Exclusive ordinance violation jurisdiction

Jury trials

Justice of the Peace Court
(835 courts) *835 judges*

CSP case types:
- Tort, contract, real property rights ($0/$5,000), small claims ($5,000), and mental health
- Misdemeanor
- Moving traffic, parking, and miscellaneous traffic

Jury trials

Courts of limited jurisdiction

FIGURE 9-3 Texas Court Structure

Note: Some municipal and justice of the peace courts may appeal to the district court.
Source: Adapted with permission from *The State Court Caseload Statistics, 2004* (Williamsburg, VA: National Center for State Courts, 2005), p. 52.

If a defendant appeals a lower court's decision on the grounds that he or she was denied due process, the case is then heard in a general trial court. If a transcript of the original hearing is available, the general trial court reviews it and renders a decision; if no transcript was made, the general trial court retries the case.

Appellate Jurisdiction

A defendant may appeal the decision of a general trial court in a criminal case to an appellate court. Thirty-eight states have intermediate or first-level courts of appeal. These intermediate appellate courts are located between the general trial courts and the state court of last resort or supreme court. Unlike the general trial courts, which may retry a case after a decision made by a lower court, appellate courts do not retry cases, nor may they consider evidence not presented in the original trial. Instead, they examine the transcript of the case, read written briefs submitted by attorneys for both sides, and hear oral arguments. Generally, the decision of the appellate court is final. However, if the case involves a constitutional issue unresolved by the appellate court, it may then be appealed to the state's supreme court.

Intermediate appellate courts were created to reduce the caseload of state supreme courts. Based on the idea that every defendant has a right to one appellate review of the original trial court's decision, intermediate appellate courts provide an initial appeal screening. Given the dramatic differences in states' populations, however, there is wide variation in the structure of these appellate courts within the larger state court system. Some states have only one intermediate appellate court, whereas others operate appellate courts in different regions of the state. For example, California has six intermediate appellate courts, Illinois has five, New York has four, and Iowa has one. California further divides the appellate districts into 18 divisions, with nearly 90 appellate judges available to hear cases.

Courts of Last Resort

Courts of last resort are the final appellate courts within the state court system. In most states, the court of last resort is referred to as the State Supreme Court, although it is called the Court of Criminal Appeals in Oklahoma and Texas, the Court of Appeals in New York, and the Supreme Judicial Court in Maine and Massachusetts. Oklahoma and Texas also have

supreme courts, which function as courts of last resort for civil cases.

Appealed criminal cases may reach state supreme courts either by proceeding there directly from trial court or after going through intermediate appellate courts. Only certain kinds of cases (depending on state mandates) may be appealed directly to the supreme court, such as cases in which the death penalty was imposed at sentencing or in which a speedy hearing is otherwise required. Some states do not have intermediate appellate courts, in which case all appeals go directly from a general trial court to a supreme court. Nevertheless, most appeals in criminal cases reach state supreme courts through certification by an appellate court.

If an appellate court determines that the case is urgent or that the matter involves a significant constitutional issue, it may refer the case directly to the state's court of last resort. Some state court systems are structured in such a way that the state supreme court has initial control over all appeals. In these states, all appeals are initially filed with the supreme court, which then screens the cases, retaining some and referring others to the intermediate appellate court for review.[15] Only when a criminal case involves constitutional issues of due process that extend to the federal constitution—such as the denial of bail, the right to a speedy trial, the right to a jury of one's peers, or the introduction of illegally seized evidence at trial—can a case be appealed from the state supreme court to the U.S. Supreme Court.

LINK If decisions made before a trial begins are purported to violate a defendant's due process protections—for example, in the setting of bail or the selection of jurors, as discussed in Chapter 10—the appellate courts will likely be asked to review the case. More typically, appellate courts review cases involving issues raised during pretrial motions or at trial. For example, as Chapter 11 explores, the denial of a motion for a change of venue or the introduction of illegally obtained evidence is likely to be appealed.

The Federal Court System

The Constitution calls for a Supreme Court and "such inferior Courts as the Congress may from time to time ordain and establish" (see Appendix). The contemporary structure of the U.S federal court system includes lower trial courts, appellate courts, and a court of last resort in addition to a few special federal judicial entities outside the federal court system FIGURE 9-4 .

The International Criminal Court

American citizens may be subject to the jurisdiction of courts outside of the U.S. court system. In 1998, an International Criminal Court was established by the Rome Statute of the International Court. The Rome Statute, because it is an international treaty, is binding only on those countries that have agreed to comply with its provisions. Today, at least 102 countries have signed the treaty.

The International Criminal Court (ICC) was given jurisdiction limited to the most serious crimes of concern to the international community as a whole. It has jurisdiction only over the following crimes: genocide, crimes against humanity, war crimes, and crimes of aggression. These crimes include such actions as:

- Killing numbers of persons based on their national, ethnic, racial, or religious identity
- Torture
- Depriving a prisoner of war of the right to a fair trial
- Unlawful transfer or confinement of persons
- Committing outrages upon personal dignity (including humiliating and degrading treatment)
- Planning, preparing, initiating, and carrying out a war of aggression

The Rome Statute went into effect in 2002 after the required 60 nations signed on and agreed to be bound by the law; thus only crimes committed after 2002 can be brought before the ICC.

The first case to be prosecuted in the ICC began November 9, 2006, with initial hearings in the case of Thomas Lubanga Dyilo, a former Congolese warlord charged with coercing and kidnapping tens of thousands of children into his powerful and violent militia. Most of these children—some as young as age 10—were trained to be killers; others were required to be cooks, carriers, or sex slaves. Many of the children killed and were killed in the fighting in the Ituri region of the eastern Congo.

In November 2006, Germany's top prosecutor filed documents with the ICC to initiate charges against U.S. Defense Secretary Donald Rumsfeld, U.S. Attorney General Alberto Gonzales, former CIA director George Tenet, and a number of other U.S. civilians and military officers. Their alleged offenses are related to their roles in the abuses that occurred at the Abu Ghraib prison in Iraq and allegations of abuse at the U.S. detention facility at Guantanamo Bay, Cuba.

The United States is not currently a signatory member of the Rome Statute and opposes the court for several reasons. First, the ICC has little accountability to governments legitimately empowered to represent the people's interests. Second, the ICC is incompatible with U.S. standards of justice and due process and could potentially deny American citizens a trial that protects their Constitutional rights to a jury of peers guided by rules of evidence and definitions of crimes under U.S. law. Third, the United States believes that the ICC might pursue politically motivated investigations and prosecutions, putting government leaders and American soldiers at risk—which is exactly what happened in the threatened prosecutions of Rumsfeld and other Americans.

Sources: *Rome Statute of the International Criminal Court,* available at http://www.icc-cpi.int.html; Marlise Simons, "Congo Warlord's Case Is First for International Criminal Court," *New York Times,* November 10, 2006, available at http://www.nytimes.com/2006/11/10/world/europe/10hague.htm, accessed May 2, 2007. U.S. Department of State, "U.S. Restates Objections to International Criminal Court," October 15, 2002, available at http://usinfo.state.gov/dhr/Archive/2003/Oct/15-434709.html, accessed July 17, 2007; Adam Zagorin, "Exclusive: Charges Sought Against Rumsfeld Over Prison Abuse," *Time,* November 10, 2006, available at http://www.time/nation/article/0,8599,1557842,00.html, accessed May 2, 2007.

U.S. Magistrate Judges

In 1968, Congress passed the Federal Magistrates Act, which provided for the appointment of judicial officers for each federal district court by federal district court judges. There are currently 429 full-time and 150 part-time magistrate judges. Full-time magistrate judges are appointed for eight-year terms, whereas part-time magistrate judges are appointed for only four years. Either type of magistrate may be reappointed. Magistrates do not have all the powers of a judge, but rather generally handle criminal misdemeanor cases, such as failure to pay legal child support obligations and possession of a firearm after being convicted of a "qualifying" domestic violence offense. Each year, magistrate judges hear cases involving nearly 160,000 defendants.[16] In criminal cases, magistrates conduct pretrial hearings and, if both sides agree, try misdemeanors.

In 1989, in *Gomez v. United States,* the U.S. Supreme Court ruled that jury selection in felony cases

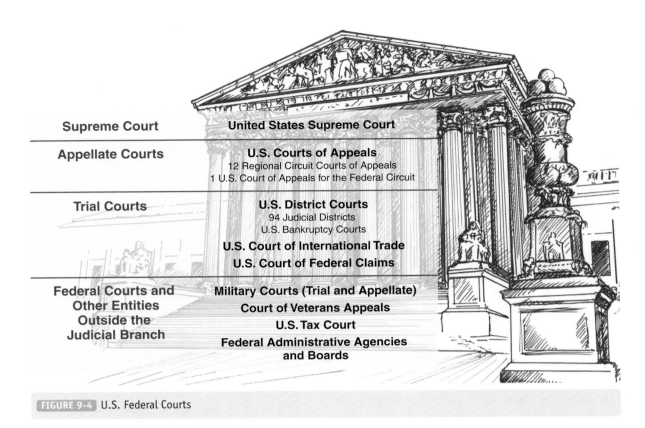

Supreme Court	**United States Supreme Court**
Appellate Courts	**U.S. Courts of Appeals** 12 Regional Circuit Courts of Appeals 1 U.S. Court of Appeals for the Federal Circuit
Trial Courts	**U.S. District Courts** 94 Judicial Districts U.S. Bankruptcy Courts **U.S. Court of International Trade** **U.S. Court of Federal Claims**
Federal Courts and Other Entities Outside the Judicial Branch	**Military Courts (Trial and Appellate)** **Court of Veterans Appeals** **U.S. Tax Court** **Federal Administrative Agencies and Boards**

FIGURE 9-4 U.S. Federal Courts

could not be presided over by a magistrate without the defendant's permission, thereby further formalizing the "lower court" nature of the magistrate's office.[17] In addition to criminal cases, magistrates may hear civil cases in which the contested amount is less than $10,000. In both criminal and civil cases, decisions by U.S. magistrate judges may be appealed either to the district court or directly to the court of appeals.

U.S. District Courts

The federal system includes a single level of general trial courts—the 94 U.S. District Courts. Each state has at least one district, as do the District of Columbia, the Commonwealth of Puerto Rico, Guam, the Virgin Islands, and the Northern Mariana Islands; California, New York, and Texas have four districts each. The number of judges in each district is determined by the population and caseloads in the district; this number ranges from only one judge to 28 judges in the densely populated Southern District of New York.

U.S. District Courts have both original and exclusive jurisdiction in all cases involving federal law or disputes over treaties. In addition to civil suits and criminal cases, these courts hear cases dealing with job discrimination and violations of equal opportunity, citizenship matters, environmental protection, and educational discrimination. In 2005, U.S. District Courts heard more than 70,000 criminal cases.[18] U.S.

District Courts also have concurrent jurisdiction with state trial courts in certain criminal matters (for example, bank robbery and kidnapping, which violate both state and federal laws) and civil disputes between people of different states where damages exceed a designated amount.

Cases in U.S. District Courts are generally heard by a single judge, as they are in state trial courts, and they may or may not be tried before a jury. Felony cases are heard before juries unless the defendant waives that right.

U.S. Courts of Appeals

Before 1891, each member of the U.S. Supreme Court was assigned to one of the federal circuits, charged with hearing appeals from district courts within that circuit. Because of a dramatic growth in appeals, however, in 1891 Congress decided to establish a separate Circuit Courts of Appeals structure. In 1948, the name was changed to the U.S. Court of Appeals.

This middle tier of the federal court system consists of 13 U.S. Courts of Appeals, including 11 numbered circuits based on geographic regions and containing three or more states, one court for the District of Columbia, and a separate Court of Appeals for the Federal Circuit **FIGURE 9-5**. The jurisdiction of the Court of Appeals for the Federal Circuit includes appeals from the 94 district courts arising from

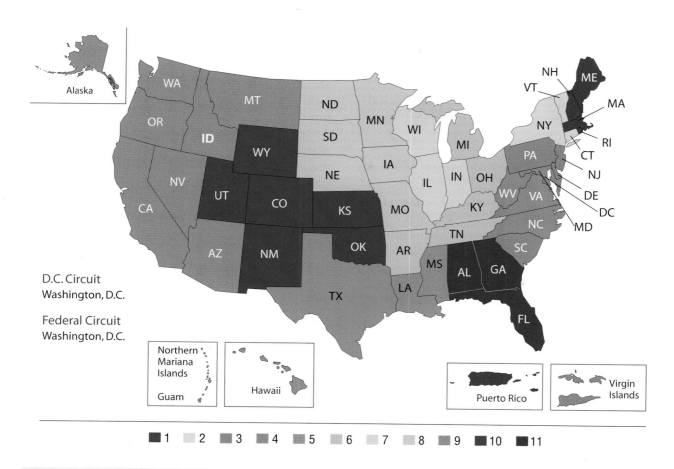

D.C. Circuit
Washington, D.C.

Federal Circuit
Washington, D.C.

Alaska

Northern Mariana Islands

Guam

Hawaii

Puerto Rico

Virgin Islands

■ 1 ■ 2 ■ 3 ■ 4 ■ 5 ■ 6 ■ 7 ■ 8 ■ 9 ■ 10 ■ 11

FIGURE 9-5 Federal Judicial Circuits

The nine Justices of the U.S. Supreme Court are the final arbiters of questions dealing with interpretation of the U.S. Constitution.

? Why is the nomination process for Supreme Court Justices so contentious? Should Supreme Court justices have life-time appointments?

cases involving patent law violations or suits against the federal government, as well as cases appealed from the Court of International Trade and Claims Court. In 2005, there were nearly 47,000 appeals from U.S. District Courts to the U.S. Courts of Appeals, including more than 32,000 civil and 14,000 criminal cases.[19]

U.S. Supreme Court

Like the various state supreme courts, the U.S. Supreme Court is the court of last resort for the federal system. In addition to its power of judicial review over Congressional legislation and acts by the executive branch, its jurisdiction extends to all cases in which a substantial federal question is involved. In other words, the Supreme Court is empowered to review any federal appellate court decision or a decision from a state supreme court in which the decision raises an issue related to federal law. The final arbitrator in cases of state law would be the appropriate state supreme court.

A case reaches the Supreme Court following an appeal from the party who wants it to review a decision of a federal or state court. To do so, the dissatisfied party files a petition asking that the case be heard by the Supreme Court. This initial petition to the Supreme Court requests that it issue a writ of certiorari, or order to the lower court to certify the court record and send it up to the higher court. The order includes a request to provide

- The lower court's decision;
- A statement of the federal legal question(s) involved in the decision; and
- A statement of the reasons why the Supreme Court should review the case.

If the Court grants the writ, the case is scheduled for the filing of briefs and for oral argument. If it denies the request for a writ, the decision by the lower case stands.

In rare instances, a case may reach the Supreme Court through a request for certification. Such a request by a lower federal court asks the Supreme Court for a ruling on a very specific legal question that the lower court has been unable to resolve in a pending case. If the Court accepts the request for certification, then its ruling is applied by the lower court in the pending case.

During the 2005 term of the Supreme Court, a total of 8521 cases were on the docket (compared to only 5268 during the 1987 term). Only 87 cases were actually argued, however, and 82 cases were disposed

of in 69 signed opinions, compared to 167 cases decided by the Court in 1987. Those cases scheduled for argument but not heard by the Supreme Court are typically held over for argument in the following term.[20] Although the Court is facing an increasing number of cases each year, it is handing down fewer decisions. Because the Supreme Court does not give reasons for why it decides to hear or not hear a case, the reason for the decline in the number of cases set for argument and resulting in decisions is not known.

In addition to sitting on the Supreme Court to review cases, each judge of the Court continues to function as a circuit justice for one or more of the federal circuits. In this role, the justice responds to special appeals from within the circuit, such as requests for a stay of execution by an inmate on death row. For example, on February 19, 1992 (the day of his scheduled execution), Leonel Herrera, a Texas death row inmate, applied to Justice Antonin Scalia through the 5th Federal Circuit for a stay of execution. Herrera claimed that newly discovered evidence would establish his innocence and, therefore, that his execution would constitute cruel and unusual punishment. Justice Scalia referred the case to the Supreme Court; that same day, the Court agreed to the stay and scheduled Herrera's case for argument before the full court. Less than a year later, the Supreme Court, in *Herrera v. Collins,* ruled that belated evidence of innocence does not ordinarily entitle an applicant to a new hearing and turned down Herrera's request that his death sentence be overturned.[21]

When its decision is published, the Court's ruling on a case becomes a precedent for the lower courts. The rulings in some cases, however, may have a very narrow scope and leave open other important questions. For example, in *Thompson v. Oklahoma,* the Court held that it was unconstitutional to execute a person who was under age 16 at the time he or she committed a capital crime but did not rule in that case as to whether juveniles ages 16 or 17 could be executed.[22] A decision on that issue was not handed down until the next year, when the Court held in two combined cases that executions of people who were 16 or 17 at the time they committed their crimes were constitutional.[23] The precedent set in that case was relatively short-lived, and juveniles are now exempt from the death penalty. In 2005, the Court ruled in *Roper v. Simmons* that "the death penalty is disproportionate punishment for offenders under the age of 18" and, therefore, is a violation of the Eighth Amendment's prohibition against cruel and unusual punishment.[24]

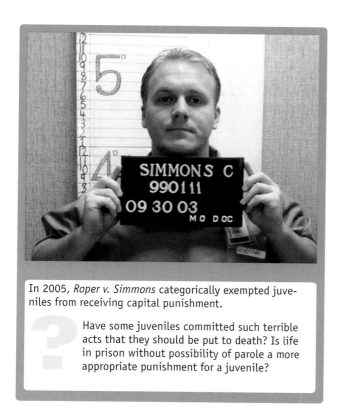

In 2005, *Roper v. Simmons* categorically exempted juveniles from receiving capital punishment.

? Have some juveniles committed such terrible acts that they should be put to death? Is life in prison without possibility of parole a more appropriate punishment for a juvenile?

LINK Between 1985 and 2003, 22 people were executed who were juveniles at the time they committed their crimes. These cases and the juvenile justice system are examined in Chapter 16.

Court Participants

Courts and judicial systems are merely the structures in which the quest for justice takes place. It is easy to describe their structures, but simple descriptions fail to accurately depict life within the court. Ultimately, the people who work in these structures define the nature of what happens there. Regular participants in the court give life and meaning to concepts such as due process and law and order. Many other participants—such as victims, defendants, witnesses, jurors, and the media—are not discussed here because none of them routinely work in the court (although some members of the news media are assigned to cover crime news and make regular visits to the court). For the most part, these people are relatively passive participants who are only temporarily in the court and are not necessarily familiar with court proceedings—only their particular case.

Court Staff

Although the court participants who receive nearly all of the media attention are the judges, prosecu-

tors, and defense attorneys, each of whom will be discussed shortly, the daily operation of a court relies on the activities of its staff.

Administrators

Larger court systems typically have court administrators. These individuals are essentially managers. They are responsible for maintaining the court's budget; hiring personnel; overseeing case flow, space, and office equipment; managing the jury system in the court; creating and maintaining uniform court record systems; managing public relations, general information, and research; and serving as liaisons with other people in the judicial or local government system.

Clerks

The clerk of the court performs the daily clerical work of the court; as a consequence, the clerk is one of the most central figures in the court. Nearly everything that goes before the judge must pass through the clerk first. Prosecutors and defense attorneys bring to the clerk motions, pleadings, and other matters to be acted upon by the judge. In addition, clerks are responsible for keeping court records, maintaining evidence submitted to the court, issuing subpoenas for jury duty, administering the jury selection system, maintaining the court docket and calendar, and gathering annual statistics for official court reports. Clerks in the federal system are appointed to their positions by federal judges, whereas clerks of state general trial courts may be either appointed or elected. Currently, about one-third of the states provide for the popular election of court clerks.

Bailiffs

The bailiff of the court (also known as the court deputy) is responsible for maintaining security and order within the courtroom and judge's chamber. In some jurisdictions, bailiffs are assigned to the court by the sheriffs' department; in other jurisdictions, they are appointed by the judge. In addition to providing security, bailiffs are responsible for keeping custody of defendants, court summons, and witnesses as well as looking after the needs of jurors during trial proceedings.

Reporters

Court reporters are responsible for the official recording of the proceedings as well as any depositions taken outside the courtroom. A deposition is the testimony of a witness taken under oath outside the court through oral questioning or written interrogation. Because testimony and depositions are part of

Brian Nichols killed four people during his escape from the courtroom where he was on trial for rape.

? Even with today's security measures, are some defendants too dangerous to appear in court?

the official court proceedings and may be required for review in an appeal of the case, it is important that they be preserved accurately.

Interpreters

Another court participant who is required in federal courts and increasingly in state courts is the court interpreter. The interpreter is responsible for providing accurate interpretation and translation for non-English-speaking defendants and witnesses.

Judges

To members of the public, judges are probably the most easily distinguished participants in the court system. They are highly visible, because they sit positioned above the center of the courtroom in their symbolic black robes. Judges appear to have tremendous power and authority in the courts. Also, because many judges are elected officials, they are often viewed as major players in local politics. Out of fear or respect, judges are able to command great deference.

Role and Responsibilities of Judges

The judicial role in criminal cases begins even before a suspect is arrested and continues past the sentencing decision. Judges in general trial courts are responsible for the following tasks:

- Determining whether there is probable cause to issue search or arrest warrants

- Determining whether there is sufficient cause to hold a suspect for prosecution
- Determining whether a defendant should be released on bail and, if so, how much bail is required
- Ruling on pretrial motions submitted to the court by prosecutors and defense attorneys
- Deciding whether to accept plea agreements

If the case goes to trial, the judge has the following duties:

- Acting as a referee to ensure that both sides "play fair"
- Deciding on the admissibility of evidence and testimony
- Deciding whether there is sufficient evidence for a jury to make a decision or whether the case should be dismissed

If the defendant is found guilty, the judge has three additional responsibilities:

- Deciding on an appropriate sentence
- Determining whether a request for a reduction in the defendant's sentence is warranted
- Deciding whether to grant requests from prisoners for early release

Nearly all of these responsibilities allow judges to use their discretionary authority; consequently, they have enormous influence in determining how any given case will proceed. Their decisions to issue warrants, determine probable cause, set bail, rule on motions, accept plea agreements, and impose sentences contribute to the widely held perception of judges as the central figures in the processing of criminal cases.

LINK In appellate courts, the judge's role is significantly more restricted. For the most part, these judges decide which cases to hear and whether to deny an appeal or to support the appeal, thus overturning a lower court's decision. See Chapters 10 and 11 for further discussion of the judge's role.

Background of Judges

Although some judges in the inferior courts have little or no legal training, judges in general trial and appellate courts at the state and federal levels must hold law degrees. Federal judges are likely to have been members of large private law firms and to have been active in state or local politics before their appointment to the court, providing them with greater visibility and the necessary political connections.[25] General trial judges in state courts are likely to have previously been prosecutors. For many judges in both state and federal courts, holding a government position in the Department of Justice in the past is not uncommon. Additionally, being involved in the right political party appears to be a requirement for a federal judicial appointment; historically, between 85 and 90 percent of all appointments have shared the political affiliation of the U.S. president who was in office at the time of the appointments.[26]

Federal judges tend to come from wealthier backgrounds than state court judges and are more likely to have been educated at top-tier schools. Judges in state-level lower courts are more likely to come from working-class or low-salaried families and to have attended state universities.[27] Not many decades ago, both women and minorities were rarely found in the judiciary. More recently, however, the gender and racial composition of the bench has changed noticeably FIGURE 9-6 , FIGURE 9-7 . The U.S. Supreme Court added its first female justice, Sandra Day O'Connor, appointed by President Ronald Reagan in 1981. A second woman, Ruth Bader Ginsburg, was appointed to the Court by President Bill Clinton in 1993.

Selection of Judges

Judges for the state courts are selected using one of three methods:

- Merit selection, with appointment by the chief executive from a certified list of qualified candidates (which is typically prepared by a state judicial merit committee)
- Direct appointment by the governor or election by the legislature
- Popular election

Terms of office vary from state to state, ranging from four or six years, to longer terms of 12 to 15 years, and, in some cases, for life.[28] Only a limited number of states use either governor appointment or legislative election of judges. Instead, by far the most popular method for selection of judges today is through popular election. Some states require that candidates note their party affiliations on ballots, whereas other states have nonpartisan elections in which no party labels are attached and no partisan campaigning is allowed.

The merit selection process, also known as the Missouri Plan, was first used in 1940 and was aimed at reducing the role of politics in the selection of judges. The popular election of judges typically requires that judges cater to party bosses and sometimes-petty politicians and bid for the support of various special-interest groups.

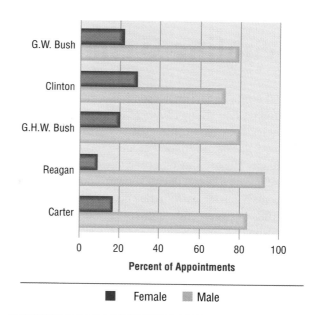

FIGURE 9-6 Gender of Presidential Appointments to the Federal District Courts

Sources: Sheldon Goldman, Elliot Slotnick, Gerard Gryski, and Sara Schavoni, "W. Bush's Judiciary: The First Term Record," *Judicature* 88:244–274 (2005); Sheldon Goldman, "Bush's Judicial Legacy: The Final Imprint," *Judicature* 76:282–297 (1993).

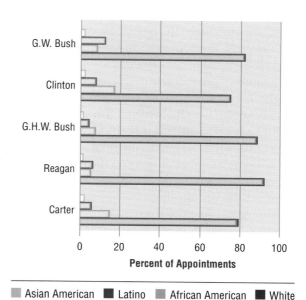

FIGURE 9-7 Race and Ethnicity of Presidential Appointments to the Federal District Courts

Sources: Sheldon Goldman, Elliot Slotnick, Gerard Gryski, and Sara Schavoni, "W. Bush's Judiciary: The First Term Record," *Judicature* 88:244–274 (2005); Sheldon Goldman, "Bush's Judicial Legacy: The Final Imprint," *Judicature* 76:282–297 (1993).

District Judge Donald Thompson was indicted on four felony counts of indecent exposure based on allegations that he had exposed himself at least 15 times to the court recorder and repeatedly used a sexual device known as a penis pump for masturbation during trials between 2001 and 2003. Thompson had spent nearly 23 years on the bench before retiring in 2004, following an investigation into his violations of the code of judicial ethics.

At his trial, a witness testified that Thompson used a penis pump almost daily during a murder trial in which a man was charged with murdering a young child. An audiotape of the trial included frequent whooshing sounds similar to air being pumped or released from a blood pressure cuff. A police officer testified that while he was testifying on the stand in a murder trial in 2003, he noticed a piece of plastic tubing disappear under the judge's robe. During the lunch break, he and another officer took photographs of the pump under the desk. Thompson denied using the pump entirely, insisting that it had been given to him as a gag gift by a longtime friend. However, investigators found semen stains on the judge's chair and surrounding carpet, as well as on his judicial robes. The jury found Thompson guilty, and he was sentenced to four years in prison.

Sources: "Oklahoma Judge Accused of Using Sexual Device in Court," *USA Today*, June 25, 2004, available at http://www.usatoday.com/news/nation/2004-06-25-judge-conduct_x.htm, accessed May 3, 2007. "Penis Pump Judge Gets 4-Year Jail Term," *USA Today*, August 18, 2006, available at htttp://www.usatoday.com/news/nation/2006-08-18-judge-sentenced_x.htm, accessed May 3, 2007.

In an attempt to eliminate partisan politics, states adopting the Missouri Plan have established nominating commissions composed of lawyers, non-lawyers, and sometimes sitting judges. When a judicial vacancy occurs, the commission solicits recommendations from the state bar association and the public, and even allows prospective candidates to nominate themselves. The commission then produces a list of names of people (usually between three and five) believed to be best qualified for the position and gives it to the governor, who makes the final appointment. After a specified number of years on the bench, judges must face reelection. In that election, a judge has no opposition; the public simply votes to retain or remove him or her from office.

The process of selecting judges for the federal courts is quite different. All federal judges—with the exception of Federal Magistrates, who are appointed by U.S. District Court judges—are nominated by the President of the United States and then confirmed by the U.S. Senate, with only a simple majority of senators present at the voting required to confirm the nomination.

Federal judges, unlike state and local judges, have lifetime appointments. As noted earlier, judges at the state and local levels must be reelected to retain their positions on the bench. Sometimes judges are removed from office not because they fail to get reelected, but because they engage in improper behavior (see "Headline Crime").

The judge is the centerpiece between the state and the individual.

Is it problematic that judges are political entities? Can justice be achieved through political appointees?

Prosecutors

Although the public generally perceives the judge to be the most central figure in the court, the prosecutor actually has far greater discretion in determining the fate of people who are arrested for crimes. In the adversarial system, prosecutors represent the state. Although they are expected to win cases, there is a presumption that they will be fair and impartial in the pursuit of justice. Prosecutors work with the police and sheriffs' departments, as well as with staff investigators, to build cases that can be won. Although prosecutors are responsible for obtaining convictions in cases brought before the court, they have an ethical duty not to prosecute people who they believe are probably innocent.

Role and Responsibilities of Prosecutors

The primary responsibilities of the prosecutor are to decide which cases are to be prosecuted, evaluate evidence, determine which charges are most likely to end in convictions, and prosecute the case before the court. In addition, the prosecutor often acts as a legal advisor to the police, giving advice about law enforcement practices that will withstand court challenges, ensuring appropriate gathering of evidence, and interviewing witnesses.

While police officers make arrests and take suspects into custody, the prosecutor must decide whether the case should proceed. Prosecutors may decide to decline prosecution, to dismiss the case at any point, to negotiate and accept a plea of guilty, or to go ahead with the prosecution. However, increased police activity—especially in terms of the enforcement of drug laws and computer-related crime—has brought an increasing number of cases to the doorsteps of prosecutors. For example, in 2005, 60 percent of all prosecutors' offices reported prosecuting computer-related crimes such as credit card fraud, bank card fraud, cyberstalking, and theft of intellectual property TABLE 9-1.

LINK Cybercrimes and identity theft affect millions of Americans each year and cause billions of dollars in damage. These emerging crimes are explored in Chapter 17.

Selection of Prosecutors

The Judiciary Act of 1789, in addition to creating the federal court system, established the process for appointing federal prosecutors, known as U.S. Attorneys. As is the case for federal judges, the president nominates these attorneys for each of the federal circuits, and their nominations must be confirmed by the U.S.

TABLE 9-1

Computer-Related Crimes Prosecuted by State Prosecutors' Offices

Offense	Percentage of All Offices
Credit card fraud	80
Bank card fraud	71
Identity theft	69
Transmitting child pornography	67
Computer forgery	40
Cyberstalking	36
Unauthorized access	23
Computer sabotage	5
Theft of intellectual property	5
Unauthorized copying	4

Source: Steven Perry, *Prosecutors in State Courts, 2005* (Washington, DC: U.S. Department of Justice, 2006), p. 5.

Senate. With the creation of the Department of Justice in 1870, the Attorney General was given responsibility for supervision of all U.S. Attorneys and assistant attorneys.

In 2005, 2344 prosecutors' offices were responsible for the prosecution of criminal cases in state courts. These offices employed about 78,000 attorneys, investigators, victim advocates, and support staff.[29] Each office is headed by a chief prosecutor, who goes by a variety of titles: district attorney, county attorney, prosecuting attorney, state's attorney, and commonwealth attorney. Chief prosecutors are typically elected at the local level, with some exceptions. In Alaska, Connecticut, and Delaware, the state attorney general is responsible for all criminal prosecutions; in New Jersey, the governor appoints a prosecutor for each county; and in the District of Columbia, the U.S. Attorney for the District of Columbia prosecutes both federal and local crimes.[30]

Prosecutors may work full- or part-time, based on the needs of their community. They generally have assistants, including several secretaries, clerks, and investigators from the police or sheriff's department. In 2005, about three-fourths of all prosecutors' offices reported having a full-time chief prosecutor, and 70 percent employed at least one full-time assistant prosecutor. More than half of all full-time prosecutors in the United States had at least one assistant prosecutor. Of these assistants, approximately 90 percent worked on a full-time basis.[31]

The perception that prosecutors and police work hand-in-hand may overlook potential conflicts between these two groups. For example, prosecutors are rarely willing to prosecute a weak case. Thus, if the police have failed to gather sufficient evidence to support a conviction, the prosecutor will likely dismiss the case—which may result in resentment by the police who worked on the case. In addition, a prosecutor may initiate a prosecution, only to later determine that the case should be dropped from further action, known as a *nolle prosequi* (a formal entry on the record stating that the prosecutor "will not further prosecute" the case). Such actions are consistent with the American Bar Association's *Standards for Criminal Justice*, which state that the "duty of the prosecutor is to seek justice, not merely to convict."[32] Furthermore, because prosecutors wish to control their own caseloads, they may inform the police that they are not interested in prosecuting certain types of cases.

Prosecutors may also hesitate to prosecute cases in which there is a very close relationship between the victim and the offender. All too often, especially in cases involving domestic violence, the case begins, but the victim then becomes uncooperative and refuses to testify after having second thoughts about the potential consequences for the offender. Without the victim's testimony, obtaining a conviction becomes much more problematic.

Prosecutorial Discretion

The discretionary power of prosecutors allows them to exert great influence over how a case proceeds and what happens to defendants during that process. Their discretion is greatest during the pretrial stages. At this point, prosecutors are in a position to evaluate the case and decide whether to dismiss it, refer the suspect to another government agency, formally charge the suspect, or, if the case is to proceed, determine the specific charges. In addition, prosecutors can use their discretionary power of charging to influence the judge's bail-setting decision by filing more serious charges, which carry higher bond amounts, or by indicating the strength of their case. They also exercise great discretion in plea bargaining negotiations. By initially overcharging (filing charges more serious than the facts of the case may warrant) or filing multiple charges or counts, prosecutors are able to increase their leverage over defendants to persuade them to enter a guilty plea.

LINK The discretion of prosecutors extends well beyond the pretrial stage. If the case goes to trial and results in a hung jury, the prosecutor can decide to file charges again or to dismiss the case, as is explained in Chapter 11. At sentencing, prosecutors may either recommend a particular sentence or stand mute and make no recommendation, as explained in Chapter 12. As Chapter 15 outlines, the prosecutor may also oppose parole at a prisoner's hearing, support it, or make no recommendation.

Despite the judge's power, the prosecutor determines whether a case will ultimately have its day in court.

? Do the courts operate like any other business? Should special attention be given to the characteristics of defendants rather than to the time and money that must be devoted to their cases?

Like the discretion exercised by the police, prosecutorial discretion is sometimes abused. Prosecutors who believe that justice or the public's interests will be served best by dismissing or prosecuting particular cases are unlikely to generate much criticism. However, if prosecutorial decisions are perceived as being based on extralegal factors such as politics, race, or gender, the public may be quick to express its outrage at the prosecutor's actions. Prosecutors are often influenced in their charging decisions by concerns for protecting and enhancing their careers.[33] Bad or messy cases, cases involving prominent suspects, and cases that will consume a disproportionate amount of resources in light of the expected outcomes are often not prosecuted; the same is true of cases that are unlikely to be won or, if won, would have an outcome that would be unpopular with the public.

All too frequently, decisions to reject or dismiss felony charges are affected by the race, gender, employment status, and even credibility of the defendant.[34] Prosecutors appear to favor female defendants over male defendants, and whites over minority defendants. More significantly, numerous studies report that the race of both offender and victim influences prosecutorial decisions in the charging of homicide cases and subsequent recommendations for the imposition of the death penalty. According to these studies, cases involving minority offenders and white victims were more likely to be charged and to have the death penalty recommended.[35]

LINK As noted in Chapter 10, sometimes charging decisions reflect the celebrity status of a case. In cases involving very wealthy or well-known defendants, the chief prosecutor may be pitted against one or more high-profile defense attorneys. Perhaps more importantly, when the charging decision is influenced by the defendant's characteristics (i.e., race, ethnicity, sex, social class, or age), the consequences become compounded at sentencing, as explored in Chapter 12.

Defense Attorneys

Although defense attorneys are the kind of lawyer most often portrayed on television, they account for only a small share of the approximately 950,000 lawyers practicing in the United States today (note that the total number of U.S. lawyers has increased approximately 150 percent since 1970).[36] Relatively few attorneys in the United States specialize in criminal law. Those who do are likely to be full-time public defenders assigned by the court to represent indigent defendants. Private attorneys who handle criminal cases typically derive most of their income from their general law practice, which often deals with civil disputes such as divorces, wills, estate planning, and torts. Most law schools deemphasize the study of criminal law because individuals who make their living practicing criminal law will likely have a lower status and income level than their corporate colleagues. Political scientist Herbert Jacob explains one of the reasons so few lawyers choose to become defense attorneys:

> A criminal law practice requires close contact with the seamy side of life. In many cases the work does not pay very well, for the clients are not wealthy. If the lawyer defends a notorious criminal, the community may misunderstand and associate the lawyer with his client. For these reasons, relatively few lawyers desire criminal cases; where at all possible they avoid taking them.[37]

Role and Responsibilities of Defense Attorneys

The role of the defense attorney is very narrow. He or she acts as the advocate for the accused and is obligated to use every lawful means to protect the interests of the defendant in court. If possible, this outcome includes achieving an acquittal of the charges or, if the defendant is convicted, minimizing the pun-

ishment. Because U.S. courts operate according to the adversarial model, the defense attorney's role is to fight for his or her client. According to Supreme Court Justice Byron White:

> [The] defense counsel has no . . . obligation to ascertain or present the truth. Our system assigns him a different mission. . . . Defense counsel need present nothing, even if he knows what the truth is. He need not furnish witnesses to the police, or reveal any confidences of his client, or furnish any other information to help the prosecution's case. If he can confuse a witness, even a truthful one, or make him appear at a disadvantage, unsure or indecisive, that will be his normal course. Our interest in not convicting the innocent permits counsel to put the State to its proof, to put the State's case in the worst possible light, regardless of what he thinks or knows to be the truth.[38]

It is the defense attorney's responsibility to probe and test every bit of evidence presented by the state to ensure that the defendant is not convicted unless the prosecution can prove the defendant guilty beyond a reasonable doubt.[39]

The Right to Counsel

Because few people have the skills and knowledge necessary to act as effective advocates on their own behalf in a court of law, defendants with sufficient resources nearly always hire private attorneys. Those who cannot afford a private attorney are assigned the services of a public defender by the court. Most state constitutions provide for the assignment of counsel to indigent defendants. The Sixth Amendment to the Constitution states, "In all criminal prosecutions, the accused shall enjoy the right . . . to have the assistance of counsel for his defense." However, just when and under what circumstances counsel must be provided was not made clear.

Until the early part of the twentieth century, the Sixth Amendment was generally interpreted to mean that a defendant had a right to an attorney if he or she could afford to hire one. During the last six decades, however, the Supreme Court's rulings in a series of cases have clarified its position on a defendant's right to counsel **TABLE 9-2**. In its first ruling on this issue in 1932 (*Powell v. Alabama*), the Court held that only an indigent person accused of a capital offense in a state court is entitled to counsel pro-

VOICE OF THE COURT · *Gideon v. Wainwright*

One of the most important landmark decisions affecting the criminal justice process was handed down by the U.S. Supreme Court in 1963. The *Gideon v. Wainwright* case established the right of a defendant to be represented by counsel in state criminal proceedings. Clarence Gideon, a 51-year-old petty thief, was arrested in Panama City, Florida, on June 4, 1961, for breaking into and entering a poolroom and stealing beer, soft drinks, and coins from a cigarette machine. At his trial, Gideon stood before Judge Robert McCrary and stated that he was unable to proceed because he did not have the assistance of an attorney. He requested that the court appoint counsel to represent him and insisted, "The United States Supreme Court says I am entitled to be represented by counsel."

The court refused to appoint an attorney to represent Gideon. Acting as his own counsel, Gideon was unable to present his defense in the manner required by law. He was found guilty by the jury and sentenced to five years in the Florida State Prison. Gideon subsequently submitted a handwritten petition to the Florida Supreme Court, which was denied without a hearing; he then filed a petition with the U.S. Supreme Court. His petition was granted on June 4, 1962, and his case was argued before the Court by an appointed attorney, Abe Fortas, who was later appointed as a justice on the Supreme Court. On March 18, 1963, Justice Hugo Black delivered the opinion of the Court:

> Put to trial before a jury, Gideon conducted his defense about as well as could be expected from a layman. He made an opening statement to the jury, cross-examined the State's witnesses, presented witnesses in his own defense, declined to testify himself, and made a short argument "emphasizing his innocence to the charge contained in the information filed in this case. . . ." Reason and reflection require us to recognize that in our adversary system of criminal justice, any person hauled into court, who is too poor to hire a lawyer, cannot be assured a fair trial unless counsel is provided for him. This seems to us to be an obvious truth. Governments, both state and federal, quite properly spend vast sums of money to establish machinery to try defendants accused of crime. Lawyers to prosecute are everywhere deemed essential to protect the public's interest in an orderly society. Similarly, there are few defendants charged with crime, few indeed, who fail to hire the best lawyers they can get to prepare and present their defenses. That government hires lawyers to prosecute and defendants who have the money to hire lawyers to defend are the strongest indications of the widespread belief that lawyers in criminal courts are necessities, not luxuries. The right of one charged with crime to counsel may not be deemed fundamental and essential to fair trials in some countries, but it is in ours. From the very beginning our state and national constitutions and laws have laid great emphasis on procedural and substantive safeguards designed to assure fair trials before impartial tribunals in which every defendant stands equal before the law. This noble ideal cannot be realized if the poor man charged with crime has to face his accusers without a lawyer to assist him.

The U.S. Supreme Court reversed the judgment and sent the case back to the Supreme Court of Florida for retrial.

Source: *Gideon v. Wainwright,* 372 U.S. 335 (1963).

TABLE 9-2

Summary of Cases Regarding a Defendant's Right to Counsel

Case	Ruling
Powell v. Alabama (1932)	An indigent person who is accused of a capital offense in a state court is entitled to counsel provided by the state.
Johnson v. Zerbst (1938)	Defendants in federal cases must be provided with counsel if they cannot afford to hire an attorney.
Betts v. Brady (1942)	States are not obligated to provide free counsel to indigent defendants in *every serious case*.
Gideon v. Wainwright (1963)	Any indigent person charged with a felony has a right to an attorney at the state's expense.
Argersinger v. Hamlin (1972)	An indigent defendant in a misdemeanor case has the right to a state-provided attorney if the penalty includes the possibility of incarceration.
Scott v. Illinois (1979)	Indigent defendants in misdemeanor cases do not have a right to free counsel in cases punishable only by fines.

vided by the state.[40] Six years later, the Court ruled that the Sixth Amendment required that defendants in federal cases be provided with attorneys if they could not afford to hire one.[41] In 1942, the Court held that states were not obligated to provide free counsel to indigent defendants in every serious case.[42] The Court reversed this position in 1963 when, in its landmark ruling in *Gideon v. Wainwright*, it held that any indigent person charged with a felony has a right to an attorney at the state's expense (see "Voice of the Court").[43] In 1972, in *Argersinger v. Hamlin*, the Court extended the right to counsel for indigent defendants to misdemeanor cases if the penalty includes the possibility of incarceration.[44] In 1979, the Court made it clear, in *Scott v. Illinois*, that the *Argersinger* decision was not meant to include cases punishable only by fines.[45]

Representing Indigent Defendants Three primary systems are used to provide counsel for indigent (i.e., poor) defendants: contract systems, assigned counsel systems, and public defender systems. Many jurisdictions use a combination of all three.

Approximately 42 percent of the largest 100 U.S. counties use contract systems, in which a private at-

torney, local bar association, or law firm is contracted to provide legal representation for indigent defendants.[46] Fees are paid to contract attorneys on a flat rate and are typically inadequate for the attorney to mount an effective defense. Consequently, the motivation to spend the necessary time and effort becomes minimal. According to Mel Tennenbaum, division chief of the Los Angeles public defender's office, under a contract system, "You don't investigate; you don't ask for continuances; you plead at the earliest possible moment."[47]

In assigned counsel systems, which are used in 89 percent of the largest 100 counties, private attorneys whose names appear on a list of volunteers are assigned to represent indigents on a case-by-case basis. Often, these attorneys are younger and less experienced and receive only minimal compensation to defend clients.

Public defender systems are used in 90 percent of the nation's largest 100 counties.[48] Most public defenders are appointed to serve on a full-time basis by either a general trial court judge or county officials, although many maintain small private practices on the side. Public defenders complain that they work hard yet receive little respect for their efforts, from either their clients or the court. Most public defenders are underpaid and overwhelmed with heavy caseloads. Many manage to spend only 30 minutes conferring with a client before their court appearance. Even in large public defender programs, attorneys' salaries are low compared to those of private attorneys. For example, the public defender program in Los Angeles, which is one of the best in the United States, offers public defenders a starting salary of $67,000—a pittance compared to the starting salaries of more than $150,000 paid to lawyers in large law firms.[49]

Each of these systems has been the subject of a number of criticisms. Both contract and assigned counsel systems are more likely to provide attorneys who have relatively little trial experience and whose primary commitment is not to criminal law. These lawyers may be well versed in civil law matters, but their knowledge of the intricacies and strategies of criminal cases is more limited.[50] Their large caseloads, low salaries, and inadequate staff and resources can place them at a disadvantage when they must go up against more experienced prosecutors, who benefit from the support provided by a staff of assistants and resources to hire investigators and expert witnesses. Of course, public defenders who have worked full-time on criminal cases for years and have extensive trial experience often enjoy an advantage in the courtroom over inexperienced assistant prosecutors who are right out of law school.

Earlier research reported that, overall, public defenders were just as effective in obtaining charge reductions, reaching plea bargains, avoiding convictions, and keeping their clients out of prison as private attorneys.[51] However, many of the studies conducted since 1980 have suggested that private counsel is now significantly more effective than public defenders.[52] The relative ineffectiveness found among public defenders, at least in terms of sentencing outcomes, may be the result of their being overworked and underpaid, and having to represent clients in the worst circumstances (i.e., serious charges and weak evidence to support the defendant).[53] Notably, these differences in effectiveness disappear when charges and strength of evidence are controlled.

Competence of Defense Attorneys

If there are relatively few differences in the effectiveness of privately retained counsel and public defenders, then the more general issue of defense attorney competence remains one way to distinguish the effectiveness of specific lawyers. Unfortunately, both private and court-appointed attorneys may provide inadequate counsel to their clients. They may refuse to meet on a regular basis with clients, conduct insufficient investigations, and fail to call important witnesses or cross-examine prosecution witnesses. They may fail to file necessary pretrial motions or post-conviction appeals in a timely manner or fail to object to the introduction of improper evidence or damaging testimony. Furthermore, they may give clients bad advice regarding a plea agreement, thereby encouraging an individual to plead guilty to a charge that clearly would not have been sustained if the case had gone to trial.

Before the 1970s, the appellate courts used a "mockery of justice" standard to determine competency of representation. In other words, a defense attorney was considered ineffective only when his or her actions reduced the trial to a farce or made it a mockery of justice.[54] This standard meant that an attorney could appear in court drunk or fall asleep during the trial and still not be found ineffective.

In 1971, the American Bar Association attempted to establish a categorical approach to defining inadequate or ineffective defense counsel by setting forth minimum requirements of effective assistance.[55] The idea of effective assistance means that "a defendant should be able to expect that his lawyer will actually

Defense attorneys are the guardians of the due process rights of the accused.

? Are defense attorneys compelled to serve because of a passion for justice or injustice?

do something to ensure fairness at the trial. In more colorful terms, the right to counsel is said to guarantee something more than 'a warm body.'"[56] Although there was some support for this approach, many judges opposed it.

In 1984, the U.S. Supreme Court provided its own definition of competence in *Strickland v. Washington*.[57] According to the Court, a defendant's claim that his or her attorney was incompetent must clearly identify acts or omissions by counsel that produced such errors so as to effectively deny the defendant the presence of counsel guaranteed by the Sixth Amendment. In addition, the defendant must show that the attorney's inadequate performance prejudiced the outcome of the trial. In other words, "the defendant must show that there is a reasonable probability that, but for counsel's unprofessional errors, the result of the proceeding would have been different."[58] Unfortunately, some courts using the *Strickland* standard have refused to overturn convictions of defendants whose lawyers were on drugs or asleep during the trial.[59]

Courtroom Workgroups

Patterns of cooperation and friendship between prosecutors and defense attorneys may contradict public perceptions of expected relationships in an adversarial system. How could prosecutors and defense attorneys be friends and cooperate when they must engage in more combative activities as they negotiate plea bargains, object to each other's motions in court, and attempt to discredit each other's witnesses at trial? According to political scientist Peter Nardulli, the criminal courts "are operated by a group of actors who are tied together by a variety of interdependencies and who share a common workplace" and, as a consequence, develop a shared courtroom subculture.[60]

LINK Courtroom subculture, like the subculture that develops in police work (discussed in Chapter 6) incorporates the vaues and perceptions of the regular members of the court community about how they ought to behave and what they can and should expect of one another.[61]

Prosecutors, judges, and defense attorneys work together to move cases through the system as efficiently as possible. In the process, they develop shared goals, norms, and expectations regarding the nature of criminal cases, defendants, and victims. Their work and social relationships, both in and out of court, create a sense of interdependency and identification with the workgroup.

Each participant learns the formal and informal expectations regulating the work activity. Prosecutors and defense attorneys learn to work together to process cases, with each side giving a little according to the demands of the workload. Especially in an adversarial system, no one likes to lose. On occasion, adversaries have to provide their opponents with opportunities to save face and make concessions.

Because defense attorneys have the lowest status within the courtroom workgroup, they are more susceptible to the largely informal rewards and sanctions of judges and prosecutors. The limited resources available to most attorneys (especially court-appointed attorneys) mean that they often must rely on prosecutors for access to police reports, names of witnesses, or other pertinent information regarding a case. Because it is not unusual for private attorneys to have trouble collecting fees from their clients, they often look to judges to grant continuances to give them more time and leverage to collect their fees. By contrast, attorneys who fail to conform to the norms and expectations of the courtroom subculture may not only fail to receive rewards, but also be subject to the group's informal sanctions. Deviant attorneys may find it difficult to get cases conveniently scheduled, or judges may drag out their trials by delaying the proceedings to attend to other business. Judges can adversely affect defense attorneys' incomes by refusing to appoint certain attorneys to represent indigent defendants or by openly criticizing the defenders in court. In addition, prosecutors may take tougher positions in plea negotiations or recommend longer than normal prison sentences.[62]

Courtroom workgroups enable court operations to run smoothly, balancing the various parties' needs within an adversarial system and allowing co-workers to develop a sense of trust within a context of conflicting roles. Like small communities outside the courtroom, a culture of common values and goals eventually emerges to foster conformity to the larger group.

WRAP UP

In many jurisdictions, the charges filed at the initial appearance are likely to originate with the police officer's complaint and may not have been reviewed by the prosecutor before the hearing. The police may file only the minimum charge relevant to the case, or they may routinely file every legally possible charge and then let the prosecutor decide which charges are most likely to result in a conviction. Understandably, charges filed by police officers are often modified by the prosecutor following a more complete review of the case.

Individual prosecutors develop general policies or models for their charging decisions. Some prosecutors give primary attention to whether the elements of the crime are sufficiently present to warrant a particular charge. Others do extensive early screening of cases to weed out the weakest cases to maximize the efficiency of the system and conserve resources. Still other prosecutors make their charging decisions based on the likelihood of obtaining a conviction at trial. Prosecutors consider the presence of necessary elements of the crime, evaluate the supporting evidence and the quality of witnesses, and determine whether any evidence might have been obtained as a result of an illegal search and seizure. If the case appears sufficiently strong, prosecutors will file charges aimed at going to trial. On other occasions, prosecutors may decide to not file charges (*nolle prosequi*) after screening for a variety of reasons—for example, if there is little concrete evidence or the victim is not credible.

In the Dominique case, police may have disregarded the allegations because many of the alleged victims were transients who admittedly went with Dominique to engage in sex or serve as prostitutes. While police and prosecutors believed that a crime occurred, they probably did not believe that the evidence or the victim accounts were credible. Dominique later admitted that he murdered his victims to avoid the possibility of going to prison for raping them.

Chapter Spotlight

- The criminal court system in the United States today is largely a product of the Anglo-Saxon and English common law courts and the rights of citizens enumerated in the Magna Carta.

- Many of the procedural safeguards of the English courts were incorporated into the early colonial legal codes. After the American Revolution, each of the new states created its own independent court system. Congress established the basic structure of the federal court system in the Judiciary Act of 1789.

- With the proliferation of state and federal courts, clashes over judicial jurisdiction questions were inevitable. The jurisdiction of courts depends on geographical boundaries, the significant characteristic or subject matter of the case, or some characteristic of the parties to the case.

- Although there are many variations in the general state court structures, all of them make similar distinctions between the courts according to their basic jurisdiction. State-level courts generally include courts of limited jurisdiction, courts of general jurisdiction, intermediate appellate courts, and courts of last resort.

- The federal court system parallels the structure of the state courts and includes U.S. Magistrates, U.S. District Courts, U.S. Courts of Appeals, and the U.S. Supreme Court.

- The criminal courts employ a number of people who are collectively responsible for scheduling cases, maintaining order, recording the actions of the courts, interpreting, and administering budgets and managing personnel.

- The judge is the most central figure in the court. He or she performs a wide variety of duties, ranging from issuing warrants, to deciding whether to accept plea agreements, to determining appropriate sentences. Today, most judges are popularly elected, although in some states they may be appointed by the governor or selected through a merit process.

- The prosecutor has far greater discretion—and, therefore, a greater influence on the fate of defendants—than the judge has. Prosecutors decide which cases to prosecute, which charges will be filed, and whether to negotiate a plea bargain. In addition, they frequently make recommendations regarding bail and sentences and may speak in support of (or in opposition to) an inmate's parole at parole hearings.

- Defense attorneys play a less central role than judges in the functioning of the courts and have relatively little discretion. Their primary role is to protect the interests of their clients and to present the best possible case before the court. The Sixth Amendment guarantees that a criminal defendant shall have the right to counsel, although the courts have only recently held that this right extends to defendants in both state and federal courts who are charged with either a felony or a misdemeanor. Indigent defendants may be represented by attorneys who are supplied through a contract, assigned counsel, or public defender system.

- Although the courts operate under an adversarial system, most develop courtroom workgroups among the participants. These informal networks and expectations regulate their members' activities and function to move cases through the system as efficiently as possible.

Putting It All Together

1. Identify some important developments in U.S. courts and legal procedure during the early colonial and postcolonial periods. How are these changes reflected in today's courts?

2. What was the importance of *Marbury v. Madison* in terms of the role of the U.S. Supreme Court?

3. Describe a criminal case that contains elements of each of the following types of jurisdiction: exclusive, concurrent, and original.

4. What methods are used for selecting judges? Which method do you believe is most likely to achieve justice?

5. Should judges be appointed to the bench for life? Why or why not?

Key Terms

appellate court Courts which hear and determine appeals from lower trial courts.

appellate jurisdiction The authority of a court to review or revise the judicial actions of a lower court.

assigned counsel systems A method for providing legal representation for indigent defendants by which the court assigns private attorneys whose names appear on a list of volunteers on a case-by-case basis.

Body of Liberties A document in the Massachusetts Bay Colony legal code outlining the provisions for protecting the rights of citizens in criminal prosecutions.

certification A request by a lower federal court asking the Supreme Court to rule on a specific legal question.

common law courts English courts established during the late twelfth century; their judgments became the law common to the entire country.

concurrent jurisdiction The authority of two or more courts to hear a particular case.

contract systems A method for providing legal representation for indigent defendants by which a local attorney, bar association, or law firm contracts with the court.

courts of general jurisdiction Courts with the authority to hear virtually any criminal or civil case.

courts of last resort In most states and in the federal court system, the final appellate court.

courts of limited jurisdiction Courts usually referred to as the lower or inferior courts, which are limited to hearing only specific kinds of cases.

deposition Testimony of a witness taken under oath outside the courtroom.

exclusive jurisdiction The authority of a court to be the only court to hear a particular case.

JBPUB.COM/ExploringCJ

judicial jurisdiction The power or authority of a court to hear a case or consider a particular legal motion.

judicial review The power of the U.S. Supreme Court to review and determine the constitutionality of acts of Congress and orders by the executive branch.

Judiciary Act of 1789 An act created by the First Congress establishing the basic structure of the federal court system.

Justice of the Peace courts Courts first established in the American colonies to hear minor criminal cases.

Magna Carta A document signed by King John in 1215 that enumerated rights and protections for the common citizens of England.

Marbury v. Madison The U.S. Supreme Court case that established the principle of judicial review.

Missouri Plan Developed in 1940, the first plan for the selection of judges based on merit.

original jurisdiction The power or authority of a court to be the first to hear a case and render a verdict.

public defender systems A method for providing legal representation for indigent defendants by which defense attorneys are appointed by the court to act as full-time defenders.

U.S. Supreme Court The highest appellate court in the U.S. judicial system; it reviews cases appealed from federal and state court systems that deal with constitutional issues.

writ of certiorari An order by the U.S. Supreme Court to a lower court to send up a certified record of the lower court decision to be reviewed.

Notes

1. Sir William Holdsworth, *A History of English Law,* 4th edition (London: Methuen, 1936).
2. Holdsworth, note 1.
3. Carl Stephenson and Frederick Marcham, *Sources of English Constitutional History,* 2nd ed (New York: Harper & Row, 1972), p. 121.
4. Herbert Johnson, *History of Criminal Justice* (Cincinnati: Anderson, 1988), pp. 82–84.
5. Johnson, note 4.
6. Roscoe Pound, *The Lawyer from Antiquity to Modern Times* (St. Paul: West, 1953).
7. Lawrence Friedman, *A History of American Law* (New York: Simon and Schuster, 1977).
8. *Virginia Constitution,* July 5, 1776.
9. *Marbury v. Madison,* 1 Cr. 137 (1803).
10. *Fletcher v. Peck,* 6 Cr. 87 (1810).
11. Robert Carp and Ronald Stidham, *The Federal Courts* (Washington, DC: Congressional Quarterly, 1985).
12. Roscoe Pound, *The Spirit of the Common Law* (Francestown, NH: Marshall Jones, 1921).
13. National Center for State Courts, *State Court Caseload Statistics, 2005* (Williamsburg, VA: National Center for State Courts, 2006).
14. National Center for State Courts, note 13.
15. Daniel Meador, *American Courts* (St. Paul: West, 1991), pp. 16–18.
16. Administrative Office of the United States Courts. *Judicial Business of the United States Courts, 2005* (Washington, DC: U.S. Government Printing Office, 2006).
17. *Gomez v. United States,* 488 U.S. 1003 (1989).
18. Administrative Office of the United States Courts, note 16.
19. Administrative Office of the United States Courts, note 16.
20. *United States Law Week, Supreme Court Proceedings* (Washington, DC: The Bureau of National Affairs), 57:3074 (1988); Supreme Court of the United States, *2006 Year-End Report on the Federal Judiciary* (Washington, DC: U.S. Supreme Court, 2007).
21. *Herrera v. Collins,* 506 U.S. 390 (1993).
22. *Thompson v. Oklahoma,* 487 U.S. 815 (1988).
23. *Stanford v. Kentucky,* 429 U.S. 361 (1989); *Wilkins v. Missouri,* 429 U.S. 361 (1989).
24. *Roper v. Simmons,* 543 U.S. 551 (2005).
25. Sheldon Goldman, "Reagan's Judicial Legacy: Completing the Puzzle and Summing Up," *Judicature* 72:318–330 (1989); John Ryan, Allan Ashman, Bruce Sales, and Sandra Shane Du Bow, *American Trial Judges: Their Work Styles and Performance* (New York: Free Press, 1980).
26. Sheldon Goldman, Elliot Slotnick, Gerard Gryski, and Sara Schavoni, "W. Bush's Judiciary: The First Term Record," *Judicature* 88:244–274 (2005); Sheldon Goldman, "Bush's Judicial Legacy: The Final Imprint," *Judicature* 76:282–297 (1993).

27. Lawrence Baum, *American Courts: Process and Policy,* 2nd edition (Boston: Houghton Mifflin, 1990), p. 140.

28. Meador, note 15, p. 57.

29. Steven Perry, *Prosecutors in State Courts, 2005* (Washington, DC: U.S. Department of Justice, 2006).

30. Perry, note 29.

31. Perry, note 29.

32. American Bar Association, *ABA Standards for Criminal Justice,* Standard 3-1.2 (Washington, DC: American Bar Association, 1992).

33. Celesta Albonetti, "Criminality, Prosecutorial Screening, and Uncertainty: Toward a Theory of Discretionary Decision Making in Felony Case Processings," *Criminology* 24:640 (1986).

34. Cassia Spohn, John Gruhl, and Susan Welch, "The Impact of the Ethnicity and Gender of Defendants on the Decision to Reject or Dismiss Felony Charges," *Criminology* 25:175 (1987); Cassia Spohn, Dawn Beichner, and Erika Davis-Frenzel, "Prosecutorial Justifications for Sexual Assault Case Rejection: Guarding the 'Gateway to Justice,'" *Social Problems* 48:206–235 (2001).

35. Nicci Lovre-Laughlin, "Lethal Decisions: Examining the Role of Prosecutorial Discretion in Capital Cases in South Dakota and the Federal Justice System," *South Dakota Law Review* 50:550–574 (2005); U.S. Department of Justice, *The Federal Death Penalty System: A Statistical Survey (1988–2000)* (Washington, DC: U.S. Department of Justice, 2000); Michael Radelet and Glenn Pierce, "Race and Prosecutorial Discretion in Homicide Cases," *Law and Society Review* 19:587–621 (1985); Raymond Paternoster, "Prosecutorial Discretion in Requesting the Death Penalty: A Case of Victim-Based Racial Discrimination," *Law and Society Review* 18:437–478 (1984).

36. American Bar Association, *Fall 2005 Enrollment Statistics,* (Chicago: American Bar Association, 2006).

37. Herbert Jacob, *Justice in America,* 4th edition (Boston: Little, Brown, 1984), pp. 78–79.

38. *United States v. Wade,* 388 U.S. 218 (1967).

39. Rodney Uphoff, "The Criminal Defense Lawyer: Zealous Advocate, Double Agent, or Beleaguered Dealer?" *Criminal Law Bulletin,* September–October, 1992, p. 420.

40. *Powell v. Alabama,* 287 U.S. 45 (1932).

41. *Johnson v. Zerbst,* 304 U.S. 458 (1938).

42. *Betts v. Brady,* 316 U.S. 455 (1942).

43. *Gideon v. Wainwright,* 372 U.S. 335 (1963).

44. *Argersinger v. Hamlin,* 407 U.S. 25 (1972).

45. *Scott v. Illinois,* 440 U.S. 367 (1979).

46. Carol DeFrances and Marika Litras, *Indigent Defense Services in Large Counties, 1999* (Washington, DC: U.S. Department of Justice, 2000).

47. Jill Smolowe, "The Trials of the Public Defender," *Time,* March 29, 1993, p. 50.

48. Smolowe, note 47, p. 50.

49. Stanford Law School, *Hiring Practices of California Public Defender Offices* (Palo Alto, CA: Stanford University, 2004); Leigh Jones, "Does It Pay to Make N.Y. Law Firms Pay?" *The National Law Journal,* June 12, 2007, available at http://www.law.com/jsp/article.jsp?id=1181552737950, accessed July 10, 2007.

50. Christopher Smith, *Courts and the Poor* (Chicago: Nelson-Hall, 1991), p. 30.

51. Roger Hanson, Brian Ostrom, William Hewitt, and Christopher Lomvardias, *Indigent Defenders: Get the Job Done and Done Well* (Williamsburg, VA: National Center for State Courts, 1992).

52. Joyce Sterling, "Retained Counsel Versus the Public Defender: The Impact of Type of Counsel on Charge Bargaining," pages 176–200 in William McDonald (ed.), *The Defense Counsel* (Thousand Oaks, CA: Sage, 1983); Dean Champion, "Private Counsels and Public Defenders: A Look at Weak Cases, Prior Records and Leniency in Bargaining," *Journal of Criminal Justice* 17:253–263 (1989); William Stuntz, "The Uneasy Relationship between Criminal Procedure and Criminal Justice," *Yale Law Journal* 107:1–76 (1997).

53. Morris Hoffman, Paul Rubin, and Joanna Shepherd, "An Empirical Study of Public Defender Effectiveness: Self-Selection by the 'Marginally Indigent,'" *Ohio State Journal of Criminal Law* 3:223–255 (2004).

54. William Erickson, "Standards of Competency for Defense Counsel in a Criminal Case, *American Criminal Law Review* 17:233, 251 (1979).

55. American Bar Association, *ABA Standards Relating to the Defense Function* (Chicago: American Bar Association, 1971).

56. Rebecca Kunkel, "Equalizing the Right to Counsel," *Georgetown Journal of Legal Ethics* 18:843–862 (2005).

57. *Strickland v. Washington,* 466 U.S. 668 (1984).

58. *Strickland v. Washington,* note 57.

59. Jeffrey Kirchmeier, "Drink, Drugs, and Drowsiness: The Constitutional Right to Effective Assistance of Counsel and the Strickland Prejudice Requirement," *Nebraska Law Review* 75:455–460 (1996).

60. Peter Nardulli, James Eisenstein, and Roy Flemming, *The Tenor of Justice: Criminal Courts and the Guilty Plea Process* (Urbana, IL: University of Illinois Press, 1988), pp. 123–124.

61. James Eisenstein, Roy Flemming, and Peter Nardulli, *The Contours of Justice: Communities and Their Courts* (Boston: Little, Brown, 1988), p. 28.

62. Note, "Breathing New Life into Prosecutorial Vindictiveness Doctrine," *Harvard Law Review* 114:2074–2097 (2001); Peter Nardulli, *The Courtroom Elite: An Organizational Perspective on Criminal Justice* (Cambridge, MA: Ballinger, 1978).

OBJECTIVES

- Understand the nature and function of the initial appearance.

- Know the reasons behind pretrial release, alternatives to money bail, and justifications for the use of preventive detention.

- Grasp the history of the grand jury, its contemporary structure and procedures, and possible problems.

- Know the major factors considered by prosecutors in selecting or dismissing charges.

- Describe the basic procedures used in preliminary hearings and the importance of discovery.

- Distinguish between the various types of pleas, understand the nature of plea bargaining, and elucidate the advantages and disadvantages of bargained justice.

Prosecution

CHAPTER 10

On July 28, 2006, Naveed Haq, a Muslim American, entered the Jewish Federation of Greater Seattle's office and shot six people, killing one. Haq told police that he was angry about the United States' cooperation with Israel and the ongoing war in Iraq. He informed a 9-1-1 dispatcher, "These are Jews, and I'm tired of getting pushed around and our people getting pushed around by the situation in the Middle East." In addition to murder and attempted murder charges, Haq was charged with multiple counts of kidnapping and malicious harassment, which was the charge under Washington's hate-crime law. King County Prosecutor Norm Maleng called the event one of the most serious crimes in Seattle history. In December 2006, Maleng announced that prosecutors would not seek the death penalty for Haq because of his long history of mental illness. Despite the fact that this crime was an attempted mass murder with hate-crime overtones, it did not result in the most serious punishment.

- Why would a prosecutor choose this course of action?
- What are the potential pros and cons of the prosecutor's decision?

Source: The Associated Press, "Prosecutors Won't Seek Death for Mentally Ill Man Accused in Jewish Center Shooting," December 20, 2006, available at http://www.courttv.com/news/2006/1220/shooting_ap.html/, accessed December 26, 2006.

Introduction

Of the more than 2.1 million people in the United States who are arrested for serious crimes each year, fewer than 45,000—about 4 percent—have their cases resolved through trials.[1] It is important to examine what happens to people who never face a jury, such as in cases in which the prosecutor decides not to charge or prosecute the person for a crime, or the defendant initially pleads guilty, or a plea bargain is reached. For many defendants, an agreement to plead guilty in exchange for some consideration may end their involvement with the courts and the formal criminal justice system. For others, a plea bargain may simply mean a shorter prison sentence or, in an extreme circumstance, avoidance of the death penalty.

Initial Appearance

The initial appearance in court is the first stage in the process of determining whether a defendant is guilty. According to the Sixth Amendment, all people who are arrested, after being booked by the police, should be taken before a judge or a magistrate without unnecessary delay and informed of the charges against them, asked to enter a plea, and informed of their rights.

Notification of Rights

At the initial appearance, the accused is informed of the charges and of three constitutionally guaranteed rights:

- right to counsel
- right to remain silent
- right to reasonable bail

In misdemeanor cases, a defendant may choose to plead guilty at the initial appearance and face immediate sentencing. In many urban courts, in which literally hundreds of minor cases are processed daily, the initial appearance may resemble assembly-line justice. In such situations, those accused of misdemeanors are often informed of the charges against them and their rights, assigned counsel, and then, after a brief discussion with counsel, plead guilty and are sentenced, usually to a fine or a short jail term. Alternatively, after a brief examination of the merits of the case, the judge may choose to dismiss the charges and release the defendant. Under more extreme time constraints, a judge may call all similarly charged defendants to the bench and handle their cases at the same time. It is not unusual for a half-dozen or more people who are accused of public intoxication, for example, to be charged collectively and then asked in turn for their pleas.

Felony charges are handled differently at the initial appearance. Defendants are rarely permitted to enter pleas and, consequently, are not sentenced at this time. Because defendants typically do not have attorneys at this stage, attempts to plead guilty to serious charges are considered ill informed and are generally not accepted by the court. Felony guilty pleas also are frequently rejected because most initial appearances are held in inferior or lower courts, which do not have general trial jurisdiction.

LINK Lower courts are not courts of record, and typically no transcripts are made of hearings that take place in such courts. Thus, as Chapter 9 discusses, if a defendant chooses to appeal his or her conviction, the absence of a transcript means that the case must be tried again in a general trial court.

Timing of the Initial Appearance

Both the states and the federal government have established guidelines that regulate the timing of the initial appearances of defendants. These guidelines are designed to protect the rights of citizens so that the government cannot let people languish in jail without knowing why, conduct secret interrogations, or coerce people into making confessions. These guidelines are often inconsistent, however, and the various levels of governments have not agreed on exceptions to the rules or the consequences of their violation. For example, the federal rule on the timing of initial appearances states only that "a person making an arrest within the United States must take the defendant *without unnecessary delay* before a magistrate judge, or before a state or local judicial officer" (emphasis added).[2] By comparison, the state of Minnesota's Rules of Criminal Procedure are more explicit in defining the maximum amount of time between arrest and initial appearance:

An arrested person . . . shall be brought before the nearest available judge of the district court of the county where the alleged offense occurred or judicial officer of such court. The defendant shall be brought before such judge or judicial officer without unnecessary delay, and in any event, *not more than 36 hours* [emphasis added] after the arrest, exclusive of the day of arrest, Sundays, and legal holidays, or as soon thereafter as such judge or judicial officer is available.[3]

Definition of "Unnecessary Delay"

McNabb v. United States, decided in 1943, was the first U.S. Supreme Court decision to address the timing of the initial appearance. In its decision, the Court ruled that confessions obtained during an unnecessary delay before a defendant was brought to an initial appearance could not be used in federal court. In this case, Benjamin McNabb had been held for six days before being brought before the court to be notified of the charges. During this time, he confessed, and the confession became the primary basis for his subsequent conviction. In declaring their strong disapproval of the police practice of secret interrogations and the coercion of confessions, the Supreme Court stated that "a conviction resting on evidence secured through such flagrant disregard of the procedure which Congress has commanded cannot be allowed to stand without making the courts themselves accomplices in willful disobedience of law."[4]

The Supreme Court extended this position in 1957 in *Mallory v. United States,* when it stated that unnecessary delays were serious violations of a defendant's basic right to due process.[5] In this case, Andrew Mallory was arrested at 2:30 P.M. but not brought before a magistrate until the next morning. During the 18-hour delay, he made a voluntary confession. Mallory was eventually convicted and sentenced to death. The Supreme Court nullified the death sentence, arguing that because the delay was not warranted, any evidence gathered by police—including a voluntary confession—was not admissible in this case.

Responding to pressure by police agencies to avoid acquittals or judicial reversals based on such technical violations, Congress included in the Omnibus Crime Control and Safe Streets Act of 1968 a modification to the ruling in *Mallory.* This act stated that voluntary confessions, if given within six hours of the arrest, were admissible even if there was a delay in bringing the defendant before a magistrate as long as the delay was not "unreasonable."

The 48-Hour Rule

In 1975, the Supreme Court ruled in *Gerstein v. Pugh* that defendants must be brought before a judge or magistrate "promptly after arrest."[6] However, this decision did not include a specific time frame. Some years later, in response to a class-action suit against a California policy allowing delays of as much as seven days before the initial appearance, the U.S. District Court of Appeals for the Ninth District imposed a 36-hour rule. The case was eventually appealed to the Supreme Court, and in 1991, in *Riverside County, California v. McLaughlin,* the Court ruled that the notion of "promptly," which was the guiding statement in *Gerstein,* did not mean "immediately upon completion of booking the defendant."[7] Justice Sandra Day O'Connor, writing for the majority, stated that the criminal justice system requires some degree of flexibility and that 48 hours provides that needed flexibility. This defined the <u>48-hour rule</u>, which allows delays of as much as 48 hours before the initial appearance.

The Court, however, presented two caveats. First, hearings that occur within the 48-hour limit may be unconstitutional if the defendant can demonstrate that the delay was unreasonable. According to Justice O'Connor, "examples of unreasonable delay are delays for the purpose of gathering additional evidence to justify the arrest, a delay motivated by ill will against the arrested individual, or delay for delay's sake."[8] Second, delays may exceed the 48-hour rule and still be considered constitutional if the state can demonstrate that the delay was caused by an emergency or other extraordinary circumstance, such as the closure of the courthouse because of a fire.

Not all people who are brought before the court to face criminal prosecution arrive there from arrest and subsequent initial appearance. Indeed, many defendants are investigated and indicted through the grand jury process, discussed later.

Pretrial Release

The belief that a person who is accused of a crime has a right to <u>pretrial release</u> (that is, being released from custody while awaiting judicial action) has been long and widely held in the United States. Generally, pretrial release has involved the posting of bail by the defendant. The Eighth Amendment states that "excessive bail shall not be required." However, this amendment merely prohibits excessive bail; it does not *guarantee* a defendant the right to pretrial release.[9] Even so, the majority of people who are arrested—even for felony offenses—are released before trial or final adjudication by the court. Not all felony defendants are equally likely to be released, however. For example, in state cases in the United States' 75 largest counties, 62 percent of those charged with assault, 42 percent of those charged with robbery,

and only 8 percent charged with murder are released FIGURE 10-1 . The percentage of federal felony defendants released prior to case disposition also varies by offense—from 22 percent for robbery defendants, to 93 percent for defendants charged with embezzlement, to 98 percent of those charged with gambling-related crimes.[10]

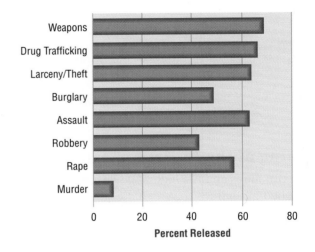

FIGURE 10-1 Pretrial Release in the 75 Largest Counties in the United States

Source: Bureau of Justice Statistics, *Felony Defendants in Large Urban Counties, 2002* (Washington, DC: U.S. Department of Justice, 2006), p. 162.

The Purpose Behind Bail

Historically, it has been believed that providing for pretrial release would do less harm to the defendant and his or her family than might occur by keeping the defendant in custody. Pretrial release has typically involved setting some <u>bail</u> amount that was large enough to ensure that the defendant would appear at subsequent hearings. This was to be accomplished by having the defendant deposit a monetary bond upon booking or at the initial appearance. To this effect, in 1951, the Supreme Court ruled in *Stack v. Boyle* that bail should be set high enough to ensure the defendant's appearance at trial.[11] However, the Court also stated that "bail set at a figure higher than an amount reasonably calculated to fulfill this purpose is 'excessive' under the Eighth Amendment." Unfortunately, the Court did not provide any standard equation for lower courts to use in determining whether a specific amount is excessive.

In many misdemeanor cases, bail is set by a police officer at the station house based on a predetermined schedule for particular offenses. In more serious cases, bail is set at the initial hearing or at a special bail hearing. If the accused is able to post the full bail amount, he or she is released. Many defendants, because they are unable to post the full amount of the bail, turn to a bail bondsman for assistance. For a fee of 10 to 15 percent of the full bond, the bondsman will guarantee payment to the court of the full fee.

Bias in Pretrial Release

For many defendants, nearly any bail amount is unaffordable. As a consequence, these individuals may wait in jail for weeks or months while the wheels of justice turn. The poor and minorities suffer disproportionately from a loss of freedom before trial because they often lack the resources needed to gain their pretrial release by posting bail. A substantial body of evidence indicates that race, ethnicity, and gender bias bail decisions. Studies that have examined the patterns of pretrial release in the United States' 75 largest counties as well as in smaller, more rural counties, for example, have consistently found that African Americans and Latinos are more likely to be denied bail than white defendants in felony cases. In addition, Latinos are more disadvantaged than African Americans in terms of both the likelihood of their pretrial release and the dollar amount required for release. Moreover, in addition to race and ethnicity, defendants with weak community ties (local residence and employment status) are less likely to be released prior to trial.[12]

Race and ethnicity are not the only extra-legal factors to significantly bias release decisions: Criminologists Stephen Demuth and Darrell Steffensmeier report that gender also affects the likelihood of release. Females across all racial groups are more likely to receive pretrial release than their male counterparts, even when controlling for legally relevant factors, such as seriousness of the crime and prior record. Even so, significant gender differences are observed within racial and ethnic groups when it comes to pretrial release patterns. For example, the gender difference is smallest for White Americans and greatest for Latinos, with African Americans falling in the middle.[13]

Even when they are not convicted of any crime, poor and minority defendants are often left to face unpleasant and punitive consequences of their condition. According to criminologist John Goldkamp:

The due process notion of presumption of innocence requires that criminally charged defendants be treated as innocent until guilt has been established. Thus, a glaring affront to due process is offered by the very existence of the practice of pretrial detention. To the defendant, pretrial detention is essentially punitive; it institutionalizes the punishment of persons who theoretically are presumed innocent.[14]

The negative impact of sitting in jail for these defendants is great. Not only do they lose income from work, but they may also lose their jobs, their families may suffer financially, and they are less able to assist in the preparation of their cases. In addition, these individuals are more likely to face conviction

Offenders who are unable to post bond and remain in jail potentially face increased legal problems.

Are defendants who are unable to be released on bail punished more severely than other offenders? Is this reasonable?

and, if convicted, they are more likely to receive a prison sentence.[15] Two important factors may account for the greater probability of detainees ultimately being convicted and incarcerated. The first factor involves negative perceptions of detainees held by judges and jurors alike: A well-dressed defendant who walks into court accompanied by his or her attorney produces a very different impression than a defendant who wears a jail uniform and is brought into the courtroom by guards. The implications are not lost on the judge or jury that the latter defendant is probably guilty or perhaps poses a greater danger to the community. The second factor involves more positive perceptions produced by released defendants who are able to demonstrate that they are trustworthy, because they have refrained from further criminal activity and maintained their jobs while awaiting trial.

Defendants who are detained in jail quickly find that it is neither a safe nor a pleasant place. People kept in custody face not only a depressing environment, but also the possibility of assault from other detainees. Because so many U.S. jails are overcrowded, few attempts are made to separate offenders based on the seriousness of their charges or their emotional or psychological condition. Nonviolent defendants awaiting trial are often housed with people who have already been convicted of violent crimes and are waiting to be transferred to prison. Even if personal safety is not of immediate concern, the time spent in jail is typically monotonous and tedious; there are few books, little time allowed for recreation, and infrequent opportunities for contact visits from family and friends.

> **LINK** Nearly 750,000 people are confined in jail on any given day, a number that accounts for approximately 95 percent of the rated capacity of the nation's jails. The *rated capacity* of a correctional facility is the number of beds or inmates assigned by a rating official to institutions within a jurisdiction. Chapter 13 provides more information about jail and prison populations.

Alternatives to Monetary Bail

Partly because of public perceptions of inequities in the use of bail, and perhaps even more because of severe jail overcrowding, attempts to reform the bail system emerged in the 1960s. In 1961, the Vera Institute of Justice in New York City began an experiment in cooperation with the city's criminal courts called the Manhattan Bail Project. The participants in this experiment sought to identify information that would allow the court to screen defendants for possible nonmonetary release (i.e., no bail required for pretrial release). A method for predicting which defendants were likely to appear at judicial hearings, even if they were not required to post bond, was created. Such factors as the current offense and likely sentence if found guilty, employment history, marital and family situation, and length of residence were considered in recommendations for <u>release on recognizance (ROR)</u>—that is, a personal promise to appear at trial. The early results of the experiment indicated its success: During the first three years, 10,000 defendants were interviewed for possible ROR, 4000 were recommended, and 2195 were released on their own recognizance. Only 15 (less than 0.7 percent) of those released on the program failed to show up in court.[16]

Duane "Dog" Chapman has become a television star based on his bail recovery business.

? Is pretrial detention a fundamental violation of the presumption of innocence?

Soon, additional alternatives to traditional monetary bail were established in courts around the United States. These measures include the following:

- *Conditional release.* In this system of pretrial release, defendants must promise to meet stated conditions that extend beyond simply promising to appear in court. Such conditions may include agreeing to maintain one's residence or employment status, remain in the jurisdiction, adhere to a curfew, report to a third party or designated release program, or participate in a drug or alcohol treatment program.

- *Unsecured bail.* No payment is required at the time of release. However, if the defendant fails to appear at required hearings, he or she becomes liable to pay the full amount.
- *Property bail.* The defendant may post evidence of real property, such as a car title, as an alternative to cash.
- *Court deposit bail.* The defendant may deposit a certain percentage of the bail amount with the court (usually 10 to 15 percent). When the defendant fulfills his or her appearance requirements, the deposit is returned. Some courts subtract a small administrative fee from this deposit.[17]

These alternatives to the traditional posting of monetary bail proved successful enough to prompt the passage of the Bail Reform Act of 1966. This act provided for the release of defendants in noncapital federal cases in which there was reason to believe that the defendant would appear at the required hearings. It was intended to change existing bail practices so as to ensure that all people, regardless of financial status, would not needlessly be detained during the pretrial or trial process.[18] As a consequence of the Bail Reform Act of 1966, it was presumed that defendants were to be considered for ROR or conditional release unless the prosecution could show strong evidence supporting the need for monetary bail.

In an attempt to reduce the apparent race and class bias in pretrial release discussed earlier, as well as to allow for needed flexibility in these decisions, criminologists John Goldkamp and Michael Gottfredson developed a system of bail guidelines for the Philadelphia courts. Using a two-dimensional grid that plots the severity of the offense against the defendant's personal characteristics (such as his or her courtroom demeanor, any physical or mental health concerns, other current or outstanding warrants, and the possibility of a mandatory sentence in the current case), the court determines whether the defendant should receive ROR or conditional release or pay a money bail.[19] Goldkamp and Gottfredson noted that the use of bail guidelines increased equity in bail decisions but resulted in no significant change in failure-to-appear rates: Only 25 percent of all released defendants failed to appear in court, and the differences between those who posted bond and those who received ROR were negligible.[20]

Preventive Detention

Another concern—the need to protect society from the additional commission of crimes by dangerous offenders awaiting trial—led the U.S. Congress to pass the Bail Reform Act of 1984. The Bail Reform Act was part of a much larger crime control package, known as the Comprehensive Crime Control Act of 1984, which was submitted to Congress by then-President Ronald Reagan.[21] This act extended the opportunity for ROR for suspects in many federal cases by mandating that defendants should not be kept in custody simply because they could not afford money bail. However, it also provided for the preventive detention of defendants who are charged with particularly serious crimes and thus are perceived to be high risks for committing further crimes or absconding before trial.

To counter a growing public fear of new crimes committed by arrested felons, the possibility of intimidation of witnesses, and the ability of some offenders in drug trafficking cases to post hundreds of thousands of dollars in bail and then flea, the Bail Reform Act of 1984 gave the government new tools to protect the public and to keep likely-to-flee suspects in custody. Temporary preventive detention without bail is permitted if a defendant is on pretrial release for another offense, is on probation, or is out on parole, or if the judicial officer finds that no conditions or combination of conditions will reasonably ensure the appearance of the person as required and the safety of any other person and the community. The burden of rebuttal is forced on the defendant; that is, the defendant must convince the court that he or she is not a risk.

The no-bail presumption contained in the Bail Reform Act of 1984 raises a crucial constitutional question: Does the presumption of future dangerousness and denial of pretrial release deny the accused of his or her due process rights by imposing punishment before actually facing trial? This question was considered by the Supreme Court in 1987 in *United States v. Salerno*, when the Court, in a 6 to 3 decision, ruled that the preventive detention provisions of the Bail Reform Act were constitutional (see "Voice of the Court").[22]

The idea of preventive detention was extended in the no-bail provision of the Immigration Nationality Act, which authorized the federal government to detain noncitizen offenders who were facing deportation after having been convicted of felonies. The U.S. Supreme Court, in *Demore v. Kim*, held that such detention was constitutional because, under the cir-

Anthony Salerno was arrested on charges of racketeering, mail fraud, and extortion. At a pretrial hearing, the federal prosecutor stated that Salerno was the head of an organized crime family and, based on the testimony of two witnesses, that Salerno had conspired to commit two murders. The judge denied bail, and Salerno appealed. The critical issue Salerno raised in his appeal was whether the court had the authority to deny bail based on its prediction of future dangerousness. Salerno also appealed on the basis that the Bail Reform Act of 1984 violated his Eighth Amendment right of protection against excessive bail.

In a 6 to 3 decision, the U.S. Supreme Court ruled that the Bail Reform Act of 1984 had a legitimate regulatory function and was not a form of punishment. Rather, it was designed to protect the community from crimes committed by people who are out on bail. The Court determined that society's need for safety outweighs an individual's right to bail. Furthermore, the Court stated that the act did not violate the Eighth Amendment protection against excessive bail and reaffirmed that this amendment only prohibits excessive bail—it does not prevent a court from denying bail in certain cases.

According to Chief Justice William Rehnquist, who delivered the opinion of the Court:

> The Bail Reform Act of 1984 allows a federal court to detain an arrestee pending trial if the government demonstrates by clear and convincing evidence after an adversary hearing that no release conditions "will reasonably assure . . . the safety of any other person and the community."

The legislative history of the Bail Reform Act clearly indicates that Congress did not formulate the pretrial detention provisions as punishment for dangerous individuals. Congress instead perceived pretrial detention as a potential solution to a pressing society problem. There is no doubt that preventing danger to the community is a legitimate regulatory goal. . . . On the other side of the scale, of course, is the individual's strong interest in liberty. We do not minimize the importance and fundamental nature of this right. But, as our cases hold, this right may, in circumstances where the government's interest is sufficiently weighty, be subordinated to the greater needs of society.

Respondents also contend that the Bail Reform Act violates the Excessive Bail Clause of the Eighth Amendment. . . . This Clause, of course, says nothing about whether bail shall be available at all. Respondents nevertheless contend that this Clause grants them a right to bail calculated solely upon considerations of flight. . . . While we agree that a primary function of bail is to safeguard the courts' role in adjudicating the guilt or innocence of defendants, we reject the proposition that the Eighth Amendment categorically prohibits the government from pursuing other admittedly compelling interests through regulation of pretrial release.

The *Salerno* decision reflected the Supreme Court's agreement with Congress's concern that communities should be protected from dangerous offenders through preventive detention of suspects awaiting trial.

Source: *United States v. Salerno*, 481 U.S. 739 (1987).

cumstances, deportable aliens presented a special risk to flee while they were awaiting deportation hearings.[23]

The Grand Jury

It is not unusual to pick up a newspaper and read that a grand jury recently indicted someone for a crime, investigated corruption in some governmental office, or declared that a public agency such as a housing authority or a prison was in violation of government regulations. The grand jury of today comprises a body of citizens who are called by a prosecutor to investigate the conduct of public officials and agencies and criminal activity in general and to determine whether probable cause exists that a crime has been committed. Grand jurors are asked to decide the following question: "Would the average person believe that the defendant committed the crime he or she is charged with, based on the evidence presented at the hearing?" It is not necessary that the evidence be sufficient to prove guilt beyond a reasonable doubt; the prosecutor need only demonstrate a *probability* that the defendant committed the crime.

In some states, all felony cases are presented to the grand jury for review and determination of whether an <u>indictment</u>—that is, a formal written accusation—should be issued. In other states, the prosecutor may bypass the grand jury when filing charges. In these cases, the formal written accusation is called an information.

Selection of Grand Jury Members

Grand jury members are usually selected through a random drawing from a list of registered voters in the jurisdiction, although procedures may vary between jurisdictions. Members are required to be age 18 or older, must be citizens of the United States, must have been residents of the jurisdiction for at least one year, and must have sufficient command of the English language. Each potential juror must commit to being in town and available for the term of the grand jury. Terms vary from state to state, but typically range from 6 to 18 months, with most lasting one year.

Once the jury is selected and impaneled (that is, the final jury list is formally designated), the judge either appoints one member to act as the foreperson or instructs the jury to elect a foreperson from among themselves. The foreman or forewoman is responsible for leading the discussion by members and communicating the jury's decision to the prosecutor.

Grand Jury Hearings

As mentioned earlier, the grand jury is asked to determine whether there is probable cause to believe that a crime was committed by the individual who is under investigation. The jury meets in relative secrecy to keep the targets of the investigation from knowing who and what are being investigated. Although the existence of specific hearings may be known, the public, the media, family and friends, and often even attorneys for witnesses are prohibited from attending grand jury sessions.

How often a grand jury meets depends on its level of activity. Grand juries in large cities usually meet on a weekly basis, whereas juries in smaller communities may meet only occasionally. Although grand juries may initiate hearings on their own, typically they proceed only at the request of the prosecutor. The prosecutor, instead of a judge, manages

the hearings, selects witnesses, presents evidence, and directs the jury in what they should consider. The judge is present only for the selection and swearing in of jury members.

Witnesses, including target witnesses (i.e., those suspected of wrongdoing), have no right to present evidence on their own behalf or to cross-examine other witnesses. In most states and in the federal system, they do not have the right to have an attorney present and may generally not raise objections.[24] States that do allow an attorney to be present restrict the attorney to giving advice to his or her client. In fact, a targeted witness does not even have the right to be present at the hearings. Jurors may ask questions of witnesses, but typically allow the prosecutor to set the stage and establish the direction of the questioning. The exclusionary rule, which prohibits illegally obtained evidence from being presented in court proceedings, does not apply in grand jury hearings.[25] Although such evidence may not be used directly in a later criminal trial, the testimony is recorded and may be used at trial to impeach the witness's statement. For example, if a targeted witness admits to a grand jury that he or she committed a particular crime, but later at trial denies doing so, the earlier testimony may be submitted to show the inconsistency of the person's statements.

LINK The exclusionary rule applies to evidence or confessions that are obtained illegally, in violation of a person's Fourth Amendment protection against unreasonable searches and seizures, as discussed in Chapter 7.

Immunity

When testifying before a grand jury, witnesses are often hesitant or unwilling to answer questions. Targeted witnesses do not wish to incriminate themselves, and other witnesses may be unsure of the real nature of the investigation and fear incriminating someone they wish to protect. Prosecutors may grant witnesses immunity to compel them to answer; if they still refuse, they may be held in contempt of court.

Two kinds of immunity are available: transactional immunity and use immunity. A granting of <u>transactional immunity</u> provides a witness with a blanket protection against prosecution for any crimes the witness may testify about while under immunity. The granting of <u>use immunity</u> is more restricted, prohibiting specific information given by the witness from being used against him or her in a later prosecution.

Former mob henchman Sammy "The Bull" Gravano received immunity and was not prosecuted for 19 homicides he committed.

 Should prosecutors be allowed to use immunity given that it erases the chance of prosecuting known crimes? Is immunity the price the system is willing to pay to prosecute larger criminals?

True Bills and Presentments

Once they hear all of the testimony, the members of the grand jury retire to deliberate and vote to determine whether there is a sufficient basis for believing that the crime was committed by the targeted witness. As mentioned earlier, unlike trial juries, grand juries are not required to make this determination beyond a reasonable doubt. Rather, they need to determine whether there is probable cause to believe that the person is guilty. Based on such a determination, the grand jury then issues a bill of indictment, which is referred to as a true bill. If the jury determines that the evidence is insufficient to charge a suspect, it returns a *no bill*. When the grand jury acts on its own to investigate criminal activity, without the prosecutor having called it together, the accusation it may issue is called a *presentment*.

Criticisms of the Grand Jury System

Many criticisms have been leveled at the concept of the grand jury. Legal scholars Marvin Frankel and Gary Naftalis have raised a number of concerns about their operation. For example, the majority of the states and the federal system restrict the participants in a grand jury hearing to witnesses, prosecutors, court reporters, and, when necessary, translators.

Typically, these persons are prohibited from disclosing any testimony presented to the jury.

Grand jury members also tend to be less diverse than members of juries in criminal trials. Some jurisdictions still impanel grand juries by nomination rather than by selecting candidates from a representative jury pool reflecting a cross section of the community.

Moreover, prosecutors are often viewed as running the show, arbitrarily selecting which witnesses will testify and what evidence will be presented. In essence, the grand jury becomes a "rubber stamp" for the prosecutor.

The protection of the rights of witnesses in grand jury investigations is also of concern, because witnesses generally do not have the right to have counsel with them. In addition, witnesses are often not informed before or during the hearings that they are targets of the investigation. Although witnesses have a right to forgo self-incrimination, there is no requirement that prosecutors or the grand jury warn them that they may invoke their Fifth Amendment right to not testify.

Finally, because the exclusionary rule does not apply in grand jury investigations, illegally obtained evidence and hearsay testimony are allowed and may influence the jury, even though their inclusion would be prohibited in any subsequent trial.[26]

The Prosecutor's Charging Decision

In many jurisdictions, the charges filed at the initial appearance are likely to originate with the police officer's complaint and may not have been reviewed by the prosecutor before that hearing. The police may file only the minimum charge relevant to the case, or they may routinely file every legally possible charge (a process called creative charging) and then let the prosecutor determine which charges are most likely to result in a conviction. Understandably, charges filed by the police are often modified by the prosecutor when a more complete review of the case is conducted. In addition, certain cases may be selected for priority or fast-track prosecution owing to the special nature of the offense or the offender (see "Focus on Criminal Justice").

In jurisdictions in which the prosecutor is responsible for all charging decisions, either horizontal overcharging or vertical overcharging may initially

Priority and Fast Track Prosecution of Offenders

In just two years, Jerry Edwards had accumulated six felony convictions for assault, robbery, and attempted rape in three different states. Because of his pattern of prior offenses and his current charge for armed robbery, he was identified by the prosecutor's office as a high-rate, dangerous offender, and his case was given top priority for prosecution in the hopes of obtaining a conviction that would take him off the streets. The prosecutor's strategy paid off: Edwards was convicted and sentenced to 30 years in prison.

Prosecutors must deal with a wide range of offenders who are brought to the courts: first-time, minor offenders, repeat drug-users, prostitutes, and murderers. Because the cases are so different, and because not all cases are equally important or winnable, prosecutors must make decisions about their priorities in charging and prosecuting such a diverse array of defendants.

Sometimes prosecutors organize their priorities into formally established categories, assigning certain cases to high-priority programs for special prosecution. One common priority prosecution program is known as career-criminal prosecution. Three types of career criminals are typically identified for high-priority prosecution:

- *Persistent offenders:* individuals who commit crimes over a long period of time
- *High-rate offenders:* individuals who commit numerous crimes per year
- *Dangerous offenders:* individuals who commit crimes of violence, often injuring their victims

Many recent career-criminal prosecution programs have begun to focus on offenders who are both high rate and dangerous. The first step in priority prosecution is to select appropriate cases. Some cases are so obvious that the prosecutor gives little attention to the selection decisions. For example, a defendant who is charged with 10 or 12 eyewitnessed armed robberies clearly qualifies as high rate and dangerous. Most cases, however, are not so clear-cut. The task for prosecutors is to identify offenders for priority prosecution from inferences they are able to make about actual behavior based on very limited data. Often, at initial screen-

ing, the prosecutor has only a brief case report containing the current charge.

To improve screening, the prosecutor can review the offender's rap sheet (record of prior arrests and convictions), the arrest report, and the investigating officer's report for information indicating whether the suspect is a high-rate and dangerous offender. Prosecutors may also screen individual cases in which the offender is charged, asking two sequential questions: Has the offender committed crimes at a high rate, and has the offender committed crimes considered dangerous?

The Fast Track Drug Prosecution Project in Lexington, Kentucky, was designed to prosecute drug offenses as quickly, consistently, and effectively as possible. Each morning, members of the Fast Track team review all incoming cases, searching for those that include a violation of the state's drug statutes. These cases are flagged for presentation to the grand jury within 28 days, rather than the usual 60-day limit, and documentation is forwarded to the Fast Track prosecutor for scheduling action. These documents include a summary of the facts, any lab results, a list of assets confiscated, and a preliminary recommendation.

Federal and local prosecutors are also collaborating to fight drug-, gang-, and violence-related crime in larger cities in the United States. Project Achilles and Mobile Enforcement Teams, for example, are programs that are designed to produce priority arrests and prosecutions of armed violent offenders and drug-oriented violent crime by federal and local law enforcement and prosecutors. By combining agencies, personnel, funding, and information, these programs target offenders for arrest, prosecution, and removal from the city streets.

Sources: Fayette County Prosecutor's Office, *Fast Track Drug Prosecution Project* (Lexington, KY: Fayette County Prosecutor's Office, 2006); Malcolm Russell-Einhorn, *Fighting Urban Crime: The Evolution of Federal–Local Collaboration* (Washington, DC: U.S. Department of Justice, 2003); Marcia Chaiken and Jan Chaiken, "Priority Prosecution of High-Rate Dangerous Offenders," *National Institute of Justice, Research in Action* (Washington, DC: U.S. Department of Justice, March 1991).

occur. <u>Horizontal overcharging</u> is the filing of a number of related charges or a number of separate counts of the same basic charge. For example, a person who burglarized three motel rooms might be charged with three counts of burglary, three counts of theft, and three counts of trespassing. <u>Vertical overcharging</u> involves filing the most serious possible charge appropriate to the criminal act even though the known circumstances do not appear to support the charge— for example, filing a charge of first-degree murder in a homicide case that is likely to produce a conviction on a charge of involuntary manslaughter.[27] Horizontal and vertical overcharging also provide greater discretion to prosecutors in plea bargaining (discussed later in this chapter).

Samuel Walker has suggested that the differential treatment of cases is similar to the structure of a

When juries return not guilty verdicts of celebrities like rapper Sean Combs (known as "Puff Daddy"), regardless of the seriousness of the charges, the public may believe that celebrities are above the law.

? Are celebrities treated differently in the U.S. Criminal Justice System?

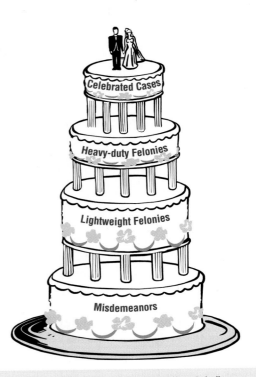

FIGURE 10-2 The Criminal Justice "Wedding Cake"

Source: Adapted from Samuel Walker, *Sense and Nonsense About Crime and Drugs: A Policy Guide,* 6th ed. (Belmont, CA: Thomson/Wadsworth, 2006), p. 36.

wedding cake **FIGURE 10-2**. Each layer is composed of cases that are perceived as being more or less important than those in the surrounding layers and are consequently given differential treatment by the courts as well as differential attention by the media and the public.[28]

- *Celebrated cases.* These are the relatively few cases that receive the greatest attention from the criminal justice system and the media. They are likely to involve wealthy or well-known people as victims or offenders—for example, Scott Peterson, who was charged with the murder of his pregnant wife, Laci; Michael Jackson, who was charged with child molestation; and Sean "Puffy" Combs, who was charged with weapons possession and bribery in connection with a nightclub shootout. These cases are likely to result in highly publicized trials involving high-profile defense attorneys and prosecutors. Through the media, these cases reinforce the public perception that the criminal justice system handles all cases similarly; in reality, few cases go to trial, and few involve high-profile private defense attorneys.

- *Heavy-duty felonies.* These cases include serious felonies such as murder, rape, robbery, and burglary—crimes that often are violent, involve strangers, and are committed by offenders with lengthy criminal records. They are generally given higher priority than less serious cases and typically result in more severe sentences for convicted offenders.

- *Lightweight felonies.* These cases are considered less serious felonies. Offenders are younger, have no prior criminal records, and are more likely to have some prior relationship with their victims. Because they are viewed as less serious, these cases have a lower priority for prosecution and typically result in plea bargains and quick disposition.

- *Misdemeanors.* Misdemeanors (about 90 percent of all criminal cases) account for nearly all cases in this layer. They include disturbing the peace, shoplifting, public intoxication, prostitution, and minor drug offenses. In these cases, guilty pleas are common and trials are rare. Convicted offenders generally receive sentences of fines, probation, restitution, or short jail terms.

Walker's wedding cake analogy suggests that prosecutors' decisions about charging, and subsequent prosecution of cases, is a product of a limited number of factors: the seriousness of the crime, the social importance of the victim or offender, and the amount of public and media interest in the crime.

Models of Prosecutor Charging

Criminologist Joan Jacoby has offered a slightly different view of prosecutor charging decisions.[29] Jacoby suggests that individual prosecutors have developed four types of general policies or models for their charging decisions. The first model is based on legal sufficiency. In this model, if the prosecutor believes that the elements of the crime are sufficiently present, it is his or her responsibility to prosecute the case.

> LINK   The key elements of a crimes—*mens rea* (criminal intent) and *actus reus* (criminal act)—are necessary for an act to be considered criminal and are presented in Chapter 2.

Conversely, a system efficiency policy leads prosecutors to obtain "speedy and early disposition of cases by any means possible."[30] In this model, prosecutors do extensive early screening of cases to weed out the weakest cases. Those cases eventually selected for prosecution are likely to be charged at a lower level to maximize the efficiency of the system and to conserve resources.

The third charging model focuses on the potential for defendant rehabilitation and leads prosecutors to accept pleas to lesser charges so that defendants may obtain treatment outside the criminal justice system. As noted earlier, such a policy depends on the availability of resources within the community.

The final charging model suggested by Jacoby involves trial sufficiency. Here, the charging decision is made based on the likelihood of obtaining a conviction at trial. In addition to the presence of necessary elements of the crime, prosecutors consider the supporting evidence, the quality of witnesses, and any constitutional issues that may arise, such as the chance of illegal search and seizure. If the case appears strong enough, prosecutors will file charges aimed at going to trial.

Dismissing Cases

Prosecutor offices in different jurisdictions vary in their felony case dismissal rates. For example, in 1999 about 10 percent of the felony cases in Atlanta were dismissed, whereas 12 percent were dismissed in Houston, nearly 5 percent in Brooklyn, and 38 percent in Washington, D.C.[31] Cases may be dismissed after screening for a variety of reasons:

- The prosecutor determines that the case is weak (there is little evidence or the victim or witness is not credible).

- The defendant has no prior arrests and the current offense is minor.

- The accused is willing to testify against another defendant in a more serious case.

In some jurisdictions, both a preliminary hearing and a grand jury indictment are required before a case can be transferred to the felony court. By the time a case reaches the felony trial court, then, the prosecutor has already determined that the case has sufficient evidence or that, even though there is marginal evidence, the seriousness of the case warrants prosecution. Consequently, few cases are dismissed once they reach the felony trial court.

In 2003, U.S. Attorneys filed criminal charges against 55 percent of the cases they received. They declined to prosecute 27 percent of the cases, and referred 11 percent to U.S. Magistrates for disposition. The decision to charge and fully prosecute suspects varied according to the nature of the charge. For example, 76 percent of the suspects in drug-related

Prosecuting Terrorists

In 2002, federal prosecutors filed terrorism-related charges against more than 1200 individuals. This number represents only about two-thirds of all terrorism cases referred to prosecutors, however. Among the most notable terrorism-related cases were the prosecutions of the 1992 World Trade Center bombers, Oklahoma City bombers Timothy McVeigh and Terry Nichols, the bombers of the U.S. embassies in Kenya and Tanzania, and the plea bargain convictions of John Walker Lindh (the "American Taliban"), Richard Reid (the "shoe bomber"), and the Lackawana Six (an alleged "sleeper" cell in New York state).

Between 2001 and 2003, 45 states amended or created more than 60 new antiterrorism laws, although the primary responsibility for prosecuting such cases lies with the federal Department of Justice. The new and existing statutes generally fall within one of the following categories:

- Laws that criminalize precursor crimes (i.e., identity theft, money laundering, sale of illegal drugs)
- Laws dealing with threats, hoaxes, and false reports of terrorist activities
- Laws designed to punish specific incidents of terrorism
- Laws enhancing the ability of agencies to investigate and prosecute acts of terrorism

One of the more controversial laws is the federal material support statute, 18 U.S.C. 2339B, which states that it is unlawful to "knowingly provide material support or resources to a foreign terrorist organization." This statute has been on the books since 1996, but it was not used until 2002. Since then, at least 20 criminal cases have been filed in which material support was one of the charges. One of the most notable cases involved the so-called Lackawanna Six. These six American Yemenis traveled from New York to Kandahar, Pakistan, where they listened to lectures on jihad and justifications for suicide bombings. They then traveled to the Al Farooq training camp, where they received training in firearms and more specialized skills such as mountain climbing; they also met with Osama bin Laden. Eventually, after returning to the United States, the six were arrested and federal prosecutors charged them with providing material support to Al Qaeda. All of the defendants pled guilty and were sentenced to prison for terms of seven to ten years.

Approximately 90 percent of all federal criminal prosecutions are resolved through plea bargains, but this is not the case for prosecutions of alleged terrorists. Between 1996 and 2002, fewer than 40 percent of alleged terrorist defendants were convicted as a result of plea bargains. It is possible that terrorists are unwilling to plead guilty because they want to use the trial as an opportunity to express their ideology and air their grievances. It is also possible that prosecutors refuse to offer "acceptable" bargains to defendants or that they, like the defendants, wish to use these trials (which will likely draw national attention) as stages to gain political attention.

Sources: Ravi Satkalmi, "Material Support: The United States v. the Lackawanna Six," *Studies in Conflict and Terrorism* 28:193–199 (2005); M. Elaine Nugent, James Johnson, Brad Bartholomew, and Delene Bromirski, *Local Prosecutors' Response to Terrorism* (Alexandria, VA: American Prosecutors Research Institute, 2005); Nora Demleitner, "How Many Terrorists Are There? The Escalation in So-Called Terrorism Prosecution," *Federal Sentencing Reporter* 16:38–42 (2003); Brent Smith and Kelly Damphousse, *American Terrorism Study: Patterns of Behavior, Investigation, and Prosecution of American Terrorists, Final Report* (Fayetteville, AR: Terrorism Research Center, University of Arkansas, 2002).

Limiting the Discretion of Federal Prosecutors

In 2003, then-Attorney General John Ashcroft issued a memo to all 94 U.S. Attorneys' offices stating that plea bargains should be offered only in very limited, specific circumstances. Ashcroft directed federal prosecutors to generally file the most severe charges and to seek the most severe penalties, rather than to plea bargain with reduced charges and reduced sentences just to obtain convictions.

Ashcroft's memo included six specific exceptions to the expectation that only the most severe charges should be filed:

1. If a defendant agrees to provide "substantial assistance" in an investigation, lesser charges may be filed. According to Ashcroft,

"if defendants will cooperate, the green light is on for negotiation."

2. Certain cases are to be assigned to fast-track programs in which defendants charged with relatively common crimes, such as drug violations and immigration violations in the Southwest, receive preset charges with sentences shorter than those called for the federal sentencing guidelines.

3. Reduced charges may be filed if a prosecutor determines that the original charge would likely be too difficult to prove in court as a result of witness problems, evidence being suppressed, or for other reasons.

4. If a reduced charge would have no impact on the likely sentence, then lesser charges may be filed.

5. If a prosecutor determines that a longer sentence resulting from the filing of "enhancements," such as multiple charges related to the main crime, would remove the incentive for a defendant to plead guilty, then prosecutors could drop one or more of the related charges.

6. Supervisors would still be able to allow exceptions on a case-by-case basis.

Source: Curt Anderson, "Ashcroft Limits Prosecutor Discretion," *Associated Press*, September 23, 2003.

cases and 54 percent of the suspects in violent crimes and property offenses were prosecuted, but only 29 percent of those accused of public order offenses.[32] In late 2003, U.S. Attorneys had some of their discretion regarding the filing of charges limited by the U.S. Attorney General, who indicated that federal prosecutors should offer reduced charges less often.

Reasons for which federal attorneys decline to prosecute cases are similar to the reasons cited by state attorneys for failure to prosecute. In 22 percent of such cases, federal attorneys determined the evidence was too weak to support a conviction. They declined another 24 percent of cases because of the lack of a prosecutable offense (either no federal law had been violated or no proof of criminal intent was found). They declined another 4 percent because the federal attorneys had minimal interest in prosecuting them—the cases were simply not serious enough to warrant the time and effort. Finally, 5 percent of the cases were subject to some alternative resolution, such as pretrial diversion (such as a requirement that a defendant participate in a drug or alcohol treatment program, or civil action).[33]

Preliminary Hearings

When cases are not initially dismissed or probable cause is determined by a grand jury, the defendant faces a preliminary hearing. About half of all U.S. states now use a preliminary hearing to review charges, determine probable cause, and set bail. In these jurisdictions, the prosecutor decides which route to take. Typically, if the defendant has been arrested, is in custody, and is charged with a fairly common felony (such as burglary or possession of stolen property), the prosecutor is likely to choose the preliminary hearing over the grand jury.

The purpose of the court appearance at this stage is to determine probable cause. That is, the prosecutor must produce sufficient evidence to show the judge that the facts and circumstances would lead a reasonable person to believe that a crime was committed and the defendant committed it. If the proof indicates that the defendant is probably guilty, the prosecutor files an information stating the charges and the accused is bound over for trial. If the judge believes that the evidence submitted by the prosecutor is not sufficient, however, the charge or charges are dismissed and the defendant is released.

The preliminary hearing goes beyond the mere investigation of charges. The following critical actions are taken during preliminary hearings:

- Review of the charges
- Review of bail for possible reduction
- Initial presentation of witnesses
- Cross-examination of witnesses
- Discovery (the right of the defense to be informed of the basic facts and evidence that the state will introduce at trial)

Suspects have a legal right to a preliminary hearing within a reasonable number of days following their initial appearance in court. This hearing typically must occur no later than 10 days following the initial appearance if the defendant is held in custody and no later than 20 days if the defendant has been released from custody. A defendant may waive his or her right to a preliminary hearing if the judge determines that the decision was made knowingly and intelligently. Defendants in about 50 percent of all cases waive their right to the preliminary hearing at the initial appearance.[34] This course is typically taken when the charging complaint stating probable cause is supported by affidavits (sworn signed statements of fact) and the defense believes that there would be no advantage to the defendant's participation in a preliminary hearing.

Format of Preliminary Hearings

In some ways, the preliminary hearing resembles a mini-trial, although no jury is involved at this stage. The case is presented before a judge, who evaluates the evidence and initial testimony to determine whether probable cause exists to hold the defendant for trial. Because the burden of proof is on the prosecutor, he or she must present evidence and call witnesses to support the state's case. As in grand jury investigations, because the prosecutor must prove only that there is probable cause—not that the defendant is guilty beyond a reasonable doubt—the prosecutor is not likely to call all witnesses or present all available evidence at this hearing.

A record or transcript of the hearing is kept for a variety of reasons. First, the prosecutor is able to document, for use in subsequent proceedings, the testimony of a witness who may not be available at a later time or who may be vulnerable to intimidation. Second, this record preserves initial statements by people who may have a memory loss or who may be confused about specific times or events at a later date. Finally, testimony by the defendant or a witness may be introduced later at trial to impeach statements that differ from those given at the preliminary hearing.

As in a trial, the defendant has the right to be represented by counsel during a preliminary hearing, to cross-examine witnesses for the state, and to challenge any evidence submitted by the prosecutor. The defense may also call witnesses to testify, although this practice is very uncommon because it enables the prosecutor to examine those witnesses and exposes the defense's strategy to the state.

Discovery

Perhaps the greatest advantage to a defendant during the preliminary hearing is the initial opportunity for discovery, or the legal motion that allows the defense to discover the basis for the prosecutor's case. The defense can find out which theories or evidence will be used in trying the case, is given access to statements and physical evidence, and is informed which witnesses will be called. Discovery is often viewed as providing a level playing field for both parties in the adversarial system and as helping the defendant make an informed judgment as to which plea to enter.

Not all states recognize the defendant's right to pretrial discovery. Moreover, at least two-thirds of the states permit the prosecution to discover the defense's case. Discovery by the defense typically includes the following kinds of information:

- Learning the prosecutor's intention to use evidence obtained through a search, wiretap, or line-up
- Access to evidence in possession of the prosecutor for inspection or copying
- Names, addresses, and statements of witnesses known to the prosecution who may be relevant to the case
- Papers, reports, objects, or statements obtained from the defendant or from others that pertain to the case

Pretrial discovery by prosecutors may include nearly all evidence that a defendant might plan to introduce at trial:

- Names, addresses, and statements of defense witnesses
- Scientific, psychiatric, and medical reports
- Notice of defenses, including alibi, insanity, excuses, or exemptions

Exclusionary sanctions may be imposed on the prosecution or the defense if either party fails to dis-

In France, violations of the criminal law fall into one of three categories:

- Minor offenses (*contraventions*), which are tried in police courts
- Intermediate offenses (*délits*), which are tried in correctional courts
- Serious offenses (*crimes*), which are tried in assize courts

Pretrial screening of serious offenses is very rigorous, because once a prosecutor makes the initial decision to refer a case to the assize court, the prosecutor loses any further control over the case.

French prosecutors, as part of the judiciary, are obligated by the law to provide judges with just or correct solutions to cases rather than pursuing convictions in all cases brought to them by the police. Prosecutors may choose to not prosecute a case even when there appears to be sufficient evidence to establish guilt. The majority of cases (between 50 and 80 percent) in France do not proceed to trial and are dismissed prior to formal presentation to the court.

Prosecutors are not required to engage in pretrial screening of minor and intermediate cases to be tried in either police courts or correctional courts. For such offenses, prosecutors have full discretion to decide whether to further prosecute a case or to dismiss it.

The pretrial investigation begins immediately after the suspect is arrested. Witnesses for both sides are called, their testimony is taken, and the presiding magistrate and prosecutor carefully question the defendant about the crime before deciding whether to take the case to trial. The pretrial investigation is conducted in secrecy by an examining magistrate, with the assistance of the prosecutor and police investigators. The purpose of this secrecy is to avoid unnecessary trials and to protect the accused from any adverse pub-

licity before the state has determined that it has a sufficiently strong case to justify prosecution.

French law requires that defendants be represented by attorneys during pretrial investigations, and the accused may not refuse assistance of counsel. Although defendants also have the right to remain silent during the investigation, there is a presumption by the court that the accused will cooperate in the investigation by answering all questions and raising all points that might aid the defense.

While the law requires prosecutors to prosecute all serious offenses in the assize court, they have the discretion to reduce a charge (a practice known as correctionalizing) so that it may be tried in one of the lower courts. Prosecutors may not drop criminal charges once they are filed with the court unless they obtain the approval of the court magistrate. Thus the tendency to overcharge a case is diminished: If there is not sufficient evidence to support the charges, the judge may correctionalize the charges or dismiss the case entirely.

Suspects in the French criminal justice system may not simply plead guilty to a reduced charge; rather, they must be proved guilty through the entire process of pretrial investigation, development of a dossier (a complete record of the investigation), and a trial. If the pretrial investigation reveals that a suspect is likely innocent, the prosecution ends. Conversely, if charges are finally filed, there is no need for plea bargaining; the case goes to trial and the certainty of its outcome is rather predictable.

Sources: Yue Ma, "Prosecutorial Discretion and Plea Bargaining in the United States, France, Germany, and Italy: A Comparative Perspective," *International Criminal Justice Review* 12:22–52 (2002); Erika Fairchild, *Comparative Criminal Justice Systems* (Belmont, CA: Wadsworth, 1993); Mauro Cappelletti and William Cohen, *Comparative Constitutional Law* (New York: Bobbs-Merrill, 1979).

close all relevant material under discovery. The typical sanction would be the exclusion of the use of the evidence or testimony of the witness at trial.[35]

Entering Pleas

Jurisdictions vary in their organization and the sequence of judicial procedures for dealing with peo-

ple charged with crimes. One of these variations is the time at which a defendant may enter a plea—the individual's response to a criminal charge. In misdemeanor cases, a person may enter a plea at the initial appearance in response to the first reading of the criminal charge. In felony cases, following either a filing of an information by the prosecutor or a grand jury indictment, the suspect enters a plea at an arraignment. In federal court, the defendant is asked to enter a plea at arraignment immediately after the reading of the indictment or information. In addition to standard pleas of guilty, not guilty, and *nolo contendere*, a person may enter a plea of not guilty by reason of insanity.

Guilty Plea

A guilty plea is an admission of guilt to the crime as charged. In 1969 in *Boykin v. Alabama,* the U.S. Supreme Court stated, "A plea of guilty is more than a confession which admits that the accused did various acts; it is itself a conviction; nothing remains but to give judgment and determine punishment."[36] A guilty plea must be made in open court, and the judge must inform the defendant of the nature of the charge or charges, the mandatory minimum sentence (if any), and other consequences of such a plea. As a result of the decision in *Boykin,* defendants must also be informed that to plead guilty means that they waive their rights to avoid self-incrimination, have a trial by jury, and confront their accusers.

A defendant has no absolute right to withdraw a guilty plea once it has been entered.[37] Nevertheless,

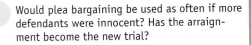

Arraignment is an important legal step because approximately 95 percent of criminal cases reach a disposition from a plea bargain.

? Would plea bargaining be used as often if more defendants were innocent? Has the arraignment become the new trial?

as long as a sentence has not yet been imposed, a defendant may petition the court to permit him or her to withdraw the guilty plea and enter a new plea. In federal courts, a guilty plea may be withdrawn upon showing a "fair and just reason" before sentencing.[38]

Once sentencing has occurred, it is unlikely that the court will reconsider the guilty plea, but some exceptions have been made. In *Henderson v. Morgan,* the U.S. Supreme Court ruled that a guilty plea could be withdrawn if the original plea had been made involuntarily.[39] In this case, the defendant had pleaded guilty to a charge of second-degree murder. However, he had not been informed that intent was a required element of such a charge, nor had he been informed of the minimum sentence he would receive upon pleading guilty. The Court ruled that his due process rights had been violated and allowed the defendant to withdraw the guilty plea.

Nolo Contendere Plea

A plea of *nolo contendere* (literally, "no contest") means that the defendant neither admits nor denies the charge. It is similar to a guilty plea in that the defendant is still subject to any sentence imposed by the court; unlike a guilty plea, however, a *nolo contendere* plea may not be used later against a defendant in any civil suit based on the same act. For instance, in an assault or rape case in which the evidence is overwhelming, the defendant may plead *nolo contendere* to save the expense of the criminal trial and to avoid creating an official record of admission of guilt that may be used as evidence in a later civil suit brought by the victim or the victim's family.

Not Guilty Plea

Very few people who are accused of felonies initially plead guilty or *nolo contendere* and then move directly to sentencing. Instead, most initially plead not guilty, and then later plead guilty after negotiating a plea bargain. Although many minor misdemeanor cases handled in the lower courts are resolved through guilty pleas, the standard plea in felony cases is not guilty. When a defendant refuses to enter any plea (stands mute), is obviously incompetent, or is not represented by counsel, the judge will enter a plea of not guilty on his or her behalf. After a not guilty plea is entered, a trial date is set. In some jurisdictions, if a not guilty plea is entered at the initial appearance, the case must be sent on either to a grand jury or to a preliminary hearing before a trial date is set.

Plea of Not Guilty by Reason of Insanity

A defendant may also enter a plea of <u>not guilty by reason of insanity</u>. In this situation, the defendant does not deny committing the crime but does deny that he or she is criminally responsible because of insanity at the time of the offense. The federal courts and many states require that a defendant who intends to enter this plea do so in a pretrial motion (a written or oral request to the judge before the beginning of the trial) to allow the government time to have a psychiatrist examine the defendant. Contrary to public perceptions, pleas of not guilty by reason of insanity are rarely entered (in fewer than 5 percent of all criminal cases); when such a plea is entered, it is even more rarely successful (less than 1 percent of those pleas entered).[40]

LINK As a result of the Insanity Defense Reform Act of 1984, the burden of proof has shifted from the prosecutor to the defense, such that a defendant now bears the burden of proving he or she is insane by presenting clear and convincing evidence supporting that contention. The various legal tests for insanity are discussed in Chapter 2.

Plea Bargaining

Plea bargaining, sometimes referred to as plea negotiation, continues to be one of the most controversial and misunderstood facets of the criminal justice system. <u>Plea bargaining</u> is an interactive process that involves the prosecutor and the defense, who attempt to arrive at a mutually satisfactory disposition of a case without going to trial. In some jurisdictions, it requires the approval of the victim, and in all jurisdictions it requires the approval of the judge. In contrast to the popular conception based on fictional courtroom dramas, approximately 95 percent of all criminal cases resulting in convictions in state courts are disposed of by guilty pleas, while only 2 percent are resolved by jury trials and 3 percent by bench trials.[41] Defendants in federal criminal cases are also overwhelmingly more likely to enter guilty pleas rather than go to trial (FIGURE 10-3).

LINK Plea bargaining allows defendants to resolve their cases relatively quickly compared to those who choose to go to trial. The latter defendants must often wait in jail during lengthy pretrial motions, selection of a jury, and a potentially time-consuming trial, as explained in Chapter 11.

Bruce Green, professor of law and director of the Louis Stein Center for Law and Ethics at Fordham

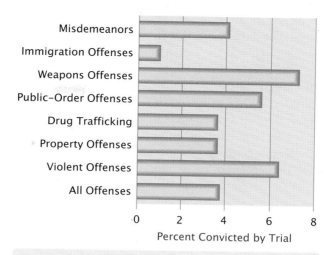

FIGURE 10-3 Percentage of Convicted Defendants in Federal District Courts Found Guilty by Trial

Source: Bureau of Justice Statistics, *Federal Criminal Case Processing, 2002* (Washington, DC: U.S. Department of Justice, 2005), Table 5, p. 11.

University, argues that plea bargains are rarely the outcome of real bargaining between the defense and the prosecutor, despite the impression given by television programs. According to Green, "Sometimes the notion of plea negotiations suggests that real negotiations are going on, that there's some give and take, the way you might negotiate it if you were buying something in the market." This is not the case. Generally, as Green describes, "Prosecutors say, 'Here's the offer: If you plead guilty, you get time served, you get probation, you get three years, you get five years, take it or leave it.'"[42]

The concept of "bargained justice," the fairness of plea bargaining and sentencing, and the role of these negotiations in the modern criminal justice system are difficult to understand without examining the historical development of plea bargaining.

History of Plea Bargaining

Plea bargaining can be traced to the legal systems in France and Spain during the Inquisition of the thirteenth and fourteenth centuries, in which defendants were allowed to plead guilty to avoid torture.[43] By the sixteenth century, English common law allowed a defendant to stand mute before the court as an alternative to pleading guilty. Because there was no formal entry of a guilty plea, the court was prohibited from imposing the most severe sanctions on a defendant who chose to stand mute, but the defendant was required to forfeit all property. English court records indicate that by the end of the sixteenth century, defendants were able to plead guilty to reduced charges to avoid the death penalty.[44]

The first recorded evidence of plea bargaining in the United States dates back to 1749 in colonial Massachusetts, when a prosecutor reduced an initial charge of burglary to simple theft in return for guilty pleas by the three defendants in the case. By the end of the nineteenth century, guilty pleas accounted for more than 85 percent of convictions in New York City.[45]

Plea bargaining was also common in Oakland, California, during the last few decades of the nineteenth century. Legal historians Lawrence Friedman and Robert Percival report that at least 14 percent of the defendants charged in this city altered their pleas from not guilty to guilty, and nearly half entered pleas of guilty as charged.[46] In <u>implicit plea bargaining</u>, defendants enter guilty pleas without negotiating with the prosecutor with the expectation that they will receive some kind of consideration by the court in return.

According to Milton Heumann, many guilty pleas during the nineteenth century reflected implicit plea bargaining in situations in which no actual negotiations took place between the prosecution and the defense, but the defendant nevertheless perceived that it would be better to plead guilty than to lose at trial.[47] Although some defendants may have pleaded guilty out of feelings of remorse, shame, guilt, or self-hate, Heumann notes that most consciously worked through their options. If a defendant perceived that he or she was likely to be given a lesser sentence as a reward for pleading guilty, there was little need for prompting by the prosecutor or defense counsel to enter a guilty plea.[48]

By the 1920s, guilty pleas—both implicit and explicit (i.e., bargains formally agreed to by both prosecution and defense and signed off on by the judge)—accounted for far more convictions than either bench or jury trials. For example, such pleas accounted for 70 percent of all convictions in Dallas, 85 percent in Chicago, 86 percent in Cleveland, and 88 percent in New York City.[49] Initial pleas of guilty continued to be very common until the early 1950s, when they began to decline markedly and formally bargained pleas became more common. By the 1970s, the initial plea entered by most defendants was not guilty; later, usually through their attorneys, they typically negotiated with the prosecutor for a lessened sentence or charge.[50] Today, nearly all cases are settled without going to trial.

Deciding to Bargain

Prosecutors take many issues into consideration when making a decision to offer a plea bargain to a defendant:

- Defendant's willingness to cooperate in the investigation or prosecution of others
- Defendant's history with respect to criminal activity
- Nature and seriousness of the offense(s) charged
- Defendant's remorse or contrition and his or her willingness to assume responsibility for his or her conduct
- Desirability of prompt and certain disposition of the case
- Likelihood of obtaining a conviction at trial
- Probable effect on witnesses
- Probable sentence or other consequences if the defendant is convicted
- Public interest in having the case tried rather than disposed of by a guilty plea
- Expense of the trial and any subsequent appeal
- Need to avoid delay in the disposition of other pending cases
- Effect on the victim's right to restitution

The vast majority of criminal cases are resolved through a plea agreement or plea bargain.

 Would the system criminal justice system grind to a halt if more cases went to trial? Are plea bargains effective because most defendants committed some crime?

Types of Plea Bargains

There are five types of plea bargains:

1. Reducing the seriousness of the charge
2. Reducing the number of charges or counts
3. Reducing the sentence
4. No recommendation regarding sentence
5. Altering charges with negative labels

Reducing the Seriousness of the Charge

Reducing the seriousness of the charge involves pleading guilty to a charge that is less serious than the initial charge. Such a plea bargain most often occurs under one of two circumstances.

The first involves a reevaluation of the strength of the case. Once the probable cause hearing has been completed and both the prosecutor and the defense have had an opportunity to evaluate the quality of the case (that is, to determine whether it is winnable), the prosecutor may decide that obtaining a conviction on the original charge is unlikely. Consequently, the prosecutor may offer the defense an opportunity to plead guilty to a less serious charge rather than go to trial.

The second circumstance occurs when the prosecutor intentionally engages in vertical overcharging—the practice of filing initial charges that are more serious than the case warrants in hopes of intimidating the defendant into pleading guilty to a reduced charge. For example, a defendant with a prior felony conviction for drug distribution may face a mandatory prison sentence if found guilty on the current felony charge. By contrast, conviction on a misdemeanor, such as simple possession of marijuana, would make a probationary sentence possible. Many defense attorneys approach the prosecutor with offers for the defendant to plead guilty to a lesser charge in an attempt to avoid potentially more serious punishments.

Reducing the Number of Charges or Counts

Defendants often face a number of different charges or counts stemming from a single criminal event. As noted earlier, this often happens when prosecutors or the police engage in horizontal overcharging. In many jurisdictions, the sole purpose behind this pattern of charging is to overwhelm the defendant with a sense of futility in fighting all of the charges or counts. In addition, defendants are likely to consider the accumulation of sentences that might occur if they are convicted on each charge. Consequently, either side may suggest dropping some of the charges or counts in exchange for a guilty plea.

Reducing the Sentence

A third form of plea bargaining occurs when the defendant enters a plea of guilty to the charge in exchange for the prosecutor's agreement to support a reduced sentence. This agreement may involve a lengthy prison sentence rather than the death penalty, a shorter rather than longer prison sentence, or probation rather than prison. Because the judge has the final say in sentencing, the prosecutor cannot guarantee that the sentence will be applied, although most judges accept their recommendations.[51]

No Recommendation Regarding Sentence

Sometimes, the best a defendant can hope for is an agreement that the prosecutor will stand mute at sentencing or not make a sentence recommendation at all in exchange for the guilty plea. This is more likely to occur when the evidence pertaining to the case is substantial and any recommendation by the prosecutor would carry great weight in the decision by the judge. Under this kind of plea agreement, when a judge asks for the prosecutor's recommendation, the prosecutor states that he or she will make no recommendation, and the defense is then allowed to present its case for leniency without objection.

Altering Charges with Negative Labels

Certain criminal charges carry much greater negative stigma than others. For example, defendants may be subjected to potential harm (either in prison or in the community) or their rehabilitation made more difficult if they are convicted of rape instead of sexual assault, child molestation instead of child neglect, or indecent exposure instead of disorderly conduct. Assuming that the impact of the crime on the victim was not great, defendants may be protected from excessive stigma through this kind of bargain.

The Judge's Role in Plea Bargains

Although plea bargains are generally worked out between the prosecutor and the defense attorney, the judge often plays a major role in shaping the terms of the bargain. The judge's role in plea bargaining is similar to that of the police officer who attempts to maintain order on the streets. Like town watchmen, judges frequently take a more reactive—rather than proactive—approach to enforcing local courtroom norms in the bargaining process.[52] When a pattern of going rates for pleas has been established (that is, an understood relationship between the charge, prior criminal record, and sentence), it is the judge's responsibility to keep violations of the expectations regarding bargains in check. As one judge put it:

> I'm a pragmatist. . . . If a plea fits within a scale of what I would say is reasonable, I will accept it. I do reject pleas. And I reject them both ways, both from the standpoint that the prosecutor wants too much and the defense wants too little.[53]

Some states prohibit judges from any direct involvement in plea bargains; in those states, judges are limited to either accepting or rejecting the plea. In these circumstances, it becomes very important for prosecutors and defense attorneys to anticipate judges' understandings of the going rates and to gauge how they are likely to sentence cases. In jurisdictions that allow judicial participation, judges may reject pleas if they believe those agreements fall outside the going rates, or they may refuse to be bound by recommendations attached to pleas, believing that sentence recommendations infringe on their prerogatives in sentencing. In illustrating this point, one judge stated:

> Here's how I run a plea bargain. . . . I want them to understand that what they recommend to me is only that, a recommendation. The ultimate sentencing prerogative and duty are mine. That's what I get paid for. I'm here as a judge and I'm supposed to be the person who makes that ultimate determination.[54]

Although some judges refuse to get involved in the actual negotiations and simply ratify the terms of the pleas brought to them, most judges play a more active role. Judges can influence pleas by requiring that cases be reviewed in pretrial conferences. In these conferences, the prosecutor and the defense attorney meet with the judge to discuss the strengths and weaknesses of particular cases in an attempt to eliminate the scheduling of both cases that are likely to end up as guilty pleas and cases for which trials are agreed to be unnecessary, thereby freeing up court time for more complicated cases. Judges also are able to influence pleas by controlling the court docket and timing of trials. If the judge sets an early trial date, defense attorneys have less time to prepare for trial and are therefore encouraged to pressure the defendant to plead guilty. Finally, judges can exert their influence on court-appointed defense attorneys by offering incentives, such as by frequently appointing attorneys who demonstrate an ability to move their cases to early dispositions through plea bargains.[55]

The Victim's Role in Plea Bargains

Victims are essential in reporting crime, assisting the police investigate and clear cases through arrest, and providing testimony at trial. But what if there is no trial? Historically, when prosecutors and defense attorneys engaged in plea bargaining, the victim all too often sat on the sidelines waiting to learn what bargain was struck. More recently, reforms establishing or expanding victims' rights have led to increasing involvement of victims in the plea bargaining process. Today, more than 30 states allow for some form of victim input into plea agreements.

LINK As noted in Chapter 3, crime is not just about criminals—it is also very much about the victims of crime, and victims may play an important role in the criminal justice process.

When victims have been permitted to provide input into plea agreements, this right has typically been granted at two stages of the criminal justice process. The first is when conferring with the prosecutor during plea bargaining; the second is when addressing the court, either orally or in writing, before the entry of the plea.

Most states provide victims with some level of consultation with the prosecutor about the negotiated plea agreement, but the extent of their participation varies widely from state to state. In a number of states, victims are given a general right to confer with the prosecutor. In other states, the obligation to confer is typically limited to notifying, informing, or advising victims of a plea bargain or agreement that has already been reached before presenting the proposed agreement to the court.

In at least 22 states, the victim's right to confer with the prosecutor requires a prosecutor to obtain the victim's views concerning the proposed plea. One-third of the states permit the victim to be heard at the time the plea is being entered. In a few states, a written impact statement may be submitted early in the criminal justice process and used by the court when the plea agreement is presented. For example, victims in Rhode Island have the right to prepare a written impact statement for insertion into the prosecutor's case files. The statement is then submitted to the court for review or the victim is given a chance to address the court before the plea is entered.

Although victims' rights have increased in recent years, crime victims often remain the forgotten players in the "criminals' justice system."

? Do criminal defendants have too many rights? Do crime victims have too few rights?

As an alternative to—or, in some states, in addition to—permitting the victim to address the court or submit a victim impact statement, the prosecutor must inform the court of the victim's position on the plea agreement. For example, in Minnesota, when a victim is not present to express his or her opinion regarding the plea agreement, the prosecutor must bring to the attention of the court any known objections expressed by the victim.

The right to confer with the prosecutor does not give a victim the right to veto a decision to plea bargain rather than go to trial, nor does it give the victim a veto over a plea agreement. Victims merely have an opportunity to be heard—giving them a voice, not a veto.[56]

Are Plea Bargains Constitutional?

Throughout much of the twentieth century, the public viewed plea bargaining as unethical, if not downright illegal. The full extent of its use remained largely hidden from public view. Prosecutors and defense attorneys would often negotiate in secret and then give the appearance in open court of a voluntary plea on the part of the defendant to an altered charge. No mention was made of a bargain, and the public was unaware of the negotiation. Consequently, the acceptance of the plea to a reduced charge created public perceptions that something shady had taken place.

Although there was little doubt about the illegitimacy of plea bargains if defendants were coerced into deals (for example, by the threat of severe punishment for not accepting a bargain), the constitutionality of plea bargains remained unsettled until 1970, when the U.S. Supreme Court ruled in *Brady v. United States*.[57] In this case, the Court found that plea bargains were a legitimate part of the criminal justice process as long as they were fully voluntary and the defendant had been informed of all the consequences of pleading guilty. One year later, the Court, in *Santobello v. New York*, reaffirmed that plea bargaining was an essential component of the administration of justice.[58] Furthermore, the Supreme Court stated that a prosecutor must live up to the terms of the plea bargain. Sixteen years later, in *Ricketts v. Adamson*, the Court ruled that defendants must also live up to the terms of the bargain to receive promised leniency.[59]

While plea bargaining has become a standard feature of the criminal courts, prosecutors continue to have a greater advantage in the process and are able to use plea bargaining as a means to put pressure on defendants. For example, prosecutors are not required to engage in plea bargaining and may refuse to bargain with a defendant unless the defendant agrees that any statements made by him or her during the bargaining process can be used against him or her in a possible trial.[60]

Just how far can a prosecutor go in pressuring a defendant to plead guilty? Is a threat to reindict and prosecute on a more serious charge coercive and unconstitutional? In 1978, the Supreme Court, in *Bordenkircher v. Hayes,* determined that it was not. According to the Court, the ability of prosecutors to use their bargaining chips in the negotiation process is an essential part of plea bargaining.[61] In 1987, Congress added a rule to the Federal Rules of Procedure to ensure that plea bargains negotiated in federal criminal cases would meet the requirements of both *Brady* and *Santobello.*

Who Wins with Bargained Justice?

Adversarial systems of criminal justice are grounded in the notion that one side wins and the other side loses. Each side presents its case as well as possible and then lets the jury decide guilt. Ideally, if the defendant is guilty, the state wins through conviction; if the defendant is innocent, the defendant wins with an acquittal. Of course, cases often arise in which defendants are guilty of either the crime charged or

In agreeing to hear *Santobello v. New York,* the U.S. Supreme Court was to determine whether the state's failure to keep a commitment concerning the sentence recommendation on a guilty plea required a new trial. In this case, the prosecutor agreed to make no recommendation regarding the defendant's sentence in exchange for a plea of guilty to a misdemeanor gambling charge by the defendant. Conviction of the misdemeanor would have carried a maximum one-year sentence, but Rudolph Santobello was hoping that with the prosecutor standing mute, he had a good chance at a suspended sentence. The court accepted the plea and set a date for sentencing.

At Santobello's sentencing, however, a different prosecutor represented the State of New York and recommended a one-year prison sentence. The defense counsel objected on the grounds that the plea had been entered with the promise that the prosecutor would make no recommendation regarding sentence. The second prosecutor, who was unaware of the prior commitment, argued that nothing in the record supported the claim that the promise had been made. After discussion ended, the judge sentenced the defendant to the maximum sentence of one year.

In the Supreme Court decision, Chief Justice Warren Burger stated:

> On this record, petitioner "bargained" and negotiated for a particular plea in order to secure dismissal of more serious charges, but also on condition that no sentence recommendation would be made by the prosecutor. It is now conceded that the promise to abstain from a recommendation was made. . . . That the breach of agreement was inadvertent does not lessen its impact.
>
> [The sentencing judge] stated that the prosecutor's recommendation did not influence him and we have no reason to doubt that. Nevertheless, we conclude that the interests of justice and appropriate recognition of the duties of the prosecution in relation to promises made in the negotiation of pleas of guilty will be best served by remanding the case to the state courts for further consideration. The ultimate relief to which petitioner is entitled we leave to the discretion of the state court. To which case petitioner should be resentenced by a different judge, or whether, in the view of the state court, the circumstances require granting the relief sought by petitioner, i.e., the opportunity to withdraw his plea of guilty.

The Court was making a clear statement in this case: Promises made by prosecutors in plea bargain agreements must be kept, and any breaking of the promise violates the defendant's due process rights.

Source: *Santobello v. New York,* 404 U.S. 257 (1971).

a related offense, but the courts are unable to try all cases in an adversarial confrontation. Is it possible to achieve an acceptable degree of justice with all parties winning? Plea bargaining has largely developed as an alternative to the win-lose trial system.

Plea bargaining may allow all parties to benefit from a negotiated plea. Defendants are most likely to benefit directly from plea bargains: They spend less time in custody, incur fewer costs (such as loss of income and attorney expenses), and are likely to receive more lenient sentences. A study of felony plea bargaining over a five-year period in a large urban Ohio county, for example, found that defendants who were convicted at trial were likely to receive prison sentences 14 months longer than defendants who were convicted through a plea bargain, even when controlling for other relevant factors.[62] The going rates for defendants with similar prior criminal records and facing similar charges in many criminal courts are fairly well known to defense attorneys who are established in the courtroom workgroup. According to one attorney, "The court will never state that there are penalties for going to trial. . . . They're unstated, but they exist."[63]

One of the most significant benefits of plea bargaining for prosecutors is the increased predictability they offer. Trials produce uncertainty, and the longer a case takes, the greater the uncertainty and the likelihood that a defendant will be acquitted. Through plea bargains, prosecutors increase their confidence that people they believe are probably guilty will be convicted.[64] Prosecutors also benefit because they free up more time to prepare other cases that must go to trial, they improve their conviction rates, they obtain convictions in weak cases, they ensure some form of punishment for offenders even when the seriousness of the original charge was reduced, and they save time, money, and other resources of the department.

Likewise, defense attorneys benefit from plea bargains. They gain the appreciation of clients who receive lighter sentences. Like prosecutors, they are able to spend more time on cases that must go to trial. Because public defenders are typically paid an annual salary rather than receiving compensation on a case-by-case basis and because court-appointed attorneys may receive no fee for their service, disposing of many cases through plea bargains allows them to allocate time and resources to other income-producing parts of their work.

Many other parties involved in the case may also benefit from plea bargaining. Victims may be spared the stress of a trial and gain the satisfaction of knowing that the offender was convicted of something. Witnesses and prospective jurors benefit because they do not need to leave their jobs or arrange for child care to attend trial. The courts realize improved efficiency as caseloads thin out and the overall cost of prosecutions is reduced. The public also benefits from plea bargaining because the tax burden is reduced by a more efficient criminal justice system.

A final advantage in plea bargaining relates to its ability to individualize justice when legislatures, attempting to respond to public pressure to "get tough on crime," create more severe sentences or mandatory prison statutes that restrict the sentencing options of judges. Unnecessarily severe punishments can be softened by allowing a defendant to plead to a less serious crime that carries a shorter sentence or the possibility of probation.

LINK   As Chapter 12 examines, plea bargaining may help to individualize justice, although the prosecutorial discretion that goes with bargaining may also allow for disparities in sentencing based on race, ethnicity, sex, and socioeconomic status.

Who Loses with Bargained Justice?

Although defendants stand to win through plea bargains, they may also be the first to lose under such agreements. Most notably, defendants give up a number of constitutionally protected due process rights as part of a plea bargain. Innocent defendants may be enticed or coerced into pleading guilty to a minor charge to avoid the costs of a larger defense or the possibility of a severe punishment.

Victims of crime may also lose in plea bargains. In many jurisdictions, they are not involved in any stage of the negotiation process and are told of the outcome only after it has been finalized. Victims frequently perceive plea bargains as failing to punish the offender appropriately for the crime committed.

As a result of excessive plea bargaining, the criminal court system itself may lose. When the public perceives that judges are being soft on criminals, that prosecutors are lazy, or that defense attorneys are willing to sell out clients to move cases through the system more quickly, the judicial system can lose valuable public support. Public support may also be eroded when the emphasis on efficiency means that poor and minority defendants are more likely to receive an abbreviated form of due process, or when victims perceive that the court is ignoring their needs to serve the needs of the court. At a minimum, such loss of support for the courts is demoralizing to citizens and diminishes their willingness to cooperate; in extreme situations, it can result in violence.

Reducing the sentence is a form of plea bargaining that occurs when the defendant enters a plea of guilty to the charge in exchange for the prosecutor's agreement to support a reduced sentence. This may involve a lengthy prison sentence rather than the death penalty, a shorter rather than longer prison sentence, or probation rather than prison. Because the judge has the final say in sentencing, the prosecutor cannot guarantee the sentence, although most judges accept their recommendations.

In taking the death penalty off the table, King County Prosecutor Norm Maleng provided an incentive to the defendant whose heinous crime was clearly deserving of death based on the facts of the case. At an earlier preliminary hearing, Naveed Haq attempted to plead guilty to the charges; thus the prosecutor was confident that an enticement of life imprisonment would bring another guilty plea. For the prosecutor, this plea bargain saved time, money, and resources, and quickly resolved a heinous crime with a severe punishment (albeit not death).

Of course, the prosecutor's failure to pursue the death penalty in this case likely angered Seattle residents who wanted Haq to face the most severe legal punishment. Also, citizens might be skeptical of the defendant's claims of mental illness and instead view Haq's behavior as a hate-inspired murder. If Haq's mental instability is clearly substantiated, then choosing to not pursue the death penalty reflects well on the prosecutor who showed restraint and compassion while still securing a conviction and harsh punishment.

Finally, this case demonstrates another important component of the decision to pursue the death penalty: politics. In a progressive city like Seattle, pursuing the death penalty against an obviously mentally ill defendant (if Haq is shown to be) might carry negative political ramifications. Because prosecutors are elected, they must pay attention to the wishes of the general public and walk a fine line in balancing crime control and due process.

Chapter Spotlight

- At the initial appearance, defendants are informed of the charge(s) against them and their right to counsel, right to remain silent, and right to reasonable bail.

- The Eighth Amendment prohibits the setting of excessive bail. During the past few decades, a number of alternatives to monetary bail have been created to allow less affluent defendants to be released from custody. Demands for reforming pretrial release, however, led to the Bail Reform Act of 1984, which provided for preventive detention.

- In its early incarnation in the United States, the grand jury was intended to protect citizens against arbitrary use of the criminal courts by a potentially coercive state. Over much of the twentieth century, the rigid control of the grand jury by prosecutors, the denial of important rights to witnesses who are called to testify before grand juries, and the secrecy surrounding grand juries have contributed to a decline in their use.

- The prosecutor's charging discretion includes options ranging from the decision to drop the case to the decision to file additional charges. A number of factors influence the charging decision, such as the suspect's prior record, the characteristics of the victim, the chances of obtaining a conviction, and the visibility and seriousness of the charge.

- The preliminary hearing is an opportunity to present the outlines of the state's case to determine whether to continue prosecution. It is also the first opportunity for the defense to discover the strength of the prosecution's case.

- Once arrested and charged, a suspect is asked to enter a plea. In some jurisdictions, the plea is entered at the initial appearance when the charges are read; in other jurisdictions, it is entered later, at the preliminary hearing or arraignment.

- Nearly 95 percent of criminal cases are resolved by guilty pleas. Although many defendants initially plead guilty (especially in misdemeanor cases), many others negotiate a plea bargain. Advocates defend plea bargaining as benefiting all people involved, whereas critics suggest that it corrupts the very notion of justice.

Putting It All Together

1. The U.S. Supreme Court has stated that defendants should be brought before a magistrate within 48 hours of being arrested. Is this a reasonable period?

2. In *United States v. Salerno,* the Supreme Court ruled that the preventive detention of dangerous defendants was constitutional. Do you agree or disagree with this decision?

3. Should the grand jury be abolished? Why or why not?

4. Which factors should be considered by a prosecutor when deciding which charge or charges to file against a defendant? Which factors should not enter into this decision?

5. Should plea bargaining be abolished? Why or why not?

Key Terms

48-hour rule Supreme Court ruling in *Riverside County, California v. McLaughlin,* that a defendant must be brought before a magistrate within 48 hours of his or her arrest.

arraignment A hearing at which felony defendants are informed of the charges and their rights and given an opportunity to enter a plea.

assembly-line justice The mechanical disposition of misdemeanor cases to move them swiftly through the courts.

bail Money or a cash bond deposited with the court or bail bondsman allowing the defendant to be released on the assurance that he or she will appear in court at the proper time.

bail bondsman A person who guarantees court payment of the full bail amount if the defendant fails to appear.

bail guidelines Use of a grid to plot a defendant's personal and offense characteristics to determine probability of appearance.

Bail Reform Act of 1966 Act providing for release on recognizance (ROR) in noncapital federal cases when it is likely that the defendant will appear in court at required hearings.

Bail Reform Act of 1984 Act extending the opportunity for release on recognizance (ROR) in many federal cases but also providing for preventive detention without bail of dangerous suspects.

defendant rehabilitation A model of charging decisions in which the prosecutor accepts pleas to lesser charges to enable the defendant to obtain treatment outside the criminal justice system.

discovery Legal motion to reveal to the defense the basis of the prosecutor's case.

grand jury A group of citizens who are called upon to investigate the conduct of public officials and agencies and criminal activity in general and to determine whether probable cause exists to issue indictments.

guilty plea An admission of guilt to the crime charged.

horizontal overcharging The practice of filing of a number of related charges or a number of separate counts related to the same basic charge.

implicit plea bargaining The entering of a guilty plea with the expectation that the defendant will be looked upon favorably by the court at sentencing.

indictment A formal criminal charge filed by the prosecutor.

initial appearance A defendant's first appearance in court, at which the charge is read, bail is set, and the defendant is informed of his or her rights.

legal sufficiency A model of charging decisions in which cases are prosecuted if the prosecutor believes that the elements of the crime are sufficiently present to warrant bringing the case to trial.

nolo contendere A plea of no contest; essentially the same as a guilty plea except that the defendant neither admits nor denies the charge.

not guilty A plea denying guilt.

not guilty by reason of insanity A plea in which the defendant does not deny committing the crime but claims that he or she was insane at the time of the offense and, therefore, is not criminally responsible.

plea A defendant's response to a criminal charge.

plea bargaining The negotiation between a prosecutor and a defense attorney in which they seek to arrive at a mutually satisfactory disposition of a case without going to trial.

preliminary hearing An early hearing to review charges, set bail, present witnesses, and determine probable cause.

pretrial release Release of defendant from custody while he or she is awaiting trial.

preventive detention The practice of holding a defendant in custody without bail if he or she is deemed likely to abscond or commit further offenses if released.

release on recognizance (ROR) A personal promise by the defendant to appear in court; does not require a monetary bail.

Santobello v. New York U.S. Supreme Court decision that plea bargaining is an essential component of the criminal justice system and that prosecutors must honor the terms of a plea bargain.

system efficiency A model of charging decisions in which the prosecutor pursues only those cases that are most likely to achieve efficiency in the system by speedy disposition.

transactional immunity A blanket protection against prosecution for crimes a witness may testify about to a grand jury while under immunity.

trial sufficiency A model of charging decisions in which the decision to prosecute is based on the ability to obtain a conviction at trial.

true bill An indictment issued by a grand jury charging a person with a crime; similar to a prosecutor's filing of an information.

United States v. Salerno U.S. Supreme Court ruling that the preventive detention provisions of the Bail Reform Act of 1984 were constitutional.

use immunity Protection that prohibits specific information given during grand jury testimony from being used against the witness.

vertical overcharging The practice of filing the most serious possible charge appropriate to a criminal act even though the known circumstances do not support the charge.

Notes

1. Federal Bureau of Investigation, *Crime in the United States, 2006* (Washington, DC: U.S. Department of Justice, 2007); Brian Ostrom, Shauna Stricklaund, and Paula Hannaford-Agor, "Examining Trial Trends in State Courts: 1976–2002," *Journal of Empirical Legal Studies* 1:755–782 (2004).
2. *Federal Rules of Criminal Procedure* (Washington, DC: U.S. Government Printing Office, 2005).
3. *Minnesota Rules of Criminal Procedure* (2005).
4. *McNabb v. United States,* 318 U.S. 332 (1943).
5. *Mallory v. United States,* 354 U.S. 449 (1957).
6. *Gerstein v. Pugh,* 420 U.S. 103 (1975).
7. *Riverside County, California v. McLaughlin,* 500 U.S. 44 (1991).
8. *Riverside County, California v. McLaughlin,* note 7.
9. *United States v. Salerno,* 481 U.S. 739 (1987).
10. *Compendium of Federal Justice Statistics, 2001* (Washington, DC: U.S. Department of Justice, 2003), pp. 42, 44.
11. *Stack v. Boyle,* 342 U.S. 1 (1951).
12. Traci Schlesinger, "Racial and Ethnic Disparity in Pretrial Criminal Processing," *Justice Quarterly* 22:170–193 (2005); K. B. Turner and James Johnson, "A Comparison of Bail Amounts for Hispanics, Whites, and African Americans: A Single County Analysis," *American Journal of Criminal Justice* 30:35–56 (2005); Stephen Demuth, "Racial and Ethnic Differences in Pretrial Release Decisions and Outcomes: A Comparison of Hispanic, Black, and White Felony Arrests," *Criminology* 41:873–908 (2003); Cassia Spohn, *Offender Race and Case Outcomes: Do Crime Seriousness and Strength of Evidence Matter? Final Activities Report* (Omaha, NE: University of Nebraska at Omaha, 2000).
13. Stephen Demuth and Darrell Steffensmeier, "The Impact of Gender and Race-Ethnicity in the Pretrial Release Process," *Social Problems* 51:222–242 (2004).
14. John Goldkamp, "Philadelphia Revisited: An Examination of Bail and Detention Two Decades After *Foote*," *Crime & Delinquency* 26:183 (1980).
15. William Landes, "Legality and Reality: Some Evidence in Criminal Procedure," *Journal of Legal Studies* 3:287–337 (1974); Paul Wice, *Freedom for Sale: A National Study of Pretrial Release* (Lexington, MA: Lexington Books, 1974); Stevens Clarke and Gary Koch, "The Influence of Income and Other Factors on Whether Criminal Defendants Go to Prison," *Law and Society Review* 11:57–92 (1976); Malcolm Feeley, *The Process Is the Punishment* (New York: Russell Sage Foundation, 1979).
16. Charles Ares, Anne Rankin, and Herbert Sturtz, "The Manhattan Bail Project," *New York University Law Review* 38:68 (1963).

17. Andy Hall, *Pretrial Release Program Options* (Washington, DC: National Institute of Justice, 1984), pp. 32–33.
18. Pub. Law 89-465, 89th Cong. S. 1357, 80 Stat. 214.
19. John Goldkamp and Michael Gottfredson, *Judicial Guidelines for Bail: The Philadelphia Experiment* (Washington, DC: U.S. Department of Justice, 1984); John Goldkamp and Michael White, *Restoring Accountability in Pretrial Release: The Philadelphia Pretrial Release Supervision Experiments, Final Report* (Philadelphia: Crime and Justice Research Institute, 2001).
20. Goldkamp and Gottfredson, note 19.
21. *Comprehensive Crime Control Act of 1984,* Public Law 98-473, October 12, 1984.
22. *United States v. Salerno,* 481 U.S. 739 (1987).
23. *Demore v. Kim,* 538 U.S. 510 (2003).
24. Niki Kuckes, "The Useful, Dangerous Fiction of Grand Jury Independence," *The American Criminal Law Review* 41:1–66 (2004).
25. *United States v. Calandra,* 414 U.S. 338 (1974).
26. Marvin Frankel and Gary Naftalis, *The Grand Jury: An Institution on Trial* (New York: Hill and Wang, 1977), pp. 52–115; Michael Vitiello and J. Clark Kelso, *Grand Jury Background Study* (Stockton, CA: McGeorge School of Law, University of the Pacific, 2001).
27. Michael Cox, "Discretion: A Twentieth Century Mutation," *Oklahoma Law Review* 28:311–322 (1975).
28. Samuel Walker, *Sense and Nonsense about Crime and Drugs: A Policy Guide,* 6th edition (Belmont, CA: Thomson/Wadsworth, 2006), pp. 22–27.
29. Joan Jacoby, "The Charging Policies of Prosecutors," in William McDonald (ed.), *The Prosecutor* (Beverly Hills, CA: Sage Publications, 1979), pp. 75–97.
30. Jacoby, note 29, p. 83.
31. Martin Schneider, Amina Rana, and Richard Amberg, "D.C. Prosecutors Drop 38% of Felonies," *The Washington Times,* August 23, 1999:A1; Kings County District Attorney's Office, *Annual Report* (Brooklyn, 2006).
32. Bureau of Justice Statistics, *Federal Criminal Case Processing, 1980–90* (Washington, DC: U.S. Department of Justice, 1992).
33. Bureau of Justice Statistics, *Compendium of Federal Justice Statistics, 1989* (Washington, DC: U.S. Department of Justice, 1992), pp. 9–10.
34. Ronald Carlson, *Criminal Justice Procedure,* 4th edition (Cincinnati: Anderson Publishing, 1991), p. 65.
35. Eric Blumenson, "Constitutional Limitations on Prosecutorial Discovery," *Harvard Civil Rights-Civil Liberties Law Review* 18:123–156 (1983).
36. *Boykin v. Alabama,* 395 U.S. 238 (1969).
37. *United States v. Buckles,* 490 U.S. 1099 (1989).
38. *United States v. Hyde,* 520 U.S. 670 (1997).
39. *Henderson v. Morgan,* 426 U.S. 637 (1976).
40. Henry Steadman and Jeraldine Braff, "Defendants Not Guilty by Reason of Insanity," in John Mohanan and Henry Steadman (eds.), *Mentally Disordered Offenders* (New York: Plenum, 1983), pp. 109–129.
41. Matthew Durose and Patrick Langan, *Felony Sentences in State Courts, 2002* (Washington, DC: U.S. Department of Justice, 2004), p. 8.
42. "Interview Bruce Green," available at http://www.pbs.org/wgbh/pages/frontline/shows/plea/interviews/green.html, January 29, 2004, accessed July 2, 2007.
43. Joseph Sanborn, Jr., "A Historical Sketch of Plea Bargaining," *Justice Quarterly* 3:113–116 (1986).
44. Sanborn, note 43, p. 43.
45. Raymond Moley, "The Vanishing Jury," *Southern California Law Review* 2:97–127 (1928–1929).
46. Lawrence Friedman and Robert Percival, *The Roots of Justice: Crime and Punishment in Alameda County, California, 1870–1910* (Chapel Hill, NC: University of North Carolina Press, 1981), p. 176.
47. Milton Heumann, *Plea Bargaining: The Experiences of Prosecutors, Judges, and Defense Attorneys* (Chicago: University of Chicago Press, 1978), p. 158.
48. Milton Heumann, "A Note on Plea Bargaining and Case Pressure," *Law and Society Review* 9:515 (1975).
49. Lawrence Friedman, "Plea Bargaining in Historical Perspective," *Law and Society Review* 13:255 (1979).
50. Friedman, note 49, p. 49.
51. John Hagan, John D. Hewitt, and Duane Alwin, "Ceremonial Justice: Crime and Punishment in a Loosely Coupled System," *Social Forces* 58:506–527 (1979).
52. Roy Flemming, Peter Nardulli, and James Eisenstein, *The Craft of Justice: Politics and Work in Criminal Court Communities* (Philadelphia: University of Pennsylvania Press, 1992).
53. Flemming et al., note 52, p. 115.
54. Flemming et al., note 52, p. 117.
55. Flemming et al., note 52.
56. Office for Victims of Crime, *Victim Input into Plea Agreements* (Washington, DC: U.S. Department of Justice, 2002).
57. *Brady v. United States,* 397 U.S. 742 (1970).
58. *Santobello v. New York,* 404 U.S. 257 (1971).
59. *Ricketts v. Adamson,* 483 U.S. 1 (1987).
60. *United States v. Mezzanatto,* 513 U.S. 196 (1995).
61. *Bordenkircher v. Hayes,* 434 U.S. 357 (1978).
62. Anthony Walsh, "Standing Trial Versus Copping a Plea: Is There a Penalty?" *Journal of Contemporary Criminal Justice* 6:226–236 (1990).
63. Flemming et al., note 52, p. 172.
64. Scott Baker and Claudio Mezzetti, "Prosecutorial Resources, Plea Bargaining, and the Decision to Go to Trial," *Journal of Law, Economics, and Organization* 17:149–167 (2001); William McDonald, *Plea Bargaining: Critical Issues and Common Practices* (Washington, DC: U.S. Government Printing Office, 1985); Celesta Albonetti, "Criminality, Prosecutorial Screening, and Uncertainty: Toward a Theory of Discretionary Decision Making in Felony Case Processing," *Criminology* 24:623–644 (1986).

JBPUB.COM/ExploringCJ

Interactives

Key Term Explorer

Web Links

Trial

CHAPTER

11

Across the United States, prosecutors are now using rap music to make their case. Several defendants who have been charged with serious violent felonies also happen to write rap lyrics that often glamorize and depict their own criminal behavior. For example:

In New York, a defendant who referred to himself as "Rated R" challenged his enemies to wear bullet-proof vests and boasted of leaving .45-caliber slugs in their heads. The lyrics were introduced as evidence that ultimately led to his conviction for murdering two undercover police officers. He was sentenced to death in 2007.

In California, a jury convicted a reputed gang member of murdering a 17-year-old boy who was an alleged gang rival. The defendant had written a rap where he warned that rivals would "die looking at my barrel with your very last breath."

In Alabama, a jury sentenced a man to death for his role in the murder of three Birmingham police officers after prosecutors showed the jury rap lyrics and drawings that the defendant had in his jail cell. The lyrics glorified explicit details of the murders.

- Why would prosecutors be especially interested if the defendant wrote violent rap lyrics?

- What legal issues are raised by introducing a defendant's personal rap lyrics as evidence in a trial?

Source: Associated Press, "More Rap Lyrics Are Showing Up in Court as Evidence," December 21, 2006, available at http://www.courttv.com/news/2006/1221/rap_ap.html, accessed July 5, 2007.

Introduction

The first live, nationally televised criminal trial aired in the United States in 1991. It involved the trial of William Kennedy Smith, a nephew of Senator Edward Kennedy, who was charged with the rape of Patricia Bowman. Before then, the public relied on news accounts or brief taped footage of testimony to learn about the events at a trial. Now, the public was able to watch the jury selection process, presentation of evidence, cross-examination, instructions to the jury, and reading of the verdict.

Smith's trial did not resemble the highly dramatized events previously aired in television programs such as *Law and Order* and films such as *The Firm, To Kill a Mockingbird, The Verdict,* and *American Tragedy.* These fictionalized trials presented a much more polished—and ultimately less realistic—process than occurred in actual trials. With the goal of entertainment and excitement, these programs and films exaggerated the questioning of witnesses and

their testimony, the introduction of surprise witnesses at the last moment, and the ability of defense attorneys to outwit prosecutors.[1] Prior to the Smith trial, the public's beliefs about the trial process had been based on such fictionalized presentations and were far from the reality of what actually occurs as defendants go through the stages of a trial from pretrial motions to jury selection, trial, and, for many, appeal of conviction FIGURE 11-1.

Pretrial Motions

Although the vast majority of cases are resolved via the plea bargaining process, some defendants choose to have their cases decided through a trial. Once a case has been set for trial, the prosecutor may reevaluate the strength of the case based on the outcome of various pretrial motions raised by the defense. Pretrial motions are written or oral requests to the judge to make a ruling or to order that action be taken

William Kennedy Smith's 1991 rape case was the first live, nationally televised trial in the United States.

? Should trials be aired on television? Does media attention lead to greater opportunities for justice to be compromised?

in favor of the applicant. By definition, they are made before any opening statements or presentation of evidence and usually even before jury selection, although the timing of these motions varies widely among the states.

Pretrial motions are important to both the defense and the prosecution in moving a case toward trial. Most motions are made by the defense and are motivated by a desire to gain some legal advantage. For example, if the defense can persuade the judge to suppress certain evidence, the prosecutor may no longer have a winnable case and may move to dismiss the charges.

Motion for Dismissal of Charges

A motion to dismiss is often one of the first motions made by the defense before trial and is based on the defense's claim that the indictment is not sufficient to justify a trial. The prosecutor may also file a motion to dismiss the case by entering a *nolle prosequi* (a Latin term meaning that the prosecutor "will not further prosecute" the case). The reasons that prosecutors cite to dismiss cases vary but often result from a discovery that the original allegation is unfounded, that evidence produced by the police was unlawfully obtained, or that critical witnesses are no longer available.

If a judge accepts the motion for dismissal of charges prior to the beginning of the trial, the prosecutor may legally reinstate the case later if, for example, new incriminating evidence is found. The Fifth Amendment prohibition against double jeopardy does not go into effect until the trial actually

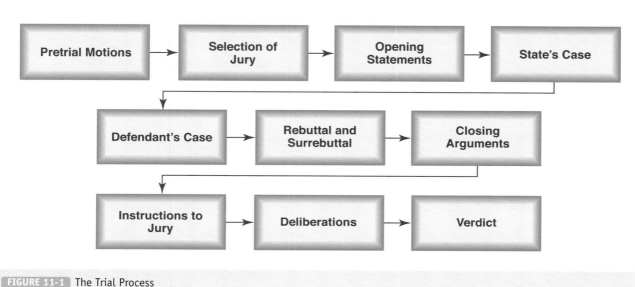

FIGURE 11-1 The Trial Process

begins. Jeopardy attaches in a jury trial once the jury has been impaneled and sworn in, and in a bench trial once the judge begins to hear evidence.[2]

Motion for Change of Venue

Typically, criminal cases are tried in the county (or, in federal cases, the district) in which the crime was committed. Sometimes, however, the defense may move for a change of venue if there has been excessive pretrial publicity about the case or if there is reason to believe that substantial prejudice exists that would deny the defendant a fair trial. If the trial judge agrees to this motion, the trial is moved to another county or district. The prosecution frequently objects to motions for a change of venue because of the difficulty and expense in transporting staff and witnesses to another city.

Motion for Severance of Defendants

When two or more defendants are jointly charged with the same offense, most states require that they be tried together. The savings in time and money from conducting a single trial are obvious, and, in many cases, a joint trial generally presents no problems to the defendants. On some occasions, a defendant may file a motion for severance of defendants if it is in his or her best interest to do so. For example, evidence presented against a co-defendant may not apply to the defendant, and the jury may have difficulty in keeping the cases separate in their minds.

Motion for Severance of Charges

Sometimes, multiple charges are filed against one defendant. It is to the prosecutor's advantage to try the defendant on all of these charges at the same time: It saves time and effort, and the collective weight of the charges may have a greater negative impact on the jury. Because the defense may wish to use different strategies to deal with each charge, and to avoid the prejudicial effect on the jury owing to the existence of multiple charges, the defense may file a motion for a severance of charges in such a case, requesting a separate trial for each charge.

Motion for Discovery

In many jurisdictions, the defense is given an opportunity for discovery of the state's case at a preliminary hearing. At this hearing, the prosecutor is required to provide the basic evidence against the defendant and identify the witnesses who will testify about the crime. In jurisdictions that do not allow discovery at the preliminary hearing or in cases where the defendant was indicted by a grand jury, it is critical for the defense to file a pretrial motion for discovery to request access to evidence and the list of witnesses who the prosecutor plans to present at trial.

Motion for a Bill of Particulars

Similar to a motion for discovery, a motion for a bill of particulars asks for a written statement from the prosecutor revealing the details of the charge(s), including the time, place, manner, and means of commission of the crime. Having this information allows the defense to prepare a more accurate defense, set limits on the evidence that the prosecutor may present at trial, avoid surprise claims of criminal acts, and establish the basis for a claim of double jeopardy if the defendant has already been tried for the same crime.

Motion for Suppression of Evidence

The exclusionary rule prohibits the introduction at trial of any evidence obtained illegally. Through motions for discovery, the defense may become aware of evidence that the prosecutor plans to introduce. If the defense believes that this evidence was unlawfully acquired, it may file a motion to suppress the evidence. In response to such a motion, the court may hold an evidentiary or suppression hearing to determine whether the evidence has been tainted by any illegal search or seizure. Even though such a hearing takes place, the judge may still reserve judgment on the introduction of the evidence until the issue actually comes up at trial, when the defense may again move to suppress the evidence. At this point, the judge will excuse the jury from the courtroom and hear arguments from the prosecutor and defense on the merits of the evidence. The judge may determine the evidence to be appropriate and allow it to be introduced, or find it inappropriate and disallow it.

LINK The exclusionary rule was originally applicable only to federal cases. It was not until 1961, in *Mapp v. Ohio*, that the rule was applied to state cases, as outlined in Chapter 7.

Motion of Intention to Provide an Alibi

The use of an alibi defense typically requires the filing of a pretrial motion indicating the intent to use this de-

The number of motions that can be filed may make trials very lengthy and drawn out.

? Should the number of motions that counsel can file be seriously curtailed? Which motions are attempts to extend the trial to better serve the client's interests?

fense. A motion of intention to provide an <u>alibi</u> (an assertion that the defendant was somewhere else at the time the crime was committed) states that the defendant's attorney plans to present a defense based on this notion. The motion must place the defendant in a location different from the crime scene at the time of the crime in such a way that it would have been impossible for him or her to be the guilty party. The names and addresses of witnesses to the defendant's location at the time of the crime must be disclosed to the prosecutor before the trial. In federal prosecutions, government attorneys may request that the defense notify the government if there is to be an alibi defense. Within 10 days of the request, the defendant must indicate the intent to provide an alibi defense and must provide the specific place where he or she claims to have been at the time of the alleged offense as well as the name, address, and telephone number of each alibi witness the defense intends to call.[3]

Motion for Determination of Competency

If the defense counsel questions his or her client's sanity while preparing for trial, or at any time dur-

ing the trial, the attorney should file a motion for <u>determination of competency</u>. A defendant who lacks the capacity to understand the charge or the possible penalties if convicted, assist or confer with counsel, or understand the nature of the court proceedings is considered not competent to stand trial.[4]

A competency hearing is not designed to establish the defendant's guilt or innocence, but rather to determine his or her present mental competence. Prior to the hearing, the court and the defense typically select two or three psychiatrists or psychologists to evaluate the defendant. Additional witnesses may be called, and evidence is presented by both sides. If the defendant is found not competent to stand trial, he or she is ordered to be confined to a mental institution until judged by the staff at the institution to be competent. At that time, unless the charge has been dismissed, the defendant will be brought back to court to resume the prosecution. If, as a result of the hearing, the defendant is found competent to stand trial, other pretrial motions will be considered by the judge and the trial date will be set.

Motion for Continuance

One of the most frequently filed pretrial motions is the motion for continuance, or postponement or adjournment of the trial until a later date. This kind of request is typically filed by the defense. Why would the defense choose to delay a trial, especially if the defendant is confined in jail during this process? The defense is likely to claim the need to prepare for the trial, but requests for a continuance are often part of a defense strategy to wear down victims and to allow for the possibility that prosecution witnesses may move away, be unable to be located, forget details of the crime, or even die by the time the trial begins. In addition, when the trial is delayed, public outrage over a particularly heinous crime may have subsided.

The Right to a Speedy, Public, and Fair Trial

The Sixth Amendment to the Constitution states that "In all criminal prosecutions, the accused shall enjoy the right to a speedy and public trial, by an impartial jury." This statement raises several issues: How soon after a person is arrested must a trial be held? How much access to pretrial hearings and the trial itself should the public have? How do we ensure that

Successful Use of Alibi in Robert Blake Murder Trial

Bonny Bakley, wife of actor Robert Blake, was shot in the head as she waited in the car outside an Italian restaurant where she and her husband had just finished dinner. Blake told authorities that he had returned to the restaurant to retrieve a revolver he had accidentally left behind after it had fallen under the table. Blake maintained that he was inside the restaurant at the time his wife was shot. Despite Blake's claim of innocence, the 71-year-old star of the 1970s television show *Baretta* was tried for first-degree solicitation for murder. After nearly 35 hours of deliberation, the jury acquitted Blake, indicating a lack of sufficient evidence to convict beyond a reasonable doubt, even though no witnesses had testified to seeing him actually return to his table in the restaurant.

Source: Lisa Sweetingham, "Actor Robert Blake Acquitted of His Wife's Murder," March 17, 2005, available at http://www.courttv.com/trials/blake/031605_verdict_ctv.html, accessed July 1, 2007.

---

the jury is untainted by prejudice? Unfortunately, the Constitution does not provide even broad outlines of how we should guarantee these rights. Those issues have been left to the courts and legislative bodies to decide.

The Right to a Speedy Trial

Few people would argue with the proposition that "Justice delayed is justice denied." The plight of a defendant who languishes in jail for months while the state prepares its case is clearly objectionable. Nevertheless, many defendants must wait long periods of time to have their day in court. How long they wait from arrest to sentencing varies greatly across jurisdictions and even from court to court within particular jurisdictions. However, the greatest influence on case processing is the seriousness of the charge. For example, for cases involving motor vehicle theft, the median number of days between arrest and sentencing is only 99; by comparison, it is 222 days for robbery and 412 days for murder FIGURE 11-2 .

Some of the delay may be due to court congestion—too many cases and too few judges. Other delays may result from poor management of court calendars, the number of jury trials scheduled, excessive motions for continuance by the prosecution or defense, or the seriousness of the case.[5] However, some evidence suggests that although courts with the highest rates of jury trials are generally the slowest to process cases, some courts with relatively high

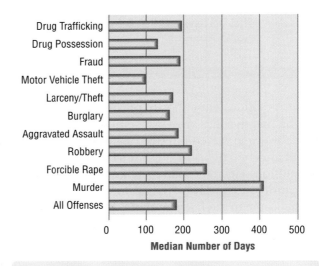

FIGURE 11-2 Median Number of Days from Arrest to Sentencing

Source: Matthew Durose and Patrick Langan, *Felony Sentences in State Courts*, 2002 (Washington, DC: U.S. Department of Justice, 2004).

jury trial rates are also among the fastest to bring defendants to trial.[6] Whatever the cause, any unnecessary delay in moving a case to its conclusion causes hardship for both defendants, especially those who are detained in jail, and victims, who often wait in anguish for justice to be done.

Defining the Limits of a Speedy Trial

The Sixth Amendment does not mean that a defendant may demand an immediate trial following his or her arrest. Both the prosecutor and the defendant have a right to prepare their cases before the trial begins. In attempting to define the meaning of "speedy trial," both federal and state governments have established rules limiting the time between arrest and trial. For example, in Illinois, the defendant must be brought to trial within 120 days if in custody or within 160 days if free on bond. By comparison, a defendant in California must be brought to trial within 60 days for either a felony or a misdemeanor, and in federal cases the trial must begin within 70 days. If the prosecutor exceeds that limit, the defendant may file a motion to have the charge dismissed.

In *Barker v. Wingo,* the U.S. Supreme Court held that a defendant's right to a speedy trial is not necessarily infringed upon by prosecutorial requests for continuances; rather, such decisions should be made on a case-by-case basis.[7] If the defense files a motion for continuance, it is considered a waiver of the right to a speedy trial. If neither the defendant nor the defense counsel has done anything to cause the delay, and the trial has not begun within the designated number of days, the judge may dismiss the case.

Federal Speedy Trial Act of 1974

Congress passed the Speedy Trial Act of 1974, later amended in 1979, to guide the prosecution of federal cases. According to this act, charges must be filed within 30 days of the suspect's arrest or after the defendant has received a summons on the charge. The trial should begin within 70 days of the filing date or after the defendant's initial appearance in court, whichever date is later.[8] In addition, because both sides need time to prepare for trial, the Speedy Trial Act provides that the trial may not begin less than 30 days from the initial appearance unless the defendant waives that right. Defendants sometimes waive this right if they are detained in jail and believe they are likely to be acquitted or sentenced to probation, thereby obtaining an earlier release from confinement.

The Courts, the Public, and the Press

The Sixth Amendment to the Constitution also guarantees defendants the right to a public trial. This right was established to ensure that people who were accused of crimes were treated fairly by the government by allowing the public full access to information regarding the proceedings. In 1948, the Supreme Court stated, in *In re Oliver,*

Whatever other benefits the guarantee to an accused that his trial be conducted in public may confer upon our society, the guarantee has always been recognized as a safeguard against any attempt to employ our courts as instruments of persecution.[9]

In 1965, in *Estes v. Texas,* the Supreme Court held that

Clearly the openness of the proceedings provides other benefits as well: it arguably improves the quality of testimony, it may induce unknown witnesses to come forward with relevant testimony, it may move all the trial participants to perform their duties conscientiously, and it gives the public the opportunity to observe the courts in the performance of their duties and to determine whether they are performing adequately.[10]

The Constitution does not tell us what makes a trial "public," how many people must be allowed to view a trial, or how they might view it, nor does it say whether a defendant may waive this right and request a private trial. Furthermore, there has been extensive debate over the conflict between the press's First Amendment right to report information about a case, the defendant's Sixth Amendment right to an impartial jury, and the right of victims and witnesses to privacy, as implied by the Fourth Amendment.

Over the years, a common-sense definition of a public trial has come to mean one that the public is free to attend, and the limit on the number of people allowed or required has been determined by the size of normal courtrooms. Other questions have not been so easily resolved.

The Public's Right to Attend

A number of U.S. Supreme Court decisions have dealt with the question of just how open trials must be. In *Gannett Co., Inc. v. DePasquale,* the Court held that the public and press could be barred from pretrial hearings.[11] Because many pretrial hearings involve sensitive matters such as determining the admissibility of evidence, access of the public and press to these hearings could pose special risks of unfairness. In 1986, the Supreme Court ruled that the press and public should not be excluded from the jury selection stage of the trial, even though the decision of the lower court in question was based on a desire to protect the privacy of potential jurors and to promote their candor during questioning.[12]

In *Richmond Newspapers, Inc. v. Virginia,* the Supreme Court held that trials must be open to the public, even if both the defendant and the prosecutor agree to closing them.[13] Chief Justice Warren Burger stated:

> We hold the right to attend criminal trials is implicit in the guarantees of the First Amendment; without the freedom to attend such trials, which people have exercised for centuries, important aspects of freedom of speech and of the press could be eviscerated.[14]

Some criminal cases involve very awkward and sensitive testimony by victims or require young children to testify about traumatic sex crimes. May the court close such portions of the trial to the public? The State of Massachusetts passed legislation barring the public and press during testimony in any trial involving specified sex offenses against people younger than age 18. After being excluded from the trial of a defendant charged with the rape of three teenage girls, a Massachusetts newspaper appealed this state law, claiming that it violated the press's constitutional right of access to the courtroom. In *Globe Newspapers Co. v. Superior Court for the County of Norfolk,* the Supreme Court stated that although "the right of access to criminal trials is of constitutional stature, it is not absolute," and upheld the Massachusetts state law.[15]

Freedom of the Press and Pretrial Publicity

In 2004, Santa Barbara County Superior Court Judge Rodney Melville issued a blanket gag order that prohibited singer Michael Jackson, his accuser, and attorneys on both sides from publicly commenting on Jackson's child molestation case. Representatives of the media and Jackson's attorneys appealed to the California Supreme Court to have the gag order lifted, but the judge refused to drop the order.

Judges in high-profile cases have frequently issued gag orders to prohibit the reporting of highly inflammatory information based on comments from attorneys outside the court or testimony during a trial. Until recently, the U.S. Supreme Court had not clearly defined the limits of gag orders designed to prevent extensive or prejudicial pretrial publicity that could deny a defendant a fair trial. In 1976, in *Nebraska Press Association v. Stuart,* the Supreme Court finally set forth guidelines for the use of prior restraint by lower trial courts.[16] The Court held that prior restraint should be used only when absolutely

Despite several accusations and financial settlements with alleged victims, Michael Jackson has never been convicted of child molestation.

 In what ways can an accusation in the "court of public opinion" result in the same damage as a conviction in a trial? What does this suggest about the public belief in the presumption of innocence?

necessary and only when the court can show all of the following:

- There is a clear threat to the fairness of trial.
- Such a threat is posed by the actual publicity to be restrained.
- No less restrictive alternatives are available.

Unfortunately, the issue regarding gag orders was not clearly settled in that case, and numerous challenges to gag orders have since been filed by the media, resulting in mixed rulings. For example, the Second, Fourth, Fifth, and Tenth Federal Circuit Courts of Appeal have held that a trial court may gag participants if the court determines that their comments would create a "reasonable likelihood" or a "substantial likelihood" of prejudicing a fair trial.[17] Four other Federal Circuit Courts of Appeal (the Third, Sixth, Seventh, and Ninth) have required a higher standard. According to these courts, gag orders may not be imposed unless there is a "clear and present danger" or a "serious and imminent threat" of prejudicing a fair trial.[18]

Some evidence suggests that general pretrial publicity may, indeed, influence jurors. As they consider the testimony at hand, jurors may remember what they have heard about similar but unrelated cases and may draw upon this knowledge to evaluate the case before them.[19] For example, jurors who have seen news coverage of rape cases and heard defendants' claims that the victims consented to have sex may apply their understanding of those cases to the case they are hearing, thereby influencing their evaluation of the testimony.

With the exception of a change of venue, the regulation of general pretrial publicity is beyond the court's control. Attempts to reduce the impact of such publicity can be made only at the time of jury selection by asking potential jurors about their knowledge of the case and inquiring whether that knowledge has created any bias in their minds or made it impossible for them to judge the case on its merits. Those who express a current or probable bias may be excused from the pool of prospective jurors.

Cameras in the Courtroom

Prior to 1977, if the news media wished to report on a criminal trial, they had to assign a reporter to attend the trial, take notes and perhaps make drawings, and then write the news story. Newspapers, radio, and television could then report the story. That situation changed in 1977, when television cameras were first admitted into the Florida courts on a regular basis. Since then, all 50 states have developed rules permitting cameras into their courts under certain circumstances.

The 1977 Florida law that pioneered the introduction of cameras into the courtroom presumes the right of news media to present live or taped coverage of trials unless the court finds compelling reasons to ban such a broadcast. This law was tested in 1981 in *Chandler v. Florida*.[20] In that case, the U.S. Supreme Court held that the risk of prejudice does not justify an absolute constitutional ban on all broadcast coverage, and that to demonstrate prejudice in a specific case a defendant must show that the media's coverage compromised the ability of the jury to judge the case fairly.

Rules permitting cameras in the courts remained only experimental in some states. In the state of New York, that experiment ended when the legislature once again banned televising criminal trials. The Court TV network sued to overturn this ban, but the New York Court of Appeals in 2005 held that "the press has no greater right of access to court proceedings than the general public under either the state or federal Constitutions." The New York court also stated that

[T]he governmental interests of the right of a defendant to have a fair trial and for the trial court to maintain the integrity of the courtroom out-

weigh any absolute First Amendment or Article 1, Section 8 right of the press or the public to have access to trials.[21]

The federal courts, which had banned cameras from their courtrooms since the sensationalized coverage of the Lindbergh baby kidnapping trial in 1932, briefly experimented with allowing cameras in between 1991 and 1994. Congress is currently considering legislation that would permit cameras in all federal courts, including the Supreme Court. In addition, the U.S. Court of Military Appeals has permitted television coverage of a number of important cases since 1989.

Bench Trial Versus Jury Trial

One of the more critical decisions a defendant must make along the path to trial is whether to ask for a trial by jury. Of all criminal defendants convicted at trial, about 40 percent are convicted by juries; the other 60 percent are tried and convicted before the judge alone in a <u>bench trial</u>.[22]

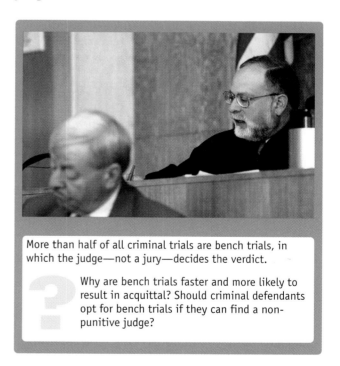

More than half of all criminal trials are bench trials, in which the judge—not a jury—decides the verdict.

? Why are bench trials faster and more likely to result in acquittal? Should criminal defendants opt for bench trials if they can find a non-punitive judge?

There are several reasons why a defendant might choose to waive his or her right to a jury trial and request a trial before a judge:

- The crime may be so heinous that it could generate a greater emotional reaction in jurors, who are generally less familiar with such crimes than are judges.

- A defendant's unusual appearance may create a bias in the minds of jurors.
- Excessive media coverage of the crime may make it difficult to ensure a fair and impartial jury.
- The case may be too complex for a jury to understand.
- Attorney fees may be lower because a bench trial generally involves less total attorney time than a jury trial.[23]

Other factors may also influence a defendant's decision to select a bench trial. Because defendants learn who the judge will be before they must make a decision, they may waive the right to a jury trial if the judge is known for being lenient in similar cases. Additionally—and perhaps most importantly—the rate of acquittals is much higher in bench trials than it is in jury trials.[24] In the federal courts, only 16 percent of jury trials end in acquittals, whereas 45 percent of defendants are acquitted in bench trials.[25] Lastly, because they avoid the jury selection and deliberation stages, bench trials generally take less time to complete. For example, the average number of days between arrest and conviction in felony cases is significantly greater for jury trials than for bench trials—9 months for jury trials versus 6 months for bench trials.[26]

Trial by Jury

During the late 1600s, Baptist preacher John Bunyan was imprisoned for more than 12 years for his political crimes against the Church of England. After being convicted and imprisoned a second time, Bunyan drafted his allegorical novel *Pilgrim's Progress,* published in 1678, in which he wrote:

> Then went the jury out, whose names were Mr. Blind-man, Mr. No-good, Mr. Malice, Mr. Love-lust, Mr. Live-loose, Mr. Heady, Mr. High-mind, Mr. Enmity, Mr. Liar, Mr. Cruelty, Mr. Hate-light, and Mr. Implacable, who every one gave in his private verdict against him among themselves, and afterwards unanimously concluded to bring him in guilty before the judge.[27]

Clearly, Bunyan was suggesting that jurors are unable to get past their own self-interests and that they are prejudiced by past experiences.

Bunyan wrote *Pilgrim's Progress* more than three centuries ago, but is it fair to suggest that today's juries are still filled with such self-interested people? Many people—especially minorities and the poor—believe that jurors are selected specifically because of identifiable self-interests, and there is much evidence to bear out this perception. Even so, most people today consider the right to be tried before a jury of one's peers one of the United States' most deeply held values. It is noteworthy that the right to a jury trial is found only in the handful of countries with criminal justice systems operating under common law procedures, such as the United Kingdom, Canada, and Australia. Today, the United States accounts for nearly 80 percent of all jury trials that take place around the world.[28]

Constitutional Right to a Trial by Jury

The Sixth Amendment guarantees not only the right to a jury trial, but also the right to a trial by an impartial jury. However, early Supreme Court interpretations of this provision limited those rights to federal cases. In 1937, the Court held, in *Palko v. Connecticut,* that only those rights "so rooted in the traditions and conscience of our people as to be ranked as fundamental"[29] and considered essential to the "principle of justice" were to be applied to the states through the due process clause of the Fourteenth Amendment.[30] Jury trials were not deemed to be a fundamental right. According to Justice Benjamin Cardozo, "The right to a trial by jury and the immunity from prosecution except as the result of an indictment may have value and importance. Even so, they are not of the very essence of a scheme of ordered liberty."[31]

Although most states provided for the right to a jury trial in their own constitutions, federal appellate courts did not view defendants in state criminal cases as having legitimate constitutional claims when they argued that their right to an impartial jury trial had been denied. In 1968, the Court reversed that interpretation and held that jury trials were fundamental to the criminal justice process "to prevent oppression by the government." Any serious crime that would qualify for a jury trial in federal court entitles a defendant the right to jury trial in state court.[32] In 1989, however, the Supreme Court held that defendants do not have the right to jury trial in minor criminal cases that carry punishments of less than six months in prison.[33] Furthermore, the right to a jury trial is not considered a fundamental right in juvenile proceedings.[34]

LINK Chapter 16 discusses the 1971 case of *McKeiver v. Pennsylvania,* in which the U.S. Supreme Court held that a jury trial is not necessary to have a fair hearing and that to require jury trials in the juvenile courts would make juvenile proceedings fully adversarial.

Size of the Jury

Most states and the federal courts use 12-person juries. However, the Sixth Amendment does not establish a required size for juries. In 1970, the Supreme Court ruled on the constitutionality of juries with fewer than 12 people.[35] The justices held that the number 12 was only a "historical accident, unnecessary to effect the purposes of the jury system and wholly without significance." They determined that juries should be "large enough to promote group deliberations and . . . to provide a fair possibility for obtaining a representative cross section of the community."

The size of juries is determined by each state, and juries with as few as six people have been held to be constitutional, except in death penalty cases. To reaffirm the Supreme Court's belief that at least six people were needed to achieve reasonable deliberations, in 1978 the Court, in *Ballew v. Georgia,* struck down a conviction by a five-person jury.[36]

Jury Selection

While the Sixth Amendment guarantees trial by an impartial jury, neither the prosecutor nor the defense really desires a fully impartial jury. That is, each party generally attempts to select people who they believe will favor their side. The complex jury selection process is designed to eliminate certain people from serving while retaining others. Both the prosecutor and the defense attorney, and sometimes the judge, ask questions of potential jurors. Once a sufficient number of jurors are found acceptable by both sides, the jury is established.

Eligibility for Serving on Juries

There are few requirements for jury service. Minor variations exist among the states, but the basic qualifications are the same. A prospective juror must meet the following requirements:
- United States citizenship
- 18 years of age or older
- Minimum residence in the jurisdiction in which the trial is being held (generally one year, but may be one month in some states)

Jury duty is one of the primary ways in which citizens engage in civic activities.

? Why do people often attempt to get out of jury duty? What does this trend indicate about persons who actually serve on juries?

- Possession of natural faculties (be able to see, hear, talk, feel, and smell and be relatively mobile) to be able to evaluate evidence
- Ordinary intelligence
- Knowledge of the English language sufficient to understand the proceedings and communicate with the other jurors

State rules for excluding people from serving on juries vary. Generally, those excluded are people who have been convicted of a felony or illegal act while in public office or those who have served on a jury during the preceding year.

Many jurisdictions also exempt some people from jury duty because of their occupation or duties they perform within the community. For example, attorneys, clergy members, teachers, physicians, firefighters, law enforcement and correctional officers, military personnel, many public officials, caregivers of young children, and those with proven hardship are generally exempt from jury service, though they may serve if they wish to do so.

Selection Process

The actual selection of people to serve on a jury involves a number of steps FIGURE 11-3 . The first step is the construction of a master list, called the jury pool, which contains approximately 1000 names of citizens randomly selected from the community. In many communities, jury pools are compiled from

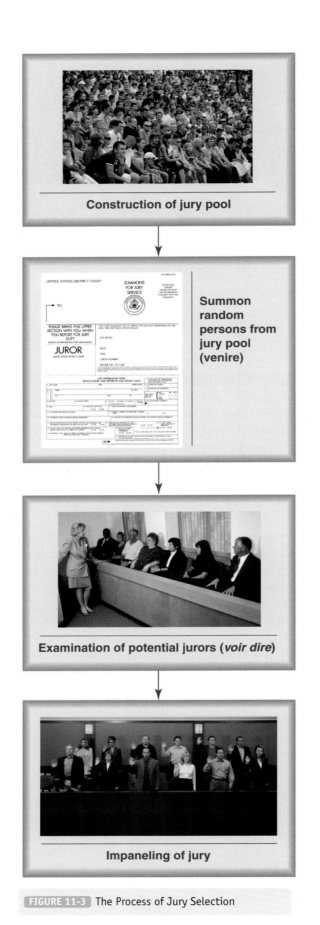

Construction of jury pool

Summon random persons from jury pool (venire)

Examination of potential jurors (*voir dire*)

Impaneling of jury

FIGURE 11-3 The Process of Jury Selection

voter registration lists. This process can produce very biased jury pools because certain groups of people—especially minorities and the poor—are much less likely than others to register to vote.[37] In an attempt to ensure the broadest inclusion of people, most communities now construct jury pools from several sources, such as voter and welfare lists, property tax rolls, lists of licensed drivers, telephone directories, and even utility records.

Each person in the jury pool receives a questionnaire that asks basic questions about residency, occupation, ability to understand English, prior felony convictions, and physical impairments that might automatically exclude the person from jury duty. In addition, the questionnaire is designed to elicit information about the respondent's possible biases in a criminal trial, such as prior jury duty, involvement in lawsuits, or familiarity with police officers in the community.

The second step is to randomly select a group of people from the jury pool to establish the jury panel, called the venire. The <u>venire</u> (from the Latin meaning "to come") is composed of people who are sent a summons (a legal document notifying them to report for jury duty). The number of potential jurors summoned is usually determined by the expected difficulty in obtaining a qualified jury. These people then become the panel from which the final jury is selected.

Unfortunately, not all potential jurors actually show up for jury selection. In many jurisdictions, as few as 40 percent of the venire report to the courthouse.[38] Many people are exempted from jury duty if they do not meet the eligibility requirements outlined previously; others may be exempted because they will be out of town during the term of possible service. Many others simply skip the responsibility because they do not want to take the time or effort to participate, and few of them are ever sanctioned by the court. Consequently, it is not unusual for juries to have disproportionate numbers of homemakers with grown children, retirees, and the unemployed.

Once assembled in the courtroom, the venire is sworn in and the *voir dire* begins. The <u>*voir dire*</u> (derived from the French, meaning "to speak the truth") is a preliminary examination of potential jurors. It seeks to answer two questions:

1. Is this person qualified to serve on the jury?
2. Is he or she capable of making a determination of guilt or innocence without prejudice?

During the *voir dire*, the prosecutor, the defense counsel, and often the judge ask questions to elicit information about the person's familiarity with the case and possible biases that might affect his or her judgment and willingness to listen impartially to all the evidence before making a decision. There are a few limits on the kind of questions that may be asked during this process, however. For example, potential jurors may not be asked specific questions about their sexual orientation or their religious affiliation and beliefs.

After the *voir dire*, prosecutors, defense attorneys, or the judge may excuse people from jury duty using one of two methods—challenge for cause or peremptory challenge.

Challenge for Cause

A challenge for cause is a call by the prosecutor or the defense to dismiss a person from a jury panel for a legitimate cause. Any number of people may be challenged for cause and dismissed from the panel. For example, if a prospective juror has vision or hearing problems that could interfere with observing or understanding the proceedings; if he or she indicates an existing bias toward one of the parties in the case; if the individual is worried about a sick family member; or if the person has knowledge of the defendant, witnesses, or attorneys involved in the case, then, with the agreement of the judge, that person is excused.

In addition, a person may be excluded for cause in a capital case if he or she could never vote for a guilty verdict if it carried the death penalty.[39] However, the Supreme Court has held that people could not be excluded from serving on a jury "simply because they voiced general objections to the death penalty or expressed conscientious or religious scruples against its infliction."[40]

Peremptory Challenge

Peremptory challenges allow attorneys to eliminate people from the jury who they believe are not likely to be sympathetic to their arguments. Because both the prosecutor and the defense attorney may challenge a limited number of prospective jurors without giving any reasons for doing so, the use of peremptory challenges has become a very controversial part of the jury selection process.

Inasmuch as the process of excusing people based on personal characteristics or views introduces a form of bias into the jury and may deny a defendant or the state a representative jury, government statutes limit the number of peremptory challenges allowed to each side.[41] The minimum in felony cases is usually three, but both sides may be allowed as many as 20 in highly publicized murder cases or cases in which several defendants are being tried together.

When a sufficient number of jurors (usually 12) and one or two alternates have been accepted by the defense and prosecution, they are impaneled—that is, sworn in.

Bias in Jury Selection

Can a person be excused from serving on a jury simply because of his or her race or gender? Does a defendant have a right to a jury composed of people with particular racial, ethnic, or gender characteristics? Must the jury be a cross section of the community? The Supreme Court has tried to answer these questions. In 1947, the Court ruled as follows:

> There is no constitutional right to a jury drawn from a group of uneducated and unintelligent persons. Nor is there any right to a jury chosen solely from those at the lower end of the economic and social scale. But there is a constitutional right to a jury drawn from a group which represents a cross section of the community.[42]

The requirement of a representative cross section of the community is met only if there is no systematic exclusion of any particular group of people. Of course, owing to peremptory challenges, when no reason for the challenge is given, it may be difficult to determine whether a particular race or gender is being deliberately excluded from the jury.

Race

In 1986, the Supreme Court held, in *Batson v. Kentucky,* that the Fourteenth Amendment's equal protection clause "forbids the prosecutor to challenge potential jurors solely on account of their race or on the presumption that black jurors as a group will be unable impartially to consider the State's case against a black defendant."[43] The Court further stated that for a defendant to establish that his or her constitutional right was violated, the defendant must show "that he is a member of a cognizable racial group . . . and that the prosecutor has exercised peremptory challenges to remove from the venire members of the defendant's race." Excluding members of a certain racial group would deny the defendant the right to be tried by an impartial jury representative of a cross section of the local community. In 1991, the Court, in *Hernandez v.*

New York, extended the *Batson* rule to the ethnicity of defendants.[44]

Six years later, in *Georgia v. McCollum,* the Supreme Court held that the Constitution prohibits a criminal defendant from engaging in purposeful discrimination on the ground of race in the exercise of peremptory challenges.[45] While noting that peremptory challenges are not constitutionally protected fundamental rights, but rather a state-created means of assisting in providing an impartial jury and fair trial, the Court stated that

A criminal defendant's racially discriminatory exercise of peremptory challenges inflicts the harms addressed by *Batson.* Regardless of whether it is the State or the defense who invokes them, discriminatory challenges harm the individual juror by subjecting him to open and public racial discrimination and harm the community by undermining public confidence in this country's system of justice.[46]

In 1990 in *Holland v. Illinois,* a white defendant objected to the prosecutor's exclusion of all African Americans from the jury, and the Supreme Court held that the Sixth Amendment requirement of a fair cross section on the venire is a means of ensuring only an impartial jury—not necessarily a representative jury.[47] In other words, defendants are not entitled to a jury of any particular composition; there is no constitutional right to have African Americans, White Americans, the poor, or the rich on the jury. The Court held that the cross-section requirement referred only to who is excluded from juries, not to who must be included. However, in *Powers v. Ohio* (1991), the Supreme Court held that all citizens have an equal opportunity to serve on juries and that a "defendant is injured by exclusion of racial groups even if he is not a member of those groups."[48] In other words, the systematic exclusion of particular racial groups denies a defendant a fair and impartial trial.

Religion Even though jurors may not be asked specific questions about their religious affiliation or beliefs during *voir dire,* indirect indicators or responses during questioning may provide information about the religion of potential jurors. The constitutionality of using peremptory challenges based on a potential juror's religion has not yet been resolved. For example, Connecticut has recently prohibited religion-based peremptory challenges. In contrast, the Texas appellate court has held that *Batson* does not apply to such challenges.[49] Finally, a Florida appellate court may have opened the door to prohibiting religion-based challenges in that state based on the "free exercise of religion" protection in the First Amendment to the U.S. Constitution.[50]

Scientific Jury Selection

In an attempt to make effective decisions about potential jurors, many attorneys are turning to psychological and sociological studies designed to correlate background characteristics, personality profiles, courtroom behaviors, body language, and facial expressions with particular biases. Such studies include community surveys conducted before a trial to identify characteristics of people who are more inclined to be sympathetic toward either the prosecution or the defense. People on the venire exhibiting those characteristics are more likely to be either challenged or accepted as attorneys try to build a favorable jury.

Sometimes potential biases are identified during *voir dire.* In addition, studies on group dynamics have been conducted to help predict which jurors are most likely to play leadership roles in the jury.[51] However, legal scholar Shari Diamond cautions that the effects of scientific jury selection on final juror decisions are modest at best and likely to remove little of the uncertainty once the jury retires to deliberate.[52]

The Trial

Although jury trials account for only 5 percent of criminal prosecutions today, each one provides an occasion for public scrutiny of the judicial process and, in many ways, symbolizes the entire criminal justice system itself. Whether the trial is televised live or simply reported in the media, members of the public are given an opportunity to observe and comment on how the wheels of justice turn. Trials today are generally less entertaining than those of many decades ago, when flamboyant judges and attorneys used them as stages for oration. Most trials now are rather routine and highly regulated in terms of procedure, and are typically completed in a matter of hours or days. Only the most complex and celebrated cases take a week or more to conclude. **TABLE 11-1** shows the average length of jury trials.

VOICE OF THE COURT — Gender Bias in Jury Selection

In 1994, in *J.E.B. v. Alabama,* the U.S. Supreme Court ruled that peremptory challenges based on gender were unconstitutional. The Court, in a 6-3 decision, held that "gender, like race, is an unconstitutional proxy for juror competence and impartiality." The justices ruled that:

Intentional discrimination on the basis of gender by state actors violates the Equal Protection Clause, particularly where, as here, the discrimination serves to ratify and perpetuate invidious, archaic, and overbroad stereotypes about the relative abilities of men and women.

Source: J.E.B. v. Alabama, 511 U.S. 127 (1994).

TABLE 11-1
Average Length of Jury Trials

Court	Average Time[a]
Elizabeth, New Jersey	6:20
Paterson, New Jersey	7:24
Golden, Colorado	8:10
Monterey, California	9:27
Denver, Colorado	10:50
Jersey City, New Jersey	12:09
Marin County, California	17:44
Oakland, California	23:16

[a]In hours and minutes. Does not include jury deliberation time.

Source: Adapted from Dale Anne Sipes, *On Trial: The Length of Civil and Criminal Trials* (Williamsburg, VA: The National Center for State Courts, 1988), p. 17.

LINK Chapter 9 presents an overview of the U.S. court system and the adversarial system of justice, which is designed to arrive at the truth of a criminal case by having opponents (the prosecutor and the defense attorney) present their evidence and scrutinize each other's evidence. Both sides carefully build what they believe to be an accurate representation of the reality in the case, and then let the jury determine which side is most convincing.

Opening Statements

Once the jury has been sworn in and seated and the formal charges against the defendant have been read, the prosecutor presents an opening statement to the jury. (The prosecutor goes first because it is the state's burden to prove the defendant's guilt.) The length of the statement depends on the complexity of the case, but its purpose is always to provide a factual outline of the case the prosecutor intends to prove. The opening statement may include a restatement of the charges, a general overview of why the prosecutor

believes the defendant is guilty, and a brief listing of the witnesses the prosecutor intends to call and what each will testify to.

Nothing said by the prosecutor in the opening statements may be considered by the jury as evidence or facts in the case. Rather, it is commentary designed to help jurors follow the case as it develops. Prosecutors are not allowed to make statements considered inflammatory or prejudicial against the defendant, such as commenting on evidence already known to be inadmissible. Such statements constitute prejudicial error—an error substantially affecting the defendant's rights—and may result in either a mistrial or a reversal on appeal if the defendant is eventually convicted.

The defense is then given an opportunity to make its opening statement. Often, the defense may choose to delay its opening statement until after the prosecution's case has been fully presented. In any event, the defense will usually stress during this statement that the prosecutor must prove his or her case beyond a reasonable doubt, given the presumption of innocence.

Presentation of Evidence

In criminal cases, the prosecutor always presents evidence first. Evidence is presented by witnesses, not by the attorneys. The prosecutor and the defense attorney call various witnesses to present different kinds of evidence. Ideally, each witness will present evidence that lays a foundation for subsequent witnesses.

Types of Evidence
All evidence submitted must be relevant, competent, and material. That is, it must relate directly to the issue at hand and to the material elements of the

Real Evidence: Fingerprints and DNA

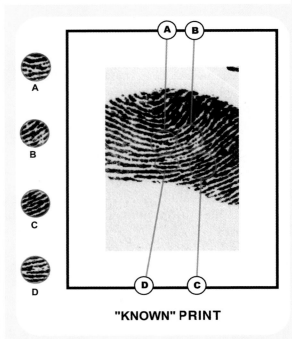

"KNOWN" PRINT

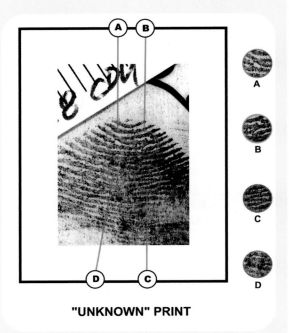

"UNKNOWN" PRINT

Rick Jackson was arrested in Upper Darby, Pennsylvania, for a gruesome murder and told that the police had solid evidence against him—photographs of his bloody fingerprints taken from the crime scene. Even though experts agreed that the fingerprints were a match, Jackson insisted that they couldn't be his. He was found guilty and sentenced to a life sentence without possibility of parole. Two years later, other fingerprint experts testified that the prosecution had been wrong and that the prints, in fact, did not belong to Jackson. He was released after spending two years in prison.

While fingerprints have long been considered the "gold standard" of identification at trial and frequently are the key evidence used to obtain convictions, their reliability is increasingly being questioned, especially with the growing refinement of DNA technologies. DNA samples are often retrieved from crime scenes and used by prosecutors in minor property crimes. These samples may consist of small amounts of blood on broken glass in burglaries, sweat on a hat or mask, or skin cells left behind on a drinking glass, cigarette butt, chewing gum, or food. All of these materials are real evidence and are admissible in court.

The DNA of property offenders collected at crime scenes is matched against the nearly 2 million DNA samples contained in the Combined DNA Index System (CODIS) housed by the Federal Bureau of Investigation (FBI). The CODIS database includes both state and federal DNA samples collected from convicted felons, prison inmates, and adults and juveniles charged with serious crimes. One such DNA match helped solve a "cold case" homicide of an 11-year-old girl murdered in 1986 in Fort Worth, Texas, as well as 9 rapes and 22 homicides in Kansas City, Missouri. However, DNA samples have also been used to establish that individuals were wrongfully convicted when reexamination of their cases revealed mishandling of evidence.

Experts claim that a DNA match is conclusive, with less than one in 200 million matches likely to be faulty. Nearly all states now admit DNA evidence at trial, although the specific standards for analysis and testing vary. Both prosecutors and defense attorneys look for DNA evidence to strengthen their cases.

As DNA evidence gains popularity, fingerprint evidence has developed a more questionable reputation. In 2002, a federal judge, in *United States v. Plaza*, first ruled that fingerprint evidence is unreliable and declared that he would no longer allow fingerprint experts to testify with certainty that prints from a defendant match or do not match those found at a crime scene. Two months later, the same judge overruled himself, stating that the methodology used by the FBI is generally accepted and reliable enough to be presented to juries.

But the debate about the reliability of fingerprint matches continues. Matches identified when known or rolled (sometimes called "inked") prints—those fingerprints obtained from individuals by inking fingers and purposefully rolling fingertips onto paper—are compared to other known prints are generally considered valid. There is much less certainty about matches made by comparing known prints to latent prints—those obtained from surfaces of objects at the crime scene. Latent prints are typically invisible under normal viewing conditions and are rendered visible only with the help of special technologies, such as the use of powders or other chemicals. Unfortunately, latent prints are generally incomplete, representing only a little more than 20 percent of the average size of a known print, and are frequently smudged or distorted. Comparisons of latent prints to known prints are considered subjective and subject to error. Indeed, there

is currently no agreed-upon standard in the United States regarding exactly how many points in prints are required to declare a match. Fingerprint examiners in Sweden require 7 points to declare a match, while Australia requires 12, and Brazil requires 30.

Many critics argue that testimony regarding fingerprint analysis is only opinion, with little or no empirical research proving the reliability of known-to-latent print matches. The use of computer fingerprint databases produces only a narrowed-down list of possible matches; fingerprint examiners then must visually—and subjectively—make the comparison and draw conclusions. When DNA evidence is not available, fingerprints are still important evidence and will continue to have a role at trial.

Sources: Lesley Stahl, "Fingerprints: Infallible Evidence?", *CBS News 60 Minutes,* June 6, 2004, available at http://www.cbsnews.com/stories/2003/07/16/60minutes/main563607.shtml, accessed July 16, 2007; Office of Justice Programs, *DNA in "Minor" Crimes Yields Major Benefits in Public Safety* (Washington, DC: U.S. Department of Justice, 2004); Donna Lyons, "Capturing DNA's Crime Fighting Potential," *State Legislatures* 32:16–18 (2006); Lisa Kreeger and Danielle Weiss, *Forensic DNA Fundamentals for the Prosecutor* (Alexandria, VA: American Prosecutors Research Institute, 2003); *United States v. Plaza,* Eastern District of Pennsylvania, No. 98-362 (2002); Tamara Lawson, "Can Fingerprints Lie?: Re-weighing Fingerprint Evidence in Criminal Jury Trials," *American Journal of Criminal Law* 31:1–67 (2003); Sandy Zabell, "Fingerprint Evidence," *Journal of Law and Policy* 13:143–179 (2005).

crime and be provided by someone considered competent or qualified to testify. Although there is often overlap, most evidence falls into one of the following categories:

- Real evidence includes physical objects such as fingerprints, clothing, weapons, stolen property, documents, confiscated drugs, and genetic material. Sometimes, the original real evidence is not convenient to present to the jury, so photographs or reconstructions may be used.
- Testimonial evidence includes the testimony of witnesses who are qualified to speak about specific real evidence. For instance, an expert witness such as a forensic chemist may be called to testify to the fact that drugs confiscated by the police are what they are purported to be. Typically, potential expert witnesses are questioned by the prosecutor and the defense attorney about their job experience, years of education, or special training and knowledge. Based on these credentials, the court certifies them as expert witnesses.

- Direct evidence is provided by eyewitnesses to the crime about what they directly observed.
- Circumstantial evidence requires that the jury draw a reasonable inference from the testimony. A witness may testify that he or she heard a scream followed by a gunshot coming from a victim's apartment; moments later, the witness saw the defendant leave the apartment. Although the witness did not directly observe the shooting, absent any other suspects in the apartment, it is reasonable to infer that the defendant was involved in the crime. In the absence of eyewitnesses, or to bolster inferences from circumstantial evidence, some courts now admit filmed reenactments to demonstrate graphically to the jury how the incident may have occurred.
- Hearsay evidence includes testimony based on something that the witness does not have direct knowledge of but has heard or been told. Generally, hearsay evidence is considered inadmissible. For example, if John were to testify, "Jerry told me that Cheryl sold Chris a kilo of

marijuana," it would be considered hearsay evidence because John had no direct knowledge of the alleged sale. However, several exceptions to this rule exist:

1. *Dying declarations.* Because it is assumed that dying people will not lie when they believe themselves close to death, such testimony may be admitted for the jury's consideration.

2. *Statements made by victims of child abuse.* Such statements made to a caseworker, teacher, or doctor may be admitted.

3. *Admission of a criminal act by the defendant to a witness.* Because both the witness and the defendant are in court and the defendant may rebut the statement, it may be admitted as evidence.

Evidence about the Sexual Past of a Victim

One of the most controversial issues in the admission of evidence has to do with the sexual past of a victim. Until recently, defense attorneys were allowed to present evidence about a rape victim's past sexual history in an attempt to discredit his or her claim that the victim had not consented to sexual intercourse. Since 1974, when Michigan passed the first rape shield law, and 1978, when the federal rape shield law was introduced, rape victims have become relatively protected from the presentation of such evidence.[53] These laws exclude an accuser's prior "sexual behavior," although some exceptions are made. For example, the Federal Rules of Procedure and laws in most states include exceptions when the evidence is used to

- Show that someone other than the accused was the source of the physical evidence (i.e., semen or injury)
- Prove consent was based on the accuser's prior sexual behavior with the defendant
- Establish that the exclusion of sexual evidence regarding the accuser would violate the defendant's constitutional rights

If an exception is made, it may quickly bring an end to the case, because few accusers are willing to face a vigorous examination of their sexual history and practices when being cross-examined by a defense attorney.

The Prosecution's Case

The prosecutor presents the state's case through a succession of witnesses and evidence. Questioning is usually straightforward, with each witness being asked to discuss what he or she knows to be the facts of the case. The defense attorney may object to a question on several grounds:

- The question is irrelevant (immaterial) or incompetent (not admissible).
- The question calls for speculation on the part of the witness (only expert witnesses may offer opinions).
- The prosecutor is leading the witness's response (presenting the desired response in the question itself).

The judge either sustains (consents to) or overrules the objection. If the objection is sustained, the question must be rephrased or replaced; if it is overruled, the witness is asked to answer the question.

When the prosecutor is finished questioning a witness, the defense may cross-examine the witness but may cover only those issues raised in the prosecutor's direct examination. The cross-examination is designed to discredit the witness by identifying inconsistencies or contradictions in the testimony. If the defense attorney believes that the witness was telling the truth and that any further questioning might strengthen the state's case, he or she may waive the right to cross-examine a particular witness.

Once cross-examination is complete, the prosecutor may ask additional questions through redirect examination. The purpose of redirect examination is to clarify any issue brought out in response to questions posed by the defense. If the prosecutor chooses to redirect, the defense is given a final opportunity to ask clarifying questions based on the redirect examination through recross-examination. Thus each side has a maximum of two opportunities to question a witness.

Once the prosecutor has presented the state's evidence, the prosecution rests. At this point, the defense may ask for a judgment of acquittal, claiming that the state failed to establish that a crime was committed or that the defendant committed it. In most states, if the judge believes that the prosecutor has not established a sufficient case, he or she can take the case out of the hands of the jury and enter a directed verdict—a judgment of acquittal—which bars any further prosecution of the defendant on the crime charged. In some states, a directed verdict means that the judge directs the jury to return a not guilty verdict; however, a few states allow the jury to disregard the instruction and return a guilty verdict.

The Defense's Case

In the United States, a defendant is innocent until proven guilty. Thus the prosecution must prove its case to win a conviction. The defense is not required to present any case at all, although this is rarely done. If the defense does decide to present its case, it calls its own witnesses to testify on the defendant's behalf after the prosecution rests.

The defendant is not required to testify. This right is protected by the Fifth Amendment, which states that no one "shall be compelled in any criminal case to be a witness against himself." There are many reasons why a defendant might decline to testify, such as a prior criminal record or the desire not to give the jury a bad impression. In any case, a prosecutor may not comment on the refusal of a defendant to testify.[54]

The defense must decide what kind of defense to present. Typical defenses include alibi, self-defense, insanity, entrapment, and intoxication. In most cases, the defense simply attempts to present sufficient evidence to cast doubt on the prosecutor's case, thereby creating reasonable doubt about guilt in the minds of the jurors.

LINK Defenses involving justification or excuse are discussed in detail in Chapter 2.

Rebuttal and Surrebuttal

Once the defense has completed its case and the cross-examination and redirection of defense witnesses are complete, the prosecutor may present additional evidence in a rebuttal to issues raised in the defense's case. Witnesses called by the prosecutor may then be questioned by the defense and additional new evidence presented in surrebuttal.

Closing Arguments

After the defense rests its case, both sides present their summation, also called closing arguments. This statement gives each side an opportunity to review and summarize the facts of the case, highlight the significant weaknesses in the opposing case, and, if necessary, make an emotional appeal to the jury in a final attempt to win them to their side.

Typically, the prosecution presents its closing argument first, arguing that guilt has been established and emphasizing to the jury the wrongness of the crime, the impact of the crime on the victim, and the jury's responsibility to return a guilty verdict. The defense then presents its closing argument, in which it summarizes facts presented in direct and cross-examination, and likely argues that it is the prosecutor's responsibility to prove the guilt of the defendant. After the defense makes its closing argument, the prosecution is allowed to make a final rebuttal.

The give-and-take in the examination of witnesses and the often antagonistic positions taken by the prosecution and the defense in their opening statements and closing arguments are critical parts of the adversarial process in criminal trials. Although this adversarial nature of trials is taken for granted by most Americans as the only logical way of discovering the truth about a case and achieving justice, the adversarial system is not universally used. Many countries use trial procedures similar to the sixteenth-century inquisitorial system, which is based on a European tradition different from the Anglo-Saxon system that formed the basis of U.S. courts. In contemporary inquisitorial systems, the prosecutor and the defense attorney assist the judge in his or her inquiry and play much less significant roles.

Judicial Instruction

The judge's instructions, or charge to the jury, provide the members of the jury with guidelines for making their decision. Most states have developed standard jury instructions covering the issues of standards of proof, the responsibility of the state to prove guilt, the rights of the defendant, the elements of the crime that must be proven, possible verdicts, restrictions on communicating with others during jury deliberations, and suggestions for determining the credibility of witnesses.

In addition, the judge may give special instructions regarding the nature of the offense and any lesser included charges. Often, the prosecutor and the defense attorney will confer with the judge before the charging of the jury to discuss additional instructions that they wish to include.

Many studies suggest that a large percentage of jurors have little comprehension of the judicial instructions.[55] Other studies suggest that jurors who receive standard instructions comprehend their responsibilities no better than jurors who receive no instructions at all—both groups appear to make similar decisions and to raise similar questions of the judge after beginning to deliberate.[56]

If jurors are unable to fully comprehend and apply the judge's instructions, then the fairness of

Inquisitorial Court Procedures: Germany and England

In 1924, German courts replaced juries of citizens with panels of lay and professional judges who were responsible for determining guilt in criminal cases. This system remains in place today. Typically, three or more judges hear each case, and a two-thirds majority is required to acquit or convict and determine an appropriate sentence.

Once a suspect is apprehended, the German Code of Criminal Procedure requires prosecutors to prosecute all defendants on each charge for which sufficient evidence exists. Furthermore, the prosecutor is prohibited from entering into any plea bargain with the defendant regarding charges or sentence. Whether the accused pleads guilty or not guilty, the case goes to trial.

In the German system, judges call and question witnesses, decide on the order for the presentation of evidence, and rule on the admissibility of such evidence. Expert witnesses are not called by the prosecutor or the defense; rather, they are selected and paid by the government, thereby avoiding the spectacle of "dueling experts." Unlike in U.S. courts, the prosecutor and the defense attorney play relatively minor roles in German trials and may examine witnesses only after the judge has questioned them. The defendant is questioned by the judge at the beginning of the trial; his or her appearance is then followed by testimony from other witnesses, who are examined and cross-examined by the judge, prosecutor, and defense attorney. Moreover, victims of particular serious crimes (or relatives in the case of a murder) have the right to ask questions of the defendant and may formally act as adjunct prosecutors.

In England, the system is somewhat different, resembling the adversarial system in place in the United States. However, in 1994, a key element found in inquisitorial systems was reintroduced into English criminal procedure. The Criminal Justice and Public Order Act of 1994 modified the right of a defendant to remain silent in criminal trials. It allowed the court to inform jurors that they may draw adverse inferences from a defendant's refusal to testify. The English Court of Appeal has approved the following jury directions regarding a defendant's silence:

The defendant has not given evidence. This is his right. He is entitled to remain silent and require the prosecution to prove its case. You must not assume he is guilty just because he has not given evidence because failure to give evidence cannot, on its own, prove guilt. However . . . you are entitled, when deciding whether the defendant is guilty of the offence(s) charged, to draw such inferences from his failure to give evidence as you think proper. . . .

If, in your judgment, the only sensible reason for his decision not to give evidence is that he had no explanation or answer to give, or none that could have stood up to cross-examination, then it would be open to you to hold against him his failure to give evidence, that is take into account as some additional support for the prosecution's case.

Critics of this law suggest the inquisitorial limit on the right to remain silent at trial could greatly erode or eliminate the more general right to silence, thereby shifting the burden of proof from the state to the accused. They also believe that it may shift the accusatorial system's focus on establishing proof by witnesses to the inquisitorial system's focus on questioning of suspects. Finally, critics are concerned that limits on silence will lower the government's level of proof. That is, if the prosecution is able to establish even a *prima facie* case (submitting evidence that is sufficient, if not rebutted, to prove a particular proposition or fact), yet falls short of proof beyond a reasonable doubt, the defendant will likely have to testify: If he or she refuses to testify, the government's case becomes strengthened by silent inference of the defendant's guilt.

Sources: Erika Fairchild, *Comparative Criminal Justice Systems* (Belmont, CA: Wadsworth, 1993); Jonathan Doak, "Victims' Rights in Criminal Trials: Prospects for Participation," *Journal of Law and Society* 32:294–316 (2005); *Criminal Justice and Public Order Act 1994*, Section 35 (Eng); Gregory O'Reilly, "England Limits the Right to Silence and Moves Towards an Inquisitorial System of Justice," *Journal of Criminal Law and Criminology* 85:402–452 (1994).

their deliberations may be called into question. According to linguistics professor Bethany Dumas, standard jury instructions tend to be "written in dense, complex, lawyer's language. . . . They are delivered orally, often monotone, to the jurors under instruction, who, as average Americans, have an average reading level of sixth to eighth grade."[57] The problems of readability and comprehension have led many who practice in the courts to call for plain-English jury instructions. Indeed, some states, such as Michigan and California, have developed such guides for incorporating plain, simplified, and straightforward language to help ensure jurors understand the law and apply it appropriately.[58]

Jury Nullification

In recent years, juries have acquitted a number of defendants in very high-profile criminal cases, including O. J. Simpson, Lorena Bobbitt, Robert Blake, Michael Jackson, and the police officers accused of beating Rodney King of major charges, despite widespread perceptions that the evidence against the defendants was overwhelming.

In the judge's instructions to the jury, jurors are typically told that when they were sworn onto the jury, they took an oath to uphold the law and they are only to be finders of the facts, not interpreters of the law. Historically, however, juries have been allowed to make a decision based on the fairness of a particular law, known as jury nullification. Rooted in common law rights that were established in Europe and ultimately transported to America, juries, until recently, were able to acquit a defendant because they believed the particular law in the case was wrong. By acquitting the defendant, they nullified the law in that particular case.

Because jury deliberations are held in secret and many jurisdictions prohibit testimony based on what went on in the jury room in an appeal, actual instances of jury nullification are rarely discovered. A number of beliefs held by jurors may lead to jury nullification:[59]

- The law is unfair.
- The law is fair but is being unfairly applied to the defendant in this particular case due to exceptional circumstances (i.e., the defendant is too old or lacking in mental capacity to understand his or her actions).
- The entire criminal justice system is unfair (i.e., biased against minorities).

- The prosecutor acted unfairly (i.e., unscrupulously).

Although the vast majority of trial and appellate courts refuse to instruct jurors about the right of nullification, a number of state legislatures have recently considered new statutes or amendments to their constitutions that would require courts to inform jurors of their right to extend mercy to clearly guilty defendants through jury nullification.[60] For example, the state constitutions of Indiana and Maryland include provisions requiring a judge to tell jurors that his or her instructions about the law are only advisory and that they may interpret the law as well.

The Presumption of Innocence and Reasonable Doubt

One of the most important instructions a judge gives to jury members is to remind them that a defendant is considered innocent and that the state must prove its case beyond a reasonable doubt. The presumption of innocence means that a person is presumed to be innocent until the state proves beyond a reasonable doubt that he or she is guilty of the crime charged. This is the principle under which the process of determining guilt operates.

Many people believe that because there was probable cause to arrest and charge a suspect, the suspect

The courtroom conduct of many of O. J. Simpson's attorneys garnered headlines and resulted in an acquittal.

 Are high-profile cases more likely to result in acquittals? How do televised trials affect jury nullification?

must be guilty. In reality, probable cause means only that there are reasonable grounds to believe the accused committed certain acts. A fine line exists between the grounds necessary to make an arrest and those necessary to convict.

The requirement that the accused be judged guilty beyond a reasonable doubt means that the jury (or the judge in the case of a bench trial) must find the evidence entirely convincing and must be satisfied beyond a moral certainty of the defendant's guilt to convict him or her. Fanciful or imagined doubt, or doubt that arises in the face of the unpleasant task of convicting or acquitting a defendant, is not sufficient. This standard does not require that the case be proved beyond *all* doubt—a situation generally unlikely given human nature.

The Supreme Court has held that, although the Bill of Rights does not include a provision guaranteeing the standard of proof beyond a reasonable doubt, this standard is still constitutionally required. In *In re Winship*, it ruled:

> A society that values the good name and freedom of every individual should not condemn a man for commission of a crime when there is reasonable doubt about his guilt. . . . There is always in litigation a margin of error, representing error in fact finding, which both parties must take into account.[61]

This requirement protects defendants from being convicted of crimes when the case against them is not very strong. In essence, the Supreme Court was saying that it is better to let a guilty person go free as a result of less than adequate proof than to convict an innocent person solely on the basis of the probability of guilt. When much of the evidence presented against a defendant is circumstantial, or when testimony by both prosecution and defense witnesses appears reasonable and yet contradictory, it is better for the jury to err on the side of setting the defendant free than to send him or her to prison.

Jury Deliberations

After hearing the judge's instructions, the jury retires to the jury room to begin deliberations. If the case has generated much public attention or deliberations are likely to take some time, the judge may request that members of the jury be sequestered (segregated from all outside contact). In some highly publicized trials, juries have been sequestered from the point at which they were impaneled. The jury members are kept together for the duration of their deliberations, sometimes receiving temporary housing and meals, to protect them from outside influences. In most cases, such precautions are not needed and jurors simply take an oath to stay together during breaks and to refrain from talking to others about the case.

The jury selects a jury foreperson to act as the leader. It is this juror's responsibility to conduct the voting, to communicate with the judge regarding requests for clarification or additional instructions, and to read the verdict once it has been made. Studies report that not all members of the jury are equally likely to be selected as foreperson or to exert equal influence in deliberations. Typically, better educated and more affluent males with higher socioeconomic status are more likely to be chosen as leaders and to take a more dominant role in the debate and negotiation process.[62] Just as not all jurors are equally likely to be selected as foreperson, neither are all jurors equally involved in jury deliberations. Studies reveal that three jurors on average are responsible for approximately 50 percent of the discussion by the group, jurors seated at the head or end of the table provide about one-third of the discussions, and males, jurors with more education and higher-status occupations, and jury forepersons all contribute significantly more to the deliberations.[63]

Judgment by an impartial jury is a cornerstone of U.S. law and criminal justice.

 Should jurors be shielded from the media during deliberations? Is common sense ever sacrificed because juries are protected from public scrutiny?

The Judge's Instructions to the Jury

After reading a statement of the formal charges against the defendant, the judge defines the charge(s) according to state (or federal) statute and indicates the elements that must be proven to the jury. The judge then delivers instructions regarding the finding of guilt or innocence.

The following instructions were given to a jury regarding the issues of reasonable doubt and weighing of evidence in a felony drug case involving the illegal distribution of cocaine. While university students may find these instructions easy to understand, they are likely to be difficult for many average citizens serving on juries to fully comprehend. According to the Flesch-Kincaid Grade Level scale, the instructions are written at the twelfth-grade level.

If you are convinced beyond a reasonable doubt that the defendant is guilty of the offense charged, or one of the lesser included offenses, but have a reasonable doubt as to which of such offenses he is guilty, then it is your duty to resolve such doubt in the defendant's favor, and you can only convict him of the least serious offense.

The defendant is presumed innocent until proven guilty beyond a reasonable doubt, and this presumption prevails until the conclusion of the trial, and you should weigh the evidence in the light of this presumption of innocence and it should be your endeavor to reconcile all the evidence with this presumption of innocence if you can, but if this cannot be done, and the evidence so strongly tends to establish the guilt of the defendant as to remove all reasonable doubt of the guilt of such defendant from the mind of each juror, then it is the duty of the jury to convict. A "reasonable doubt" is a fair, actual, and logical doubt that arises in your mind after an impartial consideration of all of the evidence and circumstances in the case. It should be doubt based upon reason and common sense and not a doubt based upon imagination or speculation.

If, after considering all of the evidence, you have reached such a firm belief in the guilt of the defendant that you would feel safe to act upon that conviction, without hesitation, in a matter of the highest concern and importance to you, when you are not required to act at all, then you will have reached that degree of certainty which excludes reasonable doubt and authorizes conviction.

If, after careful consideration of all the evidence in this case, you are left with two different theories: one consistent with the defendant's innocence, and the other with his guilt, both reasonable; and you are not able to choose between the two, you must find the defendant not guilty.

You are the exclusive judges of the evidence, the credibility of the witnesses and of the weight to be given to the testimony of each of them. In considering the testimony of any witness, you may take into account his or her ability and opportunity to observe; the manner and conduct of the witness while testifying; any interest, bias, or prejudice the witness may have; any relationship with other witnesses or interested parties; and the reasonableness of the testimony of the witness considered in the light of all the evidence in the case.

In considering this case you will no doubt meet with conflicts in the evidence. It is a matter of common knowledge that witnesses to an event rarely see or hear all the circumstances alike. Whenever you meet with such conflicts, reconcile them, if you can, on the assumption that each witness has testified truthfully. If you cannot so reconcile the conflicts, it is for you to determine what you will believe and what you will not believe.

Sources: Extracted from Juror Instruction Sheet, Muncie, Indiana, 2006. Robert Flesch, "A new readability yardstick," *Journal of Applied Psychology* 32:3:221–233 (1948).

Generally, before the members of the jury begin deliberating, they take a straw vote (an unofficial vote) to get a sense of how each person feels about the case. With few exceptions, the initial vote is indicative of the final vote.[64] Mock jury experiments confirm these findings, showing that jurors reversed the vote only 6 percent of the time.[65]

Hung Jury

Because a jury must reach a unanimous verdict, if the jury is split (even if it is 11 to 1) and no verdict can be reached, it is a <u>hung jury</u>. Hung juries do not occur frequently; the National Center for State Courts reports the average hung jury rate in state courts is only 6.2 percent (although it ranges from less than 1 percent to nearly 15 percent, depending on jurisdiction). The average hung jury rate in the federal courts is much lower, averaging only 2.5 percent.[66] There is often disagreement among jurors at the early stages of deliberations. Indeed, it is not unusual for the judge to call the jury back to the courtroom if deliberations have taken more than a few days (or longer in complex cases) to ask whether the foreperson believes that a verdict can be reached. However, the judge may not intervene to push the jury for a verdict. Because a hung jury is considered by law to be a legitimate outcome in a trial, any attempt to put undue pressure on the jury to reach a verdict is viewed as inappropriate. It is considered coercive and improper for judges to emphasize the expense and inconvenience that will result from a retrial or to require a jury to continue deliberations for more than 24 hours without sleep.[67]

If only a few jurors are dissenting from the majority's position, the judge may instruct the jury to continue deliberations. If the jury again reports that it cannot reach a verdict and that further deliberation would be futile, the judge may once more instruct them to continue. Generally speaking, the longer a judge keeps the jury deliberating, the more likely it becomes that the jury will reach a verdict.[68]

In some jurisdictions, the judge may give the jury so-called Allen instructions, which are designed to push them to reach a verdict. Such instructions are based on the 1896 Supreme Court decision in *Allen v. United States*.[69] The judge may lecture dissenting jurors about the importance of listening to other jurors' opinions and considering whether the doubt in his or her own mind is reasonable. The purpose of such instructions is to encourage a compromise that will allow the jury to arrive at a verdict.

Research examining the issue of hung juries has looked at both state and federal courts and found the following to be among the most important factors resulting in a hung jury:

- Weak or ambiguous evidence
- Evidence or nuances of law that are difficult for jurors to understand
- Lack of thorough discussion of evidence by jurors
- Cases involving many charges (with many charges, the jury is more likely to be hung on at least one charge)
- Conflict among jurors
- Domination of the jury by one or two jurors
- A large presence of "unreasonable" people on the jury[70]

Judgment or Verdict

The federal courts and most states require unanimous verdicts. Two states allow for nonunanimous decisions in non-death-penalty felony cases: Louisiana allows convictions to be based on agreement of 10 of 12 jurors for less serious felonies, and Oregon accepts an agreement of 10 of 12 furors in all felony cases except for murder. The U.S. Supreme Court has held that such verdicts are constitutional.[71]

Upon arriving at a verdict, the jury returns to the courtroom. The jury foreperson gives the signed verdict to the bailiff, who then gives it to the judge. The judge glances at the verdict and gives it to the clerk, who reads it to the court. Next, the judge asks the prosecutor or the defense attorney whether they would like the jury polled; if requested, the judge asks each juror to state his or her vote on the verdict. Sometimes jurors feel pressured to go along with the majority, even though they truly believe the defendant to be innocent. If a juror states that he or she did not really agree with the verdict, the judge will send the jury back to the jury room to deliberate again in an attempt to reach a unanimous verdict.

Appeal of the Verdict

If the verdict is "not guilty," the defendant is acquitted and released. The state may not appeal a verdict of not guilty because the Fifth Amendment guarantees that no person "shall be subject for the same offense to be twice put in jeopardy of life or limb." This prohibition against double jeopardy means that once a person has been tried and acquitted of a criminal charge, he or she may not be recharged and retried for the same offense. By contrast, if the verdict is "guilty," the defendant has the right to appeal and may or may not be released on bail while awaiting the outcome of that appeal.

An appeal of the verdict is a request to the state appellate court to correct mistakes or injustices that occurred in the trial process, such as a judge's error in permitting certain evidence to be introduced, misconduct by the prosecutor, or jury tampering. Even though all 50 states provide for an appeal process, the question of the right of a defendant to appeal is not clear. In 1984, in *McKane v. Durston*, the U.S. Supreme Court held that states were not required to provide a right to an appellate review or even access to the appellate courts.[72] Although not required to do so, the courts have allowed criminal defendants this right as a matter of legal tradition.

The U.S. Supreme Court has held that indigent defendants have the right to be represented by counsel on appeal.[73] This right is limited to the defendant's first appeal only. The Court has also held that a defendant loses the right to counsel in a first appeal if he or she delays in filing the appeal with the court.[74] Until 2005, Michigan prohibited the appointment of appellate counsel to indigent defendants who pleaded *nolo contendere* (with very few exceptions), but the Supreme Court ruled in *Halbert v. Michigan* that this law was unconstitutional.[75]

The issues raised in an appeal must be based on objections raised by the defense in pretrial motions or at the time of the trial, such as a motion for a change of venue or a motion to suppress evidence. With two exceptions, issues not raised in pretrial motions or during the trial may not be considered. Issues that are the result of plain error and those that affect substantial rights of the defendant may be appealed—for example, a claim that the court lacked legal jurisdiction to hear the case is grounds for an appeal. An appeal may also be based on the claim that the defendant's attorney was incompetent.

Appeals are sometimes rejected because they are based on harmless error—an error, defect, irregularity, or variance that does not affect substantial rights of the defendant. Harmless errors include the following:

- Technical errors having no bearing on the outcome of the trial
- Errors corrected during trial (for example, when testimony was allowed but then ordered stricken and the jury admonished to ignore it)
- Errors resulting in a ruling in the appellant's favor

- Situations in which the appeals court believes that even without the error, the appealing party would not have won at trial

In 1967, *Chapman v. California,* the Supreme Court held that defendants were not necessarily entitled to a new trial even if certain constitutional violations had occurred at trial.[76] According to this ruling, appellate courts should dismiss arguments about particular constitutional errors when these were "so unimportant and insignificant that they may, consistent with the Federal Constitution, be deemed harmless, not requiring automatic reversal of a conviction." The Supreme Court, however, stated that the appellate court must be certain beyond a reasonable doubt that the error did not affect the outcome of the case.

Very few appeals result in an overturned conviction; most produce only minor modifications, leaving the conviction itself undisturbed.[77] Nevertheless, if the appellate court finds that a significant error did occur, it may reverse the conviction, thus setting aside the guilty verdict and sending the case back to the trial court. If the conviction is reversed, the prosecutor may appeal the decision of the appellate court.

Contrary to widely held perceptions, few criminals are eventually set free as a result of errors in the original trial. Approximately half of all offenders who are retried after appeal are convicted again.[78]

Other Post-Conviction Remedies

Once a defendant has exhausted all appeals through the state appellate courts and is incarcerated, he or she may still seek post-conviction relief by filing a petition for a writ of habeas corpus with the Federal Courts of Appeals or directly with the U.S. Supreme Court. A petition of habeas corpus asks the federal court to release the defendant from an alleged illegal imprisonment or confinement by the state.

Habeas corpus petitions differ from appeals in several ways. For example, they may be filed only by people who are actually confined, and they must raise constitutional issues rather than issues of error. They may be broader in scope than appeals; for example, an inmate may claim that the current conditions in his or her prison constitute cruel and unusual punishment. Since the 1960s, courts have granted writs of habeas corpus in several instances:

- To release defendants on bail when the bail amount was considered excessive

- To release inmates from prison when the sentence was considered excessive
- To overturn capital punishment sentences
- To release inmates who claimed their attorneys did not provide competent counsel

In 1963, in *Fay v. Noia,* the Supreme Court required federal courts to consider habeas corpus petitions even from inmates who had failed for some reason to appeal their case properly in the state courts, as long as the inmate had not deliberately bypassed the state's appeal process and the allegation of newly discovered evidence was not irrelevant, frivolous, or incredible.[79] By the early 1990s, with nearly 10,000 habeas corpus petitions being filed annually and the justices having a decidedly more conservative bent, the Supreme Court placed severe restrictions on the conditions under which federal courts would hear such petitions. Today, only a small fraction of petitions are granted. Most petitions are viewed as frivolous, and federal judges dismiss the majority of them without a hearing.

One of the first attempts to reduce the burden placed on federal courts by habeas corpus petitions came in 1991, in *Coleman v. Thompson.*[80] In this case, the defendant's lawyer filed the habeas corpus petition with the Virginia Supreme Court three days after the 30-day time limit had expired. The Virginia court held that, although there was "no doubt an inadvertent error . . . the petitioner must bear the risk of attorney error" and barred any further review in the federal courts. The next year, in *Keeney v. Tamayo-Reyes,* the U.S. Supreme Court ruled that the federal courts are not obligated to grant a hearing on a state prisoner's challenge to his or her conviction, even if the prisoner can show that the defense attorney did not properly present crucial facts in a state-court appeal.[81] In 1993, in *Herrera v. Collins,* the U.S. Supreme Court ruled that a Texas court's rejection of a defendant's late claim of innocence based on new evidence did not necessarily violate the petitioner's right to due process.[82] Texas, as well as several other states, requires that such a claim be filed within 30 days; other states allow up to a few years for such an appeal, and nine states have no time limit for filings based on new evidence.

The implications of these decisions are far-reaching, affecting the lives of inmates who may have been incorrectly convicted. At the same time that the Supreme Court has tightened the reins on the numerous frivolous petitions from prisoners, it has also significantly reduced the rights of prisoners to post-conviction relief from judicial errors and wrongful imprisonment.

VOICE OF THE COURT

Herrera v. Collins

In early 1993, the U.S. Supreme Court ruled that a death-row inmate whose case had been appealed to the Court based on a belated claim of innocence after having exhausted his appeals on procedural grounds was not entitled to a new hearing. Leonel Herrera was convicted of murdering two police officers and was sentenced to death in January 1982. After an unsuccessful challenge of his conviction on direct appeal in the Texas state courts, as well as an unsuccessful habeas corpus petition to the Supreme Court, Herrera filed a second writ of habeas corpus with the Court, claiming that he was innocent of the murder for which he was sentenced to death and that he could provide testimony establishing that his broker (who had since died) was actually the killer.

Texas law requires that a motion for a new trial based on undiscovered evidence must be filed within 30 days after a conviction. Herrera's new evidence was uncovered 10 years after his conviction. Herrera claimed that the 30-day limit was fundamentally unfair, violating the Eighth Amendment's prohibition against cruel and unusual punishment and the Fourteenth Amendment's guarantee of due process of law. In a 6-3 decision, the Supreme Court disagreed. Chief Justice William Rehnquist wrote the majority opinion:

> Once a defendant has been afforded a fair trial and convicted of the offense for which he was charged, the presumption of innocence disappears. . . .

Here, it is not disputed that the State met its burden of proving at trial that petitioner was guilty of the capital murder . . . beyond a reasonable doubt. Thus, in the eyes of the law, petitioner does not come before the Court as one who is innocent, but on the contrary as one who has been convicted by due process of law of two brutal murders.

Claims of actual innocence based on newly discovered evidence have never been held to state a ground for Federal habeas corpus relief absent an independent constitutional violation occurring in the underlying state criminal proceeding. . . . This rule is grounded in the principle that Federal habeas courts sit to ensure that individuals are not imprisoned in violation of the Constitution, not to correct errors of fact.

The Court added that Herrera was not left without a forum in which to raise his claim of innocence, noting that under Texas law, Herrera could file a request for executive clemency (that is, a motion asking the governor to commute the death sentence to life imprisonment). According to Justice Rehnquist, "[C]lemency is deeply rooted in our Anglo-American tradition of law and is the historic remedy for preventing miscarriages of justice where judicial process has been exhausted." No clemency was granted, and Leonel Herrera was executed on May 12, 1993.

Source: Herrera v. Collins, 506 U.S. 390 (1993).

WRAP UP

The use of rap lyrics as evidence in a criminal trial is controversial and could lead the defense to file a motion to suppress. Given that the prosecution must prove the defendant's guilt beyond a reasonable doubt, rap lyrics would have great probative value if they documented details of the crimes that only the perpetrator would have known. Because the character of the defendant is an implicit part of every trial, presenting graphic, violent, and crude rap lyrics helps to portray the defendant in a negative light. For the defense, the rap lyrics would be considered prejudicial for the same reasons. Defense attorneys would claim that rap lyrics represent nothing more than the creative writings of a defendant and merely serve to inflame the decent sensibilities of the jury. For both prosecution and defense, rap lyrics symbolize who the defendant is—graphic thug or creative poet.

Chapter Spotlight

- Few events draw more public attention than a criminal trial. As an alternative to the less public plea bargaining method of obtaining justice, the trial epitomizes the adversarial process in which the prosecution and the defense present evidence and arguments in their attempts to convince a jury of their side of a case.

- The prosecutor and the defense attorney typically submit pretrial motions to the judge before the beginning of the trial. Some of these motions must be decided before the start of the trial, but the judge may choose to rule on other motions, such as those to suppress evidence, later during the trial when the evidence is actually submitted.

- The Sixth Amendment guarantees defendants the right to a speedy, public, and fair trial. State time guidelines differ widely, but the federal Speedy Trial Act of 1974 requires that charges be filed within 30 days of a suspect's arrest and that the trial begin within 70 days of the filing of charges. Although trials must be open to the public, just how open and at which stages a trial must be open have been the subject of much debate and various rulings by the Supreme Court.

- Approximately 40 percent of all cases that go to trial are tried before juries. The remaining 60 percent involve bench trials.

- Defendants charged with either a felony or a misdemeanor are entitled to a trial by jury. Most states, as well as the federal government, use juries composed of 12 people, although juries with as few as six members have been deemed constitutional except in death penalty cases. Jury members are drawn from lists of local citizens.

- The prosecutor and the defense attorney ask questions during the *voir dire* to determine which people will make the best jurors.

- During the trial itself, the prosecution and the defense alternate in presenting evidence, questioning the evidence of the opposing side, and then submitting the evidence to the jury. After both sides present their closing arguments, the judge gives his or her charge (instructions) to the jury.

- After receiving its charge, the jury retires to the jury room to begin deliberations. The federal courts and most states require unanimous verdicts, although the Supreme Court has held that nonunanimous verdicts are constitutional in non-death-penalty cases. A jury that cannot arrive at a verdict is known as a hung jury.

- If the verdict is not guilty, the defendant is acquitted and released and may not be retried on the same charge (double jeopardy). The defendant may appeal a guilty verdict to an appropriate appellate court based on objections raised by the defense in pretrial motions or at trial.

- Defendants who have exhausted the appeal process and are incarcerated may still seek post-conviction relief by filing a writ of habeas corpus. The Supreme Court has restricted the grounds on which such petitions may be filed.

Putting It All Together

1. Should the defense or the prosecutor be limited in the number of continuances requested? Why?
2. Should the public and the press be allowed to attend criminal trials? Why or why not?
3. How much control should the court exercise over the media in reporting of trials?
4. What are the benefits of peremptory challenges of potential jurors? Why do some critics argue that they should be abolished?
5. Should all jury verdicts be unanimous? Why or why not?

Key Terms

alibi An assertion that the defendant was somewhere else at the time the crime was committed.

bench trial A trial before a judge alone, as an alternative to a jury trial.

beyond a reasonable doubt The requirement that the jury (or the judge in the case of a bench trial) must find the evidence entirely convincing and must be satisfied beyond a moral certainty of the defendant's guilt before returning a conviction.

bill of particulars A written statement from the prosecutor revealing the details of the charge(s), including the time, place, manner, and means of commission.

challenge for cause A challenge by the prosecutor or the defense to dismiss a person from a jury panel for a legitimate cause.

charge to the jury The judge's instructions to the jury, which are intended to guide their deliberations.

circumstantial evidence Testimony by a witness that requires jurors to draw a reasonable inference.

closing arguments The final presentation of arguments to the jury.

cross-examination Questioning of a witness by counsel after questions have been asked by the opposing counsel.

determination of competency A determination as to whether the defendant lacks the capacity to understand the charge or possible penalties if convicted, assist or confer with counsel, or understand the nature of the court proceedings.

direct evidence Testimony by an eyewitness to the crime.

habeas corpus A judicial order to bring a person immediately before the court to determine the legality of his or her detention.

harmless error An error, defect, irregularity, or variance that does not affect substantial rights of the defendant.

hearsay evidence Testimony involving information the witness was told but has no direct knowledge of.

hung jury A jury that is deadlocked and cannot reach a verdict. As a result, the judge may declare a mistrial.

judgment of acquittal A defense motion for dismissal of a case based on the claim that the prosecution failed to establish that a crime was committed or that the defendant committed it.

jury nullification The right of a jury to interpret and negate the law in a case.

jury pool The master list of community members who are eligible to be called for jury duty.

mistrial An invalid trial due to some unusual event, such as the death of a juror or an attorney, a prejudicial error, or inability of the jury to reach a verdict.

opening statement The initial presentation of the outline of the prosecution's and the defense's cases to the jury.

peremptory challenge A challenge by the defense or the prosecution to excuse a person from a jury panel without having to give a reason.

prejudicial error Inflammatory or biasing statements made by an attorney to the jury.

presumption of innocence The notion that a person is presumed to be innocent unless proved guilty beyond a reasonable doubt.

pretrial motion A written or oral request to a judge for a ruling or action before the beginning of trial.

real evidence Physical evidence introduced at the trial.

rebuttal The presentation of additional witnesses and evidence by the prosecutor in response to issues raised in the defense's presentation of witnesses.

severance of charges A motion requesting a separate trial for each charge.

Speedy Trial Act of 1974 Act of Congress requiring that a federal trial must begin within 70 days of the filing of charges or the defendant's initial appearance.

surrebuttal Questioning by the defense of witnesses who were presented by the prosecutor during rebuttal.

testimonial evidence Sworn testimony of witnesses who are qualified to speak about specific real evidence.

venire A group of people who are selected from the jury pool and notified to report for jury duty.

voir dire Preliminary examination by the prosecution and defense of potential jurors.

Notes

1. Donna Burchfield, "Appearance v. Reality: 'L.A. Law,'" *Phi Kappa Phi Journal* (Fall 1991), p. 20.
2. *Downum v. United States*, 372 U.S. 734 (1963); *Serfass v. United States*, 420 U.S. 377 (1975).
3. *Federal Rules of Criminal Procedure* (Washington, DC: U.S. Government Printing Office, 2005).
4. *Dusky v. United States*, 362 U.S. 402 (1960).
5. Thomas Church, Jr., *Justice Delayed: The Pace of Litigation in Urban Trial Courts* (Williamsburg, VA: National Center for State Courts, 1978); Barry Mahoney, *Changing Times in Trial Courts* (Williamsburg, VA: National Center for State Courts, 1988); Bureau of Justice Statistics, *Felony Sentences in State Courts, 2002* (Washington, DC: U.S. Department of Justice, 2004).
6. Church, note 5, pp. 31–36.
7. *Barker v. Wingo*, 407 U.S. 514 (1972).
8. *Federal Rules of Criminal Procedure*, note 3.
9. *In re Oliver*, 333 U.S. 257 (1948).
10. *Estes v. Texas*, 381 U.S. 532 (1965).
11. *Gannett Co., Inc. v. DePasquale*, 443 U.S. 368 (1979).
12. *Press Enterprise Co. v. Superior Court of California*, 478 U.S. 1 (1986).
13. *Richmond Newspapers, Inc. v. Virginia*, 448 U.S. 555 (1980).
14. *Richmond Newspapers, Inc. v. Virginia*, note 13.
15. *Globe Newspapers Co. v. Superior Court for the County of Norfolk*, 457 U.S. 596 (1982).
16. *Nebraska Press Association v. Stuart*, 427 U.S. 539 (1976).
17. *In re Dow Jones & Co., Inc.*, 842 F.2d 603 (2nd Cir. 1988); *In re Russell*, 726 F.2d 1007 (4th Cir. 1984); *U.S. v. Brown*, 218 F.3d 415 (5th Cir. 2000); *U.S. v. Tijerina*, 412 F.2d 661 (10th Cir. 1969).
18. *Bailey v. Systems Innovation, Inc.*, 852 F.2d 93 (3rd Cir. 1988); *Chicago Council of Lawyers v. Bauer*, 552 F.2d 242 (7th Cir. 1975; *Levine v. U.S. Dist. Ct.*, 764 F.2d 590 (9th Cir. 1985); *U.S. v. Ford*, 830 F.2d 596 (6th Cir. 1987).
19. Edith Greene, "Media Effects on Jurors," *Law and Human Behavior* 14:442 (1990).
20. *Chandler v. Florida*, 449 U.S. 560 (1981).
21. *Courtroom Tel. Network LLC v. State of New York*, NYSlipOp 05386 (2004).
22. Brian Ostrom, Shauna Strickland, and Paula Hannaford-Agor, *Court Statistics Project* (Williamsburg, VA: National Center for State Courts, 2004).
23. Orville Richardson, "Jury or Bench Trial? Considerations," *Trial* 19:58–63 (1983); Harry Kalven and Hans Zeisel, *The American Jury* (Boston: Little, Brown, 1966), pp. 56–60.
24. Uzi Segal and Alex Stein, "Ambiguity Aversion and the Criminal Process, *Notre Dame Law Review* 81:1–68 (2006); Kalven and Zeisel, note 23, pp. 56–60.
25. Andrew Leipold, "Why Are Federal Judges So Acquittal Prone?" *Washington University Law Quarterly* 83:151–177 (2005).
26. Bureau of Justice Statistics, *State Court Sentencing of Convicted Felons, 1996* (Washington, DC: U.S. Department of Justice, 2000).
27. John Bunyan, *Pilgrim's Progress* (Philadelphia: Lippincott, 1939).
28. Valerie Hans and Neil Vidmar, *Judging the Jury* (New York: Plenum, 1986), p. 109.
29. *Snyder v. Massachusetts*, 291 U.S. 97 (1934).
30. *Palko v. Connecticut*, 302 U.S. 319 (1937).
31. *Palko v. Connecticut*, note 30.
32. *Duncan v. Louisiana*, 391 U.S. 145 (1968).
33. *Blanton v. North Las Vegas*, 489 U.S. 538 (1989).
34. *McKeiver v. Pennsylvania*, 403 U.S. 528 (1971).
35. *Williams v. Florida*, 399 U.S. 78 (1970).
36. *Ballew v. Georgia*, 435 U.S. 223 (1978).
37. James Levine, *Juries and Politics* (Pacific Grove, CA: Brooks/Cole, 1992), p. 43.
38. Robert Boatright, "Why Citizens Don't Respond to Jury Summonses and What Courts Can Do about It," *Judicature* 82:156–163 (1999); Nancy King, "Juror Delinquency in Criminal Trials in America, 1796–1996," *Michigan Law Review* 94:2673–2715 (1996); Levine, note 37, p. 44.
39. *Lockhart v. McCree*, 476 U.S. 162 (1986).
40. *Witherspoon v. Illinois*, 391 U.S. 510 (1968).
41. Amanda L. Kutz, "A Jury of One's Peers: Virginia's Restoration of Rights Process and Its Disproportionate Effect on the African American Community," *William and Mary Law Review*

46:2109–2162 (2005); Marvin Steinberg, "The Case for Eliminating Peremptory Challenges," *Criminal Law Bulletin* 27:216–229 (1991).

42. *Fay v. New York*, 332 U.S. 261 (1947).

43. *Batson v. Kentucky*, 476 U.S. 79 (1986).

44. *Hernandez v. New York*, 500 U.S. 352 (1991).

45. *Georgia v. McCollum*, 505 U.S. 42 (1992).

46. *Georgia v. McCollum*, note 45.

47. *Holland v. Illinois*, 493 U.S. 474 (1990).

48. *Powers v. Ohio*, 499 U.S. 400 (1991).

49. Thomas Scheffey, "Connecticut Outlaws Religion-Based Juror Challenges," *Connecticut Law Tribune*, April 5, 1999; *Casarez v. State*, 913 S.W.2d 468 (Tex. Crim. App. 1995).

50. Kelly Kuljol, "Where Did Florida Go Wrong? Why Religion-Based Peremptory Challenges Withstand Constitutional Scrutiny," *Stetson Law Review* 32:171–203 (2002).

51. Jay Schulman, Phillip Shaver, Robert Coleman, Barbara Emrich, and Richard Christie, "Recipe for a Jury," *Psychology Today* 37:37–44 (1973); Richard Christie, "Probability v. Precedence: The Social Psychology of Jury Selection," in G. Bermant, C. Nemeth, and N. Vidmar (eds.), *Psychology and the Law* (Lexington, MA: Lexington Books, 1976); Robert Buckhout, *Studies in Systematic Jury Selection: An Inside View* (Brooklyn, NY: Center for Responsive Psychology, 1978).

52. Shari Diamond, "Scientific Jury Selection: What Social Scientists Know and Do Not Know," *Judicature* 74:178–183 (1990).

53. Tom Lininger, "Bearing the Cross," *Fordham Law Review* 74:1353–1423 (2005); Lorraine Dusky, "Bryant Lawyer Is Using Dirty Tactics," *Newsday*, March 2, 2004, p. A51; J. Tanford and A. Bocchino, "Rape Victim Shield Laws and the Sixth Amendment," *University of Pennsylvania Law Review* 128:544–602 (1980).

54. *Griffin v. California*, 380 U.S. 609 (1965).

55. Amiram Elwork, Bruce Sales, and James Alfini, "Juridic Decisions: In Ignorance of the Law or in Light of It?" *Law and Human Behavior* 1:163–178 (1977); R. Reid Hastie, Steven Penrod, and Nancy Pennington, *Inside the Jury* (Clark, NJ: Lawbook Exchange, 2002); David Strawn and Raymond Buchanan, "Jury Confusion: A Threat to Justice," *Judicature* 59:478–483 (1976); Laurence Severance, Edith Greene, and Elizabeth Loftus, "Toward Criminal Jury Instructions That Jurors Can Understand," *Journal of Criminal Law and Criminology* 75:198–233 (1984).

56. Amiram Elwork, Bruce Sales, and James Alfini, *Making Jury Instructions Understandable* (Charlottesville, VA: Michie, 1982); Walter Steele, Jr., and Elizabeth Thornburg, "Jury Instructions: A Persistent Failure to Communicate," *Judicature* 74:249–254 (1991).

57. Bethany Dumas, "Jury Trials: Lay Jurors, Pattern Jury Instructions, and Comprehension Issues," *Tennessee Law Review* 67:701–742 (2000).

58. Lynn Holton, "New Plain-English Jury Instructions Adopted to Assist Jurors in California Courts," *Judicial Council of California News* 42:1–2 (2003); Joseph Kimble, "The Route to Clear Jury Instructions," *Scribes Journal of Legal Writing* 6:163–165 (1997).

59. Otis Grant, "Rational Choice or Wrongful Discrimination: The Law and Economics of Jury Nullification," *George Mason University Civil Rights Law Journal* 14:145–187 (2004).

60. David Brody, "*Sparf* and *Dougherty* Revisited: Why the Court Should Instruct the Jury of Its Nullification Right," *American Criminal Law Review* 33:89–122 (1995).

61. *In re Winship*, 397 U.S. 358 (1970).

62. Fred Strodtbeck, Rita James, and C. Hawkins, "Social Status in Jury Deliberations," *American Sociological Review* 22:713–718 (1957); Michael Saks, *Jury Verdicts: The Role of Group Size and Social Decision Rule* (Lexington, MA: D.C. Heath, 1977); Fred Strodtbeck and Richard Lipinski, "Becoming First Among Equals: Moral Considerations in Jury Foreman Selection," *Journal of Personality and Social Psychology* 49:927–936 (1985).

63. Hastie et al., note 55; Penny Darbyshire, Andy Maughan, and Angus Stewart, "What Can the English Legal System Learn From Jury Research Published up to 2001?", April 14, 2006, available at http://www.kingston. ac.uk/~ku00596/elsres01.pdf, accessed July 17, 2007.

64. Kalven and Zeisel, note 23, p. 488.

65. Sarah Tanford and Steven Penrod, "Jury Deliberations: Discussion Content and Influence Processes in Jury Decision Making," *Journal of Applied Social Psychology* 16:322–347 (1986).

66. Paula Hannaford-Agor, Valerie Hans, Nicole Mott, and Thomas Munsterman, *Are Hung Juries a Problem?* (Williamsburg, VA: National Center for State Courts, 2002).

67. Bennett Gershman, "Judicial Misconduct During Jury Deliberations," *Criminal Law Bulletin* 27:291–314 (1991).

68. Levine, note 37, p. 164.

69. *Allen v. United States*, 164 U.S. 492 (1896).

70. Hannaford-Agor et al., note 66.

71. *Johnson v. Louisiana*, 406 U.S. 356 (1972); *Apodaca v. Oregon*, 406 U.S. 404 (1972).

72. *McKane v. Durston*, 153 U.S. 684 (1894).

73. *Douglas v. California*, 372 U.S. 353 (1963).

74. *United States v. MacCollom*, 426 U.S. 317 (1976).

75. *Halbert v. Michigan*, 545 U.S. 605 (2005).

76. *Chapman v. California*, 386 U.S. 18 (1967).

77. Joy Chapper and Roger Hanson, "Understanding Reversible Error in Criminal Appeals," *State Court Journal* 14:16–18, 24 (1990); David Neubauer, "Winners and Losers before the Louisiana Supreme Court: The Case of Criminal Appeals," *Justice Quarterly* 8:85–106 (1991).

78. Robert Roper and Albert Melone, "Does Procedural Due Process Make a Difference? A Study of Second Trials," *Judicature* 65:136–141 (1981).

79. *Fay v. Noia*, 372 U.S. 391 (1963).

80. *Coleman v. Thompson*, 501 U.S. 722 (1991).

81. *Keeney v. Tamayo-Reyes*, 504 U.S. 1 (1992).

82. *Herrera v. Collins*, 506 U.S. 390 (1993).

Sentencing

In December 2005, Sharen and Michael Gravelle were accused of forcing some of their 11 adopted special-needs children to sleep in chicken-wire cages. Both members of the Ohio couple were convicted of four felony counts of child endangerment, two misdemeanor counts of child endangerment, and five misdemeanor counts of child abuse, and both were acquitted of 13 other charges. The defendants were sentenced to two years in prison on each of the four felony counts.

During the three-week trial, the adoptive parents argued that the children needed to be caged for their own protection. Prosecutors suggested the couple wanted so many special-needs children for financial reasons—namely, to take advantage of the adoption and foster care subsidies that accompanied them. The sheriff's deputies who made the arrest testified that the cages were stained with urine and lacked pillows and bedding. The children testified to various abuses at the hands of the parents. They were removed from the Gravelles' home and placed in foster care.

- What sentence would you recommend to achieve the following goals: rehabilitation, deterrence, retribution, incapacitation, and proportionality?
- How would you determine which punishment goal is most appropriate?

Source: M. R. Kropko, "Caged kids' parents still hoping to get children back," July 11, 2007, available at http://news.aol.com/story/_a/caged_kids_parents_still_hoping_to_get/N2007071120120999003, accessed July 17, 2007.

Introduction

A number of theories have been put forth regarding ways to deal with offenders and the goals of sentencing. According to contemporary American political scientist James Q. Wilson, "Wicked people exist. Nothing avails except to set them apart from innocent people."[1] If he is right, then the best way to deal with people who are convicted of serious crimes may be to set them apart from "good" people through imprisonment. Wilson's position represents a commonly held belief: Criminals are basically bad people who must be held accountable for their crimes and face incapacitation (incarceration) to prevent them from committing future crimes. If the rate of imprisonment is viewed as an indicator of the amount of punishment in a society, the United States certainly punishes a substantial portion of its population every year. Even so, crime remains widespread in this country.

An alternative view, which was held by seventeenth-century English philosopher Thomas Hobbes, suggests that society should punish offenders solely to achieve future goodness in people. Hobbes argued that "in revenges and punishments men ought not to look at the greatness of evil past, but at the greatness of the good to follow."[2] That is, sentencing should be imposed solely to deter offenders (past, present, and future) from committing crimes.

LINK Thomas Hobbes's ideas helped to lay the groundwork for the classical school of criminology, discussed in Chapter 4, which held that people choose their behavior.

The Goals of Sentencing

The history of sentencing reflects a variety of reasons for imposing sanctions, some of which are contradictory:

- Retribution
- Deterrence
- Incapacitation
- Rehabilitation
- Proportionality

These goals are not mutually exclusive; indeed, they often overlap. For example, the wish to express moral outrage over a crime and extract a degree of revenge (retribution) by sentencing the offender to life in prison may correspond with the desire that

LINK Chapter 15 discusses the concept of restorative justice, a goal of sentencing that focuses on restoring or repairing social balance and those relationships disrupted by crime. This method of sentencing holds offenders accountable by requiring restitution to victims, promoting offender competency and responsibility, and balancing the needs of the community, the victim, and the offender through involvement in the restoration process.

Gary Ridgway, the most prolific serial killer in U.S. history, pled guilty to 48 counts of first-degree murder.

? Does the concept of evil have criminological relevance today? Why do most criminologists not embrace evil as an explanation for crime?

the harsh sentence will be viewed by others as a warning not to commit similar crimes (deterrence).

Retribution

Retribution reflects society's moral outrage or disapproval of a crime. It is a moral statement that the act was fundamentally wrong and must be punished. Here the focus is on the crime rather than on the individual who committed it; there is virtually no concern with deterring future crime or changing the individual through punishment or rehabilitation. Furthermore, retribution calls for sentences to be proportionate to the crime and thus is designed to balance the harm caused by the crime. According to criminologists Joel Feinberg and Hyman Gross, "It is morally fitting that a person who does wrong should suffer in proportion to his wrongdoing. That a criminal should be punished follows from his guilt, and

the severity of the appropriate punishment depends on the depravity of the act."[3]

The idea of a proportionate punishment is probably the oldest rationale in the history of criminal sanctions. The Bible, for example, proclaims "an eye for an eye."[4] Early retributionists, such as eighteenth-century philosopher Immanuel Kant, believed that retribution should include the principle of *lex talionis* (the law of retaliation). In other words, "the punishment should fit the crime"—not only being in proportion to the seriousness of the crime, but also reflecting the nature of the crime (such as fines for theft, beatings for assault, and castration for rape). Today, many people believe that criminals should "get what they asked for" in the form of retributive punishment.

Under this theory of sentencing, the exact proportion and manner of punishment are very problematic. Criminologists Harold Pepinsky and Paul Jesilow argue that it is difficult to determine how many years of a person's life in prison are equal to the value of a lost television set or whether the theft of a car from a wealthy person should be treated in the same way as the theft of a car from someone who is poor.[5]

Contemporary retributionists, such as Andrew von Hirsch, see retribution as not only a just punishment for the offender but also a means of deterring those who might consider committing criminal acts. Through punishment, offenders receive their just desserts, and potential offenders are made aware of the consequences of crime. As a result, the total misery produced by punishing the few who commit crimes is outweighed by the total greater reduction in misery that would be created by the failure to punish. According to von Hirsch, through retributive punishment, "Fewer innocent persons will be victimized by crimes, while those less deserving—the victimizers—will be made to suffer instead."[6]

Deterrence

The writings of utilitarian philosophers in Europe during the seventeenth and eighteenth centuries gave rise to deterrence theories. These philosophers believed that people are rational beings with free will who prefer pleasure over pain; therefore, people weigh the benefits and costs of future actions before deciding to act. Utilitarians such as Cesare Beccaria and Jeremy Bentham believed that people chose crime when they believed that the benefits resulting from the criminal behavior would exceed its costs.[7] This

They May Not Be Cruel, but Are They Unusual Punishments?

Although the Eighth Amendment prohibits cruel and unusual punishment, the Supreme Court has yet to rule on what constitutes an unusual punishment. As local judges search for effective means to punish offenders and to achieve a degree of retribution, they sometimes arrive at sentences many people might consider unusual and often humiliating. During the past few years, the following sentences have been imposed:

- An Ohio judge ordered a man to spend two hours on a city sidewalk next to a 350-pound pig and a sign reading, "This is not a police officer" for disorderly conduct in a confrontation in which he called a police officer a pig (amid several other obscenities).

- A Nebraska judge ordered a woman who was convicted of making a false accusation of rape to apologize in half-page advertisements in four newspapers and 10 spot announcements on two radio stations.

- A Delaware judge ordered a man who twice exposed himself to a 10-year-old girl at his work-

place to wear a T-shirt with the words "I am a registered sex offender" in bold letters.

- An Illinois judge ordered a woman to quit smoking for a year or face imprisonment for stealing cigarettes.

- A Tennessee woman who was convicted of molesting her sons agreed to be sterilized as an alternative to going to prison.

- A Wisconsin judge ordered a woman who had stolen money from her employer to donate her family's Green Bay Packers seats to the Make-a-Wish Foundation.

- In Georgia, a judge suspended most of a man's sentence for cocaine possession and driving under the influence, with the requirement that he buy a casket and keep it in his house to remind him of the costs of drug addiction.

- In New Mexico, men convicted of domestic abuse were required to attend meditation classes with scented candles and herbal teas.

- An Ohio judge ordered a woman to spend the night in the woods without water, food, or entertainment as part of her punishment for abandoning 35 kittens.

Sources: "Pigheadedness lands man in pigpen," February 14, 2002, available at http://www.courttv.com/news/scm/scm_021402.html, accessed August 21, 2007; "Man Ordered to Wear 'Sex Offender' T-Shirt," Reuters, November 6, 2006, available at http://today.reuters.com/misc/2006-11-06, accessed November 6, 2006; "Smoking Deal: Judge Snuffs Shoplifter's Habit," Arizona Republic, January 17, 1991, p. A13; "Sentence for Lie on Rape Charge Creates Debate," New York Times, July 7, 1990, p. 8; "Woman Who Molested Sons Agrees to Sterilization," New York Times, January 31, 1993, p. A13; "Shame on You," Washington Post, September 18, 2005, p. B3; "Woman Ordered to Spend Night in Woods for Abandoning Kittens," ABC News, November 23, 2005, available at http://abcnews.go.com/GMA/LegalCenter/story?id=1322751, accessed July 17, 2007.

theory viewed punishment as justified as a means to deter (prevent) people from committing crime. According to utilitarian theory, punishment should be designed to achieve the greatest good or to do the least harm to those who are affected. Although deterrence theory emerged more than 250 years ago, its support declined during the late nineteenth and early twentieth centuries. In the late twentieth and early twenty-first centuries, it has reemerged to become the dominant justification for punishment.

Specific deterrence is aimed at individual offenders. According to this theory, criminals are punished for past crimes in an effort to make them afraid (or unable, in the case of the death penalty) to commit new crimes. General deterrence, by contrast, aims to dissuade potential offenders through the punishment of convicted criminals. That is, if people are made aware of punishments received by criminals, then they may consider the likely consequences of crime and ultimately be deterred from engaging in criminal activity. For example, a number of communities have published in local newspapers the names of people who were arrested for soliciting prostitutes in the hope that such public humiliation would deter others from doing the same.

Research on the actual effectiveness of deterrence has produced mixed findings. For example, Steven Levitt reports that, based on an analysis of annual crime data for the largest 59 U.S. cities over a 23-year period, deterrence appears to have a strong effect; he notes that the amount of violent crime dropped dramatically in both Washington and California following the introduction of significantly more severe sentencing statutes.[8] Other research suggests that the deterrent impact of punishment varies by both the type of crime and the individual offender.[9] Rational crimes (those involving planning and anticipation of potential costs and benefits) are more easily deterred than are crimes of passion (those motivated by emotion), and some individuals are more responsive to the threat of punishment than others (for example, elderly individuals and females are more easily deterred than are younger persons and males). This may account for the different conclusions about deterrence in studies on the impact of the severity of sentence on re-offending by spouse abusers. For example, Lawrence Sherman and Richard Berk report that harsher sentences lead to significant reductions in spousal abuse,[10] while Robert Davis found that sentence severity had no difference in the likelihood of reoffending.[11]

Incapacitation

Incapacitation is the removal of offenders from the community through imprisonment or banishment (e.g., deportation). Offenders who commit crimes while in prison can be further incapacitated by being placed in solitary confinement. Although banishment has been practiced throughout the ages, the incapacitation of offenders through imprisonment did not become popular until the nineteenth century.

Incapacitation is a form of specific deterrence in that it is designed to prevent individual offenders from committing future crimes. Research on the effectiveness of incapacitation, like studies of the deterrent effect of punishment, has yielded inconclusive results. Studies by criminologists David Greenberg and Stephan Van Dine suggest that increasing the average prison sentence or making prison terms mandatory for all people who are convicted of felonies would have little overall effect on reducing serious crime,

because much crime goes unreported and only a small percentage of reported crime results in arrest and conviction.[12] Other studies suggest that the selective incapacitation of repeat offenders who are convicted of serious crimes would have a much greater impact. Because a small number of offenders are responsible for a large proportion of all serious crimes, imposing mandatory sentences or longer prison terms on those high-rate offenders could significantly alter crime patterns.[13]

It also appears that incapacitation affects different offenders differently. For offenders who have few ties to the community and more extensive arrest records, longer terms of incarceration seem to lead to a longer time until rearrest. By contrast, the incarceration of first-time offenders may expose them to criminal values in the prison and actually increase their likelihood of rearrest after release.[14]

Rehabilitation

Without a doubt, rehabilitation—reforming an offender to become a productive member of society through treatment, education, or counseling—was the dominant philosophical orientation guiding sentencing practices in the United States from the 1940s through the 1970s. During this period, most state legislatures included statements in their sentencing

Because of multiple risk factors, prisoners often have difficulty reintegrating into mainstream society.

? Can prisons undo the damage caused by the lifelong problems that many inmates have? Is rehabilitation possible?

laws that established rehabilitation as the primary goal of punishment.

This theory holds that the causes of crime are located within individuals or in their immediate social environment. Once these causes are identified, appropriate treatment can change the internal make-up of offenders or alter the way they respond to their environment. Thus the offender can be successfully reintegrated into the community. From a rehabilitative perspective, punishments should be designed to help the offender change his or her behavior.

Treatment of offenders does not necessarily address psychological or emotional disabilities but more broadly assumes that offenders can benefit from drug or alcohol counseling, education, and vocational skill development, whether provided within institutions or through community-based correctional programs. Because individuals differ in their capacity to be rehabilitated, different types or lengths of sentences are justified. Of course, the exact length of time necessary to treat an offender cannot be known in advance, so the use of an indeterminate prison sentence (in which minimum and maximum numbers of years to be served are set) allows correctional officials to determine when offenders are ready to be released.

A flurry of evaluations of rehabilitation programs conducted during the 1960s and 1970s questioned the effectiveness of the rehabilitative approach in corrections. The general consensus from these studies was that—at best—only some forms of treatments work for certain types of offenders in various settings and there was little reason to be enthusiastic about future rehabilitation efforts.[15] The rehabilitation movement was set back by these reports, though more recent studies have provided strong support to those who advocate rehabilitation over strict punishment.

These newer studies conclude that certain forms of rehabilitation are effective in reducing recidivism.[16] The most effective treatment approaches have the following characteristics:

- They take place outside formal correctional settings and institutions.
- They provide service for higher-risk cases.
- They attend to extrapersonal circumstances of the offender (such as family and peers).
- They provide specialized academic programming and intensive structured skill training.
- They occur within behaviorally oriented individual counseling, group counseling, or structured milieu systems.[17]

Proportionality

As with all aspects of criminal justice, one of the goals of sentencing is fairness, which is often interpreted as proportionality—making the punishment fit the crime. How do we determine what kind of punishment is appropriate or what extent of punishment is proportional to a particular offense?

While capital punishment is often reserved for those convicted of first-degree or felony murder, South Carolina lawmakers voted in 2006 to allow the death penalty for sex offenders who were convicted a second time of raping children younger than 11 years old. Three other states—Louisiana, Florida, and Montana—allow the death penalty for particular sex crimes, although executions for such crimes have not been carried out in more than three decades.[18]

The U.S. Supreme Court ultimately determines whether these kinds of sentencing guidelines are proportional. For example, in 1910, in *Weems v. United States,* the Court ruled that a sentence involving 12 years at hard labor while wearing leg irons, a heavy fine, and loss of all political rights during imprisonment for falsifying official records of the U.S. Coast Guard violated the defendant's Eighth Amendment protection against cruel and unusual punishment.[19] The Court held that what constitutes cruel and unusual punishment is not static and stated that what may have been an acceptable sentence at one point in time may be considered too severe at another point in time. Nearly five decades later, the Court, in *Trop v. Dulles,* further clarified this point by noting that an interpretation of "cruel and unusual" must draw its meaning from the evolving standards of decency that reflect a maturing society.[20]

Ultimately, the idea of proportionality remains elusive. For example, in 1991, in *Harmelin v. Michigan,* the Supreme Court upheld the mandatory sentence of life in prison without possibility of parole for a defendant who was convicted of possession of more than 650 grams of cocaine.[21] The actual amount of cocaine, the Court reasoned, would have had a potential yield of between 32,500 and 65,000 doses; it decided that possession of such a large amount "is momentous enough to warrant the deterrence and retribution of a life sentence without parole." Although the defendant argued that the sentence was clearly disproportionate to his offense, the Court held that neither the Eighth Amendment nor previous cases (i.e., *Weems* and *Trop*) established a constitutional proportionality requirement or guaranteed a standard concerning the type or length of sentence for a specific crime.

AROUND THE GLOBE — Islamic Crimes and Punishments

Islamic law retains an extensive use of corporal punishments based on the Qur'an (the Holy Book of Islam) and the teachings of the Prophet Mohammed.

Offense	Punishment
Adultery	*Married person:* Stoning to death. Convict is taken to a barren site. Stones are thrown first by witnesses, then by the *qadi* (a religious judge), and finally by the rest of the community. For a woman, a grave is dug to receive the body; no grave is dug for the body of a man. *Unmarried person (party to adultery):* 100 lashes.
Highway robbery	*With homicide:* Death by beheading. The body is then displayed in a crucifixion-like form. *Without homicide:* Amputation of right hand and left foot. *If arrested before commission:* Imprisonment until repentance.
Use of alcohol	*Free person:* 80 lashes. *Slave:* 40 lashes.
Theft	*First offense:* Amputation of hand at wrist. *Second offense:* Amputation of second hand at wrist. *Third offense:* Amputation of foot at ankle or imprisonment until repentance.
Willful murder with weapon	Death by retaliation by victim's family, compensation, exclusion from inheritance, family, compensation, exclusion from inheritance, or pardon.
Sodomy	Death by sword followed by incineration, live burial, or being thrown from a high building and stoned.
Embezzlement, false testimony	Public disclosure and stigmatization (display of offender in various sections of the city, announcement of the offense and the punishment) or fines.

Frank Vogel reports that in Saudi Arabia in the early 1990s, nearly 40 persons were publicly executed for robberies and similar property crimes. At least two of the executions involved crucifixion after stoning. Floggings in many Muslim countries are also public events. One Egyptian man was sentenced to 4000 lashes for robbery in 1990, and more than 170 youths had been whipped in Riyadh, Saudi Arabia, for harassing girls. Imprisonment is actually used much less frequently in Islamic countries than in Western nations. For example, while the U.S. incarceration rate is 486 per 100,000, the rate is 250 per 100,000 in Tunisia, 156 per 100,000 in Iran, 95 per 100,000 in Turkey, 50 per 100,000 in Pakistan, and only 45 per 100,000 in Saudi Arabia.

Sources: Terance Miethe and Hong Lu, *Punishment: A Comparative Historical Perspective* (New York: Cambridge University Press, 2005), pp. 180–184; Frank Vogel, *Islamic Law and Legal System* (Boston: Brill, 2000), p. 370; Matthew Lippman, Sean McConville, and Mordechai Yerushalmi, *Islamic Criminal Law and Procedure* (New York: Praeger, 1988), pp. 42–45.

Types of Sentences

The history of punishment reflects the creativity of humans, who have responded to behavior defined as wrong or evil in myriad ways. From the imposition of the death penalty by stoning, beheading, gassing, hanging, shooting, electrocution, or lethal rejection, to banishment, imprisonment, whipping, caning, branding, house arrest, fines, counseling, restitution, and community service, society has attempted to find the best way to respond to crime and to the offender.

Sentences today are primarily linked to particular offenses and are specified by the criminal codes established by the state and federal governments. Judges typically have a wide range of sentencing options available to them as they seek to determine the most appropriate sentence for an offender. To help ensure proportionality, the kinds and lengths of sentences that may be imposed for misdemeanors differ from the punishments for felonies, reflecting the severity of the crime. Misdemeanors typically do not carry prison terms of a year or longer, and the death penalty is unlikely to be imposed in state criminal cases that

CHAPTER 12 Sentencing **319**

do not involve homicide or a very limited number of other extremely serious offenses (i.e., treason).

Sentences also vary for different groups of offenders. Juveniles who are found to be delinquent by juvenile courts may not be sentenced to adult prisons, for example. Likewise, first-time offenders generally have access to sentencing alternatives that are not available to repeat offenders (such as probation or placement in a community treatment facility).

Intermediate Sentences

Incarceration and probation are common sentences for people who are convicted of felonies and major misdemeanors. Prison overcrowding and excessively large probation caseloads, however, have recently led to the creation of a wider range of intermediate, or community-based, sentencing options:

- Intermittent incarceration allows offenders to serve relatively short sentences in jail on specified days. For example, a 30-day sentence could be served over a period of 15 successive weekends. The advantage to offenders is that they are able to retain their jobs or stay involved with their families. Such sentences are typically used when the offense warrants some period of incarceration but not full-time confinement.

- Intensive probation supervision is intended to ensure frequent and close contact between probation officers and offenders. With small caseloads, probation officers are able to give greater attention to offenders assigned to house arrest, electronic monitoring, drug or alcohol treatment, or psychological counseling.

- Fines, which are probably the most widely used form of sanction by lower courts, have been used for traffic offenses, local ordinance violations, and misdemeanors. The inability of poor defendants to pay even modest fines and the minimal impact of fines on wealthier offenders may be partly responsible for the limited use of fines in felony cases.[22]

- Community service requires offenders to make reparation to the community. Typically, it involves contributing less than 100 hours of time to a local government agency or volunteer organization.

- Restitution is one of the oldest forms of punishment. It generally takes one of two forms: direct financial payment to victims to compensate them for their losses or indirect restitution by payment into a victim-assistance program. In addition, restitution may require that the offender provide some service to the victim or the community.

- In forfeiture, offenders give up property that they obtained through criminal activity. The federal Drug Enforcement Administration (DEA) has used forfeiture extensively in its war on drugs. It is not always necessary for a person to be convicted of criminal activity to have property forfeited; in some states, simply being charged with a crime may result in confiscation of property believed to be connected to the offense.

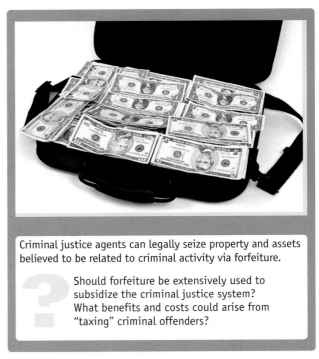

Criminal justice agents can legally seize property and assets believed to be related to criminal activity via forfeiture.

? Should forfeiture be extensively used to subsidize the criminal justice system? What benefits and costs could arise from "taxing" criminal offenders?

Intermediate punishments are rarely used alone because each, by itself, is generally considered by the courts to be an insufficient sanction. More often, several such punishments or the combination of an intermediate punishment and either probation or a short term of incarceration is imposed.

LINK Chapter 15 offers a more detailed discussion of intermediate sentencing options, such as fines, house arrest, and community service.

Indeterminate Prison Sentences

An indeterminate sentence establishes a minimum number and a maximum number of years to be served. The actual time for release is determined by a parole board after the offender has served a portion of the minimum sentence. For example, a burglar might receive a 2- to 5-year sentence and a robber might receive a 10- to 20-year sentence. Although states vary in terms of the number of years that must be served before an offender can be considered for parole, at least one-third of the minimum term is the typical point at which parole becomes available. In 2002, approximately one-fourth of all prison releases resulted from parole board decisions (as opposed to the inmate completing the maximum sentence).[23]

LINK Prison releases and parole decision making are examined in Chapter 15.

Determinate or Structured Prison Sentences

Determinate sentences, which feature fixed terms of imprisonment, emerged as a result of the sentencing reform movement in the early 1970s. Reformers argued that rehabilitation did not work and that indeterminate sentences led to gross disparities and unfairness.[24] Determinate sentencing systems built upon retributive and deterrence theories of punishment were advocated, with the following goals:

- To eliminate the disparities created by the sentencing discretion of judges
- To create a system of relatively uniform sentences for offenders convicted of similar crimes and with similar records
- To redistribute time served in prison (so that less serious offenders spent less time in prison and more serious felons spent more time) without significantly increasing the total person-years served or the associated costs of incarceration[25]

The adoption of determinate sentencing did not entirely eliminate judicial discretion, nor did it guarantee that inmates who were convicted of similar offenses would actually serve similar amounts of time in prison. For example, in many sentencing systems (such as presumptive systems, which are discussed below), judges retain the discretionary authority to impose or suspend a prison sentence, although the length of the sentence is regulated. In addition, determinate sentencing systems are often linked to correctional policies that allow inmates to earn early release from prison by accumulating good time. Consequently, the actual time spent in prison continues to vary dramatically.

The determinate sentencing systems that were developed over the past two decades have generally taken one of two forms: presumptive sentencing or sentencing guidelines systems.

Presumptive Sentencing

Presumptive sentencing relies on a range of minimum and maximum terms of incarceration that have been established by the legislature for particular categories or classes of crimes, with the judge determining the specific number of years to be served within this range. If aggravating or mitigating factors exist, the judge may depart from the presumptive sentence by adding or subtracting time, within limits **TABLE 12-1**. Any departure from the presumptive sentence typically requires the judge to submit a written explanation for the decision.

TABLE 12-1

Aggravating and Mitigating Factors

Aggravating Factors

- The offender has a criminal history.
- The offender has recently violated the terms of his or her probation or parole.
- A reduced or suspended sentence would make the crime appear less serious than it is.
- The victim is older than age 65 or is mentally or physically infirm.
- The offender is in need of correctional or rehabilitative treatment that can best be provided by commitment to a penal facility.

Mitigating Factors

- The circumstances of the crime are unlikely to recur.
- The offender acted under strong provocation.
- The victim of the crime induced or facilitated the offense.
- The crime did not cause or threaten serious harm to people or property.
- Grounds exist to justify or excuse the crime.
- The offender has no prior criminal record or has led a law-abiding life for a substantial period of time before committing the crime.
- The character and attitude of the offender indicate that he or she is not likely to commit another crime.
- The offender has made or will make restitution to the victim.
- The offender is likely to respond affirmatively to probation or short-term imprisonment.

Source: Todd Clear, John D. Hewitt, and Robert M. Regoli, "Discretion and the Determinate Sentence: Its Distribution, Control, and Effect on Time Served," *Crime & Delinquency* 24:428–445 (1978).

While presumptive sentencing might appear to provide increased fairness and consistency in sentencing, many state and federal lawmakers believed that it did not go far enough in regulating sentences. Consequently, formal sentencing guidelines with even greater structure and predictability were developed.

Sentencing Guidelines

Alternatives to presumptive sentencing schemes, known as sentencing guidelines, were developed to limit judges' discretion, which many people believed resulted in widely disparate sentences for similar offenders. Like presumptive sentences, sentencing guidelines were intended to deemphasize rehabilitation as a primary goal in sentencing.

Guidelines like the one used in Minnesota **TABLE 12-2** direct the judge to arrive at a sentence by plotting the criminal history score of an offender against the severity of the current offense. Using these guidelines, judges are allowed to exercise some degree of discretion by selecting a specific number of months within a specified range. As in other presumptive sentencing systems, judges must write a statement explaining any departures from sentencing guidelines.

By 1994, the U.S. District Courts and 18 states were using some form of sentencing guidelines. Some states developed advisory or nonbinding guidelines that judges could consult for suggested—but not mandatory—sentences. Other states as well as the federal courts developed prescriptive or mandatory guidelines for sentencing. In these systems, judges are required to follow the guidelines and impose only sentences that fall within the guideline ranges.

Criticisms quickly arose regarding sentencing guidelines. Opponents of the systems argued that modern-day guidelines developed by using past patterns of sentencing might incorporate past race and class biases and historical disparities.[26] For example, federal guidelines established much longer sentences for persons convicted of possession of crack cocaine than for persons convicted of possession of powdered cocaine—and African Americans were disproportionately more likely to be arrested for crack-related offenses. Other critics suggested that guidelines that try to achieve consistency in sentences may actually foster disparity because they fail to consider significant material differences between cases.[27]

Of greater concern was the possibility of significant departures from guidelines. In 2004, in *Blakely v. Washington,* the U.S. Supreme Court ruled that the state of Washington's guidelines were unconstitutional because they violated a defendant's Sixth Amendment protections.[28] In this case, Ralph Blakely pleaded guilty to the kidnapping of his estranged wife. The facts admitted in his plea supported a maximum sentence of 53 months. At his sentencing, the judge rejected the prosecutor's recommendation based on the guidelines and instead imposed a sentence of 90 months, justifying the sentence on the grounds that Blakely had acted with "deliberate cruelty." The Supreme Court held that "the facts supporting the petitioner's exceptional sentence were neither admitted by petitioner nor found by a jury," meaning that the sentence violated Blakely's Sixth Amendment right to trial by jury and to be found guilty beyond a reasonable doubt.

One year later, in *United States v. Booker,* the Supreme Court ruled that the federal sentencing guidelines were unconstitutional and reduced them to an advisory status.[29] Although the federal guidelines are not prescriptive, judges continue to use them as a starting point in deciding a sentence and must provide a written justification for any departure from the guidelines.

The *Blakely* case affected 12 states that had used guidelines similar to those established in Washington. Since this ruling, six of these states have either adopted or are in the process of adopting revised guideline procedures that will use jury fact-finding to determine aggravating circumstances that might increase the presumptive sentence. At least four states have decided not to adopt jury fact-finding. In three of these states—Colorado, Indiana, and Tennessee—recommended sentences are now voluntary rather than presumptive.[30] In California, the U.S. Supreme Court struck down a law in 2007 that allowed judges to add years to a sentence based on their own fact-finding of aggravating circumstances.[31]

Mandatory Sentences

Many states have established mandatory sentences, requiring imprisonment of offenders who are convicted of certain types of serious crimes. Mandatory sentences prohibit judges from suspending prison sentences and placing offenders on probation. For example, some states have gun laws mandating prison sentences for persons who are convicted of using firearms in the commission of crimes, and a number of states and the federal government have adopted mandatory minimum sentences for certain violent crimes and drug offenses.[32] Some critics believe that much of the national increase in both incarceration rates and sentence length is attributable to such laws.[33]

Habitual Offender Statutes

In the past few years, a number of states have passed habitual or persistent offender legislation, also known as three-strikes laws (as in, "Three strikes and you're out"). Most such statutes require lengthy prison sentences or even life in prison without parole for offenders who are convicted of a third violent felony,

TABLE 12-2

Minnesota Sentencing Guidelines Grid (Presumptive Sentence Lengths in Months)

Italicized numbers within the grid denote the range within which a judge may sentence without the sentence being deemed a departure. Offenders with nonimprisonment felony sentences are subject to jail time according to law.

Severity Level of Conviction Offense		Criminal History Score						
		0	1	2	3	4	5	6 or More
Murder, Second Degree (intentional)	XI	306 *61–367*	326 *276–391*	346 *295–415*	366 *312–439*	386 *329–463*	406 *346–480²*	426 *353–480²*
Murder, Third Degree Murder, Second Degree (unintentional)	X	150 *128–180*	165 *141–198*	180 *153–216*	195 *166–234*	210 *179–252*	225 *192–270*	240 *204–288*
Assault, First Degree Controlled Substance Crime, First Degree	IX	86 *74–103*	98 *84–117*	110 *94–132*	122 *104–146*	134 *114–160*	146 *125–175*	158 *135–189*
Aggravated Robbery, First Degree Controlled Substance Crime, Second Degree	VIII	48 *41–57*	58 *50–69*	68 *58–81*	78 *67–93*	88 *75–105*	98 *84–117*	108 *92–129*
Felony Driving While Intoxicated	VII	36	42	48	54 *46–64*	60 *51–72*	66 *57–79*	72 *72–86*
Assault, Second Degree Felon in Possession of a Firearm	VI	21	27	33	39 *34–46*	45 *39–54*	51 *44–61*	57 *49–68*
Residential Burglary Simple Robbery	V	18	23	28	33 *29–39*	38 *33–45*	43 *37–51*	48 *41–57*
Nonresidential Burglary	IV	12¹	15	18	21	24 *21–28*	27 *23–32*	30 *26–36*
Theft Crimes (more than $2500)	III	12¹	13	15	17	19 *17–22*	21 *18–25*	23 *20–27*
Theft Crimes ($2500 or less)	II	12¹	12¹	13	15	17 *17–22*	19 *18–25*	21 *18–25*
Sale of Simulated Controlled Substance	I	12¹	12¹	12¹	13	15	17	19 *17–22*

Presumptive commitment to state imprisonment. First-degree murder is excluded from the guidelines by law and continues to have a mandatory life sentence.

Presumptive stayed sentence; at the discretion of the judge, up to a year in jail and/or other non-jail sanctions can be imposed as conditions of probation. However, certain offenses in this section of the grid always carry a presumptive commitment to state prison.

1. One year and one day

2. M.S. § 244.09 requires the Sentencing Guidelines to provide a range of 15% downward and 20% upward from the presumptive sentence. However, because the statutory maximum sentence for these offenses is no more than 40 years, the range is capped at that number.

Source: Minnesota Sentencing Guidelines Commission, effective August 1, 2006.

After lengthy deliberations and numerous votes, a single holdout juror kept the jury from handing down a death sentence for Zacarias Moussaoui for his involvement in the September 11, 2001, terrorist attacks.

U.S. Judge Leonie Brinkema followed the jury's recommendation and sentenced Moussaoui to six life sentences, to be divided into two terms to run consecutively. In response to the sentences, Moussaoui declared, "God save Osama bin Laden—you will never get him!" and "America, you lost . . . [and] I won." Judge Brinkema told Moussaoui, "Mr. Moussaoui, you came here to be a martyr in a great big bang of glory but, to paraphrase the poet T. S. Eliot, instead you will die with a whimper."

A couple of days after his sentencing, Moussaoui filed a motion to withdraw his guilty plea, saying he had lied on the witness stand about his involvement in the terrorist plot. Judge Brinkema told Moussaoui, "You do not have a right to appeal your convictions, as was explained to you when you pled guilty. You waived that right." The judge added that he did have a right to appeal his sentence, but noted, "I believe it would be an act of futility."

Sources: *Courttvnews*, "Moussaoui Formally Sentenced to Life," available at http://www.courttv.com/trials/moussaoui/050406_sentencing_ap.html, accessed July 2, 2007; *Courttvnews*, "Moussaoui Says He Wants to Withdraw Guilty Plea," available at http://www.courttv.com/trials/moussaoui/050806_ap.html, accessed July 3, 2007; *Courttvnews*, "Report: Lone Holdout Juror Kept Moussaoui Alive," available at http://www.courttv.com/trials/moussaoui/051206_ap.html, accessed July 6, 2007.

---

although some states do not require any of the felonies to be violent crimes. Currently, 25 states and the federal government have three-strikes laws. Some states, such as Georgia, have passed even tougher laws mandating life-without-parole sentences for offenders who are convicted twice of particular violent crimes such as murder, armed robbery, kidnapping, and rape.[34]

Research has shown that three-strikes provisions have not led to any significant overall reduction in violent crime or felonies. In many states with three-strikes laws, the crime rates were already dropping prior to the introduction of the new legislation, so it is impossible to separate out the effects of these laws. Furthermore, crime rates have dropped as much or more in states without such laws.[35]

Three-strikes laws have also had some unintended consequences. Recent studies have found a positive association between such laws and homicide rates; that is, homicide rates appear to have actually increased after the passage of the laws.[36] In addition, three-strikes laws appear to have disproportionately harsh effects on African Americans. In their study of the impact of California's three-strikes laws, Franklin Zimring and his colleagues report that African Americans, who make up less than 10 percent of the state's population, account for nearly two-thirds of third-strike eligible arrests.[37] Moreover,

The effectiveness of three-strikes laws has been hotly debated, and some politicians have even advocated two-strikes laws that would impose life imprisonment after a person commits a second serious felony.

 Do crimes such as murder, rape, and kidnapping warrant mandatory life sentences? Why are so many criminals able to amass dozens of serious felony convictions, and what does this indicate about the justice system?

African Americans account for slightly more than 30 percent of the state prison population but fully 44 percent of inmates serving three-strikes sentences.[38]

Truth in Sentencing

The wave of sentencing reforms implemented during the past two decades also included legislation to establish a much closer correspondence between judicially imposed sentences and the actual time offenders serve in prison. A number of states have moved to create truth-in-sentencing (TIS) laws, which require that offenders—and especially violent offenders—serve at least 85 percent of their sentences. The federal Violent Crime Control and Law Enforcement Act of 1994, as amended in 1996, provided federal grants to states to expand their prison capacity if they imposed TIS requirements of 85 percent time served for violent offenders. A number of states have acceded to the federal government's push for increased sentences; by 2005, at least 40 states had some form of TIS laws.

Most research suggests that, as a consequence of TIS laws, there is now a more truthful relationship between the imposed sentence and the time actually served. Moreover, although critics of TIS laws argued that this kind of legislation would likely significantly increase already historically large prison populations, most research to date suggests that it has played only a minor role in the increase. Existing patterns of crime and other social and economic forces are believed to have had much greater effect in the growth of the prison population.[39]

Concurrent and Consecutive Sentences

Offenders are often convicted on more than one charge. In such a case, the judge must first determine the appropriate prison sentence for each conviction charge. Then, whether in an indeterminate or determinate sentencing system, the judge must decide whether to impose these sentences concurrently or consecutively. Concurrent sentences are served at the same time. That is, a person with two convictions, each carrying a 10-year sentence, would serve the sentences together and still be eligible for release in 10 years, assuming he or she served the full maximum term. Consecutive sentences, by contrast, require that the offender serve one sentence before the next. Thus an offender who is facing two 10-year prison sentences might not be released until the full 20 years has been served.

Arriving at an Appropriate Sentence

Despite determinate sentencing reforms in many jurisdictions, sentencing remains a difficult decision for judges. They must consider three interrelated questions:

- What are the appropriate goals of sentencing, and how should they be weighed in light of the facts of the current case?
- How can the goals of sentencing be achieved under the circumstances of the case?
- What specific sentence is justified given the facts of the case?[40]

Judges seek to answer these questions as they try to determine the most appropriate sentence for a given case, using any sentencing guidelines or case-specific information contained in the presentence investigation report.

The Presentence Investigation Report

The presentence investigation (PSI), which is widely believed to be the cornerstone of the sentencing decision, is a comprehensive report on the background of the offender, the offense, and any other information that the judge considers relevant to deciding an appropriate sentence, including aggravating and mitigating circumstances. The PSI report helps the court understand the nature of the crime within the context of the offender's life. In jurisdictions that rely on sentencing guidelines, such individualized considerations are viewed as less relevant and consequently have less impact on the sentence that is ultimately imposed.

The PSI report is based on interviews with the offender, family members, employers, and friends, as well as reviews of police reports and victim statements concerning the crime. It includes information about the offender's background, such as his or her educational, employment, medical, sexual, and military history; past juvenile and adult criminal records; evidence of alcohol and drug use; and any psychological or psychiatric evaluations. It may also include a sentence recommendation from a probation officer, although this practice varies by jurisdiction and even among judges within local court systems.

Research indicates that judges often closely follow the sentence recommended by the probation officer.[41] In reality, this high correlation between the probation officer's recommendation and the judge's sentence may be a result of the probation officer's anticipation of the judge's decision. In many jurisdictions, probation officers are appointed by the court and serve at the will of the judge. Even in jurisdictions where they are more independent, probation officers may still perceive a need to please their immediate supervisors. In many jurisdictions, the probation officer's role in sentencing may be little more than ceremonial, designed to create the appearance of individualized justice.[42]

The Sentencing Hearing

Sentencing in felony cases typically occurs at a sentencing hearing, which is often scheduled three to six weeks after conviction (except in lower court sentencing of minor misdemeanors and ordinance violations, where offenders may be sentenced immediately upon conviction). At this hearing, the offender has an opportunity to deny, explain, or add to information contained in the PSI report. In addition, the prosecutor may submit a written statement or make an oral recommendation regarding the sentence. In many plea agreements, the prosecutor may agree to support a reduced sentence or concurrent sentences, or to make no recommendation about the sentence. Judges typically make the final sentencing decisions in felony cases, whereas juries must make the sentencing decision in capital cases.

The sentencing of offenders who are convicted of capital crimes where the death penalty may be imposed requires a bifurcated trial, which includes a separate jury hearing after a guilty verdict to determine the offender's penalty. At sentencing, the jury considers mitigating and aggravating circumstances and then recommends either the death penalty or a prison sentence.

Prior to 2002, judges typically viewed jury recommendations regarding death sentences as merely that—recommendations—with the judge making the final determination. However, the U.S. Supreme Court, in *Ring v. Arizona,* held that juries—not judges—must make the critical fact-finding of aggravating circumstances that would lead to a death sentence.[43] In 2003, the Ninth Circuit Court of Appeals, in *Summerlin v. Stewart,* applied the *Ring* decision retroactively and vacated the death sentences of nearly 100 inmates on death row in Arizona, Idaho, and Montana who had been sentenced to death by judges rather than juries.[44] This decision of the

Until 2002, juries in six states (Arkansas, Kentucky, Missouri, Oklahoma, Texas, and Virginia) decided sentences in all felony cases except those involving the death penalty.

? Should death penalty cases be an exception to standard sentencing procedures? Why should juries play a role in sentencing capital offenders?

Ninth Circuit was reversed by the U.S. Supreme Court in 2004, in *Schriro v. Summerlin,*[45] when it held that the *Ring* decision does not apply retroactively. The original death sentences were, therefore, reinstated.

The Role of the Victim in Sentencing

Before the development of common law in England, when crimes often led to blood feuds, victims played a central part in the prosecution and punishment of offenders. Over the years, the role of the victim has declined, being replaced by the bureaucratic machinery of the state. In the late twentieth century, with the revival of demands that offenders receive their just desserts (the retributive model of justice), the role of the victim in sentencing began to expand once again. Demands that victims be heard at sentencing stimulated legislative reforms in many states.[46] Today, all but five states specifically permit victims to be heard at sentencing TABLE 12-3.

Many jurisdictions now allow victims to prepare written or oral victim impact statements (VIS) to be given to the court during the plea bargaining stage or at the sentencing hearing.[47] The VIS details the impact of the crime on the victim (including financial, social, physical, and psychological effects) and describes the victim's feelings about the crime, the offender, and the proposed sentence.

TABLE 12-3

Admissibility of Victim Impact Evidence by State

Undecided	Limited Admissibility	Admissible	
Connecticut	Indiana	Alabama	Nevada
Montana	Mississippi	Arizona	New Jersey
New Hampshire		Arkansas	New Mexico
New York		California	North Carolina
Wyoming		Colorado	Ohio
		Delaware	Oklahoma
		Florida	Oregon
		Georgia	Pennsylvania
		Idaho	South Carolina
		Illinois	South Dakota
		Kansas	Tennessee
		Kentucky	Texas
		Louisiana	Utah
		Maryland	Virginia
		Missouri	Washington
		Nebraska	

Source: National Center for Victims of Crime, *The 1996 Victims' Rights Sourcebook: A Compilation and Comparison of Victims' Rights Laws* (Arlington, VA, 1996).

In reality, the use of a VIS at sentencing may have relatively little effect on sentencing, especially when juries decide sentences, because victims seldom avail themselves of the opportunities to make a statement. In California, for example, victims appear or submit statements in less than 3 percent of all cases.

Criminologist Anthony Walsh has examined the role of the VIS in 417 sexual assault cases in a metropolitan Ohio county.[48] Ohio law provides that in all felony cases involving victims, a VIS may be included as part of the proceedings. Nevertheless, Walsh found, only slightly more than half of the victims submitted a VIS and only 6 percent spoke at the sentencing hearing. He concluded that the VIS had little impact on the judge's choice between probation and prison. Although there was greater overall agreement than disagreement between the victim recommendation and the imposed sentence, Walsh noted that the outcome was primarily explained by legal constraints rather than by the VIS.

According to sociologist Edna Erez, "There is also no evidence that including victims in the criminal justice process results in punitiveness or that requests by victims for harsh sanctions influence the sentences." Instead, she argues, victims' rights may correctly be viewed as a type of "placebo justice."[49]

According to Erez and her colleagues, a self-fulfilling prophecy may be at work. Victims may participate infrequently in sentencing because they believe their input will have little effect on the outcome; in turn, lack of participation by victims undermines attempts at reintegrating the victim into the criminal justice process.[50]

At least one recent study did find that the VIS could have a significant effect on jury sentencing decisions. Using an experimental design, Bryan Myers and Jack Arbuthnot presented two separate mock juries with trial transcripts of a murder case. One of the juries read the VIS of the grandmother of the murder victim; the other jury did not. Two-thirds of the jurors who read the VIS voted to impose the death penalty compared to less than one-third of those jurors who had not read the VIS.[51]

Disparity and Discrimination in Sentencing

The use of indeterminate sentencing demanded sensitivity to the differences between offenders. Believing in the rehabilitative ideal, judges attempted to fit the sentence to the unique characteristics and circumstances in each case. Little attention was paid to the possibility that extralegal factors—such as the offender's race, ethnicity, gender, socioeconomic status, or age, or even victim characteristics—might produce unintentional disparities in sentencing or cause judges to consciously or unconsciously discriminate against particular groups of defendants. Determinant sentencing systems—and especially sentencing guidelines—have reduced but not eliminated disparities in sentencing, although the range of differences is more limited. For the most part, while limiting the sentencing discretion of judges, discretion has simply been shifted to prosecutors.

Sentencing disparities occur when similar cases are sentenced differently.[52] Disparities are statistical differences in sentencing that reflect some characteristic of interest, such as the race or ethnicity of the offender. Disparities may or may not reflect intentional or unintentional bias or discriminatory sentencing practices; they simply indicate that differences in sentences appear to be related to certain case-related factors.

When judges are allowed to have some degree of discretion in making their sentencing decisions, they may take into consideration a wide variety of case factors relating to the offense and the offender.[53] For example, if judges are honestly and fairly trying to fit the best sentence to the defendant, similar case factors may be weighed differently for different people. A judge may sentence a white, middle-class college student who is convicted of possession of marijuana to probation and require that the student participate in a drug treatment program provided by the university. The same judge may sentence a young, unemployed African American youth convicted of possession of crack cocaine to a short prison term. On the surface, the difference in sentencing might appear to reflect racial prejudice on the part of the judge. However, it is also possible that the judge was weighing differences in circumstances and available options in both cases. If, as in many communities, there are no drug treatment programs available for inner-city drug users, the judge may believe that the best available option is short-term incarceration in an institution that can provide drug counseling. The lack of drug treatment programs in the inner city may reflect discriminatory policies by the government in budgetary decisions, but would not necessarily be an indicator of discrimination by the judge in sentencing.

In addition, even the decision by a defendant to request a jury trial may affect sentencing. For exam-

ple, when Jeffery Ulmer and Mindy Bradley examined sentencing data from county criminal courts in Pennsylvania between 1997 and 2000, they found that defendants who were convicted after a jury trial (as opposed to a bench trial or a guilty plea) were significantly more likely to receive more severe sentences for more serious violent offenses.[54]

Sentencing discrimination exists when illegitimate, morally objectionable, or extralegal factors are taken into account in the sentencing decision. Discrimination often leads to disparities in sentences based on extralegal factors. It occurs when a judge sentences certain offenders to prison or to longer terms while sentencing other offenders convicted of similar offenses and with prior records to probation or shorter sentences solely on the basis of the defendant's race, socioeconomic status, age, or gender, or the characteristics of the victim.

Most studies of sentencing disparity and discrimination have examined the effects of four extralegal factors tied to the offender: race/ethnicity, gender, socioeconomic status, and age. These factors are important because they appear to be related to decisions made at nearly every stage in the criminal justice process.

Race and Ethnicity

In 2005, African Americans accounted for 40 percent of the U.S. prison population even though they represented only 12 percent of the U.S. population. Latinos accounted for an additional 20 percent of inmates, and whites made up another 35 percent.[55] Compared to whites, African Americans are more likely to receive prison sentences instead of jail or probation regardless of the offense of which they are convicted FIGURE 12-1 , FIGURE 12-2 , FIGURE 12-3 . It is not clear whether these disparities are a reflection of clear racial discrimination, factors unrelated to race, or more subtle institutional racism.

Many studies find only a small—if not entirely absent—effect of race and ethnicity on the decision to incarcerate or the length of prison sentence. Generally, race and ethnicity appear to play minor roles compared to other legally relevant factors such as the offense committed and the offender's prior record.[56]

In truth, race may actually have a more complicated and less obvious impact on sentencing. For example, James Gibson, in his study of sentencing patterns in Atlanta, found that African American and white defendants received similar sentences, even when controlling for the effect of the current offense and the offender's prior record. When Gibson exam-

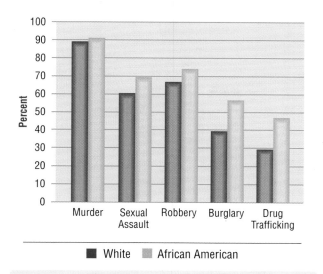

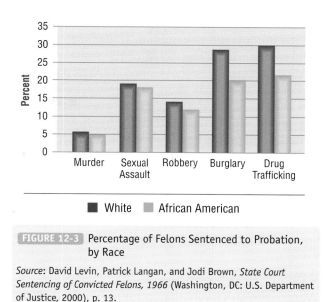

FIGURE 12-3 Percentage of Felons Sentenced to Probation, by Race

Source: David Levin, Patrick Langan, and Jodi Brown, *State Court Sentencing of Convicted Felons, 1966* (Washington, DC: U.S. Department of Justice, 2000), p. 13.

FIGURE 12-1 Percentage of Felons Sentenced to Prison, by Race

Source: David Levin, Patrick Langan, and Jodi Brown, *State Court Sentencing of Convicted Felons, 1966* (Washington, DC: U.S. Department of Justice, 2000), p. 13.

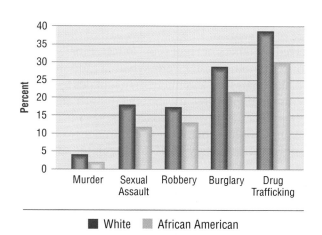

FIGURE 12-2 Percentage of Felons Sentenced to Jail, by Race

Source: David Levin, Patrick Langan, and Jodi Brown, *State Court Sentencing of Convicted Felons, 1966* (Washington, DC: U.S. Department of Justice, 2000), p. 13.

ined the sentencing patterns of individual judges, however, he found that the race of the defendant was clearly a factor in the decisions of some judges. Some judges were clearly biased against African Americans, some judges were biased in favor of African Americans, and other judges appeared not to consider race at all.[57]

Kathleen Daly and Michael Tonry note that while most sentencing studies do not find statistical advantages for whites at sentencing, such findings are likely to reflect institutional discrimination masking the real effects of race and ethnicity.[58] For example, African Americans are more likely to be held in jail awaiting trial and to have court-appointed attorneys. Perhaps, as Cassia Spohn suggests, African Americans are more likely than whites to be involved in more ambiguous or borderline cases where the seriousness of crime, the strength of the evidence, and the defendant's degree of culpability are less clear. In such instances, ambiguity tends to increase judicial discretion and enhance the potential for extralegal factors, such as race, to enter into the sentencing decision.[59]

Studies have also examined whether the effects of race and ethnicity are reduced when sentencing guidelines are used. Researchers in Maryland and Pennsylvania, for example, found that race and ethnicity have an effect on upward and downward departures within the sentencing grids. Shawn Bushway and Anne Piehl found that African Americans, on average, received 20 percent longer sentences than whites who were similarly situated on the sentencing guidelines grid, although this disparity does not necessarily reflect discrimination.[60] In Pennsylvania, according to Brian Johnson, African American and Latino defendants are significantly less likely to have downward departures but are more likely to receive upward departures in sentence length compared to similarly situated white defendants.[61] Finally, Darrell

Steffensmeier and Stephen Demuth report that while federal judges tend to hand out similar sentences for similar defendants convicted of the same offenses, those defendants who have been convicted of more serious offenses, have more extensive prior records, and have been convicted at trial (rather than pleading guilty) are significantly more likely both to be sentenced to prison and to receive longer sentences. Steffensmeier and Demuth also found evidence that Latino defendants were sentenced somewhat more harshly than white defendants, while African Americans fell in between the two extremes, possibly due to differences in the seriousness of the offenses committed and the offenders' prior records.[62]

Gender

Women are less likely than men to be arrested, denied bail, or prosecuted. As they go through the criminal justice process, women find increasing leniency at every stage. They account for only 7 percent of all offenders incarcerated in state and federal correctional institutions in 2005, and they are less likely to be sentenced to prison and more likely to be sentenced to jail or probation than males who commit similar offenses **FIGURE 12-4** , **FIGURE 12-5** , **FIGURE 12-6** .[63]

Generally, even when the severity of the offense, prior record, and other legally relevant factors are held constant, women are still treated noticeably more leniently at sentencing than men. In a study of murder cases in the 75 largest counties in the United States, women convicted of murder were found to be treated more leniently than men who committed the same offense.[64] More than 50 percent of the men who were convicted of murder were given life sen-

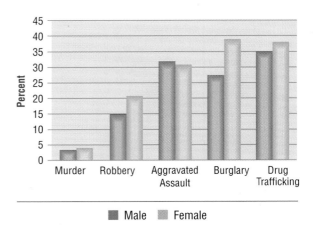

FIGURE 12-5 Percentage of Felons Sentenced to Jail, by Sex

Source: David Levin, Patrick Langan, and Jodi Brown, *State Court Sentencing of Convicted Felons, 1966* (Washington, DC: U.S. Department of Justice, 2000), p. 12.

tences, compared with only 42 percent of the women; 13 percent of the men were sentenced to death, but none of the women received this punishment. Women who were convicted of murder also received shorter sentences than men: Half of the men received a sentence of 17 years or less, whereas half of the women received sentences of eight years or less. Finally, analysis of sentencing under Pennsylvania guidelines reports that males are less likely than females to receive downward departures in sentence length and more likely to receive upward departures.[65]

What might explain these differences? One factor, according to Daly and Tonry, relates to differ-

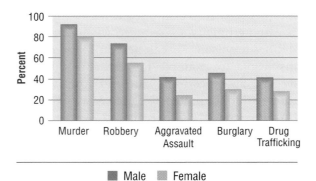

FIGURE 12-4 Percentage of Felons Sentenced to Prison, by Sex

Source: David Levin, Patrick Langan, and Jodi Brown, *State Court Sentencing of Convicted Felons, 1966* (Washington, DC: U.S. Department of Justice, 2000), p. 12.

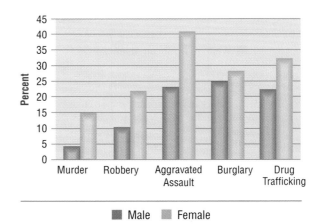

FIGURE 12-6 Percentage of Felons Sentenced to Probation, by Sex

Source: David Levin, Patrick Langan, and Jodi Brown, *State Court Sentencing of Convicted Felons, 1966* (Washington, DC: U.S. Department of Justice, 2000), p. 12.

ences in defendants' ties and responsibilities to others, especially within family relationships.[66] Having ties and responsibilities in the day-to-day care of others is often cited as a mitigating circumstance at sentencing.[67] Although men with dependents tend to receive less severe sentences than men without dependents, women—especially married ones with dependents—receive more lenient sentences overall.[68]

Another factor might have to do with gender-linked stereotyped beliefs in the greater potential for women to be reformed. Moreover, according to Daly and Tonry, "court officials assume that women are more easily deterred than men and that women will make greater efforts to reform themselves."[69]

Socioeconomic Status

Because poor defendants are less likely to be able to post bail and afford their own counsel, concern exists that they may be at a disadvantage in sentencing. Some research suggests defendants who remain in custody while awaiting trial or who have court-appointed counsel are more likely to be sentenced to prison and to receive longer sentences. Conflict theorists suggest that this trend reflects a social class bias against the poor in the courts.[70]

The actual relationship between socioeconomic status and sentence severity may be obscured because many studies fail to consider the seriousness of the current offense and the prior conviction record of the defendant. Researchers John Hagan, Theodore Chiricos, and Gordon Waldo found no difference in sentences for defendants based on their economic status in a study of more than 10,000 inmates in three Southeastern states.[71] Other studies report leniency toward "relatively disadvantaged persons and even some suggestion that relative advantage fostered punitiveness in sentencing."[72] The same studies also found that white-collar offenders received disproportionately longer sentences, at least in those counties with greater income inequality.

Age

Research on the impact of age on sentence suggests only a minor relationship between the two. Myers and Talarico found that the effect of age on sentence varied depending on the level of unemployment within the county in which the offender was sentenced. In counties with low unemployment, age had little effect; in those counties with serious unemployment, younger offenders were more likely than older people to be incarcerated.[73] Although these researchers did not spec-

ulate on the possible reasons for this relationship, it may be that the courts in high-unemployment counties considered younger offenders to be less able to obtain employment, and consequently to be at greater risk for involvement in future crimes.

Research on sentencing patterns in Kentucky, Virginia, and Tennessee found that elderly offenders (over age 60) were more likely to receive probation or shorter incarceration sentences than younger people who were convicted of the same or similar crimes.[74] This difference may stem partly from the fact that jails and prisons are not well suited for the elderly. Finally, a recent study of departures in Pennsylvania's sentencing guidelines found that younger defendants are more likely to have upward departures and less likely to downward departures than older defendants.[75]

Capital Punishment

Between 1930 and 2006, more than 4900 executions were carried out in the United States. Excluding the 10 years between 1968 and 1977, when there was a moratorium on executions because of legal challenges brought before the Supreme Court, an average of 104 executions took place each year. However, only 310 persons were executed between 1977 and 2006, or an average of slightly fewer than 36 per year. Executions peaked in 1999, when 98 persons were put to death FIGURE 12-7 .[76]

Supreme Court Decisions

The brief moratorium and subsequent reinstatement of the death penalty resulted from a series of Supreme Court decisions that attempted to clarify particular constitutional issues surrounding capital punishment. In 1968, in *Witherspoon v. Illinois*, the Court suspended all scheduled executions (at that time, there were more than 500 inmates on death row).[77] In *Witherspoon*, the defendant had originally been convicted and sentenced to death by a jury explicitly selected from among people who had expressed that they had no opposition to capital punishment. The Court held that excluding all persons opposed to capital punishment from a jury, thus making it a death-qualified jury, was unconstitutional and overturned Witherspoon's sentence. However, the Supreme Court did not hold that "conviction-prone" juries were unconstitutional, and in 1986, in *Lockhart v. McCree*, the Court stated that the "conviction-prone" orientation

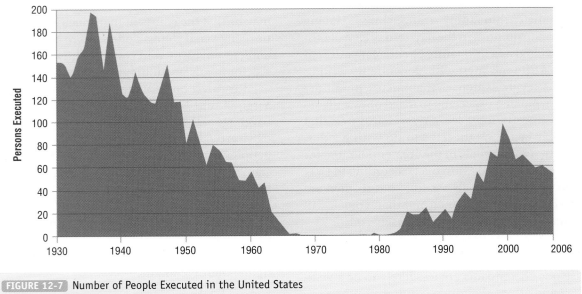

FIGURE 12-7 Number of People Executed in the United States

Source: Tracy Snell, *Capital Punishment, 2006* (Washington, DC: U.S. Department of Justice, 2007), p. 10.

of a jury supporting the death penalty is not in itself a violation of any constitutional protections.[78] According to the Court, the death-qualification process in jury selection does not necessarily mean the jury cannot be impartial, and there is no constitutional requirement for a jury to be composed of a balance of particular viewpoints or attitudes.

LINK A person who could never vote to convict a defendant facing the death penalty can be excluded from the jury based on a challenge for cause, as explained in Chapter 11.

In its 1972 landmark decision in *Furman v. Georgia*, the Supreme Court ruled 5 to 4 that the death penalty, as it was applied, was unconstitutional, thus overturning death penalty statutes in 37 states.[79] The *Furman* case actually consisted of three unrelated cases, each involving an African American defendant. In all three cases, the defendants had been sentenced to death by a jury that was allowed complete discretion in the decision to impose a sentence of imprisonment or execution. The Court argued that the death penalty was being imposed in an arbitrary, infrequent, and discriminatory manner and held that it was unconstitutional for juries to exercise total discretion in imposing the sentence of death. The *Furman* decision suggested that states might rewrite statutes to reduce inequities in the application of the death penalty by removing judicial and jury discretion in sentencing. It was also suggested

that this goal might be achieved by specifying mandatory death sentences for all capital crimes or by developing statutes that guide the decision making of judges and juries.

During the next few years, 35 states developed new capital punishment statutes. In 1976, the Supreme Court handed down a critical decision in *Gregg v. Georgia*.[80] Troy Gregg had been convicted of robbing and murdering two men. After he was found guilty by the trial jury, a penalty hearing was held before the same jury, where aggravating and mitigating circumstances surrounding the case were presented. The judge carefully instructed the jury that the sentence of death would not be authorized unless the jury found beyond a reasonable doubt that one of several specific aggravating circumstances listed in the death penalty statute was present. The jury determined that two of the circumstances were present and returned a sentence of death. The Supreme Court, in a 7 to 2 decision, upheld Gregg's sentence, ruling that the Georgia statute providing for a bifurcated process and consideration of aggravating and mitigating circumstances was constitutional.

In a group of related cases in 1976, the Supreme Court struck down statutes that provided for mandatory death sentences because they failed to consider broadly enough the circumstances surrounding the crime.[81] A little more than a decade later, the Court ruled in *Sumner v. Shuman* that a mandatory death

Furman v. Georgia (1972) rendered capital punishment unconstitutional, until this penalty was reinstated in 1976.

? Is it likely that capital punishment will again be held unconstitutional? Why is this law so controversial when so few offenders are executed?

the defendant that the imposition of the death penalty had been influenced by racial considerations and was, therefore, unconstitutional. Warren McCleskey, an African American, had been convicted of two counts of armed robbery and the murder of a white police officer. Despite the introduction of extensive social science research demonstrating that African American offenders convicted of killing white victims were significantly more likely to receive a death sentence than were whites who killed whites, the Court held that there was no evidence to suggest that racial discrimination entered into the sentencing decision in McCleskey's case.

The Supreme Court has also ruled on whether a mentally retarded person can be executed. In 1989, in *Penry v. Lynaugh,* it held that a mentally retarded person may be sentenced to death.[85] The Court noted there was little consensus among the states to bar the execution of mentally retarded offenders and that only two states prohibited such executions at the time. Thirteen years later, in 2002, the Court reversed its decision allowing mentally retarded persons to be executed. In *Atkins v. Virginia,* the Court held that much had changed since the reasoning in *Penry* and that now "a significant number of states have concluded that death is not a suitable punishment for a mentally retarded criminal," and that to execute such a person would be a violation of the Eighth Amendment's prohibition of cruel and unusual punishment.[86]

In 1991, in *Payne v. Tennessee,* the Supreme Court overturned two prior decisions[87] that prevented the prosecutor from introducing evidence about the murder victim's character and the impact of the crime on the victim's family.[88] Chief Justice William Rehnquist, in writing the 6 to 3 decision that upheld the conviction of a Tennessee murderer, stated that barring victim impact statements in cases involving the death penalty "unfairly weighted the scales in a capital trial" in favor of the defendant, who is allowed to present all possible mitigating evidence about the crime and his or her own life circumstances.[89]

It is not unusual for a prosecutor to present evidence of the defendant's future dangerousness to the sentencing jury in an effort to persuade the jury members to apply a death sentence as an alternative to life imprisonment. Many people also believe that offenders who are serving life sentences are very likely to become eligible for parole. As a consequence, jurors may be more likely to vote for a death sentence if they are considering the offender's future dangerousness. Many states, however, have enacted life-

sentence was unconstitutional for a prisoner who was serving a life sentence and was convicted of murder while in prison, because mitigating factors must be considered.[82]

In 1977, the Supreme Court overturned a lower court's death sentence for rape in *Coker v. Georgia*.[83] In this case, the Court held that "a sentence of death is grossly disproportionate and excessive punishment for the crime of rape and is therefore forbidden by the Eighth Amendment as cruel and unusual punishment." However, the Court did not stipulate whether other nonhomicide offenses may be prohibited from punishment by the death penalty, and this controversial issue may appear before the Supreme Court in the near future.

One of the more controversial death penalty cases since the *Furman* decision in 1972 was *McCleskey v. Kemp*, which the Supreme Court ruled on in 1986.[84] In a 5 to 4 decision, the Court rejected the claim by

without-parole laws that would theoretically protect the public as effectively as capital punishment. In 1994, the Supreme Court, in *Simmons v. South Carolina,* ruled that when life-without-parole is an alternative sentence to the death penalty, and if the defendant's future dangerousness is an issue, the trial court must inform the jury that the defendant is ineligible for parole.[90]

Perhaps the most controversial death penalty decision by the Supreme Court in recent years was that handed down in 2005 in *Roper v. Simmons.*[91] This ruling overturned a 1989 Supreme Court decision in *Stanford v. Kentucky,* which allowed the execution of persons who were age 16 or 17 at the time they committed their crimes.[92] In *Roper,* the Court held that the execution of a person under the age of 18 is disproportionate punishment under the Eighth Amendment and, therefore, is cruel and unusual punishment.

The Death Penalty Today

Today, 38 states and the federal government have statutes authorizing the death penalty. However, since 1977, executions have been carried out in only 32 states. Five states—Texas, Florida, Oklahoma, Missouri, and Virginia—account for 66 percent of all executions since 1977, with Texas performing more than one-third of all executions in the United States.[93] Given that more than 3300 people are currently on death row, the potential for large numbers of executions in the future remains very high.

In 2006, 3366 prisoners were under a sentence of death in the United States. The majority were white, had never been married, had less than a high school education, and were overwhelmingly male **TABLE 12-4**. Of the 1057 prisoners executed between 1977 and 2006, 58 percent were white, 34 percent were African American, and 7 percent were Latino. During this period, only 11 women were executed, accounting for only 1 percent of all executions.[94]

Methods of Execution

Several methods of execution are used in the United States, including lethal injection, electrocution, lethal gas, hanging, and firing squad **TABLE 12-5**. Since 1977, 82 percent of all executions have been by lethal injection; indeed, all of the executions in 2006 used this method. Today 37 states use lethal injection as the exclusive or primary method of execution. Seventeen states authorize the use of more than one method; in such a case, the condemned prisoner is generally given the choice of method. The method

TABLE 12-4		
Demographic Profile of Prisoners Under Sentence of Death		
Characteristic	**Year-End 2005**	**2005 New Admissions**
Total number under sentence of death	3254	128
Sex		
Male	98.4%	96.1%
Female	1.6	3.9
Race		
White	55.5%	54.7%
African American	42.2	40.6
Other	2.4	4.7
Ethnicity		
Latino	12.7%	15.5%
Non-Latino	87.3	84.5
Median Age	28 years	
	(at time of arrest)	
Education		
Eighth grade or less	14.3%	9.9%
Ninth to eleventh grade	36.9	31.7
Twelfth grade	39.6	48.5
Any college	9.2	9.9
Median education	Eleventh grade	Twelfth grade
Marital Status		
Married	22.2%	17.6%
Divorced/separated	20.5	19.6
Widowed	2.9	3.9
Never married	54.4	57.0

Source: Tracy Snell, *Capital Punishment, 2005* (Washington, DC: U.S. Department of Justice, 2006).

of execution for federal prisoners is exclusively lethal injection.

Although the popularity of lethal injection may reflect public perceptions that it is more humane, legal challenges to this method of execution have recently come before the Supreme Court. In 2006, the Court heard an appeal by Clarence Hill, a Florida inmate who was convicted of murdering a police officer in 1982. Hill's appeal claimed that the three chemicals used in the lethal injection process would cause unnecessary pain in the execution and that the risk of inflicting unnecessary and wanton infliction of pain is contrary to contemporary standards of decency, thus constituting cruel and unusual punishment. During questioning, Justice Antonin Scalia noted that "lethal injection is much less painful than hanging, which has sometimes resulted in decapitation or slow strangulation."[95] The Court denied Hill's stay and he was executed on September 21, 2006, by lethal injection.

Christopher Simmons was 17 years old when he brutally murdered Shirley Crook on a September day in 1993. He was tried, convicted, and sentenced to death. His case was eventually appealed to the U.S. Supreme Court after the Missouri Supreme Court set Simmons's death sentence aside based on his claim that the death penalty was unconstitutional.

The majority in this 5 to 4 decision argued that 16- and 17-year-old murderers must be "categorically exempted" from capital punishment because they "cannot with reliability be classified among the worst offenders." The argument was built around three perceived differences between adults (age 18 and older) and juveniles:

1. Juveniles generally lack maturity and responsibility and are significantly more reckless than adults.

2. Juveniles are more vulnerable to outside influences, especially peer influences, because they have less control over their surroundings.

3. A "juvenile's character is not as fully formed as that of an adult."

According to the Supreme Court, "these differences render suspect any conclusion that a juvenile falls among the worst offenders" and that "from a moral standpoint it would be misguided to equate the failings of a minor with those of an adult."

In addition, the Court expressed its belief that juvenile murderers could be reformed: "Only a relatively small proportion of adolescents who experiment in risky or illegal activities develop entrenched patterns of problem behavior that persist into adulthood." Finally, the Court argued that the evolving standards of decency and perceived reduction in public support for the application of the juvenile death penalty, combined with the "the overwhelming weight of international opinion against the juvenile death penalty," required the Court to conclude that the execution of juvenile offenders was no longer constitutional.

Justice O'Connor and Justice Scalia wrote dissenting opinions. O'Connor argued that "no evidence impeaching the seemingly reasonable conclusion reached by many state legislatures: that at least some 17-year-old murderers are sufficiently mature to deserve the death penalty in an appropriate case" had been presented by the justices in the majority. She also noted that their analysis was "premised on differences in the aggregate between juveniles and adults, which frequently do not hold true when comparing individuals."

Scalia stated that the majority provided no evidence of a national consensus opposing the juvenile death penalty and that, in fact, "a number of legislatures and voters have expressly affirmed their support for capital punishment of 16- and 17-year-old offenders." In response to the majority drawing on international opinion to help form their decision, Scalia wrote that "the basic premise of the Court's argument—that American law should conform to the laws of the rest of the world—ought to be rejected out of hand."

Source: *Roper v. Simmons*, 543 U.S. 551 (2005).

Earlier in 2006, an inmate in California scheduled to be executed by lethal injection received an indefinite stay of execution because the state could not find any medical professional willing to attend the execution, as required by state law. In North Carolina and Tennessee, state courts have recently dismissed challenges to the use of lethal injection.[96]

The Capital Punishment Debate

Arguments for Capital Punishment

Contemporary arguments in support of capital punishment have largely focused on the deterrent impact of the penalty and the need to express moral condemnation of heinous crimes. Advocates of capital punishment are quick to point out its specific deterrent effect. According to James Q. Wilson, "Whatever else may be said about the death penalty, it is certain that it incapacitates."[97] In other words, people who are executed do not kill again. Advocates also argue that the death penalty carries a strong general deterrent: The very threat of the death penalty is thought to deter potential killers. In 1783, writer Samuel Johnson wrote that "executions are intended to draw spectators. If they do not draw spectators, they don't answer their purposes."[98] For the death penalty to deter murder, the potential offender must know of the penalty's application and believe that the certainty of its application is sufficient to make the possibility of incurring it an unacceptable risk.

TABLE 12-5
Method of Execution

Lethal Injection		Electrocution	Lethal Gas	Hanging	Firing Squad
Alabama	Montana	Alabama	Arizona	Delaware	Idaho
Arizona	Nevada	Arkansas	California	New Hampshire	Oklahoma
Arkansas	New Hampshire	Florida	Missouri	Washington	Utah
California	New Jersey	Kentucky	Wyoming		
Colorado	New Mexico	Nebraska			
Connecticut	New York	Oklahoma			
Delaware	North Carolina	South Carolina			
Florida	Ohio	Tennessee			
Georgia	Oklahoma	Virginia			
Idaho	Oregon				
Illinois	Pennsylvania				
Indiana	South Carolina				
Kansas	South Dakota				
Kentucky	Tennessee				
Louisiana	Texas				
Maryland	Utah				
Mississippi	Virginia				
Missouri	Washington				
	Wyoming				

Source: Tracy Snell, *Capital Punishment, 2005* (Washington, DC: U.S. Department of Justice, 2006), p. 4.

A large number of studies appear to support the general deterrence hypothesis. Economist Isaac Ehrlich's analysis of homicide between 1934 and 1969 led him to conclude that an additional execution per year would prevent seven or eight other murders annually.[99] When Steven Layson examined homicides and executions from 1936 to 1977, he concluded that 18 murders are prevented by each execution. Paul Zimmerman's analysis of executions and subsequent homicides between 1978 and 1997 found that each execution deters 14 murders per year on average.[100] Finally, a study by Hashem Dezhbakhsh and his colleagues of data from 2054 U.S. counties between 1977 and 1996 found that the murder rate was significantly reduced by both death sentences and executions and that, on average, each execution produced 18 fewer murders.[101]

Another justification for capital punishment rests on the principle of retribution. Retribution may involve a base desire for revenge or represent a general statement by society that the most extreme crimes must be dealt with by the most extreme punishment. Society's moral outrage at murder requires that the act be condemned by an equally severe response. The Supreme Court, in its justification of capital punishment in the *Gregg* decision, noted:

Despite his argument that lethal injection is cruel and unusual punishment, Clarence Hill was executed by lethal injection in 2006.

? Why is each of the three drugs in lethal injections administered in a fatal dose? How can critics contend that inmates suffer cruel and unusual punishment when they are rendered unconscious in seconds?

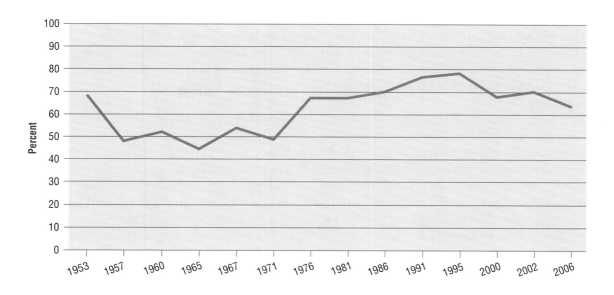

FIGURE 12-8 Public Support for the Death Penalty in the United States, 1953–2006

Source: Gallup Organization, *The Gallup Poll* June 1, 2006, available at http://www.galluppoll.com/content/?ci=1606&pg1, accessed January 10, 2007.

Indeed, the decision that capital punishment may be the appropriate sanction in extreme cases is an expression of the community's belief that certain crimes are themselves so grievous an affront to humanity that the only adequate response may be the penalty of death.[102]

Some scholars believe that capital punishment is actually morally required as a response to murder. For example, Cass Sunstein and Adrian Vermeule say that although many people believe that it is morally wrong for the state to intentionally take the life of a person, the state may be morally obligated to maintain capital punishment to save the lives of innocent people.[103] Sunstein and Vermeule suggest that if the death penalty is the only or the most effective means to prevent some larger number of murders from occurring, then failing to use capital punishment would also result in a failure to protect life.

In the United States, there continues to be broad public support for the death penalty. Since the mid-1970s, between 60 percent and 70 percent of Americans surveyed have maintained that they are in favor of the death penalty for a person convicted of murder **FIGURE 12-8** . Support declines, however, if people are asked if they would still support capital punishment if there was a viable alternative, such as life in prison without parole. When given such a

choice, support for the death penalty drops to about 52 percent. Americans also are less supportive of capital punishment if the defendant is mentally retarded or mentally ill, female, or under age 18.[104]

Arguments against Capital Punishment

In 2006, more than 10,000 people were arrested for homicide and non-negligent manslaughter.[105] During the same year, 114 offenders received the death penalty, but only 53 were actually executed. Although a large number of people who were arrested for murder did not commit death-eligible homicides (crimes that by law may carry the death penalty), it is clear that throughout the criminal justice process, decisions are made that result in very few prosecutions on death-eligible charges and even fewer death sentences.

Critics of capital punishment point to apparent-discrimination in the application of the death penalty. They argue that many prosecutors are more likely to select homicide cases for prosecution that have African American defendants and white victims, that juries are more likely to convict African Americans, and that death sentences are disproportionately imposed on minority defendants. The race of both offender and victim appear to be significant factors in this selective process. A number of studies have shown that African Americans who are charged with the murder of whites have a much greater risk of receiving a death sentence[106] and also are less likely to have their death sentences commuted to a lesser sentence[107] than whites who kill African Americans or people who are involved in intraracial killings.

Studies have shown that, regardless of the race of offender, the race of the victim alone appears to have a significant effect on death sentencing.[108] When Glenn Pierce and Michael Radelet examined all homicides and related death sentences in California from 1990 through 1999, they found that convicted murderers were significantly more likely to receive the death penalty if their victims were white.[109] Those who murdered whites were over three times more likely to be sentenced to death than were those who murdered African Americans and over four times more likely to be sentenced to death than those who murdered Latinos.

The evidence also suggests that gender is an important factor in application of the penalty. Women who are convicted of homicides that are eligible for the death penalty are less likely to be sentenced to death than men, although women who kill intimates (i.e., family members or lovers) are more likely to receive death sentences than men who kill intimates.[110] However, studies on how the race and gender of the victim interact to influence death sentencing outcomes have yielded mixed findings. While Marian Williams and Jefferson Holcomb[111] report that killers of white women are at greater risk of receiving a death sentence in Ohio, Amy Stauffer and her colleagues report no effect of race and gender in their analysis of death sentences in North Carolina.[112]

Critics also claim that the pursuit of retribution through capital punishment produces a brutalizing effect on society, ultimately increasing—rather than decreasing—homicide rates. The argument here is that even low levels of executions "brutalize" members of society, leading people to devalue human life and to see some degree of legitimacy in the use of retaliatory violence.[113] Research provides some support for the claim that violence may increase as a result of executions. For example, sociologist William Bowers examined the relationship between executions and violence and reported that a clear short-term brutalizing effect occurs within the first month or two following an execution.[114] Bowers believes that a small number of people existing in a state of emotional turmoil are ready to kill and are pushed over the edge by the message they perceive from executions—namely, that it is appropriate to kill those who betray, disgrace, or dishonor them.

In addition, other studies suggest that capital punishment fails to deter murder more effectively than long-term imprisonment.[115] It is unknown whether this trend arises because murderers believe that they are unlikely to receive the death penalty owing to its infrequent use or because they are unlikely to be apprehended. Also, most people who commit murder do so under the influence of alcohol, other drugs, or severe emotional turmoil, so they may not be able to make rational judgments about the penalties attached to the crime.

Additionally, opponents of the death penalty point to the fact that the criminal justice system is not infallible. Truly innocent people are sometimes arrested and even convicted of crimes they did not commit. Often, such errors are corrected on appeal, but no appeal is possible once the defendant has been executed. In 2000, Illinois Governor George Ryan suspended executions and then three years later commuted the death sentences of all inmates on death row, arguing that the possibility of a death-row inmate being innocent was simply too great.[116]

While we cannot be certain that a truly innocent person has been executed in recent decades, strong evidence exists that a number of persons on death row were not guilty of the crimes that resulted in their death sentences. According to the Innocence Project (a national litigation and public policy organization), 200 persons have been exonerated by DNA evidence in the United States since the Innocence Project began its work in 1994.[117] While most of the 200 exonerations did not involve death-row inmates, at least 13 inmates facing execution have been exonerated on the basis of DNA tests and more than 60 have been exonerated largely on the basis of either new evidence establishing the prisoner's innocence or the exclusion at retrial of inappropriately admitted evidence from the initial trial. It is very possible that the discovery of these false convictions was a result of the much greater attention given to the review of death penalty cases after convictions. Alternatively, false convictions may be more likely to occur in capital murder cases simply because there is greater room for error—the victims are dead and unavailable as witnesses, there is an incentive for the real killers to frame others to avoid punishment, police face enormous pressure to clear murders, and prosecutors want to obtain a conviction and help bring closure to the victim's family.[118]

Appeal of Sentence

LINK Unlike the post-conviction appeal of errors that may have occurred in a trial (discussed in Chapter 11), appellate review of sentencing is more restricted.

Grounds for Appellate Review of Sentence

Five types of sentences typically lead to an appeal:

- Sentences that are in violation of the Eighth Amendment prohibition against cruel and unusual punishment
- Sentences that are disproportionate to the seriousness of the offense
- Sentences based on an abuse of discretionary power of the sentencing judge
- Sentences that fall outside statutory sentencing guidelines
- Sentences imposed by a court not having jurisdictional authority to impose the sentence

All cases that meet one or more of these criteria may be appealed to a higher court and follow the same process of appeal.

LINK Appeals and habeas corpus petitions differ in several ways, as discussed in Chapter 11.

The Appeal Process

As in general post-conviction appeals, a notice of appeal of a sentence must be filed with the court (usually within 30 to 90 days following the defendant's conviction). Following the filing of the notice, the defendant must file a written brief identifying the error in sentencing. The state is then required to respond by filing a brief in the case, and the appellate court schedules both briefs for review. If the appellate court believes that an error was made, the case will be returned to the lower court for resentencing.

Since the *Furman v. Georgia* decision in 1972, all death penalty sentences have been required to undergo automatic review by an intermediate appellate court or the state court of last resort.[119] Death sentences also may be appealed directly to the Supreme Court through a petition by the offender for a writ of habeas corpus.

Recent decisions by the Supreme Court restricting habeas corpus petitions by prisoners included those filed by inmates under death sentences. For example, in *McCleskey v. Kemp*, Warren McCleskey's first petition for a writ of habeas corpus was accepted by the Court and resulted in the 1986 decision discussed previously. Five days before his scheduled execution in July 1987, McCleskey appealed a second time, this time claiming that his Sixth Amendment right to counsel had been violated after it was discovered that prosecutors had planted an informant in an adjoining cell. The trial judge stayed McCleskey's execution and ordered a new trial. Later, a federal appeals court reinstated McCleskey's conviction on the grounds that the defendant should have raised the informant issue in his first appeal, despite the fact that neither McCleskey nor his lawyers had any way of knowing of the existence of the informant. Once again, McCleskey petitioned for a writ of habeas corpus from the Supreme Court. The Court, in *McCleskey v. Zant*, held that McCleskey had no legitimate excuse for failing to bring up his claim.[120] Reflecting the Court's growing irritation with seemingly endless numbers of habeas corpus petitions by prisoners, Justice Anthony Kennedy, in writing the 6 to 3 majority opinion, stated, "Perpetual disrespect for the finality of convictions disparages the entire criminal justice system." Now, after exhausting state appeals, most prisoners are allowed only one appeal in the federal courts.

WRAP UP

Several punishment philosophies are at the heart of criminal sentencing. Rehabilitation aims to reform an offender; thus a sentence of three years probation with conditions would provide ample opportunities for the defendants to understand what they did wrong, resolve the problems or issues that led to their criminal behavior, and provide an incentive (avoiding confinement) to work toward their treatment.

By contrast, deterrence aims to send a message to offenders to reduce crime. Because these defendants had little to no prior criminal record, it is doubtful that they are a risk to re-offend. A sentence of community service and a $10,000 fine would likely deter future offending.

Retribution expresses society's moral outrage or disapproval of a crime. On the surface, it is the punishment philosophy that most immediately comes to mind with severe child abuse cases because the details of the crime are reprehensible. A sentence of 11 years in prison—that is, one year for each of the 11 victims—would express society's disapproval.

Finally, incapacitation means to remove an offender from society. The prison terms imposed on the couple would limit their ability to commit child abuse.

Chapter Spotlight

- There are five general goals in sentencing: retribution, deterrence, incapacitation, rehabilitation, and proportionality.

- Sentencing is primarily linked to specific offenses or categories of offenses, with more serious offenses generally carrying more severe punishments.

- Intermediate sentences offer alternatives to the traditional sentences of imprisonment and probation. They include intermittent sentences, intensive probation, fines, community service, restitution, and forfeiture.

- Indeterminate prison sentences have been largely replaced by determinate sentencing systems, including presumptive sentencing, sentencing guidelines, mandatory sentences, and three-strikes laws, in an attempt to reduce the discretion of judges and make sentences more uniform.

- Multiple incarceration sentences may run concurrently (at the same time) or consecutively (one after the other).

- At the sentencing hearing, a judge reviews the presentence investigation (PSI) report, the offender is given an opportunity to speak and make a case for a lenient sentence, the prosecutor may make a recommendation regarding the sentence, and the victim is allowed to submit a victim impact statement or statement of opinion.

- The discretion permitted in the judge's sentencing decision in many jurisdictions often results in sentencing disparities (differences in sentences for people who are convicted of similar offenses). Sentencing discrimination may also occur when judges abuse their discretion and consider illegitimate, morally objectionable, or extralegal factors in determining sentences.

- Many proponents of capital punishment stress its deterrent effect; others support capital punishment as a form of retribution. Critics of the death penalty argue that it is discriminatory, has a brutalizing effect on society, fails to deter homicides more effectively than long-term imprisonment, and has resulted in innocent people being put to death.

- An offender may appeal what he or she considers to be an unjust or faulty sentence. If the appellate court finds that an error was made in sentencing, the case is returned to the lower court for resentencing.

Putting It All Together

1. Which goals of punishment (e.g., retribution, deterrence, incapacitation, rehabilitation, and propotionality) should guide sentencing?

2. To what extent should the background characteristics of offenders (i.e., family, education, prior military service) play a role in determining a sentence?

3. Do structured sentencing systems provide greater justice in sentencing? Why or why not?

4. Is the death penalty a positive or negative factor in dealing with crime in society?

5. Are there crimes other than first-degree murder that might justify the use of the death penalty? If so, what are they?

bifurcated trial A two-stage trial: the first stage determines guilt, and the second stage determines the sentence.

Coker v. Georgia The Supreme Court ruling that a death sentence that is grossly disproportionate to the crime is unconstitutional.

concurrent sentences Two or more prison sentences to be served at the same time.

consecutive sentences Two or more prison sentences to be served one after the other.

determinate sentence A prison sentence with a fixed term of imprisonment.

deterrence A punishment philosophy based on the belief that punishing offenders will deter crime.

Furman v. Georgia The Supreme Court ruling that the death penalty, as applied at that time, was unconstitutional.

general deterrence Punishing offenders to discourage others from committing crimes.

Gregg v. Georgia The Supreme Court ruling that the death penalty was constitutional under a state statute requiring the judge and the jury to consider both aggravating and mitigating circumstances.

incapacitation A punishment aimed at removing offenders from the community through imprisonment or banishment.

indeterminate sentence A prison sentence that identifies a minimum and a maximum number of years to be served by the offender; the actual release date is set by a parole board or the institution.

mandatory sentence A requirement that an offender must be sentenced to prison.

McCleskey v. Kemp The Supreme Court ruling that rejected the claim that the death penalty law in

Georgia was unconstitutional because it promoted racial discrimination.

Payne v. Tennessee The Supreme Court ruling that statutes that bar the introduction of victim impact statements in death penalty cases are unconstitutional.

presentence investigation (PSI) A comprehensive report including information on the offender's background and offense and any other information the judge desires to determine an appropriate sentence.

presumptive sentencing The use of ranges—that is, minimum and maximum number of years of incarceration—set for types of particular crimes. The judge determines the number of years to be served from within this range.

proportionality A punishment philosophy based on the belief that the severity of the punishment should fit the seriousness of the crime.

rehabilitation A sentencing objective aimed at reforming an offender through treatment, education, or counseling.

retribution A punishment philosophy based on society's moral outrage or disapproval of a crime.

sentencing discrimination Differences in sentencing outcomes based on illegitimate, morally objectionable, or extralegal factors.

sentencing disparities Differences in sentencing outcomes in cases with similar case attributes.

sentencing guidelines Sentencing schemes that limit judicial discretion; the offender's criminal background and severity of current offense are plotted on a grid to determine the sentence.

sentencing hearing A court hearing to determine an appropriate sentence, which is typically scheduled within three to six weeks after the offender's conviction.

JBPUB.COM/ExploringCJ

specific deterrence Punishing offenders to prevent them from committing new crimes.

three-strikes laws Laws that provide a mandatory sentence of incarceration for persons who are convicted of a third separate serious criminal offense.

truth-in-sentencing Laws that require offenders, especially violent offenders, to serve at least 85 percent of their sentences.

victim impact statement (VIS) A statement informing the sentencing judge of the physical, financial, and emotional harm suffered by the crime victim or his or her family.

Notes

1. James Q. Wilson, *Thinking about Crime,* revised edition (New York: Vintage Books, 1985), p. 260.
2. Thomas Hobbes, cited in Graeme Newman, *The Punishment Response* (Albany, NY: Harrow and Heston, 1985), p. 201.
3. Joel Feinberg and Hyman Gross, *Philosophy of Law,* 2nd edition (Belmont, CA: Wadsworth, 1980), p. 515.
4. Exodus 21:24, *New American Standard Bible: Reference Edition* (New York: Cambridge, 1977).
5. Harold Pepinsky and Paul Jesilow, *Myths That Cause Crime* (Cabin John, MD: Seven Locks Press, 1984).
6. Andrew von Hirsch, *Doing Justice: The Choice of Punishments* (New York: Hill and Wang, 1976), p. 54.
7. Cesare Beccaria, *On Crimes and Punishments,* trans. by Henry Paolucci (New York: Bobbs-Merrill, 1963); J. H. Burns and H. L. A. Hart (eds.), *The Collected Works of Jeremy Bentham: An Introduction to the Principles of Morals and Legislation* (London: Athlone Press, 1970).
8. Steven Levitt, "Why Do Increased Arrest Rates Appear to Reduce Crime: Deterrence, Incapacitation, or Measurement Error?" *Economic Inquiry* 36:353–373 (1998).
9. Ronald Clarke and Derek Cornish, "Modeling Offender's Decisions: A Framework for Research and Policy," in Michael Tonry and Norval Morris (eds.), *Crime and Justice,* Volume 7:147–186 (Chicago: University of Chicago Press, 1985).
10. Lawrence Sherman and Richard Berk, "The Specific Deterrent Effects of Arrest for Domestic Assault," *American Sociological Review* 49:261–272 (1984).
11. Robert Davis, Barbara Smith, and Laura Nickles, "The Deterrent Effect of Prosecuting Domestic Violence Misdemeanors," *Crime & Delinquency* 44:434–442 (1998).
12. David Greenberg, "The Incapacitative Effect of Imprisonment: Some Estimates," *Law and Society Review* 9:541–580 (1975); Stephan Van Dine, Simon Dinitz, and John Conrad, "The Incapacitation of the Dangerous Offender: A Statistical Experiment," *Journal of Research in Crime and Delinquency* 14:22–34 (1977).
13. Levitt, note 8; Don Gottfredson, *Effects of Judges' Sentencing Decisions on Decisions on Criminal Careers* (Washington, DC: National Institute of Justice, 1999).
14. Christina DeJong, "Survival Analysis and Specific Deterrence: Integrating Theoretical and Empirical Models of Recidivism," *Criminology* 35:561–576 (1997).
15. Robert Martinson, "What Works? Questions and Answers about Prison Reform," *Public Interest* 35:22–54 (1974); Douglas Lipton, Robert Martinson, and Judith Wilks, *The Effectiveness of Correctional Treatment: A Survey of Evaluation Studies* (New York: Praeger, 1975); William Bailey, "Correctional Outcome: An Evaluation of 100 Reports," *Journal of Criminal Law, Criminology, and Police Science* 57:153–160 (1966); Hans Eysenck, "The Effects of Psychotherapy," *International Journal of Psychiatry* 1:99–144 (1965).
16. D. A. Andrews, Ivan Zinger, Robert Hoge, James Bonta, Paul Gendreau, and Francis Cullen, "Does Correctional Treatment Work? A Clinically Relevant and Psychologically Informed Meta-Analysis," *Criminology* 28:369–404 (1990); Lawrence W. Sherman, Denise Gottfredson, Doris MacKenzie, John Eck, Peter Reuter, and Shawn Bushway, *Preventing Crime: What Works, What Doesn't, What's Promising* (Washington, DC: National Institute of Justice, 1998); Lawrence Sherman, David Farrington, Doris MacKenzie, Brandon Walsh, Denise Gottfredson, John Eck, Shawn Bushway, and Peter Reuter, *Evidence-Based Crime Prevention* (London: Routledge and Kegan Paul, 2002).
17. Andrews et al., note 16; Mark Lipsey, *The Efficacy of Intervention for Juvenile Delinquency: Results from 400 Studies,* paper presented at the American Society of Criminology annual meeting, Reno, NV (1989).
18. "Louisiana Supreme Court Upholds Death Penality in Child Rape Case," available at http://www.associatedcontent.com/article/256067/louisiana_supreme_court_upholds_death.html, accessed July 17, 2007.
19. *Weems v. United States,* 217 U.S. 349 (1910).
20. *Trop v. Dulles,* 356 U.S. 86 (1958).
21. *Harmelin v. Michigan,* 501 U.S. 957 (1991).
22. Sally Hillsman, Barry Mahoney, George Cole, and Bernard Auchter, *Fines as Criminal Sanctions* (Washington, DC: National Institute of Justice, 1987).

23. Matthew Durose and Patrick Langan, *Felony Sentences in State Courts, 2002* (Washington, DC: U.S. Department of Justice, 2004).

24. Todd Clear, John D. Hewitt, and Robert M. Regoli, "Discretion and the Determinate Sentence: Its Distribution, Control, and Effect on Time Served," *Crime & Delinquency* 24:428–445 (1978).

25. David Fogel, ". . . We Are the Living Proof . . . ," in *The Justice Model for Corrections* (Cincinnati: Anderson, 1975), p. 310.

26. John D. Hewitt, Robert M. Regoli, and Todd Clear, "Evaluating the Cook County Sentencing Guidelines," *Law and Policy Quarterly* 4:260 (1982).

27. Austin Lovegrove, *Judicial Decision Making, Sentencing Policy, and Numerical Guidance* (London: Springer-Verlag, 1989), p. 21.

28. *Blakely v. Washington,* 542 U.S. 296 (2004).

29. *United States v. Booker,* 543 U.S. 220 (2005).

30. Don Stemen and Daniel Wilhelm, "After *Blakely v. Washington,* States Seek to Protect or Revise their Sentencing Structures," *State News,* June/July 2005. pp. 20–25.

31. *Cunningham v. California*, 05-6551 (2007).

32. Milton Heumann and Colin Loftin, "Mandatory Sentencing and the Abolition of Plea Bargaining: The Michigan Felony Firearm Statute," *Law and Society Review* 13:393–430 (1979); U.S. Department of Justice, "Mandatory Sentencing: The Experience of Two States," *Policy Briefs* (Washington, DC: U.S. Government Printing Office, 1982).

33. Barbara Meierhoefer, *The General Effect of Mandatory Minimum Prison Terms* (Washington, DC: Federal Judicial Center, 1992).

34. Larry Rohter, "In Wave of Anticrime Fervor, States Rush to Adopt Laws," *New York Times,* May 10, 1994, p. A19.

35. Tomislav Kovandzic, John Sloan, and Lynne Vieraitis, "'Striking Out' as Crime Reduction Policy: The Impact of 'Three Strikes' Laws on Crime Rates in U.S. Cities," *Justice Quarterly* 21:207–240 (2004); Kevin Meehan, "California's Three-Strikes Law: The First Six Years," *Corrections Management Quarterly* 4:22–34 (2000); Walter Dickey and Pam Hollenhorst, "Three-Strikes Laws: Five Years Later," *Corrections Management Quarterly* 3:1–19 (1999).

36. Katherine Rosich and Kamala Kane, "Truth in Sentencing and State Sentencing Practices," *NIJ Journal* 252:18–21 (2005).

37. Massachusetts Sentencing Commission, *Survey of Sentencing Practices: Truth-in-Sentencing Reform in Massachusetts* (Boston: Massachusetts Sentencing Commission, 2000); William Sabol, Katherine Rosich, Kamala Kane, David Kirk, and Glenn Dubin, *The Influences of Truth-in-Sentencing Reforms on Changes in States' Sentencing Practices and Prison Populations* (Washington, DC: Urban Institute, 2002).

38. Kovandzic et al., note 35; Tomislav Kovandzic, John Sloan, and Lynne Vieraitis, "Unintended Consequences of Politically Popular Sentencing Policy: The Homicide Promoting Effects of 'Three-Strikes' Laws in U.S. Cities," *Criminology and Public Policy* 1:399–424 (2002); Thomas Marvell and Carlisle Moody, "The Lethal Effects of Three Strikes Laws," *Journal of Legal Studies* 30:89–106 (2001); Franklin Zimring, Sam Kamin, and Gordon Hawkins, *Crime and Punishment in California* (Berkeley, CA: Institute of Governmental Studies Press, 1999).

39. Dickey and Hollenhorst, note 35.

40. Lovegrove, note 27, p. 9.

41. Robert Carter and Leslie Wilkins, "Some Factors in Sentencing Policy," *Journal of Criminal Law, Criminology, and Police Science* 58:503–514 (1967); Curtis Campbell, Candace McCoy, and Chimezie Osigweh, "The Influence of Probation Recommendations on Sentencing Decisions and Their Predictive Accuracy," *Federal Probation* 54:13–21 (1990); John Rosencrance, "Maintaining the Myth of Individualized Justice: Probation Presentence Reports," *Justice Quarterly* 5:235–256 (1988).

42. John Hagan, John D. Hewitt, and Duane Alwin, "Ceremonial Justice: Crime and Punishment in a Loosely Coupled System," *Social Forces* 58:506–527 (1979).

43. *Ring v. Arizona,* 536 U.S. 584 (2002).

44. *Summerlin v. Stewart,* 341 F.3d 1082 (2003).

45. *Schriro v. Summerlin,* 542 U.S. 348 (2004).

46. Maureen McLeod, "Victim Participation at Sentencing," *Criminal Law Bulletin* 22:501–517 (1986).

47. McLeod, note 46, pp. 503–504.

48. Anthony Walsh, "Placebo Justice: Victim Recommendations and Offender Sentences in Sexual Assault Cases," *Journal of Criminal Law and Criminology* 77:1126–1141 (1986).

49. Edna Erez, "Victim Participation in Sentencing: Rhetoric and Reality," *Journal of Criminal Justice* 18:29 (1990); Robert Davis and Barbara Smith, "Victim Impact Statements and Victim Satisfaction: An Unfulfilled Promise," *Journal of Criminal Justice* 22:1–12 (1994).

50. Edna Erez and Pamela Tontodonato, "The Effect of Victim Participation in Sentencing on Sentence Outcome," *Criminology* 28:455 (1990).

51. Bryan Myers and Jack Arbuthnot, "The Effects of Victim Impact Evidence on the Verdicts of Sentencing Judgments of Mock Jurors," *Journal of Offender Rehabilitation* 29:95–112 (1999).

52. Alfred Blumstein, Jacqueline Cohen, Susan Martin, and Michael Tonry, *Research on Sentencing: The Search for Reform,* Volume I (Washington, DC: National Academy Press, 1983), p. 8.

53. Lovegrove, note 27, p. 4.

54. Jeffery Ulmer and Mindy Bradley, "Variation in Trial Penalties among Serious Violent Offenses," *Criminology* 44:631–670 (2006).

55. Paige Harrison and Allen Beck, *Prisoners in 2005* (Washington, DC: U.S. Department of Justice, 2006), p. 8.

56. Gary Kleck, "Racial Discrimination in Sentencing: A Critical Evaluation of the Evidence with Additional Evidence on the Death Penalty," *American Sociological Review* 46:783–805 (1981); Martha Myers and Susette Talarico, *The Social Contexts of Criminal Sentencing* (New York: Springer-Verlag, 1987); Cassia Spohn, "The Sentencing Decisions of Black and White Judges: Expected and Unexpected Similarities," *Law and Society Review* 24:1197–1216 (1990); Stephen Klein, Joan Petersilia, and Susan Turner, "Race and Imprisonment Decisions in California," *Science* 247:812–816 (1990); John Dawson and Barbara Boland, *Murder in Large Urban Counties, 1988* (Washington, DC: U.S. Department of Justice, 1993).

57. James Gibson, "Race as a Determinant of Criminal Sentences: A Methodological Critique and a Case Study," *Law and Society Review* 12:455–478 (1978).

58. Kathleen Daly and Michael Tonry, "Gender, Race, and Sentencing," *Crime and Justice* 22:201–252 (1997).

59. Cassia Spohn, *Offender Race and Case Outcomes: Do Crime Seriousness and Strength of Evidence Matter? Final Activities Report Submitted to the National Institute of Justice* (Washington, DC: National Institute of Justice, 2000).

60. Shawn Bushway and Anne Piehl, "Judging Judicial Discretion: Legal Factors and Racial Discrimination in Sentencing," *Law and Society Review* 35:733–764 (2001).

61. Brian Johnson, "Racial and Ethnic Disparities in Sentencing Departures Across Modes of Sentencing," *Criminology* 41:449–490 (2003).

62. Darrell Steffensmeier and Stephen Demuth, "Ethnicity and Sentencing Outcomes in U.S. Federal Courts: Who Is Punished More Harshly?" *American Sociological Review* 65:705–729 (2000).

63. Harrison and Beck, note 55, p. 4.

64. Dawson and Boland, note 56.

65. Johnson, note 61.

66. Daly and Tonry, note 58.

67. Roy Lotz and John D. Hewitt, "The Influence of Legally Irrelevant Factors on Felony Sentencing," *Sociological Inquiry,* 47:39–48 (1977); John D. Hewitt, "The Effects of Individual Resources in Judicial Sentencing," *Review of Public Data Use* 5:30–51 (1977); Darrell Steffensmeier, John Kramer, and Cathy Streifel, "Gender and Imprisonment Decisions," *Criminology* 31:411–446 (1993).

68. Kathleen Daly, "Discrimination in the Criminal Courts: Family, Gender, and the Problem of Equal Treatment," *Social Forces* 66:154 (1987).

69. Daly and Tonry, note 58.

70. David Jacobs, "Inequality and the Legal Order: An Ecological Test of the Conflict Model," *Social Problems* 25:515–525 (1978); Alan Lizotte, "Extra-Legal Factors in Chicago's Criminal Courts: Testing the Conflict Model of Criminal Justice," *Social Problems* 25:564–580 (1978).

71. Theodore Chiricos and Gordon Waldo, "Socioeconomic Status and Criminal Sentencing: An Empirical Assessment of a Conflict Proposition," *American Sociological Review* 40:753–772 (1975).

72. Myers and Talarico, note 56, p. 81.

73. Myers and Talarico, note 56, p. 63.

74. Gerri Turner and Dean Champion, "The Elderly Offender and Sentencing Leniency," *Journal of Offender Counseling, Services, and Rehabilitation* 13:125–140 (1988/1989).

75. Johnson, note 61.

76. Tracy Snell, *Capital Punishment, 2006* (Washington, DC: U.S. Department of Justice, 2007).

77. *Witherspoon v. Illinois,* 391 U.S. 510 (1968).

78. *Lockhart v. McCree,* 476 U.S. 162 (1986).

79. *Furman v. Georgia,* 408 U.S. 238 (1972).

80. *Gregg v. Georgia,* 428 U.S. 153 (1976).

81. *Roberts v. Louisiana,* 428 U.S. 325 (1976); *Woodson v. North Carolina,* 428 U.S. 280 (1976).

82. *Sumner v. Shuman,* 483 U.S. 66 (1987).

83. *Coker v. Georgia,* 433 U.S. 584 (1977).

84. *McCleskey v. Kemp,* 481 U.S. 279 (1986).

85. *Penry v. Lynaugh,* 492 U.S. 302 (1989).

86. *Atkins v. Virginia,* 536 U.S. 304 (2002).

87. *Booth v. Maryland,* 482 U.S. 496 (1987); *South Carolina v. Gathers,* 490 U.S. 805 (1989).

88. *Payne v. Tennessee,* 498 U.S. 1076 (1991).

89. *Payne v. Tennessee,* note 88.

90. *Simmons v. South Carolina,* 114 S. Ct. 2187 (1994).

91. *Roper v. Simmons,* 543 U.S. 551 (2005).

92. *Stanford v. Kentucky,* 492 U.S. 361 (1989).

93. Snell, note 76.

94. Snell, note 76.

95. *Hill v. Florida,* No. SC06-2 (2006).

96. Ron Word, "High Court to Hear Lethal Injection Case," *Mercury News,* April 23, 2006, p. A1; Bill Mears, "Justices: Does Lethal Injection Hurt?" April 26, 2006, available at http://www.cnn.com.2006/LAW/04/26/scotus.injection/index.html, accessed July 17, 2006.

97. Wilson, note 1, p. 178.

98. James Boswell, *Boswell's Life of Johnson* (Chicago: Scott, Foresman, 1923).

99. Isaac Ehrlich, "The Deterrent Effect of Capital Punishment: A Question of Life and Death," *American Economic Review* 65:397–417 (1976).

100. Paul Zimmerman, "State Executions, Deterrence, and the Incidence of Murder," *Journal of Applied Economics* 7:163–193 (2004).

101. Hashem Dezhbakhsh, Paul Rubin, and Joanna Shepherd, "Does Capital Punishment Have a Deterrent Effect? Evidence from Postmoratorium Panel Data," *American Law and Economics Review* 5:344–376 (2003).

102. *Gregg v. Georgia,* note 80.

103. Cass Sunstein and Adrian Vermeule, "Is Capital Punishment Morally Required? Acts, Omissions, and Life-Life Tradeoffs," *Stanford Law Review* 58:703–750 (2005).

104. Pamela Paul, "The Death Penalty," *American Demographics* 23:22–23 (2001).

105. Federal Bureau of Investigation, *Crime in the United States, 2006* (Washington, DC: U.S. Department of Justice, 2007).

106. Thomas Keil and Gennaro Vito, "Race and the Death Penalty in Kentucky Murder Trials: An Analysis of Post-*Gregg* Outcomes," *Justice Quarterly* 7:189–206 (1990); Samuel Gross and Robert Mauro, *Death and Discrimination: Racial Disparities in Capital Sentencing* (Boston: Northeastern University Press, 1989); Raymond Paternoster, "Prosecutorial Discretion in Requesting the Death Penalty: A Case of Victim-Based Racial Discrimination," *Law and Society Review* 18:437–478 (1984); Michael Radelet and Glenn Pierce, "Race and Prosecutorial Discretion in Homicide Cases," *Law and Society Review* 19:587–621 (1985); Hans Zeisel, "Race Bias in the Administration of the Death Penalty: The Florida Experience," *Harvard Law Review* 95:456–468 (1981); Kleck, note 56.

107. Marvin Wolfgang, Arlene Kelly, and Hans Nolde, "Comparison of the Executed and Commuted Among Admissions to Death Row," *Journal of Criminal Law and Criminology* 53:301–310 (1962).

108. David Baldus, Charles Pulaski, and George Woodworth, "Comparative Review of Death Sentences: An Empirical Study of the Georgia Experience," *Journal of Criminal Law and Criminology* 74:661–685 (1983); David Baldus and George Woodworth, "Race Discrimination in the Administration of the Death Penalty: An Overview of the Empirical Evidence with Special Emphasis on the Post-1990 Research," *Criminal Law Bulletin* 39:194–227 (2003).

109. Glenn Pierce and Michael Radelet, "The Impact of Legally Inappropriate Factors on Death Sentencing for California Homicides, 1990–1999," *Santa Clara Law Review* 46:1–47 (2005).

110. Elizabeth Rapaport, "The Death Penalty and Gender Discrimination," *Law and Society Review* 25:367–383 (1991); Dawson and Boland, note 56.

111. Marian Williams and Jefferson Holcomb, "The Interactive Effects of Victim Race and Gender on Death Sentence Disparity Findings," *Homicide Studies* 8:350–376 (2004).

112. Amy Stauffer, Dwayne Smith, John Cochran, Sondra Fogel, and Beth Bjerregaard, "The Interaction Between Victim Race and Gender on Sentencing Outcomes in Capital Murder Trials," *Homicide Studies* 10:98–117 (2006).

113. Joanna Shepherd, "Deterrence Versus Brutalization: Capital Punishment's Differing Impacts Among the States," *Michigan Law Review* 104:203–255 (2005); Carol Steiker, "No, Capital Punishment Is Not Morally Required: Deterrence, Deontology, and the Death Penalty," *Stanford Law Review* 58:751-790 (2005).

114. William Bowers, "The Effect of Executions Is Brutalization, Not Deterrence," in Kenneth Haas and James Inciardi (eds.), *Challenging Capital Punishment: Legal and Social Science Approaches* (Newbury Park, CA: Sage, 1988), pp. 49–89. See also William Bowers and Glenn Pierce, "Deterrence or Brutalization? What Is the Effect of Executions?" *Crime & Delinquency* 26:453–484 (1980).

115. William Bailey, "Imprisonment vs. the Death Penalty as a Deterrent to Murder," *Law and Human Behavior* 1:239–260 (1977); Ruth Peterson and William Bailey, "Felony Murder and Capital Punishment: An Examination of the Deterrence Question," *Criminology* 29:367–395 (1991); Lawrence Katz, Steven Levitt, and Ellen Shustorovich, "Prison Conditions, Capital Punishment, and Deterrence," *American Law and Economics Review* 5:318–343 (2003); Richard Berk, "New Claims about Executions and General Deterrence: Déjà Vu All Over Again?" *Journal of Empirical Legal Studies* 2:303–330 (2005); John Donohoe III and Justin Wolfers, "Uses and Abuses of Empirical Evidence in the Death Penalty Debate," *Stanford Law Review* 58:791–846 (2005).

116. Michael Radelet and Marian Borg, "The Changing Nature of Death Penalty Debates," *Annual Review of Sociology* 26:43–61 (2000); Donohoe III and Wolfers, note 115.

117. Richard Willing, "DNA to Clear 200th Person," *USA Today,* April 23, 2007, p. 1A.

118. Samuel Gross, Kristen Jacoby, Daniel Matheson, Nicholas Montgomery, and Sujata Patil, "Exonerations in the United States, 1989 Through 2003," *Journal of Criminal Law and Criminology* 95:523–560 (2005).

119. *Furman v. Georgia,* note 79.

120. *McCleskey v. Zant,* 499 U.S. 467 (1991).

JBPUB.COM/ExploringCJ

Interactives

Key Term Explorer

Web Links

Corrections

SECTION

4

O nce offenders have been convicted and sentenced, they enter the corrections segment of the criminal justice system. Corrections include those programs or institutions designed to punish, confine, or rehabilitate people convicted of crimes.

The contemporary corrections system has evolved dramatically from its colonial origins when the death penalty was liberally applied for a wide range of offenses and corporal punishment was commonplace. The birth of the prison did not occur until the nineteenth century. People's perceptions of the causes of crime and what should be done with offenders had changed over the past five centuries, resulting in the creation of an expansive, and very expensive, system of corrections at both the state and federal levels. The composition of prison populations has also evolved because of changes in the nature of crime and in the racial, ethnic, and gender characteristics of those people convicted of crimes.

Section 4 examines the history of U.S. corrections, the various reform movements responsible for its transformation, the structure of modern corrections and the contemporary prison, and the movement toward greater use of community corrections. The history of corrections, the birth (and rebirth) of the prison, and the structure of state and federal prison systems today are explored in Chapter 13. Chapter 14 examines the contemporary institution of the prison, characteristics of prisoners, prison management, life in prison for male and female inmates, and the impact of the prisoners' rights movement. Lastly, Chapter 15 looks at community corrections and its various components.

CHAPTER 13

Corrections History and Structure

CHAPTER 14

Prisons

CHAPTER 15

Corrections in the Community

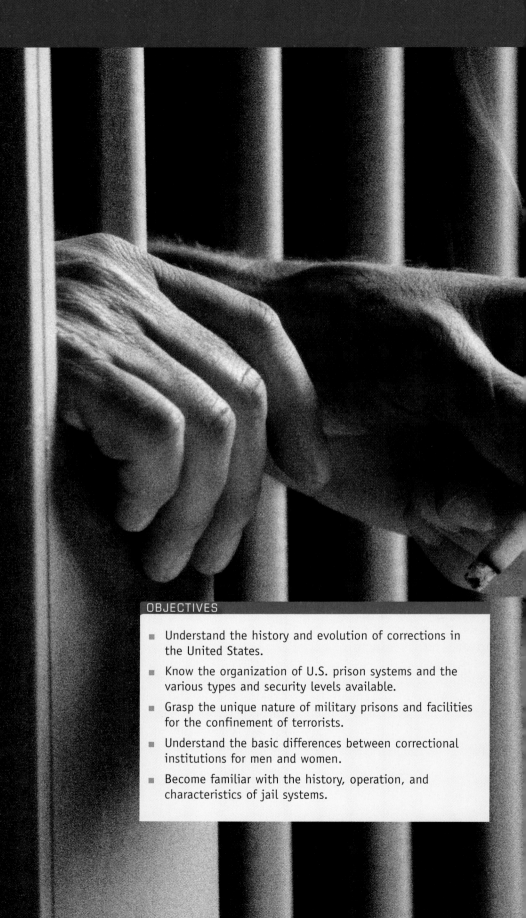

OBJECTIVES

- Understand the history and evolution of corrections in the United States.

- Know the organization of U.S. prison systems and the various types and security levels available.

- Grasp the unique nature of military prisons and facilities for the confinement of terrorists.

- Understand the basic differences between correctional institutions for men and women.

- Become familiar with the history, operation, and characteristics of jail systems.

CHAPTER 13

Corrections History and Structure

FEATURES

THINKING ABOUT CRIME AND JUSTICE

AROUND **THE GLOBE**

FOCUS **ON CRIMINAL JUSTICE**

𝕳𝖾𝖆𝖉𝖑𝖎𝖓𝖊 𝕮𝖗𝖎𝖒𝖊

VOICE **OF THE COURT**

LINK

WRAP UP

THINKING ABOUT CRIME AND JUSTICE: CONCLUSION

Chapter Spotlight

Putting It All Together

Key Terms

Notes

JBPUB.COM/ExploringCJ

Interactives

Key Term Explorer

Web Links

One of the most infamous criminals in U.S. history, and one whose crimes were directly responsible for improved criminal justice policies, was Kenneth McDuff. McDuff was a chronic juvenile delinquent who was imprisoned for burglary in the 1960s. In 1966, McDuff and his accomplice abducted three teenagers. They shot the two boys in the face and raped and strangled the girl. McDuff was sentenced to death, but as a result of a 1972 Supreme Court decision that invalidated the death penalty for juvenile offenders, his death sentences were vacated and converted to life imprisonment without parole.

In 1989, McDuff was released on parole because of prison overcrowding and his advanced age. After his release from the Texas prison, he was soon rearrested and convicted for menacing a group of African American children with a weapon. He was returned to prison and released again within a few months.

Over the next several years, McDuff again preyed on women and murdered at least five victims before his arrest. Authorities estimate that McDuff is responsible for roughly 15 homicides. The only person to be sentenced to death on two separate occasions for different crimes, McDuff was executed in Texas in 1999. Outrage over his crime spree contributed to toughened correctional policies restricting the parole of violent offenders in Texas, now making it one of the most punitive states in the nation.

- What factors may have contributed to McDuff's sentence?
- Why does the correctional system sometimes enable serious offenders like McDuff?

Source: Matt DeLisi and Ed Munoz, "Future Dangerousness Revisited," *Criminal Justice Policy Review* 14:287–305 (2003).

Introduction

In the United States, once offenders have been prosecuted, convicted, and sentenced, they enter the <u>correctional system</u>. This system consists of institutions such as jails, reformatories, and prisons and the correctional practices of parole, intermediate sanctions, and probation. It employs more than 700,000 people and has total costs of nearly $61 billion each year.[1]

Within the correctional system, three major settings focus on incarceration: jails, reformatories, and prisons. Although these terms are often used interchangeably, they describe different types of facilities.

- <u>Jails</u> are used to confine people who are awaiting trial and people who have been sentenced to short-term incarceration (one year or less), typically those convicted of misdemeanors or petty offenses.

- <u>Reformatories</u> typically confine younger, first-time offenders between the ages of 16 and 30 who have been tried as adults and convicted of felonies.

- <u>Prisons</u> are used for long-term confinement (more than one year) for serious or repeat felony offenders.

> **LINK** Correctional practices outside of these settings, such as parole, intermediate sanctions, and probation, are examined in Chapter 15.

The modern American corrections system differs significantly from its early form in the colonial period, reflecting changes in how people perceive crime, criminals, and victims. When perceptions about these phenomena change, society's reactions to them also change. Thus it is not surprising that state and federal corrections have undergone numerous reforms that parallel changes in society's perceptions of crime.

Colonial Punishments

The early colonial settlements were tightly governed, with the goal of establishing a pious and God-fearing citizenry. Crime was defined as sin and was not publicly tolerated; when discovered, it was swiftly brought to public reckoning through ridicule, labeling, and corporal punishment. Control of criminal behavior was enthusiastically embraced by devout Christians, who gladly reported offenders to the local constable. In a nineteenth-century account of punishments, the temper of the colonial period is succinctly described:

> In these barbarous methods of degrading criminals, the colonists in America copied the laws of the fatherland. Our ancestors were not squeamish. The sight of a man lopped of his ears, slit of his nose, or with a seared brand or great gash in his forehead or cheek could not affect the stout stomachs that cheerfully and eagerly gathered around the bloody whipping post and the gallows.[2]

These practices derived from the colonists' perception of the offender as "willful, a sinner, immoral, a captive of the devil, simply pauperized or defective."[3] The colonists did not, according to criminologist David Fogel,

> believe that a jail could rehabilitate, or intimidate or detain the offender. They placed little faith in the possibility of reform. Prevailing Calvinist doctrines that stressed the natural depravity of man and the powers of the devil hardly allowed such optimism. Since temptations to misconduct were not only omnipresent but practically irresistible, rehabilitation could not serve as a basis for a prison program.[4]

The Death Penalty

In the colonies, the death penalty was applied for more than 160 offenses, including murder, arson, absence from church, and disrespect toward one's parents.[5] Lesser offenses brought corporal punishments, except in the case of repeat offenders, who were publicly executed.[6]

LINK   The modern death penalty and controversies surrounding it were presented in depth in Chapter 12.

Fines and Corporal Punishment

Offenses that did not merit the death penalty were punished by fines, corporal punishment, or a combination of the two. Wealthy offenders usually paid fines; poor ones were publicly whipped. In addition,

Corporal punishment was once a standard mode of punishment in the criminal justice system.

? Why is the United States reluctant to employ corporal forms of punishment? Do physically painful punishments serve the public good?

other corporal punishments were implemented, because it was widely believed that public humiliation was an effective reformative tool. Great efforts were taken to ensure that offenders were degraded before the local community. The most popular corporal punishments included flogging, the pillory, the dunking stool, stocks, and branding.

Flogging

Flogging, or whipping, was the favored corporal punishment, principally because it was cheap and swift. Flogging was meted out for a wide range of offenses, including stealing a loaf of bread, shooting fowl on the Sabbath, and swearing.[7] Both men and women were whipped. Offenders received a set number of lashes, depending on their offenses. Two types of whips were used for such punishments: the cat-o'-nine-tails consisted of nine knotted thongs of rawhide attached to a handle; more cruel was the Russian knout—a lash constructed of many dried and hardened thongs of rawhide with wires and hooks on their ends that would tear flesh open upon contact with the offender's body.[8]

Pillories, Stocks, and Dunking Stools

The pillory (or "stretch-neck"), a wooden frame with holes for a person's head, hands, and feet, tightly held the offender in public view at all times. Documents dating from the colonial period chronicle many instances of both men and women being confined in a pillory for various offenses. For example, a man named John Hawkins stood an hour in the pillory for forgery, and a woman named Mariam Fitch was sentenced to the pillory for one hour for being a cheat.[9] While confined, offenders were routinely whipped or pelted with rocks, stones, and eggs; oftentimes, their ears were nailed to the pillory's beams. To extricate themselves, offenders had to rip their ears loose from the nails or have them cut off by an official.[10]

Offenders subjected to the dunking stool were strapped to a chair fastened to a long lever. The offender was hoisted over a river and dipped into water (a river or pond) for approximately one-half minute, for as many times as it took for him or her to publicly repent.[11] This punishment was generally reserved for village scolds and gossips, although records show that brewers of bad beer and bakers of bad bread were dunked as well. Sometimes quarrelsome married couples would be tied back-to-back and set upon the stool and dunked.

Colonial magistrates also maintained a supply of stocks, which held the prisoner, sitting down, with his or her feet and hands fastened in a locked frame. Stocks were used primarily to punish lower-class men who were gamblers, Sabbath-breakers, drunkards, and traveling musicians. The accused person usually made a public confession in church and was then placed in the stock to receive public ridicule.

Branding

One particularly humiliating punishment, branding, was used in America until the end of the eighteenth century. A brand was impressed somewhere on the offender's body, corresponding to the crime committed. For example, the letter "T" was burned on the left hand of thieves, and the letter "F" was burned into the forehead of forgers. For other offenses, a large cloth letter (made famous in Nathaniel Hawthorne's novel, *The Scarlet Letter*) was sewn on the offender's clothing, such as "A" for adultery.[12]

Early American Prisons

When the Revolutionary War began in 1776, there were no prisons for the long-term incarceration of convicted felons. The few jails that existed held people awaiting trial or offenders convicted of petty offenses. After the war, the writings of Cesare Beccaria and Jeremy Bentham began to take hold in the United States. Their more optimistic view of human nature led to the belief that crime stemmed from the individual's environment. Beccaria's and Bentham's writings inspired considerable penal reform throughout the Western world, resulting in a reduction in the death penalty and corporal punishment as well as the creation of the first prisons to punish offenders through the loss of liberty.

LINK Chapter 4 discussed the philosophical arguments of Beccaria and Bentham and supporting reform efforts of their theories.

This new way of thinking was consistent with the Quaker belief that all people could obtain God's grace, but whether they did so depended on how they treated others. For those individuals who had already fallen from grace (criminal offenders), the best way to improve was through penance and silent contemplation. Toward that end, the Quakers believed institutions should be built where offenders could spend uninterrupted time contemplating their sins.

In 1787, the Philadelphia Society for Alleviating the Miseries of Public Prisons was established to bring about changes in how offenders were treated. Led by social reformers Benjamin Rush and Benjamin Franklin, the Society sought to replace capital and corporal punishments with incarceration. It proposed the establishment of a prison program that would accomplish the following goals:

* Classify prisoners by gender and type of offense
* Provide labor for individual inmates to perform
* Include gardens to provide food and outdoor areas for recreation
* Classify convicts according to a judgment about the nature of the crime (whether it arose out of passion, habit, temptation, or mental illness)
* Impose indeterminate periods of confinement based on the convicts' reformative progress[13]

In 1790, the Pennsylvania state legislature established a wing of the Walnut Street Jail as a <u>penitentiary house</u>—that is, as a place of penitence and repentance for all "hardened and atrocious offenders."[14] The Walnut Street Jail had been open since 1776 to house criminals of minor offenses. While the jail continued to house all offenders together, regardless of their gender, age, or seriousness of criminal offense, it did implement an inmate classification system, prison labor program, and a policy of firmness and kindness instead of punishment.[15] However, overcrowding, lack of productive work for inmates, incompetent personnel, and public apathy led to the jail's collapse in 1835.

The Philadelphia Prison Society eventually put together a plan for a new type of prison, calling for "complete solitude with labor in the cells and recreation in a private yard adjacent to each cell."[16] While waiting for the Pennsylvania state legislature to act, prison reformers in New York developed a different strategy for handling criminals. The divergence in the two states' approaches ultimately resulted in two distinctly different prison models. Despite their differences, both models were grounded in the idea that criminals were victims of their environment and could be reformed.

The Pennsylvania Model

In the early nineteenth century, the Pennsylvania state legislature ordered the construction of two prisons, the Western State Penitentiary in Pittsburgh and the Eastern State Penitentiary in Philadelphia. In both prisons, inmates were kept in isolation. Separation was built into the physical structure of the prison; inmates were confined in single cells where they ate, slept, prayed, and worked. Separate confinement had the following aims:

- To provide opportunities for hard work and selective forms of suffering without vengeful treatment
- To prevent the prison from becoming a corrupting influence

Eastern State Penitentiary was once the most luxurious prison in the United States.

Were the Quakers on the right track with their vision of the prison? Has prison ceased being a place for contemplation and redemption?

- To allow time for reflection and repentance
- To provide punishing discipline by denying social contact
- To lessen the amount of time necessary to benefit from the penitential experience
- To ease the financial burden of imprisonment by minimizing the number of prison keepers required[17]

The Western and Eastern State Penitentiaries were the first institutions of their kind in the world. As a consequence, there were no existing architectural models for these facilities, and different architects were selected to design each prison. Their different architectural designs produced different outcomes. The design of the Western prison, an octagonal monstrosity with small dark cells that provided solitary confinement and no labor, proved ineffective. In fact, seven years after it opened in 1833, the Pennsylvania legislature had it demolished and rebuilt.

In contrast, the Eastern prison (known as Cherry Hill because its site had once been a cherry orchard) became well established and one of the most famous prisons in the world. Cherry Hill looked like a square wheel with seven wings, each containing 76 cells, radiating from a central hub where prison officials were stationed. Each inmate was placed in a solitary cell with a private exercise yard, which the inmate was permitted to use one hour per day.

To eliminate distractions, Cherry Hill prisoners spent their entire sentences in their cells, where they worked at handcrafts such as shoemaking, spinning, and weaving. Inmates wore blindfolds whenever they were taken to or from their cells and were not allowed to speak to each other. Even the guards who brought them food were not allowed to talk to them. Inmates were referred to only by numbers (rather than by their names) for the duration of their sentences. Visits with relatives were forbidden, and inmates were not allowed to receive mail. At the compulsory chapel services, prisoners were seated in chairs resembling upended coffins to prevent them from seeing each other.[18]

Despite their efforts and public expectations, both prisons reached the same conclusion: Solitary cells did not create penitent offenders. Complete isolation did more to damage the sanity and health of the inmates than it did to instill sorrow for wrongdoings.[19] The very modest prison labor carried out in individual cells and the requirement that all inmates were to be fed in their cells meant that the institution was also extremely costly. Nevertheless, the philosophy underlying the Pennsylvania model continues to influence contemporary corrections. The belief that prisoners can be transformed into law-abiding citizens remains popular with many modern penologists.

The New York Model

The New York state legislature took a different approach with the construction of the Auburn Prison, which opened in 1829 in central New York. Rather than a system of separate confinement, this prison adopted a paramilitary program called the congregate system (or silent system). In this model, prisoners were isolated during the night but worked together during the day under a strict rule of silence.

The cells in the Auburn Prison were much smaller than those at Cherry Hill, and no light could penetrate them, which made them damp in the summer and cold in the winter. Some of the trademarks of early prisons—including black-and-white uniforms and lock-step marches—originated at the Auburn Prison in an effort to treat all prisoners equally, make supervision easier, and prevent conversation between inmates.

Prisoners who challenged the strict regime of hard labor, silence, and unquestioning obedience received severe punishments. Auburn's first warden, Captain Elam Lynds, was a strict disciplinarian who believed that reformation began once the inmate's spirit was broken. To break the spirit, Lynds used the whip. In fact, some inmates received as many as 500 lashes at one time.[20]

Criticisms of the Two Models

Ultimately, neither the Pennsylvania nor the New York prison rehabilitated inmates very well. Critics of the Pennsylvania model held that these kinds of prisons were too expensive to build and operate and that separate confinement led to widespread insanity within the prison population. Opponents of the New York model argued that the system was too cruel and inhumane to affect people's lives in a positive way.[21]

The New York model became more popular in the United States, whereas the Pennsylvania model was preferred by European philosophers such as Gustav de Beaumont and Alexis de Tocqueville. After inspecting both types of prisons, they concluded that, "communication between [inmates] renders their moral reformation impossible, and becomes even for them the inevitable cause of alarming corruption. . . . No salutary system can possibly exist without the

The Auburn system was seen as an advance in penology but was characterized by brutality.

? Are U.S. prisons in a constant state of development and a quest for decency?

separation of the criminals."[22] Over time, the Pennsylvania model was widely imitated throughout the rest of the world. England, Belgium, Spain, and Germany all used the Eastern State Prison as their model for prison building, as did Japan and China. Germany, Italy, France, and the Netherlands adopted the separate system, in which inmates wore face masks when outside their cells, exercised in individual yards, and sat in individual stalls in schools and for religious services.[23]

In the end, the New York model was more widely adopted in the United States for one principal reason: It reinforced the emerging industrial philosophy that allowed states to use convict labor to defray prison costs.[24]

Reformation of Penology

As the inmate population increased, prison reformers started to question the future of a prison system that had become overcrowded, understaffed, and financially floundering. Corrupt administrators had failed to provide proper discipline, and regardless of the prison's design, these institutions had become unproductive and brutish, with little direction. As an alternative to large, sterile, forbidding prisons designed to isolate and punish offenders, a new treatment philosophy began to emerge—one built on the

The Prison Discipline Society, founded in Boston in 1825, championed the Auburn Prison. The following excerpt from the Society's first annual report illustrates the daily routine at Auburn Prison:

The whole establishment from the gate to the sewer is a specimen of neatness. The unremitted industry, the entire subordination, and subdued feeling among the convicts, has probably no parallel among an equal number of criminals. In their solitary cells, they spend the night, with no other book but the *Bible*; and at sunrise, they proceed in military order, under the eye of the turnkeys [guards], in solid columns, with the lock march, to their workshops, thence in the same order at the hour of breakfast, to the common hall, where they partake of their wholesome and frugal meal in silence. Not even a whisper is heard; though the silence is such that a whisper might be heard through the whole apartment. The convicts are seated in single file, at narrow tables, with their backs towards the centre, so that there can be no interchange of signs. If one has more food than he wants, he raises his left hand; and if another has less, he raises his right hand, and the waiter changes it.

When they have done eating, at the ringing of a little bell, of the softest sound, they rise from table, form in solid columns, and return under the eye of their turnkeys to the workshops. From one end of the shops to the other, it is the testimony of many witnesses that they have passed more than three hundred convicts without seeing one leave his work, or turn his head to gaze at them. There is the most perfect attention to business from morning till night, interrupted only by the time necessary to dine; and never by the fact that the whole body of prisoners have done their tasks, and the time is now their own, and they can do as they please.

At the close of the day, a little before sunset, the work is all laid aside at once, and the convicts return in military order to the solitary cells, where they partake of the frugal meal, which they are permitted to take from the kitchen, where it is furnished for them, as they returned from the shop. After supper, they can, if they choose, read scriptures, undisturbed and then reflect in silence on the errors of their lives. They must not disturb their fellow prisoners by even a whisper. The feelings which the convicts exhibit to their religious teacher, as he passes from one cell to another are generally subdued feelings. . . . The men attend to their business from the rising to the setting of the sun, and spend the night in solitude.

Source: Prison Discipline Society, *First Annual Report, 1826.* Reprinted in *Reports of the Prison Discipline Society of Boston, 1826–1854* (Montclair, NJ: Patterson Smith, 1972), pp. 36–37. Reprinted with permission.

idea that offenders could be reformed when provided with the right incentives. This reformatory movement was initiated through the efforts of people such as Captain Alexander Maconochie in Australia, Sir Walter Crofton in Ireland, and Enoch Wines and Zebulon Brockway in the United States.[25]

The Mark System

For nearly two centuries, from 1597 until 1776, England had transported its worst criminals to the United States. The exact number of offenders England sent to America is unknown; estimates vary between 15,000 and 100,000.[26] This practice ended with the American Revolution, after which England began transporting its criminals to Australia.

In 1840, Captain Alexander Maconochie was placed in charge of one of the worst British penal colonies, located about 1000 miles off Australia's coast on Norfolk Island. This site was where twice-condemned criminals were sent—offenders who had committed felonies in England, been transported to Australia, and committed additional crimes there. In response to the brutal treatment that had been instituted on Norfolk Island before his arrival, Maconochie introduced humane reforms that would give prisoners some degree of hope for their future. He proposed the following changes, known as the mark system:

- Criminal sentences should not be for a specific period of time; rather, release would be based on the performance of a specified quantity of labor.

Tens of thousands of British convicts were transported to the United States and Australia.

 Is prison incapacitation the modern version of transportation and banishment? Do similar rationales underlie these sanctions?

- The quantity of labor that prisoners must perform should be expressed in a number of marks that must be earned before their release.
- While in prison, inmates should earn everything they receive; all sustenance and indulgences should be added to their debts of mark.
- Prisoners would be required to work in groups of six or seven people, and the entire group should be held accountable for the behavior of each of its members.
- Prisoners, while still obliged to earn their daily tally of marks, should be given proprietary interest in their own labor and be subject to a less rigorous discipline so as to prepare them for release into society.[27]

Maconochie did not have the support of his superiors in England for his progressive ideas and, before he could implement them, Maconochie was recalled home to England. Nonetheless, his ideas persisted and later influenced the work of Sir Walter Crofton in Ireland.

The Irish System

Sir Walter Crofton of Ireland was impressed with Maconochie's idea of preparing convicts for their eventual return to society. He proposed a similar set of reforms in a program called the Irish system or indeterminate system, based on the idea "that criminals can be reformed, but only through employment in a free community where they are subjected to ordinary temptations."[28]

Under Crofton's plan, an inmate's release was based on sustained good behavior. Prisoners moved through a series of stages, where each stage was characterized by increasing freedom and responsibilities, and were ultimately placed in a work environment outside the prison as part of their conditional release. This program was referred to as a ticket-of-leave for inmates who agreed to live up to the conditions attached to their release and was the forerunner to our modern system of parole.

LINK Today, more than 760,000 persons are on parole. Details of the parole system are examined in Chapter 15.

The Elmira Model

At the 1870 meeting of the National Prison Association (now the American Correctional Association), prison reformers had gathered to discuss and resolve the principal differences between the Pennsylvania and Auburn models. What emerged from their discussions was the reformatory—a new type of institution especially designed for young offenders.

Based on what he saw in the Irish system, American reformer Enoch Wines believed that some criminals—specifically younger ones—could be successfully rehabilitated.[29] This idea was adopted as part of the National Prison Association's Declaration of Principles—a document consisting of 37 proposals advocating a philosophy of reformation that included the classification of inmates based on a mark system, indeterminate sentencing, and the cultivation of the inmate's self-respect. The first institution in the United States to implement this new ideology was the Elmira Reformatory, built in Elmira, New York.

In 1876, the Elmira Reformatory began receiving inmates between the ages of 10 and 30. Elmira was built much like the Auburn Prison, with inside cell blocks for solitary confinement at night and communal workshops. However, some of its cells were built with outside courtyards, similar to those at Cherry Hill. This modified design allowed natural light to penetrate the building. Elmira also used more artificial light than the Auburn and Cherry Hill prisons and featured more modern sanitary facilities.

Inmates at the Elmira Reformatory were subject to indeterminate sentencing, much like their counterparts in the Irish system. Prisoners received a maximum sentence but could win early release on parole if they exhibited good behavior. At entry, all prisoners were placed in the second grade. After six months of good conduct, they were promoted to the first grade. Six months of continued good conduct entitled them to parole. Prisoners who misbehaved were demoted to the third level, where a month's good conduct was required to restore them to the second grade. Inmates who regularly misbehaved were obliged to serve their maximum sentence.

Zebulon Brockway, Elmira's first superintendent, believed that prisoners could be reformed only in an atmosphere conducive to rehabilitation, and insisted that the prison include the following features:

- Uniforms were not degrading and represented the respective grades of the inmates.
- The liberal prison diet was designed to promote vigor.
- The gymnasium was equipped with baths, exercise equipment, and facilities for field athletics.
- Facilities were provided for vocational training (such as mechanical and freehand drawing, woodworking, metalworking, clay modeling, cabinet making, and iron molding).
- Trade or vocational instruction was offered based on the needs and capacities of individual prisoners.
- An educational curriculum, adapted for all levels, was implemented.
- Prisoners had access to a library.
- A weekly institutional newspaper (in lieu of all outside newspapers) was edited and published by prisoners.
- Opportunities for inmate recreation and entertainment were provided.
- Nondenominational religious opportunities were available.[30]

Over the next 25 years, the Elmira model was adopted in correctional systems across the United States TABLE 13-1. However, the reforms proved to be ineffective and, at Brockway's retirement in 1900, crime was still widespread. One of the most crucial problems inherent in the Elmira system was the fact that guards were unwilling to adjust to a correctional philosophy that provided inmates with autonomy. Emphasis on security remained their first priority, especially given the increasing populations of the institutions. Administrators were forced to create holding areas for more violent offenders, thus making reform programs available only to a few.

The enlightened concepts of the reformers gave way once more to a more control-oriented approach to corrections. At Elmira and elsewhere, simple custody reemerged as the primary goal and punishment as the method for controlling prisoners.[31]

TABLE 13-1

Early Reformatories Adopting the Elmira Model

State	Location	Date Opened
New York	Elmira	1876
Michigan	Ionia	1877
Massachusetts	Concord	1884
Pennsylvania	Huntingdon	1889
Minnesota	St. Cloud	1889
Colorado	Buena Vista	1890
Illinois	Pontiac	1891
Kansas	Hutchinson	1895
Ohio	Mansfield	1896
Indiana	Jeffersonville	1897
Wisconsin	Green Bay	1898
New Jersey	Rahway	1901
Washington	Monroe	1908
Oklahoma	Granite	1910
Maine	South Windham	1912
Wyoming	Worland	1912
Nebraska	Lincoln	1912
Connecticut	Cheshire	1913

Source: Wayne Morse, *The Attorney General's Survey of Release Procedures* (Washington, DC: U.S. Government Printing Office, 1940).

Prisons for Women

More than 107,000 women, accounting for about 7 percent of all prisoners, are confined in state and federal prisons, and an additional 95,000 women are held in local jails.[32] Although new studies are beginning to shed light on the early treatment of women in prison, relatively little has been written about the imprisonment of women during the nineteenth century or the overall contributions of women to corrections.

English Servitude

Much of the dismal treatment of women in early American prisons can be attributed to gender discrimination based on English common law, ignorance, and religious zealotry. Early on, some women were forced into sexual servitude. Penal reformers interested in improving conditions for incarcerated women voiced their concerns in England. Elizabeth Fry, an English Quaker, was one of the most outspoken reformers. Through her efforts, living conditions at London's Newgate prison were improved to allow for separate facilities for women and staffing by female wardens.[33]

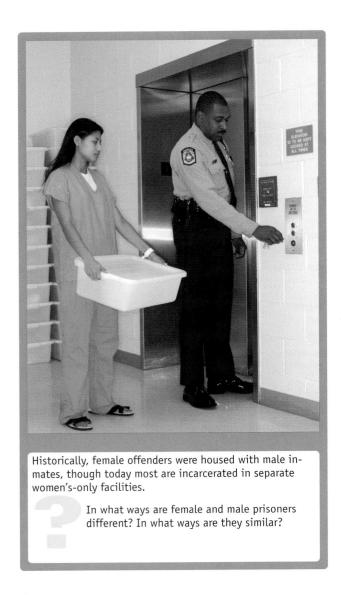

Historically, female offenders were housed with male inmates, though today most are incarcerated in separate women's-only facilities.

? In what ways are female and male prisoners different? In what ways are they similar?

Colonial Treatment

In America, female inmates received differential treatment from the very earliest colonial period.[34] Some early lawmakers believed female prisoners threatened the social order by sinning and by removing the "moral constraints on men." They assumed that, by their nature, women were more chaste and virtuous than men. Consequently, it was thought that incarcerated women had fallen even further than men because it was assumed that they were initially more pure.[35]

Some early reformers viewed incarcerated women—and particularly prostitutes—as a "serious eugenic danger to society."[36] They feared that female criminals might produce defective offspring, and that perceived threat to society prompted some states to pass laws providing for the sterilization of female inmates. In addition to imprisonment, women were subject to harsh punishments, including whipping

and death by hanging, because it was widely held that some women, due to some sort of inherent deficiency, could not be reformed.

Reform Efforts

American reformers lobbied for female wardens to govern female inmates, and in 1822, Maryland became the first state to hire a female jail keeper. Nearly from the beginning, female offenders were confined in the same penitentiaries as men. Generally, women were placed in separate cells or a small wing of the facility. For example, the first women admitted to the Auburn Prison in 1825 were placed in the third-floor attic. As the number of female inmates increased, however, separate buildings were built at the prisons to house women. The Mount Pleasant Female Prison—the first facility exclusively devoted to female offenders—opened in 1835.[37]

Motivations for separating women from men in prison were not always based on concern for the humanitarian treatment of female offenders. For example, in 1835, women incarcerated in the penitentiary in Alton, Illinois, were considered a negative influence on maintaining good order and discipline of male inmates. The Illinois Committee of Inspectors reported in 1845 that

> when it is known that while convicts of both sexes are confined in the same yard, it is impossible for them to be restrained within the bounds of propriety, or their morals reformed; and from past experience, not only in our own State, but in others, one female prisoner is of more trouble than twenty males.[38]

Although women were separated from male offenders, they still endured many physical hardships. In 1843, Sing Sing penitentiary in New York, best known for the long-term incarceration of hardened male prisoners, included a small number of female inmates. This institution housed mothers with children, pregnant women, and other females in small, crowded, and unsanitary rooms, where they were often beaten for minor rule violations.[39]

As an alternative to incarcerating women in prisons, a number of houses and shelters sprang up in the latter part of the nineteenth century. In several cities, these homes, whose founders were motivated by religious fervor to save "fallen" women from prostitution, opened their doors as alternatives to prison. Many of the homes were run exclusively by women and emphasized religion, education, and domestic skills.[40]

Most of the women incarcerated in reformatories were younger, white, and not considered dangerous; they were typically convicted of misdemeanors. African American females were segregated into separate cottages. Although a few reformatories emphasized outdoor labor, most offered little more than domestic training, and women were rewarded for maintaining a feminine appearance.[41]

By 1900, female prison administrators, in contrast to their counterparts in reformatories, were more concerned with prison management than with inmate spirituality. Moreover, the Great Depression of the 1930s had a dramatic effect on prison populations. Reformatories and prisons started receiving larger numbers of female felons, and as additional institutions were built, an increasing number of female administrators began to manage the institutions.

LINK Today there are nearly 110,000 women in prison in the United States, accounting for nearly 7 percent of the total prison population. See Chapter 14 for a discussion of the treatment of incarcerated women.

Rebirth of the Prison

By the early twentieth century, crime appeared to have reached epidemic proportions in the United States. Prison populations were swelling, and demands for new prisons increased. As public support for the reformatory system declined, penologists blamed crime on bad people, not on bad laws; they saw criminals as defective people, rather than as the victims of an arbitrary and capricious criminal justice system. This line of thinking set the stage for the emergence of a new ideology that perceived prisoners as sick people who were in need of treatment. This philosophy exerted a profound influence on corrections; it laid the groundwork for a new model with changes in prison design and operations. Prison reform came to mean the addition of libraries, recreation facilities, schools, and vocational programs. Particularly interesting among these innovations were the vocational programs. Like the hard labor programs that preceded them, vocational programs were grounded in the idea that inmates needed to work. The new philosophy, however, emphasized the idea that inmates need to work so they would be prepared with a trade and good work habits upon their release so they could find employment.

Prison Industry

The vocational programs developed in the reform movement meant that inmates learned skilled labor

and could produce goods that were sold by the state to defray the costs of prison operations. Perhaps not surprisingly, these industrial prisons faced stiff opposition from private industry, which could not compete with the low cost of prison-made goods made possible by the low inmate wages. Responding to pressure from organized labor, between 1929 and 1940, Congress passed three federal statutes that brought an end to the industrial prison.

With the passage of the Hawes–Cooper Act in 1929, prison-made products became subject to the laws of the state to which they were being shipped. In 1935, Congress passed the Ashurst–Sumners Act, which prohibited the transportation of prison goods to any state whose laws forbade it. This act was later amended in 1940 to exclude almost all prison-made products from interstate commerce.

The industrial prison was temporarily revived during World War II when inmates in both state and federal institutions manufactured war goods. After the war ended, however, inmates resumed a more idle lifestyle, playing cards, lifting weights, and mulling about in the yard.[42] Prisons continued to follow this pattern until 1979, when Congress passed the Percy Amendment, which allowed states to sell prison-made goods across state lines as long as they complied with strict rules, such as making sure unions were consulted and preventing manufacturers from undercutting existing wage structures.[43]

As a result of the Percy Amendment, prison industry rapidly expanded and engaged an increasing number of inmates in meaningful work experiences. Although some states continue to prohibit the sale of prison-made goods, most states provide opportunities for private-sector employment of inmates, and some prison-made goods are even sold over the Internet to buyers around the world (see "Focus on Criminal Justice"). The sale of these goods, which range from furniture to clothing, helps to offset the cost of incarceration and contributes funds to victim compensation and restitution programs.

Emergence of the Medical Model

Although the seeds for the medical model were planted in the late nineteenth century with the work of psychoanalyst Sigmund Freud, it was not until the 1930s that the medical model began to take shape. The medical model viewed crime as symptomatic of pathology—a biological or psychological defect in offenders that, once identified, could be treated.[44] During the 1940s and 1950s, this model was widely implemented, and the goal of corrections once again became reha-

Made on the Inside to Be Worn on the Outside

In 1997, voters in Oregon overwhelmingly passed Ballot Measure 49, which required the Oregon Department of Corrections to create inmate work programs to defray the costs of incarceration. Approximately 80 percent of inmates' wages is held by the institution and applied toward the cost of room and board, victim restitution, child support, and taxes. The remaining 20 percent is invested in an inmate savings account, which becomes available to inmates at the time of their release.

One result of Measure 49 was the creation of a line of clothing called Prison Blues, which is produced at the Eastern Oregon Correctional Facility in Pendleton, Oregon. Prisoners are paid the prevailing industry wages to produce a clothing line that includes blue jeans, denim jackets, sweatshirts, T-shirts, and yard coats.

Initially sold only through a limited number of retail stores, Prison Blues has expanded into markets in Asia and Europe via an arrangement with a catalog distributor and established an Internet-based store (www.prisonblues.com). In 2002, Rob Waibel, 20-time world champion in lumberjack sports competitions including tree climbing, ax throwing, log rolling, wood chopping, and cross-cut sawing, endorsed the Prison Blues brand; he wears the clothing as he competes. According to Waibel, "Prison Blues makes the toughest, U.S.A.-made work jean I have been able to find. In addition, it is a company with a purpose and objective I believe in."

Sources: "Prison Blues Launches Online Retail Store," available at http://www.prisonblues.com, accessed June 17, 2006; Robert Goldfield, "Prison Blues Hoping to Find Favor in Germany," *Portland Business Journal* (April 13, 2001), available at http://portland.bizjournals.com/portland/stories/2001/04/16/newscolumn5.html, accessed July 31, 2007.

bilitation. Thinking of offenders as "sick" and in need of individualized treatment contributed to the renewed emphasis on rehabilitation and led to an increased emphasis on inmate classification. This perspective reflected the medical model notion that before hospitals could treat ill people, they needed a system for identifying the diseases from which their patients suffered. Under such a system, prisoners were assigned to units according to security level and treatment needs.

LINK   Inmate classification, as discussed in Chapter 14, is unfortunately often based on whether beds are available and which jobs and programs need to be done or have openings, rather than inmates being placed in the job or program best suited for them.

Other Correctional Innovations

While many of the reform efforts were aimed at changing conditions within prisons, correctional innovations outside of prisons were also being developed during this time. These advances included the emergence of community-based corrections and calls to deinstitutionalize inmates by removing them from prisons. Advocates for community-based programs argued that rehabilitation could be best achieved by allowing inmates to interact directly with members of the community to which they would be released. Toward this end, halfway houses were opened to allow inmates to move back into society gradually by living in a secure house while working in the community.

In 1961, the first prerelease guidance centers were opened in Los Angeles and Chicago by the Federal Bureau of Prisons (BOP) assist juvenile offenders reintegrating into their communities after their release from custody. In these unlocked facilities, inmates had much more freedom to come and go as they pleased, access to jobs and education, and less stigma than that associated with prisoners in traditional guarded structures.

LINK   Chapter 15 examines a wide variety of community-based corrections currently being used in the United States and other countries.

Work Release

Work release was authorized by the federal government through the Federal Prisoner Rehabilitation Act of 1965. This act permitted prisoners (usually those in minimum-security prisons) to work outside the institution during the day and return after work. Inmates in work release programs typically are expected to contribute some portion of their wages for their room and board, but they also have greater access to drug or alcohol treatment programs in the community and are able to draw upon a wide variety of community resources to increase the likelihood of their success upon reentry into society.

Conjugal Visits

<u>Conjugal visits</u>—private visits between inmates and their spouses intended to help them maintain interpersonal and sexual relationships and strengthen the family unit—became widely used in many states in the 1960s. In 1968, the California Correctional Institute at Tehachapi permitted inmates nearing parole a three-day-per-month family visitation pass that brought an inmate's spouse (and children, when appropriate) to an apartment within the prison compound. Generally, the inmate and his or her spouse and children stayed in the apartment until the visit was over. This community-based concept was endorsed by the National Advisory Commission on Criminal Justice Standards. Today, in part because of increased concerns with security, only six states—California, Connecticut, Mississippi, New Mexico, New York, and Washington—permit conjugal visits.[45]

Modern Prisons

In 2005, 1,446,269 men and women were incarcerated in state and federal prisons in the United States; an additional 747,529 were held in local jails. Between 1995 and 2005, the number of inmates held in state and federal prisons increased by approximately 33 percent.[46] The imprisonment rate in the United States continues to outpace the rest of the world, although Russia is close behind. Most other countries have much lower rates **FIGURE 13-1**. Even within the United States, there is substantial variation in rates of imprisonment. **TABLE 13-2** compares the states with the highest and lowest total incarceration rates.

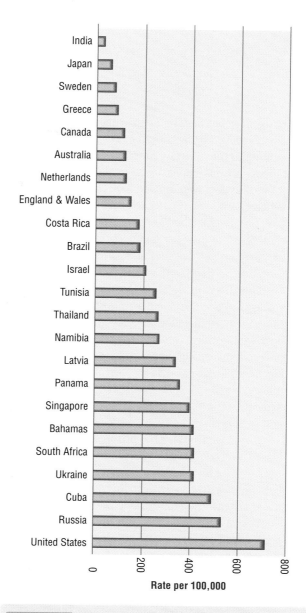

FIGURE 13-1 Imprisonment Rates Around the World

Source: Roy Walmsley, *World Prison Population List* (London: International Centre for Prison Studies, 2005).

TABLE 13-2

States with the Highest and Lowest Incarceration Rates

State	Rate per 100,000
States with the Highest Incarceration Rates	
Louisiana	797
Texas	691
Mississippi	660
Oklahoma	652
Alabama	591
States with the Lowest Incarceration Rates	
North Dakota	208
New Hampshire	192
Rhode Island	189
Minnesota	180
Maine	144

Source: Paige Harrison and Allen Beck, *Prisoners in 2005* (Washington, DC: U.S. Department of Justice, 2006), p. 1.

Prison Security Levels

The federal government and all state governments operate a system of correctional institutions. These institutions are classified according to their level of security—that is, the amount of security required to safely confine inmates with different levels of potential for violence or escape:

- Minimum
- Low
- Medium

China does not classify prisons according to their security levels, but rather organizes prisons largely according to their primary functions. For example, many Chinese prisons emphasize education as a means to reform offenders who are serving relatively short sentences. Other prisons, known as *Laogai* (meaning "reform through labor") emphasize labor and have little interest in prisoner reform. Some of these prisons confine tens of thousands of inmates serving very long sentences in large work camps.

Inmates who violate prison rules may find themselves in the "little number" punishment box, a solitary confinement cell where they are shackled and fed only sporadically. In some *Laogai*, prisoners are fed only sorghum and ground corn cobs. Inmates often become too weak to work in the iron mines to which they are assigned. Such inmates are labeled as "work avoiders," and their already meager food is decreased and sometimes cut off.

Those who survive the prison experience are "released" and relocated to a Forced Job Placement Camp. Here, the ex-prisoners have a few privileges, such as being allowed to be with their families, but they all remain at the camp for life, and their children grow up to become workers on the farm. Inmates rise at 5 A.M. and work all day, often until late at night. Interspersed through the day are political study sessions, which are essentially aimed at re-educating (i.e., brainwashing) inmates so that they will conform to China's political and economic policies. While most inmates in the regular prisons have been convicted of criminal offenses, many—if not most—of those in *Laogai* have been sentenced for "re-education through labor." People may be sent to labor camps for as long as four years without trial or formal sentencing for such "offenses" as public or private political dissent, belonging to unauthorized religious groups, or separatist-minded ethnic minorities.

One explanation for this system is that prison labor has become a mainstay of the Chinese economy. In China today, between 2 million and 20 million people are estimated to be confined in *Laogai* and forced labor camps. There is virtually no product that is not manufactured inside them. The prison-made products are sold both domestically and abroad. The most popular export products are coffee, tea, spices, salt, cement, plastic, rubber, textiles, ceramics, glass, iron, steel, copper, lead, nuclear reactors, boilers, machinery, sound equipment, footballs, medical gloves, high-grade optical equipment, and vehicles. Every one of these products is produced by inmate labor under appalling conditions.

Sources: Hongda Harry Wu, *Laogai: The Chinese Gulag,* translated by Ted Slingerland (Boulder, CO: Westview Press, 1992); Violet Gwynne, "Nineteen Years in the Chinese Gulag," *The Times* (London), April 22, 1993, p. 41; Harry Wu, "China's Gulag: Suppressing Dissent through the *Laogai*," *Harvard International Review* 20:20–23 (1997/1998); Tim Luard, "China's 'Reforming' Work Programme," *BBC News,* available at http://news.bbc.co.uk/go/pr/fr/-/1/hi/world/asia-pacific/4515197.stm, accessed July 31, 2007.

- Maximum
- Super-maximum

Each security category has its own unique characteristics that determine the nature of the prison environment.

There are approximately 814 minimum-security prisons in operation around the country, including work camps and farms. These institutions do not have high walls or armed guards in towers but often have fences around their perimeters. Most inmates in these institutions are serving relatively short sentences for less serious offenses such as property and drug crimes. Work furloughs and education programs are widely available, and inmates receive more privileges and are given more personal time for recreational activities than in more secure prisons. This security level is designed for nondangerous, stable offenders, who are given the opportunity to avoid the stress and violence found in more secure facilities.

Low-security prisons, established by the BOP, operate between medium and minimum security levels. They typically house nonviolent offenders with drug or alcohol problems. Low-security facilities have double-fenced perimeters, mostly dormitory or cubicle housing, and strong work and program components. The staff-to-inmate ratio in these institutions is higher than in minimum-security facilities.

More than 800 minimum-security prisons supervising low-risk inmates operate in the United States.

? Given that inmates in minimum-security facilities are low risk, should they be incarcerated at all? Should low-risk offenders be supervised only in the community?

In medium-security prisons, correctional officers typically are not armed and there are no high walls. In fact, many of the 522 medium-security prisons in operation today have been built using a campus design, allowing inmates to live in single rooms that have a window or in dormitory-style housing rather than in cells (which have three solid walls and an obtrusive steel security door). In these facilities, inmates are typically a mix of violent and property offenders serving less than life sentences. Medium-security prisons may give the impression that inmates are not under constant surveillance, but they are actually closely supervised by unobtrusive surveillance equipment such as hidden cameras.

Nationally, 332 maximum-security prisons collectively house more than 245,000 prisoners.[47] They are designed to prevent escape and to exert maximum control over inmates. Many resemble Auburn Prison, with high walls, watchtowers, and barbed wire or electronic fences to deter escape attempts. These fortress-like structures are home to many of society's most violent offenders. Inmate privacy is limited in these facilities, which include open bathrooms and showers; inmates may be subjected to full-body searches, especially after seeing visitors.

Some states and the federal government have recently constructed super-maximum-security prisons, where the most predatory or dangerous criminals are

confined. Only 20 years ago, the only super-max prison was a federal facility located in Marion, Illinois. Today, there are at least 57 super-max institutions located in more than 40 states that serve a total inmate population of about 20,000, though these facilities are somewhat controversial.[48] According to the National Institute of Corrections, super-max prisons seek to provide maximum control over inmates who exhibited violent or seriously disruptive behavior while incarcerated and are viewed as a security threat in standard correctional settings.[49] In super-maximum facilities, inmates are separated, their movement is restricted, and they have only limited direct access to staff and other inmates. These facilities are the most expensive institutions in the correctional system, sometimes having twice the per inmate cost of maximum-security prisons.[50]

Prison Farms and Camps

A number of state correctional systems operate minimum-security prison farms and forestry camps. The inmates who are placed in these facilities are typically serving short sentences and are considered very-low-risk offenders. Prison farms produce much of the livestock, dairy products, and vegetables used to feed inmates in the state prisons. For example, the Bolduc Correctional Facility in Maine, which was built in the early 1930s, became one of the largest dairy and beef farms in the state; it currently produces potatoes and dried beans for Maine correctional facilities.[51]

The Iowa Department of Corrections operates farms at all except two of its prisons. These farms operate with no state appropriations and must generate a net profit to remain in operation. Inmates in the Iowa farms not only raise crops and livestock, but also operate a nursery, mend fences, repair dikes, and repair and construct farm buildings.[52]

Prison forestry camps, sometimes known as conservation camps, provide labor for the maintenance of state parks, tree planting and thinning, wildlife care, maintenance of fish hatcheries, and cleanup of roads and highways. Some camps, such as the South Fork Forest Camp in Oregon and Sugar Pine Conservation Camp in California, provide a labor source for fighting forest fires. These inmate crews are paid $1 per hour (which is significantly more than their standard pay while incarcerated in the camp) to provide emergency firefighting assistance as well as to respond to floods, search and rescues, and earthquakes.[53]

Life in a Super-Max Prison

In most super-maximum-security prisons, inmates spend 23 hours a day locked in their 7 x 14 foot cells. A light remains on in the cell at all times, though it is sometimes dimmed. Cells have solid metal doors with metal strips along their sides and bottoms that prevent conversation or communication with other inmates. All meals are taken alone in the inmate's cell instead of in a common eating area.

Super-max inmates around the country experience additional security when they venture outside their cells. For example, at California's Pelican Bay State Prison, during the hour inmates are not in their cells, they are typically strip-searched, shackled, and then transported by guards to the exercise room or shower, where the chains are removed. To ensure the safety of the prison staff, contact between inmates and staff members is kept to a minimum. In some super-max prisons, inmates are not transported by guards to shower or exercise. Instead, an inmate is moved from one location to another by the use of intercoms, automatic doors, and surveillance cameras controlled by guards in a separate location. Some super-max prisons have been designed for constant monitoring of all inmates.

At the Illinois Closed Maximum Security Correctional Center, for example, inmates are housed in a number of pods (similar to those in the new-generation jails, discussed later in this chapter) with 60 cells surrounding a central control station.

Most politicians and citizens are supportive of super-max prisons, believing that they provide the greatest safety for both inmates and staff. Critics argue that near-total isolation of inmates likely produces serious psychological and emotional problems. Inmates may begin to exhibit symptoms of psychiatric decomposition, depression, psychological withdrawal, and heightened anxiety, and some even lose touch with reality. Critics also charge that the extreme deprivations in super-max prisons may violate inmates' Eighth Amendment protection against cruel and unusual punishment.

Sources: Charles Pettigrew, "Technology and the Eighth Amendment: The Problem of Supermax Prisons," *North Carolina Journal of Law and Technology* 4:191–215 (2002); Robert Sheppard, Jeffrey Geiger, and George Welborn, "Closed Maximum Security: The Illinois Supermax," *Corrections Today* 58:84–87, 106 (1996); Terry Kupers, *Prison Madness: The Mental Health Crisis Behind Bars and What We Must Do about It* (San Francisco: Jossey-Bass, 1999); Hans Toch, "The Future of Supermax Confinement," *The Prison Journal* 81:376–388 (2001).

Boot Camps

Boot camps are highly structured and regimented correctional facilities where inmates undergo rigorous physical conditioning and discipline. Boot camps were established in 27 states and the federal system in the early 1980s and quickly gained favor with politicians and the public looking for high-profile, "get-tough" responses to crime. Boot camps evolved through three distinct stages:

1. First-generation camps stressed military-style discipline, physical training, and hard work.

2. Second-generation camps emphasized rehabilitation by adding such components as alcohol and drug treatment and prosocial skills training.

3. Third-generation camps dropped the military-style training, implementing programs stressing educational and vocational skills instead.

Offenders who are eligible for boot camps are generally young adults (younger than 30 years of

Initially a military-style, get-tough initiative, boot camps today stress educational and vocational training.

 Do you believe the "boot camp" experience changes the hearts and minds of offenders and makes them want to desist from crime?

Wilkinson v. Austin

In 2005, in *Wilkinson v. Austin*, the U.S. Supreme Court examined a case brought by a group of inmates against administrators of the super-maximum-security Ohio State Penitentiary (OSP). The inmates alleged that the procedures for assignment to OSP and subsequent inmate requests for consideration for movement to a lower security classification violated their due process rights based on the Fourteenth Amendment.

Inmates had claimed that assignment to OSP was arbitrary and that, once confined in OSP, their requests for reclassification and transfer to a lower-security facility were often ignored. Because of the severe nature of the confinement, they argued, reclassification and transfer should not occur without appropriate due process. In other words, inmates had a significant "liberty interest" in avoiding reclassification.

Placement at OSP results in severe deprivations similar to those found in other super-max facilities, so prisoners strongly wish to avoid transfer to the institution. If placed in OSP, they face solitary confinement in cells for at least 23 hours a day. Inmates are shackled whenever they are outside their cells. An in-

mate on leaving his cellblock is placed in a cage and strip-searched. Inmates are prohibited contact with other prisoners and have limited phone and visitation privileges. All medical or psychiatric interviews are conducted through a narrow slot in the door, which requires inmates to discuss their problems publicly.

After considering the procedures for transfer of inmates to OSP and reclassification once an inmate was there, the Supreme Court delivered a unanimous decision in response to the inmates' suit. The Court found the review procedure to provide a sufficient level of due process and suggested that risks of erroneous placement at OSP are very unlikely given the multiple levels of review and the possibility of a recommendation being overturned at any level. It also noted that the state's first obligation must be "to ensure the safety of guards and prison personnel, the public, and the prisoners themselves," recognizing that "prolonged confinement in Supermax may be the State's only option for the control of some inmates."

Source: Wilkinson v. Austin, 545 U.S. 209 (2005).

age) serving their first prison term. They are typically given a choice of a boot camp or a traditional prison. If they elect the boot camp option, their sentence will be shorter (90 to 180 days), but what is expected of them will be very demanding. Offenders who successfully complete boot camp are released to community supervision. Inmates who fail boot camp (for example, those who drop out or break rules) are returned to traditional prisons, where they serve out their complete sentences.

The goals of boot camps include reducing recidivism and shortening the length of prison sentences, which should translate into a reduction in both the size of the prison population and the costs of corrections. However, a national study by the National Institute of Justice (NIJ) looking at boot camps over a 10-year period concluded that these goals were not being achieved. Boot camps did not reduce recidivism, possibly owing to offenders' relatively short stays in the camp environment, the lack of effective

post-release programs, and the absence of a strong underlying treatment model. Other research on boot camps in Georgia and Illinois found no difference in recidivism compared to matched groups; an evaluation of the Work Ethic Camp in Washington actually found higher rates of recidivism among participants. The NIJ study also found that boot camps achieved only a very small reduction in prison populations. On a more positive note, boot camp inmates did seem to experience more prosocial attitudes, increased self-esteem, and improved coping skills.[54]

Critics of boot camps have raised additional concerns. Many offenders selected for boot camps do not complete the program; they either request transfer to a regular prison because they do not like the rigor of the boot camp or are determined to be unfit for the program. More importantly, critics worry about the potential for abuse of offenders during their stay in the program. Guards are generally given wide discretion in dealing with inmates and are often relatively untrained or insensitive to the impact of such regimented discipline. Sometimes the physical abuse results in serious injuries or even death. For example, in April 2006, the state of Florida shut down all of its juvenile boot camps after 14-year-old Martin Lee Anderson died soon after being roughed up by guards at the Panama City boot camp. At one time, Florida operated nine juvenile boot camps; by 2006, however, it operated only five, with fewer than 140 offenders in the camps.[55]

Despite their growing popularity in the 1990s, at which point more than 7000 inmates were housed in approximately 75 adult state and federal boot camps around the nation, one-third of the state-run camps had closed by 2000. The BOP closed all of its boot camps in 2005.[56]

Military Prisons

The military penal system differs significantly from the state and federal systems. First, the military has its own criminal code, called the *Uniform Code of Military Justice*. Many of its rules are peculiar to the armed services. For example, civilians can quit their jobs and walk away, but military personnel who walk away from their jobs are court-marshaled. Similarly, civilians who disobey an order from their bosses may be fired, but they cannot be prosecuted. By contrast, military personnel who disobey a lawful order from a superior are usually prosecuted in a military court. If convicted, they are sentenced to a military prison.

The military prison system operates 59 facilities, which currently hold about 2400 inmates. The most common offense among military inmates is rape, followed by drug possession and trafficking. Only a relatively small percentage (about 12 percent) of the inmates have been convicted of desertion, malingering, absence without leave, or other specifically military offenses.[57]

Military prisons are divided into three tiers:

- Tier One (includes Navy and Marine brigs, Army detention cells, and Air Force correctional facilities)—most offenders are either pretrial and post-trial detainees or inmates serving sentences of up to one year
- Tier Two—regional correctional centers for inmates with sentences up to seven years but with minimum-security requirements
- Tier Three—maximum-security facilities

The United States Disciplinary Barracks, located in a portion of Fort Leavenworth Prison in Kansas,

is the only maximum-security prison operated by the Department of Defense. It houses nearly 1400 prisoners from all branches of the military. An additional Tier Three facility, the Miramar Naval Consolidated Brig, is designated for female personnel only.

The Navy also operates a unique set of correctional facilities aboard ships at sea to supplement its waterfront brigs at major naval bases. On any given day, approximately 150 sailors are confined in onboard brigs around the world.[58] The military also operates 11 overseas correctional Army and Air Force facilities; in 2002, these facilities confined 142 inmates.[59]

Camp Delta: A Prison for Terrorists

The Pentagon runs one of the most unusual prisons in the U.S. correctional system: Camp Delta, which is located at the Guantanamo Naval Base in Cuba and operated by U.S. Military Police. Shortly after the terrorist attacks on September 11, 2001, President George W. Bush issued a military order for the "Detention, Treatment, and Trial of Certain Non-Citizens in the War against Terrorism." This order authorized the Secretary of Defense to detain any persons subject to the order. In early 2002, the first Al Qaeda and Taliban inmates from Afghanistan arrived at the temporary Guantanamo detention facility known as Camp X-Ray. A few months later, Camp Delta, a permanent facility for confinement of suspected terrorists, opened to hold more than 300 detainees.[60] According to camp officials, the detainees held at Camp Delta are viewed as very dangerous and have expressed a commitment to kill Americans or American allies if released. They are not common criminals, but "enemy combatants" detained because of their acts of war against the United States.[61]

Camp Delta actually comprises at least five detention camps. The first three camps are maximum-security facilities designed to house a maximum of 800 detainees in solitary confinement. Camp 4 is considered a medium-security facility in which detainees are held in podular complexes, with a communal living room and larger cells with private toilet and shower. Camp 5 is a concrete and steel supermaximum facility designed to hold the most dangerous detainees or those considered to have the most valuable information.

Another lower-security facility, Camp Iguana, holds detainees between the ages of 13 and 15. The living quarters are air-conditioned and have a small living room, bathroom, and kitchen. Detainees are provided counseling, are permitted to watch television, and are tutored in math and geography.[62]

Controversy has surrounded Camp Delta since it was opened. Although much of the debate has centered on the legality of detaining terrorists who were captured during the war in Afghanistan, recent concerns have focused on attempted suicides by detainees. Numerous detainees have also engaged in hunger strikes, and military officials have responded by force-feeding detainees. While medical personnel were able to stop many attempted suicides, three detainees successfully hanged themselves in a coordinated suicide effort in 2006. According to Rear Admiral Harry Harris, Jr., commander of Camp Delta, each of the three detainees had very close ties to terrorist organizations and their suicides were not acts of desperation but rather acts of "asymmetric warfare" against the United States intended to stir up increased international criticism of the camp.[63]

Since April 28, 2002, Camp Delta in Guantanamo Bay, Cuba, has detained Al Qaeda and Taliban enemy combatants.

? Should enemy combatants be provided with all due process rights in accord with the U.S. Constitution?

Co-correctional Prisons

One approach to reducing some of the pains of imprisonment for U.S. inmates is reflected in the relatively small number of co-correctional prisons, where men and women are confined in the same institution. Until recently, men and women were often housed together in the same institutions, often even the same cells. The contemporary approach has been to create sexually integrated institutions, where inmates are housed separately but interact in normal institutional activities.

LINK The unique pains of imprisonment experienced by offenders serving time, including the absence of normal heterosexual contact, is discussed in greater detail in Chapter 14.

There are currently 21 co-correctional prisons located throughout the United States.[64] Most of these are small, minimum-security prisons, although there are some medium- and maximum-security coed institutions. In these prisons, inmates live in sexually segregated housing units—either in different buildings, in walled-off wings of buildings, or in separate cottages. During the day and evening hours, men and women come together in prison shops, educational or vocational programs, and recreational activities.

One goal of coed prisons is to create a more normal prison environment by allowing male and female inmates to interact in daily prison routines. It is believed that this environment enriches life in prison and reduces many prison problems. For example, women may gain greater access to vocational and educational programs than are typically available in women-only prisons, and male inmates report that women bring a more humanizing influence to the institution, resulting in a reduction of assaults and homosexual rapes.[65]

However, the mingling of men and women in prison does present other problems as prison officials confront the natural attraction between the sexes. A major concern in coed prisons is how to regulate physical contact between inmates, and the rules governing such contact vary greatly between institutions. In some prisons, all touching is prohibited; in others, inmates are allowed to hold hands or walk arm-in-arm. Moreover, there is increasing evidence that male and female prisoners may not generally benefit equally from coed prisons. Criminologists John Smykla and Jimmy Williams report that while coed prisons increase economic, educational, vocational, and social opportunities for male prisoners, female inmates do not realize equivalent benefits from these arrangements.[66] Perhaps owing to the lack of significant advantages demonstrated by these prisons and to their opponents' preference for the get-tough approach to punishment, many co-correctional facilities have closed in recent years.[67]

The Federal Prison System

Although most of the penal institutions in United States are operated by the various states, cities, and counties, the federal government has its own prison system that is operated by the U.S. Bureau of Prisons, a division of the Department of Justice.

In 1891, Congress authorized the purchase of land for three federal prisons. Leavenworth, Kansas, was selected as the site of the first prison because the Justice Department had taken over a military facility there (Fort Leavenworth) from the War Department. The existing facility proved to be inadequate, however, and in 1896, Congress authorized the construction of a maximum-security prison on the Fort Leavenworth grounds. The prison was built by inmates at Fort Leavenworth, and on February 1, 1906, they were transferred to the new facility. Soon, two more federal prisons were constructed: one in Atlanta, Georgia, and one on McNeil Island, Washington. These facilities were modeled, in both architecture and philosophy, after the prison in Auburn. The first federal facility for women, also based on the Auburn model, opened in 1927 in Alderson, West Virginia.

From its inception, the federal prison system was plagued by overcrowding and inconsistent administration. It soon became clear that the system needed to be overhauled. In 1930, President Herbert Hoover signed a law creating the BOP. It called for the

> proper classification and segregation of federal prisoners according to their character, the nature of the crimes they have committed, their mental condition, and such other factors as should be taken into consideration in providing an individualized system of discipline, care, and treatment.[68]

Within four years, federal facilities were divided into penitentiaries, reformatories, prison camps, drug treatment facilities, and a hospital. All inmates were processed through a classification system to determine their needs and abilities.

Since the 1930s, the federal system has expanded significantly. Today, more than 106 federal correctional institutions employ about 35,000 people across seven regional offices. The BOP operates numerous facilities, including the following:

- *U.S. penitentiaries.* The six maximum-security U.S. penitentiaries are designed to house male inmates who have been convicted of serious crimes and are serving long sentences.

- *Federal correctional institutions.* These 40 institutions are lower-security facilities designed to house male and female prisoners who are serving terms of two to five years in dormitory-style housing.

- *Metropolitan correctional centers.* There are six such centers housing prisoners who are serving short sentences and offenders of any security level who are awaiting trial or sentencing.

- *Medical centers.* The BOP operates four regional medical centers for patients from any institution in the federal prison system who require medical, surgical, or psychiatric care.

- *Federal prison camps.* The 14 minimum-security federal prison camps are primarily work-oriented facilities for minor federal offenders. These camps use dormitory-style housing and are often located adjacent to larger institutions.

- *Federal detention center.* The single detention center, located in Oakdale, Louisiana, houses pretrial detainees or noncitizens convicted of crimes and awaiting deportation.

In total, more than 170,000 people are being held by the BOP. More than half are white, 93 percent are male, and the majority are incarcerated on drug offenses FIGURE 13-2 .

Privatization of Correctional Facilities

Private prisons have become increasingly popular, largely as a response to the rapidly growing prison population. Privatization is the process in which state and federal governments contract with the private sector to help construct, finance, and operate correctional facilities for agreed-upon fees.

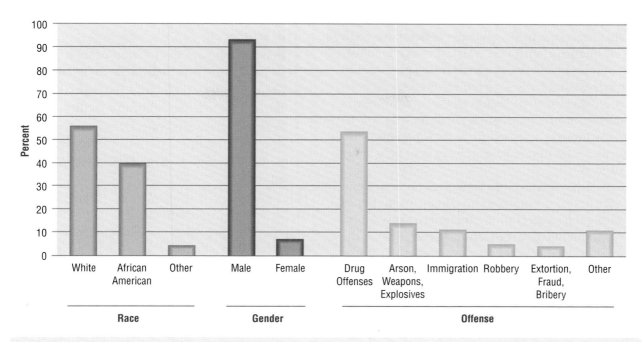

FIGURE 13-2 Selected Characteristics of Prison Inmates

Source: Federal Bureau of Prisons, available at http://www.bop.gov/about/facts.jsp#4, accessed May 28, 2007.

Since the early 1800s, five types of private prison labor systems have been dominant in the United States:

- *Contract labor system.* Private contractors provide prisons with machinery and raw materials in exchange for the inmate labor to produce finished products.

- *Piece-price system.* Contractors give raw materials to prisons, which use convict labor to produce finished products. Once the goods are manufactured, they are sold by the piece to the contractor, which resells them on the open market.

- *Lease system.* Contractors bid against one another to own the rights to inmate labor. Inmates work outside the prison facility, under the supervision of a private contractor, which is responsible for the inmates' food, shelter, and clothing.[69]

- *Public account system.* The state retains control of inmate labor and provides convicts with the machinery and raw materials to produce finished products. The state sells these products on the open market and uses the profits to defray the cost of prison operations.

- *State-use system.* Prison labor is used to produce goods for state-supported institutions (e.g., schools and hospitals).

Privatization continues to be a mainstay of penal philosophy. In the 1980s, with encouragement from President Ronald Reagan, who wanted to involve private enterprise in both state and federal government operations, privatization of prisons began to flourish.[70] Today, state and federal correction agencies contract with private companies that provide a wide range of services, including drug and psychiatric treatment; high school, college, and vocational education; physicians' services; staff ironing; and even the full operation of correctional facilities.[71]

Approximately 100,000 inmates (slightly less than 7 percent of the total prison population) are confined in more than 150 privately contracted prisons. Since the 1980s, several criminal justice agencies at the federal, state, and local levels have contracted with private agencies. In 2004, nearly 25,000 inmates (an increase of 60 percent since 2000) from the BOP, the Immigration and Naturalization Service, and the U.S. Marshal Service were confined in federally contracted facilities.[72]

Privatization is a controversial issue. Proponents argue that private construction is faster and cheaper and that the resulting facilities are less expensive to operate. Costs can be kept down owing to these prisons' lower labor costs (e.g., through use of non-unionized workers and elimination of overtime and employee benefits), less bureaucratic red tape, ability to negotiate item costs, and bulk purchasing.[73] Private firms claim to show greater concern for quality and quantity, with less waste. Private management may also be better at transforming prisons into "factories within fences" to reduce inmate idleness. Research suggests that recidivism rates for inmates released from private prisons are at least as low as those for inmates released from state prisons, with the former individuals' new offenses generally being less serious.[74]

Critics, however, voice a variety of concerns about privatization. Foremost is the constitutionality of delegating the incarceration function to a private entity.[75] Opponents of privatization also argue that the apparent efficiency and cost-effectiveness of privatization will eventually produce less humane treatment for prisoners because private firms have no incentive to reduce overcrowding, because they are paid on a per-prisoner basis. Furthermore, a recent review of 33 cost evaluations comparing private and public prisons found "no overall significant pattern of cost savings for private over public prisons."[76]

Another concern is that personnel will be of the "rent-a-cop" variety—that is, poorly trained, unprofessional, undereducated, and willing to accept low wages. In addition, the liability costs of regulating private industry could possibly negate any real savings.[77] What would happen, for example, if a state or federal government contracted with a private agency to detain inmates and a guard employed by the private contractor shot and killed an inmate who was trying to escape? Such an incident occurred in the 1980s, and a U.S. Court of Appeals ruled that the government could be held legally responsible for the inmate's safety in *Medina v. O'Neill*.[78] In 1997, the Supreme Court, in *Richardson v. McKnight*, held that individual correctional officers employed in private prisons do not have full immunity from civil suits brought by inmates as do officers in public prisons.[79] Four years later, in *Correctional Services Corporation v. Malesko*, the Court further defined the rights of inmates in private prisons when it ruled that although an inmate could not sue the prison itself in a civil rights claim, the inmate may file suit against the individual employee of the prison.[80]

Jail Systems

Jails confine the following types of individuals:
- Persons serving short sentences, typically less than one year

- Persons awaiting arraignment, trial, conviction, or sentencing
- Persons who are in violation of their probation or parole or who absconded while out on bail
- Mentally ill persons and individuals waiting movement to a mental health facility
- Persons who are in protective custody, contempt of court, or witness protection programs
- Federal or state inmates from overcrowded facilities

Jails in America have historically differed from prisons in several important ways:
- *They are more numerous.* There are approximately three times as many jails as there are prisons.
- *They have more inmates.* The number of people confined in jails in any given year is much greater than the number held in prisons.
- *They are more likely to have inferior facilities.* Criminologists who have studied jails have come away calling them the "ultimate ghetto," "human warehouses," "brutal," "filthy," and "cesspools of crime."[81]
- *They are typically more poorly managed.* The sordid conditions of jails—a result of years of neglect, insufficient funding, and poor management—are a heavy burden on inmates that may adversely affect them for the rest of their lives.[82]

Colonial Jails

The origin of the modern jail can be traced to medieval England when, in 1166, Henry II ordered the sheriff of each county to establish a jail to detain suspects awaiting trial.[83] Over time, the jail came to be used for the confinement of criminal offenders and vagrants. When the colonists came to America, they brought the English concept of the jail with them.

The earliest jails in America were built in the 1650s and typically consisted of a house-like structure with stocks, a pillory, and a whipping post. Colonial jails had only small rooms, where 20 or 30 people might be housed closely together with no heat. Inmates were required to buy their food from the jailer; destitute inmates were forced to rely on their families, friends, and the goodwill of others for assistance.

Modern Jails

The adage "The more things change, the more they stay the same" applies to modern jails: They have changed relatively little in the past 200 years. The

In Phoenix, Arizona, Sheriff Joe Arpaio implemented round-the-clock broadcasts of his "Jail Cam." Four cameras feed live footage over the Internet of the intake, holding, and searching cells, as well as the hallway of the Maricopa County jail. The cameras provide views of inmates being searched, sleeping in small bunk beds or on the concrete floor, and generally milling about. According to Sheriff Arpaio, "We get people booked in for murder all the way down to prostitution. . . . When these johns are arrested, they can wave to their wives on the camera." He also noted that the webcasts are educational: "Kids can tune in and see what it's like in the jail. Maybe they'll learn something."

Although the Jail Cam website received between 3 and 10 million hits each day, critics believe the sheriff crossed over the fine line between education and exploitation. In 2001, inmates filed a lawsuit charging that the Jail Cam violated detainees' basic right to privacy, because detainees have only been arrested but not yet convicted. In 2004, the 9th Circuit Court of Appeals affirmed a lower court decision ruling against the use of the Jail Cam. In its majority opinion, Judge Richard Paez wrote that the Jail Cam broadcasts amounted to little more than a "reality show" and went beyond what would be considered a reasonable deterrent to crime.

According to Judge Paez,

> Exposure to millions of complete strangers, not to mention friends, loved ones, co-workers and employers, as one is booked, fingerprinted, and generally processed as an arrestee, and as one sits, stands, or lies in a holding cell, constitutes a level of humiliation that almost anyone would regard as profoundly undesirable.

Sources: Heather Haddon, "Sneak Peek: 'Jail Cam' Raises Hackles and a Lawsuit When Links to Porno Sites are Discovered," *Village Voice*, June 6, 2001, available at http://www.village voice.com/news/0123.haddon.25345.html, accessed September 11, 2007; *Demery v. Arpaio*, 378F. 3d 754 (9th Cir. 2004).

majority of the nearly 3400 jails in the United States are located in rural areas.[84] Jails still serve the same basic functions they did in the colonial era: They detain people awaiting trial and provide short-term confinement for petty offenders. All jail inmates, whether they have been convicted or not, are housed together. In fact, approximately 60 percent of all people being held in jail are awaiting trial. Jails are also a refuge for what criminologist John Irwin calls "rabble"—people society finds disreputable and disruptive (i.e., the uneducated, unemployed, chronic alcoholics, and homeless).[85]

At the end of 2005, 747,529 people were being held in U.S. jails—nearly 5 percent more than in the previous year. The average daily occupancy was 95 percent of the rated capacity of the nation's jails—up from 65 percent in 1978. As with prison populations, the characteristics of jail inmates reflect a disproportionate number of minorities and males being held in local jails. Unlike prison inmates, however, nearly two-thirds of the incarcerated jail population is unconvicted FIGURE 13-3 .

Jails in urban areas are often overcrowded; by comparison, rural jails typically operate under capacity, but are generally older and poorly staffed.[86] Many jails, whether urban or rural, provide inmates with relatively little treatment, education, or recreation programs. These poor conditions can have serious consequences. For example, jail overcrowding has been found to contribute to violence, rape, and sickness. Some research indicates that prolonged exposure to overcrowding may reduce life expectancy.[87] Rural jails, which are poorly staffed compared to urban jails, have less supervision of inmates and report higher inmate homicide and suicide rates.[88] To combat this problem, many new, regional jails have been built to serve multiple counties.

New-Generation Jails

Historically, jail standards have been criticized by both the public and the private sectors as being unacceptably low in many jurisdictions. Responding in part to mounting public pressure, the federal government has stepped in to reform jail conditions. In 1974, three new federal Metropolitan Correctional Centers were opened in New York, Chicago, and San Diego. These federal jails were the first intentionally planned and designed to implement a new approach to jail administration, known as functional unit management, a precursor to the current model of direct supervision.

Other changes to improve jail conditions have focused on jail design and inmate supervision. In older jails, cells are in rows, and staff members walk down

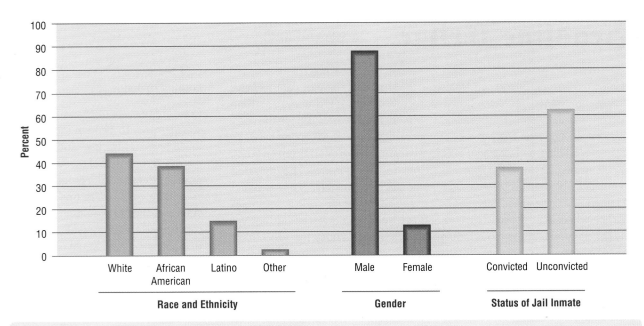

FIGURE 13-3 Selected Characteristics of Jail Inmates

Source: Paige Harrison and Allen Beck, *Prison and Jail Inmates at Midyear 2005* (Washington, DC: U.S. Department of Justice, 2006).

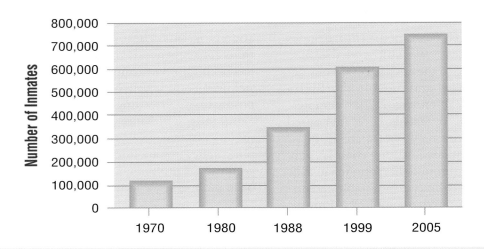

FIGURE 13-4 Annual Number of Jail Inmates

Sources: Paige Harrison and Allen Beck, *Prisoners in 2005* (Washington, DC: U.S. Department of Justice, 2006); James Stephan, *Census of Jails, 1999* (Washington, DC: U.S. Department of Justice, 2001).

a central corridor to observe the behavior of inmates through bars. In contrast, most <u>new-generation jails</u> are designed to increase staff interaction with inmates by placing the staff inside the inmate housing unit. This new layout increases direct surveillance so that staff can help actively change inmate behavior patterns rather than just reacting to them.[89]

New-generation jails are built on a podular design with small living units for each inmate situated around the perimeter of either a circle or triangle to permit staff to view all cells from the center.

Are podular designs more humane than traditional linear designs? Do inmates feel empowered when they can interact with other inmates and staff?

Additional innovations at the management level extended the reach of these reforms. Federal jails, for example, sought to redefine the role and responsibilities of guards by changing their job title to "Correctional Officer" and promoting more communication and counseling skills in dealing with inmates. These officers moved out of control rooms and into the living areas of inmates to encourage more direct interaction.[90]

Research examining differences in staff and inmate attitudes in traditional and new-generation jails demonstrates the value of direct supervision. A number of studies looking at new-generation jails have reported that staff and inmates felt more positively about the overall jail environment, including living and working conditions, than did staff or inmates in traditional jails. They have also found that direct supervision reduced the number of inmate-on-inmate assaults and lowered the overall costs of operation.[91]

Direct supervision may also prevent inmate suicides. Males continue to be at much greater risk of committing suicide in jail, being about 56 percent more likely to commit suicide than female jail inmates. In addition, white jail inmates are about six times more likely to commit suicide than African American inmates and three times more likely to commit suicide than Latino inmates.[92]

Despite the variety of reform efforts over the decades, jails have experienced rapid population growth in recent years, with their populations growing by more than one-third in less than a decade FIGURE 13-4.

WRAP UP

Along with approximately 600 condemned offenders nationwide, McDuff's death sentences were commuted to life imprisonment without parole following the Supreme Court's ruling in *Furman v. Georgia* (1972). In this case, the Court held that the manner in which the death penalty was imposed and executed was cruel and unusual and, therefore, was a violation of the Eighth Amendment.

In most states, life imprisonment without parole does not really mean that the offender will never be released. Instead, the sentence is carried out in accordance with state statutes that specify some lengthy but not indefinite term to be served (e.g., 25 years) before the offender is eligible for parole. McDuff was eligible for parole because he was fairly compliant while in custody, had already served 23 years in custody, and was considered advanced in age. Owing to these considerations and overcrowded prison conditions, McDuff was released.

Prison is perceived as a "revolving door" by the general public because many serious offenders are arrested, convicted, sentenced to prison, and released, only to repeat the process. Most correctional systems are overcrowded and face resource limitations. For this reason, parole boards seek to release those inmates who appear to offer the greatest potential for rehabilitation or who appear to have "aged out" of crime. Unfortunately, parole systems are often overburdened and cannot always provide careful supervision of parolees. As such, offenders often commit crimes while on parole. Although McDuff's case is exceptional given the number of homicides and death sentences, his tale of recidivism in an overtaxed prison system is commonplace.

Chapter Spotlight

- Colonial correctional philosophy was largely guided by strong religious beliefs. As a consequence, people used severe sanctions to deter both serious and minor violations.

- The first U.S. prisons were built after the American Revolution. Two competing models of prisons emerged: the Pennsylvania model, based on a philosophy of separate confinement, and the New York model, based on a congregate system.

- Between 1865 and 1900, correctional philosophy underwent a major shift, reflected in the reformatory movement. The Elmira Reformatory for younger, first-time offenders opened in 1876.

- The first correctional facility exclusively devoted to female offenders was established in 1835. Before that time, women were confined in houses of refuge, shelters, or local jails.

- Prisons are classified by their level of security: minimum-security, low-security, medium-security, maximum-security, and super-maximum-security.

- The military penal system includes special facilities for confining offenders from each branch of the armed forces. In addition, the military is responsible for managing Camp Delta, a special prison for captured terrorists located at the Guantanamo Naval Base in Cuba.

- The Federal Bureau of Prisons operates U.S. penitentiaries, federal correctional institutions, metropolitan correctional centers, medical centers, minimum-security prison camps, and a federal detention center.

- Nearly 100,000 persons are incarcerated in the more than 150 privately contracted prisons. While privatization is controversial, it does appear to cost less and often provides higher-quality services than public prisons.

- Jails have their origins in medieval England, where each county was ordered to establish a jail to confine offenders. Contemporary U.S. jails provide a combination of high security with greater treatment opportunities for those confined in them.

Putting It All Together

1. Which system was probably better for prisoners—the Pennsylvania model or the New York model? Why?
2. Should inmates have the right to conjugal visits? Should such visits be limited to an inmate's spouse, or should they be expanded to include any intimate of the offender?
3. What kinds of issues do private business ventures in prison raise, and how might those issues be overcome?
4. Who benefits from confinement in co-correctional institutions? Should the use of these facilities be expanded?
5. Should prisons be privatized, or should the function of incarcerating and punishing offenders be the exclusive role of the state?

Key Terms

Ashurst–Sumners Act Legislation passed by Congress in 1935 and amended in 1940 that prohibited interstate transportation of prison goods.

boot camp A highly regimented correctional facility where inmates undergo extensive physical conditioning and discipline.

Camp Delta A facility at the Guantanamo Naval Base in Cuba that is used for the confinement of suspected terrorists.

co-correctional prison An institution where men and women are confined together.

congregate system A nineteenth-century model that held prisoners in isolation during the night, allowing them to work together during the day in silence; it was implemented at New York's Auburn Prison.

conjugal visit A private visit that some prison systems allow between inmates and their spouses to help them maintain sexual and interpersonal relationships.

correctional system Programs, services, and institutions designed to manage people accused or convicted of crimes.

Hawes–Cooper Act Legislation passed by Congress in 1929 requiring that prison products be subject to the laws of the state to which they were shipped.

jail An institution to hold pretrial detainees and people convicted of less serious crimes.

low-security prison An institution that operates between the medium and minimum security levels.

mark system System by which prisoners earned "marks" for good behavior to achieve an early release from prison.

maximum-security prison The most secure prison facility, having high walls, gun towers, and barbed wire or electronic fences.

medical model A treatment approach popular between 1930 and 1960 that attributed criminality to a biological or psychological defect of the offender.

medium-security prison A middle-level prison facility, which has more relaxed security measures and fewer inmates than a maximum-security prison.

minimum-security prison A prison facility with the lowest level of security that houses nondangerous, stable offenders.

new-generation jails Jails that are designed to increase staff interaction with inmates by placing the staff inside the inmate housing unit.

penitentiary house An eighteenth-century place of penitence for all convicted felons except those sentenced to death.

Percy Amendment Law that allowed states to sell prison-made goods across state lines as long as they complied with strict rules to make sure unions were consulted and to prevent manufacturers from undercutting existing wage structures.

prison An institution for the confinement of people who have been convicted of serious crimes.

prison farms Correctional institutions that produce much of the livestock, dairy products, and vegetables used to feed inmates in the state prisons.

prison forestry camps Correctional institutions that provide labor for the maintenance of state parks, tree planting and thinning, wildlife care, maintenance of fish hatcheries, and cleanup of roads and highways.

privatization The process in which state and federal governments contract with the private sector to help finance and manage correctional facilities.

reformatory A penal institution generally used to confine first-time offenders between the ages of 16 and 30.

separate confinement A nineteenth-century model of prison that separated inmates; it was implemented in Pennsylvania's Western and Eastern penitentiaries.

super-maximum-security prison A prison where the most predatory and dangerous criminals are confined.

work release A program allowing the inmate to leave the institution during the day to work at a job.

Notes

1. Bureau of Justice Statistics, Expenditure and Employment Statistics, available at http://www.ojp.usdoj.gov/bjs/glance/exptyp.htm, accessed May 22, 2007.
2. Alice Morse Earle, *Curious Punishments of Bygone Days* (Chicago: Herbert S. Stone and Company, 1896), p. 138.
3. David Fogel, *We Are the Living Proof . . . ,* 2nd edition (Cincinnati: Anderson, 1979), p. 10.
4. Fogel, note 3, p. 10.
5. David Rothman, *The Discovery of the Asylum* (Boston: Little, Brown, 1971), p. 15.
6. Fogel, note 3, p. 8.
7. Earle, note 2.
8. Robert Caldwell, *Red Hannah* (Philadelphia: University of Pennsylvania Press, 1947).
9. Earle, note 2.
10. William Andrews, *Bygone Punishments* (London: William Andrews, 1899); Earle, note 2.
11. Earle, note 2.
12. Earle, note 2.
13. Wayne Morse, "The Attorney General's Survey of Release Procedures," in George Killinger and Paul Cromwell, Jr. (Eds.), *Penology* (St. Paul, MN: West, 1973), pp. 23–53.
14. Norman Johnston, "The World's Most Influential Prison: Success or Failure?" *The Prison Journal* 84:20S–40S (2004).
15. Harry Elmer Barnes, *The Evolution of Penology in Pennsylvania* (Indianapolis: Bobbs-Merrill, 1928).
16. Morse, note 13, p. 36.
17. Thorsten Sellin, "The Origin of the Pennsylvania System of Prison Discipline," *The Prison Journal* 50:15–17 (1970).
18. Harry Elmer Barnes, *The Story of Punishment* (Boston: Stratford, 1930).

19. Larry Sullivan, *The Prison Reform Movement: Forlorn Hope* (Boston: Twayne Publishers, 1990), p. 13.
20. Margaret Wilson, *The Crime of Punishment* (New York: Harcourt Brace, 1931).
21. Harry Elmer Barnes and Negley Teeters, *New Horizons in Criminology,* 3rd edition (Englewood Cliffs, NJ: Prentice Hall, 1959).
22. Gustav de Beaumont and Alexis de Tocqueville, *On the Penitentiary System in the United States and Its Application in France* (Carbondale, IL: Southern Illinois University Press, 1964).
23. Johnston, note 14.
24. Rothman, note 5.
25. John Vincent Barry, "Captain Alexander Maconochie," *Victorian Historical Magazine,* June 27, 1957:5.
26. Wilson, note 20.
27. John Vincent Barry, "Alexander Maconochie," *Journal of Criminal Law, Criminology and Police Science* 47:145–161 (1956).
28. Barnes and Teeters, note 21, p. 423; also see Torsten Eriksson, *The Reformers,* translated by Catherine Djurklou (New York: Elsevier, 1976), pp. 91–97.
29. *Twenty-Sixth Annual Report of the Executive Committee of the Prison Association of New York and Accompanying Documents for the Year 1870* (Albany, NY: State of New York, 1871).
30. Zebulon Brockway, *Fifty Years of Prison Service* (Montclair, NJ: Patterson Smith, 1912/1969).
31. Harry Elmer Barnes and Negley Teeters, *New Horizons in Criminology,* revised ed. (New York: Prentice Hall, 1942/1945), pp. 555–556.
32. Paige Harrison and Allen Beck, *Prisoners in 2005* (Washington, DC: U.S. Department of Justice, 2006).
33. Russell Dobash, Rebecca Emerson Dobash, and Sue Gutteridge, *The Imprisonment of Women* (New York: Basil Blackwell, 1986).
34. Joycelyn Pollock, *Women, Prison, and Crime,* 2nd edition (Pacific Grove, CA: Wadsworth, 2001); Dobash et al., note 33, p. 33.
35. Estelle Freedman, "Their Sisters' Keepers," *Feminist Studies* 2:77–95 (1974).
36. Richard Hofstadter, *The Age of Reform* (New York: Vintage Books, 1955); Philip Klein, *Prison Methods in New York State* (New York: Longmans Green, 1920).
37. Kay Harris, "Women's Imprisonment in the United States: A Historical Analysis of Female Offenders Through the Early 20th Century," *Corrections Today* 60:74–80 (1998).
38. Mara Dodge, "'One Female Prisoner Is of More Trouble Than Twenty Males': Women Convicts in Illinois Prisons, 1835–1896," *Journal of Social History* 32:907–930 (1999).
39. Clarice Feinman, "An Historical Overview of the Treatment of Incarcerated Women," *The Prison Journal* 63:12–26 (1984).
40. Estelle Freedman, *Their Sisters' Keepers* (Ann Arbor, MI: University of Michigan Press, 1981).
41. Harris, note 37; Nicole Rafter, *Partial Justice: Women in State Prisons, 1800–1935* (Boston: Northeastern University Press, 1985); Sheryl Nicole Rafter, "Gender and Justice," in Lynne Goodstein and Doris MacKenzie (Eds.), *The American Prison* (New York: Plenum, 1989), pp. 89–109.
42. Harry Elmer Barnes, *Prisoners in Wartime* (Washington, DC: War Production Board, 1944).
43. Public Law 96-157, 827, codified as 18 U.S. Code 1761(c).
44. Rudolph Alexander, *Counseling, Treatment, and Intervention Methods with Juvenile and Adult Offenders* (Belmont, CA: Wadsworth, 2000).
45. *National Advisory Commission on Criminal Justice Standards and Goals, Standard 2.17, Part 2c* (Washington, DC: U.S. Department of Justice, 1974).
46. Harrison and Beck, note 32.
47. James Stephan and Jennifer Karberg, *Census of State and Federal Correctional Facilities, 2000* (Washington, DC: U.S. Department of Justice, 2003).

48. Daniel Mears and Jamie Watson, "Toward a Fair and Balanced Assessment of Supermax Prisons," *Justice Quarterly* 23:232–270 (2006).

49. National Institute of Corrections, *Supermax Housing: A Survey of Current Practices* (Longmont, CO: National Institute of Corrections Information Center, 1997).

50. Jesenia Pizarro and Vanja Stenius, "Supermax Prisons: Their Rise, Current Practices, and Effects on Inmates," *The Prison Journal* 84:248–264 (2004).

51. Bulduc Correctional Facility, Maine Department of Corrections (2006), available at http://www.maine.gov/corrections/Facilities/bcf/index.htm, accessed May 28, 2007.

52. Dennis Prouty, "Prison Farms," *Issue Review* (Des Moines, IA: Iowa Legislative Fiscal Bureau, 2002).

53. South Fork Forest Camp, Oregon Department of Corrections, available at http://www.oregon.gov/DOC/OPS/PRISON/sffc.shtml, accessed May 28, 2007; Sugar Pine Conservation Camp, California Department of Forestry and Fire Protection, available at http://www.fire.ca.gov/php/fire_er_consrvncamp.php, accessed May 28, 2007.

54. "Federal Prisons to Eliminate Boot Camps," *Corrections Today* 67:13 (2005); National Institute of Justice, *Correctional Boot Camps: Lessons from a Decade of Research* (Washington, DC: U.S. Department of Justice, 2003).

55. "Florida Closes Youth Boot Camps," *CNN.com,* April 26, 2006, available at http://www.cnn.com/2006/US/04/26/boot.camp.ap/index.html, accessed May 28, 2007.

56. "Federal Prisons to Eliminate Boot Camps," note 54.

57. David Haasenritter, "The Military Correctional System: An Overview," *Corrections Today* 65:58–61 (2003).

58. William Peck and Timothy Purcell, "U.S. Navy Corrections: Confining Sailors Both at Sea and on Land," *Corrections Today* 65:66–70 (2003).

59. Haasenritter, note 57.

60. John Crank and Patricia Gregor, *Counter-Terrorism after 9/11: Justice, Security and Ethics Reconsidered* (Cincinnati: LexisNexis, 2005).

61. Josh White, "Three Detainees Commit Suicide at Guantanamo," *Washington Post,* June 11, 2006, p. A01.

62. "Guantanamo Bay—Camp Delta" (June 2006), available at http://globalsecurity.org/military/facility/guantanamo-bay_delta.htm, accessed May 28, 2007.

63. White, note 61.

64. American Correctional Association, *2006 Directory: Adult and Juvenile Correctional Departments, Institutions, Agencies, and Probation and Parole Authorities* (Lanham, MD: American Correctional Association, 2006).

65. Barry Ruback, "The Sexually Integrated Prison," in John Smykla (Ed.), *Coed Prison* (New York: Human Sciences Press, 1980), pp. 43–44.

66. James Ross, Esther Heffernan, James Sevick, and Ford Johnson, "Characteristics of Co-correctional Institutions," in John Smykla (Ed.), *Coed Prison* (New York: Human Sciences Press, 1980), pp. 77–78; John Ortiz Smykla and Jimmy Williams, "Co-corrections in the United States of America, 1970–1990: Two decades of disadvantages for women prisoners," *Women and Criminal Justice* 8:61–76 (1996).

67. Dede Short, "Illinois Correctional Policy-Makers Initiate Historical Changes," *Corrections Today* 64:102–106 (2002).

68. 18 U.S. Code 1762, 1988 edition (Washington, DC: U.S. Government Printing Office).

69. J.C. Powell, *The American Siberia* (Chicago: H. J. Smith, 1891); "Prison Labor in 1936," *Monthly Labor Review* 47:251 (1938).

70. Philip Ethridge and James Marquart, "Private Prisons in Texas," *Justice Quarterly* 10:31–50 (1993).

71. Linda Calvert Hanson, "The Privatization of Corrections Movement," *Journal of Contemporary Criminal Justice* 7:1–28

(1991); also see Charles Thomas, "Does the Private Sector Have a Role in American Corrections?", paper presented at A Critical Look at Privatization conference, Indianapolis, IN, 1988; John Dilulio, Jr., *Private Prisons* (Washington, DC: U.S. Department of Justice, 1988); T. Don Hutto, "The Privatization of Prisons," in John Murphy and Jack Dison (Eds.), *Are Prisons Any Better?* (Newbury Park, CA: Sage Publications, 1990), pp. 111–127.

72. Paige Harrison and Allen Beck, *Prisoners in 2004* (Washington, DC: U.S. Department of Justice, 2005).

73. Dina Perrone and Travis Pratt, "Comparing the Quality of Confinement and Cost-Effectiveness of Public Versus Private Prisons: What We Know, Why We Do Not Know More, and Where to Go from Here," *The Prison Journal* 83:301–322 (2003).

74. William Bales, Laura Bedard, Susan Quinn, David Ensley, and Glen Holley, "Recidivism of Public and Private State Prison Inmates in Florida," *Criminology and Public Policy* 4:57–82 (2005); Charles Thomas, "Recidivism and Public and Private State Prison Inmates in Florida: Issues and Unanswered Questions," *Criminology and Public Policy* 4:89–99 (2005).

75. Ira Robbins, *The Legal Dimensions of Private Incarceration* (Washington, DC: American Bar Association, 1988).

76. Travis Pratt and Jeff Maahs, "Are Private Prisons More Cost-Effective Than Public Prisons? A Meta-Analysis of Evaluation Research Studies," *Crime & Delinquency* 45:358–371 (1999).

77. Dilulio, note 71.

78. *Medina v. O'Neill,* 589 F. Supp. 1028 (1984).

79. *Richardson v. McKnight,* 521 U.S. 399 (1997).

80. *Correctional Services Corporation v. Malesko,* 534 U.S. 61 (2001).

81. Michael Charles, Sesha Kethineni, and Jeffrey Thompson, "The State of Jails in America," *Federal Probation* 56:56–62 (1992).

82. John Irwin, *Jails* (Berkeley: University of California Press, 1985), p. 7.

83. J. M. Moynahan and Earle Stewart, "The Origin of the American Jail," *Federal Probation* 42:41–50 (1978).

84. Ralph Weisheit, David Falcone, and L. Edward Wells, *Rural Crime and Rural Policing: An Overview of Selected Issues* (Normal, IL: Rural Police Project, 1994).

85. Irwin, note 82.

86. G. Larry Mays and Joel Thompson, "Mayberry Revisited: The Characteristics and Operations of America's Small Jails," *Justice Quarterly* 5:421–440 (1988).

87. Weisheit et al., note 84.

88. Paul Paulus, Garvin McCain, and Verne Cox, "Prison Standards: Some Pertinent Data on Crowding," *Federal Probation* 45:48–54 (1981).

89. Lois Spears and Don Taylor, "Coping with Our Jam-Packed Jails," *Corrections Today* 52:20 (June 1990).

90. Richard Wener, "The Invention of Direct Supervision," *Corrections Compendium* 30:4–7, 32–34 (2005).

91. Richard Wener, "Effectiveness of the Direct Supervision System of Correctional Design and Management: A Review of the Literature," *Criminal Justice and Behavior* 33:392–410 (2006); Linda Zupan, *Jails: Reform and the New Generation Philosophy* (Cincinnati: Anderson, 1991); Linda Zupan and Ben Menke, "Implementing Organizational Change: From Traditional to New Generation Jail Operations," *Policy Studies Review* 7:615–625 (1988); Linda Zupan and M. Stohr-Gillmore, "Doing Time in the New Generation Jail," *Policy Studies Review* 7:626–640 (1988).

92. Christopher Mumola, *Suicide and Homicide in State Prisons and Local Jails* (Washington, DC: U.S. Department of Justice, 2005).

OBJECTIVES

- Comprehend the characteristics of persons currently confined in prison and understand how they have changed in recent years.

- Become familiar with the variety of prison programs provided to inmates.

- Understand how prisons are managed and how discipline is maintained.

- Examine special prison populations and their needs.

- Grasp the nature of prison life, the ways in which inmates adapt to prison, and the problems facing inmates and staff.

- Know the current status of prisoner rights and discover how these rights evolved over time.

CHAPTER

14

Prisons

In January 2006, Clarence Ray Allen was executed in California for three murders. Allen had been on death row for 23 years and, at 76 years of age, was one of the oldest prisoners ever executed in U.S. history. He had an extensive criminal history and had been incarcerated throughout much of the twentieth century.

During his life, Allen witnessed major changes in the inmate subculture—changes that, in the eyes of many old-school convicts, were not for the better. Some penologists (people who study prison management and criminal rehabilitation) argue that the old honor code among prison inmates and their solidarity against guards and administration created a safer environment than modern prisons, which are now populated by younger inmates and gang members. Others argue that new systems of inmate organization and strengthened security have made modern prisons much safer than the one Allen walked into, despite the shift in prison subculture.

- How has inmate subculture changed in the past 50 years?
- What promoted these changes within the prison system?

Source: Bill Lockyer, *"People v. Clarence Ray Allen,"* January 2006, Office of Victims Services, Office of the Attorney General, California Department of Justice, January 2006, available at http://caag.state.ca.us/victimservices/pdf/CAllenPressPackE2.pdf, accessed March 20, 2007.

Introduction

A female inmate at the federal women's prison at Alderson, West Virginia, wrote:

> When I came through the gate, I said to myself: "This is a prison?" All the trees and flowers—I couldn't believe it. It looked like a college with the buildings, the trees, and all the flowers. But after you're here a while—and it don't take too long—you know it's a prison.[1]

This inmate's reaction to entering prison reflects some of the same misconceptions about prisons many people have today—that prisons are fortress-like structures with high stone walls, guard towers, and a foreboding sense of isolation and danger. Despite the trees and flowers, this facility was still a prison, and this inmate soon discovered it would confine her in close quarters with other convicted women, deprive her of privacy, and place restrictions on when and how often she could be visited by friends and relatives. This difference between the appearance and reality of prisons illustrates the ambivalence of

many Americans about what prison is and what it should be.

LINK Chapter 13 examined the history of American corrections and explored how correctional philosophies and prison structures have evolved over the past century.

Certainly the early prisons, such as Auburn Prison in New York, were cold, dark fortresses where inmates were removed from society for extended periods of time. The bleakness of most prisons up to the early twentieth century reflected the widely held belief that prisoners should be confined in structures with few of the amenities of the outside world. This notion of <u>less eligibility</u> posited that prisoners should always live in conditions that were less desirable than even the worst conditions in free society.

Over the past 100 years, both changes in correctional philosophy and Supreme Court decisions concerning prisoners' constitutional rights have altered the general appearance of prisons and the conditions under which inmates are confined. Many prisons today have flowers and trees, and some have tennis courts and exercise facilities, high school and college classrooms, rooms for watching television (or even TV sets in cells), law libraries, and state-of-the-art vocational training. Even so, they are still prisons: Their primary function is to house people who have been convicted of crimes. The addition of amenities has not changed the emphasis on security in prison, nor has it made life in prison more pleasant than life on the outside, although it has clearly improved living conditions for inmates compared to the conditions faced by their counterparts from earlier decades.

Prisons and Prisoners

Sociologist Erving Goffman characterized prisons (along with mental hospitals and the military) as total institutions, where "all aspects of life are conducted in the same place and under the same single authority."[2] All persons there are treated alike, and activities are tightly scheduled, with the institution providing for all the basic needs within an environment closed off from the outside world. The large group of people within a total institution is divided into two subgroups: those who exercise control and those who are controlled. In prison, this means that all aspects of inmates' lives, from the time they awake each morning until they go to sleep at night, are governed by institutional rules enforced by prison staff. When inmates arrive at the prison, they are stripped of their personal clothing and possessions and given institutional clothes and a set of rules that govern how they will interact with the staff and with other inmates. Their contact with the outside world is limited, and many normal social activities are curtailed. Phone calls and visits with family members are minimal, and most inmates are deprived of heterosexual contact.

Interestingly, the lives of staff are also regulated: Their interaction with inmates is strictly controlled by institutional rules, and the functions and relationships of each segment of the staff are formalized according to a military style of organization. A large portion of the staff is also relatively isolated from the outside world. Many prison wardens live in quarters located on the prison grounds outside the walled security area, and many counselors, teachers, and correctional officers spend their entire work days locked inside the institutions along with the inmates.

Despite these characteristics, some scholars question whether prisons are truly total institutions. Sociologist Keith Farrington, for instance, suggests that prisons might better be characterized as "not-so-total" institutions.[3] He argues that prisons are supplied with goods, services, and materials from the outside world; employees maintain lives outside the prison; inmates receive visits from relatives and friends; and many inmates are granted brief furloughs to return home as a reward for good behavior. As a consequence, prisons—and those who work and live in them—may not be as isolated as Goffman's concept of the total institution suggests.

Classification of Prisoners

Few inmates arrive at correctional institutions directly from the courts that sentenced them. Instead, they usually go from the court to a reception or diagnostic center for orientation and initial classification, where the type of custody and treatment appropriate to their needs is determined. In some correctional systems, the classification system simply sorts inmates according to their age, severity of offense or prior incarceration record, and work experience. Classification allows administrators to assign inmates to institutions or housing units appropriate to their security level, thereby separating more aggressive or violent inmates from those considered to be more vulnerable to assault or exploitation (e.g., because they are younger, more timid, or smaller) as well as members of conflicting gangs; inmates may also be separated according to their work or program assignments.[4]

The classification process usually takes 60 to 90 days (which count toward completion of the inmate's sentence). During this time inmates undergo a variety of psychological and physical evaluations. Classification staff review the following issues for each offender:[5]

- Criminal and behavioral history
- Length of sentence
- Mental health needs
- Inmate records (such as the presentence investigation report)
- Americans with Disabilities Act requirements
- Bed availability in facilities

The classification process, however, does not stop once an inmate arrives at his or her prison destination. Inmates are often returned to the reception and diagnostic center for additional evaluation and may be reclassified and transferred to another institution. Sometimes this return to the center results from requests made by inmates, who may want to be

According to the Sentencing Project, nearly 10 percent of all state prisoners are serving a life sentence.

? Given the "nothing to lose" mentality of some inmates sentenced to life imprisonment, should prisons be segregated by type of offense? Should all inmates serving life sentences be placed together in one institution for safety reasons?

placed in institutions closer to their home communities or facilities where they may participate in vocational or educational programs that are not available in their current institutions. Inmates may also be reclassified or transferred as a result of behavior problems, involvement with prison gangs, or vulnerability to assaults (often caused by snitching or accumulating debts to other inmates). In addition, in many states, inmates who are within one year to three months of their release dates are brought to the center for prerelease evaluation and possible reclassification, which could make it possible for them to be assigned to a work-release center in a nearby community.

The Changing Prison Population

Largely as a result of the federal government's drug-enforcement policies, the number of prisoners has skyrocketed in the past decade. In 2005, 1,446,269 prisoners were incarcerated in state and federal prisons—an increase of more than 360 percent since 1980 FIGURE 14-1.[6] The increasing number of prisoners has required both the states and the federal government to build new prisons and employ more people to operate them. The type of inmate has changed as well: Prisons have become populated by younger and more violent offenders. As a result, correctional institutions have been forced to increase the sizes of their staffs. In 1958, state prisons employed one correctional officer for every nine inmates, on average; in 2001, there was one officer for every 4.5 inmates. As shown in FIGURE 14-2, however, this number varies greatly from state to state. In federal prisons, which have smaller proportions of violent offenders, the ratio of inmates to officers in 1990 was roughly double the average state ratio, with one officer for every nine inmates.[7]

While most inmates in state and federal prisons are male (nearly 93 percent of the total prison population), the number of female inmates has increased at a much higher rate than the number of male inmates in recent years (a 4.6 percent increase for females versus a 3 percent increase for males).[8] As shown in TABLE 14-1, the majority of inmates are members of racial and ethnic minorities. Approximately half of all state prison inmates were sentenced for violent crimes, and nearly half of all federal prisoners were sentenced for drug offenses.[9]

A large number of inmates are also serving lengthy sentences, and a substantial number are serving life sentences. According to the Sentencing Project, slightly less than 10 percent of all prisoners (more than 127,000 inmates) are serving life sentences—an increase of 83 percent since 1992. Of those serving life, one-fourth are serving sentences of life without parole.[10]

TABLE 14-1

Characteristics of State and Federal Prison Inmates

Characteristic	Percentage of Prison Inmates
Gender	
Females	7.0
Males	93.0
Race and Ethnicity	
White, non-Latino	34.3
African American	40.7
Latino	19.2
Other	5.8
Gender and Race	
Females	
White	44.2
African American	33.4
Latina	15.6
Males	
White	33.6
African American	41.2
Latino	19.5

Source: Paige Harrison and Allen Beck, *Prisoners in 2004* (Washington, DC: U.S. Department of Justice, 2005).

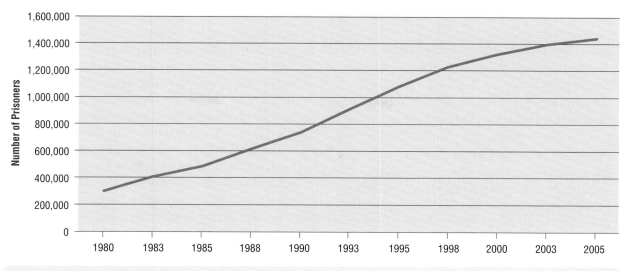

FIGURE 14-1 State and Federal Prison Populations

Sources: Paige Harrison and Allen Beck, *Prisoners in 2005* (Washington, DC: U.S. Department of Justice, 2006); *Bureau of Justice Statistics, National Prisoners Statistics—1* (Washington, DC: U.S. Department of Justice, 1994).

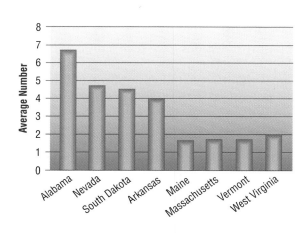

FIGURE 14-2 Average Number of Inmates per Guard

Source: James Stephan, *State Prison Expenditures, 2001* (Washington DC: U.S. Department of Justice, 2004), p. 5.

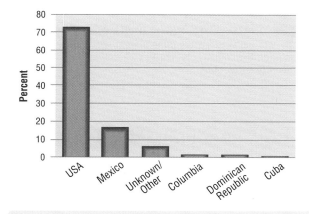

FIGURE 14-3 Federal Prison Population by Citizenship

Source: "Inmate Breakdown by Citizenship," Federal Bureau of Prisons (Washington, DC: 2006), available at http://www.bop.gov/news/quick.jsp, accessed August 2, 2007.

Many of those incarcerated in state and federal prisons today are not U.S. citizens FIGURE 14-3 and are subject to deportation after they finish serving their sentences. For example, noncitizen prisoners account for more than 10 percent of the prison populations of Arizona, New York, Nevada, and California, and nearly 20 percent of the federal prison population.[11] These changes in the inmate population present prisons with new and more difficult challenges.

Prison Crowding

The growth in prison populations has resulted in greater crowding in most facilities. Federal prisons are currently at 134 percent of their rated capacity (the optimal number of beds for the facility), while state prisons are at 101 percent of their capacity.[12] Consequently, an increasing number of prisoners have to share living space. While many inmates are housed in single-person cells, most are forced to adjust to sharing a 6-foot by 9-foot cell; increasingly, many live in more open, but less secure, dormitories.

Several methods to relieve prison crowding have been suggested:

- *Build more prisons.* Unfortunately, the economic constraints faced by many states may not permit this approach.
- *Contract with private corporations.* An increasing number of states have begun to contract with private corporations to build and manage prisons as for-profit ventures.
- *Change sentencing provisions.* If crimes that carry mandatory or long sentences were punished with shorter sentences, it would divert people from prison or keep them in for shorter periods of time. Many states saw their prison populations soar after they implemented harsher sentencing codes.
- *Expand community-based corrections.* An increasing number of states are expanding their use of probation, electronic monitoring, and restorative justice as alternatives to institutional incarceration.

LINK   Only about 20 percent of all offenders under some form of correctional supervision are in prison; the rest are being supervised in the community. Chapter 15 explores the concept of community corrections.

Prison Programs

It was not until the middle of the twentieth century that prisons began expanding beyond mere custody to include a variety of programs geared toward meeting both inmate and institutional needs. Today, with the exception of a few super-maximum-security institutions, nearly all prisons offer counseling, health and medical services, academic and vocational education, and prison industries.

LINK   As discussed in Chapter 13, early American prisons provided a few basic programs for inmates, such as industrial labor for men and sewing and cooking for women, and very limited medical care or counseling for prisoners of either gender. In those days, the primary purpose of the prison was to maintain secure custody of prisoners.

The nature and extent of each kind of program vary among institutions, depending on the size of the institution, the state funding available, and the composition of the inmate population (i.e., prisoners' gender, age, security levels, or psychological and medical needs). Inmates are typically assigned to academic or work programs as a result of the classification process or by request once they arrive at the institution. If prisoners participate in education or industry programs, they will receive a basic—albeit very low—wage. They may spend these earnings on incidentals such as candy and cigarettes, save them in their prison account, or send a portion home to their family. Some states require that a percentage of an inmate's pay be contributed to make restitution or be sent to a victim compensation program. Inmates who are assigned to work in private prison industries may earn significantly higher wages based either on federal minimum wage standards or the prevailing wages in the industry.[13]

In 2006, the 5th Circuit Court of Appeals rejected a suit brought by an inmate in Texas who argued that his job as a drying machine operator qualified him for protection under the Fair Labor Standards Act and, therefore, required that he receive the then-minimum wage of $5.15 per hour. The Court held that prisoners are not employees and thus are not entitled to minimum wages.[14]

Counseling

The past two decades have seen a dramatic decline in the emphasis on inmate counseling and rehabilitation. A number of studies published in the 1970s called into question the ability of prisons to achieve

Many inmates have psychiatric problems and benefit from mental health treatment while detained.

? Should psychiatric counseling be mandatory for all inmates? Are mental health problems at the root of criminality, or are they only one of many problems that offenders face?

any significant rehabilitation of inmates. Although critics called for spending less money and effort on treatment, they were also quick to suggest placing a much greater emphasis on security and control in institutions. Some states went so far as to change the stated goal in their sentencing code from rehabilitation to punishment.

Even so, most prisons continue to provide some degree of counseling and treatment for inmates, with programs aimed at "the elimination of criminal behavior and the establishment of prosocial behavior, both during imprisonment and subsequent release."[15] Psychological treatment typically includes a variety of individual counseling techniques:

- *Reality therapy,* in which the therapist attempts to improve the inmate's willingness or ability to behave more responsibly by focusing on the consequences of acting inappropriately
- *Group psychotherapy,* which involves bringing inmates into confrontational group interactions with other inmates as they discuss shared problems and call each other on their attempts to manipulate or rationalize their behaviors
- *Behavior therapy,* in which behavior modification techniques, such as rewards and punishments, are used to alter the behaviors of inmates

While there is little empirical evidence that psychological treatment in prison has significant or lasting positive effects on inmates, some approaches do seem to produce positive results, at least for some inmates. These strategies include the teaching of interpersonal skills, a stress on moral reasoning, individual counseling utilizing cognitive-behavioral therapy, and the integration of in-prison treatment with systematic follow-up treatment once the inmate is released into the community.[16]

Health and Medical Services

Because most prisons are large institutions in which hundreds of inmates live in close quarters for years at a time, the normal medical and health concerns of prisoners become a major responsibility of the prison. In addition, inmates frequently bring with them a history of inadequate medical attention, poor diets, and drug and alcohol abuse. Unfortunately, the funding for most correctional institutions is not large enough to provide the same kinds of medical and health care that many civilians receive.

Every prison has some sort of medical and health services, but few institutions have anything close to state-of-the-art facilities and many do not have a full-

time physician. Typically, nurses and paraprofessionals meet the daily or routine needs of inmates; the prison contracts with the local medical community to provide weekly visits by doctors and dentists. Most prison hospitals are more like small clinics where inmates stay when ill, receive medication, and have cuts and broken bones treated. Inmates are transported to a community hospital for more serious injuries and illnesses or surgery. Many of the nation's larger prisons are also faced with a growing number of human immunodeficiency virus (HIV)-positive inmates (discussed later in this chapter), presenting new medical demands for care and treatment.

Academic and Vocational Education

Roughly two-thirds of all inmates have not finished high school by the time they enter prison. About one-third of all men and more than half of all women in prison were not employed at the time of their arrest.[17] Their lack of education and vocational skills may have contributed to the problems inmates faced in the outside world before coming to prison, and it exacerbates the difficulties they confront once they are released. As part of the rehabilitative goals of correctional institutions, therefore, academic and vocational education programs have become a standard part of the program services offered.

All federal prisons and roughly 90 percent of state prisons provide educational programs to their inmates. These programs include the following:

- Adult basic education
- Adult secondary education
- Special education
- College coursework
- Vocational education
- Study release

Secondary education programs designed to prepare inmates for the graduate equivalency degree (GED) exam are the most common type of courses offered in prison. Next most common are basic classes in math and reading. Only 27 percent of state prisons offer college classes.[18]

Slightly more than half of all state and federal prisoners participate in educational programs. Among those who participate, most enroll in vocational training or GED preparation coursework. A relatively large proportion of inmates also participate in college-level coursework, including 10 percent of state inmates and 13 percent of federal prisoners. These numbers have declined since the early 1990s, because federal legislation in 1994 made inmates ineligible for Pell grants.[19]

Much of the prison-based course work involves individualized instruction because of the constant turnover of students. Traditional semester-long courses are nearly impossible to conduct, because new inmates arrive and are assigned to school programs, inmates get transferred to other institutions, and still other inmates are released from prison. Inmates doing college course work typically enroll in correspondence courses or attend in-prison classes taught by instructors from nearby colleges or universities.

A number of studies consistently show that inmates who participate in educational programs are less likely to be rearrested, reconvicted, or returned to prison compared to inmates who do not participate in such activities. Generally, inmates who participate in education programs are approximately one-third as likely to be reincarcerated as nonparticipants.[20]

Slightly less than one-third of inmates in state and federal prisons participate in vocational education. Prisoners who complete a vocational training program are awarded certificates and may be given the opportunity to take state exams for licensing or entry into certain apprentice programs on the outside. These kinds of vocational education opportunities are more widely available in men's prisons than in institutions for women. They typically include auto mechanics and body work, plumbing, welding, air conditioner and small appliance repair, dental and optical technician training, computer programming, and carpentry and cabinet making. Unfortunately, much prison vocational education is limited by the use of outmoded equipment and difficulties in hiring qualified instructors: Few well-qualified teachers are attracted by the thought of teaching convicted felons within a locked institution. In addition, the training offered often focuses on careers that are prohibited to people with felony convictions. For example, many plumbing, electrical, and masonry unions bar membership to convicted felons, and some states still deny barber licenses to people trained in their own state prisons' barber schools.

Prison Industries

All institutions must be maintained, and prisons are no exception. Within the prison, the clothes must be laundered, lawns mowed, meals prepared, and plumbing and electrical systems serviced. Most of these tasks are performed by inmates. Nearly half of

Despite often checkered work histories, inmates are expected to have gainful employment upon release.

 Is it unrealistic to expect that prisoners will find suitable work upon release? What does this suggest about the likelihood of recidivism?

all state and federal inmates are assigned to maintaining facilities, while another 20 percent are assigned to prison industries such as farming, repair of office equipment, food processing, or the manufacturing of goods such as license plates, road signs, and prison clothing.[21]

The goals of prison industry include the reduction of idleness and the development of inmate work skills to enhance inmates' chances of getting reasonable employment once released; however, there are major obstacles to these goals. Because of the outmoded equipment or institution-specific nature of the work, most prison industry actually does little to develop valuable work skills in inmates. In addition, there are many more inmates than available work assignments, and even fewer opportunities to approximate outside employment conditions.[22]

Although most prison industries are designed to produce products for use by the correctional system itself or by other government agencies, a small number of states have established partnerships with private industries to run factories either within or attached to prisons, with the products being sold on the open market. For example, companies such as Victoria's Secret and Boeing subcontract with companies that use prison labor to manufacture lingerie and aircraft components; prisoners in a South Carolina prison make graduation gowns for South Carolina Cap and Gown, Inc.; and TWA has a contract with the California Youth and Correctional Agency to use inmate labor for making airline reservations.[23]

The Tihar Prison Complex located outside of Delhi holds more than 14,000 inmates in seven prisons—more than double the rated capacity. Approximately 60 percent of the inmates are younger than age 30, and almost 4 percent are women.

The Tihar prisons provide inmates with a wide variety of programs, including educational and vocational classes, various approaches to rehabilitation (including yoga, meditation, and creative arts therapy), and access to legal aid. Students at the University of Delhi Law School periodically visit the prisoners and help them prepare for filing their appeals, petitions, and other legal matters.

Many inmates involved in vocational training develop skills in pen manufacturing, book binding, screen printing, envelope making, tailoring, and shoemaking. Nearly 600 inmates are employed in one of the production centers in the prison factory producing finished materials contracted by businesses for sale outside the prison. The weaving section manufactures cloth for uni-

forms and cotton dresses; the carpentry section manufactures desks for schools in and around Delhi; the chemical section produces mustard oil both for consumption within the prison and for sale; and the paper section produces hand-made paper as well as carry-bags, file covers, envelopes, and cardboard paper.

An unusual feature of the Tihar Prison is that prisoners are encouraged to participate in the management of their welfare activities. Prisoner groups, called *panchayats,* are organized to help prison administration in the areas of education, vocational training, and legal counseling, among others. Prisoners and administrators meet once a year to discuss grievances and ways of improving programs that affect the prisoners. In addition to giving prisoners a sense of pride, this approach enables inmates to let off steam, which helps to diffuse emotions that might otherwise lead to serious consequences.

Source: "Tihar Prisons," June 2006, available at http://tiharprisons. nic.in/, accessed June 1, 2007.

Managing Prisons

Prisons are highly structured organizations. According to criminologist Donald Cressey, the organization of prisons varies depending on the emphasis placed on custody and security versus rehabilitation.[24] Institutions that emphasize custody and control are highly regimented, with relationships between staff and inmates being strictly regulated. Such institutions emphasize strict obedience to prison rules, and staff members spend much of their time supervising inmates. By contrast, institutions that emphasize rehabilitation are less highly regulated and may even encourage the development of informal relationships between staff and inmates in an attempt to reinforce the treatment goals of the prison. Most prisons operate with some combination of these organizational styles.

Prison Personnel

Each prison is managed as a hierarchy. At the top of this hierarchy is a warden or superintendent, followed by deputy wardens, and then an administrative staff.

- Wardens are primarily responsible for the overall operation of the prison and report to the central office of the state Department of Corrections or Federal Bureau of Prisons. Deputy wardens are responsible for overseeing the various functions of the prison, such as custody, programs, and industry.

- Line personnel are employees who have direct contact with inmates and whose jobs are intended to achieve the custodial and rehabilitative goals of the prison. These employees include custody staff (guards), teachers, counselors, classification officers, medical personnel, and religious and recreational staff.

- Staff personnel provide support services to the line personnel and administrators. They include clerks, secretaries, training officers, and research, budget, and accounting personnel.

The responsibilities of these various groups of employees are quite diverse and often come into conflict. For example, custody and treatment staff frequently find themselves working at cross-purposes: Guards believe that inmates should be highly controlled and punished for infractions of prison rules, whereas counselors believe that inmates should have greater freedom and more responsibility for making individual decisions. Despite these conflicts, all employees are expected to work toward the primary goals of the institution: the control and rehabilitation of inmates.

Approximately two-thirds of all prison employees are correctional officers **FIGURE 14-4**. As a result of civil rights legislation and court decisions regarding affirmative action and equal opportunity hiring practices, the racial, ethnic, and gender composition of prison employees has changed dramatically in the past few decades. Before the 1980s, most prison employees were white males. Since then, however, women and minorities (especially Latinos) have been employed in increasing numbers: Between 1990 and 2000, the number of women employed in prisons increased by 135 percent, and the number of Latinos jumped by 161 percent **TABLE 14-2**.

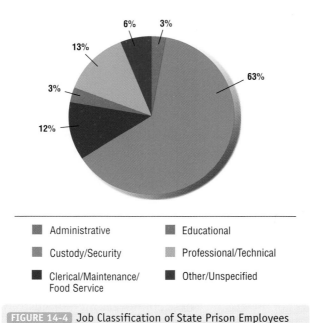

- Administrative
- Custody/Security
- Clerical/Maintenance/ Food Service
- Educational
- Professional/Technical
- Other/Unspecified

FIGURE 14-4 Job Classification of State Prison Employees

Source: James Stephan and Jennifer Karberg, *Census of State and Federal Correctional Facilities, 2000* (Washington, DC: U.S. Department of Justice, 2003), p. 13.

Custody Staff

<u>Correctional officers</u> (guards) have the most access to and potential influence over inmates. Their primary functions are to supervise inmates throughout the institution, conduct periodic counts of prisoners, search for contraband, patrol the grounds and the walls, and transport inmates to court hearings and to other institutions. They may also sit on dis-

TABLE 14-2

Gender, Race, and Ethnicity of State and Federal Prison Employees

	Number of Employees		
	1990	2000	Percent Change
Male	173,976	278,725	+60
Female	57,339	135,189	+135
White, non-Latino	172,046	264,900	+54
African American	44,076	78,739	+79
Latino	11,549	30,105	+161
Other	3644	7601	+108

Sources: Bureau of Justice Statistics, *Prisons and Prisoners in the United States* (Washington DC: U.S. Department of Justice, 1992), p. 10; James Stephan and Jennifer Karberg, *Census of State and Federal Correctional Facilities, 2000* (Washington DC: U.S. Department of Justice, 2003), p. 13.

ciplinary hearing committees and classification boards within the institution.

Officers are often hired more as a result of availability than suitability. In some states, they receive very little formal training for the job. Other states provide standardized training through corrections academies, which operate similarly to police academies.[25] In addition, some states and the Federal Bureau of Prisons provide continuing education credits for officers. Having no felony convictions or history of drug use and being in reasonably good physical condition are the most important criteria for employ-

Correctional officers perform a thankless and dangerous job.

? Which characteristics should a good correctional officer possess?

ment. In addition, more states are establishing minimum education requirements for correctional officers. However, fewer than half of all states use psychological tests to screen potential officers and predict their work performance.[26]

Officer-Inmate Relations

Owing to the high degree of officer-inmate contact, these relationships are often strained. The institution expects guards to provide prisoners' basic needs: food, shelter, clothing, and medical attention. Officers may even be reproached by co-workers for trying to help inmates resolve problems.[27] If they go beyond the basics, officers may find themselves being manipulated by inmates. For example, once a guard provides an inmate with special privileges, the inmate may pressure the guard to continue such activities under threat of being reported to the administration.

Many officers find personal security by adhering closely to rules, policies, and procedures. As they adjust to their roles as keepers and caregivers, however, they learn the importance of discretionary control. By ignoring minor infractions, the officer is able to maintain a reasonable level of social order, much like police officers who seek to maintain order by choosing to ignore minor criminal violations. Research also shows that officers who try to be of service to inmates are better able to cope with the tensions and stress within the prison setting and express more job satisfaction than other officers.[28]

LINK Job-related stress is not limited to correctional officers. Police officers experience similar problems, as noted in Chapter 8.

Gender-Related Issues

Until the 1970s, women were generally prohibited from working in the inner confines of men's prisons. Today, females account for 15 to 20 percent of all correctional officers. Some debate has arisen about women guarding men in all-male institutions, however, and there is a growing concern about men working as correctional officers in women's prisons.

Male Prisons Female correctional officers in all-male prisons, like female police officers who have to deal disproportionately with male suspects, face a variety of special problems and barriers in their work. Because of their generally smaller size, lesser strength, and gender stereotypes, female officers are perceived by many prison administrators and male officers as be-

ing more vulnerable to inmate assaults or more likely to be conned or manipulated by inmates. In reality, this may not be the case: Research suggests that female officers are assaulted only about 28 percent as often as male officers, and that male inmates generally accept female guards, find them easier to interact with, and believe that they are less likely to be punitive than male officers.[29] Some studies have actually found that female officers are significantly less likely than male officers to be assault victims. This difference may largely reflect the less aggressive style of many female officers, who tend to use communication rather than threats to gain cooperation and rely more on established disciplinary rules when dealing with inmate problems.[30]

Another issue is the potential violation of inmates' rights to privacy. Guards are often expected to supervise inmates in the shower and toilet areas; to conduct pat-down searches of inmates, including the genital area; and to occasionally conduct visual body cavity searches. In general, the courts have held that female officers' rights to equal employment override the privacy rights of inmates.

In 1985, in *Grummett v. Rushen,* a U.S. District Court held that the observation of male inmates in showers and cells by female guards was "infrequent and casual . . . or from a distance" and that no privacy rights had been violated.[31] The same court also held that pat-down searches violated no rights of inmates, as long as the searches were conducted in a professional manner. In 1991, in *Timm v. Gunter,* another federal district court ruled that whatever privacy rights inmates may have are outweighed by the interests of the institution in maintaining security.[32]

Female Prisons Male correctional officers in female prisons raise a special issue. Lawsuits have been filed by female prisoners in a number of states over sexual harassment, abuse, and even sexual assaults by male correctional officers. In 1997, Michigan was one of two states sued by the federal government over recurring sexual abuse of female inmates; in only a little more than 15 years, at least 30 male officers had been convicted of sexual assaults against female prisoners.[33]

Maintaining Discipline

More than half of all inmates are charged with violating prison rules at least once per sentence, and at least one-third commit multiple infractions. Such rule violations range from serious infractions, such

as carrying a weapon or assault, to minor violations, such as destroying property or disobeying orders. Most violations are committed by young inmates with long histories of crime and drug involvement. A higher percentage of rule violations occur in larger prisons or maximum-security institutions than in smaller or minimum-security facilities; recidivists are more prone to break rules than first-timers.[34] Most rule violators are charged with administrative infractions, followed by possession of contraband and violence without injury. Only about 10 percent are involved in incidents of injury, threat, or escape.[35]

No guard enforces all of the rules all of the time or enforces all rules equally. As a result of the discretion exercised by correctional officers, the official institutional data on rule-breaking in prisons greatly underestimate the extent of the problem. A study of inmate self-reported rule-breaking in a federal correctional facility in Fort Worth, Texas, found that inmates admitted to many more rule violations than were reported and that guards actually reported very few of the violations they did observe.[36] In their desire to achieve a smoothly running shift, many guards attempted to gain inmate acquiescence with the carrot rather than the stick. Furthermore, the same study found that guards were less concerned with reporting violations of prison rules and more focused on preventing infractions that might come to the attention of their superiors.[37]

The Disciplinary Process

Prison staff are bound by a variety of legal rules and judicial mandates set forth by the U.S. Supreme Court in 1974 in *Wolff v. McDonnell*.[38] Before the *Wolff* decision, few procedural rules existed to control prison staff's discretionary power in applying disciplinary sanctions. Inmates could be placed in administrative segregation (often referred to as solitary confinement) without a hearing for an indefinite time. In *Wolff,* the Court rejected the right of inmates to counsel and to confront or cross-examine witnesses in disciplinary hearings, noting that to exercise such rights could threaten the security of the institution, though reprisals may be taken by accused inmates against staff members or inmates who brought the charges. It also held that administrators must follow these steps in the disciplinary process:

- Write and file an incident report
- Provide a formal hearing
- Give a written 24-hour notification of the charge(s) and supporting evidence
- Allow inmates to call witnesses on their own behalf unless it would jeopardize institutional safety
- Provide inmates with the assistance of a staff member or another inmate in presenting their defense if the inmate is illiterate or the issue is unusually complex

LINK The formal disciplinary hearing is conducted in a manner similar to preliminary hearings in criminal courts in misdemeanor cases, as discussed in Chapter 10.

At the disciplinary hearing, the disciplinary committee informs the inmate of the charge and gives the inmate and the person who wrote the incident report an opportunity to testify regarding the incident. In addition, the inmate may present witnesses on his or her behalf, although inmates are not entitled to cross-examine witnesses. Although inmates may not be represented by attorneys at disciplinary hearings, many institutions allow them to be represented by an inmate advocate. If the inmate is found guilty of the charge, the hearing officer or committee may then impose a sanction. Most state correctional systems allow inmates to appeal disciplinary hearing decisions to the warden, to the superintendent of the institution, or to the commissioner of the state department of corrections.

Sanctions for Rule Violations

Rule violators generally face one of three forms of sanction: administrative segregation, loss of good time, and loss of privileges.

When in administrative segregation, an inmate is placed in a single-person cell in a high-security section of the institution for a specified period of time. Inmates placed in segregation usually spend 22 to 23 hours a day in their cells.[39] This punishment typically is applied in cases involving more serious rule violations, such as assault, but may also be used for inmates who are repeatedly insubordinate. Most states limit the amount of time an inmate may be in segregation to a maximum of 20 days.

Good time is the practice of reducing the days of an inmate's sentence for maintaining a record of good behavior, for participating in educational, vocational, or treatment programs, or for providing some sort of special service to the institution (such as assisting a guard who is in trouble). In some jurisdictions, for each good day served, an inmate's sentence is reduced by one day; in other jurisdictions, an inmate may be required to earn seven good days for each day to be reduced from his or her sentence. Inmates who earn the maximum amount of good time can dramatically reduce the total amount of time they serve in prison.

The loss of good time for an inmate and the resulting longer stay in prison is a severe punishment for rule infractions. Thus guards and prison administrators wield great discretionary power over inmates by threatening to extend their stay in prison. Theoretically, an inmate close to release who violates an institutional rule could lose all accumulated good time and be required to remain in prison for many years longer. Most states have attempted to make the loss of good time a more reasonable sanction, however, and have limited or otherwise regulated the amount of good time an inmate may lose. For exam-

Inmates can significantly reduce their sentence through the use of "good time" for compliant behavior.

? Does good time make correctional officers' jobs easier? Does it help to rehabilitate inmates?

ple, any disciplinary board hearing that results in a loss of good time must be supported by at least a modicum of evidence so as to meet due process requirements.[40] In some states, an inmate's good time (or a portion of it) is vested after a certain number of days (for example, 90 or 120 days) have been accumulated. Good time that is vested cannot be taken away from the inmate.

Inmates who violate institutional rules may also face loss of privileges. Correctional institutions are places with few amenities, but inmates do have limited privileges. Because they are considered privileges, rather than guaranteed rights, these benefits can be taken away. For example, prisoners can lose privileges such as receiving visits, receiving or sending mail, participating in recreation, and having access to the commissary.

Other sanctions for rule violations may include reprimands, assignment of extra duties during leisure hours, confinement to quarters, change in job assignment, or change in housing (for example, being moved from a dormitory housing unit, which prisoners see as providing greater freedom, to a single-person cell).

Managing Special Populations

While most inmates are capable of being managed with standard procedures in traditional institutional settings, some groups have special needs or present unique challenges for prison administrators.

Inmates with HIV/AIDS

HIV infection and acquired immune deficiency syndrome (AIDS) have become major concerns for today's prison administrators. Because many inmates have histories that include high-risk behavior (such as intravenous drug use and male homosexual activity), prison and public health officials are now taking a closer look at the rate of HIV/AIDS in prison populations. According to the Centers for Disease Control and Prevention (CDC), residents of prisons have the highest rate of HIV infection among all residents of public institutions. In 2003, 2 percent of state prison inmates and 1.1 percent of federal prison inmates tested positive for HIV. Of those nearly 24,000 inmates testing positive for HIV, 5643 inmates were confirmed as having AIDS. During the same year, 268 inmates in state prisons died from AIDS-related illnesses.

HIV-infected inmates tend to be concentrated in a small number of states. New York, Florida, and Texas have traditionally held the largest number of HIV-positive inmates: These three states housed nearly half (48 percent) of all HIV-infected inmates in state prisons and also had the largest number of confirmed AIDS cases in 2003.

According to the CDC data, female inmates (2.8 percent) were more likely than male inmates (1.9 percent) to test positive for HIV in 2003. However, male inmates with AIDS were nearly twice as likely as females to die from AIDS-related causes. African American inmates were about 3.5 times more likely than whites and almost 2.5 times more likely than Latinos to die of AIDS-related causes.[41]

Testing inmates for HIV has become controversial. Those who advocate mass testing argue that correctional institutions need to identify any HIV-infected inmates so as to prevent the spread of the virus among prisoners. In addition, advocates believe that the staff needs to know which inmates have HIV so that they can protect themselves against infection.

Critics of such testing argue that mass screening is likely to produce a class of outcasts within the prison: Inmates who are identified as being infected with HIV would be vulnerable to harassment, discrimination, and possible assault by other inmates.[42] Critics are also concerned that HIV-positive inmates—even those who may not have AIDS yet—would be subject to segregation from the general prisoner population. Some states, for example, require that HIV-infected inmates be housed separately from uninfected inmates if they are believed to be sexually active or predatory and if the prison administration believes that separate confinement is in the best interests of the department of corrections and the inmate population.[43]

Inmates with Mental Disorders

LINK   Guilty, but mentally ill statutes, discussed in Chapter 2, recognize that some offenders are not exempt from criminal responsibility, even if they do suffer from mental impairment.

Many severely mentally disordered offenders are routinely sent to prison, as are much larger numbers of inmates who have less problematic mental or emotional problems. Surveys of state correctional departments have found that approximately 16 percent of inmates (nearly 300,000 individuals) are either mentally disordered or deficient.[44] Prisons frequently receive inmates with histories of mental disorders or institutionalization for mental health treatment. In addition, stress can exacerbate an inmate's mental disorder once he or she is confined to a prison.[45]

Special services for mentally disordered offenders are typically limited, and only about one-fourth of disordered inmates see psychiatrists or other licensed therapists while in prison. Many of these inmates are housed in segregated units for the mentally disordered to provide protection for themselves or others, including staff and other inmates.[46] Because some mentally disordered inmates are unpredictably violent or aggressive, in 1990 the Supreme Court, in *Washington v. Harper,* agreed that such inmates could be forced to take antipsychotic drugs for their benefit or if they are a danger to themselves or others.[47]

Elderly Inmates

Older inmates are the fastest-growing population in state prisons. Today, approximately 67,000 prisoners are age 55 or older; this is double the number of older persons who were imprisoned just a decade earlier and accounts for nearly 5 percent of all inmates.[48] The same baby boom that produced the dramatic rise in crime in the late 1960s and early 1970s is also generating new concerns for prison administrators around the country. According to the National Center on Institutions and Alternatives, elderly inmates cost two to three times more to incarcerate than younger inmates.[49]

The number of elderly inmates will increase in coming decades given the recent changes in sentencing policies.

? Should older inmates be released given the skyrocketing medical costs faced by prisons?

TABLE 14-3

Comparison of Older Inmates' Chronic Illnesses

Health Condition	State			
	Iowa	Michigan	Mississippi	Tennessee
Arthritis	45.4%	42.8%	36.2%	15.3%
Hypertension	39.7	28.8	42.1	49.6
Stomach ulcers	21.6	13.8	19.5	11.9
Prostate problems	21.0	21.1	16.4	3.1
Myocardial infarction	19.0	29.1	27.4	33.9
Emphysema	21.2	22.5	15.5	15.0
Ulcers	*	13.8	17.6	11.8
Diabetes	11.2	5.1	15.6	16.6
Cancer	6.9	*	10.7	4.4

*Not reported.

Source: Adapted from Ronald Aday, *Aging Prisoners: Crisis in American Corrections* (Westport, CT: Praeger, 2003), p. 94.

mates have been in prisons for decades and have already made the necessary adjustments to prison.

One consequence of the graying of U.S. prisons is the development of new housing strategies. Since the late 1980s, a number of states have built special-needs facilities or separate housing units for their elderly inmates. Some of the units (often designated as aged/infirm, medical/geriatric, or disabled) house a mix of older inmates with younger inmates who have disabilities or long-term illnesses. The use of consolidated housing for inmates with similar health-care needs allows specialized staff members to supervise and treat elderly inmates as their health problems change.[52]

Inmates with Disabilities

An estimated 10 percent of inmates in state prisons have disabilities, ranging from mental retardation and learning disabilities to drug dependency and a variety of physical disabilities.[53] Although prisons generally have programs or separate housing for inmates with certain disabilities, few prisons have kept pace with the needs of the physically disabled. For example, few prisons have made accommodations for inmates in wheelchairs, and it is rare for prisons to have elevators, ramps, special toilets, or handles or bars to assist physically disabled inmates while showering.

According to the Americans with Disabilities Act of 1990, all public accommodations are prohibited from discriminating against persons with physical

One of the main reasons for the greater cost of caring for elderly inmates is the range of debilitating health conditions they bring into prison or acquire as a result of growing older in prison. Common conditions faced by these individuals include arthritis, hypertension, myocardial infarction (heart attack), and diabetes **TABLE 14-3** .[50]

Older inmates entering prison for the first time face unusually difficult adjustments and are especially vulnerable to intimidation and victimization by younger and stronger inmates. Such inmates are no longer able to have a life tied to lifelong routines and patterns of coping; instead, they must adjust to a system characterized by regimentation and a severe lack of privacy.[51] Of course, many elderly in-

disabilities, including prisoners. In 1998, the Supreme Court reaffirmed that prisons were public entities and that inmates with disabilities must be appropriately accommodated within the facilities and given reasonable access to most prison programs.[54]

Juvenile Offenders

Approximately 2266 persons younger than age 18 were in custody in adult prisons in 44 states in 2005, with slightly more than 6700 being housed temporarily in adult jails. Juveniles accounted for slightly less than 0.2 percent of all inmates in state prisons.[55]

Approximately 96 percent of all juveniles in prison are males, 59 percent are African American, and 79 percent were age 17 at the time they entered prison. Compared to young adult inmates ages 18 to 24, juvenile inmates were more likely to have been convicted of violent crimes (primarily robbery and assault) and less likely to have been convicted of drug offenses. In general, juvenile inmates have less formal education and are more likely to have been convicted of more serious crimes than their adult counterparts.[56]

LINK Juveniles generally are much more vulnerable than older inmates and consequently are more likely to be victimized by other predatory inmates. Chapter 16 discusses the special problems associated with housing juveniles with adults in prison in greater detail.

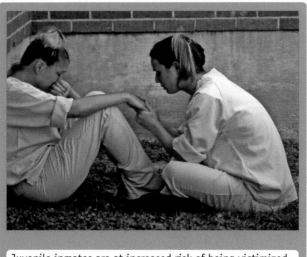

Juvenile inmates are at increased risk of being victimized while incarcerated.

? Should juvenile inmates be mixed with adult prisoners?

In most states, juvenile inmates are housed in correctional facilities as part of the general population. That is, they are assigned to the same cell blocks as adults and may even be placed in cells with adults. Of those states with juvenile inmates, only 13 maintain separate facilities or units for juvenile offenders.[57] Only six states (Arizona, Hawaii, Kentucky, Montana, Tennessee, and West Virginia) require separate housing for all inmates younger than age 18.

While some states permit the incarceration of juveniles as young as age 13 in adult institutions, no offender younger than age 16 can be placed in an adult prison in North Dakota and California. A total of 12 states use graduated incarceration, in which all inmates younger than age 18 are required to begin their sentences in juvenile facilities. Once they reach a certain age (typically 18), they are then transferred to adult facilities to serve the remainder of their sentences. Eight states use segregated incarceration as part of their youthful offender programs. In these instances, juvenile offenders are assigned to specific facilities based on their age and specific programming needs.[58]

It is not surprising that juveniles incarcerated in the general prison population in adult prisons are more likely to be victims of violence than they would have been if they had been incarcerated in juvenile institutions. According to Martin Forst, slightly more than one-third of all youths who are assigned to juvenile facilities become victims of violence, compared to nearly half of the juveniles housed in adult facilities. Moreover, juveniles in prison are five times more likely to be victims of sexual assault than are youths in juvenile institutions.[59]

Female Offenders

Although their numbers are increasing, relatively few women are sentenced to prison. In 2005, slightly more than 107,000 women were under the jurisdiction of state and federal prisons, or 7 percent of the total inmate population.[60] Those women who are incarcerated are more likely to be placed in smaller institutions, face a less formal and bureaucratic administration, receive greater freedom of movement within the institution, and face less violence or intimidation by other inmates than their male counterparts.

Most female inmates are young, unmarried, have less than a high school education, and have children younger than age 18 **TABLE 14-4**. More than 29 percent of all women sentenced to prison in 2005 were drug offenders **FIGURE 14-5**.

TABLE 14-4

Characteristics of Female Inmates in State Prisons

Characteristic	Percentage of All Female Prisoners
Race and Ethnicity	
White, non-Latina	44.2
African American	33.4
Latina	15.6
Other	6.8
Age	
24 or younger	13.9
25–34	31.4
35–44	37.7
45–54	14.0
55 or older	2.9
Marital Status	
Married	17.3
Widowed	5.8
Divorced	19.9
Separated	9.8
Never married	47.1
Education	
Eighth grade or less	8.4
Some high school	55.5
High school graduate	21.7
Some college or more	14.4
Prearrest Employment	
Employed	50.7
Unemployed	49.3
Have Children Younger Than Age 18	
Yes	65.8
No	34.2
Ever Physically or Sexually Abused Before Incarceration?	
Yes	57.2
No	42.8

Sources: Paige Harrison and Allen Beck, *Prisoners in 2004* (Washington, DC: U.S. Department of Justice, 2005); U.S. General Accounting Office, *Women in Prison: Issues and Challenges Confronting U.S. Correctional Systems* (Washington, DC: United States General Accounting Office, 1999).

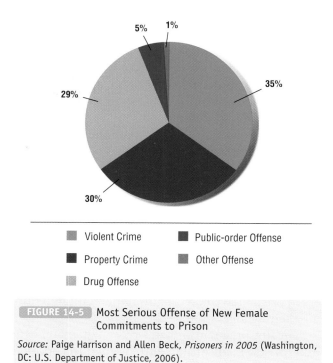

- Violent Crime
- Property Crime
- Drug Offense
- Public-order Offense
- Other Offense

FIGURE 14-5 Most Serious Offense of New Female Commitments to Prison

Source: Paige Harrison and Allen Beck, *Prisoners in 2005* (Washington, DC: U.S. Department of Justice, 2006).

Social Order of Women's Prisons

Female inmates, like their male counterparts, must make adjustments to prison life. Women in prison are generally in greater need of social support than are male inmates, however. Males tend either to be more anomic and isolated or to join prison gangs, whereas females are more likely to seek out companionship and social support through the development of interpersonal relationships with other inmates. Imprisoned women sometimes establish homosexual relationships, but more often platonic friendships, and they sometimes become members of surrogate families.[61] Surrogate families, sometimes referred to as fictive families or pseudofamilies, involve a number of inmates coming together to establish a set of relationships based on traditional family roles. These kinship groups perform most of the normal functions of a family and provide stability, warmth, security, and social bonding for women seeking primary group relationships.[62]

Women adjusting to prison face some of same problems of doing time as male inmates. One of these problems for women is finding ways to stay out of "the mix." "The mix," according to Barbara Owen, is an ambiguous concept involving continuing the behavior that got the woman into prison; this behavior can bring conflict or trouble with other prisoners or with staff—such as involvement in drugs, fights, and just "being messy" (i.e., not minding your own business). It becomes important for women prisoners to develop "prison smarts" as a means for avoiding "the mix." To do so, a female inmate may construct a personal world inside prison that involves the woman's living situation, job, education, or treatment programs, and relationships with others that are satisfying and removed from the more aggravating elements of prison life.[63]

Mothers in Prison

Nearly 80 percent of the women in prison have children, compared to only 60 percent of the men in prison. Between 60 and 70 percent of these women have one or more children younger than age 18.[64] Most of these inmate-mothers report that at least one of their dependent children lives with a grandparent, father, or other relative/friend; approximately 10 percent have children in foster care or in other institutions.[65]

The special pains of imprisonment for these women are often expressed in feelings of helplessness, guilt, and anxiety about their children's care, education, and health. In addition to the anguish of explaining to their children why they have been sent to prison and concerns about their well-being, inmate-mothers often fear the possibility of losing custody. Additionally, because most women's prisons are some distance from inmates' homes, female inmates do not see their children often; more than half of all mothers in prison have no visits from their children during their incarceration.[66]

Ideally, pregnant inmates receive prenatal care from prison medical staff and are then transported to a local community hospital for the delivery. In some instances, however, inmates are offered little prenatal care and deliver the baby in the prison hospital. Most states require that the baby and mother be separated soon after birth; arrangements are made for a custody hearing for placement of the child.[67] There are no consistent state policies regarding newborn infants and whether, and for how long, they are allowed to stay with their incarcerated mothers. Two states, Nebraska and New York, operate nurseries for infants and toddlers in their women's prisons. These programs provide parenting classes, stressing appropriate child-rearing skills. In Nebraska, older children are permitted overnight visits with their mothers. The Federal Bureau of Prisons operates Mothers and Infants in Transition, a program that allows selected pregnant inmates to live in a special community-based unit for a period from shortly before giving birth to a specified period following the birth.[68]

Many states offer programs intended to facilitate and strengthen mother–child relationships among female inmates. At least 11 states participate in a program known as Mothers and Their Children, which focuses on improving visiting procedures and maintaining children's centers for mothers and their children to play and learn together. Perhaps one of the most widespread programs is the Girl Scouts Beyond Bars, a program begun in 1992. In cooperation with local Girl Scout troops, special troops of girls with mothers in prison have been formed. In addition to

Family disruption is one of the many collateral consequences of imprisonment.

? Do children born in prison have any chance of success in U.S. society? What special risk factors do they face?

their regular meetings, twice a month the meetings are held at the women's prison where mothers and daughters participate in a variety of activities and projects.[69]

Inmate Subculture

Prison is a place that requires order, whether it is imposed by institutional staff or created informally by inmates. Many people believe that order is maintained by correctional officers backed up by authority and, sometimes, the barrel of a gun. In reality, the social order of prisons is largely a product of an informal subculture that inmates create to help them cope with their imprisonment. Although debate has arisen regarding the origins of the inmate subculture, scholars tend to support one of two explanations for its development.

The first explanation, which is known as the <u>deprivation model</u>, views inmate behavior as a matter of adaptation to the unique demands and conditions of the prison environment and the inmate's response to the loss of amenities and freedoms such as possessions, dignity, autonomy, security, and heterosexual relationships. Because all inmates share these deprivations, or <u>pains of imprisonment</u>, they create bonds to help them cope.[70] These bonds form a social system that permits inmates to maintain self-esteem and a sense of dignity by rejecting the formal norms of the prison. The inmate social system is also responsible for maintaining an underground economy in which inmates are able to obtain and distribute valued and scarce items—both legal and illegal—such as cigarettes, snack food, magazines, shaving cream, drugs, alcohol, and sex. These items' availability and cost, which is often measured in packs of cigarettes, can vary according to amount of attention the institution gives to controlling the economy. Many inmates seek out particular job assignments so that they will have greater access to goods. These inmates then sell or trade the goods to other inmates.

The second model suggested as an explanation for the development of inmate subculture is the importation model. Proponents of this theory contend that inmates bring to prison their previously established values and patterns of behavior.[71] They adapt these beliefs, identities, and group allegiances from existing street subcultures to the prison context. In addition to a well-developed set of attitudes and values, many inmates may enter prison armed with a deep hostility to authority, which causes them to reject the attitudes and opinions of prison staff and administration.[72]

Coping in Prison

The process by which inmates adjust to or become assimilated into the prison subculture is called prisonization, a term coined by sociologist Donald Clemmer.[73] Once in prison, inmates adjust to prison routines and the inmate subculture. Clemmer suggests that individuals adapt to prison life by taking on the inmate role little by little and then remaining "cons" until release. However, sociologist Charles Tittle argues that prisoner socialization follows more of a bell-shaped curve in terms of individuals' adaptation to the prison subculture. According to Tittle, most inmates are more likely to express pro-staff beliefs (including an acceptance of the legitimacy of institutional rules and emphasis on treatment and rehabilitation) during the first and last six months of their imprisonment, whereas inmates express much more pro-prisoner beliefs in the middle of their sentences.[74]

Regardless of which form prisonization takes, it is clear that even inmates who adapt to the subculture vary a great deal in how they cope with prison life, largely because imprisonment produces different psychological responses in different people.[75] Most inmates' maladaptation to prison, as reflected by disciplinary reports or other indications of prob-

lem behavior, peaks fairly early during confinement and then diminishes over time. Much of the adjustment to prison is seen as a direct result of aging or normal maturation.[76] In other words, the majority of prisoners appear to cope well with living in prison, although this adaptation does not mean that they are making progress toward rehabilitation. They learn to adjust to the formal and informal demands of institutional life, serve their time, and are then eventually released. Prison appears to have neither a great negative nor positive impact on them.[77]

The Inmate Code

In 1960, criminologists Gresham Sykes and Sheldon Messinger described the inmate code, or system of informal norms regulating inmate behavior, which help to order relationships among inmates and between inmates and prison employees. The major tenets of the code include the following norms:

- *Don't exploit inmates.* This maxim demands that inmates do not break their word, steal from other inmates, or fail to pay their debts.
- *Don't lose your head.* Inmates are told to "play it cool" and "do their own time," meaning that they should not mess with other inmates. Inmates should not argue with one another and should minimize emotional frictions by ignoring the irritants of daily life in prison.
- *Don't weaken.* The ideal inmate should be tough and never back down from a fight. According to the code, this involves showing courage, strength, and integrity when faced with deprivations or threats from other inmates or guards.
- *Don't interfere with inmate interests.* Inmates should not be too curious about the activities or interests of other inmates; inmates should mind their own business. In addition, inmates are told, "Never rat on a con" and "Don't put another inmate on the spot." Instead, inmates should always be loyal to their own kind.
- *Don't be a sucker.* Inmates should always regard prison officials with suspicion and distrust. In any conflict between inmates and prison employees, the employees should always be considered in the wrong.[78]

The code is violated to some degree by most inmates. Those who adhere strictly to the code are considered to be "right guys" and are accorded a certain amount of esteem by other inmates. Those who blatantly violate the code suffer the consequences: ostracism, beatings, or even death. However, many criminologists believe that this traditional inmate code has lost much of its influence over inmate behavior during recent decades as a consequence of the increasing number of racial and ethnic prison gangs.

Prison Gangs

A number of corrections specialists believe that the traditional inmate code and subculture have given way to new patterns of social organization reflecting the changing racial and ethnic composition of inmate populations and the emergence of prison gangs. Partly as a result of the growing influence of the Black Power movement during the 1960s and 1970s and the rapid growth in the African American inmate population in the 1970s and 1980s, the old inmate social order crumbled as African American inmates challenged the white inmate power structure. At the same time, growing cultural awareness by Latino inmates gave rise to the emergence of Latino gangs in institutions with sizable Latino populations. Racial and ethnic divisions became hardened, and the solidarity among prisoners broke down as inmates shifted their allegiance from the larger inmate subculture to their racial or ethnic group.

Gangs develop in prison for a variety of reasons, including solidarity, power, and self-preservation. One of the first sociological studies of the rise of prison gangs was conducted in 1977 by James Jacobs

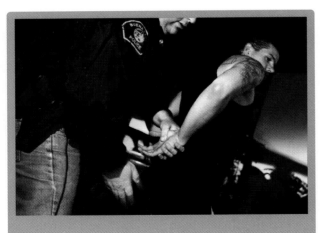

The Aryan Brotherhood is unsurpassed among prison gangs in terms of its violence, organization, and criminal enterprise.

 Should prison gangs be segregated or forced to integrate? Why are hate crime statutes not applied in cases involving racially motivated inmate violence?

at the Stateville Prison in Illinois.[79] Jacobs noted that the street gangs of Chicago, such as the Vice Lords, El-Rukns, Latin Kings, and Disciples, brought their organizational structure and ideologies along when members were sentenced to prison. Jacobs argues that the gangs placed the old inmate order in jeopardy, as inmates began to relate to one another on the basis of group affiliation rather than as individuals. The sheer numbers and solidarity of the gangs also made it much more difficult for staff to manage or control individual inmates who were affiliated with gangs. Formal prison control weakened as established relations between inmates and staff were replaced by gang–staff relations, prison rackets were taken over by gangs, and gang leaders gained informal control over the distribution of prison jobs. New inmates belonging to Chicago gangs were met by gang associates as they entered prison, for example, and were quickly absorbed into the prison gang structure, thereby providing them with an immediate sense of identity and belonging.

Today, prison administrators around the country, as well as the Federal Bureau of Prisons, face significant prison gang problems. A National Major Gang Task Force survey of gang activity in the nation's prisons conducted in 2002 reported that six major prison gangs are in operation and pose serious security threats to the institutions, staff, and inmates:[80]

- The *Mexican Mafia* (also known as La Eme), a Mexican American/Latino gang formed in the late 1950s in California Youth Correctional Facility at Duel, is considered to be one of the most dangerous gangs both state- and nationwide. It uses extremely gruesome killings to establish fear and intimidation.

- *La Nuestra Familia* established itself in Soledad prison in California in the mid-1960s to protect younger, rural Mexican Americans from other inmates and the gang's chief rival—the Mexican Mafia. This gang is involved in drug trafficking, extortion, and pressure rackets.

- *Neta* originated in 1970 in a Puerto Rican prison and is composed of Puerto Rican Americans and Latinos. Members are involved in drug activity, extortion, and the use of violence to deal with disrespect to the gang.

- The *Black Guerrilla Family* was founded in San Quentin by former Black Panther leader George Jackson in 1966. Members of this African American supremacist group are highly political and hostile to government and prison officials.

- The *Texas Syndicate* was formed in the early 1970s in California's Folsom prison as a direct response to other prison gangs, especially the Mexican Mafia and Aryan Brotherhood.
- The *Aryan Brotherhood*, also known as the Brand, was formed in 1967 in the prison at San Quentin in California. This white supremacist group organized to serve the interests of white inmates and has since spread to prisons around the country. Much of its activities have been centered on drug trafficking, prostitution, extortion, and murder, especially of African American inmates.

Sex in Prison

Most inmates serve their sentences in same-sex institutions with no heterosexual outlet, producing a number of problems for inmates and prison staff. The normal need for sexual release by inmates leads to various adaptations. Some inmates engage in homosexual activity; others resort to prostitution, engaging in homosexual sex while maintaining a heterosexual identity.

LINK A few states (e.g., California, Mississippi, and New York) allow conjugal visits with spouses, as discussed in Chapter 13.

Many male inmates experience intense sexual harassment included unwanted and offensive sexual advances, which frequently produce an acute fear of sexual assault. Within prison, sexual harassment involves both physical incidents, such as kissing, touching, or fondling, and verbal harassment, including statements that feminize an inmate, sexual propositions, and sexual extortion.[81]

Male Inmates

Sexual harassment of inmates is relatively common in men's prisons. Daniel Lockwood's study of two New York prisons found nearly 30 percent of inmates reported being sexually targeted, 33 percent propositioned, and 7 percent touched or grabbed.[82] When Peter Nacci and Thomas Kane looked at inmate sex

---

Headline Crime   The Aryan Brotherhood

On July 28, 2006, a jury in Santa Ana, California, convicted four leaders of the Aryan Brotherhood prison gang— Barry Mills, Tyler Bingham, Edgar Hevle, and Christopher Gibson—under the Racketeer Influenced and Corrupt Organization law on charges of using murder and intimidation to protect their drug trafficking activities in prison. The four had been charged with 32 murders and attempted murders involving members of the Aryan Brotherhood over nearly three decades. Mills and Bingham were found guilty of ordering the murder of African American inmates from their cells at the super-max prison at Pelican Bay.

Over the years, most of the gang's violence was directed at African American inmates. The bloodiest attack occurred on August 28, 1997, when gang members at the Lewisburg federal prison armed themselves with shivs (homemade knife-like weapons) and stabbed six African American inmates, killing two.

While originally formed to strike at African American inmates belonging to the Black Guerrilla Family, the Aryan Brotherhood soon evolved into a full-fledged criminal enterprise involving extortion, drug trafficking, and the sale of "punks" (inmates forced into prostitution). To be accepted into the gang, a recruit had to "make his bones," which often required killing another inmate. Although only a few hundred inmates belong to the gang, its impact has been far-reaching. For example, the Aryan Brotherhood accounts for most of the drug distribution that occurs in state and federal prisons.

The conviction of the four gang leaders suggests an attempt by prison officials to break up one of the most violent prison gangs and to restore a greater control over prison gangs, even in super-max prisons.

Sources: Greg Risling, "Jury Convicts 4 White-Supremacists," *Washington Post* (July 28, 2006), available at http://www.washingtonpost.com/wp-dyn/content/article/2006/07/28/AR2006072801079.html, accessed June 1, 2007; Associated Press, "4 Aryan Brotherhood Leaders Convicted," *CBSNews* (July 28, 2006), available at http://www.cbsnews.com/stories/2006/07/28/national/main1847262.shtml, accessed August 1, 2007; David Grann, "The Brand: How the Aryan Brotherhood Became the Most Murderous Prison Gang in America," *New Yorker* 80:156–171 (2004).

in federal prisons, they found that 29 percent of the inmates had been propositioned, 12 percent had engaged in sexual acts while in prison, and 7 percent had been "seduced" into a sexual act in exchange for gifts and favors.[83] Finally, Wayne Wooden and Jay Parker's study of a medium-security prison in California reported that 53 percent of the inmates had been frequently victimized by sexual innuendo, sexual harassment, and verbal or physical threats of a sexual nature.[84] All three studies report that among male prisoners, white, non-Latino, younger males were the most likely to be targeted.

Sometimes sexual coercion becomes significantly more serious, involving aggressive sexual assault and rape. A national survey of state and federal prisons as well as local jails found that more than 6241 acts of sexual violence were reported in 2005, up from 5386 incidents in 2004. Nearly 40 percent of allegations involved staff sexual misconduct; 35 percent involved inmate-on-inmate nonconsensual sexual acts; 17 percent involved staff sexual harassment; and 10 percent involved inmate-on-inmate abusive sexual contact.[85] While prison rape is frequently motivated by a desire for sexual release, more often it is the result of an inmate's attempt to gain power and control over another inmate. Male inmates most likely to become rape victims tend to have the following characteristics:

- First-time offenders
- Young, small, and slight of stature
- Feminine in appearance or manners
- Exhibiting an appearance of being weak
- With a history of sexual abuse
- Mentally challenged (often "persuaded" rather than forced to perform sexual acts)
- Child molesters (often targeted in a form of "jailhouse justice")[86]

Female Inmates

Female inmates also have to deal with sexual coercion and assaults by both staff and by other inmates. While rates of sexual coercion are much higher in men's prisons, recent research shows that women are sexually victimized more frequently than previously believed.[87] For example, in their study of three Midwestern women's prisons, Cindy and David Struckman-Johnson reported rates of sexual coercion ranging from a low of 8 percent at one of the prisons to 19 percent at a second facility. The prison with higher rates was more racially and ethnically diverse, and nearly half of its inmates had committed violent crimes. Female targets of sexual coercion were most likely to be heterosexual and white, but women from all racial groups reported victimization. Approximately half of the sexual incidents, ranging from sexual grabs to rapes, were carried out by female inmates; the other half were perpetrated by prison staff.[88]

Prison Violence

In 2002, 48 state prison inmates were victims of homicide and another 168 inmates committed suicide.[89] However, both the prison homicide rate and the suicide rate have dropped dramatically since 1980 even as prison populations have soared FIGURE 14-6 .

Unfortunately, violence directed by inmates against prison staff, especially correctional officers, has not declined in similar fashion. For example, two inmates at a Maryland State maximum security prison escaped from their cells on July 26, 2006, and attacked David McGuinn, a 42-year-old correctional officer, repeatedly stabbing him in the upper torso before returning to their cells. McGuinn was the second Maryland correctional officer to be killed during the first half of the year, and the rate of assaults against correctional officers had almost doubled between 2004 and 2005.[90] Individual acts of violence in prison, however, rarely catch the attention of the public or news media. Instead, interest is largely focused on the explosive, collective violence produced in prison riots.

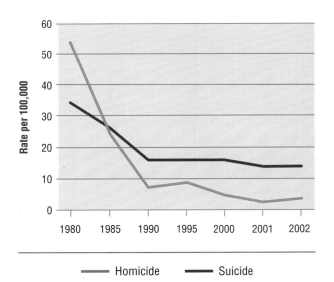

FIGURE 14-6 State Prison Homicide and Suicide Rates

Source: Christopher Mumola, *Suicide and Homicide in State Prisons and Local Jails* (Washington, DC: U.S. Department of Justice, 2005), p. 2.

On September 4, 2003, President George W. Bush signed the Prison Rape Elimination Act into law. The unanimous passage of this legislation reflects a national effort to reduce sexual violence in prison and to protect the rights of inmates. The purposes of the Act are to:

- Establish a zero-tolerance standard for the incidence of prison rape

- Make the prevention of prison rape a top priority in each prison system

- Develop and implement national standards for the detection, prevention, reduction, and punishment of prison rape

- Increase the available data and information on the incidence of prison rape, thus improving the management and administration of correctional facilities

- Standardize the definitions used for collecting data on the incidence of prison rape

- Increase the accountability of prison officials who fail to detect, prevent, reduce, and punish prison rape

- Protect the Eighth Amendment rights of federal, state, and local prisoners

- Increase the efficiency and effectiveness of federal expenditures through grant programs dealing with social and public health issues

- Reduce the costs that prison rape imposes on interstate commerce.

The Prison Rape Elimination Act also provides $60 million for a two-year study of the actual extent of prison rape and recommendations for its elimination.

Source: Prison Rape Elimination Act of 2003, Public Law 108-79 (2003).

Prison Riots

The past few decades have experienced some of the nation's most explosive and bloody prison riots. In a 1971 riot at Attica State Prison in New York, 39 people died as a direct result of the riot and another 4 later died whose deaths were attributed to the riot.[91] Only nine years later, the New Mexico State Penitentiary turned into a battlefield as inmates went on a rampage that left 33 dead before police and the National Guard regained control of the prison.

Although the New Mexico riot was not the most deadly prison riot, it was one of the most grotesquely violent. All of the inmates killed in the New Mexico riot died at the hands of other inmates. Twelve guards were taken hostage and were seriously beaten and, in some cases, sodomized by inmates. The rioters who burned and gutted the prison facility were composed of about a dozen inmates referred to as the death squad. They moved about the prison torturing and killing inmates who had reputations as snitches, psychos, or weaklings or who simply got in the way. Even inmates on the hit list who were housed in protection units were not safe from the death squad, which burned their cell doors off with an acetylene torch.[92] The impetus for this riot originated four years earlier, when prison administrators

reacted severely to a strike by inmates. After the strike was broken, strike leaders were either placed in administrative segregation or transferred to other institutions, and an extensive snitch (informant) system was put into place by prison officials. The new get-

Ultimately, prison violence is caused by a myriad of factors that are not yet well understood.

Given the conditions in prison and the attributes of inmates, correctional personnel, and the institutional structure, is prison violence to be expected? Would more complete and credible inmate classification systems effectively separate problematic inmates and problematic officers from one another?

tough policy also produced a power vacuum, resulting in violent struggles for power among inmates that sometimes take the form of rioting.

Prisons have continued to have to deal with riot situations:

- In 1984, a group of 182 inmates at Attica State Prison went on an eight-hour rampage to protest the shooting of an inmate by guards.
- In 1987, Cuban inmates awaiting deportation in federal prisons in Atlanta and Oakdale, Louisiana, took hostages and seized control over portions of the institutions.
- In 1989, in a riot at the Pennsylvania State Correctional Institution at Camp Hill, 123 employees and inmates were injured.
- In 1991, an uprising by inmates at the Montana State Prison at Deer Lodge left five inmates dead, all of whom were killed by other inmates.
- In 1993, the longest prison riot occurred at the Lucasville Prison in Ohio. After an 11-day standoff, nine inmates and one correctional officer had been murdered by rioters.
- In 2007, a riot involving about 500 inmates broke out at the New Castle Correctional facility in Indiana. Two staff members and seven inmates were injured.

The New Castle riot was unique because it involved large groups of inmates from two different states: Indiana and Arizona. The riot occurred six weeks after a group of 600 inmates from Arizona had been transferred to the Indiana facility in a move to relieve overcrowding in Arizona's prisons and fill empty cells in the privately run prison at New Castle. The Indiana and Arizona prisoners lived in dormitories separated by a fence. Both groups ate and exercised in a common area. According to officials, the riot likely stemmed from the Arizona inmates' displeasure with a smoking ban, limitations on what they could have in their cells, and the fact that they were required to eat meals after the Indiana inmates and had to wait for their time in the exercise yard until after the Indiana inmates were done, sometimes not getting to exercise until after midnight. After the riot, 220 inmates were moved to other facilities and plans to transfer additional inmates from Arizona were put on hold.[93]

Causes of Prison Violence

When more inmates are confined in a small space, conflicts among inmates or between inmates and staff intensify.[94] Sociologists have also suggested that riots often occur as a result of the breakdown in the traditional inmate subculture and attempts by prison administrators to establish formal control of the prison by strict rule enforcement. When administrators shake things up by imposing new rules and prison routines and attempting to run the prison "by the book," they alter the informal inmate-staff relations that have developed over time. This change may spark disorder and violent outbursts.[95]

Criminologists Burt Useem and Jack Goldstone offer alternative explanations of prison riots:[96]

- *External pressures.* New or increased demands on prison administrators that are not accompanied by provision of resources needed to meet those demands are likely to result in staff layoffs and cuts in inmate services.
- *Internal pressures from corrections staff.* New policies or reforms can produce intense opposition from staff. When correctional officers become sufficiently alienated, absenteeism and turnover increase, officers are more likely to not follow prison routines, and groups of staff are more likely to confront administrators, further increasing the general sense of unease.
- *Internal pressures from inmates.* Inmates are likely to place demands on administrators when they perceive arbitrary rule enforcement, excessive use of force by staff, reduction in inmate services, and lack of inmate safety, creating an environment of greater violence.
- *Inmate ideologies that justify revolt.* Radical inmate ideologies can establish a framework for opposition to prison administration and justify rebellion. Inmate ideologies claiming that prison conditions are not just bad, but actually wrong, can lead to a goal of rioting to attract media attention to the unjust conditions.
- *Unjust prison administration actions.* Administrators may act in ways that are perceived to be ineffective or unjust. Rioting may result when administrators are weak in responding to assaults on officers or other inmates, when they mishandle inmates' attempts to escape, or when they bungle attempts to conduct security operations, such as lockdowns and cellblock searches.

Prisoners' Rights

Many people believe that persons who have been convicted of serious felonies and sent to prison should lose all of their basic rights and privileges. Most prisoners are citizens of the United States, however, and thus subject to the protections provided by the Bill of Rights. Nonetheless, as discussed in previous chapters, the courts have not always been very clear on just which rights should be extended to certain people under particular circumstances. This is also true when looking at prisoner rights.

In many states, people who are convicted of serious felonies forfeit certain civil rights and privileges. This forfeiture is known as loss of civil liberties and includes the loss of the right to vote, hold public office, enter into contracts, sit on juries, obtain certain jobs or occupational licenses, marry, keep or adopt children, and even procreate. For example, a federal appeals court recently held that while inmates do not have a fundamental right to procreate via a conjugal visit with their spouse, they may exercise that right through artificial insemination.[97]

Perhaps the most controversial forfeiture is the loss of the right to vote. All but two states—Maine and Vermont—deny prisoners the right to vote while they are in prison. In addition, 36 states prevent parolees from voting, and 31 states deny convicted felons on probation the right to vote TABLE 14-5 .

Historically, the federal courts deferred to prison administrators to make decisions about how prisons should be run, including the treatment of prisoners. The hands-off doctrine essentially stated that the courts do not have the expertise to supervise prison administrators or to interfere with ordinary prison rules and regulations.[98] For the most part, the courts confined themselves to matters of writs of habeas corpus, which challenge the legality of confinement of certain inmates.

LINK As noted in Chapter 11, a writ of habeas corpus is an order by a court requiring the government to bring a confined person to court so that the prisoner's claims can be heard.

The 1960s brought a new focus upon civil rights, not just for minorities in society but also for prison inmates.[99] Over the next three decades, prisoners successfully challenged a variety of policies and practices involving rights of religion, privacy, mail and access to the media, access to the courts and legal services, protection against cruel and unusual punishment, medical care, and general prison conditions.[100]

Freedom of Religion

As the number of African American Muslim prisoners increased, they began to demand the same reli-

TABLE 14-5

Disenfranchisement of Convicted Felons

State	Prison	Probation	Parole
Alabama	✓	✓	✓
Alaska	✓	✓	✓
Arizona	✓	✓	✓
Arkansas	✓	✓	✓
California	✓		✓
Colorado	✓		✓
Connecticut	✓		✓
Delaware	✓	✓	✓
District of Columbia	✓		
Florida	✓	✓	✓
Georgia	✓	✓	✓
Hawaii	✓		
Idaho	✓	✓	✓
Illinois	✓		
Indiana	✓		
Iowa	✓	✓	✓
Kansas	✓	✓	✓
Kentucky	✓	✓	✓
Louisiana	✓	✓	✓
Maine			
Maryland	✓	✓	✓
Massachusetts	✓		
Michigan	✓		
Minnesota	✓	✓	✓
Mississippi	✓	✓	✓
Missouri	✓	✓	✓
Montana	✓		
Nebraska	✓	✓	✓
Nevada	✓	✓	✓
New Hampshire	✓		
New Jersey	✓	✓	✓
New Mexico	✓	✓	✓
New York	✓		✓
North Carolina	✓	✓	✓
North Dakota	✓		
Ohio	✓		
Oklahoma	✓	✓	✓
Oregon	✓		
Pennsylvania	✓		
Rhode Island	✓	✓	✓
South Carolina	✓	✓	✓
South Dakota	✓		✓
Tennessee	✓	✓	✓
Texas	✓	✓	✓
Utah	✓		
Vermont			
Virginia	✓	✓	✓
Washington	✓	✓	✓
West Virginia	✓	✓	✓
Wisconsin	✓	✓	✓
Wyoming	✓	✓	✓
U.S. Total	49	31	36

Source: The Sentencing Project, *Felony Disenfranchisement Laws in the United States* (Washington, DC: The Sentencing Project, 2006).

gious rights as were accorded to Christian inmates. Many prison officials, however, believed that Muslim inmates were revolutionaries who were organizing to challenge prison authority. In 1962, the U.S. District Court for the District of Columbia ruled, in *Fulwood v. Clemmer*, that Muslims have the same constitutional right to practice their religion and to hold worship services as inmates of other faiths.[101] This decision opened the door for further expansion and clarification of religious rights of prisoners in several subsequent court cases:

- *Gittlemacker v. Prasse* (1970): Inmates must be given the opportunity to practice their religion, but the state is not required to provide a member of the clergy to conduct services.[102]
- *Cruz v. Beto* (1972): Buddhist inmates must be given a reasonable opportunity to worship in their own manner, just like other inmates who participate in more conventional religious practices.[103]
- *Kahane v. Carlson* (1975): Orthodox Jewish inmates may not be denied a kosher diet unless prison officials can show cause for why such a diet cannot be provided.[104]
- *Gallahan v. Hollyfield* (1982): Native American inmates may wear long hair if it is a requirement of their sincere religious beliefs.[105]
- *Childs v. Duckworth* (1983): Prison officials may place limits on the practice of Satanism.[106]
- *Abdullah v. Kinnison* (1985): Muslim inmates may be prohibited from wearing white robes in their cells if prison officials believe that it interferes with prison security.[107]
- *O'Lone v. Estate of Shabazz* (1987): Inmates may be assigned to work schedules that deny them the opportunity to attend religious services if no reasonable alternative schedule exists.[108]
- *Cutter v. Wilkinson* (2005): Inmates may practice their own religion, even if Wiccan or Satanist, unless the practices undermine prison security and safety, according to the Religious Land Use and Institutionalized Persons Act of 2000.[109]
- *Americans United for Separation of Church and State v. Prison Fellowship et al.* (2006): U.S. District Court ruled that the Bible-based Prison Fellowship program operating within an Iowa prison violated the separation of church and state doctrine by using state funds to promote Christianity to inmates.[110]

Right to Privacy

How much privacy should prisoners have? Given the legitimate concerns of prisons related to the safety of both staff and inmates, it is understandable that the privacy of inmates is constrained. In a series of cases, the federal courts ruled that inmates have little right to privacy:

- *United States v. Hitchcock* (1973): Search warrants are not required for prison officials to search an inmate's cell.[111]
- *Bell v. Wolfish* (1979): Strip searches of inmates, including body cavity searches after contact visits, may be conducted when the security needs of the institution outweigh the inmate's personal right to privacy.[112]
- *Hudson v. Palmer* (1984): Prison officials may search cells and confiscate any contraband found there. Inmates do not have a reasonable expectation of privacy during their period of incarceration.[113]
- *Block v. Rutherford* (1984): Inmates do not have the right to be present during cell searches.[114]

Mail and Access to the Media

If inmates and their cells may be searched, do such policies extend to limiting, opening, reading, and censoring inmate correspondence? Do inmates have the right to have access to the media? The courts have held that limits may be placed on inmates' mail and media access when necessary to maintain prison security:

- *Nolan v. Fitzpatrick* (1971): Inmates may correspond with newspapers unless their correspondence contains discussion of escape plans, contraband, or objectionable material.[115]
- *Procunier v. Martinez* (1974): Censorship of inmates' mail is constitutional if it is necessary to maintain order or to facilitate an inmate's rehabilitation. Even then, such censorship may be no greater than necessary to protect legitimate government interests. Censorship may not be practiced simply to eliminate unflattering or unwelcome opinions or factually inaccurate statements.[116]
- *Ramos v. Lamm* (1980): Policies prohibiting the delivery of mail written in a language other than English is unconstitutional.[117]
- *Shaw v. Murphy* (2001): Inmates do not have a constitutional right to correspond with other inmates

if prison administrators believe that such correspondence might undermine prison security.[118]

- *Banks v. Beard* (2005): The U.S. Court of Appeals for the Third Circuit ruled that prison administrators in Pennsylvania could not prohibit magazines, photographs, and newspapers sent through the mail to be received by even the most violent inmates.[119]

Access to Courts and Legal Services

Do inmates have a right to legal assistance in petitioning the courts? Must prisons provide inmates with legal access? Legal procedures are complex and formal, and very few inmates have the necessary legal knowledge or resources to take advantage of such access. Until recently, in fact, it was not unusual for prison administrators to discipline inmates who petitioned the courts.

- *Ex Parte Hull* (1941): Access to the courts is a prisoner's most fundamental right.[120]
- *Johnson v. Avery* (1969): Prison officials must provide some legal assistance to inmates and cannot prohibit inmates from assisting one another in preparing legal materials. (This ruling allowed jailhouse lawyers—inmates who are knowledgeable in the legal process—to help other inmates who wanted to submit habeas corpus petitions to the courts. In this case, however, the court held that jailhouse lawyers could be used only if institutions provided no reasonable alternative.)[121]
- *Younger v. Gilmore* (1971): Prisons must maintain adequate law libraries for inmate use.[122]
- *Wolff v. McDonnell* (1974): Jailhouse lawyers may assist inmates wishing to file legal actions against prison officials for civil rights violations.[123]
- *Procunier v. Martinez* (1974): Law students and paraprofessionals cannot be prohibited from assisting inmates.[124]
- *Bounds v. Smith* (1977): State prisons must provide inmates with law libraries or assistance by people trained in the law. According to the court, such people may include inmates who are trained as paralegals and working under a lawyer's supervision, paraprofessionals and law students, and volunteer lawyers.[125]
- *DeMallory v. Cullen* (1988): The use of jailhouse lawyers is not an acceptable alternative to providing access to a law library in prison.[126]

- *McCleskey v. Zant* (1991): Habeas corpus petitions by death row inmates may be denied to inmates who have abused the appeals process "by failing to raise a claim through inexcusable neglect."[127]

> **LINK** Since the *McCleskey* case, discussed in greater detail in Chapter 12, most prisoners who have exhausted their state appeals now have only one appeal in the federal courts.

Protection Against Cruel and Unusual Punishment

According to the Eighth Amendment, inmates are not to be subjected to cruel and unusual punishments. The issue of cruel and unusual punishment was raised in 1979, in *Bell v. Wolfish*.[128] In this case, the court ruled that the pretrial detention of a group of suspects at the Metropolitan Correctional Center in Manhattan did not constitute cruel and unusual punishment, because they were detained for a valid government purpose. Placing unruly inmates in the "hole" (administrative segregation) for extended periods of time, however, was found to be unnecessary and beyond acceptable standards.

Correctional officers are also restricted in the amount of force they may use to obtain inmate compliance with their demands. Specifically, officers may use no more force than is necessary. In an attempt to define excessive use of force by officers, the Supreme Court, in 1986 in *Whitley v. Albers*, held that inmates have a legitimate claim that excessive force was used only under the following conditions:

- The intent of the officer was to inflict harm.
- The use of force was unnecessary to achieve compliance.
- There was infliction of severe pain or injury.[129]

In early 1992, the Supreme Court broadened those guidelines in *Hudson v. McMillian*. In that decision, the court stated that "When prison officials maliciously and sadistically use force to cause harm, contemporary standards of decency always are violated."[130]

In 2002, in *Hope v. Pelzer*, the Supreme Court held that a correctional officer who handcuffed an Alabama prisoner to a hitching post for two hours for disruptive conduct was not entitled to qualified immunity. In other words, a correctional official who knowingly violates the Eighth Amendment rights of an inmate can be held liable for damages.[131]

Hudson v. McMillian

Keith Hudson, an inmate at the State Penitentiary in Angola, Louisiana, got into a physical confrontation with correctional officer Jack McMillian. Two other officers came to assist McMillian and subdued Hudson, placing him in handcuffs and shackles. They took him out of his cell and walked him toward the prison's administrative segregation area. Along the way, the officers kicked and punched Hudson in the mouth, eyes, chest, and stomach. As a result, Hudson suffered minor bruises and swelling of his face, mouth, and lip. The blows also loosened his teeth and cracked his partial dental plate. Hudson brought a civil suit against the officers, seeking compensation and damages, and won an $800 damage award. However, a federal appeals court overturned Hudson's suit, ruling that the inmate's injuries were minor. Hudson then appealed to the U.S. Supreme Court, which reversed the appellate court's ruling.

Justice Sandra Day O'Connor, writing for the majority, stated, "When prison officials maliciously and sadistically use force to cause harm, contemporary standards of decency always are violated." Justice Clarence Thomas wrote a strongly worded dissenting opinion, stating:

> A use of force that causes only insignificant harm to a prisoner may be immoral, it may be tortious [wrong], it may be criminal, and it may even be remedial under other provisions of the Federal Constitution, but it is not "cruel and unusual punishment." . . . Today's expansion of the cruel and unusual punishment clause beyond all bounds of history and precedent is, I suspect, yet another manifestation of the pervasive view that the Federal Constitution must address all ills in our society. Abusive behavior by prison guards is deplorable conduct. . . . But that does not mean that it is invariably unconstitutional.

Source: Hudson v. McMillian, 503 U.S. 1 (1992).

Right to Medical Care

Prisoners' right to medical care, including mental health treatment, has also been supported by the courts within the framework of the Eighth Amendment's prohibition against cruel and unusual punishment:

- *Estelle v. Gamble* (1976): Indifference to an inmate's medical needs constitutes cruel and unusual punishment; it is the government's obligation to provide medical care for inmates in its correctional system.[132]
- *Ruiz v. Estelle* (1980): The Texas Department of Corrections was providing inadequate medical treatment programs; the court ordered the state to provide more qualified medical staff, establish diagnostic and sick-call procedures, improve clinic facilities and recordkeeping, and discontinue assigning inmates to work as medical assistants and to dispense drugs in the pharmacy.[133]

Prison Conditions

One of the most significant developments in defining inmate rights in recent years has been the introduction of the standard of totality of conditions for judging possible constitutional violations within a single prison or an entire prison system. This standard means that a pattern of abuses—none of which by itself would justify the court's intervention, but that would meet this criterion when considered together—constitutes a violation of the Eighth Amendment's protection against cruel and unusual punishment.

The federal courts first applied the totality of conditions standard in 1970 in *Holt v. Sarver*, when it was ruled that the entire Arkansas prison system was in violation of the Eighth Amendment. Prisons in that state had traditionally authorized armed inmate trustees to supervise and punish other inmates, and they frequently resorted to violence to maintain order. In addition to inappropriate use of trustees, the court found extensive unsanitary conditions in food preparation and service, a lack of rehabilitative programs, and crowded and filthy isolation cells. In applying the standard of totality of conditions, it held that confinement in the Arkansas correctional system was unconstitutional.[134]

Six years after *Holt*, in *Pugh v. Locke*, a federal judge found the Alabama prison system to be unconstitutional. In this case, conditions were so deplorable that the court ordered 44 major reforms to bring the system into constitutional compliance. Among the reforms were a minimum of 66 square feet for cells, at least one toilet for every 15 inmates, one shower for every 20 inmates, and improvements in ventilation, heating, and lighting of cellblocks.[135]

In 1980, in *Ruiz v. Estelle*, the Texas prison system was found in violation of the Eighth Amendment by a federal district court judge because of the use of inmate trustees in security positions, patterns of inmate discipline, overcrowding, and inadequate medical care.[136]

One year later, in *Rhodes v. Chapman*, the Supreme Court, using the totality of conditions standard, reversed lower court rulings that had found overcrowding in an Ohio prison violated the Eighth Amendment prohibition against cruel and unusual punishment. In the *Rhodes* case, the Court ruled that the policy of double-bunking inmates in cells designed for a single inmate did not inflict "unnecessary and wanton pain," nor were the conditions "grossly disproportionate to the severity of the crime" for which the inmate was sentenced. According to this ruling, overcrowding per se does not violate the Eighth Amendment unless the institutional policies themselves result in deplorable living conditions; in other words, in this case the totality of conditions in the prison was held to be within constitutionally acceptable limits.[137]

Finally, in 1994, in *Farmer v. Brennan, Warden, et al.*, the Supreme Court dealt with a case in which a preoperative transsexual with feminine characteristics was incarcerated with other males in the federal prison system and was beaten and raped by another inmate after having been placed in the general population. The inmate claimed that prison officials had acted with "deliberate indifference" to the inmate's safety because they knew that (1) the prison had a violent environment and a history of inmate assaults and (2) the inmate would be particularly vulnerable to sexual attack. Although the appeals court had rejected the claim, the Supreme Court upheld it, stating that a prison official may be held liable under the Eighth Amendment for acting with "deliberate indifference" to inmate health or safety if the official knows that inmates face a substantial risk of serious harm and disregards that risk by failing to take responsible measure to abate it.[138]

WRAP UP

During the social upheaval of the 1960s, inmates became less willing to abide by the prison order created by administration and staff. During this time, militant groups such as the Black Panthers began to organize inmate society according to ideology and race. In addition, as a result of the police and prosecutors targeting street gangs, gang members and their leaders began to replace the old inmate social order, creating new relationships and stresses between inmates and prison staff. Inmates are now more likely to tie themselves to racial and ethnic groups for protection, solidarity, and power. Power arrangements were not the only changes in prisons; the nature of the prison population also changed. No longer predominantly white males, inmate populations today reflect the diversity of the larger society. In turn, the growing social, cultural, and language diversity in prison creates even greater distance between inmates and staff.

Chapter Spotlight

- As they enter prison, new inmates are classified according to their security level as well as their housing, work, or educational assignment.

- Changes in the prison population in recent decades have transformed it into a younger, more violent, but still overwhelmingly male group of inmates.

- Nearly all prisons provide counseling, health and medical services, academic and vocational programs, and prison industries.

- All correctional institutions are managed by a hierarchy of staff, with the warden appearing at the top of the bureaucratic structure.

- Custody staff (known as correctional officers or guards) supervise inmates throughout the institution and often find themselves torn between their roles as keepers and caregivers.

- Inmates charged with rule violations face a disciplinary hearing. Sanctions for rule violations include placement in administrative segregation, loss of the "good time" credit toward early release, or loss of privileges.

- Prisons must manage a variety of special populations, including inmates with HIV or AIDS, inmates classified as mentally disordered or mentally deficient, elderly inmates, inmates with disabilities, juveniles, and female offenders.

- The social order of prisons is largely maintained by an inmate subculture that prescribes expected relationships among inmates and prison employees.

- Many scholars believe that the traditional inmate subculture has given way to new social arrangements reflecting the emergence of prison gangs. Inmates are now more likely to tie themselves to racial and ethnic groups for protection, solidarity, and power.

- Inmates lose many constitutional rights when they enter prison. Conversely, as a result of a number of Supreme Court decisions over the past three decades, a number of inmate rights have been extended.

Putting It All Together

1. Who really controls prisons: staff or inmates?
2. Should women be allowed to work as correctional officers directly supervising male inmates, and should men be allowed to work as correctional officers directly supervising female inmates?
3. Should pregnant inmates be allowed to keep their babies with them in prison and, if so, for how long?
4. Do you believe inmates should have the same rights as free citizens? What limits would you place on the rights of inmates, if any?
5. Should prisoners with HIV/AIDS be segregated from the general prison population? What would you propose as an alternative?

administrative segregation The placement of an inmate in a single-person cell in a high-security area for a specified period of time; sometimes referred to as solitary confinement.

classification System for assigning inmates to levels of custody and treatment appropriate to their needs.

correctional officers Also known as guards; the lowest-ranking prison staff members, who have the primary responsibility of supervising inmates.

deprivation model An explanation of the inmate subculture as an adaptation to loss of amenities and freedoms within prison.

disciplinary hearing A hearing before a disciplinary board to determine whether the charge against an inmate has merit and to determine the sanction if the charge is sustained.

Estelle v. Gamble The Supreme Court decision that indifference to inmates' medical needs constitutes cruel and unusual punishment.

Fulwood v. Clemmer The U.S. District Court decision that African American Muslim inmates have the same constitutional rights to practice their religion and to hold worship services as inmates of other faiths.

good time The practice of reducing an inmate's sentence for good behavior.

hands-off doctrine The position taken by the Supreme Court that it will not interfere with states' administration of prisons.

Holt v. Sarver The Supreme Court decision that applied the "totality of conditions" principle to find the Arkansas prison system in violation of the Eighth Amendment.

Hudson v. McMillian The Supreme Court decision that the use of excessive force by correctional officers may constitute cruel and unusual punishment, even if the inmate does not suffer serious injury.

importation model A view of the inmate subculture as a reflection of the values and norms inmates bring with them when they enter prison.

inmate code A system of informal norms created by prisoners to regulate inmate behavior.

jailhouse lawyers Inmates who have studied legal proceedings and who help other inmates in making petitions to the courts.

less eligibility The belief that prisoners should always reside in worse conditions than should the poorest law-abiding citizens.

line personnel Prison employees who have direct contact with inmates.

loss of privileges A disciplinary sanction involving the loss of visits, mail, recreation, and access to the prison commissary.

pains of imprisonment Deprivations—such as the loss of freedom, possessions, dignity, autonomy, security, and heterosexual relationships—shared by inmates.

prisonization The process by which inmates adjust to or become assimilated into the prison subculture.

staff personnel Prison employees who provide support services to line personnel and administrators.

surrogate families Fictive families created by female prisoners to provide stability and security.

total institutions Institutions that completely encapsulate the lives of the people who work and live in them.

totality of conditions A principle guiding federal court evaluations of prison conditions: The lack of a specific condition alone does not necessarily constitute cruel and unusual punishment.

warden The superintendent or top administrator of a prison.

Wolff v. McDonnell The Supreme Court decision that inmates facing disciplinary action must have a formal hearing, 24-hour notification of the hearing, assistance in presenting a defense, and ability to call witnesses.

1. Quoted in Rose Giallombardo, *Society of Women: A Study of a Women's Prison* (New York: John Wiley & Sons, 1966), p. 23.

2. Erving Goffman, *Asylums: Essays on the Social Situation of Mental Patients and Other Inmates* (New York: Doubleday, 1961), p. 6.

3. Keith Farrington, "The Modern Prison as Total Institution? Public Perception Versus Objective Reality," *Crime & Delinquency* 38:6–26 (1992).

4. John Irwin and James Austin, *It's about Time: America's Imprisonment Binge* (Belmont, CA: Wadsworth, 1994), p. 74.

5. Daryl Vigil, "Classification and Security Threat Group Management," *Corrections Today* 68:32–34 (2006).

6. Paige Harrison and Allen Beck, *Prisoners in 2005* (Washington, DC: U.S. Department of Justice, 2006); Bureau of Justice Statistics, *National Prisoners Statistics—1994* (Washington, DC: U.S. Department of Justice, 1994).

7. James Stephan, *State Prison Expenditures, 2001* (Washington, DC: U.S. Department of Justice, 2004); James Stephan and Jennifer Karberg, *Census of State and Federal Correctional Facilities, 2000* (Washington, DC: U.S. Department of Justice, 2004).

8. Harrison and Beck, note 6.

9. Harrison and Beck, note 6.

10. Marc Mauer, Ryan King, and Malcolm Young, *The Meaning of "Life": Long Prison Sentences in Context* (Washington, DC: The Sentencing Project, 2004).

11. Harrison and Beck, note 6.

12. Stephan and Karberg, note 7.

13. "Pennsylvania Correctional Industries" (March 2005), available at http://www.cor.state.pa.us, accessed May 31, 2007; Section 4122(b) of Title 18, United States Code; Joan Mullen, Kent Chabotar, and Deborah Carrow, *The Privatization of Corrections* (Washington, DC: National Institute of Justice, 1985).

14. *Loving v. Johnson*, No. 05-10679 (5th Cir. 2006).

15. T. Paul Louis and Jerry Sparger, "Treatment Modalities Within Prison," *Are Prisons Any Better? Twenty Years of Correctional Reform* (Newbury Park, CA: Sage, 1990), p. 151.

16. David Wilson, Leana Bouffard, and Doris Mackenzie, "A Quantitative Review of Structured, Group-Oriented, Cognitive-Behavioral Programs for Offenders," *Criminal Justice and Behavior* 32:172–204 (2005).

17. Harrison and Beck, note 6.

18. Caroline Harlow, *Education and Correctional Populations* (Washington, DC: U.S. Department of Justice, 2003); Stephen Klein, Michelle Tolbert, Rosio Bugarin, Emily Cataldi, and Gina Tauschek, *Correctional Education: Assessing the Status of Prison Programs and Information Needs* (Washington, DC: U.S. Department of Education, 2004).

19. Klein et al., note 18.

20. Stephen Steurer, Linda Smith, and Alice Tracy, *The Three State Recidivism Study* (Lanham, MD: Correctional Education Association, 2001); Mary Batiuk Karen Lahm, Matthew McKeever, Norma Wilcox, and Pamela Wilcox, "Disentangling the Effects of Correctional Education: Are Current Policies Misguided?" *Criminal Justice* 5:55–74 (2005); Michelle Fine et al., "Changing Minds: The Impact of College in a Maximum Security Prison," Graduate Center of the City University of New York (2001), available at http://www.barnard.columbia.edu/sfonline/prison/minds_01.htm, accessed July 31, 2007.

21. Stephan and Karberg, note 7.

22. Timothy Flanagan, "Prison Labor and Industry," in Lynne Goodstein and Doris MacKenzie (Eds.), *The American Prison: Issues in Research and Policy* (New York: Plenum, 1989), p. 147.

23. Julie Light, "Look for That Prison Label," *The Progressive* 64(6):21–23 (2000).

24. Donald Cressey, "Prison Organizations," in James March (Ed.), *Handbook of Organizations* (Chicago: Rand McNally, 1965), pp. 1023–1070.

25. Emily Herrick, "Number of COs Up 25 Percent in Two Years," *Corrections Compendium* 13:9–21 (1988).

26. Nancy Jurik and Russell Winn, "Describing Correctional Security Dropouts and Rejects: An Individual or Organizational Profile," *Criminal Justice and Behavior* 14:5–25 (1987).

27. Nancy Jurik and Michael Musheno, "The Internal Crisis of Corrections: Professionalization and the Work Environment," *Justice Quarterly* 3:457–480 (1986).

28. Francis Cullen, Bruce Link, Nancy Wolfe, and James Frank, "The Social Dimensions of Correctional Officer Stress," *Justice Quarterly* 2:505–533 (1985).

29. Joseph Rowan, "Who Is Safer in Male Maximum Security Prisons?" *Corrections Today* 58:186–189 (1996).

30. Richard Wortley, *Situational Prison Control: Crime Prevention in Correctional Institutions* (New York: Cambridge University Press, 2002); Mary Stohr, "'Yes, I've Paid the Price, but Look How Much I Gained': The Struggle and Status of Women Correctional Officers," in Claire Renzetti, Lynne Goodstein, and Susan Miller (Eds.), *Rethinking Gender, Crime, and Justice* (Los Angeles: Roxbury, 2006), pp. 262–277.

31. *Grummett v. Rushen,* 779 F.2d 491 (9th Cir. 1985).

32. *Timm v. Gunter,* 50l U.S. 1209 (1991).

33. Melvin Claxton, Ronald Hansen, and Norman Sinclair, "Sexual Abuse Behind Bars: Guards Assault Female Inmates," *Detroit News* (May 22, 2005), available at http://detnews.com/2005/specialreport/0505/24/A01-189215.htm, accessed May 31, 2007.

34. Bureau of Justice Statistics, *Prison Rule Violators* (Washington, DC: U.S. Department of Justice, 1989), pp. 1–8.

35. Bureau of Justice Statistics, note 34.

36. John D. Hewitt, Eric D. Poole, and Robert M. Regoli, "Self-Reported and Observed Rule-Breaking in Prison: A Look at Disciplinary Response," *Justice Quarterly* 1:437–447 (1984).

37. Hewitt et al., note 36, p. 446.

38. *Wolff v. McDonnell,* 418 U.S. 539 (1974).

39. Irwin and Austin, note 4, p. 91.

40. *Superintendent, Massachusetts Correctional Institution, Walpole v. Hill,* 472 U.S. 445 (1985).

41. Laura Maruschak, *HIV in Prisons, 2003* (Washington, DC: U.S. Department of Justice, 1993).

42. Mark Blumberg and Denny Langston, "Mandatory HIV Testing in Criminal Justice Settings," *Crime & Delinquency* 37:5–18 (1991).

43. Barbara Belbot and Rolando del Carmen, "AIDS in Prison: Legal Issues," *Crime & Delinquency* 37:144–145 (1991).

44. William Branigin and Leef Smith, "Mentally Ill Need Care, Find Prison," *Washington Post,* November 25, 2001, p. A01; Luke Birmingham, Debbie Mason, and Don Grubin, "A Follow-up Study of Mentally Disordered Men Remanded to Prison," *Criminal Behaviour and Mental Health* 8:202–214 (1998).

45. Kenneth Adams, "Former Mental Patients in a Prison and Parole System: A Study of Socially Disruptive Behavior," *Criminal Justice and Behavior* September: 358–394 (1983), Hans Toch, *Living in Prison: The Etiology of Survival* (New York: Free Press, 1977).

46. Brian McCarthy, "Mentally Ill and Mentally Retarded Offenders in Corrections: A Report of a National Survey," *New York State Department of Correctional Services, Sourcebook on the Mentally Disordered Prisoner* (Washington, DC: National Institute of Corrections, 1985), pp. 14–29.

47. *Washington v. Harper,* 494 U.S. 210 (1990).

48. Harrison and Beck, note 6; Patrick McMahon, "Aging Inmates Present Prison Crisis," *USA Today,* August 10, 2003, available at http://www.usatoday.com/news/nation/2003-08-10-prison-inside-usat_x.htm, accessed May 31, 2007.

49. Pat Shellenbarger, "Aging Inmates Bring Increased Ailments, Expenses," *Grand Rapids Press,* September 25, 2006, p. A3.

50. Ronald Aday, *Aging Prisoners: Crisis in American Corrections* (Westport, CT: Praeger, 2003).

51. Aday, note 50, pp. 114–115.

52. Aday, note 50, pp. 153, 208–209.

53. Harrison and Beck, note 6.

54. *Pennsylvania Dept. of Corrections v. Yeskey,* 118 F.3d 168 (1998).

55. Harrison and Beck, note 6.

56. Melissa Sickmund, *Juveniles in Corrections* (Washington, DC: U.S. Department of Justice, 2004), pp. 20–21.

57. Howard Snyder and Melissa Sickmund, *Juvenile Offenders and Victims: 2006* National Report (Washington, DC: U.S. Department of Justice, 2006).

58. Sickmund, note 56.

59. Martin Forst, Jeffrey Fagan, and T. Scott Vivona, "Youth in Prisons and Training Schools: Perceptions and Consequences of the Treatment–Custody Dichotomy," *Juvenile and Family Court Journal* 40:1–14 (1989).

60. Harrison and Beck, note 6.

61. Theresa Severance, "'You Know Who You Can Go to': Cooperation and Exchange Between Incarcerated Women," *The Prison Journal* 85:343–367 (2005).

62. Rose Giallombardo, "Social Roles in a Prison for Women," *Social Problems* 13:268–288 (1966); Alice Propper, "Make-Believe Families and Homosexuality Among Imprisoned Girls," *Criminology* 20:127–139 (1982); Susan Cranford and Rose Williams, "Critical Issues in Managing Female Offenders," *Corrections Today* 60:130–134 (1998); Severance, note 61.

63. Barbara Owen, *The Mix: The Culture of Imprisoned Women* (Albany, NY: State University Press of New York, 1998).

64. U.S. General Accounting Office, *Women in Prison: Issues and Challenges Confronting U.S. Correctional Systems* (Washington, DC: U.S. General Accounting Office, 1999); Judith Greene and Kevin Pranis, "Growth Trends and Recent Research," in Ann Jacobs and Sarah From (Eds.), *Hard Hit: The Growth in the Imprisonment of Women, 1977–2004* (New York: Women's Prison Association, 2006), pp. 7–28.

65. Harrison and Beck, note 6.

66. Greene and Pranis, note 64.

67. Karen Holt, "Nine Months to Life: The Law and the Pregnant Inmate," *Journal of Family Law* 20:537 (1982).

68. Joann Morton and Deborah Williams, "Mother/Child Bonding: Incarcerated Women Struggle to Maintain Meaningful Relationships with Their Children," *Corrections Today* 60:98–104 (1998).

69. Morton and Williams, note 68.

70. Gresham Sykes, *Society of Captives: A Study of a Maximum Security Prison* (New York: Random House, 1956).

71. John Irwin and Donald Cressey, "Thieves, Convicts, and the Inmate Culture," *Social Problems* 10:142–155 (1962).

72. Stanton Wheeler, "Socialization in Correctional Communities," *American Sociological Review* 26:250 (1961).

73. Donald Clemmer, *The Prison Community* (New York: Holt, Rinehart & Winston, 1940).

74. Charles Tittle, "Prison and Rehabilitation: The Inevitability of Failure," *Social Problems* 21:385–394 (1974).

75. Hans Toch, Kenneth Adams, and J. Douglas Grant, *Coping: Maladaptations in Prisons* (New Brunswick, NJ: Transaction, 1989), p. xiii.

76. Toch et al., note 75, p. 254.

77. Edward Zamble and Frank Porporino, *Coping, Behavior, and Adaptation in Prison Inmates* (New York: Springer-Verlag, 1988).

78. Gresham Sykes and Sheldon Messinger, "The Inmate Social System," in Richard Cloward et al. (Eds.), *Theoretical Studies in the Social Organization of the Prison* (New York: Social Science Research Council, 1960), pp. 5–19.

79. James Jacobs, *Stateville: The Penitentiary in Mass Society* (Chicago: University of Chicago Press, 1977), p. 207.

80. Florida Department of Corrections, "Major Prison Gangs," *Gang and Security Threat Group Awareness*, available at http://www.dc.state.fl.us/pub/gangs/prison.html, accessed August 1, 2007.

81. James Robertson, "Cruel and Unusual Punishment in United States Prisons: Sexual Harassment Among Male Inmates," *American Criminal Law Review* 36:1–51 (1999).

82. Daniel Lockwood, *Prison Sexual Violence* (New York: Elsevier, 1980).

83. Peter Nacci and Thomas Kane, "The Incidence of Sex and Sexual Aggression in Federal Prisons," *Federal Probation* 47:31–36 (1983).

84. Wayne Wooden and Jay Parker, *Men Behind Bars: Sexual Exploitation in Prison* (New York: Da Capo Press, 1982).

85. Allen Beck and Paige Harrison, *Sexual Violence Reported by Correctional Authorities, 2005* (Washington, DC: U.S. Department of Justice, 2006).

86. Barbara Owens and James Wells, *Staff Perspectives: Sexual Violence in Adult Prisons and Jails* (Washington, DC: National Institute of Corrections, 2006).

87. Cindy Struckman-Johnson and David Struckman-Johnson, "Sexual Coercion Reported by Women in Three Midwestern Prisons," *Journal of Sex Research* 39:217–227 (2202); Leanne Alarid, "Sexual Assault and Coercion Among Incarcerated Women Prisoners: Excerpts from Prison Letters," *The Prison Journal* 80:391–406 (2000); Agnes Baro, "Spheres of Consent: An Analysis of the Sexual Abuse and Sexual Exploitation of Women Incarcerated in the State of Hawaii," *Women and Criminal Justice* 8:61–84 (1997); Cindy Struckman-Johnson, David Struckman-Johnson, Lisa Rucker, Kurt Bumby, and

Stephen Donaldson, "Sexual Coercion Reported by Men and Women in Prison," *Journal of Sex Research* 33:67–76 (1996).

88. Struckman-Johnson and Struckman-Johnson, note 87.

89. Christopher Mumola, *Suicide and Homicide in State Prisons and Local Jails* (Washington, DC: U.S. Department of Justice, 2005).

90. Eric Rich, Hamil Harris, and Nelson Hernandez, "Inmates Kill Guard at Prison in Maryland," *Washington Post*, July 27, 2006, p. A01.

91. Tom Wicker, *A Time to Die: The Attica Prison Riot* (Lincoln, NE: University of Nebraska Press, 1994).

92. W. G. Stone, as told to G. Hirliman, *The Hate Factory* (Agoura, CA: Paisano Publications, 1982), pp. 76–79.

93. Associated Press, "220 Inmates Shipped Out After Riot," April 26, 2007, available at http://suntimes.com/news/nation/358684,CST-NWS-riot26.article, accessed June 1, 2007; Joy Leiker, "New Castle Prison Riot Report Assigns Blame," May 25, 2007, available at http://www.thestarpress.com/apps/pbcs.dll/article?AID=/20070525/NEWS01/705250345, accessed June 1, 2007.

94. Paul Paulus, *Prison Crowding: A Psychological Perspective* (New York: Springer-Verlag, 1988).

95. Richard Wilsnack, "Explaining Collective Violence in Prisons: Problems and Possibilities," in Albert Cohen, George Cole, and Robert Bailey (Eds.), *Prison Violence* (Lexington, MA: Lexington Books, 1976), pp. 61–78.

96. Burt Useem and Jack Goldstone, "Forging Social Order and Its Breakdown: Riot and Reform in U.S. Prisons," *American Sociological Review* 67:499–525 (2002).

97. *Gerber v. Hickman*, 264 F.3d 882 (9th Cir. 2001).

98. *Banning v. Looney*, cert. denied, 348 U.S. 859 (1954).

99. James Jacobs, "The Prisoner's Rights Movement and Its Impacts," in Norval Morris and Michael Tonry (Eds.), *Crime and Justice: An Annual Review of Research*, vol. 2 (Chicago: University of Chicago Press, 1980), pp. 557–586.

100. American Correctional Association, *Legal Responsibility and Authority of Correctional Officers* (College Park, MD: American Correctional Association, 1987), p. 8.

101. *Fulwood v. Clemmer*, 206 F. Supp. 370 (D.C. Cir. 1962).

102. *Gittlemacker v. Prasse*, 428 F.2d 1 (3d Cir. 1970).

103. *Cruz v. Beto*, 405 U.S. 319 (1972).

104. *Kahane v. Carlson*, 527 F.2d 492 (2d Cir. 1975).

105. *Gallahan v. Hollyfield*, 670 F.2d 1345 (4th Cir. 1982).

106. *Childs v. Duckworth*, 705 F.2d 915 (7th Cir. 1983).

107. *Abdullah v. Kinnison*, 769 F.2d 345 (6th Cir. 1985).

108. *O'Lone v. Estate of Shabazz*, 482 U.S. 342 (1987).

109. *Cutter v. Wilkinson*, 544 U.S. 709 (2005).

110. *Americans United for Separation of Church and State v. Prison Fellowship et al.*, 4:03-cv-90074 (2006).

111. *United States v. Hitchcock*, cert. denied, 410 U.S. 916 (1973).

112. *Bell v. Wolfish*, 441 U.S. 520 (1979).

113. *Hudson v. Palmer*, 468 U.S. 517 (1984).

114. *Block v. Rutherford*, 468 U.S. 576 (1984).

115. *Nolan v. Fitzpatrick*, 451 F.2d 545 (1st Cir. 1971).

116. *Procunier v. Martinez*, 416 U.S. 396 (1974).

117. *Ramos v. Lamm*, 639 F.2d 559 (10th Cir. 1980).

118. *Shaw v. Murphy*, 523 U.S. 233 (2001).

119. *Banks v. Beard*, No. 03-1245 (3d Cir. 2005).

120. *Ex Parte Hull*, 312 U.S. 546 (1941).

121. *Johnson v. Avery*, 393 U.S. 483 (1969).

122. *Younger v. Gilmore*, 404 U.S. 15 (1971).

123. *Wolff v. McDonnell*, note 38.

124. *Procunier v. Martinez*, note 116.

125. *Bounds v. Smith*, 430 U.S. 817 (1977).

126. *DeMallory v. Cullen*, 855 F.2d 442 (7th Cir. 1988).

127. *McCleskey v. Zant*, 499 U.S. 467 (1991).

128. *Bell v. Wolfish*, note 112.

129. *Whitley v. Albers*, 475 U.S. 312 (1986).

130. *Hudson v. McMillian*, 503 U.S. 1 (1992).

131. *Hope v. Pelzer*, 536 U.S. 730 (2002).

132. *Estelle v. Gamble*, 429 U.S. 97 (1976).

133. *Ruiz v. Estelle*, 503 F. Supp. 1265 (S.D. Tex. 1980).

134. *Holt v. Sarver*, 309 F. Supp. 362 (E.D. Ark. 1970).

135. *Pugh v. Lock*, 406 F. Supp 318 (N.D. Ala. 1976).

136. *Ruiz v. Estelle*, note 133.

137. *Rhodes v. Chapman*, 452 U.S. 337 (1981).

138. *Farmer v. Brennan, Warden, et al.*, 511 U.S. 825 (1994).

OBJECTIVES

- Understand the history of probation, including why it is used and how it currently operates to provide support for probationers and safety for the community.

- Become familiar with the research on the effectiveness of probation and parole.

- Comprehend the variety of intermediate sanctions used in community corrections.

- Grasp the technology used in electronic monitoring of offenders in the community as well as the major criticisms of the use of this technology.

- Identify the key principles of restorative justice and understand its strengths and weaknesses.

- Understand the history of parole, the other types of prison release that are possible, and the process by which decisions are made to release inmates into the community.

Corrections in the Community

CHAPTER

15

Consider the following five defendants before a judge for sentencing:

Defendant 1: An African American male, age 34, pleads guilty to his third offense of driving while intoxicated. He has no other criminal history. He is employed, owns a home, and has a daughter.

Defendant 2: A Latina female, age 52, pleads guilty to felony forgery, felony fraud, and felony theft. She has six prior convictions for larceny. She is employed, has been married for 36 years, and has seven children.

Defendant 3: A white male, age 19, pleads guilty to felony assault, possession of an illegal weapon, and possession of methamphetamine. He is on probation for drug possession stemming from an incident when he was a juvenile. He is currently receiving substance abuse and psychiatric counseling, works part-time, and lives with his parents.

Defendant 4: A white male, age 61, pleads guilty to robbery and assault. He is a chronic transient and was the victim of assault in a recent case. He has been arrested approximately 65 times, mostly for municipal violations and failure to appear warrants.

Defendant 5: A Latino male, age 25, was found guilty at trial of aggravated robbery, assault with intent to do bodily harm with a deadly weapon, and menacing with a deadly weapon. This is his first arrest since being released from prison four years ago, when he served time for drug and weapons convictions. He is married, is employed, has two children, and is linked to a local street gang.

- Which features of these cases warrant different punishment?
- Under what circumstances do they deserve the same punishment?

Introduction

Penologist Tom Murton once said that "placing a man in prison to train him for a democratic society is as ridiculous as sending him to the moon to learn how to live on earth."[1] Prison life is significantly different from living in a local community of a democratic society, and inmate adaptation may actually lead to greater conformity to the criminal norms stressed by the prison subculture than to the conventional norms preferred by the community.

Communities face several additional problems linked to corrections. First, the public's desire for get-tough policies often includes demands for more prisons, but few communities are willing to have correctional facilities built in their backyards. In addition, federal and state funds to build and maintain enough prisons to accommodate offenders are limited. Even if unlimited funds were available for this purpose, constructing endless prisons and filling them to capacity may not be the wisest course of action. The United States already imprisons a greater proportion of its citizens than any other Western nation, and there is little reason to believe that increased reliance on the use of prisons will reduce crime.

The most common alternatives to long-term incarceration—probation and parole—are community-based sentences that involve supervision of offenders. Many communities also are experimenting with intermediate sanctions such as fines, house arrest, and community service, both for offenders who are diverted from traditional incarceration and for offenders who are released from prison early. Each of these alternatives reflects the belief that offenders can (and often should) be dealt with within the community. Advocates of community corrections argue that these alternatives are less expensive and often more effective in rehabilitating offenders. Critics disagree, insisting that the shift away from incarceration places too many criminals into the community too soon.

There is little consensus among criminologists regarding how community corrections should be defined and whether it should include probation and parole. Paul Hahn defines community-based corrections as all programs that reduce the use of institutions, the duration of institutional confinement, or the social distance between the offender and the community; this definition encompasses probation, parole, educational and work release programs, and prison furloughs (discussed later in this chapter).[2]

The movement toward community corrections in the United States can be traced to the early nineteenth century, when the concept of probation first appeared. Since then, community corrections have become a popular and necessary alternative to incarceration. The prison population increased 360 percent between 1980 and 2005, and the number of persons being supervised in the community on probation or parole skyrocketed. By 2005, of the slightly more than 7 million adult offenders under the care or custody of a correctional agency (about 3.2 percent of the adult population of the United States), only 20 percent were incarcerated; the remaining 80 percent were living in the community under probation, intermediate sanctions, or parole.[3]

Probation

Probation is a sentencing option in which an offender is released into the community under the supervision of a probation officer. Many offenders are given incarceration sentences that are suspended by the courts, only to be placed on probation for a period of time specified in the probation agreement. In 2005, 59 percent of all offenders under correctional supervision were on probation. More than 2.2 million adults were admitted to probation supervision in 2005, and nearly the same number were removed from probation during the same year. Approximately 77 percent of these probationers were male, and 55 percent were white FIGURE 15-1 .

Nearly two of every three correctional clients in the United States are on probation.

? Are probation and other community corrections effective tools for assisting inmates to successfully become integrated into society?

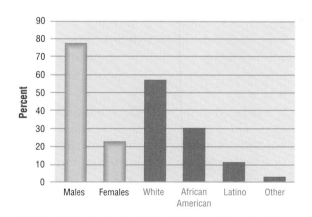

FIGURE 15-1 Probationers by Sex, Race, and Ethnicity

Source: Lauren Glaze and Thomas Bonczar, *Probation and Parole in the United States, 2005* (Washington, DC: U.S. Department of Justice, 2006), p. 6.

The use of probation varies from state to state, reflecting legislative mandates, correctional policies, the seriousness of crime in the area, and public attitudes. States that require mandatory prison sentences for people convicted of particular offenses or for habitual offenders have taken the option of probation away for such offenders. In addition, legislatures or county governments may allocate greater or lesser funding to probation according to prevailing attitudes about the importance or effectiveness of probation. For example, in 2005, New Hampshire had fewer than 460 persons on probation per 100,000 population; by contrast, Massachusetts and Rhode Island both had more than 3000 persons on probation per 100,000 population TABLE 15-1 .[4]

TABLE 15-1

State Variations in the Use of Probation

	Number on Probation	Rate per 100,000 Population
States with the Highest Rates of Persons on Probation		
Massachusetts	163,719	3350
Rhode Island	26,085	3091
Minnesota	113,121	2988
Delaware	18,725	2828
Ohio	230,758	2745
Indiana	121,675	2583
Texas	428,836	2580
Michigan	126,630	2350
Washington	111,193	2155
New Jersey	143,315	2117
States with the Lowest Rates of Persons on Probation		
New Hampshire	4285	457
West Virginia	6977	533
Utah	10,267	578
Nevada	12,645	709
Kansas	14,439	723
Maine	8907	776
Virginia	43,470	788
North Dakota	3749	791
New York	124,853	810
South Dakota	5372	889

Source: Lauren Glaze and Thomas Bonczar, *Probation and Parole in the United States, 2005* (Washington, DC: U.S. Department of Justice, 2006), p. 4.

The Emergence of Probation

The first form of probation, known as judicial reprieve, was practiced in early English courts as a temporary suspension of sentence. Recognizing that most offenders were neither evil nor dangerous, the reprieve allowed a defendant time to petition the Crown for a pardon. If the request was denied, the accused was punished, but the reprieve allowed the court to seek an alternative to the full punitive force of the law. Eventually the reprieve evolved into a suspended sentence, whereby the sentence was set aside under the condition that the offender continue to obey the law.

The concept of probation was first applied in 1841 by John Augustus, a shoemaker in Boston. As a humanitarian with both wealth and the desire to implement many of the principles emphasized by the temperance movement, Augustus was instrumental in establishing one of the first significant reforms of the U.S. criminal justice system. Working without pay, he became the nation's first probation officer. For 18 years, Augustus assisted people convicted of crimes: He investigated offenders, evalu-

ated their character, helped them find jobs, and then supervised them in the community. He built friendships with offenders and gained their trust so that they would be more responsive to his counsel about the evils of alcohol and the importance of avoiding further criminal behavior. Of the 2000 offenders whom Augustus handled personally, only 10 reneged on their agreement to obey the conditions of behavior he set forth.

In 1878, after recognizing the positive results achieved by Augustus, Boston hired probation officers as extensions of the court. These officers' activities were monitored by the superintendent of police. Probation was implemented statewide in Massachusetts by 1880. Several other states soon followed suit, albeit with some variations. For example, some states implemented a more restrictive probationary concept that excluded people convicted of serious crimes such as murder, rape, and robbery; others administered probation through social service agencies. Chicago extended probation to juvenile offenders with the Juvenile Court Act of 1899, with police officers assuming the responsibilities of a probation officer. The federal courts authorized the use of probation and hired their first probation officers in 1925. By 1956, probation was available for adult offenders in every state.[5]

The Rationale Behind Probation

Much support for probation has been based on the belief that maintaining the offender in the community has a greater rehabilitative effect than incarceration, thereby reducing recidivism (the commission of a new offense). Advocates of probation argue that probation has the following advantages:

- It provides more individualized treatment or counseling than is available in prison.
- It allows offenders greater opportunities to deal with their problems.
- It avoids subjecting offenders to the negative effects of prison, such as inmate victimization, exposure to more serious criminal role models, loss of self-esteem, and inability to support a family.
- It is no more likely to lead to recidivism than is incarceration.
- It is less expensive than incarceration.

Although the research on most of these contentions is mixed, fairly strong evidence supports the notion that the cost of probation is less than the cost of incarceration. In Tennessee, for example, the annual cost of keeping an offender in prison is slightly less than $20,000, while the cost of supervising an offender on probation is only $956 per year.[6]

Probation Administration

Several different models of probation administration are currently in use, and approximately 2000 agencies administer probation services at the federal and state levels.

- *Federal supervision.* Probation supervision at the federal level is provided by the Division of Probation in the Administrative Office of the United States Courts. Federal probation officers are appointed by judges in the federal district courts.

- *State supervision.* Administration of probation on the state level takes various forms. In more than half of the states, probation is provided by a central, statewide probation office, under the executive branch through the state department of corrections or the judicial branch of the state government. In other states, probation and parole services are organized into a single agency administered by the judicial branch of government; officers are often assigned to county judges but are ultimately responsible to, and funded by, the state agency.

- *County supervision.* About 25 percent of the states administer probation at the county or local level. This system accounts for nearly two-thirds of all offenders supervised on probation in the states.[7] Local probation officers are generally appointed by the local judge and serve at will.

Proponents of state-based probation services suggest that the greater resources available at the state level (such as funding and treatment programs) make this form of probation a more efficient and effective means of rehabilitating offenders. In addition, state-based probation provides greater uniformity in terms of the administration and provision of services.

Proponents of county-based probation services argue that judges are able to work with officers they can trust and whom they can expect to provide presentence information consistent with their own sentencing philosophies. Proponents also believe that local probation officers are more familiar with the community and, therefore, are better equipped to implement modifications in probation programs than are employees in the governor's office.

Unfortunately, local probation agencies are noted for extreme variations in caseloads. In counties with small populations, caseloads may be limited to only 15 to 20 clients per officer. In contrast, in large urban counties, probation officers often must supervise between 120 and 200 clients. Given that the national average caseload is slightly more than 100 clients per officer, it is not surprising that supervision or counseling of probationers is often inadequate.

Regardless of the system used, one consequence of large caseloads and scarce resources in many communities has been the establishment of volunteer probation programs designed to supplement local probation agencies. Although paid probation officers replaced volunteers in the late nineteenth century, increasing caseloads and limited resources have prompted a revival of the use of volunteers in recent decades. Today, hundreds of thousands of people are volunteer probation officers serving in more than 4000 probation departments around the country. For example, approximately 500 volunteers currently assist the Orange County, California, probation department.[8] Volunteers can supplement paid officers in counseling, assist in job placement, provide transportation for offenders, and perform clerical tasks, thereby freeing up the regular staff for more intensive or critical tasks that require special training or expertise.[9]

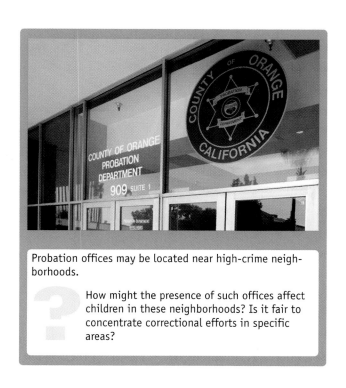

Probation offices may be located near high-crime neighborhoods.

? How might the presence of such offices affect children in these neighborhoods? Is it fair to concentrate correctional efforts in specific areas?

Functions of Probation Officers

Probation officers play an important role in the criminal justice process and have many responsibilities, including the following:

- Investigation and preparation of presentence investigation (PSI) reports
- Intake interviews with offenders to determine the potential for informal disposition of the case
- Diagnostic testing and interviews with offenders to assist in developing a treatment plan
- Risk classification of offenders to determine the type and level of supervision to be required
- Recommendation and assignment of offenders for participation in community treatment programs
- Supervision of offenders in the field
- Meeting with probationers on a regular basis (typically the most time-consuming part of a probation officer's work, as discussed later in this chapter)

In large agencies with many officers, the investigation function may be carried out by a group of officers who interview offenders, check background sources, assess the offender's potential for completion of community supervision or need for institutional placement, and prepare the PSI report for the sentencing judge. Other officers may be assigned to groups responsible for intake tasks, diagnostic testing, supervising offenders in the community, or conducting counseling sessions. In smaller agencies, officers typically perform all of these functions.

LINK As noted in Chapter 12, the presentence investigation is a valuable tool for judges as they seek to determine the most appropriate sentences.

Probation officers are usually assigned to conduct PSIs because they will be responsible for supervising the offender if he or she is placed on probation. In addition to conducting interviews with the offender and his or her family, the officer may conduct telephone interviews with schools, employers, neighbors, and sometimes military personnel in an effort to provide full and accurate information on the offender's background and current situation. The information is then prepared as a final report, with copies being given to the judge, prosecution, and defense before sentencing.

Judges often ask probation officers to include a sentencing recommendation, based on their evaluation of the offender. Other judges prefer to have only the basic report with no recommended sentence; they see the sentencing decision as their exclusive domain. Probation officers also evaluate the offender's need and potential for successful treatment in an institution or in the community. If the probation officer suggests that probation and supervision in the community would be appropriate in a particular case, he or she may also recommend a specific treatment program or programs, as well as the conditions for supervision in the community.[10]

In many jurisdictions, probation officers have taken on the added function of risk classification, which assesses the likelihood that an offender poses a continuing risk of reoffending and determines the necessary degree of probation supervision. An offender's risk score is based on a number of factors, including the seriousness of the current offense and the offender's criminal history, education, drug use, age, and employment history. The offender is then assigned to a supervisory level, ranging from intensive probation supervision, which may require as many as 20 to 30 contacts with the probation officer per month, to less frequent contact requiring only a minimum of one contact every three months.[11] Thus offenders who are convicted of more serious crimes and who present greater risks are subject to greater supervision. The philosophy behind risk classification combines the utilitarian model of sentencing, whereby offenders receive a sanction appropriate to the seriousness of their offense, with the commonly held belief that the community deserves to be protected from dangerous offenders.[12]

Eligibility for Probation

Not all offenders are eligible for probation, owing to states' imposition of mandatory sentences and sentencing guidelines. In states without such laws or guidelines, judges generally consider a number of factors when evaluating the eligibility of an offender for probation:

- The nature and seriousness of the current offense
- Whether a weapon was used and the degree of physical or emotional injury, if any, to the victim
- Whether the victim was an active or passive participant
- The length and seriousness of the offender's prior record
- The offender's previous success or failure on probation

While official misconduct is more widely known in the arena of police behavior, its occurrence among probation officers is not unknown. In 2003, a Havre, Montana, probation officer was found guilty of criminal misconduct for engaging in a consensual sexual relationship with one of his probationers. The woman had stated that she consented to have sex with the officer because "he had significant power over her" and could have recommended revocation of her probation for violating its conditions. Although Montana law forbade corrections officers to engage in consensual sexual activity with inmates, it did not specifically prohibit consensual sex between probation officers and probationers. The officer was found guilty. Following his appeal of the verdict, the Montana Supreme Court threw out the conviction, holding that while the officer's conduct was "reprehensible and op-portunistic," it did not amount to official misconduct.

In 2006, a Dakota County, Minnesota, probation officer was charged with having consensual sex on multiple occasions with one of his female probationers. The victim told police she had had sex with the officer several times while under supervision in 2005–2006. Police said that a number of witnesses saw the officer buying the woman drinks at various bars in St. Paul—in direct violation of her conditions of probation. The woman's daughters also told police that the officer visited their mother at her apartment, sometimes late at night. Unlike Montana, Minnesota law specifically prohibits sexual contact between probation officers and probationers.

Hierarchical relationships exist in a wide variety of settings (for example, businesses, universities, and the military) in which one person has au-thority over another and always have the potential for abuse. When the relationship is entered into by choice of the two parties, it is less problematic. When the two people are legally required to be in the relationship, however, the abuse of power and authority becomes a significant issue. Even the hint of coercion or the possibility that negative consequences could occur if the client does not agree to the relationship casts is inappropriate in any part of the criminal justice system.

Sources: Frederick Melo, "Probation Officer Accused of Sexual Misconduct" *Pioneer Press* (August 9, 2006), available at http://www.twin cities.com/mld/twincities/15228887.htm, accessed August 10, 2006; Jim Adams, "Dakota County Probation Officer Charged with Sexual Misconduct," *Star Tribune* (August 8, 2006), available at http://www.startribune.com/462/story/602433.html, accessed August 10, 2006; Sarah Cooke, "Misconduct Conviction Reversed" *Helena Independent Record* (December 30, 2005), available at http://www.helenair.com/articles/2005/12/30/montana/c01123005_01.prt, accessed June 5, 2007.

- The offender's prior incarcerations and success or failure on parole

Probation is not reserved for people who have been convicted of misdemeanors or minor felonies. Indeed, nearly 20 percent of persons convicted of serious violent offenses in the United States receive probation. Although only 2 percent of those convicted of murder are placed on probation, this sentence is imposed on 17 percent of all offenders convicted of rape and 23 percent of those convicted of assault FIGURE 15-2 .

LINK Some crimes carry mandatory prison sentences, as noted in Chapter 12. For example, in many states, offenders who are convicted of a second (or third) felony or of certain violent crimes or drug offenses are ineligible for probation.

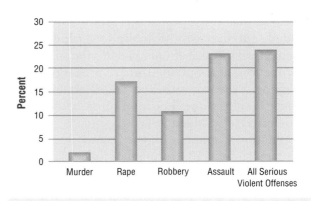

FIGURE 15-2 Percentage of Violent Felons Convicted in Large Urban Counties Receiving Probation, by Offense

Source: Brian Reaves, *Violent Felons in Large Urban Counties* (Washington, DC: U.S. Department of Justice, 2006).

Types of Probation Sentences

Some offenders eligible for probation may receive a sentence combining incarceration and probation. This typically occurs in cases in which the sentencing judge believes that the offender will be positively affected by a short period of incarceration as well as community supervision. Such sentence combinations usually take one of four forms:

- Split sentence. The court requires that the offender serve a specific period of incarceration, followed by a period of probation, with the balance of the sentence being suspended.

- Modification of sentence. An inmate may appeal to the original sentencing court for reconsideration of his or her prison sentence. Such an appeal must be made within a limited period of time. The judge may then choose to modify the original sentence and place the offender on probation.

- Shock incarceration. An offender sentenced to prison may be released after a short period of time (usually 30 to 60 days) and resentenced to probation. The brief stay in prison is believed to provide the "shock" necessary to convince an offender that prison is not a desirable place to be.

- Intermittent incarceration. An offender on probation may be required to spend a series of weekends or nights in a local jail. This is typically done when particular characteristics of the offender (such as employment or family responsibilities) or the offense (such as absence of injury) balance out the need for full-time incarceration.

About half of all probation cases involve a suspended sentence with incarceration, which is imposed only if the probationer violates the conditions of probation. In about one-third of probation cases, probation is the sole sentence; no prison term is attached. Finally, a small portion of probation cases are deferred (also called informal probation) before a conviction is entered on the record, and the offender must undergo treatment or follow strict conditions for a specified period; the prosecutor may then decline further prosecution, and the case can be dismissed.

Conditions of Probation

Probation is essentially a contract between the offender and the court, and any violations of the conditions could potentially result in revocation of probation. Most state and federal courts have established fairly uniform conditions for probation. In some states, these conditions have been created by the legislature and written into law. The conditions are established to fulfill two broad purposes:

- Treatment or rehabilitation of the offender
- Supervision to ensure law-abiding behavior

The specific conditions of probation usually depend on which purpose the probation agency emphasizes and are typically influenced by the supervisory styles of the Chief Probation Officer and his or her staff. Treatment-oriented agencies, for example, often require that the probationer attend school, undergo psychiatric treatment, or participate in drug or alcohol counseling. By contrast, agencies emphasizing control are more likely to stress rule compliance through regular and prompt reporting to the probation officer, restrictions on the offender's geographic mobility, or prohibitions against associating with particular types of people, such as ex-offenders and drug users.[13]

Most probation agencies also require probationers to pay monthly fees to help offset the expense of supervision. While these fees vary from jurisdiction to jurisdiction, a $50 per month minimum probation fee is typical.[14] However, the federal courts have ruled that lower courts may not issue any fines to help offset these costs.[15] All of the conditions of probation must be related to the purposes of probation; any imposed conditions that are substantially more severe than necessary to carry out the purposes of rehabilitation or restitution are impermissible.

The courts have also ruled on the permissibility of another condition of probation—reasonable searches. Before 2001, some states and appeals courts interpreted such searches as being limited to those focusing only on probationary issues rather than law enforcement concerns (which would require probable cause). However, the U.S. Supreme Court has recently held that reasonable suspicion is sufficient when conducting searches where submission to a search is part of the probationer order.

Supervision of Probation

The supervision of offenders in the community has always been at the heart of probation. Of course, the role of the probation officer in handling the offender in the community has changed substantially since John Augustus began the practice. Following the lead of Augustus, early probation officers took a casework

Rules of the Superior Court of the State of New Hampshire: Terms and Conditions of Probation

The terms and conditions of probation, unless otherwise prescribed, shall be as follows:

The probationer shall:

(a) Report to the probation or parole officer at such times and places as directed, comply with the probation or parole officer's instructions, and respond truthfully to all inquiries from the probation or parole officer;

(b) Comply with all orders of the Court, board of parole, or probation or parole officer, including any order for the payment of money;

(c) Obtain the probation or parole officer's permission before changing residence or employment or traveling out of State;

(d) Notify the probation or parole officer immediately of any arrest, summons, or questioning by a law enforcement officer;

(e) Diligently seek and maintain lawful employment, notify probationer's employer of his or her legal status, and support dependents to the best of his or her ability;

(f) Not receive, possess, control, or transport any weapon, explosive, or firearm, or simulated weapon, explosive, or firearm;

(g) Be of good conduct, obey all laws, and be arrest-free;

(h) Submit to reasonable searches of his or her person, property and possessions as requested by the probation or parole officer and permit the probation or parole officer to visit his or her residence at reasonable times for the purpose of examination and inspection in the enforcement of the conditions of probation or parole;

(i) Not associate with any person having a criminal record or with other individuals as directed by the probation or parole officer unless specifically authorized to do so by the probation or parole officer;

(j) Not indulge in the illegal use, sale, possession, distribution, or transportation, or be in the presence, of controlled drugs, or use alcoholic beverages to excess;

(k) Agree to waive extradition to the State of New Hampshire from any State in the United States or any other place and agree to return to New Hampshire if directed by the probation or parole officer; and

(l) Comply with such of the following, or any other, special conditions as may be imposed by the Court, the parole board or the probation or parole officer:

(1) Participate regularly in Alcoholics Anonymous to the satisfaction of the probation or parole officer;

(2) Secure written permission from the probation or parole officer prior to purchasing and/or operating a motor vehicle;

(3) Participate in and satisfactorily complete a specific designated program;

(4) Enroll and participate in mental health counseling on a regular basis to the satisfaction of the probation or parole officer;

(5) Not be in the unsupervised company of minors of one or the other sex at any time;

(6) Not leave the county without permission of the probation or parole officer;

(7) Refrain totally from the use of alcoholic beverages;

(8) Submit to breath, blood, or urine testing for abuse substances at the direction of the probation or parole officer; and

(9) Comply with designated house arrest provisions.

Source: Rules of the Superior Court of the State of New Hampshire: Terms and Conditions of Probation, available at http://www.courts.state.nh.us/rules/sror/sror-h3-107.htm, accessed August 1, 2007.

approach; that is, they became deeply involved in offenders' lives in an attempt to guide them down morally and socially acceptable paths. By the 1920s, the probation officer's role had evolved into something similar to the role of a social worker. During this era, less emphasis was placed on moral behavior and more emphasis was devoted to treatment and counseling, as well as finding solutions to problems faced by the offender such as housing, education, and job placement.

Since the early 1980s, the supervision and treatment tasks of probation officers have become more complex, often requiring specialized skills beyond the ability of many officers. For example, the increasing number of offenders with drug problems has led many agencies to turn over the treatment of drug ad-

VOICE OF THE COURT

United States v. Knights

In 1998, Mark Knights was sentenced to probation for a drug offense. The probation order that Knights signed included the following condition: that Knights would submit his "person, property, place of residence, vehicle, and personal effects to search at any time, with or without a search warrant, warrant of arrest, or reasonable cause by any probation officer or law enforcement officer."

Three days after Knights was placed on probation, law enforcement officials suspected Knights's involvement in acts of arson and vandalism. A Pacific Gas & Electric (PG&E) power transformer and an adjacent telecommunications vault were pried open and set on fire in a manner similar to more than 30 other recent acts of vandalism against PG&E. Previously, PG&E had filed a complaint of theft of services against Knights, and many of acts of vandalism against the company had coincided with Knights's court appearances.

While surveying Knights's residence, a sheriff's deputy noticed a truck in the driveway with suspicious materials, including a Molotov cocktail, explosive materials, a gasoline can, and brass padlocks that fit the description of those removed from a PG&E transformer vault. Aware of the search conditions in Knights's probation order, the deputy believed that a search warrant was not necessary and decided to conduct a search of Knights' apartment. He found a detonation cord, ammunition, liquid chemicals, bolt cutters, telephone pole climbing spurs, and a brass padlock stamped "PG&E." Knights was arrested and indicted by a federal grand jury.

The lawyer for Knights moved to suppress the evidence obtained during the search of his apartment, and the District Court granted the motion on the ground that the search was for "investigative" rather than "probationary" purposes. In its appeal to the U.S. Supreme Court, the government claimed that the search conditions in Knights's probation order did not mention anything about purpose, and that Knights had voluntarily accepted the search condition so as to be placed on probation as an alternative to going to prison. The Court agreed that, as an inherent part of probation, "probationers do not enjoy the absolute liberty to which every citizen is entitled." The Court also stated that

the very assumption of the institution of probation is that the probationer is more likely than the ordinary citizen to violate the law. The recidivism rate of probationers is significantly higher than the general crime rate . . . [a]nd probationers have even more of an incentive to conceal their criminal activities and quickly dispose of incriminating evidence than the ordinary criminal because probationers are aware that they may be subject to supervision and face revocation of probation, and possibly incarceration.

The Court held that only reasonable suspicion—not probable cause—was required to search a probationer's house:

When an officer has a reasonable suspicion that a probationer subject to a search condition is engaged in criminal activity, there is enough likelihood that criminal conduct is occurring that an intrusion on the probationer's significantly diminished privacy interests is reasonable.

Source: U.S. v. Knights, 534 U.S. 112 (2001).

dicts to private treatment centers staffed by people with greater expertise in drug rehabilitation.[16] In most probation agencies, however, officers spend the majority of their time in basic supervision, making sure that offenders abstain from drugs and alcohol, seek work or maintain employment, attend counseling sessions, and meet any other conditions of probation.

Probation supervision has historically been organized around caseloads. Thus, as crime rates and prison populations increased from the 1960s to the mid-1990s, so, too, did probation caseloads. Once caseloads in many jurisdictions reached 150 to 200 probationers per officer, experts began to question whether officers were really able to supervise their clients. Many believed that smaller caseloads were necessary to have a positive impact on offenders. Others suggested that supervision styles were the most important factor in determining officers' effectiveness. Still others argued that the whole focus on caseload was misplaced; instead of cases, probation departments should focus on "places"—that is, the communities in which probationers lived.

In an attempt to reduce caseloads, many departments hired more officers or developed probation volunteer programs. The increase in crime rates and growing dissatisfaction with the traditional social work orientation of probation that had dominated the field until the mid-1970s eventually brought about a significant change in supervision styles, with officers shifting from a service orientation to an enforcer role.[17] The model of probation officer as enforcer meant that greater attention was given to probationers' compliance with their conditions of probation as well as the application of more external controls, including drug testing, use of curfews, and increased face-to-face contacts. The demise of the social work model meant that less emphasis was placed on providing services or coordinating services and treatment with community agencies.[18]

The idea that probation officers should shift their attention from cases to places reflects the rediscovery by social scientists of the importance of neighborhood and community as critical in understanding and intervening in the lives of probationers. Because certain neighborhoods have high crime rates, which in turn leads to a large number of probationers and parolees living in those neighborhoods, it might be smart to open probation offices in the neighborhoods where most of these offenders live and develop partnerships with the local community. This would allow probation officers to concentrate their efforts on integrating probationers into community projects, such as renovating substandard housing, helping them to buy their own homes, and working with families of probationers or those with members in prison.[19]

Revocation of Probation

An offender's probationary status may be revoked if he or she commits a new crime or violates the conditions of probation (known as a technical violation). If the probationer is arrested for a new crime, the police notify the probation officer so that the probation office may take action. Violations of conditions of probation, by contrast, usually come to the direct attention of the probation officer in the course of supervision.

In either case, the probation officer has wide discretion in making the decision to initiate probation revocation proceedings. If the new crime is a relatively minor one or if the violation of conditions is deemed insignificant, the officer may choose to warn

the probationer. If the offense or violation is serious, however, the officer files a report and places the case on the court calendar for a preliminary hearing before a judge. The probationer is notified of the alleged violation and summoned to appear at the preliminary hearing.

The preliminary revocation hearing is similar to the preliminary hearing in a criminal prosecution, with a few important exceptions. First, testimony may include hearsay evidence, and the probationer cannot invoke his or her Fifth Amendment right against self-incrimination. Second, the standard of proof for determining that the probationer is guilty of a violation is simply a preponderance of the evidence (i.e., the greater weight of the evidence suggests guilt), rather than proof beyond a reasonable doubt. If no probable cause is established, the probationer is returned to probationary status. If the judge determines that there is probable cause that a violation occurred, a revocation hearing is scheduled.

After the preliminary hearing, the revocation hearing is held to determine whether probation should be revoked. Minor violations may result in a warning to the probationer and the imposition of additional conditions. If the court determines the violation to be serious, the offender's originally deferred or suspended incarceration sentence may be invoked. If no incarceration sentence was imposed at the original criminal trial, the judge may determine an appropriate prison or jail sentence during the revocation hearing.

Before 1967, probationers were not guaranteed a revocation hearing under the Constitution. Instead, they usually were returned to court for a brief hearing in which the judge could revoke probation and impose a prison sentence. In 1967, the Supreme Court held in *Mempa v. Rhay* that probationers were entitled to be represented by counsel at proceedings at which a sentence that had been deferred or suspended might be imposed.[20] As a result of this ruling, the revocation hearing came into being.

Six years later, in 1973, the Supreme Court further clarified the offender's due process rights in *Gagnon v. Scarpelli*.[21] The Court held that probationers were entitled to the following:

- Notice of the alleged violation
- A preliminary hearing to determine probable cause
- The right to present evidence
- The right to confront adverse witnesses
- An independent decision maker (not necessarily a judicial officer)

- A written report of the hearing
- A final revocation hearing

The right to be represented by counsel, however, was not held to be constitutionally guaranteed. Although the Court considered the participation of counsel undesirable (because offenders could use counsel to engage in protracted legal maneuvering) and constitutionally unnecessary at most revocation hearings (because of the straightforward nature of the process), it did foresee that fundamental fairness demands that "certain cases [will] require that the State provide at its expense counsel for indigent probationers."[22] Therefore, the need for counsel is determined on a case-by-case basis.

Effectiveness of Probation

The majority of probation cases are successful in terms of probationers getting through their period of probation without a new offense or technical violation. In 2005, 59 percent of the more than 2.2 million adults released from probation supervision in the United States successfully met the conditions of their supervision. Approximately 16 percent were discharged from supervision because of incarceration for a new offense or rule violation, 13 percent had their probation revoked without incarceration, slightly less than 3 percent absconded from supervision, and a small number were discharged to the custody of another jurisdiction.[23]

While the majority of probationers meet the conditions of their probation, a number of studies report that probation may not be significantly more effective than incarceration in preventing offenders from committing new crimes. Joan Petersilia and her colleagues tracked 1700 felony probationers for 40 months and found that within approximately three and a half years, nearly 66 percent had been rearrested, with about one-third eventually being sentenced to jail or prison.[24]

Furthermore, probation supervision with smaller caseloads, including intensive probation, proved no more effective in reducing recidivism than supervision with much larger caseloads. Ed Latessa and his colleagues report that probationers assigned to intensive probation supervision caseloads were just as likely to be rearrested and were no more likely to participate in treatment services offered by the probation department than were probationers assigned as part of regular caseloads.[25]

While not all probationers who fail to comply with the conditions of their probation or who are ar-

Probation revocation hearings are conducted with less rigorous due process rights for clients.

 Would the relaxed due process guidelines of probation revocation hearings be useful in trials? Would it be beneficial for the entire judicial system to operate on the assumption that defendants are guilty?

rested and convicted on new charges are sent to prison, the probation failure rate does affect the prison population in the United States. The incarceration rates of 20 to 40 percent for probationers with new arrests account for a significant portion of new prison admissions each year. In addition to probationers with new arrests entering prison, approximately 6 percent of state prisoners are currently in prison for technical violations of their probation.[26]

Intermediate Sanctions

The use of <u>intermediate sanctions</u> as alternatives to exclusively probation and prison has become increasingly attractive to policymakers because such sanctions are considered more severe than probation but less expensive than long-term incarceration. Intermediate sanctions include the following options:

- Intensive probation supervision
- House arrest and electronic monitoring
- Day reporting centers
- Fines
- Forfeitures
- Restitution
- Community service

Frequently, multiple sanctions are applied in combination. For example, offenders may be placed under house arrest and also required to pay fines or victim restitution.

The ability of states to provide reasonable combinations of punishment and supervision of offenders at lower costs appeals to many people. For example, a study by the Edna McConnell Clark Foundation asked hundreds of Alabama residents how they would sentence 20 convicted offenders. Nearly every respondent initially thought prison was most appropriate. After they were informed of the costs and the available alternatives, the same people "resentenced" most of the cases to intermediate sanctions.[27]

Intensive Probation Supervision

Beginning in the 1960s, many states began to experiment with intensive probation supervision (IPS) programs, in which probation officers were assigned very small caseloads (a limit of 25 cases per officer, compared to the normal 100 or more cases assigned). It was believed that IPS would increase the amount of supervision each probationer received and that the increased supervision would, in turn, lead to more positive probation outcomes, including lower recidivism rates. More broadly, IPS was seen as a way to achieve the following goals:

- Reduction in prison crowding
- Increased public protection
- Greater cost-effectiveness (owing to reduced incidents of incarceration)
- Punishment (intensive surveillance and control function as sanctions)
- Rehabilitation of the offender
- Reduction in recidivism[28]

In IPS, the probation officer's primary role is to monitor and enforce probation conditions. Some IPS programs require 15, 20, or even 30 contacts per month, compared with only three or four contacts in regular probation supervision. In reality, the average number of contacts in IPS programs is just under six per month.[29] Interestingly, smaller caseloads have not always resulted in an increase in the actual amount of supervision. An evaluation of 14 IPS programs concluded that IPS contacts of any type amounted, on average, to a total of less than 2 hours per month per offender (assuming that 20 minutes was spent per face-to-face contact). The same is true of drug testing—the average for all sites was just over two tests per month.[30]

Although IPS is used in all 50 states, its effectiveness has been questioned. Most evaluations report somewhat lower rates of new arrests but much higher rates of technical violations for IPS caseloads than for regular probationers.[31] Technical violations by offenders are more likely to be brought to the attention of supervising officers, largely as a result of closer surveillance. Of course, the primary goal of IPS is to reduce new offenses, not to catch technical violators. In addition to not being significantly more effective than traditional probation supervision, IPS is clearly more costly as a result of the small caseloads, albeit not as costly as incarceration.

House Arrest

House arrest is a sentence in which the offender is legally ordered to remain in his or her home. While under house arrest, offenders may not leave their homes except to go to work or school or to obtain medical treatment. Offenders are usually required to be in their homes during the evening and on weekends, although some may be allowed to leave their residences to perform community service.

House arrest permits punishment while offenders stay integrated into society.

 Should states convert inmates incarcerated in minimum-security prisons to clients on house arrest? Would this move save the states and federal government a tremendous amount of money while posing little risk to public safety?

House arrest has a number of advantages over incarceration. First, it is significantly less expensive than incarceration; the daily cost of monitoring an offender under house arrest ranges from $11 to $15, compared to more than $50 per day for prison confinement.[32] Second, house arrest offers a number of social advantages, in that offenders are able to keep their jobs and avoid adverse effects on families, such as economic hardship and divorce. Third, house arrest is flexible: it can be imposed at almost any point in the criminal justice process (pretrial diversion, probation, or parole) and can be used to cover particular times of the day or control particular types of offenders.

Electronic Monitoring

For more serious offenders, an electronic monitoring system may be used in conjunction with house arrest. Such programs require the offender to wear an electronic bracelet attached to the wrist or ankle. The technology involved in electronic monitoring has become increasingly sophisticated in recent years, providing a range of services depending on the degree of monitoring required:[33]

- *Radio frequency (RF) tether.* A transmitter is worn on the ankle of the offender, and a receiver is placed in the offender's home. The receiver has a preset range limiting the distance the offender may move within during the time he or she is scheduled to be at home.

- *Global Positioning Satellite (GPS).* The GPS system uses the U.S. Department of Defense's Global Positioning Satellites and provides a real-time map reporting the offender's every move. This system also allows the monitoring agency to establish "hot zones" designating areas that the offender is not allowed to enter.

- *Visual alcohol monitoring.* A visual screen and Breathalyzer are placed in the offender's home. When a randomized computer call is received, the offender is instructed to blow into the Breathalyzer and transmit his or her picture to the monitoring agency. All missed calls and positive readings are logged and reported.

- *Secure continuous remote alcohol monitoring (SCRAM).* Offenders wear an ankle bracelet that uses transdermal alcohol testing to measure the amount of alcohol that migrates through the skin, thereby determining the offender's blood alcohol content. The device constantly monitors the blood alcohol level and transmits the informa-

tion to a receiver in the home, providing 24-hour monitoring by the agency.

- *Voice verification.* The offender receives random telephone calls during the times he or she is scheduled to be home. The offender is then prompted by the computer to repeat a series of numbers. This technology identifies the offender by matching each response to the voice template that was created during the initial enrollment in the program. All missed calls and failed sessions are logged and reported to the agency.

A study examining the effectiveness of electronic monitoring using RF and GPS systems was recently conducted in Florida. After reviewing data on 75,661 offenders who were placed on home confinement between 1998 and 2002, researchers reported that both monitoring systems significantly reduced the likelihood of technical violations, reoffending, and absconding for the offenders in the sample. Furthermore, the use of electronic monitoring allowed offenders who had committed more serious crimes to be placed on house arrest with no greater likelihood of violations or reoffending than offenders who had committed less serious crimes and were on house arrest without electronic monitoring.[34]

Electronic monitoring is not foolproof, however. For example, Martha Stewart served a five-month period of house arrest during which she used a RF monitoring system with an ankle bracelet. On one occasion she was photographed at a social event without the tether, and in an interview with *Vanity Fair* magazine, Stewart stated that she knew how to remove the tether without causing an alarm. "I watched them put it on," Stewart said. "You can figure out how to get it off. It's on the Internet. I looked it up."[35]

Some critics believe that there is a point beyond which monitoring becomes counterproductive; somewhere between 60 and 120 days, offenders reach the limits of their tolerance for such monitoring.[36] Longer monitoring periods may make offenders feel that they are being unreasonably restricted, thereby creating resentment and resistance to further supervision. Critics also argue that some offenders should be incarcerated and that home monitoring is too lenient a punishment for their crimes.[37] Additionally, some critics charge that the electronic monitoring system discriminates against the poor because offenders are often required to have jobs, telephones, fixed residences, and the ability to pay daily fees to be eligible for electronic monitoring supervision.[38]

moving toward, rather than away from, becoming a 'maximum-security society.'"[39] Lastly, some people are concerned about the social and psychological stresses of using the home as a surrogate jail and the reality that people under electronic surveillance become partners in their own monitoring.[40]

Day Reporting Centers

Day reporting centers, which were initially developed to help reduce overcrowding in jails and prisons, require offenders to spend all or part of each day at a designated reporting center. Offenders may live at home and go to work or attend school, but then must report to the center to meet with a counselor. They may also be required to participate in anger management or drug or alcohol treatment programs.

An evaluation of a day reporting center for repeat driving while intoxicated offenders in Maricopa County, Arizona, examined the effectiveness of the program as an alternative to longer incarceration in prison or jail followed by probation. The researchers concluded that offenders in the day reporting center program had essentially the same statewide conviction rate for alcohol-related traffic violations as a comparison group. More importantly, they noted that the costs of providing services through the day reporting center were significantly less than the equivalent costs for jail; it cost $36.79 per day per person to keep an offender incarcerated but only $19.69 per day for the day reporting center.[41]

Fines

U.S. courts have traditionally imposed fines on offenders, often in conjunction with other sentences. Although the payment of fines appears to have few rehabilitative advantages, it does expedite the flow of offenders through the criminal justice system. The advantages of traditional fines seem clear:

- Fines can be administered to a variety of offenders.
- Fines impose minimal costs on taxpayers.
- Fines serve as a form of retribution.

Of course, traditional fines also produce inequities between rich and poor offenders. If fines are set at a level great enough to affect the wealthy, they create a proportionately much greater burden on the poor.

Structured fines, sometimes called day fines, are designed to eliminate the proportionately greater financial burden placed on poorer offenders by tying the amount of the fine to the offender's ability to pay.

Electronic monitoring and other surveillance devices are an increasing part of correctional supervision.

? Should all correctional clients wear devices that provide constant surveillance? What are the pros and cons of such a policy?

The use of electronic monitoring also raises the fear in some that the United States may be headed toward the type of society described in George Orwell's book *1984,* wherein citizens' language and movement are strictly monitored as part of the government's tools of oppression. According to sociologists Ronald Corbett and Gary Marx, "We appear to be

Structured fines are calculated in a two-step process. A probation officer or other court official initially identifies the number of "fine units" determined by the seriousness of the offense and number of prior offenses. The number of fine units is then multiplied by a proportion of the offender's net daily income using a predetermined equation. Under this scheme, an offender convicted of a minor offense and with no prior offenses would pay a proportionally smaller fine than someone who had committed a more serious offense.[42]

Forfeiture

Forfeiture is a legal procedure that permits the government to seize property used in the commission of a crime. It is not a new type of punishment. Indeed, in English common law, forfeiture was routinely directed against people and things that offended the Crown. Convicted felons forfeited their personal property to the king, and their land reverted to their lord. Nevertheless, such forfeitures were invoked only after the offender's criminal liability was clearly established. The use of forfeiture was eventually incorporated into U.S. criminal law and typically focused on narcotics, gambling, taxation, and moonshine violations, although it was rarely enforced.[43]

The recent expansion of forfeiture laws in the United States is largely a result of the federal crackdown on drugs; it represents an attempt to strike more broadly at the resources of drug traffickers. Federal prosecutors vigorously use forfeiture laws against drug traffickers by seizing their planes, vessels, homes, and vehicles, as well as cash, bank accounts, jewelry, weapons, and other possessions. Such possessions have included a horse ranch, gas stations, flower shops, a recording studio, and a bank.[44] In 1996, the Supreme Court held that a state could require the forfeiture of a person's vehicle or home even if the owner was not involved in or aware of the use of the vehicle or home in conjunction with criminal activity.[45] In many states, real estate on which marijuana is grown has been seized. Proceeds from the sale of forfeitures generally go to the state or local treasury or are divided among various law enforcement agencies. In some jurisdictions, police agencies retain some or all of the funds to continue their fight against drug traffickers. For example, seized cash may be used to buy illegal drugs in sting operations and to rent or buy buildings used as fronts in such operations.[46]

Some limitations have been placed on the use of forfeiture. Because forfeiture is a form of punishment similar to a fine, the Supreme Court has held that the Excessive Fines Clause of the Eighth Amendment "limits the government's power to extract payments, whether in cash or in kind, as punishment for some offense."[47] However, the Court has refused to establish a test for determining whether a forfeiture is constitutionally "excessive," leaving that issue for the lower courts to consider.

Restitution

The federal government and most states provide for restitution, or payment of compensation by the offender to the victim or a victim-assistance program, as an intermediate sanction imposed regardless of whether the offender is incarcerated or placed on probation. Beginning in 1972, the Minnesota Restitution Program gave prisoners convicted of property offenses the opportunity to shorten their jail stays, or avoid prison time altogether, if they went to work and turned over part of their pay as restitution to their victims. Today, many restitution programs also include payment to victims for damages resulting from violent crimes.

Courts generally rely on one of three methods to determine restitution:

- *Judicial fiat* allows the judge to specify amount of restitution based on evidence provided to the court during a sentencing hearing.
- *Insurance claim* requires that the victim provide documentation of losses (such as medical bills, property damage estimates, or evidence of missed work) so the court can determine the appropriate amount of restitution.
- *Meetings between victims and offenders* involves bringing the victim and offender together to work out a restitution agreement that is satisfactory to both parties.[48]

LINK Victims are permitted to present arguments and evidence regarding appropriate restitution through a victim-impact statement, as discussed in Chapter 12.

Community Service

Through community service, an offender provides unpaid service to the larger community. The first community service program was formally initiated in 1966 in Alameda County, California, when judges began to sentence indigents convicted of traffic violations to community service rather than to jail for failure to pay their fines. This sentencing option has several benefits:

Community service combines the interests of criminal punishment and public service.

 Is community service especially beneficial for persons with little criminal history? Are middle-class people more likely to perform community service? If so, should this opportunity be exploited to improve the common good?

- It avoids the costs of incarceration.
- It spares the offenders' families undue hardship.
- It provides needed services to the community.[49]

Today, offenders can be sentenced to voluntary work for a specific number of hours, often in charitable organizations, city park maintenance, assisting the elderly, or hospital service work.[50] Offenders assigned to community service may never repay fully the damages they caused, although symbolic or token payments may be better than no payment at all. Supporters of community service programs argue that offenders who are sitting in jail, especially if they are indigent, are unlikely to repay their victims. Thus community service represents one way of compensating victims for their losses.[51]

Restorative Justice

The two dominant paradigms in criminal justice are rehabilitation and punishment. According to Lynn Urban and her colleagues, both of these paradigms are "closed systems" that focus attention only on the offender; both victims and members of the community are essentially ignored.[52] The paradigm of restorative justice, by contrast, focuses on the community and aims to change the way people think about crime.[53] In this model, crime is viewed as more than an individual violation of criminal law—it is seen as a violation of community and relationships. Unlike the rehabilitative or punitive models, restorative justice makes the fundamental assumption that crime is a harm committed against an individual victim (more similar to a civil wrong), which is a dramatic shift from thinking of crime as an offense against the larger community or the state. According to this perspective, the government should have a more limited role and victims should be placed at the center of the criminal justice process.

The Purpose of Restorative Justice

Restorative justice aims to restore or repair relationships that have been disrupted by crime. To achieve this goal, the system holds offenders accountable by requiring restitution to victims, helping offenders accept responsibility for their acts, and attending to the various needs of the victim and the community. In its ideal form, restorative justice approaches are less formal, adversarial, punitive, stigmatizing, and costly than traditional responses to crime.[54] According to Tony Marshall, restorative justice is a process whereby "all the parties with a stake in a particular offense come together to resolve collectively how to deal with the aftermath of the offense and its implications for the future."[55]

Restorative justice takes many forms:

- *Victim–offender mediation* brings victims and offenders together with a mediator so that both parties can relate what they believe happened and why it occurred and arrive at a mutually agreed-upon reparative agreement.
- *Family group conferencing* requires that an offender admit to having committed the offense and have a representative (usually a family member) attend a conference with a person representing the victim. At this meeting, the two representatives discuss the offense and its consequences and develop a plan of action to repair the harms.
- *Circle sentencing* requires that the victim and the offender meet face-to-face with representatives from local social service agencies, law enforcement, the courts, and the community to talk about how they feel about the offense and the offender. The parties involved also determine the best ways to respond to the offense and attend to the needs of the victim and community.

- *Reparative probation* involves a special hearing board made up of local citizens who question offenders and victims and deliberate (typically in private) to develop appropriate sentencing for nonviolent offenders.[56]

Effectiveness of Restorative Justice

Perhaps due to the diversity of restorative justice programs around the country, relatively little research has been conducted to test the effectiveness of such programs. Generally, research on restorative justice programs reports modest reductions in recidivism, although a few studies have found more significant effects. For example, Paul McCold and Benjamin Wachtel examined the effects of the Bethlehem Pennsylvania Police Family Group Conferencing Project. In their study, first-time juvenile offenders who had committed moderately serious crimes were randomly assigned either to formal adjudication or to a "restorative policing" family group conferencing program. McCold and Wachtel reported that violent offenders who participated in conferences had significantly lower rearrest rates than violent offenders who declined to participate, although the same was not true for property offenders. This research also found that recidivism was probably more a function of offenders' choice to participate than the actual effects of the conferencing.[57]

In a meta-analysis of restorative justice practices, Jeff Latimer and his colleagues examined 22 unique studies of restorative justice and found that both victims and offenders who participated in the programs reported higher rates of satisfaction than did comparison groups of offenders.[58] More than two-thirds of the studies reported greater success in reducing recidivism than nonrestorative approaches to dealing with offenders. However, most of the studies in the meta-analysis likely suffered from a significant self-selection bias similar to that suggested by McCold and Wachtel. Because restorative justice is a voluntary process, offenders and victims who participate in restorative programs are more likely to be motivated to achieve positive outcomes than offenders and victims in control groups.

Criticisms of Restorative Justice

Some restorative justice advocates admit that this approach does not clearly fix blame on the offender as the traditional criminal justice process does, given that the offender's blame or responsibility is only one

Restorative justice is a new paradigm where offender, victim, and community come together to resolve a problem.

? Is restorative justice likely to help offenders change their criminal tendencies and assist victims in overcoming their fear of crime?

of many elements considered.[59] In addition, much of the attraction of restorative justice lies in its claim that punitive correctional approaches do significantly more harm than good. Critics such as Sharon Levrant, however, contend that restorative justice does not offer any realistic set of policies for the control of serious crime or methods to reduce the high rates of recidivism by offenders who continue to commit serious crimes. Levrant offers the following criticisms:

- Restorative justice approaches do not provide needed due process protections and procedural safeguards for offenders.
- Offenders are often coerced into participating in restorative programs, because they may believe any refusal to do so will result in harsher punishments.
- Policies based on restorative justice may increase punishments for offenders by requiring that they submit to both reparative conditions and traditional probation supervision.
- The current organization of probation departments and their limited resources make it unrealistic to emphasize community reintegration and expand the role of probation officers.
- Unintended race and class bias may occur because more affluent offenders may be better able to mediate or negotiate favorable sanctions.
- Restorative justice programs target low-risk, nonviolent offenders with little likelihood of recidi-

vism and have not proven to be effective for offenders who commit more serious crimes.

- Restorative justice programs are unlikely to have any long-term impact on the fundamental causes of an offender's inclination to engage in crime.[60]

As a result of these concerns, Levrant believes that restorative justice has not yet proven its worth and may actually cause more harm than good. She claims that the attractiveness of restorative justice relates "more in its humanistic sentiments than [to] any empirical evidence of its effectiveness."

Other critics, such as Adam Crawford and Todd Clear, suggest that restorative justice advocates assume community consensus without defining the boundaries of community and have not identified clear goals of restoration.[61] In addition, Crawford and Clear question whether restorative efforts in communities characterized by value and normative conflict between formal authorities and local residents can truly bring about the changes required to reduce youth crime. Moreover, many communities are too heterogeneous in terms of class, race, and ethnicity to promote consensus, and many—even those with greater homogeneity—have few of the economic and organizational resources needed to successfully bring about the restoration of victims and the reintegration of offenders.[62]

Parole and Other Releases

Parole involves the conditional release of an offender from a correctional institution after the individual serves a portion of his or her sentence. Once on parole, the offender remains under community supervision and in the legal custody of the state; if the offender violates the conditions of parole, he or she may be reincarcerated.[63]

LINK Chapter 14 discusses the dramatic rise in incarcerations during the past decade as a result of rising crime rates and get-tough policies that have, in turn, led to a growing number of offenders being placed on parole.

In 2004, more than 760,000 offenders were on parole and being supervised in the community. Overall, the U.S. parole population grew by 20,230 persons in 2005, or 1.6 percent, or slightly more than the average annual increase of 1.4 percent since 1995; some states, however, saw increases in their parole populations as high as 23 percent or decreases as much as 17 percent.[64] Thirty-eight percent of the offenders on

parole in 2004 had been convicted of drug offenses; smaller percentages had committed property crimes and violent crimes. Eighty-eight percent of adults on parole in 2004 were male. Whites accounted for 40 percent of parolees, African Americans for 41 percent, and Latinos for 18 percent TABLE 15-2 .

The Emergence of Parole

Early release from prison originated in France nearly 200 years ago when the French offered war prisoners their freedom in exchange for their promise to refrain from combat. This arrangement allowed more French soldiers to fight in the war rather than guard prisoners. Released prisoners who broke their word and rejoined the enemy forces were subject to execution.

The use of parole in the United States actually preceded the concept of probation by several decades. Until the mid-nineteenth century, nearly all prison sentences had to be served in their entirety. With the development of the indeterminate sentence, prisoners began to be released into the community when

TABLE 15-2

Characteristics of Adults on Parole

Characteristic	Percent
Sex	
Male	88
Female	12
Race	
White	40
African American	41
Latino	18
Other	1
Type of Offense	
Violent	24
Property	26
Drug	38
Other	12
Sentence Length	
Less than 1 year	5
1 year or more	95
Status of Supervision	
Active	85
Inactive	3
Absconder	7
Supervised out of state	4
Other	1

Source: Lauren Glaze and Seri Palla, *Probation and Parole in the United States, 2004* (Washington, DC: U.S. Department of Justice, 2005), p. 9.

correctional staff determined they were ready.[65] By 1900, 20 states had established parole systems. By 1932, 44 states and the federal government were using parole as a method of early prison release. Today, all jurisdictions provide for some form of early release and community supervision.

In many states, determinate sentencing laws have resulted in requirements for releasing inmates at the completion of their sentences minus good time. Good time allows prisoners to accumulate days to be subtracted from their terms for good behavior, for attending academic or training programs, or for performing meritorious service. While at least 29 states have truth-in-sentencing laws (discussed in Chapter 12) governing when inmates are to be released, four states continue to use good time rules that permit many—if not most—inmates in their prisons to serve only 50 percent of their sentences. For example, in Indiana in 2005, only 26 of about 15,000 inmates released from prison had actually reached their sentence release date; the remainder had earned sufficient good time for early release.[66] While most inmates in New Mexico's prisons receive up to 30 days credit for every 30 days served, certain "serious violent offenders" and parole violators may earn only 4 days credit for every 30 days.[67]

Other Types of Release

Not all offenders who are released from prison are placed on parole. Full release from a correctional institution may be unconditional or discretionary. In addition, many inmates are able to take advantage of temporary releases to work or attend educational programs in the community.

Unconditional Release

Unconditional release (also called mandatory release) includes the release, without supervision, of inmates who have served their maximum sentences (minus credits for good time or time served in jail before conviction). It takes two forms: commutation and clemency.

Commutation is the decision by the governor or president to reduce the severity of an inmate's sentence to allow for immediate release.

Clemency (also called pardon) is an executive decision by the governor or president to set aside an offender's punishment or exempt the offender from the punishment, and to release the individual instead. A pardon is typically granted after conviction and sentencing in a felony case, usually after the offender has been incarcerated. However, the president has the

One of the executive rights of the President of the United States is the power to commute the sentences of offenders who do not appear to deserve imprisonment.

 Should presidents be permitted to commute the sentences of offenders whose charges are political in nature, such as President Bush did with regard to Lewis "Scooter" Libby in 2007?

authority to grant a pardon before the formal filing of charges, thereby barring prosecution of an individual for a specific offense. In a famous presidential pardon, Gerald Ford, in one of his first acts as president, pardoned his predecessor, Richard Nixon, for all crimes related to the Watergate scandal and any related crimes that might be uncovered at some future time.[68] While this case made headlines, presidential pardons are actually relatively common. For example, President Bill Clinton issued a total of 395 pardons—140 of them on his last day in office.[69]

Halfway House

The Federal Bureau of Prisons and a number of state correctional systems permit eligible inmates to be transferred to a halfway house, sometimes called a community corrections center or reentry center, at a date prior to the inmate's formal release from prison. These facilities provide secure housing of inmates to ensure the public's safety while providing inmates with an opportunity to take a controlled step into the community. While at a halfway house, an inmate may attend school, work, participate in a job training program, and receive counseling. If inmates fail to return to the halfway house at the designated time of day or arrive under the influence of drugs or alcohol, they may be returned to prison to serve out the remainder of their sentences.

clemency An executive decision by the governor or president to set aside an offender's punishment and release the individual.

community corrections A correctional approach based on the belief offenders can (and often should) be dealt with within the community rather than prison.

community service A sentence in which the offender makes reparation to the community.

commutation An executive order in which a prisoner's sentence is reduced to allow for his or her immediate release.

day reporting center A center where counselors meet with offenders who are permitted to remain within the community, live at home, and go to work or attend school.

educational release A prerelease allowing the inmate to leave the institution during the day to attend classes.

electronic monitoring A sentence requiring an offender to wear an electronic device to verify his or her location.

fine A sentence in which the offender makes a cash payment to the court.

forfeiture A legal procedure that permits the government to seize property used in the commission of a crime.

furlough A prerelease allowing the inmate to have a temporary home visit, usually lasting from 24 hours to one week.

Gagnon v. Scarpelli The Supreme Court decision identifying the due process rights of an offender during a probation revocation hearing.

gate pay A small sum of cash given to an inmate upon his or her release from prison.

halfway house Facilities designed to provide secure housing of inmates and to ensure the public's safety while providing inmates with an opportunity to take a controlled step into the community.

house arrest A sentence requiring that the offender be legally confined to his or her home.

informal probation Placement of an offender on probation before a conviction is entered on the record; upon completion of the probation conditions, the offender is released and the case dismissed.

intensive probation supervision (IPS) A practice under which probation officers with very small caseloads are able to provide increased supervision.

intensive supervision parole A practice under which caseloads are very small, and officers frequently meet face to face with parolees to provide greater supervision.

intermediate sanctions Sentences that may be imposed as alternatives to traditional probation or incarceration.

intermittent incarceration A short sentence involving a series of brief periods of time in jail.

Megan's Law A law that requires both registration and community notification by sex offenders when they move into a community.

modification of sentence An alteration of an original sentence based on an appeal to the court.

Morrissey v. Brewer The Supreme Court case that spelled out the due process rights of offenders on parole.

parole The conditional release of an offender from a correctional institution after serving a portion of his or her sentence.

parole board The group of people who make the discretionary decision about a prisoner's release.

preliminary revocation hearing The first hearing to determine whether there is probable cause that the offender violated the conditions of probation or parole.

presumptive parole date A presumed date of release using a calculation of scores based on an inmate's offense and background and designed to predict his or her likelihood of successfully completing parole.

probation A sentencing option typically involving a suspended prison sentence and supervision in the community.

reentry The process in which an inmate leaves prison and returns to society.

restitution A requirement of offenders to pay money or provide services to victims.

restorative justice A punishment philosophy aimed at restoring or repairing relationships disrupted by crime, holding offenders accountable, promoting offender competency and responsibility, and balancing the needs of the community, victim, and offender through involvement in the restorative process.

revocation hearing A hearing to determine whether probation or parole should be revoked.

risk classification An assessment of an offender's likelihood of committing a new offense if granted probation.

salient factor score A guideline used to help determine the potential risk in releasing an inmate early.

shock incarceration The release of an offender from prison after a brief period (usually 30 to 60 days), with subsequent probation.

split sentence A sentence in which an offender serves a period of incarceration and is then placed on probation, with the balance of the sentence being suspended.

structured fines Fines designed to eliminate the proportionately greater financial burden placed on poorer offenders by tying the amount of the fine to the offender's ability to pay.

suspended sentence A prison sentence that is set aside.

unconditional release A release from prison that does not require additional supervision in the community.

Notes

1. Tom Murton, "Inmate Self-Government," *University of San Francisco Law Review* 6:88 (1971).
2. Paul Hahn, *Community Based Corrections and the Criminal Justice System* (Santa Cruz, CA: Davis, 1975).
3. Lauren Glaze and Thomas Bonczar, *Probation and Parole in the United States, 2005* (Washington, DC: U.S. Department of Justice, 2006).
4. Glaze and Bonczar, note 3.
5. Task Force on Corrections, *Task Force Report: Corrections* (Washington, DC: U.S. Government Printing Office, 1966).
6. Board of Probation and Parole, *Annual Report FY 2004–05* (Nashville, TN: Board of Probation and Parole, 2005).
7. Randall Guynes, "Difficult Clients, Large Caseloads Plague Probation Parole Agencies," *Research in Action* (Washington, DC: National Institute of Justice, 1988).
8. "Volunteers in Probation (VIP)," Orange County, California: Probation Department (2006), available at http://www.oc.ca.gov/Probation/programs/vip.asp, accessed June 5, 2007; National Council on Crime and Delinquency, *Volunteers in Probation: A Report on the National Survey and Questionnaire Conducted by VIP-NCCD on the Volunteer Juvenile and Criminal Justice Movement in the United States* (Royal Oaks, MI: National Council on Crime and Delinquency, 1979).

9. Belinda McCarthy and Bernard McCarthy, Jr., *Community-Based Corrections,* 2nd edition (Pacific Grove, CA: Brooks/Cole, 1991), p. 376.

10. Charles Newman, "Concepts of Treatment in Probation and Parole Supervision," *Federal Probation* 25:11 (1961).

11. Todd Clear and Vincent O'Leary, *Controlling the Offender in the Community* (Lexington, MA: Lexington Books, 1983).

12. Todd Clear, "Punishment and Control in Community Supervision," in Clayton Hartjen and Edward Rhine (Eds.), *Correctional Theory and Practice* (Chicago: Nelson Hall, 1992), pp. 31–42.

13. McCarthy and McCarthy, note 9, p. 117.

14. "Adult Probation," Arizona Superior Court in Pima County (2006), available at http://www.sc.co.pima.az.us/apo/default.htm, accessed June 5, 2007.

15. *Higdon v. United States,* 627 F.2d 893 (9th Cir. 1980).

16. "Drug Treatment Role Increasing for Probation, Parole Agencies," *Criminal Justice Newsletter* 9:6 (1988).

17. Faye Taxman, "Supervision—Exploring the Dimensions of Effectiveness," *Federal Probation* 66:14–27 (2002).

18. Taxman, note 17.

19. Todd Clear, "Places Not Cases? Re-thinking the Probation Focus," *Howard Journal* 44:172–184 (2005).

20. *Mempa v. Rhay,* 389 U.S. 128 (1967).

21. *Gagnon v. Scarpelli,* 411 U.S. 778 (1973).

22. *Gagnon v. Scarpelli,* note 21.

23. Glaze and Bonczar, note 3, p. 6.

24. Joan Petersilia, Susan Turner, James Kahan, and Joyce Peterson, *Granting Felons Probation: Public Risks and Alternatives* (Santa Monica, CA: Rand, 1985).

25. Edward Latessa, Lawrence Travis, Betsy Fulton, and Amy Stichman, *Evaluating the Prototypical ISP: Final Report* (Cincinnati, OH: University of Cincinnati and American Probation and Parole Association, 1998).

26. Stephen Cox, Kathleen Bantley, and Thomas Roscoe, "Evaluation of the Court Support Services Division's Probation Transition Program and Technical Violation Unit" (December 2005), available at http://www.jud.state.ct.us/external/news/ProbPilot.pdf, accessed June 5, 2007.

27. Michael Castle, *Alternative Sentencing: Selling It to the Public* (Washington, DC: U.S. Department of Justice, 1991), p. 2.

28. James Byrne, "The Future of Intensive Probation Supervision and the New Intermediate Sanctions." *Crime & Delinquency* 36:14 (1990); Todd Clear and Patricia Hardyman, "The New Intensive Supervision Movement." *Crime & Delinquency* 36:47 (1990).

29. Joan Petersilia and Susan Turner, *Evaluating Intensive Supervision Probation/Parole: Results of a Nationwide Experiment* (Washington, DC: U.S. Department of Justice, 1993).

30. Joan Petersilia and Susan Turner, "Comparing Intensive and Regular Supervision for High-Risk Probationers: Early Results from an Experiment in California," *Crime & Delinquency* 36:105 (1990).

31. Elizabeth Deschenes, Susan Turner, and Joan Petersilia, *Intensive Community Supervision in Minnesota: A Dual Experiment in Prison Diversion and Enhanced Supervised Release* (Washington, DC: National Institute of Justice, 1995; Benjamin Steiner, "Treatment Retention: A Theory of Post-Release Supervision for the Substance Abusing Offender," *Federal Probation* 68:24–29 (2004); Linda Burrow, Jennifer Joseph, and John Whitehead, "A Comparison of Recidivism in Intensive Supervision and Regular Probation," paper presented at the annual meeting of the Academy of Criminal Justice Sciences (Washington, DC, April 3–7, 2001).

32. National Institute of Justice, *Keeping Track of Electronic Monitoring* (Washington, DC: National Law Enforcement and Corrections Technology Center, 1999); Rene Stutzman, "Ankle Monitors Show a Higher Rate of Success," *Orlando Sentinel,* December 29, 2002, p. A15.

33. House Arrest Services, information about electronic monitoring systems, available at http://www.housearrest.com/, accessed June 6, 2007.

34. Kathy Padgett, William Bales, and Thomas Blomberg, "Under Surveillance: An Empirical Test of the Effectiveness and Consequences of Electronic Monitoring," *Criminology and Public Policy* 5:61–92 (2006).

35. Associated Press, "Martha Says House Arrest Is 'Hideous'", *FoxNews.com* (July 5, 2005), available at http://www.foxnews.com/story/0,2933,161590,00.html, accessed June 5, 2007.

36. Marc Renzema and David Skelton, *Use of Electronic Monitoring in the United States: 1989 Update* (Washington, DC: U.S. Department of Justice, 1990), p. 4.

37. Annesley Schmidt, "Electronic Monitors—Realistically, What Can Be Expected?" *Federal Probation* 55:49 (1991).

38. Stutzman, note 32.

39. Ronald Corbett and Gary Marx, "Critique: No Soul in the New Machine: Technofallacies in the Electronic Monitoring Movement," *Justice Quarterly* 8:400 (1991).

40. Ronald Ball and Robert Lilly, "The Phenomenology of Privacy and the Power of Electronic Monitoring," in Joseph Scott and Travis Hirschi (Eds.), *Controversial Issues in Crime and*

Justice (Newbury Park, CA: Sage, 1988), p. 158; Corbett and Marx, note 39, p. 401.

41. Ralph Jones and John Lacey, *Evaluation of a Day Reporting Center for Repeat DWI Offenders* (Washington, DC: U.S. Department of Transportation, 1999).

42. Barry Mahoney, Joan Green, Judith Greene, and Julie Eigler, *How to Use Structured Fines (Day Fines) as an Intermediate Sanction* (Washington, DC: Bureau of Justice Assistance, 1996).

43. Bureau of Justice Assistance, *Asset Forfeiture: Starting Forfeiture Programs, a Prosecutors' Guide* (Washington, DC: U.S. Department of Justice, 1989), pp. 7–8.

44. Bureau of Justice Assistance, *Asset Forfeiture: The Management and Disposition of Seized Assets* (Washington, DC: U.S. Department of Justice, 1988), p. 2.

45. *Bennis v. Michigan,* 516 U.S. 442 (1996).

46. NIJ Reports, "Controlling Drug Abuse and Crime: A Research Update" (Washington, DC: U.S. Department of Justice, 1987), p. 3.

47. *Alexander v. United States,* 509 U.S. 544 (1993); *Austin v. United States,* 113 S.Ct. 2801 (1993).

48. Andrew Klein, *Alternative Sentencing* (Cincinnati, OH: Anderson, 1988), pp. 156–157.

49. Douglas MacDonald, *Restitution and Community* Service (Washington, DC: U.S. Department of Justice, 1988), p. l.

50. Michael Agopian, "Targeting Juvenile Gang Offenders for Community Service," *Community Services: International Journal of Family Care* 1:99–108 (1989).

51. Anthony Doob and Diane MacFarlane, *The Community Service Order for Youthful Offenders: Perceptions and Effects* (Toronto: Centre for Criminology, University of Toronto, 1984).

52. Lynn Urban, Jenna St. Cyr, and Scott Decker, "Goal Conflict in the Juvenile Court: The Evolution of Sentencing Practices in the United States," *Journal of Contemporary Criminal Justice* 19:454–479 (2003).

53. Susan Olson and Albert Dzur, "Revisiting Informal Justice: Restorative Justice and Democratic Professionalism," *Law and Society Review* 38:139–176 (2004); John Braithwaite, *Restorative Justice and Responsive Regulation* (New York: Oxford University Press, 2002); John Perry (Ed.), *Repairing Communities through Restorative Justice* (Lanham, MD:

American Correctional Association, 2002); Gerry Johnstone, *Restorative Justice: Ideas, Values, Debates* (Portland, OR: Willan, 2002); Daniel Van Ness and Karen Strong, *Restoring Justice,* 2nd edition (Cincinnati, OH: Anderson, 2001); Shay Bilchik, Gordon Bazemore, and Mark Umbreit, *Balanced and Restorative Justice for Juveniles: A Framework for Juvenile Justice in the 21st Century* (Washington, DC: Office of Juvenile Justice and Delinquency Prevention, 1997).

54. Urban et al., note 52, p. 467.

55. Tony Marshall, "The Evolution of Restorative Justice in Britain," *European Journal of Criminal Policy and Research* 4:21–42 (1996).

56. Braithwaite, note 53; Gordon Bazemore and Curt Griffiths, "Conferences, Circles, Boards, and Mediations: The 'New Wave' of Community Justice Decisionmaking," *Federal Probation* 61:25–37 (1997).

57. Paul McCold and Benjamin Wachtel, *Restorative Policing Experiment: The Bethlehem Pennsylvania Police Family Group Conferencing Project—Summary* (Pipersville, PA: Community Service Foundation, 1998).

58. Jeff Latimer, Craig Dowden, and Danielle Muise, "The Effectiveness of Restorative Justice Practices: A Meta-Analysis," *The Prison Journal* 85:127–144 (2005).

59. David Karp and Todd Clear, "Community Justice: A Conceptual Framework," National Institute of Justice, *Criminal Justice 2000,* vol. 2 (Washington, DC: National Institute of Justice, 2000).

60. Sharon Levrant, Francis Cullen, Betsy Fulton, and John Wozniak, "Reconsidering Restorative Justice: The Corruption of Benevolence Revisited?" *Crime & Delinquency* 45:3–27 (1999).

61. Clear, note 19.

62. Adam Crawford and Todd Clear, "Community Justice: Transforming Communities through Restorative Justice?" in Gordon Bazemore and Marie Schift, *Restorative Community Justice* (Cincinnati, OH: Anderson, 2001), pp. 127–149.

63. Wayne Morse, *The Attorney General's Survey of Release Procedures* (Washington, DC: U.S. Government Printing Office, 1939).

64. Glaze and Bonczar, note 3, p. 7.

65. Harry Barnes and Negley Teeters, *New Horizons in Criminology,* 3rd edition (Englewood Cliffs, NJ: Prentice Hall, 1959).

66. Niki Kelly, "Do the Crime . . . Do Only Half the Time," *Fort Wayne Journal Gazette* (May 28, 2006), available at http://www.fortwayne.com/mld/journalgazette/14685881.htm, accessed June 5, 2007.

67. *Time Served in New Mexico Prisons* (Albuquerque, NM: State of New Mexico Criminal and Juvenile Justice Coordinating Council, 2002).

68. Michael Parenti, *Democracy for the Few,* 5th edition (New York: St. Martin's Press, 1988), p. 149.

69. "Presidential Pardons," available at http://jurist.law.pitt.edu/pardons0a.htm, accessed June 5, 2007.

70. Howard Abadinsky, *Probation and Parole: Theory and Practice,* 8th edition (Upper Saddle River, NJ: Prentice Hall, 2003).

71. Donald MacDonald and Gerald Bala, *Follow Up Study Sample of Edgecombe Work Release Participants* (Albany, NY: Division of Correctional Services, 1985).

72. Christy Visher and Jeremy Travis, "Transitions from Prison to Community: Understanding Individual Pathways," *Annual Review of Sociology* 29:89–114 (2003).

73. Visher and Travis, note 72.

74. Maureen McLeod, "Getting Free: Victim Participation in Parole Board Decisions," *Criminal Justice* 4:12–15, 41–43 (1989).

75. Connecticut Board of Parole, *Statement of Organization and Procedures,* (Hartford, CT: Connecticut Board of Parole, 1974); Park Dietz, "Hypothetical Criteria for the Prediction of Individual Criminality," in Christopher Webster, Mark Ben-Aron, and Stephen Hucker (Eds.), *Dangerousness: Probability and Prediction, Psychiatry, and Public Policy* (Cambridge, UK: Cambridge University Press, 1985), p. 32.

76. "Sex Offender Guilty of Kidnapping, Murdering Student," available at http://www.cnn.com/2006/LAW/08/30/slain.student.ap/index.html, accessed June 5, 2007.

77. Hector Castro, "Sex Offender Registry Failing," *Seattle Post-Intelligencer Reporter* (January 8, 2003), available at http://seattlepi.nwsource.com/local/103212_register08.shtml, accessed June 5, 2007.

78. John Biewen, "Hard Time: Life After Prison, Gate Money by State" (March 2003), *American Radio Works,* available at http://americanradioworks.publicradio.org/features/hardtime/, accessed June 5, 2007.

79. James Gondles, "Returning to Society," *Corrections Today* 67:6–7 (2005).

80. John Irwin, *The Felon* (Englewood Cliffs, NJ: Prentice Hall, 1970).

81. Jean Jester, *The Technologies of Probation and Parole* (Unpublished Ph.D. dissertation, State University of New York at Albany, 1981).

82. Peter Finn and Sarah Kuck, *Stress Among Probation and Parole Officers and What Can Be Done about It* (Washington, DC: National Institute of Justice, 2005).

83. *Morrissey v. Brewer,* 408 U.S. 471 (1972).

84. Marc Lifsher, "Violators of Parole Remain Free," *Fresno Bee,* April 11, 1992, pp. A1, A14.

85. Glaze and Bonczar, note 3, p. 9.

86. Patrick Langan and David Levin, *Recidivism of Prisoners Released in 1994* (Washington, DC: U.S. Department of Justice, 2002).

87. Steven Raphael and Michael Stoll, "The Effect of Prison Releases on Regional Crime Rates." in William Gale and Janet Pack (Eds.), *Brookings-Wharton Papers on Urban Affairs 2004* (Washington, DC: Brookings Institution Press, 2004), pp. 207–255.

88. Amy Solomon, Avinash Bhati, and Vera Kachnowski, *Does Parole Work? Analyzing the Impact of Post-Prison Supervision and Recidivism* (Washington, DC: The Urban Institute, 2005).

89. Jeremy Travis, *But They All Come Back: Facing the Challenges of Prisoner Reentry* (Washington, DC: Urban Institute, 2005).

90. Visher and Travis, note 72.

91. James Wilson and Robert Davis, "Good Intentions Meet Hard Realities: An Evaluation of the Project Greenlight Reentry Program," *Criminology and Public Policy* 5:303–338 (2006).

92. Peter Jones, "The Risk of Recidivism: Evaluating the Public Safety Implications of a Community Corrections Program," *Journal of Criminal Justice* 19:49–66 (1991).

SECTION

5

Special Issues

This final section explores two issues of major concern to criminologists and the public alike: juvenile offenders and the emerging crimes of terrorism and cybercrime. These issues present the criminal justice system with new and unusual demands.

During most of the twentieth century, actions related to juvenile offenders typically came under the exclusive domain of the juvenile justice system. There, offenders were treated according to a more protective philosophy than they would have had they been cast into the adult system. Today, however, an increasing number of juveniles are being transferred into adult courts and, if convicted, face adult punishments. Chapter 16 examines the history of juvenile justice, explores characteristics of juvenile offenders, and outlines the juvenile justice process from arrest to disposition. It also investigates those cases in which juveniles are transferred to the adult criminal justice system and the demise of the juvenile death penalty.

Chapter 17 explores terrorism, cybercrime, and the unique problems they pose for the criminal justice system. This chapter examines recent legislation, such as the USA Patriot Act, and government efforts to combat terrorism, including the creation of the Department of Homeland Security. It also discusses the widespread problem of identity theft and the criminal justice system's response to the unique problems posed by high-tech crime.

CHAPTER 16

The Juvenile Justice System

CHAPTER 17

Terrorism and Cybercrime

OBJECTIVES

- Know the history of the U.S. juvenile justice system.
- Highlight the distinctions between delinquency and status-offending behaviors.
- Be familiar with police procedures for arresting juveniles and the effects of legal and extralegal factors when making discretionary arrests.
- Identify each major stage of the juvenile court process, from intake to adjudication and disposition.
- Explain the differences between probation, institutional placement, and aftercare of juvenile offenders.
- Understand the reasoning and methods of waiver for juveniles to the adult criminal justice system.
- Describe what happens to juveniles once they face prosecution in criminal court.
- Outline the history of the juvenile death penalty and explain why it was abolished.

The Juvenile Justice System

For some, Nathaniel Abraham is the "poster child" of a serious juvenile offender. He first used drugs at age 6. By age 9, he was using weapons to threaten other children. By the age of 10, Abraham had amassed a record of more than 20 police contacts for arson, assault, and other charges.

In October 1997, at the tender age of 11, Abraham killed an 18-year-old youth by shooting him with a rifle. For this crime, Abraham was prosecuted as an adult and convicted of second-degree murder. The judge, however, rejected adult punishment and instead sentenced Abraham to 8 years in juvenile corrections with mandatory release at age 21. Upon his release in 2007, Michigan Rehabilitation Services footed the bill for his apartment and college tuition. Today, Abraham is in the community, and commentators on juvenile justice are eager to see how he adapts to conventional society after being incarcerated for much of his life.

- Should Abraham have been prosecuted as an adult?
- What factors should influence this decision?

Source: Jennifer Chambers, "State Pays Abraham's Housing, College Tabs," *Detroit News,* January 19, 2007, available at http://detnews.com/apps/pbcs.dll/article?AID=/20070119/METRO/701190397, accessed March 20, 2007.

Introduction

In a 1964 statement, Federal Bureau of Investigation Director J. Edgar Hoover commented on what he saw as the arrogant attitude of youth toward the criminal justice system and noted that many delinquents believed that because they were minors, the full force of the law could not touch them.[1] Most often, they were right. Many arrested offenders who are juveniles typically defined as those under age 18, though this age varies in some states) go through a separate juvenile justice system in which they receive lenient treatment and, at most, brief confinement in a detention or juvenile correctional facility. Today, most juvenile offenders have very limited contact with the adult criminal justice system, except for temporary detention in adult jails.

In 2006, juveniles accounted for approximately 15 percent of all arrests FIGURE 16-1 .[2] While this figure may be a concern to the public, lawmakers, and justice system personnel, the overall juvenile arrest rate has declined by 40 percent in the past 10 years FIGURE 16-2 . In 2006, nearly 1.3 million juveniles were arrested, with more than 293,000 being charged with serious violent or property crimes.

Numerous studies suggest that most serious juvenile crime is committed by a relatively small number of youths—as few as 6 percent of juvenile offenders have been shown to be responsible for half of all juvenile crime and nearly 70 percent of serious crimes such as homicide, rape, robbery, and aggravated assault committed by juveniles.[3] Even though juvenile crime is a minor part of the larger crime problem, the fact that many juveniles commit very serious crimes—and the fact that some offenders commit such crimes frequently—is a reality that prompts us to explore how our society responds to juvenile crime, how the juvenile justice system operates, and how we should deal with the most serious of juvenile offenders. Many of these more serious offenders will grow up to be adult offenders who subsequently engage with the criminal justice system.

History of Juvenile Justice

Before the first juvenile court was created in Illinois in 1899, juveniles who violated the law were treated in much the same way as adults. Although the legal concept of juvenile delinquency (defined as behav-

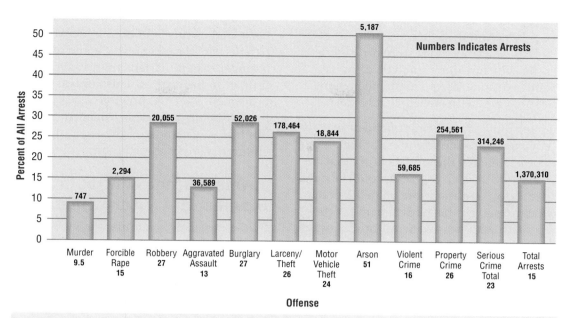

FIGURE 16-1 Arrests of Juveniles in the United States, 2006

Source: Federal Bureau of Investigation, *Crime in the United States, 2006* (Washington, DC: U.S. Department of Justice, 2007), Table 34.

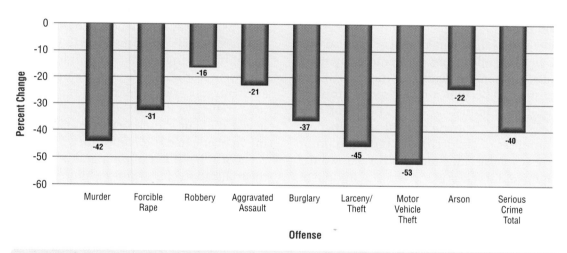

FIGURE 16-2 Percentage Change in Juvenile Arrests in the United States, 1997–2006

Source: Federal Bureau of Investigation, *Crime in the United States, 2006* (Washington, DC: U.S. Department of Justice, 2007), Table 32.

ior by a juvenile in violation of the juvenile or criminal codes) in the United States is only a little more than 100 years old, legal prohibitions against specific behaviors by juveniles have existed for centuries.

Origins of Juvenile Crime Control

One of the foundations of the modern legal system—biblical law—deals quite harshly with disobedient youth:

LINK The earliest legal code, the Code of Hammurabi (1700 B.C.), discussed in Chapter 2, set forth rules and appropriate punishments for disobedient children: If a son strikes his father, cut off his hands.

If a man have a stubborn or rebellious son, which will not obey the voice of his father, or the voice of his mother, and that, when they have chastened him, will not harken unto them. Then shall his father and mother lay hold on him, and bring him out unto the elders of his city, and unto the gate of his place; And they shall say unto the elders of his city. This our son is stubborn and rebellious, he will not obey our voice; he is a glutton, and a drunkard. And all the men of his city shall stone him with stones, that he die.[4]

Brutal punishment for juvenile offenders in early English law was also common. English King Aethelstan proclaimed in the tenth century that any thief who was more than 12 years old would be punished by death for stealing more than 12 pence, though he later reserved this punishment for those over the age of 16 who resisted arrest or ran away.[5] Even though early English law exempted children under age 12 from prosecution and punishment for criminal offenses, little distinction was made between older juveniles and adults. Juveniles age 16 or 17, who were perceived as having more substantial understanding and responsibility, were typically subject to the same punishments as adults.

Even in the Middle Ages, theft in England was considered a felony (as was murder) and carried the death penalty as an alternative to imprisonment or deportation to Australia. Most juvenile offenses involved some sort of theft, but violent crime also was common, and juveniles were frequently sentenced to prison, banishment, or death.[6] By the mid-1800s, juvenile crime had become a major social problem in England. In the industrial cities of London and Manchester, a greatly feared criminal class (including large numbers of children) was linked to problems of poverty, internal migration, and overcrowding. The emerging middle class perceived the children of the urban poor as being thieves or prostitutes, frequently employed by older criminals, often orphaned or deserted, and likely to end up in prison. Thus punishing the criminal behavior of juveniles was seen as important for preventing the spread of crime in general. Yet, English law exempted children under age 14 from criminal penalties on the assumption that they were not capable of criminal intent (*mens rea*). Even so, children under 10 were occasionally executed for crimes such as theft if they were considered to have sufficient mental maturity.[7]

In the early American colonies, it was not just the criminal activity of children that concerned colonists, but their inactivity as well. Sloth and idleness were considered sinful, corrupting influences to be prohibited by law. In 1646, the Virginia General Assembly passed legislation providing punishments for lazy or indolent children.[8] Massachusetts went one step further in 1672, requiring that "rude, stubborn, and unruly" children be separated from their parents and placed in foster homes, in the hope that stronger disciplinarians might save these children from criminal futures. This law was perhaps an improvement over the previous "stubborn child law" in place in Massachusetts, which, in accordance with biblical law, mandated the execution of stubborn children who disobeyed their parents.[9]

Juvenile Crime Control in Nineteenth-Century America

By the nineteenth century, the view of adolescents had shifted slightly. Children were viewed as corruptible, irritating, and arrogant, but they were also seen as innocent and vulnerable.[10] As such, adolescence was increasingly viewed as a unique period of life in which children needed thoughtful discipline and guidance. To ensure that children developed properly, society recognized that it must closely safeguard both children's moral and physical health in preparation for adulthood. Consequently, children needed special treatment and a period of lessened responsibility and demands.[11] To that end, new laws were created allowing the state to take responsibility for improving the lives of children from families who were seen as unable to provide appropriate guidance and supervision.[12] If parents could not or would not produce well-behaved children, then, it was believed, the state should step in to ensure proper training in obedience and conformity.

With the influx of immigrants and growing industrialization of U.S. cities, children of urban workers were often left unsupervised, largely because they no longer were able to participate in their parents' labor in the same manner as their counterparts who had resided on farms. Idleness and lack of supervision allowed many of these children to drift into immoral or criminal behavior. Concerned that these children would eventually become hardened adult criminals, reformers and philanthropists known as Child Savers sought to save them from their plight through state intervention.

Child Savers, like most other people of that time, believed in the basic goodness of children. Youthful criminality was perceived as being caused by the child's exposure to factors that made for a corrupting environment: poverty, idleness, overcrowded housing, absence of moral training, and lack of proper parental guidance. The Child Savers called for an expansion of the legal authority of the state to regulate the lives of children, including removing problem children from "bad" environments and placing them in rehabilitative milieus with a focus on constructive work programs and healthful surroundings with close supervision.[13]

Perhaps the earliest concrete expression of this desire to save children was the New York House of Refuge, opened in 1825 by the Society for the Prevention of Pauperism. The House of Refuge was designed to provide for neglected or vagrant children of the poor and immigrant members of the community. Administered almost like a prison, it required children to engage in industrial work and attend classes and subjected them to strict discipline. Children were required to march from one activity to the next, conforming to a rigid time schedule. Uncooperative or indolent children typically faced corporal punishment.[14] Children suffered much at the hands of these adults, whose mixture of hostility and benevolence produced a peculiar atmosphere.[15]

Other reformers, such as New York philanthropist Charles Loring Brace, also believed that such children posed a threat to the general well-being of society but felt that social agencies should find ways to remove wayward children from the evil and corrupting environment of the cities rather than placing them in cold, sterile, and punitive institutions. To accomplish this goal, Brace established the Children's Aid Society in 1853. The Society sought to place "unwanted" urban children in good homes in the countryside. Over the next 70 years, it ensured that nearly 250,000 children were sent west on trains and placed in foster homes or indentured as servants on farms, where it was hoped they would learn the value of hard work and the love of nature.

The effects of the "orphan trains" were decidedly mixed. There were complaints about how these children were treated, and thousands became drifters and thieves. Some ran away to return to the cities from which they were originally dispatched, whereas others lived lives of indentured servitude. Nevertheless, some found success in life, either as farmers, businesspeople, or even politicians. The Great Depression brought an end to this practice, as the need for additional laborers in the Midwest decreased.[16]

Creation of the Juvenile Justice System

Many reformers viewed the problem of juvenile crime as not being amenable to solutions that relied on

The New York House of Refuge was an institution inspired by the Child Savers' movement of 1825.

? Were houses of refuge simply glorified prisons? Why didn't the good intentions of the Child Savers' movement translate into effective correctional policy?

placing wayward children in institutions or transporting them out of cities. Instead, they believed that the United States needed to create a new juvenile legal code and court system specifically designed to handle children's cases, so that juvenile offenders could be supervised by the community without suffering severe punishment or stigma.

The first juvenile court in the United States was established in Cook County (Chicago), Illinois, in 1899. It was based on the English common law principle of *parens patriae* (literally "parents of the country"), which identified the state as the ultimate sovereign guardian of children and empowered it to act on behalf of the parents to protect the interests of children. This juvenile court was considered a civil—not criminal—court, reflecting the values and interests of the large number of social workers among the reformers. This perspective had several important implications:

- Juvenile offenders were not charged with crimes.
- Courts could impose controls without proving guilt beyond a reasonable doubt.
- Sentencing goals focused on treatment, not punishment.

In an effort to avoid the stigma associated with "criminal" behavior, violators of the juvenile code were considered "delinquent" and were placed in special institutions designed to house juveniles or supervised in the community by probation officers. Delinquency, according to the new code, included all behaviors considered crimes for adults as well as a variety of behaviors prohibited only to juveniles, known as <u>status offenses</u>. The latter included the following "undesirable" behaviors:

- Truancy from school
- Running away from home
- Curfew violations
- Incorrigible (habitually disobedient) behavior
- Use of alcoholic beverages
- Knowingly associating with thieves or other malicious or violent persons
- Vile, obscene, vulgar, or indecent language
- Indecent or immoral conduct

Status offenders were handled similarly by the juvenile court and faced the same punishments as ordinary delinquents.

Because delinquency was viewed as less serious than crime (i.e., perhaps a reflection of behavior adjustments of adolescents rather than criminal intent), delinquents were to be treated rather than punished. Thus, in theory, the new juvenile justice system focused on the special social, emotional, and developmental needs of juveniles to promote their general well-being. Other states quickly followed Cook County's lead, and by 1945, all states had juvenile codes and specialized juvenile courts.

By the mid-1960s, the number of serious crimes committed by youths had increased dramatically, and high recidivism rates led many to believe that the juvenile correctional system was incapable of successfully treating and rehabilitating young offenders. Additionally, the courts and correctional facilities were flooded with status offenders, straining the sparse resources of the fledgling juvenile justice system. To help relieve this burden, Congress passed the Juvenile Justice and Delinquency Prevention Act of 1974. Under this Act, only youths guilty of criminal offenses—and not status offenders—could be committed to secure juvenile correctional facilities. In part responding to labeling theorists' concerns, the Act was intended to reduce the stigma and negative consequences of the "delinquent" label being applied to status offenders.

In response to the Juvenile Justice and Delinquency Prevention Act, most states revised their juvenile codes to clearly distinguish between delinquent and status offenders. Today, the legal category of delinquency (in all but three states) refers only to those behaviors that violate criminal law, classifying status offenders simply as youths who need special

Juvenile justice policy has historically treated youths as both childlike innocents and young degenerates.

How does society view delinquent youth today?

treatment or state supervision. Most jurisdictions call status offenders by one of several terms:

- MINS: minor in need of supervision or services
- PINS: person in need of supervision or services
- CHINS: child in need of supervision or services

With the increasing number of serious and violent juvenile offenders who committed crimes in the 1980s, the courts took a new tack, becoming more adversarial and punitive. A number of states shifted the goals of sentencing away from treatment and toward community protection, thus holding juvenile offenders responsible for their crimes.

The Juvenile Justice System

The U.S. juvenile justice system is composed of unique judicial and correctional agencies that specialize in dealing with juvenile offenders and operate with specific policies and procedures that are intended to protect youths from the potentially stigmatizing effects of criminal courts FIGURE 16-3. Historically (at least until the early 1970s), juvenile justice systems dealt with three types of cases:

- Delinquents
- Status offenders
- Neglected, abused, or dependent (i.e., destitute, homeless, or abandoned) youths

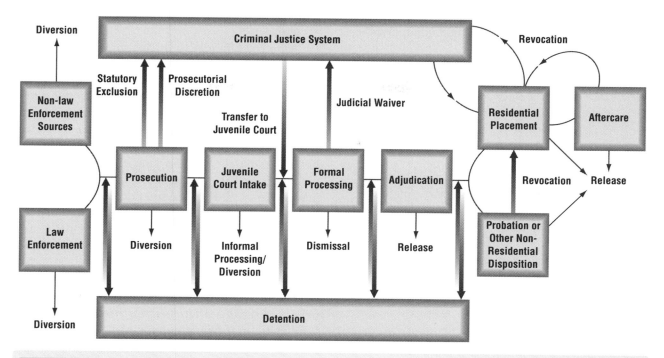

FIGURE 16-3 The U.S. Juvenile Justice System

Source: Howard Snyder and Melissa Sickmund, *Juvenile Offenders and Victims: 2006 National Report* (Washington, DC: National Center for Juvenile Justice, 2006), p. 105.

In recent years, many states have established family courts to deal with status offenders and cases of neglect, abuse, or dependency. In such jurisdictions, specialized juvenile courts focus on cases of delinquency.

The terminology used to describe the various stages and procedures associated with the juvenile justice system differs from the terminology employed in the adult criminal justice system **TABLE 16-1**. These differences are not meant to be merely linguistic; rather, they reflect the desire of the juvenile system to avoid unnecessary stigmatization of juvenile offenders.

LINK Whereas adults are "arrested," "indicted," and then "prosecuted," juvenile offenders are "taken into custody" and then a "petition is filed" before they are "adjudicated." Despite the differences in terminology, these stages in the juvenile and adult justice processes (as outlined in Chapter 1) are, in fact, quite similar.

Police

There has always been tension between police and adolescents. Many criminologists believe that this conflict stems from the beliefs held by police that separate these officers from the public, and especially from younger citizens. Many police officers are secretive, defensive, and distrustful of outsiders and see themselves as "the pragmatic guardians of the morals of the community . . . the 'thin blue line' against the forces of evil."[17] Many police view delinquent juveniles as

TABLE 16-1	
Differences in Terminology Used in Adult and Juvenile Justice Systems	
Adult Criminal Justice System	Juvenile Justice System
Crime	Delinquent act
Criminal	Delinquent
Arrest	Take into custody
Arraignment	Intake hearing
Indictment	Petition
Not guilty plea	Deny the petition
Guilty plea	Agree to an adjudication finding
Plea bargain	Adjustment
Jail	Detention facility
Trial	Adjudication hearing
Conviction	Adjudication
Presentence investigation	Social history
Sentencing	Disposition hearing
Sentence	Disposition
Incarceration	Commitment
Prison	Training or reform school, youth center
Parole	Aftercare

part of that evil force. Conversely, many youths see the police as intrusive, intimidating, and anxious to find fault.

LINK Police distrust of citizens and public animosity toward the police contribute to the development of a police subculture, which is sometimes at odds with the public, as discussed in Chapter 6.

Arrest

Although the police have a great deal of discretion in deciding when to arrest juvenile offenders, most state statutes provide some guidance for arrest procedures. The legal and extralegal factors affecting the police officer's decision to arrest a juvenile are generally similar to those affecting a decision to arrest an adult:

- Seriousness of the offense
- Time of the offense (e.g., late at night)
- Presence of evidence
- Offender's prior arrest record
- Offender's race, ethnicity, sex, and age
- Offender's attitude and demeanor
- Acceptance or denial of the allegations[18]

LINK Police discretion—the authority of officers to choose one course of action over another—plays an important role in the decision to take an offender into custody. While specific legal criteria serve as guidelines, extralegal factors such as race, ethnicity, gender, or socioeconomic class may also influence arrest decisions and create controversial disparities, as discussed in Chapter 8.

In general, the laws that govern juvenile arrests are similar to the laws that apply to adults. There is one significant difference, however: Delinquency cases do not require probable cause (a set of facts and circumstances that would lead a reasonable person to believe that a crime has been committed and that the accused committed it) prior to the juvenile's arrest. Instead, police may take any juvenile into custody if the officer has reasonable suspicion (a suspicion that creates a reasonable belief that the youth committed a delinquent act).

Once an officer arrests a juvenile, some states require the officer to notify a probation officer (or other designated official), who will then inform the youth's parents. In some jurisdictions, a juvenile who is taken into police custody goes to the police station for initial screening. Other jurisdictions give officers discretion to choose another course of action:

- Investigate the offense and the juvenile's background
- Decide to terminate the case

- Refer the offender to a community diversion program
- Send the offender to the juvenile court system

Booking

The procedure used when booking juveniles who have been arrested is essentially the same as in adult cases, with one notable exception: Some states forbid routine fingerprinting and photographing of juvenile suspects unless specifically ordered by the juvenile court.[19] When these identification techniques are used, they are intended only for temporary use and should not become part of a permanent criminal record. Advocates of fingerprinting and photographing of youth argue that these techniques provide complete records of young offenders, which are necessary for dealing with youths who refuse to reveal their identity, such as runaways, gang members, and serious offenders. Critics, however, contend that such permanent records make it more difficult for youths to be accepted by teachers or find employment.

Courts

Nearly 1 million youths are petitioned (the filing of a document setting forth the charge against the juvenile) to juvenile courts in the United States each year FIGURE 16-4 . Several key features distinguish these courts from their adult counterparts:

- Absence of legal guilt
- Nonadversarial, nonconfrontational interactions
- Focus on treatment rather than punishment
- Emphasis on the offender's background (e.g., social history, prior behavior, and clinical diagnosis)
- Absence of public scrutiny (private proceedings)
- Speed in processing cases
- Flexibility with sentencing options
- Short-term incarceration

Intake

Intake screening procedures are designed to screen out those cases that do not warrant a formal court hearing. Typically, cases that are dismissed meet one of the following criteria:

- Lack of sufficient evidence
- Minor law violations that could be handled informally (i.e., through counseling by a probation officer)
- Compensation already made to the victim

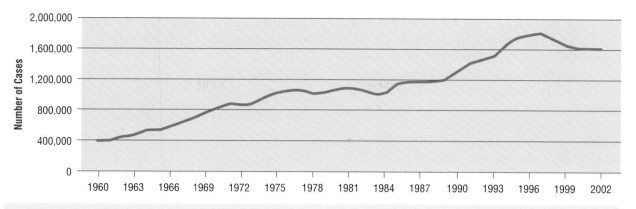

FIGURE 16-4 Delinquency Cases in U.S. Juvenile Courts

Source: Howard Snyder and Melissa Sickmund, *Juvenile Offenders and Victims: 2006 National Report* (Washington, DC: National Center for Juvenile Justice, 2006), p. 158.

- Jurisdiction inappropriate for juvenile courts (i.e., suspects found to be younger or older than the legal age for juvenile court jurisdiction)
- Circumstances that make the case more appropriate for criminal prosecution (i.e., the serious nature of the crime, the extensive criminal history of the juvenile, or a determination that the juvenile is not amenable to treatment in the juvenile system)[20]

To help make this determination, intake officers may order social background investigations or medical or psychological diagnoses. Intake officers are typically given broad discretion in determining which cases warrant formal handling and, in an effort to reduce the court's caseload, may favor informal hearings, adjudications, or probation supervision rather than referral to a judge. If intake does result in the decision for a formal hearing (which happens in approximately 60 percent of delinquency cases), the intake officer files a <u>petition</u>, which states that a delinquent act has been committed by the youth (equivalent to an indictment in criminal prosecutions).[21]

Detention

Once a juvenile has been arrested, he or she may be temporarily placed in <u>detention</u> while the court decides how to proceed. If a juvenile was placed in detention, then a petition must be filed in the case and a hearing held within 48 to 72 hours.[22] The primary goal of detention is to ensure that the youth appears at the necessary court hearings.[23]

Approximately 20 percent of all delinquents are detained at some point in the criminal justice

The juvenile court focuses on rehabilitation, unlike the criminal courts which focus on punishment.

? Does the juvenile justice system have the correct approach to responding to crime? Should the criminal justice system place greater emphasis on treatment and rehabilitation than on punishment?

process—either immediately after arrest, while awaiting a hearing, after sentencing, or before incarceration. In 2002, nearly 330,000 juveniles were held in detention for some period of time. Of these cases, 29 percent involved crimes against a person, 32 percent involved property crimes, 11 percent were related to drugs, and 27 percent involved public order offenses.[24]

Because the juvenile courts view children as more vulnerable than adults, a number of states have established limits on how long youths may be held in detention before a hearing takes place—generally within 30 days.[25] In _Schall v. Martin_ (1984), however, the U.S. Supreme Court ruled that juveniles who posed a serious risk of committing additional crimes could be held without determination of probable cause.[26] In this case, the Court reasoned that the protection of society was a sufficiently important goal in itself to justify preventive detention of juveniles.

LINK _Schall v. Martin_ expanded the ruling presented in Chapter 10 in _United States v. Salerno_ (1987), in which the Supreme Court held that the preventive detention of adult offenders who posed a likely threat to the community was constitutional.

Diversion

Even after a petition has been filed, efforts may still be taken to avoid formal hearings. Similar to plea bargaining in adult cases, the process of diversion—the early suspension or termination of the official processing of a juvenile—favors informal or unofficial alternatives. Officials may implement diversion at any of several points along the juvenile justice process in an effort to avoid the negative stigma associated with formal processing in the justice system. For example, police officers may handle delinquents informally by communicating an expectation of participation in a community recreation program, probation officers may choose to institute restitution rather than recommend a formal hearing, or judges may choose to delay sentencing while the youth is supervised on informal probation.

Good candidates for diversion programs include individuals with the following characteristics:

- First-time offenders charged with less serious offenses
- Repeat status offenders
- Offenders already participating in community-based treatment programs

These youths may then be given an opportunity to participate in various diversion efforts, including me-

diation—meetings that bring the complainant, the juvenile, and a neutral hearing officer together to reach a mutually acceptable solution.

LINK Mediation is a form of restorative justice, which has become an increasingly popular diversion approach (as discussed in Chapter 15). Restorative programs include family group conferencing, circle sentencing, and reparative probations.

Adjudication

The adjudication stage in the juvenile justice system parallels prosecution and trial in adult criminal courts. The purpose of the adjudication hearing is to determine whether the juvenile is responsible for the charges outlined in the petition.

Hearings in a juvenile court have traditionally been based more on civil—rather than criminal—proceedings. In addition, juvenile court proceedings have historically been nonadversarial. However, given the public's growing disillusionment with the ability of the courts to reduce serious juvenile crime through informal proceedings, and as a result of a series of Supreme Court decisions holding that juveniles have many of the same due process rights as adults in criminal proceedings, in recent decades juvenile courts have taken on many of the same characteristics as criminal courts. For example, juveniles may be represented by counsel, cross-examine witnesses, and invoke their Fifth Amendment protection against self-incrimination, largely as a result of the 1967 Supreme Court decision in _In re Gault_. The Supreme Court also ruled in _McKeiver v. Pennsylvania_ (1971) that juveniles are not constitutionally entitled to a trial by jury (though currently 12 states allow jury trials in serious cases if juveniles request them).[27]

Additional similarities between the juvenile and adult systems include the presumption of innocence; inadmissibility of hearsay evidence; requirement of adequate, timely, and formal notification of charges; sufficient time to formulate a response; and requirement of proof beyond a reasonable doubt to determine guilt.[28]

Disposition

At the conclusion of the adjudication hearing, the judge may either dismiss the case (equivalent to an acquittal) or sustain the petition (equivalent to a conviction). If the petition is sustained (which happens in approximately 67 percent of cases brought before the court

From its inception until the mid-1960s, the U.S. juvenile court system tolerated wide differences between the procedural rights accorded to adults and those accorded to juveniles. In practically all jurisdictions, rights granted to adults were withheld from juveniles. It was believed that juvenile court proceedings should not be adversarial or criminal. Rather, the right of the state as *parens patriae* permitted the juvenile court to act informally in the best interests of children. Consequently, juvenile proceedings were described as "civil" and, therefore, were not subject to the requirements that restrict the state when it seeks to deprive a person of his or her liberty. All of this changed in 1967, when the U.S. Supreme Court handed down its decision in what has been considered the leading constitutional case in juvenile law: *In re Gault*.

On June 8, 1964, 15-year-old Gerald Gault was arrested and taken to the Children's Detention Home in Gila County, Arizona, as a result of a verbal complaint by a neighbor, Mrs. Cook, that he had made lewd phone calls to her. No notice was given to Gault's parents that he had been taken into custody, and neither Gault nor his parents were given copies of the petition charging delinquency. At the initial hearing, Mrs. Cook did not appear and no transcript or record of the hearing was made. At a second hearing, Mrs. Cook was still not present. After this hearing, Gault was found to be delinquent and was committed to the state training school for five years.

The State of Arizona did not permit appeals by juveniles in delinquency cases, so the defense filed a writ of habeas corpus with the Arizona Supreme Court, which referred it to the Superior Court for a hearing.

The Superior Court dismissed the writ. The defense then sought review in the Arizona Supreme Court, which ruled that the juvenile court had acted appropriately. Gault appealed to the U.S. Supreme Court, arguing that the juvenile court had violated his rights of due process guaranteed by the Fourteenth Amendment.

Justice Abe Fortas delivered the opinion of the U.S. Supreme Court, stating that the basic requirements of due process and fairness must be satisfied in juvenile proceedings and that "neither the Fourteenth Amendment nor the Bill of Rights is for adults only." From this premise, he challenged the very essence of the juvenile court's operation. The court's position that its activities worked for the good of the child was shown to be suspect, and its procedure, in fact, violated juveniles' fundamental rights. According to Justice Fortas, "Under our Constitution, the condition of being a boy does not justify a kangaroo court." He further argued that the proper goal of the juvenile court would not be impaired by constitutional requirements and expressed his belief that the essentials of due process would reflect a fair and responsive attitude toward juveniles. Justice Fortas then set out the essentials of due process that should apply in juvenile delinquency proceedings, including the right to counsel, the right to confront and cross-examine one's accuser, the right against self-incrimination, and the right to timely notice of the charges.

As a result of the *Gault* decision, the operation of the juvenile court was significantly altered, making it more formal and adversarial in nature.

Source: In re Gault, 387 U.S. 1 (1967).

FIGURE 16-5, the judge may either immediately determine an appropriate disposition (equivalent to a sentence) or set a date for a disposition hearing.

A <u>disposition hearing</u> (equivalent to a sentencing hearing) is held to determine the most appropriate sentence for a delinquent. It is often an informal discussion between the following parties (although in some jurisdictions it is a more formal proceeding):

- Judge
- Probation officer
- Prosecutor
- Defense attorney
- Parents of the juvenile

These parties meet to discuss the various sentencing options:

- Probation
- Placement in a correctional facility
- Referral to a drug or alcohol treatment program
- Fines
- Restitution
- Community service

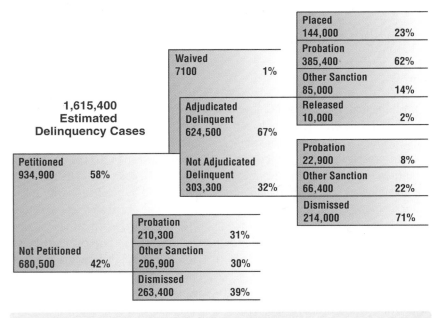

FIGURE 16-5 Processing Delinquency Cases

Source: Howard Snyder and Melissa Sickmund, *Juvenile Offenders and Victims: 2006 National Report* (Washington, DC: National Center for Juvenile Justice, 2006), p. 177.

Dispositional orders often include more than one of these options.

Corrections

The juvenile corrections system involves two main components: probation and institutional placement.

Probation

In the juvenile system, probation closely parallels probation within the adult system and involves many of the same key components:

- Conditional freedom within the community
- Close supervision by probation officers
- Rehabilitation through participation in community-based counseling programs

The philosophical foundation of probation is the belief that problem behaviors of youths are more effectively corrected within the community than in institutional settings.

Youths who are placed on probation are required to avoid further law-breaking and to comply with a variety of conditions of probation (i.e., attending school, making restitution, adhering to a curfew, receiving periodic visits from the probation officer, and submitting to random drug or alcohol tests). These conditions aim to further the goals of rehabilitation through treatment and guidance.

Occasionally, controversy arises over exactly which conditions meet this requirement. For example, courts have invalidated conditions requiring juveniles to attend church or religious school based on the notion that to do so would violate the constitutional separation of church and state, even though many people believe such attendance would serve as positive influence on a youth.

Institutional Placement

In general, placement in correctional facilities is reserved for those juveniles who have committed serious violent or property crimes **FIGURE 16-6** . The United States has approximately 3000 public and private correctional facilities for the placement of juvenile delinquents, which currently house nearly 97,000 youths.[29] These institutions include the following types of facilities:

- Training schools, reform schools, and youth centers (secure residential facilities)
- Shelter-care facilities (nonsecure housing for temporary placement of status offenders or dependent or neglected youths)
- Jails (secure facilities for holding persons who are awaiting trial or who have been convicted of misdemeanors)
- Ranches, forestry camps, and farms (nonsecure facilities for delinquents who have committed less serious crimes and that provide outdoor environments)

Blended Sentencing of Juveniles Convicted in Criminal Court

A number of states have implemented blended sentencing systems in their juvenile and criminal courts in an effort to rehabilitate juvenile offenders while still holding offenders accountable for their delinquent acts. Blended sentences expand the sentencing options available to judges when dealing with juveniles by allowing for multiple dispositions and deferral of criminal sentences. Failure to abide by the dispositional requirements can result in these criminal sentences being invoked.

Blended sentencing has been described as a "safety valve" or "emergency exit." It allows the court to extensively review the circumstances of a case and then make a more individualized decision regarding the suitability of a specific youth for juvenile or criminal treatment.

Currently, 17 states utilize blended sentencing in criminal court. Ten of these states have *exclusive blended sentencing* arrangements, in which the court imposes either a criminal sanction or a juvenile sanction. *Inclusive blend models,* which are used in seven states, allow juvenile offenders who are convicted in criminal

court to receive a combination sentence. In this model, the criminal court may suspend the adult sanction on condition of the youth's good behavior.

Some critics of blended sentencing suggest that these procedures produce a net-widening effect—that is, youths charged with somewhat less serious crimes may be transferred to criminal courts, where judges will still be able to sentence them as juveniles, placing the young offenders in juvenile correctional facilities. However, it is also possible that judges may prove more likely to revoke the probation of youths who had received blended sentences, even for technical violations, and then impose incarceration sentences.

Sources: Randi-Lynn Smallheer, "Sentence Blending and the Promise of Rehabilitation: Bring the Juvenile Justice System Full Circle," *Hofstra Law Review* 28:259–289 (1999); Howard Snyder and Melissa Sickmund, *Juvenile Offenders and Victims: 2006 National Report* (Washington, DC: Office of Juvenile Justice and Delinquency Prevention, 2006); Marcy Podkopacz and Barry Feld, "The Back-Door to Prison: Waiver Reform, 'Blended Sentencing,' and the Law of Unintended Consequences," *Journal of Criminal Law and Criminology* 91:997–1071 (2001).

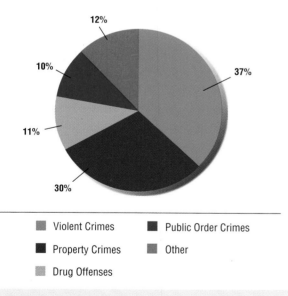

- Violent Crimes
- Property Crimes
- Drug Offenses
- Public Order Crimes
- Other

FIGURE 16-6 Offenses Committed by Juveniles Housed in U.S. Correctional Institutions

Source: Howard Snyder and Melissa Sickmund, *Juvenile Offenders and Victims: 2006 National Report* (Washington, DC: National Center for Juvenile Justice, 2006), p. 198.

These institutions aim to provide structured programming in a secure environment and further the goal of rehabilitation. In addition, a wide variety of community-based correctional programs have been established for the treatment of youth on probation.

LINK Like adult corrections discussed in Chapter 13, juvenile corrections has become increasingly privatized. Between 1997 and 2003, admissions of juveniles to private facilities increased by 3 percent, whereas admissions of juveniles to public facilities decreased by 12 percent over the same period.[30]

Aftercare

After release from juvenile correctional facilities, delinquent youths typically enter the phase of <u>aftercare</u> (equivalent to parole in the adult system)—that is, the release and subsequent community supervision of an individual from a correctional facility to ensure a more positive and effective transition back into the community. As with probation, delinquents in aftercare are subject to similar conditions and supervision requirements. Youths who violate the law or any condition of their supervision may face

Intensive aftercare programs are designed for serious, violent juvenile offenders who are released to the community.

 Are serious, violent juvenile offenders destined to become career criminals? Are they the only type of juvenile offenders who should be incarcerated?

revocation (removal from parole) and return to a correctional facility.

Nearly 100,000 juveniles enter aftercare each year. These youths share several common characteristics:

- Multiple previous commitments (40 percent of juveniles in aftercare have been held five or more times)
- A history of nonviolent offenses
- Residence in a single-parent home or dysfunctional family unit
- Relatives who have been incarcerated
- Educational setbacks when compared to peers
- Extensive time in some form of institutional placement[31]

For serious violent offenders, intensive aftercare programs (IAP) (equivalent to intensive parole

supervision) may be used to provide closer supervision. IAP officers have much smaller caseloads and are expected to conduct a number of face-to-face meetings with their parolees each week. They are also expected to establish and maintain contact with the juvenile's parents or guardians, school authorities, and, if applicable, employer on a regular basis.[32]

These relationships are also the focus of wraparound programs, which are designed to build positive relationships and support networks between youths and their families, teachers, and community agencies. Such programs typically entail a centralized coordination of services through the juvenile court, including clinical therapy, drug and alcohol treatment, special education, medical services, caregiver support, medical health care, and transportation.[33]

Juvenile Offenders in the Adult Criminal Justice System

In response to public and legislative perceptions of judicial leniency and the inability of the juvenile justice system to deal adequately with serious or repeat adolescent offenders, some delinquents—owing to their age, prior delinquent record, or seriousness of their offense—may face prosecution in criminal court and the possibility of sentencing to adult prisons.

Transfer to Criminal Courts

Waiver of jurisdiction is the judicial mechanism by which a juvenile may be transferred from juvenile court to criminal court. Waivers are usually filed in two types of cases:

- *Serious violent offenders.* Criminal courts can impose harsher punishments for serious violent offenders.[34]
- *Chronic offenders.* The criminal justice system is believed to provide more appropriate (e.g., punitive) treatment for offenders with long criminal records who have not responded positively to treatment programs.[35]

To be eligible for transfer to the adult criminal system, juvenile offenders in many states must be older than a minimum age TABLE 16-2 . In reality, most offenders waived to criminal courts are older than age 16.[36]

In *Kent v. United States* (1966), the U.S. Supreme Court held that the differences between juvenile and adult courts were so great that the transfer decision must be based on clearly established procedures de-

TABLE 16-2

Minimum Age for Judicial Transfer to Criminal Court in the United States

No minimum age specified	Alaska, Arizona, Delaware, District of Columbia, Hawaii, Idaho, Indiana, Maine, Maryland, Oklahoma, Oregon, Rhode Island, South Carolina, South Dakota, Tennessee, Washington, West Virginia
10 years	Kansas, Vermont
12 years	Colorado, Missouri
13 years	Georgia, Illinois, Mississippi, New Hampshire, North Carolina, Wyoming
14 years	Alabama, Arkansas, California, Connecticut, Florida, Iowa, Kentucky, Louisiana, Michigan, Minnesota, Nevada, New Jersey, North Dakota, Ohio, Pennsylvania, Texas, Utah, Virginia, Wisconsin

Minimum ages may not apply to all criminal offense restrictions, but represent the youngest possible age at which a juvenile may be judicially waived to criminal court.

Source: Howard Snyder and Melissa Sickmund, *Juvenile Offenders and Victims: 2006 National Report* (Washington, DC: National Center for Juvenile Justice, 2006), p. 112.

In many jurisdictions, adolescents accused of serious crimes such as murder and forcible rape are waived automatically to the adult criminal justice system. These youths may threaten public safety in the more lenient juvenile justice system, but they may be more likely to recidivate after prosecution in the adult system.

 How should the juvenile and adult criminal justice systems strike a balance between the use of juvenile and adult prosecutions for delinquents?

signed to protect the rights of juveniles.[37] Although the Court ruled that the practice of trying juveniles in criminal courts was constitutional, the juvenile may not be deprived of his or her constitutional rights. The decision to transfer offenders must include the following elements:

- A waiver hearing
- Effective counsel
- A statement of the reasons behind the decision to transfer

In its *Kent* decision, the Supreme Court also enumerated several criteria to guide judges in making transfer decisions, including the seriousness of the offense, the presence of violence, and the sophistication and maturity of the youth.[38]

Despite the existence of these guidelines, the transfer of juveniles to criminal courts remains controversial. Many citizens and correctional experts are unsure about the appropriateness of harsher punishments or the possibility of rehabilitation for juveniles in the adult correctional system. At the same time, long-term incarceration is possible only within the adult system and may be necessary to keep chronic or serious offenders off the streets. Approximately 6000 juvenile offenders are transferred to adult criminal courts in the United States each year (FIGURE 16-7), which equates to less than 1 percent of all petitioned cases.

Several procedures (or remands) are used to transfer youths between juvenile and criminal courts, though they are not all present in all states.

- **Judicial waiver**: the most common form of transfer to the criminal system. It involves a formal decision by the judge after careful consideration of the relevant issues at a transfer hearing.

- **Statutory exclusion**: the automatic transfer to the criminal system of certain juvenile offend-

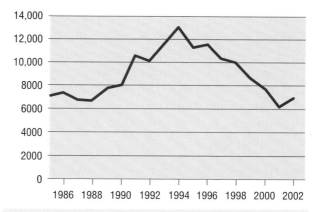

FIGURE 16-7 Juvenile Court Waivers to Criminal Court

Source: Howard Snyder and Melissa Sickmund, *Juvenile Offenders and Victims: 2006 National Report* (Washington, DC: National Center for Juvenile Justice, 2006), p. 186.

ers based on age, seriousness of offense, or prior criminal record.

- Prosecutorial waiver (or direct transfer): a form of transfer that gives the prosecutor the authority to decide whether to file a waiver of jurisdiction.
- Demand waiver: a request by a juvenile offender to be transferred to criminal court. While rare, such waivers are filed by delinquents who are seeking acquittal by jury or shorter sentences.
- Reverse waiver: a request by a juvenile who is being prosecuted in criminal court to be transferred back to the juvenile system.

To avoid situations of double jeopardy, in which the youth is tried for the same crime twice (once as a juvenile and once as an adult), the Supreme Court ruled in *Breed v. Jones* (1975) that the waiver process must begin before the evidence is presented at the adjudication hearing.[39]

Prosecution in Criminal Courts

> **LINK** When juveniles are prosecuted in criminal courts, they face the same process as adult criminal defendants (outlined in Chapter 10).

Some juvenile cases that are transferred to criminal court (approximately 16 percent) are dropped before charges are formally filed. Even when charges are filed, some cases (about 11 percent) are terminated by the prosecutor's *nolle prosequi* (decision not to prosecute).[40] If the prosecutor decides to proceed with the case, a trial may still be avoided by the process of plea bargaining.[41]

If the case does go to trial, a juvenile offender is significantly more likely than an adult to be convicted.[42] Researchers have suggested various reasons for this high conviction rate, including impaired competence to stand trial due to immature judgment and decision making. Other studies, however, have shown no significant differences between those youths who are tried in criminal courts and similar peers who are tried in the juvenile system other than the seriousness of their charges.[43]

Sentencing the Convicted Juvenile

Juveniles subject to criminal convictions have several advantages over adult criminals in obtaining lenient sentencing. First, age is a mitigating factor that is taken into consideration in determining the appropriate sentence. Second, most juveniles do not have adult criminal records, and prior juvenile records are

sometimes prohibited from being introduced in criminal hearings (primarily because the courts have traditionally believed that juvenile misbehavior reflects immaturity and should not be held against juveniles once they become adults).[44] Prior criminal records are important because many states' sentencing laws require incarceration of the offender when he or she is convicted of a second felony. Despite these considerations, research has demonstrated inconclusive findings about lenient sentencing of juveniles in general.[45]

> **LINK** Presumptive or determinant sentencing systems, discussed in Chapter 12, mean that even first-time offenders who are convicted of certain serious crimes may receive mandatory prison sentences.

Juvenile Offenders in Prison

Although only a small number of juveniles are incarcerated in state prisons (where they account for less than 0.2 percent of the total incarcerated population **FIGURE 16-8**, this population presents serious challenges for both offenders and correctional administrators.[46]

Life in adult prisons is significantly different from life in juvenile institutions, and juvenile offenders often have difficulty adjusting to prison subculture. A youth's reputation as being tough, which might have afforded him or her status on the streets or in a juvenile institution, carries little weight among older inmates. Juvenile offenders find themselves at the bottom of the status ladder, subject to both the formal authority of guards and the informal power

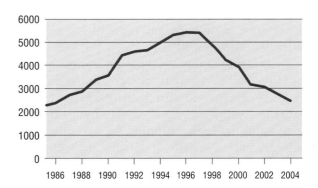

FIGURE 16-8 One-Day Count of Juveniles Held in State Prisons

Source: Howard Snyder and Melissa Sickmund, *Juvenile Offenders and Victims: 2006 National Report* (Washington, DC: National Center for Juvenile Justice, 2006), p. 238.

On October 15, 2005, 16-year-old Scott Dyleski entered the home of 52-year-old Pamela Vitale. Dyleski was planning to steal Vitale's credit cards to use in purchasing marijuana-growing equipment. During the burglary, something went wrong, and Dyleski got into a fight with Vitale. Ultimately, he bludgeoned and stabbed her to death. The brutal attack involved multiple blows to Vitale's head, dislodged teeth, broken fingers, bruises over her entire body, a stab wound to her abdomen deep enough to expose her intestines, and a symbol cut into her back.

Dyleski was a former Boy Scout who refused to eat meat or to wear leather. He lived with his mother in a communal home with two other families not far from Vitale's residence. A friend who was involved in a credit card theft scheme with Dyleski tipped off police after learning about the murder.

Jurors deliberated more than 18 hours before finding Dyleski guilty of first-degree murder with special circumstances (reflecting the unusual brutality of the murder). Dyleski was sentenced to life in prison without parole.

Source: Lisa Sweetingham, "After Guilty Verdict, Vitale Family Reflects on Their Mother's Killing and the Harrowing Trial," *CourtTVnews,* August 30, 2006, available at http://www.courttv.com/news/horowitz/082906_ctv.html, accessed August 20, 2007.

of other inmates. They may merely resent the authority or, worse, be subject to victimization. Indeed, juveniles in adult prisons are significantly more likely to become victims of violent crime or sexual assault than youths in juvenile institutions.[47]

LINK Juveniles, like adults in prison, go through the process of prisonization as they adjust to the prison subculture and learn the informal inmate code of conduct (described in Chapter 14).

Daily survival becomes the primary concern for youths in prison. For many, survival means adapting to the inmate subculture. Although this adaptation may improve their daily life in prison, it may also distract or discourage juveniles from pursuing activities that would improve their chances of getting out of prison earlier, such as participating in counseling and educational programs and conforming to institutional expectations.[48] Survival may also include forging an alliance with an older inmate who will provide protection, which all too often comes at the cost of sexual exploitation.

The characteristics of juvenile inmates also create difficulties for prison administrators. In states with very small numbers of juvenile inmates, the cost of building special housing for these juveniles (rather than integrating them into the general inmate population) becomes a serious budgetary concern. Furthermore, because juvenile inmates are younger and proportionately more violent than adult inmates,

they may have greater difficulty in adapting to institutional rules and consequently require greater supervision by institutional staff. Unlike juvenile correctional facilities, which try to foster resident and staff interactions that promote the social and personal development of youthful offenders, most adult prisons emphasize custody and control. Adult correctional facilities are not intended to cater to, nor are they typically equipped to provide for, the educational or psychological needs of juveniles. Because most states have only a handful of juvenile inmates, legislatures hesitate to allocate additional expenditures for programs or staff to give these youths specialized treatment.

The recidivism rate among juveniles released from prison also challenges criminal justice professionals. Juveniles who are paroled from prison fare no better than their adult counterparts, with approximately 60 percent returning to prison for a new offense or violation of probation in less than three years. The length of time served appears to make little difference in the recidivism rate.[49]

Abolition of the Death Penalty for Juveniles

Although juveniles have rarely been executed in the United States, at least 366 have been legally put to

Approximately 60 percent of juvenile inmates return to prison within 3 years after their release.

 Why is the recidivism rate for juveniles so high? What does this say about the rehabilitative focus of the juvenile justice system?

death since 1642. Since the 1890s, juveniles have accounted for less than 2 percent of all people executed. From the 1890s to 1930, fewer than 30 juveniles were executed in any given decade. In the 1930s and 1940s, however, an unusual increase in juvenile executions occurred, with 40 and 50 executions taking place in those two decades, respectively. Between 1965 and 1984, no juveniles were executed. However, with the execution of Charles Rumbaugh in Texas on September 11, 1985, juveniles once again faced the prospect of execution. Between 1985 and April 2003, 22 persons who were juveniles at the time when they committed their crimes were executed.[50]

Beginning in 1982, it became clear that the U.S. Supreme Court was interested in the death penalty as it applied to juveniles when it held that the youthfulness of an offender must be considered as a mitigating circumstance at sentencing, reflecting a growing ambivalence about juvenile executions among policymakers and the public alike.[51] Six years later, in *Thompson v. Oklahoma*, the Court held that the execution of a person who was under the age of 16 at the time of the commission of his or her crime was unconstitutional.[52] The next year, in *Stanford v. Kentucky*, the Supreme Court rejected an appeal that could have prohibited the execution of anyone younger than 18 at the time of his or her crime.[53]

A total of eighteen 17-year-olds and one 16-year-old were executed after the *Stanford* decision. Then, on March 1, 2005, the Supreme Court, in a divided 5-to-4 decision in *Roper v. Simmons*, ruled that "the death penalty is disproportionate punishment for offenders under the age of 18" and, therefore, is a violation of the Eighth Amendment's prohibition against cruel and unusual punishment.[54] Although the Court and the country remained deeply divided over the juvenile death penalty, it had finally come to an end.[55]

In 1993, Christopher Simmons, age 17, made a suggestion to his friends that they could literally get away with murder. Simmons and his 15-year-old friend entered the house of Shirley Crook with the intention of committing burglary and murder. The boys put duct tape over Crook's eyes and mouth, bound her hands, and drove to a state park. There, they covered her head with a towel, wrapped her hands and feet together with electrical wire, covered her face in duct tape, and threw her off a bridge. Crook drowned in the waters below. The next afternoon, her body was found by a fisherman.

Simmons, meanwhile, had been bragging about the killing. Both boys were arrested and confessed to the killing. Simmons's friend agreed to testify against him to avoid the death penalty; the friend was ultimately convicted and sentenced to prison. Simmons was convicted and sentenced to death. His case was eventually appealed to the U.S. Supreme Court, where his lawyer argued that the death penalty constituted cruel and unusual punishment for juvenile offenders.

Fifteen years earlier, the Court had ruled that the execution of 16- and 17-year-olds was constitutional. In *Roper v. Simmons,* however, the Court reversed its position. In a 5-to-4 decision, the Court argued that juveniles (those under the age of 18) should be categorically exempt from capital punishment because they "cannot with reliability be classified among the worst offenders." This conclusion was premised on three perceived differences between adults and juveniles:

1. Juveniles lack maturity and responsibility and are more reckless than adults.

2. Juveniles are more vulnerable to outside influences because they have less control over their surroundings.

3. A juvenile's character is not fully formed.

According to the Court, "these differences render suspect any conclusion that a juvenile falls among the worst offenders" and that "from a moral standpoint it would be misguided to equate the failings of a minor with those of an adult."

The Court also expressed its belief that juvenile murderers could be reformed: "Only a relatively small proportion of adolescents who experiment in risky or illegal activities develop entrenched patterns of problem behavior that persist into adulthood." Finally, the Court argued that the evolving standards of decency and a perceived reduction in support for the juvenile death penalty, combined with "the overwhelming weight of international opinion against the juvenile death penalty," led it to conclude that the executions of juvenile offenders could no longer be considered constitutional.

Both Justice Sandra Day O'Connor and Justice Antonin Scalia wrote critical dissenting opinions on behalf of the minority. O'Connor argued the majority had provided "no evidence impeaching the seemingly reasonable conclusion reached by many state legislatures—that at least some 17-year-old murderers are sufficiently mature to deserve the death penalty in an appropriate case" and the majority's analysis "is premised on differences in the aggregate between juveniles and adults, which frequently do not hold true when comparing individuals." Scalia argued that no evidence of a national consensus opposing the juvenile death penalty was presented and that, indeed, "a number of legislatures and voters have expressly affirmed their support for capital punishment of 16- and 17-year-old offenders since *Stanford.*"

Finally, both O'Connor and Scalia were critical in the Court's failing to "reprove, or even acknowledge, the Supreme Court of Missouri's unabashed refusal to follow our controlling decision in *Stanford.*" O'Connor said it was "clear error," while Scalia wrote that "allowing lower courts to reinterpret the Eighth Amendment whenever they decide enough time has passed for a new snapshot leaves this Court's decisions without any force."

Source: Roper v. Simmons, 543 U.S. (2005).

WRAP UP

In transferring an adolescent to criminal court, several points are considered. Because criminal prosecution carries more severe penalties, juvenile court officials reserve waivers for the most serious delinquents. Young offenders with multiple prior arrests (especially for violent serious offenses), prior adjudications, records of noncompliance with the juvenile justice system (e.g., probation violations), and prior commitments are most likely to be waived.

Age is also an important factor in determining whether a youth will be transferred from juvenile court to criminal court. Delinquents closer to the age of 18—especially those with lengthy criminal records—are more likely to be prosecuted as adults. In Abraham's case, the decision to prosecute was based on his history of criminal behavior and murder with a firearm. These facts suggest that Abraham was well on his way to becoming a career criminal—precisely the type of young offender for whom the waiver process was designed.

Chapter Spotlight

- Public frustration with the inability of the juvenile justice system to respond adequately to serious youth crime has led many states to revise their juvenile codes to make juvenile offenders more accountable for their criminal behavior and to make it easier to transfer juveniles to adult criminal court for prosecution.

- Before the twentieth century, juveniles who violated the law were treated much like adult criminals. However, as a result of the Child Savers' movement, a special juvenile justice system was created that included separate codes, courts, and correctional facilities for youthful offenders.

- A number of features distinguish the contemporary juvenile court from criminal court, including its definition of legal guilt, correctional philosophy, and judicial procedures.

- After a juvenile is arrested or taken into custody, a petition charging delinquency is filed with the court. An adjudication hearing is then held to determine whether the charge should be sustained. If it is sustained, a disposition hearing is held to determine the appropriate treatment or placement for the youth.

- Youths who commit serious crimes may be transferred from juvenile court to criminal court for prosecution as adults via one of three waiver procedures: judicial waiver, legislative or automatic waiver, and prosecutorial waiver. Criminal courts typically gain jurisdiction over persons who are age 18 or older, but most states have set lower minimum ages for waiver, depending on the offense charged.

- Juveniles who are prosecuted in criminal court face the same prosecutorial process that all adult defendants confront and are afforded all the same constitutional protections. Juveniles are less likely than adult defendants to receive a plea bargain, however, and are more likely to be convicted than their adult counterparts.

- In 2005, approximately 2266 juveniles were incarcerated in state and federal prisons in the United States, accounting for approximately 0.2 percent of the total incarcerated population. Most incarcerated youth are confined in the general adult prison population, which creates special problems for both youths and prison administrators: Youths often have trouble adjusting to prison rules and the inmate subculture and are more likely than older inmates to be victims of assault and exploitation by other prisoners.

- After a series of cases, the U.S. Supreme Court finally abolished the juvenile death penalty in the United States in 2005, in *Roper v. Simmons*.

Putting It All Together

1. Some critics have argued that the juvenile justice system should be abolished. Do you agree? Why or why not?

2. Should juveniles be granted the right to a jury trial? What benefits might this change afford?

3. Under what circumstances should juveniles be transferred to adult criminal courts?

4. Should juveniles who are convicted as adults be incarcerated with the general inmate population?

5. Are there any circumstances in which the death penalty should be applied to persons who were under the age of 18 at the time they committed their crimes?

Key Terms

adjudication The stage in the juvenile justice system that parallels prosecution and trial in adult criminal courts.

adjudication hearing A hearing to determine whether a juvenile committed the offense of which he or she is accused.

aftercare The release and subsequent community supervision of an individual from a correctional facility to ensure a more positive and effective transition back into the community.

Breed v. Jones The Supreme Court decision that a criminal prosecution of a child following a juvenile court hearing constitutes double jeopardy.

Child Savers A group of nineteenth-century reformers who believed that children were basically good, delinquency was the product of bad environments and that the state should remove children from such environments.

delinquent A juvenile under age 18 determined to have violated the juvenile code.

demand waiver Process by which a juvenile may request to have his or her case transferred to criminal court.

detention The temporary custody and care of juveniles pending adjudication, disposition, or implementation of disposition.

disposition hearing A hearing to determine the most appropriate placement of a juvenile adjudicated to be delinquent.

diversion The early suspension or termination of the official processing of a juvenile in favor of an informal or unofficial alternative.

In re Gault Case in which the Supreme Court held that juveniles could not be denied basic due process rights in juvenile hearings.

intake Initial screening process in the juvenile court to determine whether a case should be processed further.

intensive aftercare programs (IAP) Equivalent to intensive parole supervision; used to provide greater supervision of youths after their release from official institutions.

judicial waiver Most common waiver procedure for transferring youths to criminal court, in which the judge is the primary decision maker.

juvenile A person under the age of 18.

juvenile delinquency Behavior by a juvenile that is in violation of the juvenile or criminal codes.

Kent v. United States The Supreme Court decision requiring a formal waiver hearing before transfer of a juvenile to criminal court.

McKeiver v. Pennsylvania The Supreme Court decision that juveniles do not have a constitutional right to a jury trial in juvenile court.

New York House of Refuge The first correctional institution for children in the United States (opened in 1825), which emphasized industry, education, and strict discipline.

parens patriae A principle based on English common law that viewed the state as the ultimate sovereign and guardian of children.

petition Similar to an indictment; a written statement setting forth the specific charge that a delinquent act has been committed or that a child is dependent or neglected or needs supervision.

prosecutorial waiver Process in which the prosecutor determines whether a charge against a juvenile should be filed in criminal or juvenile court.

reverse waiver Process in which a juvenile contests a statutory exclusion or prosecutorial transfer.

Roper v. Simmons The Supreme Court decision that the death penalty for anyone who was younger than age 18 at the time of his or her crime is unconstitutional.

Schall v. Martin The Supreme Court decision authorizing the preventive detention of juveniles who are identified as "serious risks" to the community if released.

status offenses Acts prohibited to children that are not prohibited to adults (such as running away, truancy, and incorrigibility).

statutory exclusion Process established by statute that excludes certain juveniles, because of either age or offense, from juvenile court jurisdiction; charges are initially filed in criminal court.

waiver of jurisdiction A legal process to transfer a juvenile from juvenile to criminal court.

wraparound programs Programs designed to build positive relationships and support networks between youths and their families, teachers, and community agencies through coordination of services.

Notes

1. Statement by FBI Director J. Edgar Hoover, warning against "misguided policies which encourage criminal activity and often result in arrogant and defiant attitudes by youth." Quoted in Thomas Bernard, *The Cycle of Juvenile Justice* (New York: Oxford University Press, 1992), p. 34.
2. Federal Bureau of Investigation, *Crime in the United States, 2006* (Washington, DC: U.S. Department of Justice, 2007), Table 34.
3. Marvin Wolfgang, Robert Figlio, and Thorsten Sellin, *Delinquency in a Birth Cohort* (Chicago: University of Chicago Press, 1972); Marvin Wolfgang, Terence Thornberry, and Robert Figlio, *From Boy to Man, from Delinquency to Crime* (Chicago: University of Chicago Press, 1987); Simon Dinitz and John Conrad, "The Dangerous Two Percent," in David Shichor and Delos Kelly (Eds.), *Critical Issues in Juvenile Delinquency* (Lexington, MA: Lexington Books, 1980), pp. 129–155; Dora Nevares, Marvin Wolfgang, and Paul Tracy, *Delinquency in Puerto Rico: The 1970 Birth Cohort Study* (Westport, CT: Greenwood Press, 1990); Lyle Shannon, *Assessing the Relationships of Adult Criminal Careers to Juvenile Offenders: A Summary* (Washington, DC: U.S. Government Printing Office, 1982).
4. *Deuteronomy* 21:18–21, *The Thompson Chain-Reference Bible,* 5th ed. (Indianapolis, IN: B.B. Kirkbride Bible Company, 1988).
5. Fredrick Ludwig, *Youth and the Law* (New York: Foundation Press, 1955), p. 12.
6. Wiley Sanders, *Juvenile Offenders for a Thousand Years* (Chapel Hill, NC: University of North Carolina Press, 1970).
7. John Tobias, *Crime and Industrial Society in the 19th Century* (New York: Shocken Books, 1967).
8. Joseph Hawes, *Children in Urban Society: Juvenile Delinquency in Nineteenth-Century America* (New York: Oxford University Press, 1971), pp. 15–19.
9. Hawes, note 8, p. 14.
10. Phillippe Ariès, *Centuries of Childhood,* translated by Robert Baldick (New York: Knopf, 1962), pp. 411–412.
11. Ariès, note 10.
12. George Haskins, *Law and Authority in Early Massachusetts* (New York: Archon Books, 1968).
13. Anthony Platt, *The Child Savers: The Invention of Delinquency* (Chicago: University of Chicago Press, 1969).
14. Clifford Dorne, *Crimes against Children* (New York: Harrow and Heston, 1989), p. 30.
15. New York Society for the Reformation of Juvenile Delinquents, *Documents Relative to the House of Refuge* (New York: Mahlon Day, 1832).
16. Charles Brace, *The Dangerous Classes of New York and Twenty Years' Work Among Them* (New York: Wynkoop and Hellenbeck, 1880/1970); Stephen O'Connor, *Orphan Trains* (Boston: Houghton Mifflin, 2001).
17. Robert Carter, "The Police View of the Justice System," in Malcolm Klein (Ed.), *The Juvenile Justice System* (Beverly Hills, CA: Sage Publications, 1976), p. 131.
18. Robert Brown, "Black, White, and Unequal: Examining Situational Determinants of Arrest Decisions from Police–Suspect Encounters," *Criminal Justice Studies* 18:51–68 (2005); Terrence Allen, "Taking a Juvenile into Custody: Situational Factors That Influence Police Officers' Decisions," *Journal of Sociology and Social Welfare* 32:121–129 (2005); Michael Resig, John McCluskey, Stephen Mastrofski, and William Terrill, "Suspect Disrespect toward the Police," *Justice Quarterly* 21:241–268 (2004); Kenneth Novak, James Frank, Brad Smith, and Robin Engel, "Revisiting the Decision to Arrest: Comparing Beat and Community Officers," *Crime & Delinquency* 48:70–98 (2002); Robert Worden, "Situational and Attitudinal Explanations of Police Behavior: A Theoretical Reappraisal and Empirical Assessment," *Law and Society Review* 23:667–711 (1989); Irving Piliavin and Scott Briar, "Police Encounters with Juveniles," *American Journal of Sociology* 70:206–214 (1964); Geoffrey Alpert, John MacDonald, and Roger Dunham, "Police Suspicion and Discretionary Decision Making During Citizen Stops, *Criminology* 43:407–434 (2005).
19. National Advisory Committee for Juvenile Justice and Delinquency Prevention, *Juvenile Justice and Delinquency Prevention* (Washington, DC: U.S. Government Printing Office, 1976).
20. Robert M. Regoli and John D. Hewitt, *Delinquency in Society,* 6th ed. (New York: McGraw-Hill, 2006), p. 426.

21. Howard Snyder and Melissa Sickmund, *Juvenile Offenders and Victims: 2006 National Report* (Washington, DC: National Center for Juvenile Justice, 2006), p. 177.

22. Barry Feld, *Cases and Materials on Juvenile Justice Administration* (St. Paul, MN: West, 2000), p. 313.

23. James Austin, Kelly Johnson, and Ronald Weitzer, *Alternatives to Secure Detention and Confinement of Juvenile Offenders* (Washington, DC: U.S. Department of Justice, 2005).

24. Snyder and Sickmund, note 21, p. 168.

25. Barry Krisberg, Ira Schwartz, Paul Litsky, and James Austin, "The Watershed of Juvenile Justice Reform," *Crime & Delinquency* 32:5–38 (1986); Barry Krisberg, Robert DeComo, Norma Herrera, Martha Steketee, and Sharon Roberts, *Juveniles Taken into Custody: Fiscal Year 1990 Report* (San Francisco: National Council on Crime and Delinquency, 1991).

26. *Schall v. Martin,* 467 U.S. 253 (1984).

27. *McKeiver v. Pennsylvania,* 403 U.S. 528 (1971).

28. Snyder and Sickmund, note 21, p. 198.

29. Snyder and Sickmund, note 21, p. 198.

30. Snyder and Sickmund, note 21, p. 232; Howard Snyder, "An Empirical Portrait of the Youth Reentry Population," *Youth Violence and Juvenile Justice* 2:39–55 (2004).

31. Lynn Goodstein and Henry Sontheimer, "The Implementation of an Intensive Aftercare Program for Serious Juvenile Offenders," *Criminal Justice and Behavior* 24:332–359 (1997).

32. James Howell, *Preventing and Reducing Juvenile Delinquency: A Comprehensive Framework* (Thousand Oaks, CA: Sage Publications, 2003).

33. Regoli and Hewitt, note 20.

34. Barry Feld, *Bad Kids: Race and the Transformation of the Juvenile Court* (New York: Oxford University Press, 1999).

35. Daniel Mears, "A Critique of Waiver Research: Critical Next Steps in Assessing the Impacts of Laws for Transferring Juveniles to the Criminal Justice System," *Youth Violence and Juvenile Justice* 1:156–172 (2003).

36. Snyder and Sickmund, note 21.

37. *Kent v. United States,* 383 U.S. 541 (1966).

38. *Kent v. United States,* note 37.

39. *Breed v. Jones,* 421 U.S. 519 (1975).

40. Charles Thomas and Shay Bilchik, "Prosecuting Juveniles in Criminal Courts: A Legal and Empirical Analysis," *Journal of Criminal Law and Criminology* 76:439–479 (1985)

41. Cary Rudman, Eliot Hartstone, Jeffrey Fagan, and Melinda Moore, "Violent Youth in Adult Court: Process and Punishment," *Crime & Delinquency* 32:75–96 (1986); David Reed, *Needed: Serious Solutions for Serious Juvenile Crime* (Chicago: Chicago Law Enforcement Study, 1983).

42. Thomas Cohen and Brian Reaves, *Felony Defendants in Large Urban Counties, 2002* (Washington, DC: U.S. Department of Justice, 2006), p. 24; Dean Champion, *The Juvenile Justice System* (New York: Macmillan, 1992), p. 583; Rudman et al., note 41, p. 86; Donna Bishop, Charles Frazier, and John Henretta, "Prosecutorial Waiver: Case Study of a Questionable Reform," *Crime & Delinquency* 35:180 (1989).

43. Thomas Grisso and Richard Schwartz, *Youth on Trial: A Developmental Perspective on Juvenile Justice* (Chicago: University of Chicago Press, 2000); Thomas Grisso, Laurence Steinberg, Jennifer Woolard, Elizabeth Cauffman, Elizabeth Scott, Sandra Graham, et al., "Juveniles' Competence to Stand Trial: A Comparison of Adolescents' and Adults' Capacities as Trial Defendants," *Law and Human Behavior* 27:333–363 (2003); Norman Poythress, Frances Lexcen, Thomas Grisso, and Laurence Steinberg, "The Competence-Related Abilities of Adolescent Defendants in Criminal Court," *Law and Human Behavior* 30:88 (2006).

44. Dean Champion, *The Juvenile Justice System* (New York: Macmillan, 1992), p. 92.

45. Dean Champion, "Teenage Felons and Waiver Hearings: Some Recent Trends, 1980–1988," *Crime & Delinquency* 35:577–585 (1989); Mary Clement, "A Five-Year Study of Juvenile Waiver and Adult Sentences: Implications for Policy," *Criminal Justice Policy Review* 8:201–219 (1997); Thomas and Bilchik, note 40; Rudman et al., note 41; Megan Kurlychek and Brian Johnson, "The Juvenile Penalty: A Comparison of Juvenile and Young Adult Sentencing Outcomes in Criminal Court," *Criminology* 42:485–517 (2004); Jeffrey Fagan, "The Comparative Advantage of Juvenile Versus Criminal Court Sanctions on Recidivism among Adolescent Felony Offenders," *Law and Policy* 18:77–114 (1996); Carole Barnes and Randal Franze, "Questionably Adult: Determinants and Effects of the Juvenile Waiver Decision," *Justice Quarterly* 6:117–135 (1989); Gerard Rainville and Steven Smith, *Juvenile Felony Defendants in Criminal Courts* (Washington, DC: U.S. Department of Justice, 2003).

46. Paige Harrison and Allen Beck, *Prison and Jail Inmates at Midyear 2005* (Washington, DC: U.S. Department of Justice, 2006), p. 5.

47. Martin Forst, Jeffrey Fagan, and T. Scott Vivona, "Youth in Prisons and Training Schools: Perceptions and Consequences of the Treatment–Custody Dichotomy," *Juvenile and Family Court Journal* 40:1–14 (1989).

48. Zvi Eisikovits and Michael Baizerman, "'Doing Time': Violent Youth in a Juvenile Facility and in an Adult Prison," *Journal of Offender Counseling, Services and Rehabilitation* 6:10 (1982), p. 9.

49. Kathleen Heide, Erin Spencer, Andrea Thompson, and Eldra Solomon, "Who's In, Who's Out, and Who's Back: Follow-up Data on 59 Juveniles Incarcerated in Adult Prison for Murder or Attempted Murder in the Early 1980s," *Behavioral Sciences and the Law* 19:97–108 (2001).

50 . Regoli and Hewitt, note 20, pp. 484–485.

51. *Eddings v. Oklahoma,* 455 U.S. 104 (1982).

52. *Thompson v. Oklahoma,* 487 U.S. 815 (1988).

53. *Stanford v. Kentucky,* 492 U.S. 361 (1989).

54. *Roper v. Simmons,* 543 U.S. 551 (2005).

55. Regoli and Hewitt, note 20, p. 490.

Terrorism and Cybercrime

CHAPTER 17

THINKING ABOUT CRIME AND JUSTICE

Created in March 2003, Immigration and Customs Enforcement (ICE) is the largest investigative branch of the Department of Homeland Security (DHS). The agency was created after the terrorist attacks of September 11, 2001, by combining the law enforcement arms of the former Immigration and Naturalization Service and U.S. Customs Service to more effectively enforce immigration and customs laws and to protect the United States against terrorist attacks. To accomplish its mission, ICE targets illegal immigrants who support terrorism and other criminal activities. A prominent example of ICE's activity is Operation Return to Sender. Between June 2006 and January 2007, this operation was responsible for the arrest of more than 13,000 illegal immigrants across the United States, including more than 3000 inmates in local jails and state prisons. All arrestees were deported. Immigration officials estimate more than 600,000 illegal immigrants have ignored deportation orders and remain at large.

- How might ICE combine elements of criminal justice and counterterrorism?

- What types of crimes in particular does ICE target?

Sources: U.S. Immigration and Customs Enforcement, available at http://www.ice.gov/index.htm, accessed December 25, 2006; "761 Illegals Rounded Up in Southern California Sweep, One of the Largest Ever," Associated Press, January 24, 2007, available at http://www.foxnews.com/story/0,2933,246086,00.html?sPage=fnc.national/crime, accessed March 4, 2007.

Introduction

The twenty-first century truly is a new world for the criminal justice system. Developments in technology have led to new crimes, new criminal methods, and new ways to respond to crime. The foremost "new" problem that the criminal justice system must address is terrorism. Terrorism is no longer just a concern of the U.S. military and intelligence community. Indeed, criminal justice operatives—from those in policing to prosecution to corrections—now engage in counterterrorism efforts.

Although terrorism is the defining issue of the new century, other serious threats to society exist as well. Emerging crimes, such as cybercrime and identity theft, produce enormous losses and feelings of panic and uncertainty—just like terrorism. These crimes have grown considerably in the past decade, at a tremendous expense. Besides suffering from the detrimental effects of dangerous computer viruses, e-mail spam, and spyware, each year more than 10 million Americans are the victims of identity theft, a crime that has estimated annual costs of almost $50 billion.[1] These new crimes have led to innovative responses by the criminal justice system, including:

- The creation of the Department of Homeland Security

- Restructuring of the Department of Justice

- The USA Patriot Acts

- Increased communication between intelligence agencies

Terrorism

Not long ago, few Americans took the subject of terrorism too seriously. Public opinion polls indicated that terrorism was not considered a national priority compared to other concerns, such as the economy, crime, and civility/public morality.[2] When terrorist events did directly affect the United States, such as the 1993 attack on the World Trade Center and the 1995 Oklahoma City bombing, they were viewed as horrible but atypical, and were not cause for major concern. Indeed, although domestic and international terrorists have consistently been prosecuted and punished by the criminal justice system, terrorism was not even the first priority of federal and state authorities.[3]

On September 11, 2001, everything changed. Terrorists—19 Al Qaeda operatives—hijacked four commercial airliners with chilling intent, as described in the 9/11 Commission Report:

The 19 men were aboard four transcontinental flights. They were planning to hijack these planes and turn them into large guided missiles, loaded with up to 11,400 gallons of jet fuel. By 8:00 A.M.

On September 11, 2001, the United States suffered the worst terrorist attack in its history.

 How did the events of September 11 change the face of American criminal justice? How have these changes affected civil liberties in the United States?

on the morning of Tuesday, September 11, 2001, they had defeated all the security layers that America's civil aviation security system then had in place to prevent a hijacking.[4]

The result of their efforts is well known: The American Airlines and United Airlines passenger flights were turned into deadly weapons, hitting the two towers of the World Trade Center, the heart of the U.S. financial district in New York City, and the Pentagon, the heart of the country's Department of Defense. Passengers on the fourth plane attacked the hijackers, and the plane crashed into a field in southwestern Pennsylvania; the intended target was likely either the U.S. Capitol Building or the White House (the

seat of the U.S. federal government). The September 11 attacks killed almost 3000 Americans, destroyed the World Trade Center, grounded all aircraft in the United States for several days, devastated the U.S. stock market, and led to more than $17 billion in immediate costs and 200,000 lost jobs. It also placed terrorism—a known but formerly largely ignored topic—at the heart of U.S. foreign policy efforts, the federal criminal justice focus, and international debate. By 2002, nearly 75 percent of Americans were very worried that the United States would suffer another major terrorist attack.[5]

Terrorism encompasses not only extreme acts of violence, such as the September 11 attacks, but also widespread attacks that inflict massive victimization, cause financial loss, and spread fear. According to the Federal Bureau of Investigation (FBI), terrorism is the "unlawful use of force or violence to intimidate or coerce a government, the civilian population, or any segment thereof, in furtherance of political or social objectives or goals."[6] In other words, terrorism consists of politically motivated acts of violence.

Terrorism takes several forms and is differentiated according to two factors:

- The target of the attack (which often holds symbolic and practical value)
- The means by which the terror is inflicted (such as arson, bombing, assassination, or chemical attack)

Some of the most common types of terrorism are:

- Eco-terrorism (or environmental terrorism): the infliction of economic damage to those who profit from development and destruction of environmental resources (such as the logging industry, developers and construction companies, and restaurant chains that are perceived to be harmful to animals)

- Economic terrorism: the attack of banks and other financial centers that are seen as symbols of capitalistic oppression (such as those committed by the Weather Underground, whose members committed bank robberies and bombings mostly during the 1970s)

- Racial terrorism: the use of intimidation and violence against select racial and ethnic groups perpetrated by persons holding extremist views about race (such as the Ku Klux Klan (KKK), which targeted African Americans, Jewish Americans, and homosexuals as enemies of white, Christian, heterosexual Americans)

- Cyberterrorism: threats or attacks on computer or information systems
- Bioterrorism: the introduction of biological toxins (such as ricin, anthrax, or smallpox) into a food or water source or transportation hub in an attempt to inflict massive casualties

Regardless of which method they use, terrorists share many common characteristics. First, they are motivated by a profound political cause, religious belief, or social grievance that often places them in a state of inequality compared to their stated enemy. Terrorists often deeply resent this enemy, whom they blame for their problems. Second, terrorists are extremely aggressive and view the use of violence—even suicidal violence—as appropriate and justified means to strike their enemy. Third, they believe that their violence will result in social change benefiting the terrorist organization or movement; at times the social change is as simple as the destruction of the group's enemy.[7]

International terrorists tend to direct their hatred at two common enemies: the United States and Israel.[8] These two countries are targeted for long-standing reasons related to their religious and cultural differences with traditional Islam. International terrorist groups practice a radical form of Islam that views non-Muslims, or infidels, as eternal enemies.[9] In addition to extreme religious hatred, Islamic terrorist groups tend to view the political, economic, and military power of the United States and Israel as oppressive.[10] Indeed, in the wake of the events of September 11, the political motivation behind terrorism has often been conflated with religious fanaticism and Islamic fundamentalism.[11]

Although there are countless terrorist organizations, especially prominent groups in international terrorism include:

- Al Qaeda, which was formed in 1988 by Osama bin Laden and Muhammad Atef, was responsible for the September 11 attacks and is the main opposition of the United States in the War on Terror. Al Qaeda has unofficial partnerships with Egyptian Islamic Jihad, Palestinian Islamic Jihad, and Hezbollah in Lebanon.
- Hamas (the Islamic Resistance Movement), which was founded in 1987, has been engaged in a holy war, or jihad, against Israel. Its ultimate goal is to "liberate Palestine from Israeli occupation." Hamas specializes in bombings and suicide attacks and also employs a political organization that has candidates running in local elections.

Osama bin Laden is the most wanted criminal in the world.

 Will bin Laden ever be found? If he is captured, what effect would that have on world politics? Has bin Laden succeeded in spreading his terrorist message?

- The Palestinian Liberation Organization (PLO) was founded by Yasser Arafat in 1964 and is considered by many to be the terrorist organization that has achieved the greatest political currency. The PLO developed from Al Fatah, an underground terrorist organization that Arafat founded in Egypt in 1956. Today, the PLO has two offshoot organizations that are even more radical in their mission to destroy Israel: the Popular Front for the Liberation of Palestine (PFLP) and the Popular Front for the Liberation of Palestine–General Command (PFLP-GC). Although their primary target is Israel, these groups have recently established cells across the United States.[12]

Within America's borders, several homegrown, domestic terrorist groups pose significant threats:

- Left-wing (i.e., Marxist or communist) political groups, such as Macheteros, New African Freedom Fighters, and the Provisional Party of Communists, strive to destroy capitalist society and its perceived forms of oppression.
- Eco-terrorists, such as the Animal Liberation Front, Earth Liberation Front, and Earth First, use terrorist methods—especially arson and vandalism—to promote radical environmentalism.
- Islamic Fundamentalist Groups, such as Jamaat Ul Fuqra (Fuqra)—an organization created in Pakistan that consists almost entirely of African American extremists who live in communal en-

One of the most devastating acts of eco-terrorism occurred in October 1998, when the Earth Liberation Front (ELF) set a series of arson fires that destroyed eight buildings and resulted in more than $12 million in damage at the ski resort in Vail, Colorado. After the arsons, the ELF issued a statement that the fires were in retaliation for the resort's planned expansion that would destroy the last remaining habitat in Colorado for the lynx.

The ELF has three goals:

- To inflict economic damage on those who would destroy or exploit the natural environment
- To reveal and educate the public about environmental abuses committed by certain corporations
- To take all necessary precautions against harming life

While ELF members view themselves as environmentalists stridently committed to the protection of the wilderness, their actions are pure and simple terrorism.

Source: Louis J. Freeh, "Threat of Terrorism to the United States," testimony before the United States Senate Committees on Appropriations, Armed Services, and Select Committee on Intelligence, May 10, 2001, available at http://www.fbi.gov/congress/congress01/freeh051001.htm, accessed July 12, 2007.

---

vironments (called jamaats) in the United States—commit murders, bombings, white-collar crimes, cybercrime, and identity theft to serve Islam through violence.[13]

- Hate groups include the white supremacist KKK, which was founded by former Confederate soldiers after the Civil War to terrorize blacks, Jews, and even whites who sympathized with the Klan's targets.

LINK At one time, the KKK wielded considerable political influence in the United States and had approximately 5 million members.[14] Today, its legitimate power and presence on the domestic terrorism scene is minimal, with the group having fewer than 6000 members. However, as discussed in Chapter 3, the Klan continues to commit hate crimes nationwide and has become a formidable security threat group in U.S. prisons.[15]

From 1980 to 2001, 482 terrorist attacks occurred within the United States (including the September 11, 2001, attacks).[16] Most of these (67 percent) entailed bomb attacks, followed by arson (7 percent) and assassinations (4 percent) **FIGURE 17-1**. During this period, U.S. criminal justice personnel successfully prevented 133 terrorist acts or thwarted plans of known terrorist groups.[17] On an international level, the prevalence of terrorism against U.S.-related targets was somewhat lower, with 136 terrorist attacks during the same period. As with domestic terrorist acts against the United States, bombings were the most common form of international terrorism.[18]

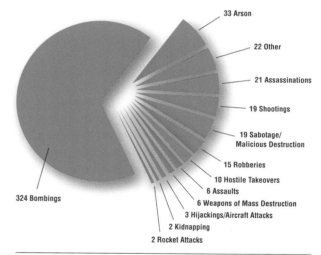

33 Arson
22 Other
21 Assassinations
19 Shootings
19 Sabotage/ Malicious Destruction
15 Robberies
10 Hostile Takeovers
6 Assaults
6 Weapons of Mass Destruction
3 Hijackings/Aircraft Attacks
2 Kidnapping
2 Rocket Attacks
324 Bombings

482 Total Incidents or Planned Attacks

FIGURE 17-1 Domestic Terrorism Events
Note: Figure includes the events of September 11, 2001, which are counted as one terrorist event.

Source: Federal Bureau of Investigation, *Terrorism 2000/2001* (Washington, DC: U.S. Department of Justice, 2004).

Timeline of Major Terrorist Events

The following timeline highlights some major terrorist events of the last 40 years, including assassinations and attempted assassinations of world leaders, airline hijackings, truck bombings, suicide attacks, kidnappings, and murders.

June 5, 1968: U.S. Presidential candidate Robert F. Kennedy is assassinated by Sirhan Sirhan, a Palestinian who was motivated by Kennedy's pro-Israel positions.

June 9, 1970: The PLO attempts to assassinate Jordanian King Hussein.

September 1970: The PFLP hijacks planes in the Netherlands, Switzerland, and Germany and kills more than 400 passengers. In response, Jordan attacks Palestinian neighborhoods, which results in more than 20,000 deaths.

September 5, 1972: Black September, an Islamic terrorist group, kills nine Israeli athletes at the Olympic Games in Munich, Germany.

November 4, 1979: Fifty-two American diplomats are taken hostage by fundamentalist Islamic students in Tehran, Iran, and held for 444 days.

May 13, 1981: Pope John Paul II is shot in an assassination attempt by a Turkish assailant who claims PFLP membership.

October 6, 1981: Egypt President Anwar al-Sadat is assassinated by Muslim extremists within the Egyptian army.

April 18, 1983: The U.S. Embassy in Beirut is destroyed in a suicide car-bombing carried out by the Radical Islamic Jihad, killing 63.

October 23, 1983: U.S. military barracks in Beirut, Lebanon, are destroyed by a truck bombing carried out by Muslims associated with Osama bin Laden's mentor, Imad Magniyah; the attack kills 241 U.S. Marines and 58 French troops.

December 21, 1988: Pan Am Flight 103 explodes over Lockerbie, Scotland, killing 259 passengers, including American students and military personnel. The Libyan government and PFLP-GC claim responsibility for the bombing.

February 26, 1993: A truck bomb explodes in the garage underneath the World Trade Center in New York City, killing 7 persons, injuring nearly 1100 people, and resulting in $500 million in damages. Ramzi Yousef and Sheik Omar Abdel Rahman, both of whom are terrorists with links to Al Qaeda, are sentenced to life imprisonment for their roles in the attack on September 5, 1996.

April 14, 1993: An assassination attempt on U.S. President George H. W. Bush is foiled in Kuwait.

October 4, 1993: Al Qaeda militants destroy U.S. helicopters in Somalia, killing 18 soldiers.

April 19, 1995: A car bomb destroys the Alfred P. Murrah Federal Building in Oklahoma City, killing 168 and wounding 600 innocent civilians. The attack is carried out by Timothy McVeigh and Terry Nichols. McVeigh is executed for his crimes on June 11, 2001; Nichols is sentenced to life imprisonment on August 9, 2004.

June 25, 1996: The Movement for Islamic Change detonates truck bombs outside the U.S. Air Force complex at Khobar Towers in Dahran, Saudi Arabia, killing 19 and wounding 515.

August 7, 1998: Al Qaeda coordinates truck bombings of the U.S. embassies in Kenya and Tanzania, killing 224 people and wounding more than 5000 African civilians.

October 12, 2000: Members of Al Qaeda ram the USS *Cole* in Aden, Yemen, with a boat laden with explosives, killing 17 and injuring 39 U.S. military personnel.

September 11, 2001: Al Qaeda operatives crash commercial airliners into the World Trade Center in New York City and the Pentagon in Washington, D.C., killing 2993 persons and injuring approximately 16,000 innocent civilians. The fourth hijacked plane was overtaken by passengers and crashed in Shanksville, Pennsylvania.

September 18, 2001: Letters containing anthrax are sent to five major U.S. media outlets, killing one civilian. A second batch of anthrax-laden letters is sent on October 9 and results in 4 deaths and 22 injuries. The FBI claims that the anthrax letters were sent by domestic terrorists, but no terrorist group took responsibility.

October 12, 2002: Al Qaeda destroys nightclubs in Bali, killing 202 persons and wounding more than 300 people.

March 11, 2004: Al Qaeda explodes bombs on three commuter trains in Madrid, Spain, killing more than 200 people and injuring 1400.

July 7, 2005: Four Al Qaeda operatives explode bombs on London Underground trains and a bus, killing 52 individuals and injuring more than 700 civilians.

August 21, 2006: An Al Qaeda plot to destroy 10 airplanes flying from England to the United States using liquid explosives is foiled by British law enforcement personnel; 21 terrorists are arrested.[19]

The War on Terror

The United States has declared a War on Terror to fight against radical Islamic terrorist groups, such as Al Qaeda, who wish to destroy the United States and its allies.[20] The DOJ has taken broad steps in this war by designating key terrorist organizations, disman-

On June 2, 2006, 12 adults and 5 juveniles (all males between the ages of 19 and 43) were arrested in Ontario, Canada, for planning to commit a series of terrorist attacks against Canadian targets in southern Ontario. Inspired by Al Qaeda, the suspected terrorists were found in possession of 6000 pounds of ammonium nitrate and other components necessary to create explosive devices. This amount was three times greater than the amount of explosives used in the 1995 Oklahoma City bombing, which resulted in 168 deaths and more than 800 injuries.

Approximately 400 intelligence and law enforcement officers participated in this effort, which was the largest counterterrorism investigation since Canada passed the Anti-Terrorism Act in the wake of the September 11, 2001, attacks on the United States. Some of the charges the suspected terrorists face include participating in terrorist group activities such as training and recruitment, the provision of property for terrorist purposes, and the commission of indictable offenses including firearms and explosives in association with a terrorist group.

Although the Canadian terrorist cell planned to destroy the Peace Tower in Canada's Parliament, the Toronto CN Tower, and the Toronto Stock Exchange, its members also allegedly met with two college students from the United States to plan terrorist strikes against U.S. oil refineries and military bases.

This was not the first time that Canada had stopped a potentially devastating terrorist attack. During the millennium celebrations, Ahmed Ressam was stopped as he attempted to enter the state of Washington from Canada with explosives and timing devices. Ressam was later convicted of conspiracy to detonate a suitcase bomb at Los Angeles International Airport.

Sources: Beth Duff-Brown, "More Arrests Possible as Canadian Terror Probe Crosses Border," *Detroit Free Press,* June 5, 2006, p. A1; Doug Struck, "Arrests Shake Image of Harmony," *Washington Post,* June 4, 2006, p. A10; Jeanne Meserve and Kevin Bohn, "Toronto Terror Plot Foiled—Canada," June 3, 1006, available at http://www.cnn.com/2006/WORLD/americas/06/03/canada.terror/index.html, accessed July 12, 2007.

---

tling terrorist threats and cells, freezing terrorist assets around the world, and killing, capturing, or otherwise incapacitating terrorist operatives. The DOJ has also significantly increased its intelligence capacity to produce information on terrorist subjects and track suspected terrorists.

Congress has worked to prevent future acts of terror against the United States and its allies by passing legislation (such as the USA Patriot Acts, discussed later in this chapter) to facilitate information sharing and cooperation among government agencies; to provide investigators with necessary tools, such as increased wiretapping and surveillance capability; to update the law to reflect new technologies and threats; and to increase the penalties for those who commit terrorist acts or who assist or harbor terrorists.

The DOJ has been successful at using the federal criminal justice system to apprehend, prosecute, convict, and incarcerate terrorists operating both within the United States and abroad. Members or associates of Al Qaeda, the Taliban (the terrorist group that seized power in Afghanistan following the Soviet–Afghan war), Hezbollah, Hamas, and dozens of other terrorist groups from Afghanistan, Colombia, England, Pakistan, and Yemen have been brought to justice. Some of the more high-profile convictions include these examples:

- Richard Reid was sentenced to life imprisonment in January 2003 for attempting to ignite a shoe bomb aboard a transatlantic flight.

- John Walker Lindh was sentenced to 20 years in prison in July 2002 for supplying services to the Taliban.

- The Lackawanna Six (New York) were sentenced to 7 to 10 years in prison in 2003 for providing material support to Al Qaeda.

- The Portland (Oregon) Cell containing seven persons was sentenced to 18 years in prison in 2003 for money laundering, conspiracy to supply goods to the Taliban, and conspiracy to commit sedition.

- Zacarias Moussaoui was sentenced to life imprisonment in May 2006 for his participation in the September 11, 2001, conspiracy.[21]

In general, Americans support the government's efforts in the War on Terror. Nevertheless, some wonder if their civil liberties are being jeopardized by the increased surveillance powers of law enforcement and the trend toward greater information sharing among intelligence, military, and criminal justice organizations. Furthermore, many people worry about violations of the constitutional rights of persons who are suspected of terrorist activity.[22]

Department of Homeland Security

One of the central problems that enabled the terrorist attacks of September 11 was the lack of information sharing between intelligence and criminal justice agencies. In the wake of these attacks, President George W. Bush proposed the creation of the <u>Department of Homeland Security (DHS)</u>, a new Cabinet-level federal unit whose primary mission is to protect the United States by coordinating intelligence efforts and communication with law enforcement. The DHS has the following mission:

- To lead the unified national effort to secure the United States
- To prevent and deter terrorist attacks
- To protect against and respond to threats and hazards to the United States
- To ensure safe and secure borders
- To welcome lawful immigrants and visitors
- To promote the free flow of commerce[23]

The DHS aims to achieve these goals by creating better transportation security systems, strengthening border security, reforming immigration processes, increasing overall preparedness, and enhancing information sharing about terrorist activities.[24] Each of these efforts is the focus of specific directorates (or components) within the DHS.

Transportation Security Administration (TSA)

The TSA was created in response to the terrorist attacks of September 11, 2001, as part of the Aviation and Transportation Security Act signed into law by President Bush on November 19, 2001. The TSA was originally part of the Department of Transportation but was moved under the aegis of the DHS in March 2003. Its mission is to protect U.S. transportation systems by ensuring the freedom of movement for people and commerce. In February 2002, the TSA assumed responsibility for security at U.S. airports and by the end of the year had deployed a federal work force to

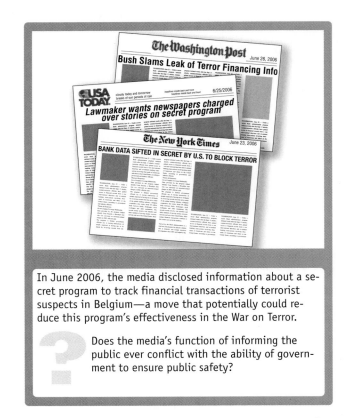

In June 2006, the media disclosed information about a secret program to track financial transactions of terrorist suspects in Belgium—a move that potentially could reduce this program's effectiveness in the War on Terror.

? Does the media's function of informing the public ever conflict with the ability of government to ensure public safety?

meet the challenging Congressional deadlines for screening all passengers and baggage.[25]

U.S. Customs and Border Protection (CBP)

As the unified border agency, the CBP combines the inspectional work forces and broad border authori-

After September 11, 2001, President George W. Bush made counterterrorism the primary focus of his presidency.

? How has the War on Terror blended the worlds of military, espionage, and criminal justice? Has this paradigm shift improved criminal justice systems?

Perhaps the most controversial criminal justice/military policy in the War on Terror has been the detention of enemy combatants (members of Al Qaeda and the Taliban), who have been held by U.S. authorities in Camp Delta in Guantanamo Bay, Cuba. According to the doctrine laid out by President George W. Bush, enemy combatants are not afforded the due process rights reserved for members of official state militaries or criminal defendants. On June 29, 2006, the U.S. Supreme Court formally disagreed with this position, when it held that terrorist suspects detained in the prison in Guantanamo Bay must receive due process provisions consistent with the military system and international standards or be released from military custody. The decision was an affront to the Bush administration's plan to try terrorists as enemy combatants in military tribunals, which do not provide the same rights for prisoners of war as the Geneva Convention does. Perhaps not surprisingly, the Bush administration was sharply critical of the Court's ruling.

In *Hamdan v. Rumsfeld*, the Court held that it had jurisdiction to rule on the matter and found that the federal government did not have authority to set up these particular special military commissions, which were ruled illegal under both the Uniform Code of Military Justice and the Geneva Convention. Although the ruling was a blow to the Bush administration's War on Terror, the Court did not categorically prohibit military commissions. In a concurring opinion with the majority, Justice Stephen Breyer wrote, "Congress has denied the President the legislative authority to cre-

ate military commissions of the kind at issue here. Nothing prevents the President from returning to Congress to seek the authority he believes necessary."

In his dissent, Justice Antonin Scalia forcefully criticized the Court and its decision. Fundamentally, Scalia argued that no court had jurisdiction to hear court requests of a detainee from Guantanamo Bay. In Justice Scalia's words:

> On December 30, 2005, Congress enacted the Detainee Treatment Act (DTA). It unambiguously provides that, as of that date, "no court, justice, or judge" shall have jurisdiction to consider the habeas application of a Guantanamo Bay detainee. Notwithstanding this plain directive, the Court today concludes that, on what it calls the statute's *most natural* reading, *every* "court, justice, or judge" before whom such a habeas application was pending on December 30 had jurisdiction to hear, consider, and render judgment on it. This conclusion is patently erroneous. And even if it were not, the jurisdiction supposedly retained should, in an exercise of sound equitable discretion, not be exercised.

It is clear that the controversy surrounding certain counterterrorism measures is hotly debated, even among Supreme Court Justices.

Sources: Hamdan v. Rumsfeld, 547 U.S. 1002 (2006); Associated Press, "Gonzales: Gitmo Ruling 'Hampered' War on Terror," available at http://www.cnn.com/2006/LAW/07/01/gonzales.gitmo/index.html, accessed July 12, 2007.

ties of U.S. Customs; the U.S. Immigration, Animal and Plant Health Inspection Service; and the U.S. Border Patrol.[26]

Immigration and Customs Enforcement (ICE)

ICE, the largest investigative branch of the DHS, was created by combining the law enforcement arms of the former Immigration and Naturalization Service and the former U.S. Customs Service. Before September 11, 2001, immigration and customs authorities were not widely recognized as being an effective counterterrorism tool in the United States. ICE changed this perception by creating a host of new

systems to better address national security threats, detect potential terrorist activities in the United States, effectively enforce immigration and customs laws, and protect against terrorist attacks. ICE accomplishes its goals by targeting illegal immigrants; the people, money, and materials that support terrorism; and other criminal activities. It is a key component of the DHS's "layered defense" approach to protecting the United States.[27]

Additional components of DHS work with local and state law enforcement and other criminal justice officials in the event of emergencies:

- Federal Law Enforcement Training Center: trains law enforcement personnel

From May 26 through June 25, 2006, ICE conducted Operation Return to Sender, a nationwide crackdown that resulted in the apprehension of 2179 criminal aliens, illegal alien gang members, fugitive aliens, and other immigration status violators. Approximately 50 percent of the arrestees had criminal records, and the roundup included 146 child molesters and 367 members of violent gangs, such as Mara Salvatrucha (MS-13). Nearly 1000 of the arrestees have already been deported and repatriated to their home countries. Operation Return to Sender is part of the Secure Border Initiative, which is the Department of Homeland Security's comprehensive, multi-year plan to secure the U.S. borders and reduce illegal immigration that contributes to crime and terrorist activity.

More than 500,000 fugitive aliens or persons have been deported and either reentered the country or never left. Operation Return to Sender is one example of the new criminal justice face of immigration and national security, which combines traditional police sweeps and crackdowns that target aliens, known criminals, and other potential threats to public safety.

Source: United States Department of Homeland Security, "ICE Apprehends More Than 2,100 Criminal Aliens, Gang Members, Fugitives, and Other Immigration Violators in Nationwide Interior Enforcement Operation," June 14, 2006, available at http://www.dhs.gov/xnews/releases/press_release_0926.shtm, accessed July 14, 2007.

- U.S. Citizenship and Immigration Services: establishes immigration services policies and priorities and manages immigration functions and adjudications

- Directorate for Preparedness: identifies threats, vulnerabilities, and targets to protect U.S. borders, seaports, bridges and highways, and information systems

- Office of Intelligence and Analysis: uses information and intelligence from multiple international, national, state, and local sources to identify and access current and future threats to the United States

- Directorate for Science and Technology: serves as the research and development arm of the DHS and assists all other components

- Directorate for Management: serves as the budgetary, appropriations and finance, and human resources arm of the DHS and assists all other components[28]

USA Patriot Act

The most important and controversial counterterrorism measure implemented since the events of September 11, 2001, is the USA Patriot Act. The Patriot Act empowers law enforcement personnel who engage in criminal justice and counterterrorism activities in several important ways. For example, it allows law enforcement to use electronic (wiretap) surveillance to monitor nonterrorism crimes that terrorists commit to build and sustain their resource infrastructure. In other words, when authorities investigate crimes such as immigration fraud, mail fraud, and passport fraud, it often leads to the discovery of more fruitful information about plans for violence.

The Patriot Act also allows "roving wiretaps" that permit the surveillance of a particular person rather than a particular telephone or communications device. Because terrorists are trained to move frequently to evade capture, the roving wiretaps give law enforcement personnel more latitude in tracking individual suspects. Similarly, the Patriot Act provides that search warrants can be obtained in any district where terrorism occurred, regardless of where the warrant will actually be executed.

Additionally, the Patriot Act permits law enforcement, under narrow circumstances, to delay for a limited time when the subject is told that a judicially approved search warrant has been executed. Such a delay gives authorities time to identify terrorist associates, eliminate immediate threats, and coordinate the arrests of multiple individuals without tipping them off beforehand.

Federal agents may, under the provisions of the Patriot Act, obtain the records of businesses that are suspected to pose risks to national security owing to their association with terrorist groups. The Patriot Act also removed the major legal barriers that prevented the law enforcement, intelligence, and national defense communities from sharing information and coordinating their efforts to ensure public safety.

The Patriot Act increased the legal penalties for an assortment of terrorism crimes, enhanced a number of conspiracy penalties, and abolished the statutes of limitations for certain terrorism crimes. It also prohibits the harboring of terrorists, attacks on mass transit systems, and bioterrorism crimes.

The Patriot Act has been criticized for its potential to infringe on civil liberties, but many media accounts fail to inform readers about the actual provisions of this legislation. In most cases, the Act simply gathered together criminal justice policies that were already in place, have been used, and have already passed constitutional muster via judicial review. Nevertheless, some view these provisions as an example of excessive and unconstitutional governmental power. For example, the American Civil Liberties Union described the Patriot Act in the following manner:

> Just 45 days after the September 11 attacks, with virtually no debate, Congress passed the USA Patriot Act. There are significant flaws in the Patriot Act, flaws that threaten your fundamental freedoms by giving the government the power to access your medical records, tax records, information about the books you buy or borrow without probable cause, and the power to break into your home and conduct secret searches without telling you for weeks, months, or indefinitely.[29]

Despite the controversy, the Patriot Act has been the major criminal justice initiative used by the United States to wage the War on Terror.

The USA Patriot Act Improvement and Reauthorization Act of 2005, which was passed by Congress in 2005, further enhanced the powers of the criminal justice system to combat crime and terrorism. It made several provisions of the original Patriot Act permanent and extended others to facilitate communication between law enforcement and national security, permitted victims of computer hackers to request law enforcement assistance in monitoring trespassers on their computers, and made several terrorism-related crimes wiretap predicates. In addition, this legislation reorganized the DOJ to create a National Security Division and clarified death penalty procedures for terrorist defendants. In an effort to reduce crime at the nation's borders, the revised Act empowered the U.S. Coast Guard to stop suspicious persons believed to be entering the country with chemical, radioactive, or nuclear materials. Sentencing enhancements have also been added for maritime crimes.[30]

After the September 11 attacks, racial profiling became a controversial issue in counterterrorism and law enforcement.

? Which issues underscore the use of race, ethnicity, or national origin as proxies for dangerousness? Are there times when public safety should override civil liberties?

Taken collectively, the Patriot Acts target the financial backing of terrorist groups in several ways: by holding U.S. banks accountable for their dealings with foreign banks, by strengthening laws on money laundering as they relate to terrorist financing, and by creating asset forfeiture laws in matters involving the funding of terrorist activities.

Foreign Intelligence Surveillance Act

The Foreign Intelligence Surveillance Act of 1978 (FISA) established a legal procedure for foreign intelligence surveillance that was separate from other types of surveillance by law enforcement.[31] FISA aims to regulate the collection of foreign intelligence information in furtherance of U.S. counterintelligence, including information in the following areas:

- Espionage and other intelligence activities
- Sabotage
- Assassinations conducted by or on behalf of foreign governments, organizations, or persons
- International terrorist activities

LINK The Foreign Intelligence Surveillance Court, discussed in Chapter 9, reviews the U.S. Attorney General's applications for authorization of electronic surveillance aimed at obtaining foreign intelligence information. The records and files of these cases are sealed and may not be revealed even to persons whose prosecutions are based on evidence obtained under FISA warrants, except to a limited degree.[32]

Police Responses to Terrorism

Federal, state, and local law enforcement agencies also play important roles in the War on Terror by turning their attention toward the white-collar crimes that help terrorist organizations finance their operations and support their infrastructure. Since September 11, 2001, more than 100 federal criminal cases have been brought against foreign and U.S. nationals who were convicted of these white-collar crime offenses for the purposes of terrorism.[33]

LINK White-collar crimes (such as money laundering, credit card theft, insurance fraud, immigration fraud, and tax evasion) preserve the anonymity of terrorists while funding their violent acts. These crimes and their victims are discussed in greater detail in Chapter 3.

Additionally, law enforcement agencies are working to respond to the needs of victims of terrorism through the government-sponsored Terrorism and International Victims Unit (TIVU). The TIVU develops programs to help victims of terrorism, mass violence, and other transnational crimes by providing medical assistance, financial aid, and legal guidance. The government is also funding terrorist-research organizations, such as the National Consortium for the Study of Terrorism and Responses to Terror, which use state-of-the-art theories, methods, and data from the social and behavioral sciences to improve understanding of the origins, dynamics, and social and psychological impacts of terrorism and enhance the resilience of U.S. society in the face of terrorist threats.[34]

Terrorism is not a traditional problem for U.S. law enforcement, so it might seem that police are poorly prepared to effectively deal with terrorist threats. In reality, though, the community policing effort has helped train police to effectively address the underlying social problems that engender crime. Additionally, local law enforcement officers are now trained and continue to participate in training throughout their careers to address and prepare for terrorism and terrorism-related criminal activity. Officers must make risk assessments of the potential damage that terrorists pose to their community, including assessments of critical infrastructure and likely targets, the seriousness of the threat, the vulnerability of the community to a terrorist attack, a calculation of casualties and other risks, and a counterterrorism response.[35]

LINK Community policing, as discussed in Chapter 5, is a collaborative effort between the police and the community (i.e., community leaders, social service agencies) to identify crimes and find solutions to these problems.

Within the DOJ, the Office of Community Oriented Policing Services (COPS) provides law enforcement and community resources to respond to terrorism threats in several important ways:

- Improving data and intelligence collection and processing
- Capitalizing on technological advancements to gather and use intelligence effectively
- Increasing communication between local police and other public safety agencies
- Working with federal agencies and local victim agencies to assist victims of terrorism[36]

Many members of "local" law enforcement are no longer doing their jobs only at the local level. That is, law enforcement agencies have increasingly established regional mutual aid agreements under which multiple criminal justice agencies are contracted to respond to specific problems of terrorism. Such mutual aid agreements have facilitated information sharing between neighboring and regional criminal justice agencies.[37]

Judicial Responses to Terrorism

With increased assistance and cooperation from federal entities, state courts are now facilitating the prosecution, deportation, and sentencing of criminal offenders who are found to be in violation of immigration law. In some cases, the defendants are involved in or suspected to be conspiring to commit terrorism; however, in the majority of cases, the defendants are involved in street crimes and immigration violations. For example, in July 2006, ICE agents arrested 154 undocumented immigrants in Columbus, Ohio, with the help of local police. These offenders were then housed in local jails, prosecuted in local courts, and deported from local airports.[38]

Local prosecutors have made concerted efforts to assist the federal government in the War on Terror, such as focusing on precursor crimes that can be used to establish the existence of organized crime or terrorist efforts. Although only 16 percent of local prosecutors have prosecuted a terrorism-related case, nearly 75 percent report that they feel actively involved in homeland security.[39] Prosecutors are part of important information-sharing networks, though many complain that federal authorities do not share enough information with local officials, which hampers prosecutors' abilities to make connections in investigating and prosecuting potential terrorists.[40]

Counterterrorism is, without question, the primary foreign policy concern of the United States today and a new facet of traditional criminal justice practice. It is an emerging crime, like cybercrime and identity theft, that has presented unique challenges for lawmakers and criminal justice personnel. It is also a development that will likely shape the future of criminal justice in the United States.

Cybercrime

Cybercrime is an umbrella term that describes various criminal behaviors that are intended to damage or destroy computer networks, fraudulently obtain money and other commodities, and disrupt normal business operations. Although cybercrime is not usually characterized by extremely violent criminal behavior, it is arguably as devastating to a society as terrorism. The global cost of cybercrime is more than $500 billion per year. Unfortunately, only approximately 10 percent of cybercriminals are reported to police, and less than 2 percent of cybercrime cases result in conviction.[41]

Cybercrime can seriously disrupt or damage travel, commerce, and the normal operating schedule of a society. It can take any of several forms:

- *Computer viruses:* hidden fragments of computer code that spread by inserting themselves into or modifying other programs
- *Cybervandalism or cybersabotage:* deliberate or malicious destruction or alteration of electronic files, webpages, data, and computer programs
- *Denial of service:* disruption or degradation of an Internet connection or e-mail service that results in an interruption of the normal flow of information
- *Hacking:* unauthorized gaining of access to computers
- *Hijacking:* installation of malicious software that effectively takes over the victim's computer, such as flooding it with obscene pop-up windows
- *Sniffing:* unlawful monitoring of data traveling over a computer network
- *Spamming:* sending of frequent, unwanted e-mail advertisements
- *Spoofing:* unauthorized gaining of access to a computer through a message from a trusted Internet address
- *Spyware:* illegally installed software that obtains personal information from a personal computer
- *Theft of proprietary information:* illegal electronic copying of copyrighted materials or other data[42]

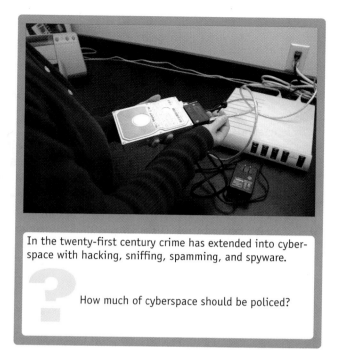

In the twenty-first century crime has extended into cyberspace with hacking, sniffing, spamming, and spyware.

? How much of cyberspace should be policed?

Some estimates suggest that cybercrime is even more costly than the preceding data would suggest. According to former U.S. Attorney General Alberto Gonzales, cybercrime results in an annual loss of $250 billion and has led to the loss of more than 750,000 jobs.[43] Nearly 75 percent of businesses claim that they have been the victims of cybercriminals.[44] In addition, nearly 1 in 5 businesses have suffered 20 or more cyberattacks in the past year **FIGURE 17-2**. In

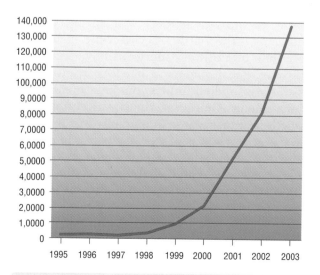

FIGURE 17-2 The Growth of Cybercrime

Source: Prasad Calyam, "Network Forensics," available at http://ftp.osc.edu/education/si/projects/forensics/index.shtml, accessed July 14, 2007.

In May 2002, David L. Smith of New Jersey was sentenced to 20 months in federal prison and fined $5000 for creating Melissa, a computer virus that caused $80 million in damages to computer networks and businesses in 1999. Smith is also liable for state fines in excess of $150,000.

Smith created the Melissa virus and disseminated it from his home computer through e-mail. The virus appeared on thousands of e-mail systems on March 26, 1999, disguised as an important message from a colleague or friend. It sent an infected e-mail to the first 50 e-mail addresses on users' mailing lists, evading antivirus software and infecting computers using Windows operating systems and other Microsoft programs. The Melissa virus was able to spread very quickly by overloading e-mail servers, which resulted in the shut-

down of networks and significant costs to repair or cleanse computer systems.

In state and federal court, Smith described how, using a stolen America Online account and his own account with a local Internet service provider, he posted an infected document on the Internet newsgroup "Alt.Sex." The posting contained a message that enticed readers to download and open the document with the hope of finding passwords to adult-content websites; it read, "Here is that document you asked for . . . don't show anyone else ;-)."

Opening and downloading the message caused the Melissa virus to infect victim computers. The virus altered Microsoft word-processing programs such that any document created using the programs would then be infected with the Melissa virus. The virus

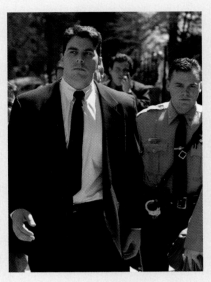

also lowered macro security settings in the word-processing programs.

Source: Department of Justice Press Release, "Creator of Melissa Computer Virus Sentenced to 20 Months in Federal Prison," May 1, 2002, available at http://www.usdoj.gov/criminal/cybercrime/melissaSent.htm, accessed August 1, 2007.

a single day, Symantec (a computer security provider) aborted 59 million attempts at cybercrime!

On an individual level, one in every three Americans is the victim of cybercrime each year.[45] More than 60 million Americans have received a "phishing" e-mail, and more than 90 percent of businesses have suffered from a computer breach from cybercriminals. Almost everyone is susceptible to cybervictimization because individuals do not control who has access to businesses' databases. For example, in 2004, a portable computer database containing personal and credit card information for almost 40,000 customers was stolen from BJ's Wholesale Club.[46] In 2006, a laptop containing sensitive data for 26 million U.S. veterans was stolen from the Veterans Administration.[47]

The U.S. government is working diligently to combat computer and intellectual property crimes worldwide. For instance, the DOJ has developed a Computer Crime and Intellectual Property Section (CCIPS) charged with combating electronic penetrations, data thefts, and cyberattacks on critical

information systems. CCIPS aims to prevent, investigate, and prosecute computer crimes by working with other government agencies, the private sector, academic institutions, and foreign counterparts.[48] In addition to establishing CCIPS, the DOJ has taken the following actions related to cybercrime:

- Created special Computer Hacker and Intellectual Property units staffed by prosecutors who specialize in the investigation of cybercrime
- Dispatched prosecutors to Asia and Europe to dismantle international organized crime rings that use cybercrime to fund other criminal activities
- Increased prosecutions of cybercriminals involved in intellectual property offenses by 98 percent
- Bolstered extradition treaties to include cybercrime
- Sponsored the Intellectual Property Protection Act, which would enhance habitual-offender penalties for certain cybercriminals[49]

LINK Habitual-offender laws, such as three-strikes provisions (presented in Chapter 12), aim to incapacitate the most active criminals by imposing harsher sentences based on their prior criminal records. These laws are effective at removing the most active offenders from society.[50]

Cyberterrorism

Cyberspace is the term used to refer to the hundreds of thousands of interconnected computers, servers, routers, switches, and fiber-optic cables that allow the U.S. infrastructure to work. Cyberspace has been referred to as the "nervous system" of the country, reflecting the fact that computers play a major part in every major area of American life, including agriculture, water, public health, defense, industry, travel, and commerce. For these reasons, controlling and protecting cyberspace is as critical as protecting the homeland from violent terrorist attack. Both terrorists and conventional criminals are keenly aware of the importance of computers to the American way of life and can inflict major damages to U.S. society without bloodshed through cyberterrorism. The bad news is that cyberspace is largely open for exploitation by cybercriminals, who have taken advantage of this vulnerability by committing new crimes such as phishing and identity theft. For instance, a nationally representative sample of 2000 U.S. households reported that nearly 30 percent had been victimized by viruses, spyware, phishing, or some other cyberattack. Because of these vulnerabilities, an estimated 2.6 million households must replace their computer systems annually. Moreover, one in five U.S. households does not have antivirus software installed to repel cyberattacks.[51]

The National Strategy to Secure Cyberspace (a DHS program created in 2002) is pursuing several goals intended to increase cyberspace security:

- Establish a public–private structure for responding to national-level cyberspace incidents
- Provide for the development of tactical and strategic analysis of cyberattacks and vulnerability assessment
- Encourage the development of a private-sector capability to protect cyberspace
- Support the role of the DHS in coordinating crisis management
- Coordinate processes for voluntary participation in cyberspace protection
- Exercise cybersecurity plans for federal systems
- Improve and enhance public–private information sharing involving cyberspace attacks, threats, and vulnerability[52]

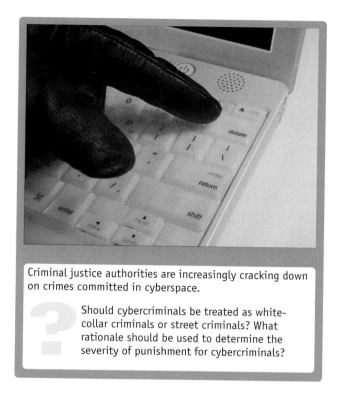

Criminal justice authorities are increasingly cracking down on crimes committed in cyberspace.

? Should cybercriminals be treated as white-collar criminals or street criminals? What rationale should be used to determine the severity of punishment for cybercriminals?

Fortunately, federal authorities and local criminal justice systems have made some headway in their response to cybercrime. The Computer Security Institute (a national organization devoted to computer security) and the FBI recently conducted the nationally representative Computer Crime and Security Survey and reported promising findings:

- Total financial losses from cyberattacks have declined dramatically, down 61 percent in the past year.
- Attacks on computer systems and misuse of systems have slowly declined.
- More organizations and businesses guard against cybercrime using antivirus software, intrusion detection systems, and server-based access control lists.
- Almost 90 percent of respondents conduct security audits to evaluate their readiness against cyberattack.[53]

Thirty states have either passed or are considering legislation that targets cybercriminals who use spyware, and in 2005 the House of Representatives passed an anti-spyware bill. In July 2005, Virginia became the first state to pass an anti-phishing bill, which America Online used to file lawsuits against phishing gangs. Other states, such as California and New York, are also pursuing new statutes to target phishers.[54]

Identity Theft

One of the fastest-growing types of cybercrime is identity theft— the use or attempted use of another person's financial account or identifying information without the owner's permission. In the 1990s, as society became increasingly dependent upon computers, cases of identity theft began to increase at a rate of more than 30 percent each year; between 1996 and 2000, identity theft fraud losses incurred by MasterCard and Visa increased by more than 43 percent. Since 2000, identity theft cases have plateaued, as consumers have become savvier at protecting themselves against identity theft.[55]

The most common types of crimes committed in tandem with identity theft are unauthorized use of existing credit cards, savings accounts, or checking accounts and the misuse of personal information to obtain new accounts or loans, or commit crimes. Many identity thieves use phishing—the creation and use of e-mails and websites designed to look like those of well-known legitimate businesses, financial institutions, or government agencies—to deceive Internet users into disclosing private financial information, such as passwords or personal identification numbers.

Although identity theft is not a widespread problem, it carries enormous costs and negative consequences for victims and consumers. The National Crime Victimization Survey estimates that nearly 3.6 million American households (or 3 percent) had at least one family member who had been victimized by identity theft in the past 6 months. The average identity theft incident resulted in $1290 in losses, and national cumulative losses total more than $3.2 billion.[56]

Identity theft presents a host of problems for the victims of this crime:

- Confrontations with debt collectors or creditors
- Banking problems, such as overdrafts of checking accounts
- Problems with credit card accounts, such as fraudulent charges resulting in damaged credit history
- Higher interest rates
- Denial of phone or utility service
- Higher insurance rates
- Civil lawsuits
- Criminal investigations

In addition to suffering billions of dollars in damages, victims of identity theft have spent an estimated 300 million hours resolving various problems stemming from the theft. Part of the reason that identity theft has thrived is that the criminal justice system has generally been unprepared to deal with the threat. There is no single database that captures all investigations and prosecutions of identity theft cases, and 75 percent of identity theft victims do not report the crime to the police.[57]

In October 1998, Congress passed the Identity Theft and Assumption Deterrence Act (Identity Theft Act) to address the problem of identity theft. Specifically, this Act made identity theft a federal crime, including knowingly transferring or using another person's identification with the intent to commit, aid, or abet any unlawful activity. In addition to the specific charge of identity theft, identity criminals may be in violation of several federal laws by committing wire fraud, credit card fraud, bank fraud, computer fraud, and others.[58] Identity theft is also a crime at the state level. Since 1998, every state except Colorado and Vermont has enacted legislation on identity theft.[59]

In 2004, Congress toughened the penalties for identity theft-related convictions with the passage of the Identity Theft Penalty Enhancement Act. This Act added two years of imprisonment onto sentences that involve the use of identity theft and an additional five years when the identity theft facilitates involvement with terrorists or terrorist organizations.

In 2006, President George W. Bush convened the Identity Theft Task Force to advise lawmakers and criminal justice systems about how to most effectively respond to the identity theft problem. Among its recommendations were increased penalties for criminal defendants who are convicted of identity theft, a universal police report to help law enforcement and victims track complaints, and victims' rights laws that would provide compensation for time lost disputing fraudulent charges on credit and debit cards.[60]

LINK The Enhancement Act is particularly tough in that it precludes probation as a sentence for identity theft and mandates consecutive—not concurrent—sentences. These types of sentences are discussed in greater detail in Chapter 12.

Victims and the criminal justice system alike were initially slow to respond to identity theft offenders because both parties viewed credit card companies as the ultimate victims of the offense. It was not until identity thieves repeatedly used various credit cards or other types of stolen financial information that the criminal justice system became involved.[61] Like all cybercrimes, identity theft cases pose problems for law enforcement because the complexity of most identity crimes makes it difficult to get an arrest warrant, and the use of the Internet makes jurisdiction difficult to establish.[62]

In 2006, the DOJ launched a National Strategy to Combat Identity Theft, administered by COPS. It includes several key components:

- *Partnerships and collaboration:* state agencies that provide crime analysis, victim assistance, and statewide investigations, and promote intelligence sharing among law enforcement agencies
- *Reporting procedures:* uniform data collection on identity theft cases for use in the Uniform Crime Reports
- *Victim assistance:* policies for responding to victims of identity theft, including written standard operating procedures to help victims mitigate the effects on their financial accounts and credit
- *Public awareness:* national public awareness campaign focusing on prevention and response techniques as well as reporting of identity theft crime
- *Legislation:* documentation of all state and federal identity theft legislation
- *Information protection:* national public education for consumers and merchants that focuses on information protection
- *Training:* requirement that police, prosecutor, victim-assistance, and private-sector organizations dealing with identity theft conduct an assessment of identity theft training needs[63]

LINK The Uniform Crime Reports, discussed in Chapter 3, collect statistics from local and state law enforcement agencies to estimate national levels of crime.

Several federal laws pertain directly to identity theft, including those dealing with identification documents, access devices, loan and credit applications, bank fraud, and Social Security number misuse. To date, law enforcement personnel with the FBI, U.S. Secret Service, and U.S. Postal Inspectors have made more than 10,000 arrests of identity theft offenders, resulting in nearly 3000 convictions secured by U.S. Attorneys.[64]

Technology enables the commission of crimes such as identity theft and technology, but it will also

Since Congress passed the Identity Theft and Assumption Deterrence Act of 1998, the criminal justice system has targeted identity thieves and increased penalties for identity theft violations.

 Would greater regulation reduce the amount of cybercrime that occurs? Should identity thieves be permanently barred from using computers as part of their sentences?

play a vital role in preventing such crimes by improving ways that individuals and organizations conduct financial transactions and by increasing authentication methods. Authentication helps verify the identity of the individual by using personal identification numbers or check verification processes. The next generation of authentication technologies will most likely capitalize on biometrics, which accurately captures an individual's unique physical attributes—such as fingerprints, voice, eyes, face, and written signature—in electronic format.[65]

WRAP UP

ICE is a key component of the DHS's layered defense approach to protect the United States. Its mission is to protect the country and ensure public safety by identifying criminal activities, eliminating vulnerabilities that pose a threat to U.S. borders, and enforcing measures intended to bolster economic, transportation, and infrastructure security. By protecting national and border security, ICE seeks to eliminate the potential threat of terrorist acts against the United States.

Before September 11, 2001, immigration and customs authorities were not widely viewed as effective counterterrorism tools in the United States. ICE changed this perception by targeting the people, money, and materials that support terrorist and criminal activities, such as money laundering, immigration violations, and fraud. The attacks of September 11 demonstrated the United States' vulnerability to undocumented immigration and the shadow financial businesses that support criminal and terrorist groups. The sharing of information between criminal justice and intelligence organizations has merged criminal justice and counterterrorism efforts in critical and unprecedented ways.

Chapter Spotlight

- Terrorism has emerged as the greatest national security threat to the United States and has resulted in massive restructuring of the way that the criminal justice system responds to a variety of crimes.

- The Department of Homeland Security reflects the new focus of the Department of Justice, which is to prevent and preempt terrorist activity.

- Terrorist groups and organizations may emerge from all over the globe, within the United States, and from extreme ideological positions on the far left and far right.

- The USA Patriot Act greatly enhanced the surveillance and investigative powers of law enforcement in the United States.

- Cyberspace is among the most important and vulnerable targets to attack from terrorists and nonterrorist criminals.

- Cyberspace is the location of an assortment of cybercrimes, most notably identity theft.

Putting It All Together

1. Should terrorist suspects be treated as prisoners of war with specific rights to due process or should they be treated as enemy combatants? Does this issue hamper the ability to fight terrorists?

2. In what ways is the War on Terror also an exercise in criminal justice?

3. What does the USA Patriot Act empower law enforcement to do? Is the controversy surrounding the Patriot Act justified?

4. Which crimes are enabled by the nature of cyberspace?

5. How do criminals obtain the information needed to commit identity theft? How common are these methods?

Al Qaeda A terrorist group led by Osama bin Laden that was responsible for the terrorist attacks of September 11, 2001, and is the most active international terrorist organization since the 1990s.

bioterrorism The use of biological toxins to inflict mass casualties.

cybercrime Criminal behaviors that are intended to damage or destroy computer networks, fraudulently obtain money and other commodities, and disrupt normal business operations.

cyberspace Interconnected computers, servers, routers, switches, and fiber-optic cables that allow the U.S. infrastructure to work.

cyberterrorism Terrorism that involves threats or attack on computer or information systems.

Department of Homeland Security (DHS) The federal agency that oversees U.S. counterterrorism efforts.

economic terrorism Terrorism directed toward banks and purported symbols of capitalist oppression.

eco-terrorism The infliction of economic damage on those who profit from development and destruction of environmental resources.

Hamas A terrorist organization also known as the Islamic Resistance Movement.

Hamdan v. Rumsfeld A Supreme Court decision that ruled that the George W. Bush administration's policy of treating terrorists as enemy combatants without due process is unconstitutional.

identity theft Any illegal act that involves the use or attempted use of another person's financial account or identifying information without the owner's permission.

Jihad Holy war of fundamentalist Islam directed against nations such as the United States and Israel.

Palestinian Liberation Organization (PLO) A terrorist organization that directs violence mostly against Israel.

phishing The creation and use of e-mails and websites designed to look like those of well-known legitimate businesses, financial institutions, or government agencies in an attempt to deceive Internet users into disclosing private financial information.

racial terrorism Terrorism motivated by extreme racial prejudice that is directed toward a specific racial or ethnic group.

terrorism The unlawful use of force or violence to intimidate or coerce a government, the civilian population, or any segment thereof, in furtherance of political or social objectives or goals.

Terrorism and International Victims Unit (TIVU) A federal agency responsible for developing programs to help victims of terrorism, mass violence, and other transnational crimes.

USA Patriot Act Federal legislation that broadened the surveillance and investigative powers of criminal justice agencies to combat terrorism and remove barriers between military, intelligence, and criminal justice entities, enabling them to share information on terrorist suspects and threats.

USA Patriot Act Improvement and Reauthorization Act of 2005 Federal legislation that revised the USA Patriot Act of 2001, enhancing the powers of the criminal justice system to combat crime and terrorism.

War on Terror The global armed conflict between radical Islamic terrorist groups, such as Al Qaeda, and the developed world, principally the United States and the G8 nations.

1. United States House of Representatives Committee on Ways and Means, "Facts and Figures: Identity Theft," available at http://waysandmeans.house.gov/media/pdf/ss/facts figures.pdf, accessed July 12, 2007.

2. Dinesh D'Souza, *What's So Great about America* (Washington, DC: Regnery, 2002).

3. Brent Smith and Kelly Damphousse, "Punishing Political Offenders: The Effect of Political Motive on Federal Sentencing Decisions," *Criminology* 34:289–321 (1996); Brent Smith, Kelly Damphousse, Freedom Jackson, and Amy Sellers, "The Prosecution and Punishment of International Terrorists in Federal Courts: 1980–1998," *Criminology and Public Policy* 1:311–338 (2002).

4. The National Commission on Terrorist Attacks Upon the United States, *The 9/11 Commission Report* (New York: W. W. Norton, 2003), p. 4.

5. The Pew Research Center for the People and the Press, *The 2004 Political Landscape: Evenly Divided and Increasingly Polarized* (Washington, DC: Pew Research Center, 2006).

6. Federal Bureau of Investigation, *Terrorism 2000/2001* (Washington, DC: U.S. Department of Justice, 2004).

7. Randy Borum, "Understanding the Terrorist Mind-Set," *FBI Law Enforcement Bulletin* 72:7–10 (2003).

8. Jonathan White, *Defending the Homeland: Domestic Intelligence, Law Enforcement, and Security* (Belmont, CA: Wadsworth, 2004).

9. Michael Hronick, "Analyzing Terror: Researchers Study the Perpetrators and the Effects of Suicide Terrorism," *National Institute of Justice Journal* 254:1–5 (2006).

10. Steven Emerson, *American Jihad: The Terrorists Living Among Us* (New York: Free Press, 2002); National Commission on Terrorist Attacks Upon the United States, note 4; Bruce Hoffman, "Combating al Qaeda and the Militant Islamic Threat," testimony presented to the House Armed Services Committee, Subcommittee on Terrorism, Unconventional Threats and Capabilities, February 16, 2006.

11. Raphael Perl, "Terrorism and National Security: Issues and Trends," *Congressional Research Service* (Washington, DC: Library of Congress, 2004); Hoffman, note 10.

12. Emerson, note 10, pp. 177–182.

13. John Kane and April Wall, *Identifying the Links between White-Collar Crime and Terrorism* (Washington, DC: U.S. Department of Justice, Office of Justice Programs, National Institute of Justice, 2005).

14. Rory McVeigh, "Structural Incentives for Conservative Mobilization: Power Devaluation and the Rise of the Ku Klux Klan, 1915–1925," *Social Forces* 77:1461–1496 (1999).

15. Gregg Etter, David McElreath, and Chester Quarles, "The Ku Klux Klan: Evolution Towards Revolution," *Journal of Gang Research* 12:1–16 (2005).

16. Federal Bureau of Investigation, note 6.

17. Federal Bureau of Investigation, note 6.

18. United States Department of State, *Patterns of Global Terrorism, 2003* (Washington, DC: U.S. Department of State, 2004).

19. Bruce Hoffman, *Inside Terrorism,* 2nd ed. (New York: Columbia University Press, 2006); Emerson, note 10; National Commission on Terrorist Attacks Upon the United States, note 10; United States Army, "Timeline of Terrorism," available at http://www.army.mil/terrorism/, accessed August 21, 2007; Robert O'Block, "Timeline of Terrorism," American Board for Certification in Homeland Security, available at http://www.acfei.com/images/PDF/Final%20Timeline%20of%20T.%20color.pdf, accessed July 13, 2007.

20. United States Department of Justice, "Waging the War on Terror," available at http://www.lifeandliberty.gov/subs/a_terr.htm, accessed August 1, 2007.

21. United States Department of Justice, "Department of Justice Examples of Terrorism Convictions Since Sept. 11, 2001," June 23, 2006, available at http://www.usdoj.gov/opa/pr/2006/June/06_crm_389.html, accessed July 13, 2007.

22. Dinesh D'Souza, *The Enemy at Home: The Cultural Left and Its Responsibility for 9/11* (New York: Broadway, 2007).

23. "Securing Our Homeland: U.S. Department of Homeland Security Strategic Plan," available at http://www.dhs.gov/xabout/strategicplan/, accessed July 14, 2007.

24. Department of Homeland Security, available at http://www.dhs.gov/index.shtm, accessed July 14, 2007.

25. Transportation Security Administration, available at http://www.tsa.gov, accessed July 14, 2007.

26. United States Customs and Border Protection, available at http://www.cbp.gov, accessed July 14, 2007.

27. Immigration and Customs Enforcement, available at http://www.ice.gov, accessed July 14, 2007.

28. United States Department of Homeland Security, available at http://www.dhs.gov, accessed July 14, 2007.

29. American Civil Liberties Union, "USA Patriot Act," November 14, 2003, available at http://www.aclu.org/safefree/resources/17343res20031114.html, accessed July 13, 2007.

30. United States Department of Justice, "Fact Sheet: USA Patriot Act Improvement and Reauthorization Act of 2005," March 2, 2006, available at http://www.usdoj.gov/opa/pr/2006/March/06_opa_113.html, accessed July 13, 2007.

31. 50 U.S.C. §§ 1801-1811, 1821-1829, 1841-1846, 1861-62.

32. Foreign Intelligence Surveillance Act, available at http://www.fas.org/irp/agency/doj/fisa/, accessed July 14, 2007.

33. Kane and Wall, note 13.

34. National Consortium for the Study of Terrorism and Responses to Terrorism (START), available at http://www.start.umd.edu/about/, accessed July 14, 2007.

35. Joel Leson, *Assessing and Managing the Terrorism Threat: New Realities* (Washington, DC: U.S. Department of Justice, 2005).

36. Robert Chapman, Shelly Baker, Veh Bezdikian, Pam Cammarata, Debra Cohen, et al., *Local Law Enforcement Responds to Terrorism: Lessons in Prevention and Preparedness* (Washington, DC: U.S. Department of Justice, 2002); David Thatcher, "The Local Role in Homeland Security," *Law and Society Review* 39:635–676 (2005).

37. Phil Lynn, *Mutual Aid: Multijurisdictional Partnerships for Meeting Regional Threats* (Washington, DC: U.S. Department of Justice, 2005).

38. Kevin Mayhood, "Ohio Sweep Nets 154," *Columbus Dispatch,* July 15, 2006, p. B1.

39. M. Elaine Nugent, James Johnson, Brad Bartholomew, and Delene Bromirski, *Local Prosecutors' Response to Terrorism* (Washington, DC: U.S. Department of Justice, 2005).

40. Nugent et al., note 39.

41. Chris Hale, "Cybercrime: Facts and Figures Concerning the Global Dilemma," *Crime and Justice International* 18:5–26 (2002).

42. Ramona Rantala, *Cybercrime Against Businesses* (Washington, DC: U.S. Department of Justice, 2004).

43. Alberto Gonzales, "Prepared Remarks to the U.S. Chamber of Commerce," June 20, 2006, available at http://www.usdoj.gov/ag/speeches/2006/ag_speech_060620.html, accessed July 14, 2007.

44. Rantala, note 42.

45. Consumer Reports National Research Center, "Cyber Insecurity: You're More Vulnerable Than You Think," *Consumer Reports,* September 2006, pp. 20–26.

46. Dennis Fisher, "Tales of Cyber-Crime Running Rampant," May 24, 2004, available at http://www.eweek.com/article2/0,1895,1597360,00.asp, accessed July 14, 2007.

47. Federal Bureau of Investigation, "Stolen Laptop and External Hard Drive Recovered," June 29, 2006, available at http://baltimore.fbi.gov/pressrel/2006/laptop_062906.htm, ac-cessed July 14, 2007.

48. Computer Crime and Intellectual Property Section, United States Department of Justice, available at http://www.cybercrime.gov, accessed August 1, 2007.

49. Gonzales, note 43.

50. Matt DeLisi, *Career Criminals in Society* (Thousand Oaks, CA: Sage Publications, 2005).

51. Consumer Reports National Research Center, note 45.

52. "The National Strategy to Secure Cyberspace," February 2003, available at http://www.whitehouse.gov/pcipb/cyberspace_strategy.pdf, accessed August 1 2007.

53. Federal Bureau of Investigation, "Calling All Business Professionals: What's the Current State of Computer Network Security?" July 25, 2005, available at http://www.fbi.gov/page2/july05/cyber072505.htm, accessed July 14, 2007.

54. Consumer Reports National Research Center, note 45.

55. United States House of Representatives Committee on Ways and Means, "Facts and Figures: Identity Theft," available at http://waysandmeans.house.gov/media/pdf/ss/factsfigures.pdf, accessed July 12, 2007.

56. Katrina Baum, *Identity Theft, 2004: First Estimates from the National Crime Victimization Survey* (Washington, DC: U.S. Department of Justice, 2006).

57. United States House of Representatives Committee on Ways and Means, note 1.

58. United States Department of Justice, Criminal Division, "Special Report on 'Phishing'," available at http://www.usdoj.gov/criminal/fraud/docs/phishing.pdf, accessed July 14, 2007.

59. Federal Trade Commission, "Identity Theft," available at http://www.ftc.gov/bcp/edu/microsites/idtheft/index.html, accessed July 14, 2007.

60. Associated Press, "Task Force Recommends Steps to Curb Identity Theft," *USA Today,* September 20, 2006, p. 5A.

61. Graeme Newman, *Identity Theft* (Washington, DC: U.S. Department of Justice, 2004).

62. Stuart Allison, Amie Schuck, and Kim Lersch, "Exploring the Crime of Identity Theft: Prevalence, Clearance Rates, and Victim/Offender Characteristics," *Journal of Criminal Justice* 33:19–29 (2005).

63. Phyllis McDonald, *National Strategy to Combat Identity Theft* (Washington, DC: U.S. Department of Justice, 2006).

64. Baum, note 56.

65. John Pollock and James May, "Authentication Technology: Identity Theft and Account Takeover," *FBI Law Enforcement Bulletin* 71:1–4 (2002).

Bill of Rights and Due Process Clause of the U.S. Constitution

First Amendment

Congress shall make no law respecting an establishment of religion, or prohibiting the free exercise thereof; or abridging the freedom of speech, or of the press; or of the people peaceably to assemble, and to petition the Government for a redress of grievances.

Second Amendment

A well-regulated militia, being necessary to the security of a free State, the right of the people to keep and bear arms, shall not be infringed.

Third Amendment

No soldier shall, in time of peace, be quartered in any house, without the consent of the owner, nor in time of war, but in a manner to be prescribed by law.

Fourth Amendment

The right of the people to be secure in their persons, houses, papers, and effects, against unreasonable searches and seizures, shall not be violated, and no warrants shall issue, but upon probable cause, supported by oath or affirmation, and particularly describing the place to be searched, and the person or things to be seized.

Fifth Amendment

No person shall be held to answer for a capital, or otherwise infamous crime, unless on a presentment or indictment of a Grand Jury, except in cases arising in the land or naval forces, or in the militia, when in actual service in time of war or public danger; nor shall any person be subject for the same offense to be twice put in jeopardy of life or limb; nor shall be compelled in any criminal case to be a witness against himself, nor be deprived of life, liberty, or property, without due process of law; nor

shall private property be taken for public use, without just compensation.

Sixth Amendment

In all criminal prosecutions, the accused shall enjoy the right to a speedy and public trial, by an impartial jury of the State and district wherein the crime shall have been committed, which district shall have been previously ascertained by law, and to be informed of the nature and cause of the accusation; to be confronted with the witnesses against him; to have compulsory process for obtaining witnesses in his favor, and to have the assistance of counsel for his defense.

Seventh Amendment

In suits at common law, where the value in controversy shall exceed twenty dollars, the right of trial by jury shall be preserved, and no fact tried by a jury shall be otherwise reexamined in any court of the United States, than according to the rules of the common law.

Eighth Amendment

Excessive bail shall not be required, nor excessive fines imposed, nor cruel and unusual punishments inflicted.

Ninth Amendment

The enumeration in the Constitution, of certain rights, shall not be construed to deny or disparage others retained by the people.

Tenth Amendment

The powers not delegated to the United States by the Constitution, nor prohibited by it to the States, are reserved for the States respectively, or to the people.

Due Process Clause— Fourteenth Amendment

Section 1. All persons born or naturalized in the United States, and subject to the jurisdiction thereof, are citizens of the United States and of the State wherein they reside. No State shall make or enforce any law which shall abridge the privileges or immunities of citizens of the United States; nor shall any State deprive any person of life, liberty, or property, without due process of law; nor deny to any person within its jurisdiction the equal protection of the laws.

GLOSSARY

48-hour rule Supreme Court ruling in *Riverside County, California v. McLaughlin,* that a defendant must be brought before a magistrate within 48 hours of his or her arrest.

abandoned property Property that is intentionally left behind or placed in a situation in which others may reasonably take the item into their possession.

actus reus Guilty act; a required material element of a crime.

adjudication The stage in the juvenile justice system that parallels prosecution and trial in adult criminal courts.

adjudication hearing A hearing to determine whether a juvenile committed the offense of which he or she is accused.

administrative segregation The placement of an inmate in a single-person cell in a high-security area for a specified period of time; sometimes referred to as solitary confinement.

affidavit for search warrant A document that outlines the evidence against the suspect and the circumstances of the crime.

aftercare The release and subsequent community supervision of an individual from a correctional facility to ensure a more positive and effective transition back into the community.

age–crime curve A curve showing that crime rates increase during preadolescence, peak in late adolescence, and steadily decline thereafter.

aging-out phenomenon The decline of participation in crime after the teenage years.

Al Qaeda A terrorist group led by Osama bin Laden that was responsible for the terrorist attacks of September 11, 2001, and is the most active international terrorist organization since the 1990s.

alibi An assertion that the defendant was somewhere else at the time the crime was committed.

anomie A social condition where the norms of society have broken down and cannot control the behavior of its members.

appellate court Courts which hear and determine appeals from lower trial courts.

appellate jurisdiction The authority of a court to review or revise the judicial actions of a lower court.

arraignment A hearing at which felony defendants are informed of the charges and their rights and given an opportunity to enter a plea.

arrest Police action of physically taking a suspect into custody on the grounds that there is probable cause that he or she committed a criminal offense.

arrest warrant A written court order instructing the police to arrest a specific person for a specific crime.

Ashurst–Sumners Act Legislation passed by Congress in 1935 and amended in 1940 that prohibited interstate transportation of prison goods.

assembly-line justice The mechanical disposition of misdemeanor cases to move them swiftly through the courts.

assigned counsel systems A method for providing legal representation for indigent defendants by which the court assigns private attorneys whose names appear on a list of volunteers on a case-by-case basis.

atavists Individuals who are throwbacks to an earlier, more primitive stage of human development, and more closely resemble their apelike ancestors in traits, abilities, and dispositions.

aversion therapy Therapy in which people are taught to connect unwanted behavior with punishment.

bail Money or a cash bond deposited with the court or bail bondsman allowing the defendant to be released on the assurance that he or she will appear in court at the proper time.

bail bondsman A person who guarantees court payment of the full bail amount if the defendant fails to appear.

bail guidelines Use of a grid to plot a defendant's personal and offense characteristics to determine probability of appearance.

Bail Reform Act of 1966 Act providing for release on recognizance (ROR) in noncapital federal cases when it is likely that the defendant will appear in court at required hearings.

Bail Reform Act of 1984 Act extending the opportunity for release on recognizance (ROR) in many federal cases but also providing for preventive detention without bail of dangerous suspects.

beat An assigned area of police patrol.

behavioral theory Theory that views behavior as a product of interactions people have with others throughout their lifetime.

bench trial A trial before a judge alone, as an alternative to a jury trial.

beyond a reasonable doubt The requirement that the jury (or the judge in the case of a bench trial) must find the evidence entirely convincing and must be satisfied beyond a moral certainty of the defendant's guilt before returning a conviction.

bifurcated trial A two-stage trial: the first stage determines guilt, and the second stage determines the sentence.

bill of particulars A written statement from the prosecutor revealing the details of the charge(s), including the time, place, manner, and means of commission.

Bill of Rights First 10 amendments to the U.S. Constitution.

bioterrorism The use of biological toxins to inflict mass casualties.

Body of Liberties A document in the Massachusetts Bay Colony legal code outlining the provisions for protecting the rights of citizens in criminal prosecutions.

booking The process of officially recording the name of the person arrested, the place and time of the arrest, the reason for the arrest, and the name of the arresting authority.

boot camp A highly regimented correctional facility where inmates undergo extensive physical conditioning and discipline.

bourgeoisie People who own the means of production.

Breed v. Jones The Supreme Court decision that a criminal prosecution of a child following a juvenile court hearing constitutes double jeopardy.

broken windows theory A theory that proposes that small signs of public disorder set in motion a downward spiral of deterioration, neighborhood decline, and increasing crime.

Brown v. Mississippi Ruling that established that involuntary confessions are inadmissible in state criminal prosecutions.

bureaucracy A model of organization in which strict and precise rules are used as a way of effectively achieving organizational goals.

Camp Delta A facility at the Guantanamo Naval Base in Cuba that is used for the confinement of suspected terrorists.

Carroll doctrine Doctrine that permits the warrantless search of vehicles whenever police have a reasonable basis for believing illegal activities are taking place.

case law Law that emerges when a court modifies a law in its application in a particular case.

certification A request by a lower federal court asking the Supreme Court to rule on a specific legal question.

chain of command A hierarchical system of authority that prescribes who communicates with (and gives orders to) whom.

challenge for cause A challenge by the prosecutor or the defense to dismiss a person from a jury panel for a legitimate cause.

charge to the jury The judge's instructions to the jury, which are intended to guide their deliberations.

child maltreatment The physical, sexual, or emotional abuse or neglect of children.

Child Savers A group of nineteenth-century reformers who believed that children were basically good, delinquency was the product of bad environments, and that the state should remove children from such environments.

Chimel v. California Ruling that established the "one arm's-length" rule, which allows police without a warrant to search suspects and, to a limited extent, the immediate area they occupy.

choice theories Theories that assume that people have free will, are rational and intelligent, and make informed decisions to commit crimes based on whether they believe they will benefit from doing so.

circumstantial evidence Testimony by a witness that requires jurors to draw a reasonable inference.

civil law A body of private law that settles disputes between two or more parties to a dispute.

classical school A school of thought that holds that criminals are rational, intelligent people who have free will and the ability to make choices.

classification System for assigning inmates to levels of custody and treatment appropriate to their needs.

clemency An executive decision by the governor or president to set aside an offender's punishment and release the individual.

closing arguments The final presentation of arguments to the jury.

co-correctional prison An institution where men and women are confined together.

Coker v. Georgia The Supreme Court ruling that a death sentence that is grossly disproportionate to the crime is unconstitutional.

common law Case decisions by judges in England that established a body of law common to the entire nation.

common law courts English courts established during the late twelfth century; their judgments became the law common to the entire country.

community corrections A correctional approach based on the belief offenders can (and often should) be dealt with within the community rather than prison.

community policing A policing model that was popular in the 1990s, in which police and citizens unite to fight crime.

community service A sentence in which the offender makes reparation to the community.

commutation An executive order in which a prisoner's sentence is reduced to allow for his or her immediate release.

competency A list of factors that reflect abilities or skills, including qualifications, test scores on promotional exams, and field performance.

concurrent jurisdiction The authority of two or more courts to hear a particular case.

concurrent sentences Two or more prison sentences to be served at the same time.

confession A voluntary declaration to another person by someone who has committed a crime in which the suspect admits to involvement in the offense.

conflict theory Theory that blames crime on inequalities in power.

congregate system A nineteenth-century model that held prisoners in isolation during the night, allowing them to work together during the day in silence; it was implemented at New York's Auburn Prison.

conjugal visit A private visit that some prison systems allow between inmates and their spouses to help them maintain sexual and interpersonal relationships.

consecutive sentences Two or more prison sentences to be served one after the other.

consent search A legal, warrantless search conducted after a person gives expressed consent to police.

constable An elected law enforcement officer in a small town without a police force.

contract systems A method for providing legal representation for indigent defendants by which a local attorney, bar association, or law firm contracts with the court.

corpus The body of the crime; the material elements of the crime that must be established in a court of law.

correctional officers Also known as guards; the lowest-ranking prison staff members, who have the primary responsibility of supervising inmates.

correctional system Programs, services, and institutions designed to manage people accused or convicted of crimes.

corruption Misuse of authority by officers for the benefit of themselves or others.

courts of general jurisdiction Courts with the authority to hear virtually any criminal or civil case.

courts of last resort In most states and in the federal court system, the final appellate court.

courts of limited jurisdiction Courts usually referred to as the lower or inferior courts, which are limited to hearing only specific kinds of cases.

crime An intentional act or omission to act, neither justified nor excused, that is in violation of criminal law and punished by the state.

crime index A statistical indicator consisting of eight offenses that is used to gauge the amount of crime reported to the police. It was discontinued in 2004.

crime mapping Computerized mapping by address of crime occurrences which helps police identify the locations and days and times of major sources of community problems.

crimes of interest The seven offenses in the National Crime Victimization Survey that people are asked whether they have been a victim of during the past year.

criminal career The progression of criminality over time or over the life-course.

criminal investigation The process of searching for evidence to assist in solving a crime.

criminal justice process The procedures that occur in the criminal justice system, from a citizen's initial contact with police to his or her potential arrest, charging, booking, prosecution, conviction, sentencing, and incarceration or placement on probation.

criminal justice system A complex set of interrelated subsystems composed of three major components—police, courts, and corrections—that operate at the federal, state, and local levels.

criminal procedure law A body of law that prescribes how the government enforces criminal law and protects citizens from overzealous police, prosecutors, and judges.

criminalists Scientists who work in crime laboratories and examine forensic evidence, which includes fingerprints, DNA analysis, bloodstains, footprints, tire tracks, and the presence of narcotics.

cross-examination Questioning of a witness by counsel after questions have been asked by the opposing counsel.

cultural deviance theory Theory that proposes that crime is the product of social and economic factors located within a neighborhood.

cultural transmission The process through which criminal values are transmitted from one generation to the next.

custody Assumed legal control of a person or object.

cybercrime Criminal behaviors that are intended to damage or destroy computer networks, fraudulently obtain money and other commodities, and disrupt normal business operations.

cyberspace Interconnected computers, servers, routers, switches, and fiber-optic cables that allow the U.S. infrastructure to work.

cyberterrorism Terrorism that involves threats or attack on computer or information systems.

dark figure of crime A term used by criminologists to describe the amount of unreported or undiscovered crime; it calls into question the reliability of UCR data.

day reporting center A center where counselors meet with offenders who are permitted to remain within the community, live at home, and go to work or attend school.

defendant rehabilitation A model of charging decisions in which the prosecutor accepts pleas to lesser charges to enable the defendant to obtain treatment outside the criminal justice system.

defense of life standard Policy mandating that officers may use deadly force only in defense of their own lives or another's life.

delegation of authority Decision making made through a chain of command in a bureaucracy.

delinquent A juvenile under age 18 determined to have violated the juvenile code.

demand waiver Process by which a juvenile may request to have his or her case transferred to criminal court.

Department of Homeland Security (DHS) The federal agency that oversees U.S. counterterrorism efforts.

deposition Testimony of a witness taken under oath outside the courtroom.

deprivation model An explanation of the inmate subculture as an adaptation to loss of amenities and freedoms within prison.

detective division A police division consisting of investigative officers, and possibly a forensic laboratory or specialized unit that focuses on specific types of crime (i.e., homicide, narcotics).

detention The temporary custody and care of juveniles pending adjudication, disposition, or implementation of disposition.

determinate sentence A prison sentence with a fixed term of imprisonment.

determination of competency A determination as to whether the defendant lacks the capacity to understand the charge or possible penalties if convicted, assist or confer with counsel, or understand the nature of the court proceedings.

deterrence A punishment philosophy based on the belief that punishing offenders will deter crime.

differential association theory Theory that explains the process by which a person becomes involved in criminality.

direct evidence Testimony by an eyewitness to the crime.

directed patrol A patrol technique in which officers are given specific instructions on how to use their patrol time.

disciplinary hearing A hearing before a disciplinary board to determine whether the charge against an inmate has merit and to determine the sanction if the charge is sustained.

discovery Legal motion to reveal to the defense the basis of the prosecutor's case.

disposition hearing A hearing to determine the most appropriate placement of a juvenile adjudicated to be delinquent.

diversion The early suspension or termination of the official processing of a juvenile in favor of an informal or unofficial alternative.

division of labor A system of assigning duties for the routine jobs completed in bureaucracies.

dizygotic twins Twins who do not share the same set of genes (fraternal twins).

double jeopardy Trying a person for the same crime more than once; it is prohibited by the Fifth Amendment.

Durham rule An insanity test that determines whether a defendant's act was a product of a mental disease or defect.

economic terrorism Terrorism directed toward banks and purported symbols of capitalist oppression.

eco-terrorism The infliction of economic damage on those who profit from development and destruction of environmental resources.

educational release A prerelease allowing the inmate to leave the institution during the day to attend classes.

ego Component of the personality that represents problem-solving dimensions.

electronic monitoring A sentence requiring an offender to wear an electronic device to verify his or her location.

entrapment The claim that a defendant was encouraged or enticed by agents of the state to engage in a criminal act.

Escobedo v. Illinois Ruling that held that suspects accused of a felony may have an attorney present during interrogation.

Estelle v. Gamble The Supreme Court decision that indifference to inmates' medical needs constitutes cruel and unusual punishment.

exclusionary rule The rule of law prohibiting the introduction of illegally obtained evidence or confessions into a trial.

exclusive jurisdiction The authority of a court to be the only court to hear a particular case.

excuses Claims based on a defendant admitting that what he or she did was wrong but arguing that, under the circumstances, he or she was not responsible for the criminal act.

extortion The use of or implicit threat of the use of violence or other criminal means to cause harm to a person, reputation, or property as a means to obtain property from someone else with his or her consent.

felony A serious crime, such as robbery or embezzlement, that is punishable by a prison term of more than one year or by death.

fine A sentence in which the offender makes a cash payment to the court.

fleeing felon doctrine Law (prior to 1985) stating that an officer could use deadly force to stop a felony suspect from fleeing.

forensic evidence Physical evidence found at a crime scene, including such things as fingerprints, DNA analysis, bloodstains, footprints, tire tracks, and the presence of narcotics.

forfeiture A legal procedure that permits the government to seize property used in the commission of a crime.

frankpledge police system An English policing system that spanned the eleventh through thirteenth centuries, in which every male older than age 12 assumed responsibility for fighting crime.

full law enforcement Law enforcement technique in which officers respond formally to all suspicious behavior.

Fulwood v. Clemmer The U.S. District Court decision that African American Muslim inmates have the same constitutional rights to practice their religion and to hold worship services as inmates of other faiths.

furlough A prerelease allowing the inmate to have a temporary home visit, usually lasting from 24 hours to one week.

Furman v. Georgia The Supreme Court ruling that the death penalty, as applied at that time, was unconstitutional.

fusion centers A police intelligence operation in which regional hubs pool information from multiple jurisdictions.

Gagnon v. Scarpelli The Supreme Court decision identifying the due process rights of an offender during a probation revocation hearing.

gate pay A small sum of cash given to an inmate upon his or her release from prison.

general deterrence Punishing offenders to discourage others from committing crimes.

geographic information systems (GIS) A system for capturing, storing, analyzing, and managing data and associated attributes that are spatially referenced to the Earth.

Gideon v. Wainwright Ruling that determined that every person who is charged with a felony has the right to appointed counsel.

global positioning system (GPS) A satellite-based radio navigation system.

good faith exception An exception to the requirement for a warrant for search and seizure; it allows evidence collected in violation of the suspect's privacy rights under the Fourth Amendment to be admitted at trial if the police had good reason to believe their actions were legal.

good time The practice of reducing an inmate's sentence for good behavior.

grand jury A group of citizens who are called upon to investigate the conduct of public officials and agencies and criminal activity in general and to determine whether probable cause exists to issue indictments.

Gregg v. Georgia The Supreme Court ruling that the death penalty was constitutional under a state statute requiring the judge and the jury to consider both aggravating and mitigating circumstances.

guilty plea An admission of guilt to the crime charged.

guilty, but mentally ill (GBMI) A substitute for traditional insanity defenses, which allows the jury to find the defendant guilty and requires psychiatric treatment during confinement. Also called guilty but insane (GBI).

habeas corpus A judicial order to bring a person immediately before the court to determine the legality of his or her detention.

halfway house Facilities designed to provide secure housing of inmates and to ensure the public's safety while providing inmates with an opportunity to take a controlled step into the community.

Hamas A terrorist organization also known as the Islamic Resistance Movement.

Hamdan v. Rumsfeld A Supreme Court decision that ruled that the George W. Bush administration's policy of treating terrorists as enemy combatants without due process is unconstitutional.

hands-off doctrine The position taken by the Supreme Court that it will not interfere with states' administration of prisons.

harmless error An error, defect, irregularity, or variance that does not affect substantial rights of the defendant.

hate crime A crime in which an offender targets a victim based on a specific characteristic (i.e., ethnicity, race, or religion), and evidence is provided that hate or personal disapproval of this characteristic prompted the offender to commit the crime.

Hawes–Cooper Act Legislation passed by Congress in 1929 requiring that prison products be subject to the laws of the state to which they were shipped.

hearsay evidence Testimony involving information the witness was told but has no direct knowledge of.

hierarchy rule A rule dictating that only the most serious crime in a multiple-offenses incident will be recorded in the Uniform Crime Reports.

Holt v. Sarver The Supreme Court decision that applied the "totality of conditions" principle to find the Arkansas prison system in violation of the Eighth Amendment.

Hopt v. Utah Ruling that established guidelines for involuntary confessions.

horizontal overcharging The practice of filing of a number of related charges or a number of separate counts related to the same basic charge.

hot spots of crime Locations characterized by high rates of crime.

house arrest A sentence requiring that the offender be legally confined to his or her home.

Hudson v. McMillian The Supreme Court decision that the use of excessive force by correctional officers may constitute cruel and unusual punishment, even if the inmate does not suffer serious injury.

hung jury A jury that is deadlocked and cannot reach a verdict. As a result, the judge may declare a mistrial.

Hurtado v. California Supreme Court decision that the Fifth Amendment guarantee of a grand jury indictment applied only to federal—not state—trials, and that not all constitutional amendments were applicable to the states.

id Component of the personality that is present at birth, and consists of blind, unreasoning, instinctual desires and motives.

identity theft Any illegal act that involves the use or attempted use of another person's financial account or identifying information without the owner's permission.

implicit plea bargaining The entering of a guilty plea with the expectation that the defendant will be looked upon favorably by the court at sentencing.

importation model A view of the inmate subculture as a reflection of the values and norms inmates bring with them when they enter prison.

In re Gault Case in which the Supreme Court held that juveniles could not be denied basic due process rights in juvenile hearings.

incapacitation A punishment aimed at removing offenders from the community through imprisonment or banishment.

incorporation The legal interpretation by the Supreme Court in which the Fourteenth Amendment applied the Bill of Rights to the states.

indeterminate sentence A prison sentence that identifies a minimum and a maximum number of years to be served by the offender; the actual release date is set by a parole board or the institution.

indictment A formal criminal charge filed by the prosecutor.

individual justice Concept that criminal law must reflect differences among people and their circumstances.

inevitable discovery rule Rule that if illegally obtained evidence would have eventually been discovered by lawful means, it is admissible regardless of how it was originally discovered.

informal probation Placement of an offender on probation before a conviction is entered on the record; upon completion of the probation conditions, the offender is released and the case dismissed.

infraction A violation of a city or county ordinance, such as cruising or noise violations.

initial appearance A defendant's first appearance in court, at which the charge is read, bail is set, and the defendant is informed of his or her rights.

inmate code A system of informal norms created by prisoners to regulate inmate behavior.

in-presence requirement A requirement that police may not make a warrantless arrest for a misdemeanor offense unless the offense is committed in their presence.

intake Initial screening process in the juvenile court to determine whether a case should be processed further.

intelligence-led policing A crime-fighting strategy driven by computer databases, intelligence gathering, and analysis.

intensive aftercare programs (IAP) Equivalent to intensive parole supervision; used to provide greater supervision of youths after their release from official institutions.

intensive probation supervision (IPS) A practice under which probation officers with very small caseloads are able to provide increased supervision.

intensive supervision parole A practice under which caseloads are very small, and officers frequently meet face to face with parolees to provide greater supervision.

intermediate sanctions Sentences that may be imposed as alternatives to traditional probation or incarceration.

intermittent incarceration A short sentence involving a series of brief periods of time in jail.

Interpol An international criminal police organization that facilitates international police cooperation.

interrogation A method police use during an interview with a suspect to obtain information that the suspect might not otherwise disclose.

intimate partner violence (IPV) Violence in intimate relationships, including that committed by current or former spouses, boyfriends, or girlfriends.

involuntary confession A confession precipitated by a promise, threat, fear, torture, or other external factor such as mental illness.

involvement crimes Crimes in which the offender directly confronts the victim, such as an armed mugging.

irresistible impulse test An insanity test that determines whether a defendant, as a result of a mental disease, temporarily lost self-control or the ability to reason sufficiently to prevent the crime.

jail An institution to hold pretrial detainees and people convicted of less serious crimes.

jailhouse lawyers Inmates who have studied legal proceedings and who help other inmates in making petitions to the courts.

Jihad Holy war of fundamentalist Islam directed against nations such as the United States and Israel.

judgment of acquittal A defense motion for dismissal of a case based on the claim that the prosecution failed to establish that a crime was committed or that the defendant committed it.

judicial jurisdiction The power or authority of a court to hear a case or consider a particular legal motion.

judicial review The power of the U.S. Supreme Court to review and determine the constitutionality of acts of Congress and orders by the executive branch.

judicial waiver Most common waiver procedure for transferring youths to criminal court, in which the judge is the primary decision maker.

Judiciary Act of 1789 An act created by the First Congress establishing the basic structure of the federal court system.

jurisdiction The territory over which a law enforcement agency has authority.

jury nullification The right of a jury to interpret and negate the law in a case.

jury pool The master list of community members who are eligible to be called for jury duty.

Justice of the Peace courts Courts first established in the American colonies to hear minor criminal cases.

justification Defense wherein a defendant admits responsibility but argues that, under the circumstances, what he or she did was right.

juvenile A person under the age of 18.

juvenile delinquency Behavior by a juvenile that is in violation of the juvenile or criminal codes.

Kansas City Preventive Patrol Experiment A study done to assess how allocating patrol at different levels of enforcement affected the crime rate and perceptions of public safety.

Kent v. United States The Supreme Court decision requiring a formal waiver hearing before transfer of a juvenile to criminal court.

kin police system An English policing system used between 400 C.E. and 500 C.E. in which each male citizen assumed responsibility for protecting his neighbor.

knock-and-announce rule Rule that requires police to announce their presence and wait about 20 seconds before entering a home.

labeling theory Theory that examines the role of societal reactions in shaping a person's behavior.

laws Formalized rules that prescribe or limit actions.

legal sufficiency A model of charging decisions in which cases are prosecuted if the prosecutor believes that the elements of the crime are sufficiently present to warrant bringing the case to trial.

less eligibility The belief that prisoners should always reside in worse conditions than should the poorest law-abiding citizens.

life-course theory Theory that explains the change in the progression of criminality over time.

lifestyle theory Theory that proposes that the way people live their lives can place them in settings with a higher or lower risk of criminal victimiztion.

line personnel Prison employees who have direct contact with inmates.

line-up A pretrial identification procedure in which several people are shown to a victim or witness of a crime, who is then asked if any of those individuals committed the crime.

loss of privileges A disciplinary sanction involving the loss of visits, mail, recreation, and access to the prison commissary.

low-security prison An institution that operates between the medium and minimum security levels.

Magna Carta A document signed by King John in 1215 that enumerated rights and protections for the common citizens of England.

mala in se Behaviors, such as murder or rape, that are considered inherently wrong or evil.

mala prohibita Behaviors, such as prostitution and gambling, that are considered wrong because they have been prohibited by criminal statutes, rather than because they are evil in themselves.

mandatory sentence A requirement that an offender must be sentenced to prison.

Mapp v. Ohio Ruling that expanded the exclusionary rule to state courts.

Marbury v. Madison The U.S. Supreme Court case that established the principle of judicial review.

mark system System by which prisoners earned "marks" for good behavior to achieve an early release from prison.

master status The status bestowed on an individual and perceived by others as a first impression.

maximum-security prison The most secure prison facility, having high walls, gun towers, and barbed wire or electronic fences.

McCleskey v. Kemp The Supreme Court ruling that rejected the claim that the death penalty law in Georgia was unconstitutional because it promoted racial discrimination.

McKeiver v. Pennsylvania The Supreme Court decision that juveniles do not have a constitutional right to a jury trial in juvenile court.

medical model A treatment approach popular between 1930 and 1960 that attributed criminality to a biological or psychological defect of the offender.

medium-security prison A middle-level prison facility, which has more relaxed security measures and fewer inmates than a maximum-security prison.

Megan's Law A law that requires both registration and community notification by sex offenders when they move into a community.

mens rea Guilty mind, or having criminal intent; a required material element of a crime.

Metropolitan Police Act The 1829 act that established the London Metropolitan Police force, which was the first salaried, uniformed police agency.

minimum-security prison A prison facility with the lowest level of security that houses nondangerous, stable offenders.

Miranda v. Arizona Ruling that established that criminal suspects must be informed of their right to consult with an attorney and their right against self-incrimination prior to questioning by police.

Miranda **warning** A warning required by law to be recited at the time of arrest, informing suspects of their constitutional right to remain silent and have an attorney present during questioning.

misdemeanor A crime that is less serious than a felony, such as petty theft or possession of a small amount of marijuana, and that is punishable by less than one year in prison.

Missouri Plan Developed in 1940, the first plan for the selection of judges based on merit.

mistrial An invalid trial due to some unusual event, such as the death of a juror or an attorney, a prejudicial error, or inability of the jury to reach a verdict.

mitigating circumstances Factors such as age or mental illness that influence the choices people make and affect a person's ability to form criminal intent.

M'Naghten rule Insanity defense claim that because of a defect of reason from a disease of the mind, the defendant was unable to distinguish right from wrong.

modification of sentence An alteration of an original sentence based on an appeal to the court.

monozygotic twins Twins who share the same set of genes (identical twins).

Morrissey v. Brewer The Supreme Court case that spelled out the due process rights of offenders on parole.

National Crime Victimization Survey (NCVS) An annual survey of criminal victimization in the United States conducted by the U.S. Bureau of Justice Statistics.

National Youth Survey (NYS) A comprehensive, nationwide self-report study of 1700 youths who reported their illegal behaviors each year for more than 30 years.

neoclassical school A school of thought that argues that there are real differences among people that must be taken into consideration when administering punishment.

New Jersey v. T.L.O. Ruling that established that school officials can conduct warrantless searches of students at school on the basis of reasonable suspicion.

New York House of Refuge The first correctional institution for children in the United States (opened in 1825), which emphasized industry, education, and strict discipline.

new-generation jails Jails that are designed to increase staff interaction with inmates by placing the staff inside the inmate housing unit.

nolo contendere A plea of no contest; essentially the same as a guilty plea except that the defendant neither admits nor denies the charge.

norms Rules and expectations by which a society guides the behavior of its members.

not guilty A plea denying guilt.

not guilty by reason of insanity A plea in which the defendant does not deny committing the crime but claims that he or she was insane at the time of the offense and, therefore, is not criminally responsible.

official crime statistics Statistics based on the aggregate records of offenders and offenses processed by police, courts, corrections agencies, and the U.S. Department of Justice.

opening statement The initial presentation of the outline of the prosecution's and the defense's cases to the jury.

operant conditioning Treatment in which rewards are used to reinforce desired behavior and punishments are used to curtail undesired behavior.

operational style The way in which police officers interact with fellow officers and the public.

original jurisdiction The power or authority of a court to be the first to hear a case and render a verdict.

pains of imprisonment Deprivations—such as the loss of freedom, possessions, dignity, autonomy, security, and heterosexual relationships—shared by inmates.

Palestinian Liberation Organization (PLO) A terrorist organization that directs violence mostly against Israel.

parens patriae A principle based on English common law that viewed the state as the ultimate sovereign and guardian of children.

parish–constable police system A police system that operated in England between 1285 and 1829, in which constables and watchmen were appointed to prevent crime.

parole The conditional release of an offender from a correctional institution after serving a portion of his or her sentence.

parole board The group of people who make the discretionary decision about a prisoner's release.

patrol Police responsibility to move through assigned areas by foot or vehicle to enforce laws, regulate traffic, control crowds, prevent crime, and arrest violators.

Payne v. Tennessee The Supreme Court ruling that statutes that bar the introduction of victim impact statements in death penalty cases are unconstitutional.

penitentiary house An eighteenth-century place of penitence for all convicted felons except those sentenced to death.

Percy Amendment Law that allowed states to sell prison-made goods across state lines as long as they complied with strict rules to make sure unions were consulted and to prevent manufacturers from undercutting existing wage structures.

peremptory challenge A challenge by the defense or the prosecution to excuse a person from a jury panel without having to give a reason.

petition Similar to an indictment; a written statement setting forth the specific charge that a delinquent act has been committed or that a child is dependent or neglected or needs supervision.

phishing The creation and use of e-mails and websites designed to look like those of well-known legitimate businesses, financial institutions, or government agencies in an attempt to deceive Internet users into disclosing private financial information.

plain view doctrine Standard that provides when police discover evidence in a place where police have a legal right to be, they have a right to seize that evidence.

plea A defendant's response to a criminal charge.

plea bargaining The negotiation between a prosecutor and a defense attorney in which they seek to arrive at a mutually satisfactory disposition of a case without going to trial.

police brutality The unlawful use of force, language, and application of the law.

police cynicism Belief of police that people are selfish and motivated by evil.

police discretion Authority of police to choose between alternative courses of action.

police operations Services that police agencies provide and the methods they use to deliver these services.

police subculture Beliefs, values, and patterns of behavior that separate officers from police administrators and the public.

police–population ratio The number of sworn officers per 1000 citizens.

positive school A school of thought that blames criminality on factors that are present before a crime is actually committed.

precinct The entire collection of police beats in a specific geographic area.

prejudicial error Inflammatory or biasing statements made by an attorney to the jury.

preliminary hearing An early hearing to review charges, set bail, present witnesses, and determine probable cause.

preliminary revocation hearing The first hearing to determine whether there is probable cause that the offender violated the conditions of probation or parole.

presentence investigation (PSI) A comprehensive report including information on the offender's background and offense and any other information the judge desires to determine an appropriate sentence.

preservation of life policy Policy mandating that police use every other means possible to maintain order before turning to deadly force.

presumption of innocence The notion that a person is presumed to be innocent unless proved guilty beyond a reasonable doubt.

presumptive parole date A presumed date of release using a calculation of scores based on an inmate's offense and background and designed to predict his or her likelihood of successfully completing parole.

presumptive sentencing The use of ranges—that is, minimum and maximum number of years of incarceration—set for types of particular crimes. The judge determines the number of years to be served from within this range.

pretrial motion A written or oral request to a judge for a ruling or action before the beginning of trial.

pretrial release Release of defendant from custody while he or she is awaiting trial.

preventive detention The practice of holding a defendant in custody without bail if he or she is deemed likely to abscond or commit further offenses if released.

preventive patrol A crime control strategy based on the idea that crime is deterred by the mere presence of police.

primary deviance The behavior that originally leads to the application of the "deviant" label.

prison An institution for the confinement of people who have been convicted of serious crimes.

prison farms Correctional institutions that produce much of the livestock, dairy products, and vegetables used to feed inmates in the state prisons.

prison forestry camps Correctional institutions that provide labor for the maintenance of state parks, tree planting and thinning, wildlife care, mainte-

nance of fish hatcheries, and cleanup of roads and highways.

prisonization The process by which inmates adjust to or become assimilated into the prison subculture.

private security police Individuals who are employed by citizens and businesses to provide security.

privatization The process in which state and federal governments contract with the private sector to help finance and manage correctional facilities.

probable cause A set of facts and circumstances that would lead a reasonable person to believe that a crime was committed and that the accused committed it.

probation A sentencing option typically involving a suspended prison sentence and supervision in the community.

procedural criminal law A body of law that specifies how crimes are to be investigated and prosecuted.

proletariat People who sell their labor to the bourgeoisie.

proportionality A punishment philosophy based on the belief that the severity of the punishment should fit the seriousness of the crime.

prosecutorial waiver Process in which the prosecutor determines whether a charge against a juvenile should be filed in criminal or juvenile court.

psychoanalytic theory Theory that unconscious mental processes developed in early childhood control the personality.

public defender systems A method for providing legal representation for indigent defendants by which defense attorneys are appointed by the court to act as full-time defenders.

racial terrorism Terrorism motivated by extreme racial prejudice that is directed toward a specific racial or ethnic group.

rational choice theory Theory in which criminals are rational people who make calculated choices regarding their actions before they act.

real evidence Physical evidence introduced at the trial.

rebuttal The presentation of additional witnesses and evidence by the prosecutor in response to issues raised in the defense's presentation of witnesses.

reentry The process in which an inmate leaves prison and returns to society.

reformatory A penal institution generally used to confine first-time offenders between the ages of 16 and 30.

rehabilitation A sentencing objective aimed at reforming an offender through treatment, education, or counseling.

release on recognizance (ROR) A personal promise by the defendant to appear in court; does not require a monetary bail.

reporting time The time lag between when a crime is committed and when the police are called.

response time The time it takes for police to respond to a call.

restitution A requirement of offenders to pay money or provide services to victims.

restorative justice A punishment philosophy aimed at restoring or repairing relationships disrupted by crime, holding offenders accountable, promoting offender competency and responsibility, and balancing the needs of the community, victim, and offender through involvement in the restorative process.

retribution A punishment philosophy based on society's moral outrage or disapproval of a crime.

reverse waiver Process in which a juvenile contests a statutory exclusion or prosecutorial transfer.

revocation hearing A hearing to determine whether probation or parole should be revoked.

risk classification An assessment of an offender's likelihood of committing a new offense if granted probation.

Roper v. Simmons The Supreme Court decision that the death penalty for anyone who was younger than age 18 at the time of his or her crime is unconstitutional.

routine activities theory Theory that examines the crime target or whatever it is the offender wants to take control of, whether it is a house to break into, a bottle of beer, or illegal music to download from the Internet.

salient factor score A guideline used to help determine the potential risk in releasing an inmate early.

Santobello v. New York U.S. Supreme Court decision that plea bargaining is an essential component of the criminal justice system and that prosecutors must honor the terms of a plea bargain.

Schall v. Martin The Supreme Court decision authorizing the preventive detention of juveniles who are identified as "serious risks" to the community if released.

search warrant A written order instructing police to examine a specific location for a certain property or persons relating to a crime, to seize the property or persons if found, and to account for the results to the judicial officer who issued the warrant.

secondary deviance Acts of deviance that occur after someone has internalized the "deviant" label and uses it as a means of defense, attack, or adjustment to the problems caused by the label.

selective law enforcement Law enforcement technique in which officers under-enforce some laws and over-enforce others.

self-control theory Theory in which people seek pleasure, are self-gratifying, and commit crimes owing to their low self-control.

self-defense Claim that a defendant acted in a lawful manner to defend himself or herself, others, or property, or to prevent a crime.

self-report surveys Surveys that ask offenders to self-report their criminal activity during a specific time period.

sentencing discrimination Differences in sentencing outcomes based on illegitimate, morally objectionable, or extralegal factors.

sentencing disparities Differences in sentencing outcomes in cases with similar case attributes.

sentencing guidelines Sentencing schemes that limit judicial discretion; the offender's criminal background and severity of current offense are plotted on a grid to determine the sentence.

sentencing hearing A court hearing to determine an appropriate sentence, which is typically scheduled within three to six weeks after the offender's conviction.

separate confinement A nineteenth-century model of prison that separated inmates; it was implemented in Pennsylvania's Western and Eastern penitentiaries.

service function Role of police to assist citizens with noncriminal matters, such as providing emergency medical assistance.

severance of charges A motion requesting a separate trial for each charge.

sheriff The principal law enforcement officer in a county.

shock incarceration The release of an offender from prison after a brief period (usually 30 to 60 days), with subsequent probation.

silver platter doctrine Doctrine that permitted officers in one jurisdiction to hand over "on a silver platter" evidence that had been illegally obtained to officers in another jurisdiction to use in court.

slave patrols Small, organized groups who controlled the slave population and outbreaks of slave revolts in pre–Civil War United States.

social bond A measure of how strongly people are connected to society.

social control theory Theory that holds that people are amoral and will break the law unless obstacles are thrown in their path.

social learning theory Theory that suggests that children learn by modeling and imitating others.

sociological theories Theories that attribute crime to a variety of social factors external to the individual, focusing on how the environment in which the person lives affects his or her behavior.

somatotype theory Theory that suggests that individuals with particular body types are likely to be inclined toward certain behaviors.

specialization The practice of dividing work among employees for it to be completed more effectively and efficiently.

specific deterrence Punishing offenders to prevent them from committing new crimes.

Speedy Trial Act of 1974 Act of Congress requiring that a federal trial must begin within 70 days of the filing of charges or the defendant's initial appearance.

split sentence A sentence in which an offender serves a period of incarceration and is then placed on probation, with the balance of the sentence being suspended.

staff personnel Prison employees who provide support services to line personnel and administrators.

stare decisis Literally, "to stand by the decision"; a policy of the courts to interpret and apply law according to precedents set in earlier cases.

state police agencies Law enforcement agencies that protect the interests of the state.

status offenses Acts prohibited to children that are not prohibited to adults (such as running away, truancy, and incorrigibility).

statute Legislation contained in written legal codes.

statute of limitations The maximum time period that can pass between a criminal act and its prosecution.

statutory exclusion Process established by statute that excludes certain juveniles, because of either age or offense, from juvenile court jurisdiction; charges are initially filed in criminal court.

stigmata Distinctive physical features.

stop-and-frisk rule Rule that police may stop, question, and frisk individuals who look suspicious.

strain theory Theory that proposes that a lack of integration between cultural goals and institutionalized means causes crime.

stress A condition that occurs in response to adverse external influences and is capable of affecting an individual's physical health.

strict liability laws Laws that provide for criminal liability without requiring either general or specific intent.

structured fines Fines designed to eliminate the proportionately greater financial burden placed on poorer offenders by tying the amount of the fine to the offender's ability to pay.

substantial capacity test An insanity test that determines whether the defendant lacked sufficient capacity to appreciate the wrongfulness of his or her conduct.

substantive criminal law A body of law that identifies behaviors harmful to society and specifies their punishments.

superego Component of the personality that develops from the ego and comprises the moral code, norms, and values the person has acquired.

super-maximum-security prison A prison where the most predatory and dangerous criminals are confined.

surrebuttal Questioning by the defense of witnesses who were presented by the prosecutor during rebuttal.

surrogate families Fictive families created by female prisoners to provide stability and security.

suspended sentence A prison sentence that is set aside.

sworn officers Officers who are empowered to arrest suspects, serve warrants, carry weapons, and use force.

system efficiency A model of charging decisions in which the prosecutor pursues only those cases that are most likely to achieve efficiency in the system by speedy disposition.

Tennessee v. Garner U.S. Supreme Court ruling that eliminated the "shoot a fleeing felon" policy and replaced it with a defense of life standard.

terrorism The unlawful use of force or violence to intimidate or coerce a government, the civilian population, or any segment thereof, in furtherance of political or social objectives or goals.

Terrorism and International Victims Unit (TIVU) A federal agency responsible for developing programs to help victims of terrorism, mass violence, and other transnational crimes.

Terry v. Ohio Ruling that determined if police observe behavior that leads them to conclude criminal activity may be in progress and the suspect is armed and dangerous, they may stop and frisk and question a suspect after identifying themselves as police officers.

testimonial evidence Sworn testimony of witnesses who are qualified to speak about specific real evidence.

theories Integrated sets of ideas that explain when and why people commit crime.

three-strikes laws Laws that provide a mandatory sentence of incarceration for persons who are convicted of a third separate serious criminal offense.

token economy A system used in penitentiaries of handing out and taking away rewards that can be exchanged for privileges such as watching television.

total institutions Institutions that completely encapsulate the lives of the people who work and live in them.

totality of conditions A principle guiding federal court evaluations of prison conditions: The lack of a specific condition alone does not necessarily constitute cruel and unusual punishment.

totality-of-the-circumstances rule Rule that requires a judge to evaluate all available information when deciding whether to issue a search warrant.

traffic enforcement Police duties related to highway and traffic safety and accident investigations.

trait theories Theories that argue that offenders commit crimes because of traits, characteristics, deficits, or psychopathologies they possess.

transactional immunity A blanket protection against prosecution for crimes a witness may testify about to a grand jury while under immunity.

trial sufficiency A model of charging decisions in which the decision to prosecute is based on the ability to obtain a conviction at trial.

true bill An indictment issued by a grand jury charging a person with a crime; similar to a prosecutor's filing of an information.

truth-in-sentencing Laws that require offenders, especially violent offenders, to serve at least 85 percent of their sentences.

U.S. Supreme Court The highest appellate court in the U.S. judicial system; it reviews cases appealed from federal and state court systems that deal with constitutional issues.

unconditional release A release from prison that does not require additional supervision in the commnity.

Uniform Crime Report (UCR) An annual publication from the Federal Bureau of Investigation that presents data on crimes reported to the police, number of arrests, and number of persons arrested.

United States v. Leon Ruling that established the good faith exception to the exclusionary rule, under which evidence that is produced in good faith and later discovered to be obtained illegally may still be admissible in court.

United States v. Salerno U.S. Supreme Court ruling that the preventive detention provisions of the Bail Reform Act of 1984 were constitutional.

unofficial crime statistics Crime statistics produced by people and agencies outside the criminal justice system, such as college professors and private organizations.

USA Patriot Act Federal legislation that broadened the surveillance and investigative powers of criminal justice agencies to combat terrorism and remove barriers between military, intelligence, and criminal justice entities, enabling them to share information on terrorist suspects and threats.

USA Patriot Act Improvement and Reauthorization Act of 2005 Federal legislation that revised the USA Patriot Act of 2001, enhancing the powers of the criminal justice system to combat crime and terrorism.

use immunity Protection that prohibits specific information given during grand jury testimony from being used against the witness.

venire A group of people who are selected from the jury pool and notified to report for jury duty.

vertical overcharging The practice of filing the most serious possible charge appropriate to a criminal act even though the known circumstances do not support the charge.

victim impact statement (VIS) A statement informing the sentencing judge of the physical, financial, and emotional harm suffered by the crime victim or his or her family.

victimization survey A method of producing crime data in which people are asked about their experiences as crime victims.

victimology The study of the characteristics of crime victims and the reasons why certain people are more likely than others to become victims of crime.

voir dire Preliminary examination by the prosecution and defense of potential jurors.

waiver of jurisdiction A legal process to transfer a juvenile from juvenile to criminal court.

War on Terror The global armed conflict between radical Islamic terrorist groups, such as Al Qaeda, and the developed world, principally the United States and the G8 nations.

warden The superintendent or top administrator of a prison.

warrantless arrest Arrest without a warrant when an officer has probable cause to believe that a crime has been or is being committed.

Weeks v. United States Ruling that established the exclusionary rule in federal cases.

Wolff v. McDonnell The Supreme Court decision that inmates facing disciplinary action must have a formal hearing, 24-hour notification of the hearing, assistance in presenting a defense, and ability to call witnesses.

work release A program allowing the inmate to leave the institution during the day to work at a job.

working personality A term that distinguishes an officer's off-the-job persona from his or her on-the-job behavior.

wraparound programs Programs designed to build positive relationships and support networks between youths and their families, teachers, and community agencies through coordination of services.

writ of certiorari An order by the U.S. Supreme Court to a lower court to send up a certified record of the lower court decision to be reviewed.

CASE INDEX

A

Abdullah v. Kinnison, 406
Albemarle Paper Co., v. Moody, 208
Allen v. United States, 304
Americans United for Separation of Church and State v. Prison Fellowship et al., 406
Argersinger v. Hamlin, 183, 243
Arizona v. Evans, 167
Arizona v. Fulminante, 181
Arizona v. Hicks, 172
Arizona v. Roberson, 180–181
Atkins v. Virginia, 333

B

Ballew v. Georgia, 291
Banks v. Beard, 407
Barker v. Wingo, 287
Bartkus v. Illinois, 49
Batson v. Kentucky, 293
Bell v. Wolfish, 406, 407
Berkemer v. McCarty, 181
Betts v. Brady, 243
Blakely v. Washington, 322
Block v. Rutherford, 406
Bordenkircher v. Hayes, 273
Bounds v. Smith, 407
Boykin v. Alabama, 268
Brady v. United States, 273
Breed v. Jones, 468
Brewer v. Williams, 179–180
Brigham City, Utah v. Stuart, 177
Brown v. Mississippi, 182
Bumper v. North Carolina, 172

C

California v. Acevedo, 171
California v. Greenwood, 168, 173–174
California v. Hodari D., 170
Carroll v. United States, 168, 171
Chambers v. Florida, 182
Chandler v. Florida, 289
Chapman v. California, 305
Childs v. Duckworth, 406
Chimel v. California, 168–169
Coker v. Georgia, 333
Coleman v. Thompson, 306
Commonwealth v. Santiago, 181
Coolidge v. New Hampshire, 172
Correctional Services Corp. v. Malesko, 370
Cruz v. Beto, 406
Cutter v. Wilkinson, 406

D

Davis v. United States, 180
Delaware v. Prouse, 171
DeMallory v. Cullen, 407
Denmore v. Kim, 257–258
Dickerson v. United States, 181
Durham v. United States, 43

E

Edwards v. Arizona, 180
Elkins v. United States, 165–166
Escobedo v. Illinois, 178, 183
Estelle v. Gamble, 408
Estes v. Texas, 287
Ex Parte Hull, 407

F

Farmer v. Brennan, Warden, et al., 409
Fay v. Noia, 306
Fletcher v. Peck, 225
Florida v. Bostick, 173
Florida v. Riley, 172
Fulwood v. Clemmer, 406
Furman v. Georgia, 332, 333, 339

G

Gagnon v. Scarpelli, 427–428
Gallahan v. Hollyfield, 406
Gannett Co., Inc. v. DePasquale, 287
Georgia v. McCollum 294
Georgia v. Randolph, 173
Gerstein v. Pugh, 253–254
Gideon v. Wainwright, 183, 242, 243
Gittlemacker v. Prasse, 406
Globe Newspapers Co. v. Superior Court for the County of Norfolk, 288
Gomez v. United States, 231–232
Gonzales v. Oregon, 42
Gregg v. Georgia, 332
Griggs v. Duke Power Co., 208
Grummett v. Rushen, 391

H

Halbert v. Michigan, 305
Hamdan v. Rumsfeld, 485
Harmelin v. Michigan, 318
Harris v. New York, 181
Harris v. United States, 168, 171
Henderson v. Morgan, 268

NAME INDEX

Lord, Carnes, 58 (n1)
Lord, Vivian, 217 (n93)
Lotz, Roy, 344 (n67)
Louis, Paul, 412 (n15)
Lovegrove, Austin, 343 (nn 27, 40, 53)
Lovre-Laughlin, Nicci, 249 (n35)
Lu, Hong, 319 (n)
Luard, Tim, 362 (n)
Ludwig, Frederick, 474 (n5)
Lundman, Richard, 214 (n34)
Lunsford, Jessica, 164, 164 (n)
Lynam, Donald, 110 (n30)
Lynch, James, 82 (nn 15, 17)
Lynds, Elam, 354
Lynn, Phil, 497 (n37)
Lyons, Donna, 297 (n)
Lyons, William, 160 (n22)

M

Ma, Yue, 267 (n)
Maahs, Jeff, 379 (n76)
MacDonald, Donald, 449 (n71)
MacDonald, Douglas, 448 (n49)
MacDonald, Heather, 116 (n)
MacDonald, John, 214 (n20), 474 (n18)
MacDowall, David, 82 (n20)
MacFarlane, Diane, 448 (n51)
MacKenzie, Doris, 342 (n16), 377 (n41), 412 (nn 16, 22)
Maconochie, Alexander, 355–356
Magniyah, Imad, 482
Maguire, Edward, 26 (n29)
Maguire, Kathleen, 12 (n), 26 (nn 17, 22)
Maher, Lisa, 109 (n13)
Mahoney, Barry, 310 (n5), 342 (n22), 448 (n42)
Mahoney, Dennis, 187 (n107)
Makarios, Matthew, 214 (n20)
Makris, John, 137 (n57)
Maleng, Norm, 252
Mallory, Andrew, 253
Mannheim, Hermann, 31, 58 (n3)
Manning, Peter, 215 (n50)
Mantovani, Andrew, 36
Mapp, Dollree, 167
Maraniss, David, 26 (n21)
March, James, 412 (n24)
Marcham, Frederick, 248 (n3)
Marquart, James, 378 (n70)
Marshall, John, 225
Marshall, Tony, 433, 448 (n55)
Martin, John, 59 (n24)
Martin, Susan, 343 (n52)
Martindale, Mike, 41 (n)
Martineau, Harriet, 109 (n19)
Martinez, Jose, 190 (n)
Martinson, Robert, 108 (n9), 342 (n15)
Maruschak, Laura, 413 (n41)
Marvell, Thomas, 343 (n38)
Marx, Gary, 431, 447 (n39), 448 (n40)
Marx, Karl, 102, 111 (n54)
Mason, Debbie, 413 (n44)

Masters, Brooke, 83 (n41)
Mastrofski, Stephen, 26 (n29), 148, 160 (nn 1, 26), 214 (n29), 214–215 (n34), 215 (nn 38, 39), 216 (nn 69, 71, 72), 474 (n18)
Matheson, Daniel, 345 (n118)
Matulia, Kenneth, 216 (n80)
Mauer, Marc, 27 (n35), 82 (n5), 412 (n10)
Maughan, Andy, 311 (n63)
Mauro, Robert, 345 (n106)
Maxwell, Christopher, 216 (nn 69, 71)
May, James, 497 (n65)
Mayhood, Kevin, 497 (n38)
Mays, Larry, 379 (n86)
McCain, Garvin, 379 (n88)
McCarthy, Belinda, 447 (nn 9, 13)
McCarthy, Bernard, 447 (nn 9, 13)
McCarthy, Brian, 413 (n46)
McCleskey, Warren, 333, 339
McCluskey, John, 215 (n38), 474 (n18)
McCold, Paul, 434, 448 (n57)
McConville, Sean, 319 (n)
McCord, Joan, 193, 214 (n22)
McCoy, Candace, 343 (n41)
McCrary, Robert, 242
McDonald, Christy, 59 (n52)
McDonald, Phyllis, 497 (n63)
McDonald, William, 249 (n52), 279 (nn 29, 64)
McDuff, Kenneth, 350
McElreath, David, 496 (n15)
McGarrell, Edmund, 161 (n46)
McGuinn, David, 402
McKay, Henry, 98, 111 (n46)
McKeever, Matthew, 412 (n20)
McLaren, Roy, 161 (n28)
McLaughlin, Eugene, 110 (n39)
McLeod, Maureen, 343 (nn 46, 47), 449 (n74)
McMahon, Patrick, 413 (n48)
McMillian, Jack, 408
McNabb, Benjamin, 253
McNaghten, Daniel, 42–43
McVeigh, Rory, 496 (n14)
McVeigh, Timothy, 154, 264, 482
Meador, Daniel, 248 (n15), 249 (n28)
Mears, Bill, 344 (n96)
Mears, Daniel, 378 (n48), 475 (n35)
Mednick, Sarnoff, 95, 109 (n26), 110 (nn 33, 38, 39)
Meehan, Kevin, 343 (n35)
Meierhoefer, Barbara, 343 (n33)
Melanson, Philip, 137 (n56)
Melo, Frederick, 423 (n)
Melone, Albert, 311 (n78)
Melville, Rodney, 288
Menke, Ben, 379 (n91)
Merton, Robert, 99, 99 (n), 111 (n48)
Meserve, Jeanne, 483 (n)
Messinger, Sheldon, 399, 414 (n78)
Messner, Steven, 82 (n20)
Mezzetti, Claudio, 279 (n64)
Miethe, Terance, 319 (n)
Miller, Susan, 413 (n30)
Miller, Todd, 109 (n13)

Thoennes, Nancy, 76 (n), 83 (nn 55, 56)
Thomas, Charles, 58 (nn 2, 4, 5), 379 (nn 71, 74), 475 (nn 40, 45)
Thomas, Clarence, 408
Thompson, Andrea, 475 (n49)
Thompson, Donald, 238
Thompson, Jeffrey, 379 (n81)
Thompson, Joel, 379 (n86)
Thompson, Martie, 83 (n54)
Thornberry, Terence, 111 (n56), 474 (n3)
Thornburg, Elizabeth, 311 (n56)
Timmendequas, Jesse, 439
Tippins, Sherrill, 110 (n36)
Tittle, Charles, 399, 414 (n74)
Tjaden, Patricia, 76 (n), 83 (nn 55, 56)
Tobias, John, 474 (n7)
Tobin, Kimberly, 111 (n56)
Toch, Hans, 216 (n76), 364 (n), 413 (n45), 414 (nn 75, 76)
Tocqueville, Alexis de, 354, 377 (n22)
Tolbert, Michelle, 412 (n18)
Tonry, Michael, 83 (n36), 108–109 (n12), 109 (n16), 215 (n50), 329, 330–331, 342 (n9), 343 (n52), 344 (nn 58, 66, 69), 415 (n99)
Tontodonato, Pamela, 343 (n50)
Townsend, Meg, 26 (n31)
Tracy, Alice, 412 (n20)
Tracy, Paul, 214 (n28), 474 (n3)
Travis, Jeremy, 449 (nn 72, 73, 89, 90)
Travis, Lawrence, 213 (n1), 447 (n25)
Triplett, Ruth, 26 (n20)
Trojanowicz, Robert, 161 (nn 31, 32, 33)
Trumbetta, Susan, 109 (n13)
Tuchman, Gary, 59 (n22)
Turner, Gerri, 344 (n74)
Turner, K. B., 278 (n12)
Turner, Susan, 344 (n56), 447 (nn 24, 29, 30, 31)
Turow, Scott, 161 (n56)

U

Uchida, Craig, 136 (nn 8, 9, 11, 18, 22)
Uggen, Christopher, 214 (n28)
Ulmer, Jeffery, 328, 344 (n54)
Umbreit, Mark, 448 (n53)
Uphoff, Rodney, 249 (n39)
Urban, Lynn, 433, 448 (nn 52, 54)
Useem, Burt, 404, 415 (n96)
Utley, Robert, 137 (n51)

V

Van den Haag, Ernest, 87, 108 (nn 1, 10)
Van Dine, Stephan, 317, 342 (n12)
Van Gogh, Theo, 175
Van Meter, Clifford, 137 (n64)
Van Ness, Daniel, 448 (n53)
Venables, Peter, 109 (n26)
Venkatesh, Sudhir, 143
Vermeule, Adrian, 337, 345 (n103)
Vidmar, Neil, 310 (n28), 311 (n51)
Vieraitis, Lynne, 343 (nn 35, 38)

Vigil, Daryl, 412 (n5)
Visher, Christy, 214 (nn 26, 27, 29), 449 (nn 72, 73, 90)
Vitale, Pamela, 469, 469 (n)
Vitiello, Michael, 279 (n26)
Vito, Gennaro, 345 (n106)
Vivona, T. Scott, 413 (n59), 475 (n47)
Vogel, Frank, 319, 319 (n)
Vollen, Lola, 161 (n56)
Vollmer, August, 121–122, 136 (n20), 148
Von Hirsch, Andrew, 315, 342 (n6)
Von Knorring, Anne-Liss, 110–111 (n39)
Voss, Harwin, 214 (n24)

W

Wachtel, Benjamin, 434, 448 (n57)
Waibel, Rob, 360
Waksler, Frances, 58 (n8)
Waldo, Gordon, 331, 344 (n71)
Walker, Samuel, 82 (n6), 135 (n3), 136 (nn 12, 13, 15, 20, 21, 22, 34, 36), 160 (nn 9, 10, 11), 214 (n13), 215 (nn 48, 54), 262–263, 279 (n28)
Wall, April, 496 (n13), 497 (n33)
Wallechinsky, David, 25 (n4)
Wallerstein, James, 82 (nn 21, 27)
Walmsley, Roy, 361 (n)
Walsh, Anthony, 82 (nn 17, 26), 109 (n18), 110 (n28), 279 (n62), 327, 343 (n48)
Walsh, Brandon, 342 (n16)
Walsh, William, 161 (n28)
Walters, Richard, 108 (n2)
Warren, Earl, 55, 178
Warren, Janet, 109 (n13)
Watson, Jamie, 378 (n48)
Watts, Coral Eugene, 62, 62 (n)
Weaver, Greg, 25 (n7)
Weber, Max, 140, 160 (n2)
Webster, Christopher, 449 (n75)
Webster-Stratton, Carolyn, 111 (n60)
Weir, Patricia Ann, 177–178
Weis, Joseph, 82 (n25)
Weisheit, Ralph, 136 (n30), 379 (nn 84, 87)
Weiss, Alexander, 161 (n46)
Weiss, Danielle, 297 (n)
Weitzer, Ronald, 475 (n23)
Welborn, George, 364 (n)
Welch, Susan, 249 (n34)
Welchans, Sarah, 75 (n), 83 (n52)
Wells, James, 414 (n86)
Wells, L. Edward, 82 (n16), 136 (n30), 379 (n84)
Wells, William, 215 (n52)
Wener, Richard, 379 (nn 90, 91)
Werthman, Carl, 214 (nn 18, 33)
Weslander, Eric, 59 (n38), 216 (n77)
Westervelt, Saundra, 161 (n56)
Westley, William, 146, 160 (n13), 213 (n1)
Wheeler, Stanton, 414 (n72)
Whitaker, Gordon, 160 (n26)
White, Byron, 241
White, Jonathan, 496 (n8)
White, Josh, 378 (nn 61, 63)

White, Michael, 279 (n19)
Whitehead, John, 447 (n31)
Wice, Paul, 278 (n15)
Wicker, Tom, 415 (n91)
Wickersham, George, 198, 215 (n60), 216 (n66)
Widom, Cathy, 214 (n22)
Wilcox, Norma, 412 (n20)
Wilcox, Pamela, 412 (n20)
Wilhelm, Daniel, 343 (n30)
Wilkins, Leslie, 343 (n41)
Wilks, Judith, 108 (n9), 342 (n15)
Willard, Brian, 206
William the Conqueror, 223
Williams, Alexander, 116
Williams, Deborah, 414 (nn 68, 69)
Williams, Jay, 82 (nn 21, 27)
Williams, Jimmy, 367–368, 378 (n66)
Williams, Marian, 338, 345 (n111)
Williams, Robert, 179–180
Williams, Rose, 414 (n62)
Willing, Richard, 345 (n117)
Wilsnack, Richard, 415 (n95)
Wilson, David, 412 (n16)
Wilson, James, 449 (n91)
Wilson, James Q., 87, 108 (n1), 124, 136 (n33), 147, 148, 156–157, 160 (nn 19, 26, 59), 161 (n44), 195, 215 (n45), 314, 335, 342 (n1), 344 (n97)
Wilson, Jeremy, 161 (n46)
Wilson, Margaret, 377 (nn 20, 26)
Wilson, Orlando W., 151, 161 (nn 28, 39)
Wilson, Ronnell, 190
Wineman, David, 111 (n41)
Wines, Enoch, 355, 356
Wingate, Keith, 161 (n54)
Winn, Russell, 412 (n26)
Wise, Daniel, 176 (n)
Witkin, Gordon, 83 (n34)
Wolfe, Nancy, 413 (n28)
Wolfers, Justin, 345 (nn 115, 116)
Wolfgang, Marvin, 109 (n20), 345 (n107), 474 (n3)

Wolfson, Warren, 183
Wooden, Wayne, 402, 414 (n84)
Woodworth, George, 345 (n108)
Woolard, Jennifer, 475 (n43)
Word, Ron, 344 (n96)
Worden, Alissa Pollitz, 217 (nn 99, 105)
Worden, Robert, 214 (n 29), 214–215 (n34), 215 (n37), 474 (n18)
Wortley, Richard, 413 (n30)
Wozniak, John, 448 (n60)
Wright, John, 27 (n36)
Wright, Richard, 109 (n13)
Wrobleski, Henry, 137 (n64), 161 (n42)
Wu, Hongda Harry, 362 (n)
Wyle, J. C., 82 (nn 21, 27)

Y

Yandrasits, Janette, 59 (n35)
Yanich, Danlo, 27 (n36)
Yates, Andrea, 15–16, 44, 59 (n36)
Yerushalmi, Mordechai, 319 (n)
Young, Malcolm, 412 (n10)
Yousef, Ramzi, 482
Yu, Roger, 149 (n)

Z

Zabell, Sandy, 297 (n)
Zagorin, Adam, 231 (n)
Zahn, Mary, 91 (n)
Zamble, Edward, 414 (n77)
Zawilski, Valerie, 217 (n95)
Zawitz, Marianne, 76 (n)
Zeisel, Hans, 310 (nn 23, 24), 311 (n64), 345 (n106)
Zimmerman, Paul, 336, 345 (n100)
Zimring, Franklin, 324, 343 (n38)
Zinger, Ivan, 342 (n16)
Zupan, Linda, 379 (n91)

SUBJECT INDEX

Constitution, U.S., 33
See also individual amendment
Bill of Rights, 33, 52–55
Constitutions, state, 33
Continuance, motion for, 285
Contract systems, 243
Copyright infringement, 36
Correctional officers (guards), 390–391
violence against, 402
Correctional systems
See also Community corrections; Jails; Prisons;
Reformatories
agencies, 14
classification of, 16
colonial, 350–352
defined, 350
juvenile, 464–466
perceptions of, 21
privatization of, 368–370
process, 18
rebirth of prisons, 359–361
reforms, 354–357
role of, 16
Corruption, 19, 121, 196
department, 197–198
investigating, 198–199
noble cause, 197
reasons for, 199
types of, 197
Costs, budgets for police departments, 126
Counsel, right to, 241–243, 252
Counseling, offered in prisons, 386–387
Counterfeiting, 130
Court deposit bail, 257
Courtroom Television network, 22
Courts
See also type of
of Appeals, 232, 234
appellate, 15, 230
bail and detention, 17
District, 232
dual system of, 227–235
federal, creation of, 225
of general jurisdiction, 228, 230
initial appearance, preliminary hearing, or arraignment, 17
juvenile, 460–464
of last resort, 230, 239
of limited jurisdiction, 227–228
lower, 15, 227–228
perceptions of, 21
plea bargaining, 17
prisoner access to, 407
sentencing and appeals, 17
state, evolution of, 224–225
trial, 17
types and levels of, 14, 15–16
Court staff
administrators, 235
bailiffs, 235
clerks, 235
defense attorneys, 241–244

judges, 236–238
interpreters, 236
prosecutors, 239–241
reporters, 235–236
workgroups, 244–245
Crime
changing efforts to fight, 8–12
costs of fighting, 8–9
defined, 33–35
elements of, 37–39
nature of, 7–8
rates (statistics), 9, 10, 12, 19, 20, 62–70, 122–123
seriousness of, 34–35
Crime Control Act (1990), 9
Crime Index, 62, 64
Crime in the United States, 63, 64
Crime mapping, 151
Crime-reporting habits, 19
Criminal career, 103
Criminal courts
emergence of, 222–223
history of American, 223–226
international, 231
jurisdiction of, 225
transfer of juveniles to, 466–468
Criminal investigation, 16, 152–153
Criminalists, 152
Criminal justice, study of, 12–13
Criminal justice process
corrections, 18
courts, 17
law enforcement, 16–17
Criminal justice system
agencies, 14
defined, 6–7
juveniles in, 466–469
levels of, 13
perceptions of, 18–23
Criminal law
codes, 31
common, 31–32
defined, 30
functions of, 35–37
origins of, 30–33
procedural, 30, 164
relationship between civil law and, 34
sources on, 32–33
stare decisis, 32
substantive, 30
Criminal Man, The (Lombroso), 92
Criminal procedure law, 30, 164
Criminals
age of, 70
socioeconomic status and, 71–72
Cross-examination, 298
Cuba, crime and due process in, 38
Cultural deviance theory, 98, 105
Cultural differences
drug policy in the Netherlands, 11
what constitutes crime, 7
Cultural transmission, 98

Homicide
  intimate partner, 75–76
  justifiable, by police, 203
Horizontal overcharging, 262
Hot spots of crime, 151
House arrest 429–430
Human Rights Watch Organization, 199
Hung jury, 17, 304

I

Id, 96
Identity theft, 492–493
Identity Theft and Assumption Deterrence Act (1998), 492, 493
Identity Theft Penalty Enhancement Act (2004), 492
Immigration and Customs Enforcement (ICE), 478, 485–486
Immigration Nationality Act, 157
Immunity, 259
Implicit plea bargaining, 270
Importation model, 399
Imprisonment rates, 361
Incapacitation, 317
Incarceration, 18
  intermittent, 424
  shock, 424
Inchoate crime, 37
Incorporation of Bill of Rights, 54–55
Indeterminate sentences, 320–321
India, prison programs in, 389
Indictment, 259
Individual justice, 89
Inevitable discovery rule, 167
Informal probation, 424
Infractions, 35
Initial appearance
  defined, 252
  notification of rights, 252–253
  timing of, 253–254
Inmate code, 399–400
Inmates. *See* Prisoners
Innocence Project, 338
In-presence requirement, 35, 177
Insanity
  death penalty and, 333
  defined, 42
  Durham rule, 43
  guilty but mentally ill, 44, 46
  irresistible impulse test, 43
  McNaghten rule, 42–43
  not guilty by reason of, 269
  state rules, 45–46
  substantial capacity test, 43
Insanity Defense Reform Act (1984), 43–44, 269
Institutional placement, 464–465
Intake, 460–461
Intelligence, criminality and, 93
Intelligence-led policing, 125–126
Intensive aftercare programs (IAPs), 466
Intensive probation supervision (IPS), 429

Intensive supervision parole, 441
Intent, general versus specific, 39
Intermediate sanctions
  community service, 432–433
  day reporting centers, 431
  defined, 428
  electronic monitoring, 430–431
  fines, 431–432
  forfeiture, 432
  house arrest, 429–430
  intensive probation supervision, 429
  restitution, 432
Intermediate sentences, 320
Intermittent incarceration, 424
International Criminal Court (ICC), 231
International Criminal Police Organization, 130–131
International imprisonment rates, 361
Interpol, 130–131
Interpretation of Dreams, The (Freud), 95
Interpreters, 236
Interrogation, 182–183
Intimate partner violence, 75–76
Intoxication, defense based on, 46
Involuntary confession, 182–183
Involvement crimes, 152
Irish system, 356
Irresistible impulse test, 43
Islamic crimes and punishments, 319
Islamic Fundamentalist Groups, 480–481
Israel, 480

J

Jailhouse lawyers, 407
Jails
  colonial, 370
  defined, 350
  differences between prisons and, 370
  modern, 370–371
  new-generation, 371, 373
  statistics, 372
  web cams, 371
Jamaat Ul Fuqra, 480–481
Japan, community policing in, 125
Jihad, 480
Judges
  background of, 236–237
  circuit-riding, 225, 226
  magistrate, 231–232
  plea bargains and role of, 271–272
  role and responsibilities of, 236
  selection of, 237–238
Judgment of acquittal, 298
Judicial instruction, 299, 301–302, 303
Judicial jurisdiction, 226
Judicial reprieve, 420
Judicial response to terrorism, 488–489
Judicial review, 225
Judicial waiver, 467
Judiciary Act (1789), 129, 225, 239
Jurisdiction, 126

Medical model, 359–360
Medical services, offered in prisons, 387
  right to, 408
Medium-security prisons, 363
Megan's Law, 439
Melissa virus, 490
Mens rea, 38–39
Mental disorders, prisoners with, 394
  See also Insanity
Mental retardation death penalty and, 333
Metropolitan Police Act (1829), 118, 119
Mexican Mafia, 400
Military prisons, 366–367
Milwaukee, crime in, 91
Minimum-security prisons, 362
Minorities
  communication with, 211
  as police officers, 145–146, 206–210
  ratio of officers in large city police departments, 127
Mint Police, 15
Miranda warning, 177–181
Misdemeanors, 35
Missouri Plan, 237–238
Mistake, 47–49
Mistrial, 295
Mitigating circumstances, 89
Mobile Enforcement Teams, 261
Model Penal Code
  duress and, 47
  intoxication and, 46
  self-defense and, 40
  Test, 43
Modification of sentence, 424
Mollen Commission, 198, 199
Monozygotic twins, 94
Mosaic law, 31
Mothers in prison, 397–398
Motor vehicle
  fatalities, 155
  high speed chases, 203–205, 206
  police patrols, 150
  searches, 171
Mount Pleasant Female Prison, 358
Movies, perceptions of criminal justice and impact of, 21
Music, perceptions of criminal justice and impact of, 21

N

National Center for Policy Analysis, 64
National Center for Women and Policing, 208
National Crime Victimization Survey (NCVS), 66–67
National Incident-Based Reporting System (NIBRS), 64, 65, 72
National Opinion Research Center, 64
National Organization for the Reform of Marijuana Laws, 11
National Prison Association, 356
National Youth Survey (NYS), 67
Necessity, 40
Neoclassical school, 88–91
Neta, 400
Netherlands, drug policy in, 11

New African Freedom Fighters, 480
New Castle Correctional riot, 404
New-generation jails, 371, 373
New Hampshire, probation in, 425
New Mexico State Penitentiary, 403–404
New Orleans, police brutality, 200
New York House of Refuge, 457
New York model, 354, 355
New York Police Department (NYPD)
  in the early 1900s, 121
  Knapp Commission, 198–199
  Mollen Commission, 198, 199
  sexual discrimination, 209
  statistics, 127
New Zealand, police moonlighting as prostitute, 196
1984 (Orwell), 37
Ninth Amendment, 53
No knock raids, 167, 201
Nolo contendere plea, 268
Norms
  conduct, 37
  defined, 7
Northwest Cyber Crime Task Force, 36
Not guilty by reason of insanity, 269
Not guilty plea, 268

O

Official crime statistics, 62
Oklahoma City bombing, 264, 482
Omnibus Crime Control and Safe Streets Act (1968), 9, 12, 174, 183, 253
On Crimes and Punishments (Beccaria), 88
One arm's length rule, 168–169
Open field searches, 172
Opening statements, 295
Operant conditioning, 98
Operational styles, 147–148
Operation D-Elite, 36
Operation Firewall, 36
Operation Return to Sender, 478, 486
Original jurisdiction, 226

P

Pains of imprisonment, 398
Palestinian Liberation Organization (PLO), 480
Pardon, 436
Parens patriae, 457
Parish-constable police system, 117
Parole, 18
  See also Release
  characteristics of adults on, 435
  conditions of, 441
  defined, 435
  effectiveness of, 442–443
  emergence of, 435–436
  revocation of, 441–442
  supervision, 441
Parole boards
  decision making, 438–439

probation and, 419
reporting of crime, 20
sentencing and, 328–330
Racial terrorism, 479
Rape
    intimate partner, 76
    prison, 402, 403
    shield laws, 298
Rational choice theory, 89–90
Real evidence, 296–297
Rebuttal, 299
Recidivism rates, 442–443, 469
Reentry, 443
Reformatories
    defined, 350
    Elmira model, 356–357
    Irish system, 356
    list of early, 357
    mark system, 355–356
Reformatory movement, 354–358
Rehabilitation, 18, 89, 317–318
Release, 18
    See also Parole
    in Canada, 437
    halfway house, 436
    of sex offenders, 439–440
    temporary, 438
    unconditional, 436
Release on recognizance (ROR), 17, 256
Religion
    freedom of, in prison, 405–406
    jury selection bias and, 294
    law and, 31
Reparative probation, 434
Reporters, court, 235–236
Reporting time, 152
Response time, 152
Restitution, 432
Restorative justice
    criticisms of, 434–435
    defined, 433
    effectiveness of, 434
    purpose of, 433–434
Retribution, 315
Reverse waiver, 468
Revocation hearing, 427
Revocation of parole, 441–442
Revocation of probation, 427–428
Right(s)
    to counsel, 241–243, 252
    juveniles and, 462, 463
    notification of, 252–253
    prisoners and, 404–409
    to privacy, 406
    to public trial, 287–289
    to reasonable bail, 252
    to remain silent, 252
    to speedy trial, 286–287
    of terrorist suspects, 485
    to trial by jury, 290–294

Riots, prison, 403–404
Risk classification, 422
Roman law, 31
Rome Statute, 231
Routine activities theory, 90
Rules, administrative, 32

S

Salient factor score, 439
School searches, 168
Scientific determinism, 89
Scooters, used by police, 149
Search and seizure
    abandoned property, 173–174
    border, 174
    consent, 172–173
    electronic surveillance, 174–175
    exclusionary rule, 165–168
    motor vehicle, 171
    one arm's length rule, 168–169
    open field, 172
    plain view doctrine, 171–172
    school, 168
    stop and frisk rule, 169–171
Search warrant, 165
    exception to, 168
Sears Roebuck Co., 132
Second Amendment, 53
Secondary deviance, 101
Secret Service, 15
Segways, 149
Selective law enforcement, 191
Self-control theory, 100–101, 105
Self-defense, 40
Self-report surveys, 67–68
Senior citizens as victims, 74–75
Sentencing, 17
    age and, 331
    appealing, 339
    blended, for juveniles, 465
    concurrent, 325
    consecutive, 325
    determinate or structured, 321
    determining appropriate, 325–327
    disparity and discrimination in, 327–331
    gender differences, 330–331
    goals of, 314–318
    guidelines, 321–322, 323
    habitual offender statutes, 322, 324–325
    hearing, 326
    indeterminate, 320–321
    intermediate, 320
    of juveniles, 468
    mandatory, 322
    presentence investigation report, 325–326
    presumptive, 321
    probation, 424
    racial differences, 328–330
    socioeconomic status and, 331

Chapter 1

Section Opener: © UPI Photo/Yuri Gripas/Landov; Chapter 1 Opener: © The Roanoke Times, Alan Kim/AP Photos; page 6: © The Roanoke Times, Matt Gentry/AP Photos; page 7: © Kevork Djansezian/AP Photos; page 8: © Dominic Dibbs/age footstock; page 9: © Lawrence Eagle Tribune/AP Photos; page 10a-b: © Reuters/HO/Landov; page 11: © david pearson/Alamy Images; page 15: © Reuters/Brett Coomer/Pool/Landov; page 18: © Reuters/Chaiwat Subprasom/Landov; page 19: © Reuters/Pool/Landov; page 21: © Reuters/Allen Fredrickson/Landov

Chapter 2

Opener: © Kenneth Lambert/AP Photos; page 30: © Mitchell Tapper/AP Photos; page 31: © Reuters/Jason Reed/Landov; page 33: © James Steidl/Shutter Stock, Inc.; page 36: © Kuzma/ShutterStock, Inc.; page 40: © Ian Thraves/Alamy Images; page 42: © Reuters/David Rae Morris/Landov; page 44: © Bob Daugherty/AP Photos; page 50 left: © Francis Specker/AP Photos; page 50 right: © Rebecca Cook/Reuters/Landov; page 51: Courtesy of Terri Jentz; page 55: Courtesy of Special Collections Department, Harvard Law School Library

Chapter 3

Opener: © Bob Child/AP Photos; page 62: © Carlos Osorio/AP Photos; page 66: © Bill Fritsch/age footstock; page 72 top: © Cobb County Police Dept./AP Photos; page 72 bottom: © Mark Green/Bloomberg News/Landov; page 73: © Eugene Richards/VII/AP Photos; page 75: © Betty Rountree/AP Photos; page 77: © Texarkan Gazette, Chris Dean/AP Photos; page 78: © tomKidd/Alamy Images; page 79: © Mark C. Ide

Chapter 4

Opener: © Iowa State Daily, Eric Rowley/AP Photos; page 86 top: © Maxim MarMur/AP Photos; page 86 bottom: © Tim Kimzey/AP Photos; page 87: © Chris O'Meara/AP Photos; page 89: © CBS/Landov; page 91: Published with permission of Journal Sentinel/Rick Wood; page 94 left: © beerkoff/ShutterStock, Inc.; page 94 right: © Design Pics /age fotostock; page 95: © Dennis MacDonald/age fotostock; page 98: © Mika Heittola/ShutterStock, Inc.; page 103: © Banana Stock/age fotostock

Chapter 5

Section Opener: © Diane Bondareff/BloombergNews/Landov; Chapter 5 Opener: © AP Photos; page 116: © Reuters/Pool/Kathy Willens/Landov; page 117: © Bettmann/Corbis; page 118: Courtesy of Perry-Castañeda Library, University of Texas at Austin; page 119: © Interfoto Pressebildagentur/Alamy Images; page 122: Courtesy of The Bancroft Library, University of California, Berkeley, used with permission; page 124: © Sascha Burkard/ShutterStock, Inc.; page 125: © Mark C. Ide; page 128: © Corbis; page 130: © Tischenko Irina/ShutterStock, Inc.; page 131: Courtesy of International Criminal Police Organization—INTERPOL

Chapter 6

Opener: © James Dawson/Image Farm Inc./Jupiterimages; page 140: Photo courtesy of the Long Beach Police Department; page 145: © NNS /The Plain Dealer/Landov; page 146 top: © The Californian, Richard D. Green/AP Photos; page 146 bottom: © Mark C. Ide; page 149: © Jeff Greenberg/age fotostock; page 150: © Kanwarjit Singh Boparai/ShutterStock, Inc.; page 157: © Kyodo/Landov

Chapter 7

Opener: © Reuters/Jason Reed/Landov; page 164: © Reuters/Brian LaPeter/Pool/Landov; page 168: © The Repository, Scott Heckel/AP Photos; page 169: © Mark C. Ide; page 171: © Mark C. Ide; page 172: © Mark C. Ide; page 174: Courtesy of Gerald L. Nino/U.S. Customs and Border Protection; page 176: © Stephen Coburn/ShutterStock, Inc.; page 177: © Mark C. Ide; page 178: © AP Photos; page 182: © John Lund/Sam Diephui/age fotostock

Chapter 8

Opener: © Mark C. Ide; page 190 © Miami Herald/MCT/Landov; page 192: © Mark C. Ide; page 194:

© Frances Roberts/Alamy Images; page 195: © UPI Photo/Billy Suratt/ Landov; page 198: © Pictorial Press Ltd/Alamy Images; page 200: © Mel Evans/AP Photos; page 204: © Kanwarjit Singh Boparai/ShutterStock, Inc.; page 204: © Jack Dagley Photography/Shutter Stock, Inc.; page 207 left: © Reuters/Brendan McDermid/Landov; page 209: © Mikael Karlsson/Alamy Images

Chapter 9

Section Opener: © Corbis/age fotostock; Chapter 9 Opener: © Elias H. Debbas II/ShutterStock, Inc.; page 222: © Terrebonne Parish Sheriff's Office-HO/AP Photos; page 225: Courtesy of Library of Congress, Prints & Photographs Division, [reproduction number LC-USZ62-3462]; page 233: © Dennis Brack/Bloomberg News/Landov; page 235: © HO/AP Photos; page 236: © Reuters/Tami Chappell/Landov; page 238 bottom: © Corbis/age fotostock; page 238 top: © Brandi Simons/AP Photos; page 240: © Banana Stock/age fotostock; page 244: © Digital Vision/age fotostock

Chapter 10

Opener: © Bill Fritsch/age fotostock; page 252: © Kevin P. Casey/AP Photos; page 255: © Dennis MacDonald/age fotostock; page 256: © Reuters/Lucy Pemoni/Landov; page 260: © Michael Ging, POOL/AP Photos; page 262: © Reuters/Carlos Barria/Landov; page 264: © David Longstreath/AP Photos; page 268: © Bill Fritsch/age fotostock; page 270: © Bill Fritsch/age fotostock; page 273: © Don Hammond/age fotostock

Chapter 11

Opener: © Corbis/age fotostock; page 282: © Reuters/Ivan Milutinovic/Landov; page 283: © Doug Jennings/AP Photos; page 285: © Brand X Pictures/Creatas; page 286: © Reuters/Pool/George Wilhelm/Los Angeles Times/Landov; page 288: © Reuters/Lucas Jackson/Landov; page 289: © Mark C. Ide; page 291: © RubberBall/Alamy Images; page 292 top: © maxstockphoto/Shutter Stock, Inc.; page 929 top middle: Courtesy of the United States District Court; page 292 middle bottom: © Dennis MacDonald/age fotostock; page 292 bottom: © PNC/age fotostock; page 296a-b: Courtesy of Cynthia D. Homer, Maine State Police Crime

Laboratory; page 301: © Reuters/Vince Bucci/Pool/Landov; page 302: © Jones and Bartlett Publishers. Photographed by Kimberly Potvin

Chapter 12

Opener: © Steve Griffin, Pool/AP Photos; page 314: © Mark Duncan, Pool/AP Photos; page 315: © UPI/Elaine Thompson/POOL/Landov; page 316: © The Plain Dealer, Chris Stephens/AP Photos; page 317: © Tim Harman/ShutterStock, Inc.; page 320: © Mark E. Stout/ShutterStock, Inc.; page 324 top: © UPI Photo/Landov; page 324 bottom: © Damian Dovarganes/AP Photos; page 326: © Pat Sullivan/AP Photos; page 333: © AP Photos; page 336: © Florida Department of Corrections/AP Photos

Chapter 13

Section Opener: © A. Ramey/PhotoEdit, Inc.; Chapter 13 Opener: © Corbis; page 350: © Brett Coomer/AP Photos; page 351: © North Wind Picture Archives/Alamy Images; page 353: Courtesy of Eastern State Penitentiary Historic Site; page 354: © Eastern Kentucky University Archives, Richmond, KY; page 356: ©Photos.com; page 358: © Ron Chapple/Thinkstock/Alamy Images; page 363: © Richard Smith/Alamy Images; page 364: © Robin Nelson/PhotoEdit, Inc.; page 367: © Reuters/Mark Wilson/Pool/Landov; page 373: © Spencer Grant/PhotoEdit, Inc.

Chapter 14

Opener: © A. Ramey/PhotoEdit, Inc.; page 382: © Reuters/HO/Landov; page 384: © Bill Fritsch/age fotostock; page 386: © Troy Maben/AP Photos; page 388: © AbleStock; page 390: © Don Hammond/age fotostock; page 392: Photos courtesy of the Wisconsin State Journal and photographer Leah L. Jones; page 393: © Ron Brown/SuperStock/age fotostock; page 395: © The Topeka Capital-Journal, Mike Shepherd/AP Photos; page 396: © Richard Lord/PhotoEdit, Inc.; page 398: © Columbus Dispatch, Renee Sauer/AP Photos; page 400: © Ann Johansson/AP Photos; page 401: © A. Ramey/PhotoEdit, Inc.; page 403: © EyePress/AP Photos

Chapter 15

Opener: © A. Ramey/PhotoEdit, Inc.; page 418: © Photos.com; page 419: © Bob Daemmrich/PhotoEdit, Inc.; page 421: © Spencer Grant/Photo Edit, Inc.; page 428: © Michael Newman/PhotoEdit, Inc.; page 429: © A. Ramey/Photo Edit, Inc.; page 431: © Madison/X17online.com; page 433: © James Shaffer/PhotoEdit, Inc.; page 434: © John Birdsall/age fotostock; page 436: © Brendan Smialowski/Bloomberg News/Landov

Chapter 16

Section Opener: Courtesy of Andrea Booher/FEMA News Photo; Chapter 16 Opener: © Reuters/Brendan McDermid/Landov; page 454: © Charles V. Tines, Pool/AP Photos; page 457: © Photos.com; page 458: © BananaStock/age fotostock; page 461: © SpencerGrant/PhotoEdit Inc.; page 466: © Spencer Grant/PhotoEdit, Inc.; page 467: © Christopher Berkey/AP Photos; page 469: © Dan Rosenstrauch, Pool/AP Photos; page 470: © Don Ryan/AP Photos

Chapter 17

Opener: © Naomi Stock/Landov; page 478: © Jason Bronis/AP Photos; page 479: © Peter Foley/Landov; page 480: © CBS/Landov; page 481: © Mark Mobley/AP Photos; page 484: Courtesy of Mate 3rd Class Joshua Karsten/U.S. Navy; page 487: © Reuters/Sue Ogrocki/Landov; page 489: © Mikael Karlsson/Alamy Images; page 490: © Daniel Hulshizer/AP Photos; page 491: © Lein de León Yong/Shutterstock, Inc.; page 493: © Peter Baxter/ShutterStock, Inc.